Advances in Solar Heating and Cooling

Related titles

Solar Energy in Buildings: Thermal Balance for Efficient Heating and Cooling
(ISBN 978-0-12-410514-0)

Solar Energy, Photovoltaics, and Domestic Hot Water: A Technical and Economic Guide for Project Planners, Builders, and Property Owners
(ISBN 978-0-12-420155-2)

Solar Energy Engineering, 2nd Edition
(ISBN 978-0-12-397270-5)

Woodhead Publishing Series in Energy:
Number 102

Advances in Solar Heating and Cooling

Edited by

R.Z. Wang and T.S. Ge

ELSEVIER

AMSTERDAM • BOSTON • CAMBRIDGE • HEIDELBERG
LONDON • NEW YORK • OXFORD • PARIS • SAN DIEGO
SAN FRANCISCO • SINGAPORE • SYDNEY • TOKYO
Woodhead Publishing is an imprint of Elsevier

WP
WOODHEAD PUBLISHING

Woodhead Publishing is an imprint of Elsevier
The Officers' Mess Business Centre, Royston Road, Duxford, CB22 4QH, UK
50 Hampshire Street, 5th Floor, Cambridge, MA 02139, USA
The Boulevard, Langford Lane, Kidlington, OX5 1GB, UK

British Library Cataloguing-in-Publication Data
A catalogue record for this book is available from the British Library

Library of Congress Cataloging-in-Publication Data
A catalog record for this book is available from the Library of Congress

ISBN: 978-0-08-100301-5 (print)
ISBN: 978-0-08-100302-2 (online)

Publisher: Joe Hayton
Acquisition Editor: Sarah Hughes
Editorial Project Manager: Alex White
Production Project Manager: Poulouse Joseph
Designer: Mark Rogers

Typeset by TNQ Books and Journals

Contents

List of contributors

C. Chang Key Laboratory of Solar Thermal Energy and Photovoltaic System, Institute of Electrical Engineering, Chinese Academy of Sciences, Beijing, China

T.T. Chow City University of Hong Kong, Kowloon Tong, Hong Kong

D.A. Chwieduk Institute of Heat Engineering, Faculty of Power and Aeronautical Engineering, Warsaw University of Technology, Warsaw, Poland

A. Duta Transilvania University of Brasov, Brasov, Romania

L. Finocchiaro Norwegian University of Science and Technology, Trondheim, Norway

T.S. Ge Institute of Refrigeration and Cryogenics, Shanghai Jiao Tong University, Shanghai, China

L. Georges Norwegian University of Science and Technology, Trondheim, Norway

J. Gong North Dakota State University, Fargo, United States

A.G. Hestnes Norwegian University of Science and Technology, Trondheim, Norway

C.A. Infante Ferreira Delft University of Technology, Delft, The Netherlands

S.A. Kalogirou Cyprus University of Technology, Limassol, Cyprus

F. Kuznik INSA Lyon, CETHIL, Villeurbanne, France

T.X. Li Institute of Refrigeration and Cryogenics, Shanghai Jiao Tong University, Shanghai, China

Y. Li Institute of Refrigeration and Cryogenics, Shanghai Jiao Tong University, Shanghai, China

G. Martinopoulos International Hellenic University, Thessaloniki, Greece

M. Moldovan Transilvania University of Brasov, Brasov, Romania

Q.W. Pan Institute of Refrigeration and Cryogenics, Shanghai Jiao Tong University, Shanghai, China

R.T.A. Prado University of Sao Paulo, Sao Paulo, Brazil

D.S. Renné Dave Renné Renewables, LLC, Boulder, CO, United States

D.S. Sowmy University of Sao Paulo, Sao Paulo, Brazil; Institute of Technological Research of Sao Paulo, Sao Paulo, Brazil

K. Sumathy North Dakota State University, Fargo, United States

R. Velraj Institute for Energy Studies, Anna University, Chennai, TamilNadu, India

I. Visa Transilvania University of Brasov, Brasov, Romania

R.Z. Wang Institute of Refrigeration and Cryogenics, Shanghai Jiao Tong University, Shanghai, China

J.C. Xu Institute of Refrigeration and Cryogenics, Shanghai Jiao Tong University, Shanghai, China

Z.Y. Xu Institute of Refrigeration and Cryogenics, Shanghai Jiao Tong University, Shanghai, China

T. Yan Institute of Refrigeration and Cryogenics, Shanghai Jiao Tong University, Shanghai, China

X. Zheng Institute of Refrigeration and Cryogenics, Shanghai Jiao Tong University, Shanghai, China

Woodhead Publishing Series in Energy

1 **Generating power at high efficiency: Combined cycle technology for sustainable energy production**
Eric Jeffs

2 **Advanced separation techniques for nuclear fuel reprocessing and radioactive waste treatment**
Edited by Kenneth L. Nash and Gregg J. Lumetta

3 **Bioalcohol production: Biochemical conversion of lignocellulosic biomass**
Edited by Keith W. Waldron

4 **Understanding and mitigating ageing in nuclear power plants: Materials and operational aspects of plant life management (PLiM)**
Edited by Philip G. Tipping

5 **Advanced power plant materials, design and technology**
Edited by Dermot Roddy

6 **Stand-alone and hybrid wind energy systems: Technology, energy storage and applications**
Edited by John K. Kaldellis

7 **Biodiesel science and technology: From soil to oil**
Jan C. J. Bart, Natale Palmeri and Stefano Cavallaro

8 **Developments and innovation in carbon dioxide (CO_2) capture and storage technology Volume 1: Carbon dioxide (CO_2) capture, transport and industrial applications**
Edited by M. Mercedes Maroto-Valer

9 **Geological repository systems for safe disposal of spent nuclear fuels and radioactive waste**
Edited by Joonhong Ahn and Michael J. Apted

10 **Wind energy systems: Optimising design and construction for safe and reliable operation**
Edited by John D. Sørensen and Jens N. Sørensen

11 **Solid oxide fuel cell technology: Principles, performance and operations**
Kevin Huang and John Bannister Goodenough

12 **Handbook of advanced radioactive waste conditioning technologies**
Edited by Michael I. Ojovan

13 **Membranes for clean and renewable power applications**
Edited by Annarosa Gugliuzza and Angelo Basile

14 **Materials for energy efficiency and thermal comfort in buildings**
Edited by Matthew R. Hall

15 **Handbook of biofuels production: Processes and technologies**
Edited by Rafael Luque, Juan Campelo and James Clark

16 **Developments and innovation in carbon dioxide (CO_2) capture and storage technology Volume 2: Carbon dioxide (CO_2) storage and utilisation**
Edited by M. Mercedes Maroto-Valer
17 **Oxy-fuel combustion for power generation and carbon dioxide (CO_2) capture**
Edited by Ligang Zheng
18 **Small and micro combined heat and power (CHP) systems: Advanced design, performance, materials and applications**
Edited by Robert Beith
19 **Advances in clean hydrocarbon fuel processing: Science and technology**
Edited by M. Rashid Khan
20 **Modern gas turbine systems: High efficiency, low emission, fuel flexible power generation**
Edited by Peter Jansohn
21 **Concentrating solar power technology: Principles, developments and applications**
Edited by Keith Lovegrove and Wes Stein
22 **Nuclear corrosion science and engineering**
Edited by Damien Féron
23 **Power plant life management and performance improvement**
Edited by John E. Oakey
24 **Electrical drives for direct drive renewable energy systems**
Edited by Markus Mueller and Henk Polinder
25 **Advanced membrane science and technology for sustainable energy and environmental applications**
Edited by Angelo Basile and Suzana Pereira Nunes
26 **Irradiation embrittlement of reactor pressure vessels (RPVs) in nuclear power plants**
Edited by Naoki Soneda
27 **High temperature superconductors (HTS) for energy applications**
Edited by Ziad Melhem
28 **Infrastructure and methodologies for the justification of nuclear power programmes**
Edited by Agustín Alonso
29 **Waste to energy conversion technology**
Edited by Naomi B. Klinghoffer and Marco J. Castaldi
30 **Polymer electrolyte membrane and direct methanol fuel cell technology Volume 1: Fundamentals and performance of low temperature fuel cells**
Edited by Christoph Hartnig and Christina Roth
31 **Polymer electrolyte membrane and direct methanol fuel cell technology Volume 2: *In situ* characterization techniques for low temperature fuel cells**
Edited by Christoph Hartnig and Christina Roth
32 **Combined cycle systems for near-zero emission power generation**
Edited by Ashok D. Rao
33 **Modern earth buildings: Materials, engineering, construction and applications**
Edited by Matthew R. Hall, Rick Lindsay and Meror Krayenhoff
34 **Metropolitan sustainability: Understanding and improving the urban environment**
Edited by Frank Zeman
35 **Functional materials for sustainable energy applications**
Edited by John A. Kilner, Stephen J. Skinner, Stuart J. C. Irvine and Peter P. Edwards

36 **Nuclear decommissioning: Planning, execution and international experience**
Edited by Michele Laraia
37 **Nuclear fuel cycle science and engineering**
Edited by Ian Crossland
38 **Electricity transmission, distribution and storage systems**
Edited by Ziad Melhem
39 **Advances in biodiesel production: Processes and technologies**
Edited by Rafael Luque and Juan A. Melero
40 **Biomass combustion science, technology and engineering**
Edited by Lasse Rosendahl
41 **Ultra-supercritical coal power plants: Materials, technologies and optimisation**
Edited by Dongke Zhang
42 **Radionuclide behaviour in the natural environment: Science, implications and lessons for the nuclear industry**
Edited by Christophe Poinssot and Horst Geckeis
43 **Calcium and chemical looping technology for power generation and carbon dioxide (CO_2) capture: Solid oxygen- and CO_2-carriers**
Paul Fennell and E. J. Anthony
44 **Materials' ageing and degradation in light water reactors: Mechanisms, and management**
Edited by K. L. Murty
45 **Structural alloys for power plants: Operational challenges and high-temperature materials**
Edited by Amir Shirzadi and Susan Jackson
46 **Biolubricants: Science and technology**
Jan C. J. Bart, Emanuele Gucciardi and Stefano Cavallaro
47 **Advances in wind turbine blade design and materials**
Edited by Povl Brøndsted and Rogier P. L. Nijssen
48 **Radioactive waste management and contaminated site clean-up: Processes, technologies and international experience**
Edited by William E. Lee, Michael I. Ojovan, Carol M. Jantzen
49 **Probabilistic safety assessment for optimum nuclear power plant life management (PLiM): Theory and application of reliability analysis methods for major power plant components**
Gennadij V. Arkadov, Alexander F. Getman and Andrei N. Rodionov
50 **The coal handbook: Towards cleaner production Volume 1: Coal production**
Edited by Dave Osborne
51 **The coal handbook: Towards cleaner production Volume 2: Coal utilisation**
Edited by Dave Osborne
52 **The biogas handbook: Science, production and applications**
Edited by Arthur Wellinger, Jerry Murphy and David Baxter
53 **Advances in biorefineries: Biomass and waste supply chain exploitation**
Edited by Keith Waldron
54 **Geological storage of carbon dioxide (CO_2): Geoscience, technologies, environmental aspects and legal frameworks**
Edited by Jon Gluyas and Simon Mathias
55 **Handbook of membrane reactors Volume 1: Fundamental materials science, design and optimisation**
Edited by Angelo Basile

56 **Handbook of membrane reactors Volume 2: Reactor types and industrial applications**
Edited by Angelo Basile
57 **Alternative fuels and advanced vehicle technologies for improved environmental performance: Towards zero carbon transportation**
Edited by Richard Folkson
58 **Handbook of microalgal bioprocess engineering**
Christopher Lan and Bei Wang
59 **Fluidized bed technologies for near-zero emission combustion and gasification**
Edited by Fabrizio Scala
60 **Managing nuclear projects: A comprehensive management resource**
Edited by Jas Devgun
61 **Handbook of Process Integration (PI): Minimisation of energy and water use, waste and emissions**
Edited by Jiří J. Klemeš
62 **Coal power plant materials and life assessment**
Edited by Ahmed Shibli
63 **Advances in hydrogen production, storage and distribution**
Edited by Ahmed Basile and Adolfo Iulianelli
64 **Handbook of small modular nuclear reactors**
Edited by Mario D. Carelli and Dan T. Ingersoll
65 **Superconductors in the power grid: Materials and applications**
Edited by Christopher Rey
66 **Advances in thermal energy storage systems: Methods and applications**
Edited by Luisa F. Cabeza
67 **Advances in batteries for medium and large-scale energy storage**
Edited by Chris Menictas, Maria Skyllas-Kazacos and Tuti Mariana Lim
68 **Palladium membrane technology for hydrogen production, carbon capture and other applications**
Edited by Aggelos Doukelis, Kyriakos Panopoulos, Antonios Koumanakos and Emmanouil Kakaras
69 **Gasification for synthetic fuel production: Fundamentals, processes and applications**
Edited by Rafael Luque and James G. Speight
70 **Renewable heating and cooling: Technologies and applications**
Edited by Gerhard Stryi-Hipp
71 **Environmental remediation and restoration of contaminated nuclear and NORM sites**
Edited by Leo van Velzen
72 **Eco-friendly innovation in electricity networks**
Edited by Jean-Luc Bessede
73 **The 2011 Fukushima nuclear power plant accident: How and why it happened**
Yotaro Hatamura, Seiji Abe, Masao Fuchigami and Naoto Kasahara. Translated by Kenji Iino
74 **Lignocellulose biorefinery engineering: Principles and applications**
Hongzhang Chen
75 **Advances in membrane technologies for water treatment: Materials, processes and applications**
Edited by Angelo Basile, Alfredo Cassano and Navin Rastogi

76 **Membrane reactors for energy applications and basic chemical production**
Edited by Angelo Basile, Luisa Di Paola, Faisal Hai and Vincenzo Piemonte

77 **Pervaporation, vapour permeation and membrane distillation: Principles and applications**
Edited by Angelo Basile, Alberto Figoli and Mohamed Khayet

78 **Safe and secure transport and storage of radioactive materials**
Edited by Ken Sorenson

79 **Reprocessing and recycling of spent nuclear fuel**
Edited by Robin Taylor

80 **Advances in battery technologies for electric vehicles**
Edited by Bruno Scrosati, Jürgen Garche and Werner Tillmetz

81 **Rechargeable lithium batteries: From fundamentals to applications**
Edited by Alejandro A. Franco

82 **Calcium and chemical looping technology for power generation and carbon dioxide (CO_2) capture**
Edited by Paul Fennell and Ben Anthony

83 **Compendium of hydrogen energy Volume 1: Hydrogen production and purificiation**
Edited by Velu Subramani, Angelo Basile and T. Nejat Veziroglu

84 **Compendium of hydrogen energy Volume 2: Hydrogen storage, transmission, transportation and infrastructure**
Edited by Ram Gupta, Angelo Basile and T. Nejat Veziroglu

85 **Compendium of hydrogen energy Volume 3: Hydrogen energy conversion**
Edited by Frano Barbir, Angelo Basile and T. Nejat Veziroglu

86 **Compendium of hydrogen energy Volume 4: Hydrogen use, safety and the hydrogen economy**
Edited by Michael Ball, Angelo Basile and T. Nejat Veziroglu

87 **Advanced district heating and cooling (DHC) systems**
Edited by Robin Wiltshire

88 **Microbial electrochemical and fuel cells: Fundamentals and applications**
Edited by Keith Scott and Eileen Hao Yu

89 **Renewable heating and cooling: Technologies and applications**
Edited by Gerhard Stryi-Hipp

90 **Small modular reactors: Nuclear power fad or future?**
Edited by Daniel T. Ingersoll

91 **Fuel flexible energy generation: Solid, liquid and gaseous fuels**
Edited by John Oakey

92 **Offshore wind farms: Technologies, design and operation**
Edited by Chong Ng and Li Ran

93 **Uranium for nuclear power: Resources, mining and transformation to fuel**
Edited by Ian Hore-Lacy

94 **Biomass supply chains for bioenergy and biorefining**
Edited by Jens Bo Holm-Nielsen and Ehiaze Augustine Ehimen

95 **Sustainable energy from salinity gradients**
Edited by Andrea Cipollina and Giorgio Micale

96 **Membrane technologies for biorefining**
Edited by Alberto Figoli, Alfredo Cassano and Angelo Basile

97 **Geothermal power generation: Developments and innovation**
Edited by Ronald DiPippo

98 **Handbook of biofuels' production: Processes and technologies (Second Edition)**
Edited by Rafael Luque, Carol Sze Ki Lin, Karen Wilson and James Clark
99 **Magnetic fusion energy: From experiments to power plants**
Edited by George H. Neilson
100 **Advances in ground-source heat pump systems**
Edited by Simon Rees
101 **Absorption-based post-combustion capture of carbon dioxide**
Edited by Paul Feron
102 **Advances in solar heating and cooling**
Edited by R.Z. Wang and T.S. Ge
103 **Handbook of generation IV nuclear power reactors**
Edited by Igor Pioro
104 **Materials for ultra-supercritical and advanced ultra-supercritical power plants**
Edited by Augusto Di Gianfrancesco
105 **The performance of photovoltaic systems: Modelling, measurement and assessment**
Edited by Nicola Pearsall
106 **Structural materials for generation IV nuclear reactors**
Edited by Pascal Yvon
107 **Organic rankine cycle (ORC) power systems: Technologies and applications**
Edited by Ennio Macchi and Marco Astolfi

Part One

Introduction

Introduction to solar heating and cooling systems

1

R.Z. Wang, Z.Y. Xu, T.S. Ge
Institute of Refrigeration and Cryogenics, Shanghai Jiao Tong University, Shanghai, China

1.1 Background

Energy and environment are two vital issues for modern society. Fossil fuels including coal, oil, and natural gas are nonrenewable and cannot provide sufficient energy sources for eternal time. In addition, utilization of these traditional energy resources has caused severe environmental problems, including global warming, air pollution, and so on. Global warming is mainly caused by carbon dioxide (CO_2) emissions, which raises the global average temperature and sea level. To solve these problems, several negotiations and conferences have been held, such as the United Nations Framework Convention on Climate Change negotiated in 1992, in which many countries participated. Conferences of the Parties have been held many times in Kyoto, Bali, Copenhagen, and Paris, in which greenhouse gas emission reduction was proposed as an important task in the world. It can be seen that to build a sustainable and green future, both energy resources and the energy-consuming systems should be reconsidered under the modern energy background.

For the energy resources, renewable energy resources including solar energy, wind power, and hydropower are among the best choices. Compared with traditional energy resources, renewable energy resources are abundant and environmentally friendly. Among the different renewable energy resources, solar energy is one of the most attractive options. It is a clean and endless power with wide distribution. In this case, there are numerous researches and businesses about solar energy and solar driven systems.

For energy-consuming systems, the heating and cooling systems take a big proportion of the entire society energy consumption. It could be as high as 30% of the total energy consumption for those developed countries. If China is taken for an example, then the energy consumption for buildings (heating, cooling, hot water supply, lighting, etc.) is greater than 10% of the total energy used. Green and energy-saving heating and cooling systems should be developed.

Considering the merits of renewable energy and high energy consumption of heating/cooling systems, the adoption of a solar energy-driven system to fulfill the heating and cooling demand is a promising solution for the aforementioned problems. Researchers all over the world have conducted innovative studies in this area. To provide a general guideline and roadmap of the solar heating and cooling systems, related technologies, including solar power, solar heating, solar cooling, solar thermal storage, and some advanced systems, will be introduced in this book.

Advances in Solar Heating and Cooling. http://dx.doi.org/10.1016/B978-0-08-100301-5.00001-1

1.2 Overview of solar heating and cooling systems

Solar energy is the primary light and heat resource of the Earth. It can provide eternal energy to maintain the atmosphere temperature and germinate plants. With technological developments, solar energy can be utilized more and more efficiently and economically.

In a solar heating and cooling system, solar energy has the potential to meet a large proportion of the heating and cooling needs of buildings and industry. There are also numerous technologies for different heat source temperatures and specific demands. To ensure steady and long-term solar utilization, heat storage is also essential. In this chapter, an overview of the solar heating and cooling technologies, including solar energy, solar heating, solar cooling, and heat storage, will be given.

1.2.1 Solar energy

Solar energy is the energy source of solar heating and cooling systems. There are mainly two modern ways to collect solar energy. One is to directly adopt the thermal energy produced by solar radiation with use of a solar collector. The solar heat gained could be then transferred to solar heating or cooling applications; this kind of system is also called a solar thermal system. The other one is to transfer solar radiation into electrical power through photovoltaic (PV) material; this kind of system is also called the solar PV system.

When solar energy is integrated with the heating and cooling systems, there are many more options for thermal-driven systems than for electrical-driven systems. In this case, the solar thermal collectors are emphasized and thermal-driven systems have been extensively researched and developed. Because of the significant price reduction of solar photovoltaics in the last 5 years, solar PV-powered systems are also becoming attractive.

There are different classifications of the solar collector. It can be classified into nonconcentrating types and concentrating types. It can also be classified into low-temperature collectors, medium-temperature collectors, and high-temperature collectors according to the working temperature. Low-, medium-, and high-temperature collectors work under 100°C, 100–200°C, and higher than 200°C, respectively. In this chapter, solar collectors are classified into nontracking solar collectors and tracking solar collectors. A brief introduction of solar PV technology is also given.

1.2.1.1 Nontracking solar collectors

This type of solar collector mainly includes the flat-plate collector (FPC), the evacuated-tube collector (ETC), and the compound parabolic concentrator (CPC). They usually work as low- and medium-temperature collectors that are suitable for space-heating and space-cooling. Water, air, or oil can be used as a thermal transport medium.

FPCs: The FPCs usually contain the glazing, absorber plate, heat transfer component, and insulation layer. FPCs are typically used for space-heating or hot water supply. It has low working temperature, but it is simple, cost-effective, and has a long lifetime. It is also easily integrated in buildings.

ETCs: When the climate is not so warm or the working temperature is high, the FPC cannot work efficiently because of heat losses, and the ETCs can be used. In the ETC, the absorber surface with selective coating (absorptivity 95%, emissivity <5%) is placed in a double-layer tube with vacuum between two layers. The vacuum surrounding the absorber can greatly reduce the convection and conduction heat losses. In this case, the efficiency can be increased.

CPCs: To increase the solar collector efficiency, concentrating collectors such as CPCs can be used. The CPC is a nonimaging concentrator with a low concentration ratio. The CPC uses a compound parabolic reflective surface to reflect and concentrate the solar radiation to the focal line. A tubular absorber is used as a receiver. In some newly developed CPC collectors, a compound parabolic surface and receiver are integrated in the evacuated tube to avoid heat losses and increase the efficiency.

1.2.1.2 Tracking solar collectors

This type of solar collector mainly includes the single-axis tracking collectors and two-axes tracking collectors. Single-axis tracking collectors include linear parabolic trough collectors (PTCs), linear Fresnel reflectors (LFRs), and cylindrical trough collector (CTCs). They have a two-dimensional concentrating effect. Two-axes tracking collectors include the parabolic dish collector and solar tower (heliostat field) collector. They have a three-dimensional concentrating effect. The tracking collectors usually work as medium- and high-temperature collectors. Water, oil, or molten salt can be used as working fluid.

PTCs: The PTC uses a parabolic trough reflector to concentrate the solar radiation. The tubular receiver integrated in the evacuated tube is placed along the focal line of the reflector. The collector needs to track the Sun along a single axis to maximize its efficiency. A higher concentration ratio than that of the CPC can be obtained. PTCs can effectively produce heat at temperatures between 50°C and 400°C. It can be used for solar thermal power generation, solar thermal energy for industry uses, and as the heat source for efficient solar cooling.

LFRs: The LFR uses several arrays of flat mirrors to reflect and concentrate the solar radiation together. Compared with PTCs, the LFR is cheaper and takes up less space. The mirror arrays are usually placed on the ground. This makes the installation easier than PTCs, especially in a large system. However, shading and blocking problems can possibly reduce its efficiency. Compact LFR technology can improve this now that it is well accepted for industry heating and solar cooling.

Parabolic dish: The parabolic dish utilizes the reflective dish to concentrate the solar radiation to one point. In this case the concentration ratio of a parabolic dish is higher than the PTC and LFR. Higher efficiency or higher working temperature can be obtained. The absorber of a parabolic dish collector is placed at the focal point. As three-dimensional concentrating is adopted, two-axes tracking is needed. Parabolic dishes have been used with power stirling engines to generate electricity.

Table 1.1 **Solar thermal collectors [1]**

Collector	Motion	Absorber type	Concentration ratio	Indicative temperature (°C)
Flat plate	Stationary	Flat	1	30–80
Evacuated tube	Stationary	Flat	1	50–200
CPC	Stationary	Tubular	1–5	60–240
PTC	Single-axis tracking	Tubular	15–45	60–300
LFR	Single-axis tracking	Tubular	10–40	60–250
Parabolic dish	Two-axes tracking	Point	100–1000	100–500
Solar tower	Two-axes tracking	Point	100–1500	150–2000

CPC, Compound parabolic concentrator; *PTC*, parabolic trough collector; *LFR*, linear Fresnel reflector.

Solar tower: The solar tower utilizes the heliostats to concentrate the solar radiation to the receiver on a tower. The heliostats are tracking mirrors spread around the tower. In this case the solar tower is also called the heliostat field or central receiver collector. Because the heliostats are individual components installed on the ground, the total reflective area and the concentration ratio can be large, which increases the system power and working efficiency. Solar tower systems have been considered as an efficient system to generate electricity from solar thermal power.

The concentrating types, tracking modes, working temperatures, and efficiencies of the mentioned collectors are given in Table 1.1. The efficiencies of solar thermal collectors are closely related to the working temperature and ambient temperature. In this case the efficiencies are not included.

1.2.1.3 Solar photovoltaics

When solar photovoltaics are used for a heating and cooling system, a conventional vapor compression system can be adopted. In a solar PV system the solar radiation can be converted into direct current electricity through the PV effect of the semiconducting materials. Solar cells could be classified as silicon cells, thin film cells, emerging solar cells, and multijunction solar cells, among which silicon and film solar cells are available on the market.

Silicon cells: Silicon-based material is the most maturely developed and commercialized PV material. It is also called “first-generation” technology. Silicon-based materials account for the biggest market share for PV products. Multicrystalline silicon and monocrystalline silicon are the most commonly used materials on the market.

Thin film cells: A thin film cell is made by depositing one or more thin layers of thin film PV material on a substrate. Its thickness varies from nanometers to tens of micrometers, which is easy for building integration. It is also called "second-generation" technology. Commercialized thin film solar cells typically use cadmium telluride, copper indium gallium selenide, and amorphous thin film silicon (a-Si). In 2014 thin film cells accounted for approximately 9% of worldwide deployment whereas the remainder comprised crystalline silicon cells [2].

Emerging solar cells: The emerging solar cells can also be called the "third-generation" solar cells. These solar cells have the potential to overcome the Shockley–Queisser limit for single bandgap solar cells [3]. They include the dye-sensitized cells and organic cells. Other available technologies include the copper zinc tin sulfide cell, perovskite cell, polymer cell, and quantum dot cell.

Multijunction cells: Traditional cells have only one p–n junction, and there is a theoretical efficiency limit. Multijunction solar cells have multiple p–n junctions made of different semiconductor materials. A theoretical efficiency up to 86.8% can be reached by infinite p–n junctions [4]. The multijunction cells vary from the junction number and material. These include the InGaP/GaAs/InGaAs cell, amorphous silicon/hydrogen alloy (a-Si)/nanocrystalline or microcrystalline silicon (nc-Si)/nc-Si thin film cell, a-Si/nc-Si thin film cell, and so on.

1.2.2 Solar heating technologies

The term *solar heating* means utilizing solar energy to fulfill space-heating and water-heating demands. The solar heating technologies are usually classified into passive and active technologies considering the use of active mechanical and electrical devices. In addition, there are also differences between space and passive water-heating systems.

1.2.2.1 Passive solar space-heating

In the passive solar space-heating system, the façade or roof are used to absorb and store the solar radiation. The stored solar energy will be transferred to heat and fulfill the space-heating demand when it is necessary. No other active mechanical and electrical devices are needed. The key point of passive solar space-heating is the building design. Available technologies include double window, Trombe wall, solar chimney, unglazed transpired solar façade, and solar roof technologies [5]. Passive solar heating can be a complementation of active solar heating.

1.2.2.2 Passive solar water-heating

In the passive solar water-heating system, solar collectors are used to heat the water. Technologies including FPCs, ETCs, integrated collector storage allied to a CPC, and the photovoltaic/thermal (PVT) system can be used. The basic elements of the system include the collector, piping, and hot water tank. The heat transfer from collector to storage tank occurs through the natural convection principle. An electrical pump is not needed.

1.2.2.3 Active solar space- and water-heating

In the active solar space- and water-heating systems, the solar collectors transfer the heat to the heating system through pumps or fans. Nontracking solar collectors are enough for these demands. Sometimes the space- and water-heating functions are integrated in one system. The heating systems can use the solar heat directly or through heat exchange processes. Water or air is used as a transport medium.

1.2.2.4 Other feasible systems

When a medium-temperature solar collector is used, a thermal-driven heat pump can be used for heating. The thermal-driven heat pump cycle usually refers to the sorption heat pump cycle. The sorption heat pump cycle contains sorption, desorption, condensation, throttling, and evaporation processes. The desorption process needs heat input whereas the sorption and condensation processes can output heat. When solar photovoltaics are used, the traditional electrical space- and water-heating technologies are all available. These include electrical heating and vapor compression heat pump systems. The condensation process releases heat output. However, these two systems are seldom seen because the low-temperature solar collector is simple, cheap, and enough for space- or water-heating.

1.2.3 Solar cooling technologies

Cooling demands mainly include refrigeration and dehumidification demands. According to the driving power and demand, solar cooling technologies can be classified into the following kinds.

1.2.3.1 Solar photovoltaic-driven refrigeration and dehumidification

Vapor compression cooling systems can be used for refrigeration and dehumidification. The refrigeration cycle includes the compression, condensation, throttling, and evaporation processes. Electrical power is transferred into mechanical power for vapor compression and then drives the cycle. The evaporation process could then output cooling.

The vapor compression air conditioner is now the most widely used refrigeration device in industrial and residential applications. The working fluids include R-134a, R-410a, R-22, R-32, R-407C, and many other organic and inorganic fluids. The cooling coefficient of performance (COP) for air conditioning under normal conditions is approximately 3.0–5.0. For refrigeration and dehumidification application, the difference lies in the evaporation temperature. The dehumidification application requires a lower evaporation temperature to cool the air down to its dew point.

Except for the solar PV system, the solar thermal power generation system can also work with a vapor compression cooling system. Such a system could be a combination of Kalina cycle and Rankine cycle. However, the solar thermal power generation is not the topic of this book and it will not be introduced here.

1.2.3.2 Solar thermal-driven refrigeration

Thermal-driven cooling technologies are different for refrigeration and dehumidification applications because of the use of a sorption working pair. For refrigeration application, the closed sorption cooling cycle can be used. The term *closed* means that the sorption working pair is isolated from the ambient. The cycle also contains the same processes with a sorption heat pump cycle, but the evaporation is the output process. The sorption cycle is built based on the sorption process of refrigerant by the binary working pair. The absorption of vapor by solution and the adsorption of vapor by solid all belong to sorption. Except for the sorption-desorption-condensation-throttling-evaporation loop for refrigerant, there is another loop of sorption-pressurizing-desorption-depressurizing for the binary working pair.

For the absorption cooling system, the common working pairs are water—lithium bromide (LiBr) and ammonia—water. For the adsorption cooling system, the common working pairs are water—zeolite, water—silica gel, ammonia-calcium chloride ($CaCl_2$), and so on. The efficiencies of the simplest single-stage sorption systems are approximately 0.5—0.8 depending on the working pair and working conditions. The most popular candidate for solar cooling is the single-effect water—LiBr absorption chiller with a COP of approximately 0.7 under a driving temperature, ambient temperature, and evaporation temperature of 90°C, 30°C, and 5°C, respectively. A higher COP can be reached with a double-effect cycle, which also requires higher driving temperature such as 140°C.

1.2.3.3 Solar thermal-driven dehumidification

To fulfill the dehumidification demand, the sorption of water vapor by the binary working pair can also be utilized. In the sorption dehumidification system, the working pair has to contact the ambient and the open sorption cycles can be used. The open sorption system is also called the desiccant cooling system. The working pair has to be related with water here. The open sorption cycle contains the sorption and desorption processes. The sorption process is used for dehumidification whereas the desorption process is used for regeneration of a sorption working pair. The desorption process needs heat input. Compared with the dehumidification completed by vapor compression cooling, the sorption desiccant dehumidification system does not need to cool the air down to dew point temperature, which is thus more energy-saving, but regeneration heat would be needed for desiccant dehumidification.

There are mainly two desiccant cooling systems, including liquid desiccant cooling and solid desiccant cooling. In a liquid desiccant cooling system, the working fluid flows between the absorber and the regenerator. In a solid desiccant cooling system, the construction is different because of the nonfluid working medium. A rotary wheel system can be adopted to ensure a continuous operation. Low-temperature solar heat can drive a desiccant cooling system.

1.2.4 Heat storage technologies

The solar power is not steady and available all day long. It varies with time, weather, and season. The instability and intermittency of solar power make high efficiency and long-term solar utilization difficult. Solar thermal storage is one of the solutions for this. There are now mainly four kinds of solar thermal storage technologies: sensible heat storage, latent heat storage, sorption heat storage, and thermochemical heat storage.

1.2.4.1 Sensible heat storage

Sensible heat storage is the simplest heat storage system. It stores the energy in sensible heat, which can be reflected by the temperature. The fluid storage media include water and oil. The solid storage media include the building fabric, metal, and rock. Take water as an example: it has heat capacity of approximately 4.2 kJ/(kgK) and a density of approximately 1000 kg/m^3, which result in an energy density of approximately 11.7 kWh/m^3 for a 10°C temperature change.

1.2.4.2 Latent heat storage

Latent heat storage stores the heat in the phase change material (PCM). Compared with sensible heat storage, its energy storage density is much higher. The research about PCM is popular because of this. The phase changing temperature is steady when the system is built. Different PCMs are needed for different energy storage temperatures.

The available PCMs include organic PCMs, inorganic PCMs, and eutectic PCMs. One of the most important groups of organic PCMs is paraffin wax. Take paraffin (*n*-docosane) with a melting temperature of 42–44°C as an example: it has a latent heat of 194.6 kJ/kg and a density of 785 kg/m^3 [6]. The energy density is 42.4 kWh/m^3. Nonparaffin organic PCMs include the fatty acids and glycols. Inorganic PCMs include salt hydrates, salts, metals, and alloys. Eutectic PCMs are a minimum-melting mixture of several different PCMs [7].

1.2.4.3 Sorption heat storage

The sorption heat storage utilizes the sorption process of the binary working pair to store the heat. The sorption heat contains both the latent heat and another part of heat released by the combination process. The stronger affinity of the working pair will result in higher specific sorption heat. Compared with PCMs, the sorption heat storage material usually has a higher energy density.

The sorption heat storage materials can be classified into absorption material, physical adsorption material, and chemical adsorption material. The absorption materials include water–LiBr, ammonia–water, water–LiCl, and water–sodium hydroxide (NaOH). The storage density of water–LiCl is approximately 253 kWh/m^3 and the storage density of water–LiBr is approximately 180–310 kWh/m^3 [8]. The physical adsorption materials include water–zeolite and water–silica gel. The heat storage density of water–zeolite can reach 124 kWh/m^3 [8]. The chemical adsorption materials, which also belong to the thermochemical heat storage, include ammonia–$CaCl_2$ and ammonia–$BaCl_2$. In addition, the novel three-phase heat storage system integrates the absorption and adsorption processes for energy storage. Water–LiCl is one of the potential materials for three-phase heat storage [9].

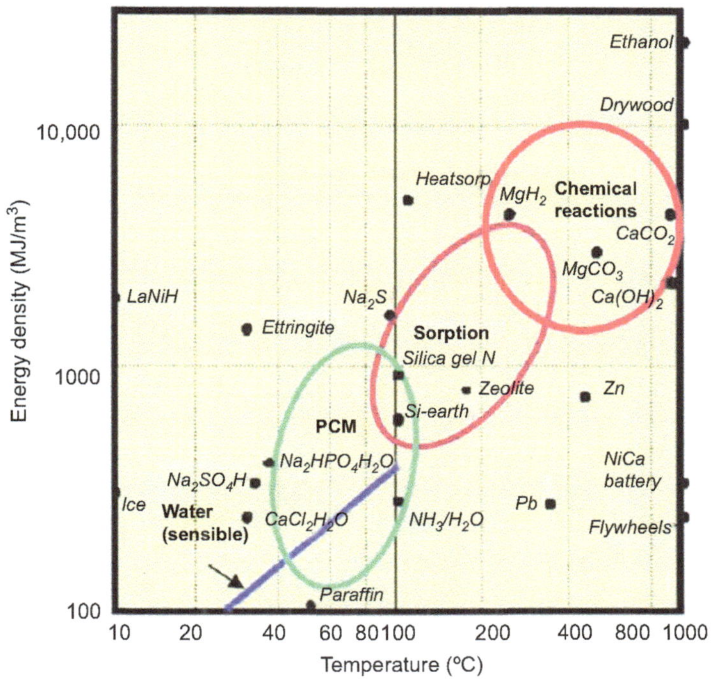

Figure 1.1 Energy density of different heat storage material [8].

1.2.4.4 *Thermochemical heat storage*

The thermochemical reaction usually has more heat release than the phase change and sorption processes. The reversible thermochemical reaction can be utilized for heat storage with high energy density. Except for the coordination reaction of ammonia mentioned in sorption heat storage, another potential reaction is the hydration reaction of salt hydrate. The materials include magnesium chloride ($MgCl_2$)/water, magnesium sulfate ($MgSO_4$)/water, and sodium sulfide (Na_2S)/water [10]. Energy densities of $MgSO_4$/water and Na_2S/water can both reach 780 kWh/m^3 [8]. Other thermochemical heat storage materials include silicon oxide, iron carbonate, iron hydroxide, and calcium sulfate [7].

To better illustrate the energy densities and working temperatures of these heat storage materials, a cited diagram about the heat storage materials is shown in Fig. 1.1.

1.3 Technology roadmap

The former sections have introduced the available technologies for solar power collection and solar-driven heating, cooling, and heat storage. To make the couplings between different technologies clearer, the contents in this chapter are summarized and a technology roadmap is given in Fig. 1.2. The energy conversion and technology

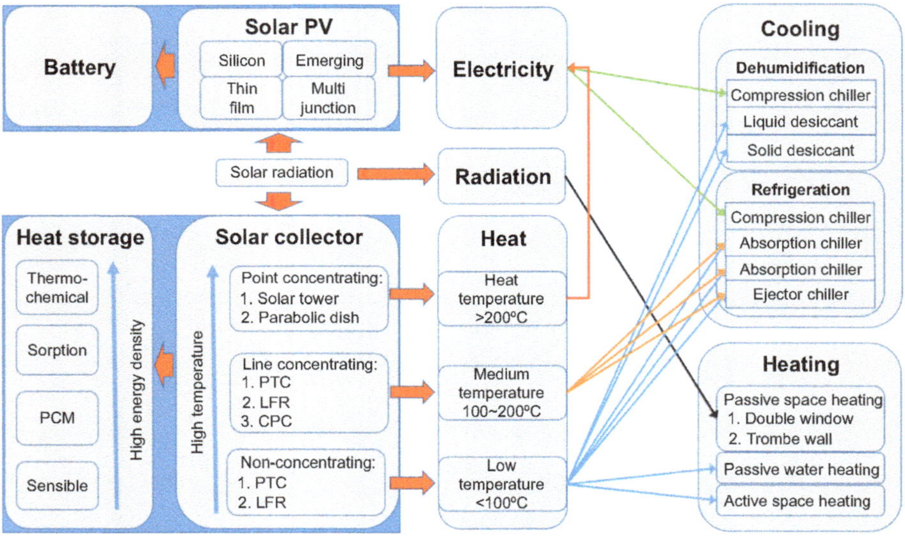

Figure 1.2 Solar heating and cooling roadmap.

combinations are shown in this figure. In the following chapters of this book, the mentioned solar heating and cooling technologies will be introduced in detail. The working principle, application, and some advanced researches will be included.

References

[1] Kalogirou SA. Solar thermal collectors and applications. Prog Energy Combust Sci 2004; 30(3):231–95.
[2] Fraunhofer ISE. Photovoltaics report. 2015.
[3] Shockley W, Queisser HJ. Detailed balance limit of efficiency of p-n junction solar cells. J Appl Phys 1961;32(3):510–9.
[4] Dimroth F, Kurtz S. High-efficiency multijunction solar cells. MRS Bull 2007;32(03): 230–5.
[5] Chan HY, Riffat SB, Zhu J. Review of passive solar heating and cooling technologies. Renewable Sustainable Energy Rev 2010;14(2):781–9.
[6] Sarı A, Karaipekli A. Thermal conductivity and latent heat thermal energy storage characteristics of paraffin/expanded graphite composite as phase change material. Appl Therm Eng 2007;27(8):1271–7.
[7] Kousksou T, Bruel P, Jamil A, et al. Energy storage: applications and challenges. Sol Energy Mater Sol Cells 2014;120:59–80.
[8] N'Tsoukpoe KE, Liu H, Le Pierrès N, et al. A review on long-term sorption solar energy storage. Renewable Sustainable Energy Rev 2009;13(9):2385–96.
[9] Yu N, Wang RZ, Lu ZS, et al. Evaluation of a three-phase sorption cycle for thermal energy storage. Energy 2014;67:468–78.
[10] Yu N, Wang RZ, Wang LW. Sorption thermal storage for solar energy. Prog Energy Combust Sci 2013;39(5):489–514.

Resource assessment and site selection for solar heating and cooling systems

2

D.S. Renné
Dave Renné Renewables, LLC, Boulder, CO, United States

2.1 Introduction

The solar resource available to solar conversion technologies is highly variable, both in time and in location. This chapter focuses on how we can obtain the best possible knowledge about this variability so that solar systems can be deployed in the optimal sites and in the most efficient configurations. Notwithstanding the nocturnal and seasonal planetary cycles that affect the solar resource reaching the top of the atmosphere, weather patterns, landform characteristics, such as coastlines, and topography all play key roles in affecting the short-term variability of the resource at the earth's surface. Although knowledge of spatial variability is important for identifying preferred sites for solar systems, it is more common that a solar system will need to be sited close to a load center for optimal system efficiency. In these cases, the knowledge of spatial variability will help determine the actual resource at a specific site, even if no measurements are available for that site, and even if the site is not at the location of the highest resource.

For certain types of solar heating systems, especially those that have some form of storage capacity, the short-term weather-related variability may have only a minor impact on system design and performance. However, for large-scale solar systems, especially those that have an impact on the load profile that must be met by the electrical grid, resource variability may have significant importance, if not for the system itself, then for the utility that must serve the net load resulting from both resource variability and electrical demand.

A clear understanding of resource characteristics is important for designing the most appropriate system for a given environment and the load requirements. Because actual ground-based measurements are often lacking at most proposed sites for solar heating and cooling systems, various tools and techniques need to be brought into play to best characterize the resource so that investment risk in the project is minimized.

Fig. 2.1 outlines the basic stages that take place in a large-scale solar project and indicates what type of solar resource information is most appropriate for each stage. In this chapter, we address the basic solar principles, including definitions, measured and modeled solar radiation data, and adaptation of solar data sets that can be applied to various stages of a project development.

Advances in Solar Heating and Cooling. http://dx.doi.org/10.1016/B978-0-08-100301-5.00002-3

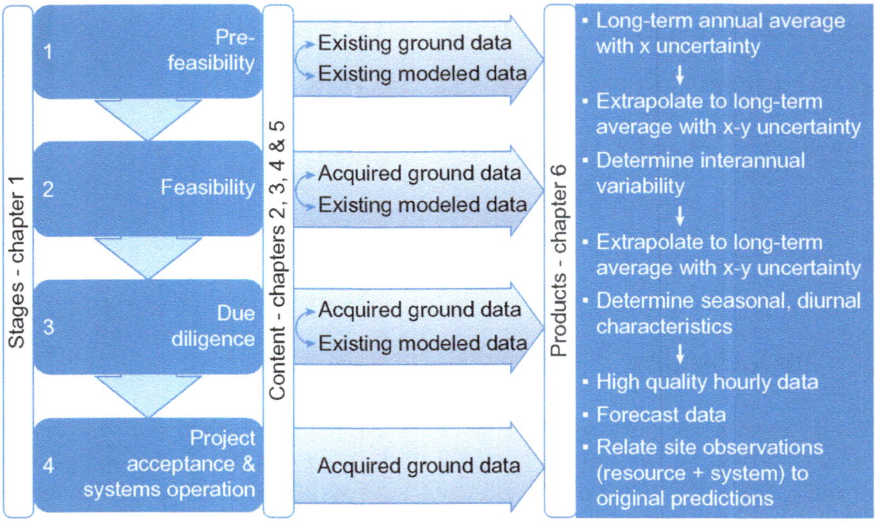

Figure 2.1 Four basic stages in a solar project development cycle, indicating basic resource data required for each cycle.
From Sengupta, M., Habte, A., Kurtz, S., Dobos, A., Wilbert, S., Lorenz, E., Stoffel, T., Renné, D., Myers, D., Wilcox, S., Blanc, P., Perez, R., February 2015. Best Practices Handbook for the Collection and Use of Solar Resource Data for Solar Energy Applications. Technical Report NREL/TP-5D00-63112. National Renewable Energy Laboratory, Golden, Colorado (USA).

The International Energy Agency's Solar Heating and Cooling Implementing Agreement has supported several solar resource assessment tasks since the program's beginning in 1977. Most recently, Task 36, "Solar Resource Knowledge Management," and Task 46, "Solar Resource Assessment and Forecasting," have brought together nearly 70 solar resource assessment experts from around the world to conduct benchmarking studies of solar resource data (measured and derived) and develop best practices in the development and use of solar resource data. Some of the information reported in this chapter has been developed through this international collaboration; further results can be found at http://task46.iea-shc.org.

2.2 Definition of solar resources

The solar radiation (or irradiance) at the top of the earth's atmosphere is called *extra-terrestrial radiation* and the current accepted value for this "solar constant" (now referred to as the *total sky irradiance*) is 1366 W/m^2. The energy emitted by the sun is virtually invariant, estimated to be about 0.1% owing primarily to variations associated with sunspot cycles. Because of the eccentricity of the earth's orbit, however, the solar constant varies by about 6.7% throughout the course of a year (Vignola et al., 2012a).

The following terms related to solar resources are often found in the literature, and therefore a definition of each term is provided here:

Solar radiation: This term is often used interchangeably with *solar energy* and *irradiation* and, less commonly, *insolation*.

Radiant energy, or **radiance**: For the purposes of this chapter, this is the amount of energy emanating from the sun, and is generally expressed in units of W/m^2 sr. The sun radiates almost as a black body, at an effective temperature of 5778 K. Radiant energy (sometimes referred to as *radiant intensity*) decreases at the rate of the inverse square of the distance from the source.

Irradiance: This is the power density of radiation incident on a surface, or the rate at which radiant energy is incident on a surface, generally expressed in units of W/m^2.

Irradiation (or **insolation**): This is the quantity of solar energy (radiation) arriving at a surface during a specified period of time, generally expressed in units of W/m^2 h or W/m^2 year, or occasionally in MJ/m^2. Thus solar energy is synonymous with solar irradiation and also represents the *solar resource* available at a location over a specified time period.

The sun emits radiation across a broad spectrum of wavelengths, as shown in Fig. 2.2, in which the dashed line shows the spectral distribution of radiant energy reaching the top of the earth's atmosphere. The portion of the solar spectrum that is visible to the human eye is generally in the range of 380–780 nm, which is also the range in which the radiance emitted by the sun is at a maximum (the absolute

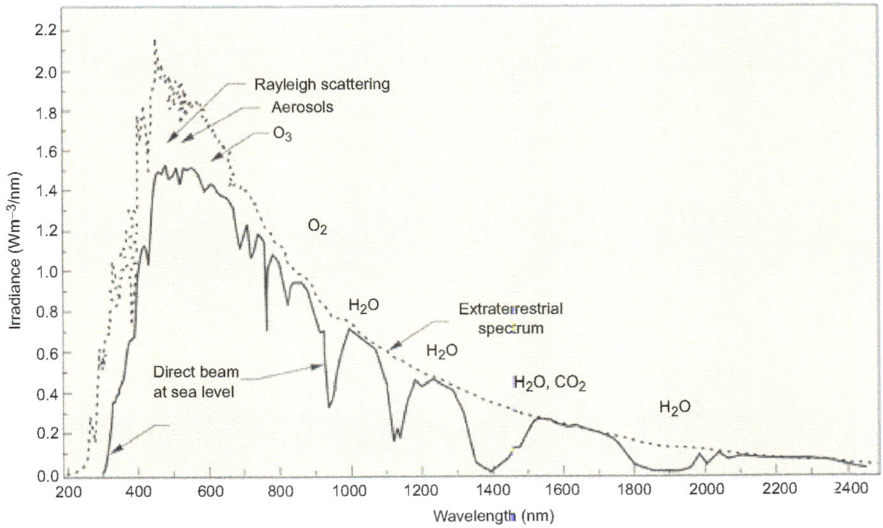

Figure 2.2 The spectral distribution of solar radiation reaching the top of the earth's atmosphere *(dashed line)* and at sea level *(solid line)*.
Image from StellarNet, Inc., http://www.stellarnet.us/popularconfigurations_radiosystems_solar.htm. See also Wehrli C. Extraterrestrial solar spectrum. Publication no. 615, Davos Dorf, Switzerland: Physikalisch-Meteorologisches Observatorium Davos, World Radiation Center, (PMOD/WRC); 1985 for further information on the solar spectrum at the top of the atmosphere.

maximum is at around 500 nm). This portion of the spectrum is called the *visible spectrum*. Radiation in wavelengths shorter than 380 nm is known as *ultraviolet* radiation, and radiation occurring in wavelengths longer than around 780 nm is called *infrared* radiation. The majority of solar radiation reaching the top of the earth's atmosphere falls in the spectral range of 300–3000 nm.

A variety of the molecular components of the atmosphere serve to absorb and scatter the incoming solar radiation at a number of discrete wavelengths, so that the spectral distribution of the radiant flux reaching the earth's surface can look quite different from the distribution at the top of the atmosphere, as shown in the solid line in Fig. 2.2. For example, Rayleigh scattering, aerosols, and ozone have a strong effect on reducing solar irradiance in the visible wavelengths at the earth's surface. Atmospheric dust and water droplets also serve to scatter radiation back to outer space. The reason the sky appears blue to the human eye is due to the preferential scattering of air molecules in the blue wavelength region of the visible spectrum (roughly 400–500 nm). Atmospheric water vapor absorbs solar energy in a number of wavelengths, especially in the infrared regions. Other trace gases in the atmosphere, such as CO_2 and methane, can also absorb solar radiation in a variety of spectral ranges (most notably in the infrared region), causing the temperature of the atmosphere to increase. The injection of additional CO_2 and methane into the atmosphere due to human activities is the well-known greenhouse gas effect, which can cause additional warming of the atmosphere due to these anthropogenic sources.

The definitions of the various *components* of the solar resource are also important. These are described in Fig. 2.3. The definitions apply primarily to the amount of radiation reaching the earth's surface in the visible spectrum.

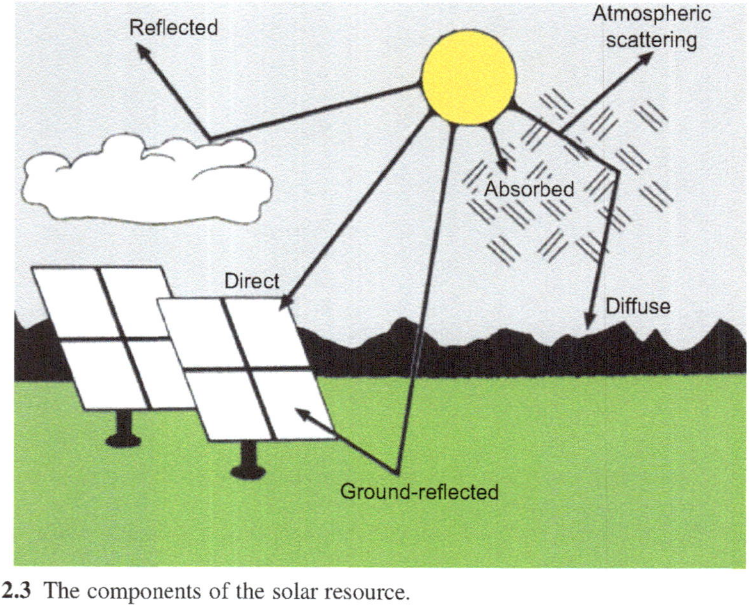

Figure 2.3 The components of the solar resource.
Image provided by NREL, courtesy of Tom Stoffel.

The amount of solar radiation from the sun reaching a flat surface at the earth's surface oriented normally to the sun's position throughout the day is known as *direct normal irradiance*, or DNI. As noted above, portions of the visible radiation reaching the top of the atmosphere can be scattered back to space or reflected toward the earth's surface by the presence of cloud droplets and aerosols. In addition, air molecules can absorb some of the radiation in the visible channels. Thus the total radiation reaching the earth's surface will be some combination of DNI and the radiation scattered back to the surface by cloud droplets and aerosols. The portion that is reflected by clouds and atmospheric aerosols is known as diffuse radiation. Measuring this component on a flat surface oriented horizontally gives the *diffuse horizontal irradiance*, or DHI. If we were to measure the total radiation falling on a flat surface horizontal to the earth's surface, we would be measuring both the solar radiation coming directly from the sun (assuming it is not obscured by a cloud) and the diffuse radiation from the clouds and sky. This, then, is the *global horizontal irradiance*, or GHI.

Thus GHI represents the total energy available from the sun and the sky on a horizontal surface and is usually the most important factor for understanding the resource available to flat plate collectors. DNI, on the other hand, is the most important component of solar radiation for understanding the energy that can be produced by concentrating collectors, such as curved mirrors used in trough technologies. Mathematically, the relationship between GHI and the other two components, DHI and DNI, is given as

$$\mathrm{GHI} = \mathrm{DNI} \times \mathrm{Cos}(Z) + \mathrm{DHI} \qquad [2.1]$$

where Z is the *solar zenith angle*, or the angle between the sun's position at any given time and the zenith, or a vertical line perpendicular to the earth's surface.

In most practical applications, flat plate collectors are generally oriented more directly toward the sun rather than on a horizontal plane. This is because a collector oriented more toward the sun will receive a higher global irradiance flux than flat surfaces for virtually all situations except for when the sun is directly overhead (at the zenith). Often, fixed flat plate collectors are oriented at an angle toward the south (in the Northern Hemisphere) equal to the latitude where the collector is installed. This is known as *latitude tilt*. In some cases flat plate collectors might even be mounted on one-axis or two-axis trackers to allow the collector to be oriented normal to the sun throughout the day and throughout the changing seasons. Even concentrating collectors will not always be oriented directly to the solar disk at all times, unless they are designed to track the sun throughout the day and throughout the seasons (two-axis tracking); in cases in which the concentrating collectors are not oriented directly toward the solar disk the direct irradiance flux will also be less than optimum.

2.3 Relationship between solar resources and solar collectors

Calculating global or diffuse radiation on a horizontal surface is useful as a global standard for being able to compare one site against another, regardless of the latitude or the solar position with respect to the zenith; however, in virtually all circumstances the total radiation falling on a flat plate collector will be highest when the collector is oriented normal to the solar disk at any given time. Therefore, it is necessary to be able to convert the GHI or DNI values to the values received by a tilted surface to depict the solar resource available to the collector at any given time. The conversion of GHI to the irradiance received on a tilted surface is called *global tilt irradiance* (GTI). Fig. 2.4(a)–(c) provides schematic depictions of a fixed collector at latitude tilt and of one-axis and two-axis flat plate collectors.

As noted in Sengupta et al. (2015), a number of models are available to convert GHI or a combination of GHI and DNI into GTI, such as Temps and Coulson (1977), Hay (1979), Kluchar (1979), Liu and Jordan (1961), Reindl et al. (1990), and Gueymard and Myers (2008). For example, the following is a description of a simple method used to convert monthly and annual average GHI to GTI for one- or two-axis trackers as part of the products developed from the US National Solar Radiation Database (NSRDB), 1961–1990, as presented by Marion and Wilcox (1994).

The irradiance I_c reaching the surface of any of these types of collectors at any given time can be estimated from the following relationship:

$$I_c = I_b \mathrm{Cos}(\theta) + I_d + I_r \quad [2.2]$$

where I_b is the incident direct beam radiation, I_d is the diffuse sky radiation, I_r is the radiation reflected from the earth's surface in front of the collector, and θ is the incident angle of the sun's rays to the collector. Algorithms developed by Menicucci and Fernandez (1988) were used for the US 1961–1990 NSRDB. The direct beam contribution, $I_b\mathrm{Cos}(\theta)$, was determined for each hour based on the DNI values reported in the NSRDB. The I_d was calculated from an anisotropic diffuse radiation model developed by Perez et al. (1990), based on DHI and DNI values from the NSRDB. The

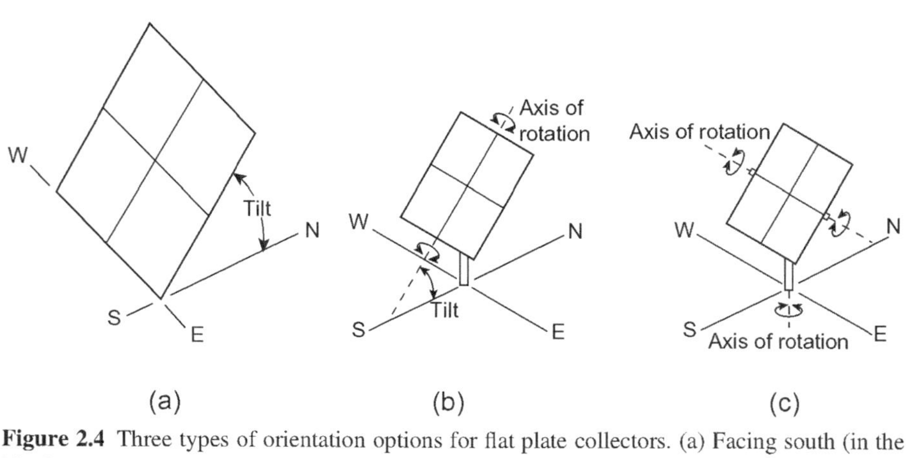

Figure 2.4 Three types of orientation options for flat plate collectors. (a) Facing south (in the Northern Hemisphere) at a fixed tilt; (b) one-axis tracking with axis of rotation oriented north–south; (c) two-axis tracking.

ground-reflected radiation received by the collector is determined from the GHI values in the NSRDB (I_h) along with the tilt of the collector from the horizontal (β) and the surface reflectivity or albedo (ρ), which is the fraction of the reflected to the incident radiation:

$$I_r = 0.5\rho I_h(1 - \cos\beta) \qquad [2.3]$$

Values of surface albedo can generally be found in meteorological textbooks. Albedos for bright surfaces, such as surfaces covered by snow or light sand, are relatively high, and typically a value of 0.6 is chosen under these conditions. For surfaces covered with heavy vegetation an albedo of 0.25 is typically chosen.

For concentrating collectors there is very little (if any) contribution from the diffuse component or from ground reflection, so that the amount of radiation received by the collector, I_c, can be determined from the following:

$$I_c = \text{DNI} \times \text{Cos}(\theta) \qquad [2.4]$$

However, particularly under hazy conditions, there could be a component of solar radiation coming from an enhanced bright aureole around the sun, called the *circumsolar radiation*. Although the circumsolar component may be factored into the DNI values based on the measurement or modeling approaches employed, the physical design of the solar collector may not be able to take advantage of this component. A detailed discussion of the extent to which the circumsolar component is captured in solar data, and the extent to which it affects a solar collector, can be found in Blanc et al. (2014).

2.4 Measuring and modeling the solar resource

For all solar thermal applications there are two primary means of obtaining solar resource data at a given location: (1) direct measurement of the solar resource using qualified instrumentation and accepted measurement practices and (2) modeling of the solar resource using "indirect" means, such as imagery of cloud characteristics observed from weather satellites, ground-based cloud cover observations, numerical weather prediction models, or sunshine duration records. This section provides an overview of best practices in measuring and modeling the solar resource for solar thermal applications.

2.4.1 Solar resource measurement techniques

Attempts to measure and quantify the amount of radiation arriving at the earth's surface from the sun date back to the early 1800s. An excellent historical perspective describing the evolution of solar measurement devices is provided in Vignola et al. (2012a). One of the earliest instruments, Pouillet's pyrheliometer (first built in 1837), was the first instrument to be called a *pyrheliometer* and was designed to provide a first estimate of the *total sky irradiance*. In the early 1900s, development started on instruments that measured the total radiation from the sun and sky (the GHI), now referred to as *pyranometers*.

In the late 1800s, the Campbell–Stokes radiometer became a very popular instrument for determining the number of hours of direct solar radiation on a daily basis. This instrument functions by focusing the direct solar beam through a glass ball onto a strip of paper; when the sun is not obscured by a cloud, the focused beam burns a line on a strip of paper calibrated in a way that relates the length of the burn line to the length of time of direct sunlight. At the end of each day the total length of the burn mark is translated into the number of direct sunlight hours for the day. Although this approach provides only a crude depiction of the solar resource, the simplicity and ease of use of the instrument made it a popular component of weather stations throughout the world. Even today many of these instruments are still operational, particularly at airport weather stations in developing countries. However, empirical relationships must be developed to convert sunshine records to solar resource assessments, and generally very high uncertainties in the conversion can result. Thus it is not recommended to use these instruments for solar resource assessment purposes.

Another approach, using *bimetallic* sensors (eg, the Robitzsch bimetallic actinograph), was developed in the early 1930s and also came into very popular use at national weather stations throughout the world by the mid-century. These instruments can still be found in wide use today.

Around the beginning of the 20th century, efforts to develop instruments that convert the incoming solar radiation to an electrical output were undertaken, and today this is the most common approach for developing precise measurements of the solar resource. Today, these types of instruments fall into two broad categories: those using *thermopile*-type detectors and those using *silicon-diode*-type detectors. A thermopile detector works on the principle of the *thermoelectric effect*, whereby a voltage is generated from the temperature difference between two dissimilar metals. Today's precision pyranometers and pyrheliometers all make use of this effect by deploying two different metals under a glass dome (or double-glass dome) and monitoring the voltage output variations due to the passage of the sun across the sky and the obscuring of the sun by clouds and haze.

Silicon photodiodes are made from crystalline silicon that has been transformed into a semiconductor, similar to the principles of a crystalline–silicon PV cell. Instruments using this technology have the advantage of much lower cost and much faster response time than a thermopile-based instrument; however, they have significant disadvantages by being limited in their spectral response characteristics (they do not respond at all to wavelengths above 1100 nm) and are therefore subject to a slightly higher uncertainty than instruments using thermopile detectors.

Examples of thermopile-based pyranometers and pyrheliometers currently on the market are shown in Fig. 2.5(a) and (b), and an example of a silicon photodiode pyranometer currently on the market is shown in Fig. 2.5(c).

The use of thermopile radiometers became common beginning in the mid-1920s owing to the innovative work of companies such as Kipp & Zonen, and later, Eppley Laboratories and EKO Instruments, which continue to develop and refine improved

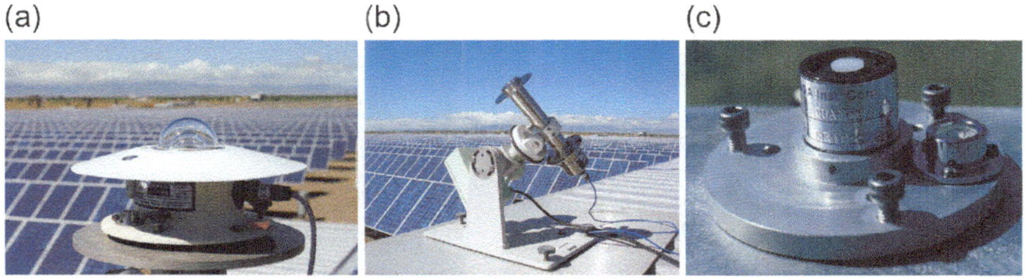

Figure 2.5 Examples of commercially available solar monitoring devices. (a) Eppley pyranometer, (b) Eppley pyrheliometer mounted on a tracker, and (c) LiCor silicon photodiode sensor.
Reproduced from NREL Image Gallery Nos. 15537, 15554, and 15483.

thermopile instruments for a variety of scientific purposes. Other companies such as Hukseflux and Yankee Environmental Systems have more recently entered thermopile-based products into the market. However, instruments based on silicon photodiode technology remain very common, because of their low cost, and can often be found in use at proposed and operational solar PV and solar thermal stations. In terms of accuracy and precision, the thermopile-type instruments and to a lesser extent the silicon photodiode–type instruments are currently the preferred choice for undertaking reliable solar resource measurement programs.

Solar resource measurement programs should be conducted in such a manner that the instrumentation used and the procedures followed in the measurement program follow globally accepted practices that ensure that the measurements can be clearly traced back to world reference standards. The globally accepted standard is the World Radiometric Reference (WRR), which is established every 5 years at the World Radiation Center in Davos, Switzerland, through an international pyrheliometer comparison (IPC). The IPC involves the use of a number of sophisticated thermopile-based precision instruments such as *absolute cavity radiometers*, operated by several global research institutions. The pyrheliometers used to establish the WRR can then be used in regional intercomparisons for purposes of identifying *reference radiometers*, which in turn can be used in the field as reference calibration sources for the instruments deployed at a given measurement site. In this way, the calibrations of the site-specific measurement can be traced back to the WRR through these intercomparison steps.

To obtain the most accurate complement of DNI, GHI, and DHI at a site, individual thermopile-based instruments must be used for each component; ie, a pyrheliometer on a tracking device for DNI and two pyranometers (one measuring GHI and one with the solar component blocked by a shading disk so that it measures only DHI). Because of the expense of such a configuration, there have been efforts under way to utilize a simpler, less costly but still reliable configuration, known as a *rotating shadowband radiometer*, or *RSR*.[1] The RSR basically consists of a silicon photodiode sensor

[1] In some cases, these instruments are referred to as Rotating Shadowband Pyranometers (RSPs) or Rotating Shadowband Irradiometers (RSIs).

Figure 2.6 Images of a rotating shadow-band radiometer (RSR). Left: Close-up of the shading arm passing over a silicon-diode sensor. Right: An RSR installed in the field.
Reproduced from NREL Image Gallery Nos. 16994 (right) and 15484 (left).

mounted on top of a box that contains a motor and a shading band (see Fig. 2.6). The motor is designed to cause the shadow band to periodically pass over the field of view of the sensor. Because of the fast response time of the sensor, as soon as the shading band completely shades the sensor, the sensor is measuring only the DHI. At all other times the sensor is measuring the GHI. Based on these two measurements, the DNI can be calculated by solving Eq. [2.5] for DNI:

$$\text{DNI} = (\text{GHI} - \text{DHI})/\text{CosZ} \qquad [2.5]$$

There has been extensive research undertaken to evaluate the uncertainties of thermopile-based pyrheliometers and pyranometers operated in the field, as well as the uncertainties in operational RSRs. For example, extensive comparative studies have been conducted at the National Renewable Energy Laboratory's (NREL's) Solar Radiation Research Laboratory in Golden, Colorado, and these studies have been published in Wilcox and Myers (2008), with an update published by Habte et al. (2014). A summary of uncertainty analyses for pyrheliometers and pyranometers is also provided in Sengupta et al. (2015). This report shows that pyrheliometers operated in the field following proper maintenance will have a subhourly uncertainty at the 95% confidence level of ±2%, whereas pyranometers will have an uncertainty of ±3% for solar zenith angles between 30° and 60°, and up to ±7−10% for solar zenith angles greater than 60°. The uncertainty of an RSR DNI measurement at the 95% confidence level is currently undergoing considerable research; in Vignola et al. (2012a) the uncertainty is reported at ±3.2%. Geuder et al. (2011), also reported in Sengupta et al. (2015), reported an uncertainty of about ±3% based on a study of 39 RSRs tested at Plataforma Solar Almería (PSA), Spain. Generally the uncertainties of an RSR are expected to be somewhat higher than with the thermopile instruments, although from the perspective of field operations these slightly higher uncertainties may be offset by their significantly lower costs. A 2015 study by Wilbert et al. (2014) provides a comprehensive overview of best practices for siting and operation of RSRs.

Other methods to develop a low-cost approach for measuring DNI are being tested and commercialized as of this writing. One such instrument is the SPN1, produced by Delta-T Devices (Fig. 2.7), which is capable of providing DNI, DHI, and GHI without the use of any moving parts. As described in Sengupta et al. (2015) this instrument consists of an array of seven thermopile detectors distributed under a glass dome in a hexagonal pattern. The sensors are located underneath a series of diffusers and a shadow mask, which is designed in such a way that throughout the day there will always be one or more detectors that are fully shaded from the sun and exposed to approximately half of the diffuse solar radiation (under overcast skies). In addition, one or more detectors are exposed to the full solar beam for all solar positions. The minimum and maximum readings of the seven detectors are then used to calculate GHI and DHI. Although at present the uncertainties of the outputs of this instrument are somewhat higher than those of the thermopile instruments or even the RSRs, based on testing at NREL and at PSA, work continues to improve the accuracy of this instrument, which can offer measurements of all three of the solar components at substantially reduced costs over the traditional suite of thermopile instruments.

2.4.2 Solar resource estimates using satellite data retrievals

In the early 1980s, when *geostationary weather satellites* (weather observation satellites that orbit at the same speed as the earth's rotation, so that they are always positioned 40,000 km above the equator at a fixed longitude) came into common use, efforts began to use the visible channel of the satellite to develop an estimate of the solar insolation at the earth's surface. Because of their fixed location above the earth's surface, these

Figure 2.7 The SPN1 pyranometer produced by Delta-T Devices.
Image provided by Delta-T, UK.

satellites have the advantage of being able to monitor cloud conditions almost continuously (every 15 min with modern-day geostationary satellites), at a visible channel resolution of about 1 km above the equator. However the "view" of these satellites is limited to the region between 60°N and 60°S. Furthermore, about five satellites spread around the world over the equator are necessary for obtaining a complete global view.

Weather services also make use of *polar-orbiting* satellites, which orbit the earth at low altitude (approx. 400 km) from pole to pole. Polar-orbiting satellites have the advantage of being able to observe the earth's clouds and surface at much higher resolution than geostationary satellites, owing to their much lower orbit, and of course, unlike geostationary satellites, they can cover the polar regions. However, because the earth is rotating under these satellites, a satellite generally passes over the same point on earth's surface just twice a day, making the time resolution of observing cloud patterns much coarser than with geostationary satellites.

Fig. 2.8 shows the current location of geostationary satellites around the world, operated by several countries. The figure also shows the typical coverage that is obtained from polar-orbiting satellites. The majority of the methods developed for solar resource assessment make use of geostationary satellite imagery, although the NASA Surface Meteorology and Solar Energy data set (https://eosweb.larc.nasa.gov/sse/) makes use of both geostationary and polar-orbiting satellites to obtain true global coverage.

The approaches developed to produce solar resource estimates from satellite observations can be categorized in three basic ways: *empirical methods*, whereby the strength of the visible signal at the satellite is correlated with various high-quality ground-based measurement stations; *semiempirical methods*, whereby the strength of the visible signal is compared with the ground albedo (as determined by monitoring

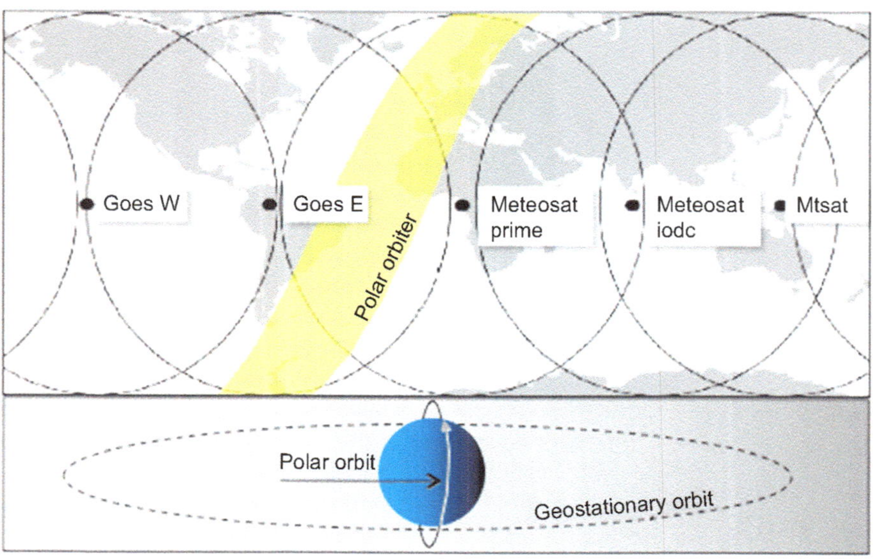

Figure 2.8 Current locations of modern-day geostationary satellites.
Image provided by Dr. Richard Perez, SUNY/Albany.

the signals at the satellite on clear days) to produce a "cloud index" from which a solar resource calculation can be made; and *physical methods*, whereby basic radiative transfer theory is applied to the satellite signals to develop a calculation of the solar irradiance at the earth's surface.

Today the semiempirical approach is the most common method used for solar resource assessments to support the solar industry, although data sets derived from both empirical methods and physical methods are also available.

One of the earliest approaches of the semiempirical method was the procedure developed by Cano et al. (1986). Many of the modern-day semiempirical approaches are based on Cano's original work. A version of this approach was developed a number of years ago by Dr. Richard Perez at the State University of New York, Albany (Perez et al., 2002) and is described below. This method has gone through a number of key improvements through the years, and currently the Perez model results are available as a SolarAnywhere product marketed by Clean Power Research. Other well-known products, such as the Heliosat Method, HelioClim, and the German Aerospace Institute's (DLR's) SOLEMI, or Solar Energy Mining product, as well as commercial products produced by companies such as GeoModel Solar and 3Tier, also have their roots from the Cano et al. approach. A summary of satellite-derived data sources can be found in Section 2.6.

The basic principles of the semiempirical approach, as reviewed by Renné et al. (1999) and illustrated in Fig. 2.9, depict the shortwave energy balance of the earth–atmosphere system.

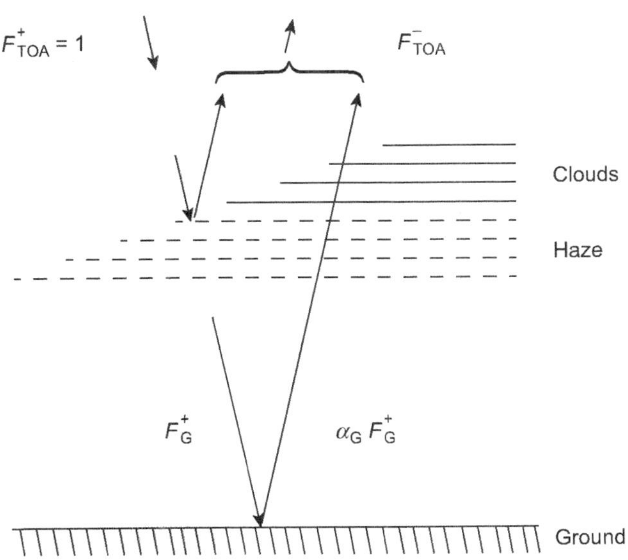

Figure 2.9 The shortwave energy balance of the earth–atmosphere system showing the relationship of the normalized incoming solar flux at the top of the atmosphere $\left(F^{+}_{TOA}\right)$, the reflected earth–atmospheric radiation measured by the satellite $\left(F^{-}_{TOA}\right)$, and the calculated ground irradiance $\left(F^{+}_{G}\right)$. α_G represents the albedo at the earth's surface.

In Fig. 2.9 we see that the net outgoing flux of shortwave radiation at the top of the atmosphere (TOA), or $F^+_{TOA} - F^-_{TOA}$ minus the amount absorbed by the atmosphere, F_A, is equal to the net flux at the surface, $F^+_G - F^-_G$ or $\alpha_G F^+_G$:

$$F^+_{TOA} - F^-_{TOA} = F^+_G(1 - \alpha_G) + F_A \qquad [2.6]$$

Equation [2.6] forms the basis of the empirical–physical as well as the physical method, in which the effort is to solve the equation for F^+_G. The main differences among the various methods are the way in which the cloud cover is characterized by the satellite's measurement of F^-_{TOA} and how the forward- and backscattering due to atmospheric particles and the absorption of solar radiation by the various trace gases in the atmosphere, such as ozone and water vapor, is addressed.

For example, under clear sky conditions an accurate determination of the radiative transfer through the atmosphere that accounts for the absorption as well as the forward- and backscattering of the shortwave flux due to haze and trace gases is required. A number of clear sky radiative transfer models have been developed since 1995 to address this fundamental issue. Gueymard (2012) provides a comprehensive review of 18 clear sky models and determines that among the most reliable is the REST2. Another model in common use is the simplified broadband version of the SOLIS clear sky model, first published by Ineichen (2008). An overview of many clear sky models can also be found in Sengupta et al. (2015).

When clouds are present the visible channel of the satellite gathering data on the earth's shortwave reflectance at the top of the atmosphere is used. Modern-day weather satellites measure this value through a relative scale based on digital counts of the reflected radiation; the higher the counts, the higher the reflectivity of the earth–atmosphere system and, therefore, the higher the cloud cover. For example the Perez et al. (2002) method has introduced the concept of the cloud index, or C_i, which is determined at the satellite level by monitoring the irradiance measured by the satellite for 15–30 days until a "clear sky" albedo at the top of the atmosphere is determined; this then represents a C_i of 0. The higher the value of the irradiance reflected from the earth–atmosphere system, the higher the C_i, ie, $C_i = f(F^-_{TOA})$. In the case of the empirical–physical models, such as that of Perez et al. (2002), the C_i is expressed as follows:

$$F^+_G / F^+_{G\text{-Clear}} = 1 - \alpha_{TOA} C_i \qquad [2.7]$$

Determination of C_i involves establishing a "dynamic range" of pixel counts received by the satellite for any given situation compared to the values determined at the lowest pixel count, which is assumed to be the clear sky value, $F_{G\text{-Clear}}$ (under which $C_i = 1$). These conditions represent the maximum dynamic range possible for determining the C_i; ie, the difference between the lowest (darkest) pixel counts observed by the satellite (lowest F^-_{TOA}, or clear sky conditions) and the highest pixel counts that occur for F^-_{TOA}, which generally occurs under completely overcast conditions. When the dynamic range is at its highest level ($C_i = 1$) clear skies are assumed to prevail ($F_G^+ = F_{G\text{-Clear}}^+$ in Eq. [2.7]). Under these conditions, only the clear sky model chosen for the satellite method comes into play. As the dynamic range becomes smaller and smaller (in other words, as cloudiness increases), the C_i decreases until $C_i = 0$, in which case it is assumed that complete overcast conditions prevail, and other models in addition to the clear sky model also come into play.

A number of factors must be addressed to improve the accuracy of this very basic approach. One is the choice of the clear sky model and its features used in the methodology. Different clear sky models invoke somewhat different approaches in addressing Rayleigh and Mie scattering, especially in the way they handle the atmospheric *aerosol optical depth*, or AOD, which is a normalized measure of the amount of dust and haze particles in the atmosphere compared to a totally clean, dust-free atmosphere. Obtaining accurate data on the AOD is quite challenging, because there are very few routine global measurements of this parameter, and the way the various models address this critical influence of the transfer of solar radiation through the atmosphere represents one of the biggest differences among the various empirical—physical and physical approaches, especially with respect to the accuracy of the DNI calculations.

The various ways in which the individual DNI, GHI, and DHI components are calculated using these satellite approaches is also important. Even under totally overcast skies, although DNI is virtually zero, there is some GHI (due to the DHI produced by forward-scattering by the clouds), so it is important that the models handle the DHI component properly, using either empirical relationships or physical radiative transfer theory.

A third important factor in the empirical—physical approach is accurately determining the ground albedo. As noted earlier the low end of the dynamic range is established by a determination of the lowest amount of reflection from the earth—atmosphere system. However, the presence of snow cover, persistent low clouds or valley fog, or the reflection from bright sand or water surfaces can result in an erroneous selection of the lower threshold of the dynamic range. Thus it is necessary to develop independent data sources to ensure when high reflective values are due to snow, fog, or bright surfaces, rather than clouds. Some of the empirical—physical models make use of the infrared channels of the satellite to obtain a clearer understanding of ground conditions as well as certain cloud characteristics.

Another important factor under research is the impact that deep, vertical clouds might have on the calculations. The three-dimensional characteristics of clouds can be determined by using the infrared channels of the satellite, because cloud temperatures can be determined using these channels. Once cloud temperatures are known, their vertical extent in the atmosphere can be determined. However, factoring in the three-dimensional characteristics of the clouds as part of a satellite retrieval method for solar resource estimates can involve time-consuming physical radiative transfer theory, so methods for a fast radiative transfer scheme also need to be developed to take advantage of this additional information on cloud characteristics. Work is currently under way at NREL, with assistance from the University of Wisconsin and Colorado State University, to address a more physical-based approach to determining surface solar radiation (Sengupta et al., 2015).

In summary, satellite-derived estimates of the solar resource are now in common use. They have the capability of providing monthly accuracies as good as 5% for GHI and 10% for DNI, and these accuracies continue to improve with new developments. Satellite estimates also give us the capability of developing solar resource information at a very high spatial resolution (1 km^2 is possible) and an acceptable temporal resolution (15 min with modern-day satellites).

Section 2.6 summarizes a variety of satellite-based data sets available through public institutions and commercial vendors.

2.4.3 Other solar resource estimation techniques

Here we briefly describe other methods that have been developed or are being researched to provide reliable estimates of the solar resource at the earth's surface. Among the very first of these methods was the Angstrom relationship, a method developed in the early 1920s to convert sunshine records such as those obtained from Campbell–Stokes radiometers to actual solar resource estimates on a monthly basis. The method involved the creation of an empirical relationship established by relating the measurements from calibrated radiometers, such as thermopile instruments, to the sunshine records obtained from the more ubiquitous Campbell–Stokes radiometer. In its simplest form the Angstrom relationship can be written as

$$Q_G/Q_{TOA} = a + b(N/N_0) \qquad [2.8]$$

where Q_G is the monthly mean global solar radiation at the earth's surface, Q_{TOA} is the monthly mean global radiation at the top of the atmosphere, N is the total observed number of monthly sunshine hours, N_0 is the monthly total possible duration of sunshine hours, and a and b are empirical coefficients that are a function of climate and geographic factors such as latitude, albedo, etc. This approach underwent a number of refinements through the years. We mention this approach here because it is referenced widely in the literature when reporting on country or regional solar resource assessments. However, satellite imagery offers a much more accurate and complete (spatial and temporal) estimate of the solar resource, and the use of solar resource information using a form of Eq. [2.8] is not recommended when reliable satellite estimates are available.

Another method involves converting cloud cover information obtained at national weather stations throughout the world into solar resource estimates using a variety of schemes involving clear sky models and empirical relationships based on the amount of cloud cover. When NREL developed the 1961–1990 NSRDB this approach was used extensively. The actual solar monitoring network using calibrated radiometers is quite limited in the United States (as in most countries), so to develop a comprehensive database for the country, it was necessary to resort to the use of cloud cover observations, which are available at virtually every weather station in the country. A model called METSTAT (described by Maxwell, 1998) was developed to convert the hourly cloud observations into hourly DNI, GHI, and DHI values at each station. Initially the database could be developed for only 239 weather stations throughout the country. However, modifications to the original approach and the application of satellite-derived estimates made it possible to derive solar resource data for 1454 stations in the NSRDB update of 1991–2005. As the reliability of satellite estimates continues to improve, and demands increase for the finer spatial resolution possible with satellite-derived data, future national databases such as the NSRDB will rely more and more on satellite methods, with very little need for the ground-based data except for access to other ancillary weather information and to maintain a long-term data record from these stations.

A third approach receiving considerable attention is to develop solar resource estimates from numerical weather prediction (NWP) models. These models allow us to simulate atmospheric processes in great detail and to predict future atmospheric conditions out to as much as 2 weeks or more. As more and more solar technologies are installed in national electricity systems, solar resource forecasting is increasing in importance, and accurate ways of converting cloud information developed in NWPs into solar irradiance values are needed. Traditionally, NWPs have not been very accurate in predicting specific cloud types and cloud amounts, so there is currently an active amount of research under way to improve NWPs to predict solar resources with acceptable accuracy for operational purposes. Such a discussion of these methods is beyond the scope of this chapter, however.

2.5 Solar resource data sets important to siting and sizing solar heating and cooling (SHC) technologies

2.5.1 *Resource variability—spatial*

Owing to the temporal and spatial variability of the solar resource, it is important to have reliable site-specific information on the resource to properly size a solar system, whether it is a thermal or a PV system. The importance of this information increases as system sizes become larger and larger, because there are critical financial implications and risk factors associated with the use of data that have high or unknown uncertainty for siting and sizing a project.

The annual spatial variability of the GHI resource is relatively low, because the diffuse component of the incoming resource can partially offset the loss of the direct component when clouds pass across the sun. In the United States, for example, based on the most recent NSRDB updates as of this writing, the annual GHI resource varies by a factor of 2 across the country, ranging from under 1200 kWh/m^2 in the cloudier regions to approximately 2400 kWh/m^2 in the sunny desert southwest. Thus in general it is possible to make use of flat plate solar collectors virtually everywhere in the United States to meet local demands; what is important is understanding the local resource in sufficient detail to size and design the system properly.

The spatial variability of the DNI resource, however, can be quite high, and for concentrating collectors it is critical to understand the resource characteristics not only for proper sizing, but also even for siting of these systems, because their output and economic feasibility will be highly dependent on the local resource. Using the United States as an example, the annual DNI resource varies by at least a factor of 3 (from around 2700 kWh/m^2 annually in the sunniest regions to about 900 kWh/m^2 in the cloudiest regions), and in addition shows much greater variability than GHI within certain regions. The value of DNI can be greatly influenced by local terrain and other effects on cloud formation.

Procedures for the optimal siting of solar systems generally involve the incorporation of spatial resource data, such as that derived from satellite methods, into Geographic Information Systems (GIS) software packages. This approach is preferred because other geospatial data, such as land use and landform features, transmission lines, highways and railways, ports, energy production facilities, load centers, protected areas, and areas excluded from development, can all be incorporated into the same software system to allow for overlaying of data sets to identify optimal sites. By establishing allowable thresholds of resource levels and then excluding all sites in these threshold regions where development cannot occur (such as protected areas), or where development might be optimal (such as proximity to load centers), the best sites for solar project development can be identified. Energy planners and government agencies use this approach to establish the technical resource potential of solar technologies and to determine optimum zones for development. This siting approach is best suited to large-scale systems requiring an ample resource, such as Concentrating Solar Power (CSP) systems, where a threshold value of, say, 5.5 kW/h day (2000 kW/h year) is established and then land use factors are taken into consideration to screen out all sites not suitable for development. An example of a regional site screening approach for CSP plants in the southwestern United States can be found in Section 6.1.2 of Sengupta et al. (2015).

2.5.2 Resource variability—temporal

The short-term temporal characteristics of the solar resource can have important effects on systems that respond immediately to irradiance fluctuations, such as PV systems. For these systems, when a cloud passes across the sun, casting a shadow on the system, a "ramp" in the solar resource occurs that results in a sudden drop in PV system output. Thus, for PV systems that have no internal built-in storage capabilities, understanding these "ramp rates" that can occur in time frames of seconds to hours is very important. However, systems that have built-in thermal storage, such as are typical of solar heating systems and now in large-scale CSP systems, are less affected by these short-term fluctuations or ramps.

The characteristics of longer-term variability, such as seasonal or interannual, are important for determining how systems might perform over the long run, whether or not they have storage capabilities. This is because most storage technologies serve only to mitigate short-term (subhourly) fluctuations and not those fluctuations that occur from day to day or season to season. Understanding interannual variability can also be important for determining a cash flow analysis of solar systems, although the interannual variability of the annual DNI resource is generally less than 10% and less than 5% for GHI.

2.5.3 Typical meteorological year data sets

Typical meteorological year (TMY) data sets represent 1 year of hourly (8760) weather data values extracted from long-term (at a minimum, 10 years) data records. TMY data sets are popular with building designers and others for modeling renewable energy systems, because they provide a relatively concise data set from which to develop system performance estimates. TMY data sets are generated by determining "typical" meteorological months from a process of weighting various weather parameters and then concatenating the months to form a typical year. TMY data sets do not provide information on extreme events and, by nature of the way they are created, do not necessarily represent actual conditions at any given time. Nevertheless they are useful for intercomparing system performance among different technology approaches without the need to ingest large amounts of data into simulation models.

There are various approaches for developing TMY data sets, but basically they are created from long-term weather records, which in many cases may require data filling. As described by Wilcox and Marion (2008), the TMY3[2] data set produced from the US NSRDB update is an empirical method that involves selecting 12 months of data from the 30-year record available in the NSRDB based on five weather parameters: GHI, DNI, dry bulb temperature, dew point temperature, and wind speed. An algorithm is created based on weighting factors applied to these five variables, so that individual months deemed "typical" of all 30 of those months would be selected, and then each month is concatenated with the next month without modification. The weighting factors assign priority to the solar radiation parameters, which means that the other weather parameters for a chosen typical month may not actually be typical.

Although TMY data sets remain in popular use today, with the convenience of being able to represent a long-term data set that may have missing data that required data filling measures, TMYs also have certain shortcomings that are important to understand, especially for financial institutions planning for the long-term cash flow expected from a project. The most important, of course, is that TMY data do not provide information on extreme events. The data incorporated into a TMY are essentially at the 50th percentile of the full distribution of possibilities; ie, the probability that a GHI value within a TMY will be exceeded is 50%. Thus a TMY data set is sometimes referred to as a P50 data set. If a developer is interested in learning what might happen to a system's performance during an unusually cloudy year, or a year in which there has been major volcanic activity causing partial obscuration of the solar resource, the TMY output will not be able to provide this information. Another important shortcoming of TMY data is that the 30-year data set from which it is derived is likely to have a number of "filled" data values, which otherwise would be missing owing to a

[2] In the United States, Sandia National Laboratories developed the original TMY data set in 1978. When NREL completed the NSRDB in the 1990s, a TMY data set was created out of this 30-year data set and designated as TMY2. An NSRDB update was produced that added in the years 1991–2005, and a new TMY data set was created, designated as TMY3, that makes use of the 1961–1990 data as well as the updated NSRDB data.

lack of observations during certain time periods, such as when a measurement device is down or a station is closed. Thus the TMY value may not represent the true long-term average value for a given parameter at a given site. Finally, as indicated in the next paragraph, the TMY data developed from the NSRDB are produced largely from modeled data sets based on the METSTAT model[3]; only about 2% of the data are actual measured data. Thus the TMY uncertainties include the uncertainties associated with the METSTAT model.

A good example of the difference between TMY data and long-term satellite-derived observations has been shown in a Clean Power Research white paper (2015). Fig. 2.10 compares TMY3 data for five local weather stations in and around the Minneapolis–St. Paul area of Minnesota in the United States with a commercial satellite-derived data set called SolarAnywhere, which is the commercial version of the Perez–SUNY/Albany model discussed in Section 2.4.2. The five stations have varying levels of data quality and completeness, resulting in differences in data quality and uncertainty across the stations. Stations indicated as class I have the lowest uncertainty, whereas class III stations have the highest uncertainty. The SolarAnywhere data cover the period 1998–2005, indicating that the TMY data are heavily weighted to solar resource estimates derived from a much earlier period than the satellite data. Fig. 2.10 shows that all of the five TMY3 stations differ from the satellite data on an annual basis, and in a few cases these differences are substantial, approaching 20%. Furthermore, the spatial pattern of the solar resource differs substantially for the two approaches, showing that spatial variability of GHI determined from the

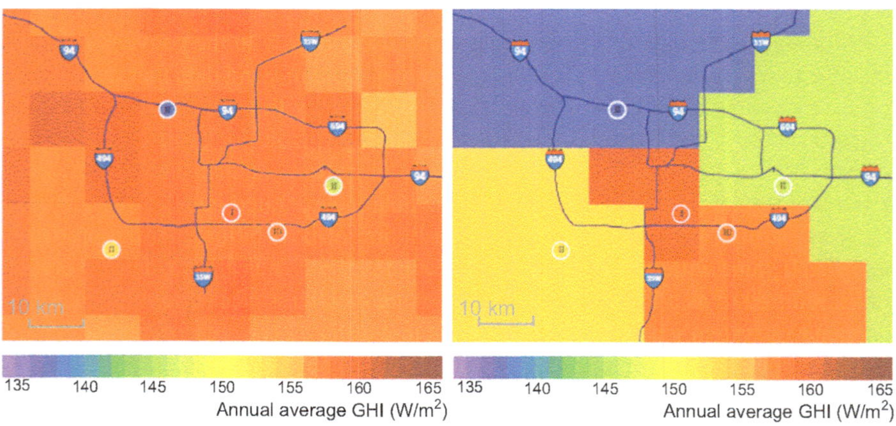

Figure 2.10 Comparison of TMY3 data with satellite-derived SolarAnywhere data for the Minneapolis–St. Paul region in the United States. The left panel shows the location of five TMY stations, including their uncertainty class (I is lowest, III is highest), with the 10-km satellite-derived data in the background. The right panel shows a gridded version of the TMY3 data based on a "nearest-neighbor" approach. It is evident from both panels that the TMY3 data can differ substantially in some cases from a more recent, validated, satellite-derived data set.

[3] Satellite data from 1998 to 2005 were used instead of the METSTAT model for the TMY3.

TMY3 stations is an unrealistically high 23 W/m^2 compared with 5 W/m^2 for the satellite-derived data. We present this information here to indicate to the reader the possible uncertainties that can result from the use of TMY data for making siting decisions and for designing and developing solar energy systems.

2.5.4 P50/P90 data sets

To avoid the concerns posed by TMY data project designers and developers and financial institutions financing these projects, data sets that provide information on *exceedance probabilities* of the solar resource should be used. For example, a P90[4] value represents the probability that a value will be exceeded 90% of the time. Exceedance values can be developed from a cumulative distribution function (CDF) of a parameter available from a large data set, such as the NSRDB. For example, Dobos et al. (2012) have looked at the NSRDB GHI and DNI data for Phoenix, Arizona, and produced CDF's of these data to illustrate the concept of P50 and P90 (Fig. 2.11). Fig. 2.11 shows that, if the annual Phoenix data were fit to a normal distribution (solid line), at CDF = 0.1 (the P90 value), an annual GHI of 2.2 MWh/m^2 will be exceeded 90% of the time (or conversely, the solar resource falls below this value 10% of the time). Similarly, for DNI, the hourly solar resource exceeds 1.96 MWh/m^2 90% of the time. However, long-term solar data do not necessarily follow a normalized distribution, so Fig. 2.11 also shows that the P90 value is somewhat lower in Phoenix when determined from an actual versus a normalized distribution.

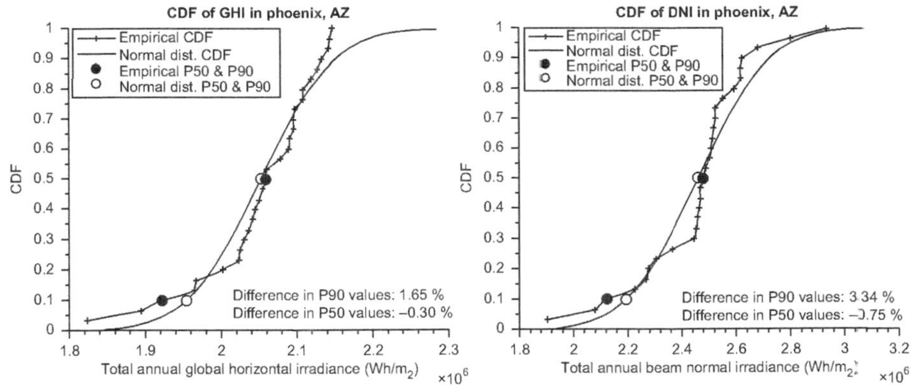

Figure 2.11 CDF of GHI (left) and DNI (right) for Phoenix, Arizona (USA) as determined from the 30-year NSRDB.
From Dobos, A., Gilman, P., Kasberg, M., 2012. P50/P90 analysis for solar energy systems using the systems analysis model. Conference Paper NREL/CP-6A20-54488, June 2012. National Renewable Energy Laboratory, Golden, Colorado. Presented at World Renewable Energy Congress, Denver, 2012.

[4] We use P90 here as an example. Other exceedance probabilities, such as P70 and P95, are also of interest to certain project designers and financial organizations.

2.5.5 The influence of data uncertainty on P90 values

A major motivation for developing exceedance probabilities for solar data sets (or for system performance parameters) is to give financial institutions information on the "bankability" of a project. Rather than just knowing what a system output might be under "typical" conditions, exceedance probabilities provide information on the amount of time a system is expected to exceed certain performance levels. For example, Dobos et al. (2012) show that if the P90 performance level of a system is 10,000 kWh-year^{-1} then this value would be exceeded in 90% of all of the years representing the total data set. Further discussion on data bankability can be found in Vignola et al. (2012b) and Gueymard (2014).

The effect of reducing the uncertainty of the true long-term solar resource at a site (which can be accomplished in a variety of ways) as a way of reducing financial risk is demonstrated in Fig. 2.12, taken from a study by Moody's Investors Services (2010). Fig. 2.12 shows that with only 1 year of data, the uncertainty of the value and the distribution of the true long-term mean is much higher than with 10 years of data. Thus, assuming that a long-term annual data set follows a Gaussian or normal distribution (which is not necessarily the case, as was shown in Fig. 2.11), the standard deviation that is assigned to a 1-year data set is much higher than for a 10-year data set, because there is more information associated with the 10-year data set. Although the median value of the distribution does not change between the 1-year and the 10-year distribution curves, the P90 value increases with the additional knowledge (lower uncertainty) associated with having a 10-year data set, which would lower the financial risk of the project.

2.5.6 Reducing uncertainty: site adaptation

As shown in Section 2.5.5, a bankable solar energy project relies on solar resource data with specified uncertainties from which investors can better understand the risk of times when performance of the system will be lower than desired. Reducing uncertainty in solar resource data is thus a key step toward producing bankable projects.

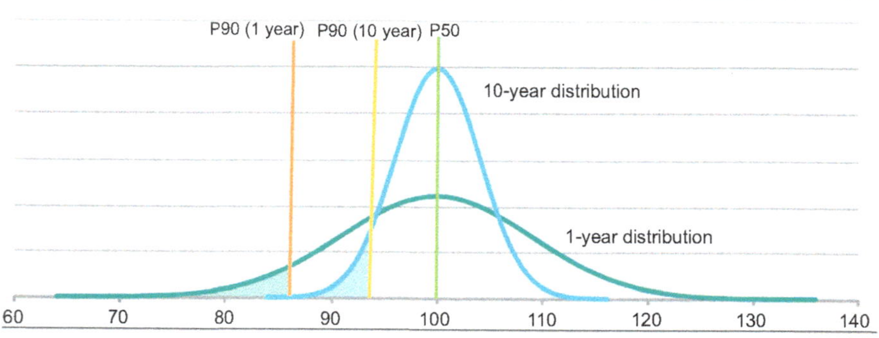

Figure 2.12 Example showing how reducing data uncertainty reduces financial risk by increasing P90 values.
From Moody's Investors Services, July 28, 2010. PV Solar Power Generation Projects. Special Comment.

There are several steps that can be taken to reduce data uncertainty and therefore project risk:

1. Collect high-quality on-site ground data for many years: this is the best way to provide the lowest risk data set to a financial organization; however, it is generally impractical owing to the time required to collect sufficient data (5−10 years);
2. Collect high-quality on-site ground data for a short time (1 year or less): this would be a more practical approach with respect to investor time scales, but, as indicated in Fig. 2.12, this poses higher risk to investors because they have limited knowledge of interannual variability and solar resource trends and therefore have high uncertainty as to whether the actual measurement period represents a high-resource or a low-resource annual value.
3. Obtain multiyear satellite-derived or other model-derived data sets: In the previous sections we have noted that we can use ancillary weather data, such as weather satellite imagery or ground observations of cloud cover, to produce long-term data sets when measurements are not available (which is the typical case). This is a relatively inexpensive way to capture enough data to produce P90 values; however, the risk to investors using these modeled data, with their inherently higher uncertainties, is higher than if actual high-quality measured data for the same time period were available.
4. Merge long-term modeled (eg, satellite-derived) data with short-term high-quality on-site measured data: this approach, known as *site adaptation*, is the most practical and cost-effective approach for developing reliable solar data for a project and offers the lowest risk to developers without causing lengthy project delays.

A detailed description of data adaptation methods is beyond the scope of this chapter; a 2015 publication by Polo et al. (2015) provides a comprehensive review of these methods. However, a few basic approaches are highlighted here. Perhaps the most straightforward approach is the *ratio method*. In this method, the ratios of selected periods of concurrent independent data sets (a short-term measurement data set concurrent with the long-term modeled data set) are calculated, and then the ratios are applied to the remaining long-term data set (Gueymard and Wilcox, 2009). However, the uncertainties in this approach increase if the biases in either data set vary from time to time. Furthermore, if there are long-term trends in the data the length of time required for the measured data set increases.

Another approach is to combine two independent data sets using weighting criteria (Meyer et al., 2008). The weighting can be determined based on data uncertainties, and additional data sets can even be included without compromising overall data uncertainty. Although this method was designed primarily for merging two different modeled data sets, measured data sets can also be included if their uncertainties are known.

A third approach is to use the CDF of ground-based data to improve satellite-derived data (Schumann et al., 2011). This approach reduces bias errors and improves Kolmogorov−Smirnov integrals (KSIs) in the satellite data. Mieslinger et al. (2014) have improved on this approach by determining both the systematic and the random deviations of the data sets based on mean bias (for systematic) and standard deviation (for random) statistics and using the KSIs to quantify the differences in the CDFs. Polynomial expressions are then used for the data adaptation. This approach is especially useful when there are large deviations between the satellite-derived and the ground-measured data. The approach requires at least 1 year of overlapping data, but preliminary results show that the corresponding biases approach zero.

2.6 Sources of solar resource information

This section provides a summary of publicly available sources of solar data, both measured and modeled, as well as key ongoing programs such as the IRENA Global Atlas and the World Bank/ESMAP resource mapping program.

Sengupta et al. (2015) provide a comprehensive summary of historical and current sources of data and data products available both through the public domain and through private companies (see Chapter 5.4 of this NREL study). Table 2.1 provides a summary of selected data sets available as of this writing, encompassing both measured and modeled data sources, derived from the NREL report.

A key development over the past few years has been the production of the International Renewable Energy Agency's (IRENA's) Global Atlas for Renewable Energy (gobalatlas.irena.org). The Atlas significantly extends an effort first started in 1999 known as the Solar and Wind Energy Resource Assessment (SWERA) project, funded by the Global Environment Facility and managed by the United Nations Environment Programme. SWERA ended in 2007, so with the formation of IRENA in 2010 the SWERA data, which resided initially with the UNEP/Global Resource Information Database in Sioux Falls, South Dakota (USA), and more recently at NREL (http://maps.nrel.gov/swera), are now incorporated into the Global Atlas. IRENA's Global Atlas provides basic foundational information on national renewable energy potentials to remove knowledge barriers and stimulate national renewable energy investments and policy formulation. The Global Atlas initially incorporated wind and solar data, but is now being expanded to include data for all renewable energy resources.

Another key development launched in October 2012 is the World Bank's Energy Sector Management Assistance Program (ESMAP) solar and wind resource mapping project (http://www.esmap.org/RE_Mapping). This project develops renewable energy resource mapping at the country level to support government planning and commercial development. The solar mapping activities include a combination of GIS-based mapping using satellite-derived techniques, with enhancements in some cases provided by NWP models. Countries that currently have very limited information on resource potential are targeted for this program, and projects are executed through the World Bank and supported by ESMAP with both funding and technical support. At the time of this writing preliminary high-resolution solar mapping has been completed for several countries (Pakistan, the Maldives, Malawi, Zambia, and Tanzania) and measurement programs have been initiated to validate the maps.

2.7 Summary

Although solar resource information has been available in a variety of formats for a number of years, the rapid growth in solar technology deployments, combined with improved ways of both measuring and modeling the solar resource, has resulted in significant activity at regional and national levels to improve and make more accessible the quality and completeness of these data. Furthermore, the many types of products

Table 2.1 **Listing of selected available solar resource data sources**

Name of data source	URL	Period of record	Coverage	Data elements
World Radiation Data Center (WRDC)	http://wrdc.mgo.rssi.ru	1964–present	Global (>1000 measurement stations)	Primarily daily total GHI; some DNI
European Solar Resource Atlas (ESRA)	http://www.soda-is.com/eng/index.html	1981–1990	Europe, 10-km resolution	Daily and monthly GHI, DHI, DNI
HelioClim (Ecole des Mines de Paris/Armines)	http://www.soda-is.com/eng/index.html	1985–present	Europe, Africa, Mediterranean Basin (5-km resolution)	Hourly, daily GHI
PVGIS (Joint Research Centre, Italy)	http://re.jrc.ec.europa.eu/pvgis/download/download.htm.	1981–1990	Europe, Mediterranean region, Africa, southeast Asia; nominally 30-km resolution	Monthly, annual GHI (Europe), daily GHI elsewhere
METEONORM 7.1.4 (Meteotest, Switzerland)	http://meteonorm.com	1991–2010	Global interpolated data from ~8325 stations	GHI, DHI, DNI, other
NASA Surface Meteorology and Solar Energy	https://eosweb.larc.nasa.gov/sse/)	July 1983–June 2005	Global, 100-km resolution	GHI, DHI, DNI
DLR-ISIS (German Aerospace Institute	http://www.pa.op.dlr.de/ISIS.	1984–2004	Global (280-km resolution)	DNI, GHI
Solar Data Warehouse	http://solardatawarehouse.com.	Up to 25 years to present	USA (~3000 measurement stations)	GHI

Continued

Table 2.1 Continued

Name of data source	URL	Period of record	Coverage	Data elements
US National Solar Radiation Data Base Updates	www.nsrdb.nrel.gov	1961–2010	USA (1454 stations, mostly modeled)	GHI, DHI, DNI, other
Baseline Surface Radiation Network	www.bsrn.awi.de/en/home/wrmc/.	1992–present	Global (~40 measurement stations)	GHI, DHI, DNI, other
3Tier by Vaisala	http://www.3tier.com/en/support/resource-maps/	1997–present	Global maps (commercial data sets available)	GHI, DNI
SolarAnywhere (Clean Power Research)	www.cleanpower.com/SolarAnywhere.	1998–present	North America (10-km, other commercial data sets available)	GHI, DHI, DNI
GeoModel Solar (Slovakia)	http://geomodel.eu/index.php	April 2004–present	Europe, Africa, and the Middle East (5-km resolution, available commercially)	GHI, DHI, DNI, other

Derived from From Sengupta, M., Habte, A., Kurtz, S., Dobos, A., Wilbert, S., Lorenz, E., Stoffel, T., Renné, D., Myers, D., Wilcox, S., Blanc, P., Perez, R., February 2015. Best Practices Handbook for the Collection and Use of Solar Resource Data for Solar Energy Applications. Technical Report NREL/TP-5D00-63112. National Renewable Energy Laboratory, Golden, Colorado (USA).

required by government policy makers, project developers, system operators, and financial institutions have been a very strong impetus to expand and improve on the type of information available. This chapter has provided a snapshot of best practices in collecting and using solar resource data for solar heating and cooling siting and performance applications. The reader is encouraged to go to the references and the URL links provided in the text to obtain more comprehensive information about solar resources and how solar resource data can be applied to all solar energy applications.

References

Blanc, P., Espinar, B., Geuder, N., Gueymard, C., Meyer, R., Pitz-Paal, R., Reinhardt, B., et al., 2014. Direct normal irradiance related definitions and applications: the circumsolar issue. Solar Energy 110, 561−577.

Cano, D., Monget, J.M., Albuisson, M., Guillard, H., Regas, N., Wald, L., 1986. A method for the determination of the global solar radiation from meteorological satellite data. Solar Energy 37 (1), 31−39.

Clean Power Research, 2015. How Misuse of Solar Resource Datasets Is Reducing Solar Industry Profits. White Paper, Clean Power Research, Kirkland, Washington.

Dobos, A., Gilman, P., Kasberg, M., 2012. P50/P90 analysis for solar energy systems using the systems analysis model. In: Conference Paper NREL/CP-6A20-54488, June 2012. National Renewable Energy Laboratory, Golden, Colorado. Presented at World Renewable Energy Congress, Denver, 2012.

Geuder, N., Hanussek, M., Halle, J., Affolter, R., Wilbert, S., 2011. Comparison of corrections and calibration procedures for rotating shadowband irradiance sensors. In: SolarPACES conference. Granada, Spain.

Gueymard, C.A., Myers, D., 2008. Validation and ranking methodologies for solar radiation models. In: Badescu, V. (Ed.), Modeling Solar Radiation at the Earth's Surface. Springer.

Gueymard, C.A., 2012. Clear-sky irradiance predictions for solar resource mapping and large-scale applications: Improved validation methodology and detailed performance analysis of 18 broadband radiative models. Solar Energy 86, 2145−2169.

Gueymard, C.A., 2014. A review of validation methodologies and statistical performance indicators for modeled solar radiation data: towards a better bankability of solar projects. Renewable and Sustainable Energy Reviews 39, 1024−1034.

Gueymard, C.A., Wilcox, S.M., 2009. Spatial and temporal variability in the solar resource: assessing the value of short-term measurements at potential solar power plant sites. In: Proceedings, Solar 2009. American Solar Energy Society Proceedings, Buffalo, New York.

Hay, J.E., 1979. Study of Shortwave Radiation on Non-Horizontal Surfaces. Report 79−12. Atmospheric Environment Service, Downsview, Ontario.

Habte, A., Wilcox, S., Stoffel, T., February 2014. Evaluation of Radiometers Deployed at the National Renewable Energy Laboratory's Solar Radiation Research Laboratory. Technical Report NREL/TP-5D00−60896. National Renewable Energy Laboratory, Golden, Colorado (USA).

Ineichen, P., 2008. A broadband simplified version of the SOLIS Clear-sky model. Solar Energy 82, 758−762.

Kluchar, T.M., 1979. Evaluation of models to predict insolation on tilted surfaces. Solar Energy 23, 111−114.

Liu, B.Y.H., Jordan, R.C., 1961. Daily insolation on surfaces tilted towards equator. ASHRAE Transactions 67, 526–541.

Marion, W., Wilcox, S., 1994. Solar Radiation Data Manual for Flat-Plate and Concentrating Collectors. NREL/TP-463–5607. National Renewable Energy Laboratory, Golden, CO.

Menicucci, D., Fernandez, J.P., 1988. User's Manual for PVFORM: a Photovoltaic System Simulation Program fro Stand-alone and Grid-Interactive Applications. SAND85-0376. Sandia National Laboratory, Albuquerque, New Mexico.

Maxwell, E.L., 1998. METSTAT – the solar radiation model used in the production of the national solar radiation data base (NSRDB). Solar Energy 62 (4), 263–279.

Moody's Investors Services, July 28, 2010. PV Solar Power Generation Projects. Special Comment.

Meyer, R., Torres Butron, J., Marquardt, G., Schwandt, M., Geuder, N., Hoyer-Klick, C., Lorenz, E., Hammer, A., Beyer, H.G., 2008. Combining solar irradiance measurements and various satellite-derived products to a site-specific best estimate. In: Proceedings, Solar-PACES Conference, March 4–7, 2008, Las Vegas, Nevada.

Mieslinger, T., Ament, F., Chhatbar, K., Meyer, R., 2014. A new method for fusion of measured and model-derived solar radiation time-series. Proceedings, SHC 2013: International Conference on Solar Heating and Cooling for Buildings and Industry, Energy Procedia 48, 1617–1626.

Perez, R., Ineichen, P., Seals, R., Michalsky, J., Stewart, R., 1990. Modeling daylight availability and irradiance components from direct and global irradiance. Solar Energy 44 (5), 271–289.

Perez, R., Ineichen, P., Moore, K., Kmiecik, M., Chain, C., George, R., Vignola, F., 2002. A new operational model for satellite-derived irradiances: description and validation. Solar Energy 73 (5), 307–317.

Polo, J., Wilbert, S., Ruiz-Arias, J.A., Meyer, R., Gueymard, C., Suri, M., Martín, L., Mieslinger, T., Blanc, P., Grant, I., Boland, J., Ineichen, P., Remund, J., Escobar, R., Troccoli, A., Sengupta, M., Nielsen, K.P., Renné, D., Geuder, N., 2015. Integration of Ground Measurements to Model-Derived Data. IEA-shc Task 46 Solar Resource Assessment and Forecasting. Final Report for Subtask B3. To be available at: http://task46.iea-shc.org/publications.

Reindl, D.T., et al., 1990. Evaluation of hourly tilted surface radiation models. Solar Energy 45, 9–17.

Renné, D.S., Perez, R., Zelenka, A., Whitlock, C., DiPasquale, R., 1999. Use of weather and climate research satellites for estimating solar resources (Chapter 5). In: Yogi Goswami, D., Böer, K.W. (Eds.), Advances in Solar Energy, vol. 13. American Solar Energy Society, Boulder, Colorado, 457 pp.

Sengupta, M., Habte, A., Kurtz, S., Dobos, A., Wilbert, S., Lorenz, E., Stoffel, T., Renné, D., Myers, D., Wilcox, S., Blanc, P., Perez, R., February 2015. Best Practices Handbook for the Collection and Use of Solar Resource Data for Solar Energy Applications. Technical Report NREL/TP-5D00-63112. National Renewable Energy Laboratory, Golden, Colorado (USA).

Schumann, K., Georg Beyer, H., Chhatbar, K., Meyer, R., 2011. Improving satellite-derived solar resource analysis with parallel ground-based measurements. In: Proceedings, Solar World Congress 2011. International Solar Energy Society, Kassel, Germany.

Temps, R.C., Coulson, K.L., 1977. Solar radiation incident up on slopes of different orientations. Solar Energy 19, 179–184.

Vignola, F., Joseph, M., Stoffel, T., 2012a. Solar and Infrared Radiation Measurements. CRC Press, Taylor & Francis Group, Boca Raton, Florida, 394 pp.

Vignola, F., Grover, C., Lemon, N., MacMahan, A., 2012b. Building a bankable solar radiation dataset. Solar Energy 86, 2218–2229.

Wehrli, C., 1985. Extraterrestrial Solar Spectrum. Publication no. 615. Physikalisch-Meteorologisches Observatorium Davos, World Radiation Center, (PMOD/WRC), Davos Dorf, Switzerland.

Wilcox, S., Myers, D., 2008. Evaluation of Radiometers in Full-Time Use at the National Renewable Energy Laboratory Solar Radiation Research Laboratory. NREL/TP-550–44627. National Renewable Energy Laboratory, Golden, CO.

Wilbert, S., Geuder, N., Schwandt, M., Kraas, B., Jessen, W., Meyer, R., Nouri, B., 2014. Best practices for solar irradiance measurements with rotating shadowband irradiometers. In: Prepared Within IEA Task 46 Subtask B1, and INS Project 1268. Available at: http://task46.iea-shc.org/publications.

Energy efficiency and environmental impact of solar heating and cooling systems

3

G. Martinopoulos
International Hellenic University, Thessaloniki, Greece

3.1 Introduction

Since the last part of the 20th century there has been a significant increase in energy demand due to three vital factors: population, per capita energy consumption, and economic growth. Most researchers agree that global energy demand will grow by more than 33% up to 2035, with developing countries like China and India as well as Middle Eastern countries being responsible for more than 60% of that increase [1]. Currently, although there is a pronounced shift toward renewable/alternative energy sources, most of the world's energy needs are met by coal, oil, natural gas, and uranium. Thus, fossil fuels remain dominant in the global energy mix [2].

Renewable energy sources compete with conventional fuels in three distinct markets: power generation (both on- and off-grid), hot water production and space heating, and transport fuels. Hot water for sanitary uses as well as space heating for millions of buildings is supplied by systems that employ biomass, solar, or geothermal energy. Biomass and geothermal energy also supply heat for the industry and agricultural sectors, whereas in transportation biofuels make small but growing contributions [3,4].

Solar heating and cooling technologies are becoming widespread, with space heating and cooling from solar energy gaining ground in several countries, although the primary application remains hot water production for sanitary uses [5].

In this chapter, a brief overview of the global solar thermal market will be given. Furthermore, the main solar thermal technologies for heating and cooling will be presented. The chapter will conclude with a description of the methods that can be used to estimate the environmental impact of various technologies and systems and their implementation on solar technologies.

3.2 Energy use in the built environment

Buildings worldwide constitute one of the biggest energy consumers, with 32% of the total final energy consumption, whereas in terms of primary energy consumption they represent around 40% in most countries according to the International Energy Agency. In the United States, commercial and residential buildings use almost 41%

Advances in Solar Heating and Cooling. http://dx.doi.org/10.1016/B978-0-08-100301-5.00003-5

of the total primary energy consumption [6]. The same trend is also evident in Europe, as the building sector accounted for 37.1% of the total final energy consumption (1157.7 million tons of oil equivalent in 2007) in the European Union (EU-27), of which 284.6 million tons of oil equivalent was used in residential buildings and 145.2 million in nonresidential buildings, with space heating representing 68% of total household consumption, followed by water heating at 12% for 2009 (Fig. 3.1) [7]. Moreover, residential buildings are the fourth largest source of CO_2 emissions in the EU and account for 9.9% of total emissions in 2007, whereas emissions from nonresidential buildings are ranked fifth and account for 3.9% of the total CO_2 emissions in EU-27 [7].

The energy used by the building sector continues to increase as the building stock steadily increases with new buildings being introduced faster than old ones are being retired. This trend will continue until buildings are designed to use energy more efficiently and furthermore to produce locally enough energy from alternative energy sources that will offset the growing energy demand of these buildings.

For these reasons, the concept of the net zero energy building (NZEB) has gained wide international attention and is now seen as the future target for the design of buildings. In the United States, the Department of Energy has established an aggressive goal to create the technology and knowledge base for cost-effective NZEBs by 2025, and the EU has implemented a series of directives that promote the use of energy alternatives for buildings, primarily for electricity, heating, cooling, and the provision of hot water, starting with Directive 2009/28/EC [8], which implied that all member states should increase the use of renewable energy sources along with energy efficiency and savings by 20% until 2020. Shortly after, the EU passed Directive 2010/31/EC [9] defining minimum rules for the performance of buildings and introducing energy certificates, taking into account the external climatic conditions and defining the NZEB. Moreover, after 2018 all newly constructed buildings that are either occupied

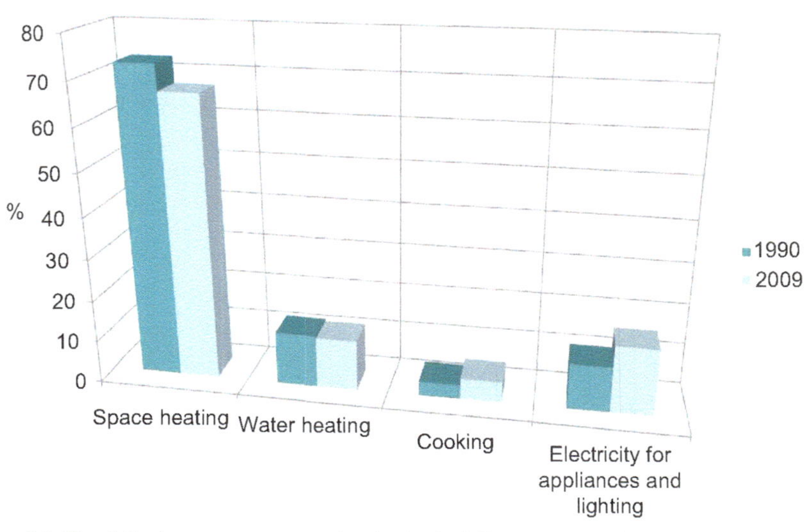

Figure 3.1 Total final energy consumption in the building sector in the European Union.

or owned by public authorities must qualify as NZEBs, with all other new buildings following suit from 2020 onward.

However, before being fully implemented in the national building codes and international standards, the NZEB concept requires a clear and consistent definition. To qualify as an NZEB, a building has "to exhibit a very high energy performance and to cover the amount of energy required to a very significant extent from renewable sources that are produced onsite or nearby" [10].

Of the various renewable energy systems that can be installed in the building sector to cover energy requirements (electrical and thermal loads), solar energy systems are currently the most widely used, mostly in the form of solar thermal and photovoltaic systems. Especially for locations with high annual solar radiation and temperatures, solar energy systems are already a viable alternative to fossil energy systems and are expected to become even more efficient and cost-competitive in the future [11].

3.3 Worldwide market penetration of solar heating and cooling systems

Solar thermal systems are considered a mature technology that has penetrated markets all over the globe with the highest total installed capacity among other renewable energy technologies. The type of solar collector, the type of system that the collectors are integrated in, and the application that they are used for vary, depending mainly on the location. According to the International Energy Agency, at the end of 2013 the cumulative installed capacity of solar thermal collectors reached almost 375 GW_{th} or 535,000,000 m^2 of collector area worldwide [12].

As presented in Fig. 3.2 more than 94% of the cumulative installed capacity is in the form of liquid collectors, whereas only 6% is in the form of air

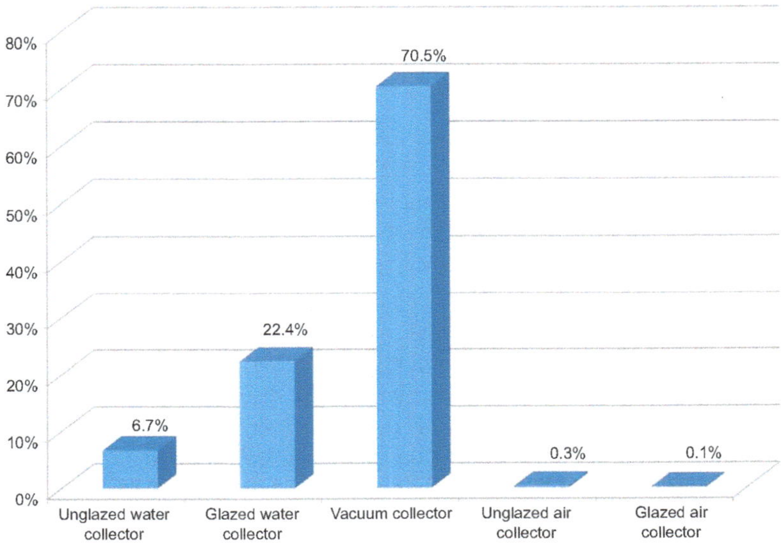

Figure 3.2 Worldwide share of various solar thermal collector types.

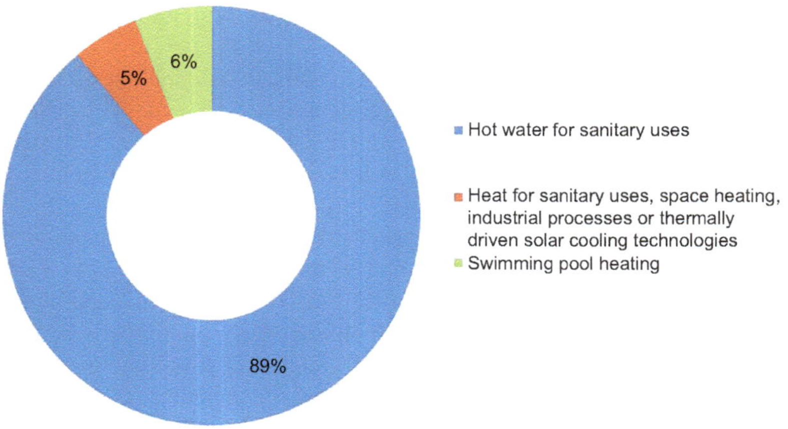

Figure 3.3 Worldwide share of solar systems by use.

collectors. The evacuated thermal collectors cover 71% of the total, mainly owing to their increased share in the world's largest market, China. Glazed liquid collectors comprise the second biggest contributor with 22% (as well as the most prominent technology in Europe) and unglazed liquid collectors cover less than 7%.

The vast majority of the installed capacity, as presented in Fig. 3.3 (almost 90%) is used to cover hot water for sanitary uses, in the form of (nonpumped) thermosiphon systems with less than a quarter being of forced circulation. Heating water for pool use is the second most used application (around 6%), mainly in the form of unglazed liquid collectors, whereas less than 5% of the total installed capacity is used for supplying heat for sanitary uses, space heating, industrial processes, or thermally driven solar cooling technologies.

The main driving force in the market, with more than 80% of the total installed capacity, is China with 262 GW_{th}, followed by Europe with 44 GW_{th} and North America, Asia (excluding China), and South America with 17, 10, and 8.7 GW_{th}, respectively [12].

In the case of Europe, according to the European Solar Thermal Industry, Germany represents almost 40% of the European market (newly installed capacity), followed by Italy (10%), Poland (9%), and France and Greece with 7% each [5]. The European market is well developed, with thermosiphon systems being employed in southern Europe, whereas more complicated forced circulation systems for combined production of hot water for sanitary uses and space heating ("combi" systems) are more common in the countries of central and northern Europe [12]. The market for solar combi systems is well established in some mature European markets such as Germany (43% of the total market) and Austria (44% of the total market) [5]. Europe is also a pioneer in solar cooling installations with around 75% of the total worldwide systems installed in Spain, Germany, and Italy [13].

3.4 Overview of technologies used for solar heating and cooling systems and their efficiency

Nonconcentrating solar thermal collectors can be utilized in a variety of systems to provide:

- Hot water for swimming pools
- Hot water for sanitary uses
- Space heating
- Space cooling
- A combination of the above

Solar heating systems for swimming pools consist of, usually, unglazed solar collectors that are connected with a filter, a circulating pump, and a hydraulic channeling system in their most simplistic form. Unglazed solar collectors typically comprise black matting or tubes made from rubber or polymer materials through which water is circulated. As this type of collector is not insulated it is used only when low temperature rises are needed for large volumes of fluid. In these systems, water from the pool is circulated by the pump, first by the filter to remove any particles and then through the solar collectors where the water is heated. The size of the system is influenced by the size of the pool (water volume), the installation site, and the operating period (yearly or seasonal). Most systems have a payback period of between 2 and 5 years depending on the parameters named previously, and in case year-round operation is needed they also incorporate a backup heater. There is a variety of published work in the literature that covers in detail the performance of solar systems employed for swimming pool heating [14–17]. In all cases, mean solar coverage of more than 80% is possible.

Solar heating systems for hot water production for sanitary use are the most common ones. A variety of systems can be employed to produce hot water, such as thermosiphon, direct and indirect circulation systems, and air systems. For the systems to be used without any issues when the air temperature is below freezing, indirect circulation systems are usually employed. Common systems consist mainly of flat plate solar collectors, a storage tank with a mounding base, and the necessary plumbing. Average annual system efficiency for the conversion of solar radiation to useful energy in the form of hot water varies between 30% and 40%, depending mainly on the type of solar collector used [18,19].

A typical liquid flat plate solar collector (Fig. 3.4) consists of a black absorber in which the absorbed solar radiation is converted to heat, which in turn is conducted to a fluid. The absorber includes the channels/pipes through which the thermal fluid flows. The back and sides of the absorber are insulated and the front side is covered by a transparent cover that allows solar radiation to reach the absorber but reduces heat losses to the atmosphere. All the above are encased in a metal housing that provides weather protection and offers structural support. Copper, steel (stainless or not), aluminum, glass, and various plastics are some of the materials used in flat plate collectors. The type of materials and the technology used in the production of solar systems are both critical in the estimation of the thermal load that the systems can cover,

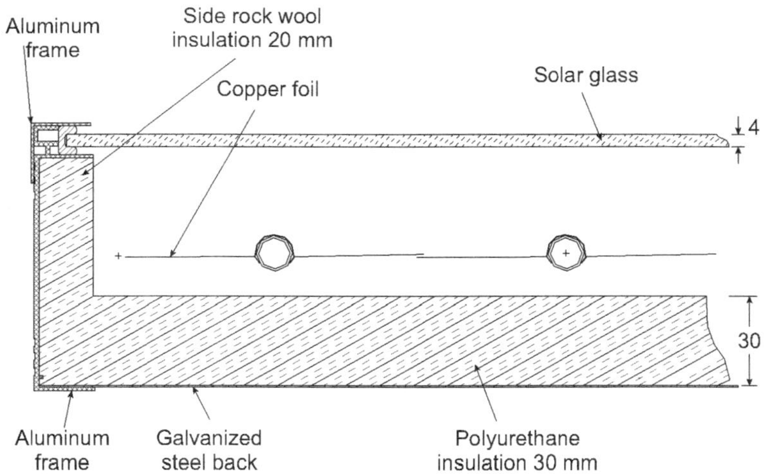

Figure 3.4 Cross-section of a typical flat plate collector.

as they influence the collectors' technical characteristics (F_RU_L, $F_R(\tau\alpha)$), and consequently system efficiency.

The absorber is the main component of the collector, containing the pipes or flow channels through which the thermal fluid circulates. The efficiency depends mainly on the absorber configuration. Usual configurations are flow channels corrugated on the body of the absorber area or pipes in contact with the absorber area or pipes welded on fins. The absorber may be coated with black powder paint to boost absorbance, whereas a selective paint is employed by the vast majority of manufacturers. To minimize heat loss to the environment, insulation is used. The materials used for insulation vary from rock wool (usually 50-mm thick) to expanded or extruded polyurethane or a combination of these. Today, most collectors use polyurethane because of its low thermal conductivity and its high moisture resistance. Low-iron tempered glass (3–5-mm thickness) is used as a glazing, whereas some collectors use an acrylic glazing. For the casing, aluminum is used for the sides and a sheet of galvanized steel for the back [20].

Hot water storage tanks are usually made of steel, with a few exceptions that use copper (Fig. 3.5). The heat exchanger is of the mantle type and the boiler is protected from the weather by an outside cylinder made of stainless steel or aluminum sheet. Between the boiler and the outside casing polyurethane is usually placed to minimize heat loss [20]. Storage tanks are placed mostly for aesthetic purposes horizontally, although a more proper mounting, owing to stratification, is vertical.

Systems that are similar to the ones described can achieve year-round coverage of as much as 80%, even for locations with harsh winters, and exhibit a payback period of as little as 3–4 years, as long as proper sizing for the thermal load needed is achieved [20–22].

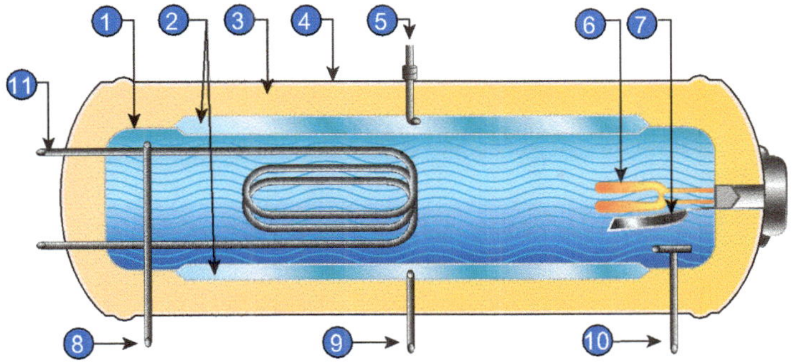

Figure 3.5 Cross-section of a typical water storage tank.

The same layout can be used to cover both space heating and hot water for sanitary uses with the only differences being the larger solar collector area and storage tank as well as the possible use of evacuated tube solar collectors (Fig. 3.6). These systems are usually called solar combis. According to a large number of studies in a variety of countries like Greece [23], Germany [24], Spain [25], and other non-European countries with climates ranging from subtropical [26] to humid continental [27], the use of a solar thermal system for space and water heating enables the minimization of energy costs, as the systems analyzed were proved to be capable of covering more than 42% of the total load and as much as 95%, while exhibiting a payback period of less than 10 and as low as 4.4 years in some cases [28].

A solar system that covers hot water needs for sanitary purposes as well as space heating and cooling is called a solar combi plus. In these systems, cooling is achieved by one of the following technologies:

- Absorption chiller, usually based on ammonia/water or water/lithium bromide.
- Absorption chiller, usually based on water/lithium chloride, water/silica gel, or water/zeolite.

In absorption chillers, solar thermal cooling systems often employ single, double, or triple cooling cycles, as the more cycles, the more efficient they are. Absorption chillers consist of a generator, a condenser, an evaporator, and an absorber. In an absorption chiller, the temperature of the chilled water is above 0°C, as it is used in air conditioning, and the refrigerant most used is typically a liquid H_2O/LiBr solution. The refrigerant (water) is evaporated in the condenser, where the cooling effect takes place, at very low pressures. The created vaporized refrigerant is absorbed by a dilute solution within the absorber. To increase the efficiency of the absorption, the process has to

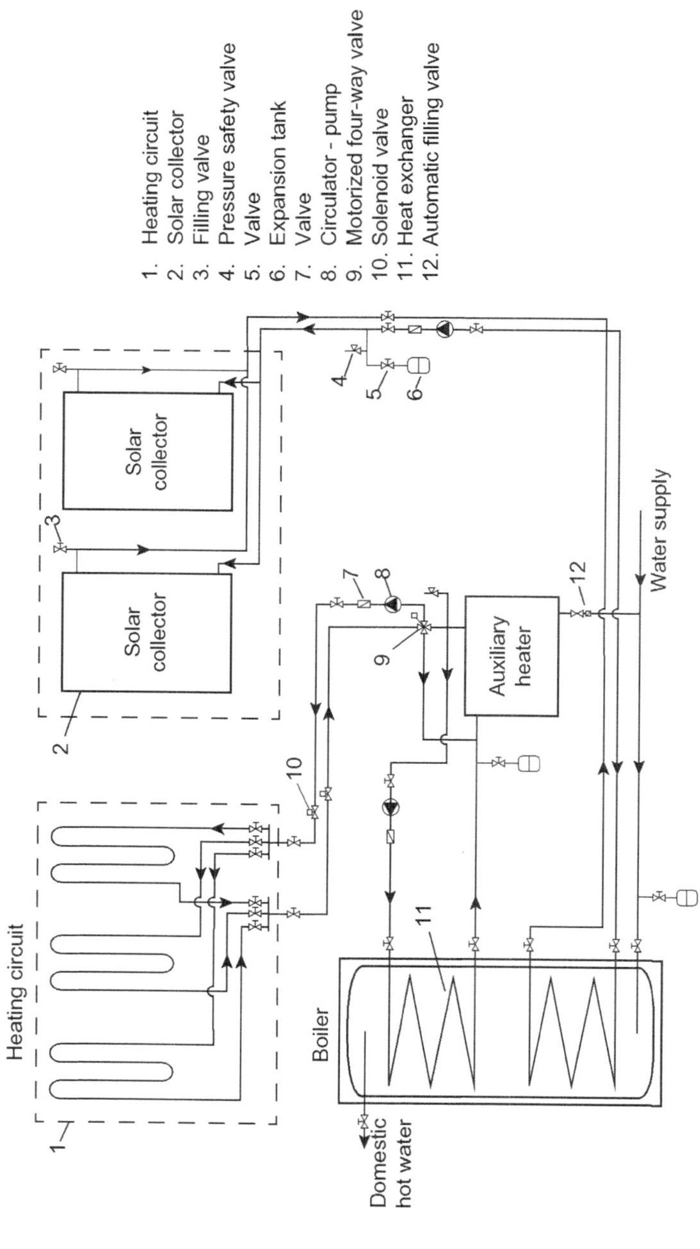

Figure 3.6 Diagram of a solar combi system.

be cooled. The solution then is pumped to the generator, where the regeneration of the solution takes place by adding heat. After the generator, the refrigerant is pumped to the condenser where it condenses through the application of cooled water and then is led with an expansion valve again into the evaporator and the procedure is repeated.

Efficient absorption chillers nominally require water of at least 85°C, making the use of common, inexpensive flat plate solar thermal collectors possible; for adsorption systems the water temperature required is usually lower than 75°C, but their coefficient of performance (COP) is also somewhat lower at 0.59. Depending on the chiller utilized the nominal area (and type) of solar collectors required differs, ranging between 0.5 and 5.5 m^2/kW of cooling power. In general NH_3/H_2O systems demand larger nominal solar fields than H_2O/LiBr systems, leading to higher initial costs. Furthermore, depending on the system the water temperature needed differs and in general the COP increases with water temperature increase. For a system to be as efficient as possible the COP is of the outmost importance [29,30].

Open systems ensure complete air conditioning by supplying cooled and dehumidified air. In the case of open systems the "refrigerant" is always water, because it is in direct contact with the atmosphere. The most common refrigeration technique in open systems is the desiccant cooling system.

In adsorption chillers solid sorption materials are used. Water is used as the refrigerant and silica gel as the sorbent. The adsorption chiller is composed of two parts, the evaporator and the condenser. The sorbent in the first compartment is regenerated using hot water from the solar collector and the sorbent in the second compartment adsorbs the water vapor that enters from the evaporator. This compartment has to be cooled to achieve a continuous adsorption. The cooling effect takes place in the evaporator where the water is transferred into the gas phase by being heated from the external water cycle [29,30]. The refrigerant used in desiccant cooling systems is water and it is in direct contact with air. The thermally driven cooling cycle is a combination of evaporative cooling with air dehumidification by a desiccant, ie, a hygroscopic material (liquid or solid). The refrigerant, after providing the cooling effect, is discarded from the system, and new refrigerant is supplied in its place in an open-ended loop. The most common technology applied today uses rotating dehumidification wheels, equipped with either silica gel or lithium chloride [5,31].

Combi plus systems have been examined by many researchers [32–35] as well as within the IEA TASK 38 (International Energy Agency Solar Heating and Cooling Programme), in which several solar cooling systems were monitored in detail. The results showed that although the electrical COP is still low, with values of up to six in the best case and below three in the worst case scenario, combi plus systems can cover between 40% and 87% of the total solar fraction depending on the location and building type. Southern locations can achieve higher solar fractions mainly because the system can cover a large percentage of the cooling load, but the situation changes as the heating load increases. The same trends can be seen when primary energy saved is considered. In general, the best performing plants are encountered where high radiation is coupled with high cooling loads and modest needs for heating and hot water preparation [36].

3.5 Environmental impact of solar heating and cooling systems

The environmental impact of products and energy utilization have gained increasing interest in recent years, with the general public becoming aware of a wide range of environmental issues. All sources of air, land, and water pollution are under constant scrutiny with an increasing number of professionals confronting pollution control-related problems and the public being interested in the impact that the consumption of products has on natural resources and environmental quality. Because many of these issues are of relatively new concern, individuals must develop proficiency and an improved understanding of technical and scientific, as well as regulatory, issues regarding pollution prevention and remediation to cope with these challenges. International efforts for the protection and restoration of the environment include interventions in the energy system, with a strong emphasis on utilization of renewable energy sources.

Many researchers have estimated the environmental impact of solar heating and cooling systems, mainly by calculating the amount of greenhouse gas (GHG) emissions from the substituted fossil fuels that are avoided. In general researchers have concluded that a typical solar thermal system for hot water production can reduce CO_2 generated on-site by 28% on average, whereas the deployment of a combined space and water heating system can reduce emissions by more than 60% [37]. For countries with a mature solar market, the environmental benefit from the use of solar thermal systems is very pronounced. In the case of Greece, one of the pioneers in the use of solar thermal systems for hot water production, a study that covered the contribution of solar thermal systems to the reduction of conventional energy and GHGs and other air pollutant emissions for a 30-year period showed that the substituted fossil fuel energy ranged from 21 GWh (0.1% of the domestic sector energy use) in 1978 to 1513 GWh (2.4%) in 2007, resulting in the abatement of 1.67 Mt of CO_2. The emissions of CO_2, NO_x, and SO_2 avoided, compared to those emitted annually in the country, were about 1.5% for CO_2, less than 1% for NO_x, and around 2.5% for SO_2 of the country's total [38].

For solar combi systems, the environmental gain is also substantial, as researchers have estimated that the use of these systems can reduce the total amount of GHGs that would otherwise be emitted to the atmosphere by up to 79.11 t for CO_2, 3.19 t for CH_4, and 0.64 t for N_2O annually on average, depending on the system size and the location of the building [28].

The same applies for solar combi plus systems, for which researchers calculated that a system employing a 10-kW $LiCl/H_2O$ absorption system installed in a detached house in Spain was able to achieve a 68% reduction in GHG emissions. The reduction in GHG emissions is 76% for cooling and reaches an aggregate value of a 68% reduction for heating, cooling, and hot water [39].

Although solar energy is considered a "clean" energy form, both the manufacture and the final disposal of solar energy systems are associated with significant environmental transactions, owing to the energy required for the raw material production and the final product formation and assembly, as well as to the final disposal of the system at the end of its life. It is considered necessary therefore to evaluate all types of solar technologies, accounting for the indirect environmental impacts caused by the systems over their whole life cycle. To minimize the effort for the estimation of these impacts, a number of software tools have been developed by various institutes and/or companies. Most of these tools are based on the implementation of the life cycle analysis (LCA) principles.

LCA is a methodology that enables quantification of environmental burdens and their potential impacts over the whole life cycle of a product, process, or activity. Although it has been used in some industrial sectors for about 20 years, LCA has received wider attention only since the beginning of the 1990s, when its relevance as an environmental management aid in both corporate and public decision-making became more evident. LCA is rather young and evolving, with its roots in research related to energy requirements in the 1960s and pollution prevention, which was initiated in the 1970s [40].

LCA methodology was initially developed by the Society of Environmental Toxicology and Chemistry and was later optimized by the International Organization for Standardization (ISO), but its real breakthrough into the business world occurred only during the 1990s. Using LCA, the environmental impacts associated with the production and utilization of solar systems can be assessed in a compatible and comparable way. This can be accomplished by recording the energy and raw materials used in the manufacturing stage and also the air, liquid, and solid pollutants emitted over the product's life cycle. Apart from obtaining a reliable assessment of the total impact, LCA can enable an existing situation to be improved by evaluating, in environmental impact terms, suggestions for manufacturing procedure modifications or substitution of materials. The results of the assessment can be used for the formation of an environmental labeling regulation [20]. A schematic diagram of the this concept is presented in Fig. 3.7.

A large number of indicators and supporting methodologies are feasible for estimating characterization factors for the various impact categories and areas of protection in LCA. Entire methodologies were compiled in addition to the stand-alone/subgroups of methods and models that were developed elsewhere for specific impact categories [40].

There are four ISO standards specifically designed for LCA application: ISO 14040, Principles and Framework; ISO 14041, Goal and Scope Definition and Inventory Analysis; ISO 14042, Life Cycle Impact Assessment; and ISO 14043, Interpretation. The ISO standards are defined in quite vague language, which makes it difficult to see if an LCA has been made according to the standard.

Unlike the 14000 standard, it is also not possible to get an official accreditation stating that an LCA, an LCA methodology, or an LCA software such as the ones used in this work have been made according to the standard. So, no software developer can claim that LCA done with a certain software tool automatically conforms to the standards.

Although all tools are based on the principles of LCA, differences between impact indicators do exist. These differences can be attributed to the fact that LCA is a comparative assessment methodology. Direct adoption of regulatory methodology and data is not always appropriate. Regulatory methods and data, again particularly in toxicological risk assessments, are not always developed for use in a comparative

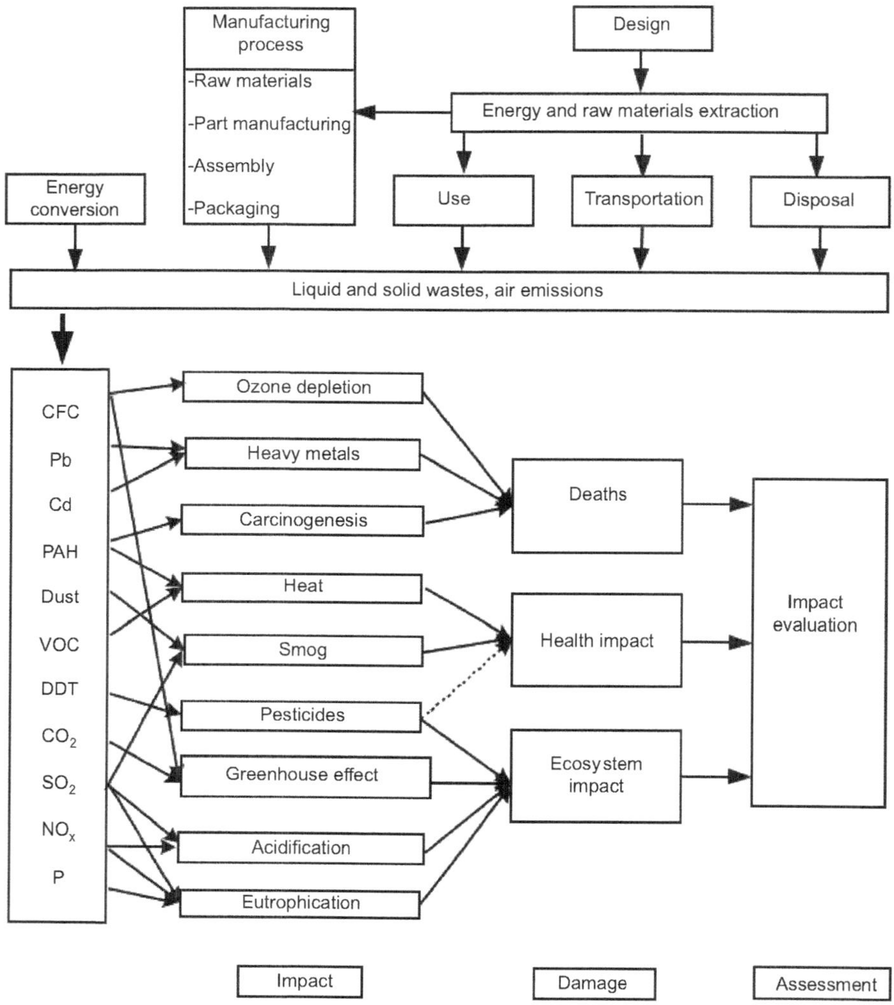

Figure 3.7 Flowchart of a product and its impact on the environment during its life cycle.

context. Best estimates are desirable in LCA, with the, often overlooked, need to account for uncertainties when making distinctions among the results [20].

Some of the best known and most popular tools among life-cycle practitioners are the ECO-it, GEMIS, SimaPro, and GaBi software. All tools claim accordance with the international standards and guidelines, although differences in the methodologies applied can be identified.

ECO-it is the simplest of all and its use is restricted solely as a tool for designers in their search for more environmentally friendly designs. A small database of materials and processes is included with the environmental impact indicators originating from the Eco-Indicator 99 methodology, in Eco-Indicator points. Standard Eco-Indicator values have been developed, as such an instrument meant for designers, a tool to be used in the search for more environmentally friendly design alternatives. Eco-Indicator 99 includes standard values for [41]:

- Materials. In determining the indicator for the production of materials, all the processes involved are included, from the extraction of the raw materials up to and including the last production stage resulting in bulk material. Transport processes along this route are also included up to the final process in the production chain.
- Production processes. This includes treatment and processing of various materials, expressed for each treatment in the unit appropriate to the particular process (eg, square meters of rolled sheet or kilograms of extruded plastic). The indicators of the production process account for the emissions not only during the manufacturing stage, but also resulting from the production of the energy needed.
- Transportation processes. These are mostly expressed in the unit tonne-kilometer. Transportation processes include the impact of emissions caused by the extraction and production of fuels and the fuel consumption during transportation. A loading efficiency for average European conditions is assumed and a possible empty return journey is accounted for.
- Energy generation processes. The energy indicators refer to the extraction and production of fuels and to the energy conversion, using average efficiencies. For the electricity score the various fuels used in Europe to generate electricity are accounted for. Different Eco-Indicators have been determined for high- and low-voltage electricity, the first intended for industrial processes and the second mainly for household and small-scale industries.
- Disposal scenarios. These are per material unit (kg), subdivided into types of material and waste-processing methods (recycling of materials, incineration, landfill, etc.). Not all products are disposed of in the same manner; therefore the most appropriate waste-processing method must be carefully considered in using indicators. In addition, scenarios have been provided for the incineration, landfill disposal, and recycling of products.

The standard Eco-Indicator 99 values can be regarded as dimensionless figures, called Eco-Indicator points (Pts). One Pt is defined as representing one one-thousandth of the yearly environmental load of an average European inhabitant [41].

GEMIS, on the other hand, was developed as a tool for the comparative assessment of environmental effects of energy and it includes an extensive database of materials and processes. GEMIS is the acronym for Global Emission Model for Integrated Systems. The model can perform complete life-cycle computations for a variety of emissions and can determine the resource use. In addition, GEMIS analyzes costs—the corresponding data of the fuels as well as cost data for energy and transport processes are included in the database.

GEMIS allows also for assessing the results of environmental and cost analyses by aggregation of emissions into so-called CO_2 equivalents, SO_2 equivalents, and tropospheric ozone precursor potential, and by a calculation of external costs. The GEMIS database offers information on energy carriers (process chains and fuel data) as well as various technologies for heat and electric power generation.

In addition to fossil energy carriers (hard coal, lignite, oil, natural gas), renewable energies, household waste, uranium, biomass (eg, fast-growing woods, rape), and hydrogen are covered. Data on various material process chains (above all for construction materials) and processes for transport services, ie, cars (gasoline, diesel, electricity, biofuels), public transport (bus, train), and airplanes, as well as processes for freight transport (trucks, light delivery vehicles, train, ships, and pipelines) are available in the database. The process data are given for a variety of countries, and a special set of data (called "generic") refer to the situation in developing countries. The results are presented in quantities of CO_2, SO_2, NO_x, etc. [42].

SimaPro stands for System for Integrated Environmental Assessment of Products. It includes several inventory databases with a large variety of materials and processes, plus the most important impact assessment methods. It can be used either as a tool for designers, like ECO-it, or as a tool for the comparative assessment of environmental effects, like GEMIS. The results are presented either in a single score (Pt) or in quantities of CO_2, SO_2, and NO_x. SimaPro incorporates the Eco-Indicator 99 methodology. The Eco-Indicator 99 methodology analyzes the following five environmental impact categories: climate change, acidification/eutrophication, ecotoxicity, fossil fuels (resources), and respiratory inorganics. The impact assessment involves three main steps: characterization or classification, normalization, and final weighted scores. There are three damage categories for the final weighted scores [40]:

- Human health. This is measured in DALY (disability adjusted life years); that is, the various disabilities caused by diseases are weighted. DALY has been developed for the World Health Organization and the World Bank. Climate change is categorized under this damage category.
- Ecosystem quality or ecotoxicity. This is measured in PDF m^2 year, which is the potentially disappeared fraction of plant species. The impact category of acidification/eutrophication is listed here. In terms of ecotoxicity, this is measured as the percentage of all species present in the environment living under toxic stress (potentially affected fraction or PAF m^2 year).
- Resources. This final damage category is measured in megajoules of surplus energy, including fossil fuels.

Furthermore, the software has the ability to calculate the quantity of emissions that were released during the life cycle of the system.

GaBi models every element of a product or system from a life-cycle perspective, equipping businesses to make the best informed decisions on the manufacture and life cycle of any product while providing a coherent database to estimate the environmental impact of sourcing and refining every raw material or processed component of a manufactured item. It can also present alternative options for manufacturing, distribution, recyclability, pollution, and sustainability [43].

All of these tools have been thoroughly used by many researchers to identify the environmental impact of various solar thermal technologies. Since 2000, there have been a large number of studies that present an LCA of solar thermal systems for hot water production; for comparison, most of them present a net energy analysis, which can also be used to produce the energy payback period of the system. For the calculations and comparisons apart from the life-cycle inventory of the systems, assumptions as well as simulation methods were also used. These studies cover mostly thermosiphon systems in various locations ranging from Australia [44] to India [45], Pakistan [46], Italy [47], Cyprus [19], and Greece [20], among others. In all cases, the energy payback period was estimated at between 6 months and 2 years depending on the location and substituted fuel, proving that the systems are clean.

Further analysis was carried out regarding the influence of various materials used in the manufacturing of the systems by analyzing 28 different systems [48]. The use of different materials for the same solar system component (solar collector, storage tank, etc.) proved to affect both the efficiency and the amount of conventional energy substituted by the system and hence the overall environmental impact of a solar thermal system for hot water production. The net environmental gain achieved by the use of these systems can differ by up to 20% in some cases, depending on the materials and techniques used. However, in all cases examined the environmental impact of the solar systems is significantly lower compared to that of the energy conserved (less than 10% in all cases). It is of interest that as thermal efficiency differs among systems their environmental performance is influenced mainly by the conventional energy substituted and to a lesser extent by the materials used for their production. Fewer studies have been carried out for solar combi systems. An NZEB in Montreal that employed a solar combi system was studied, proving that they also provide important energy savings with relatively quick energy payback times of 8–11 years and energy payback ratios of 3.6–4.8 [27].

Finally, the LCA of solar cooling systems was studied by an IEA-SHC Task Group (TASK 38) and four different cases were investigated to assess the performance of different thermally driven chillers in Palermo and Zurich with the use of the SimaPro software. The annual performance of the four systems was then compared to a vapor compression chiller and a gas boiler. The absorption chiller had a higher energy requirement and thus produced a higher global warming potential that the adsorption chiller. In all cases examined the systems provided a net environmental gain, proving that they can be a viable solution [49].

3.6 Conclusions

As the energy used by the building sector continues to increase, the use of solar thermal systems to cover heating and cooling loads seems to be a viable solution, technically, as well as from an environmental and economic point of view. In years to come, to overcome technical problems (due to the prevailing climatic conditions) apart from the use of active solar thermal systems, viable technical solutions can include hybrid solar geothermal heat pump systems or solar-assisted ground-coupled heat pump systems with satisfactory results.

References

[1] K.R. Ullah, R. Saidur, H.W. Ping, R.K. Akikur, N.H. Shuvo, A review of solar thermal refrigeration and cooling methods, Renewable Sustainable Energy Rev. 24 (2013) 499–513.
[2] F. Birol, World Energy Outlook 2012, International Energy Agency, 2012.
[3] E. Martinot, J. Sawin, Renewables Global Status Report: 2009 Update, REN21 Secretariat, Paris, France, 2009.
[4] IEA, International Energy Update 2010, International Energy Agency, 2010.

[5] ESTIF, Solar Thermal Markets in Europe – Trends and Market Statistics 2014, European Solar Thermal Industry Federation, 2015.
[6] US Energy Information Administration, Annual Energy Outlook, 2014.
[7] L. Mortensen, Consumption and the Environment – State and Outlook, European Environmental Agency, 2012.
[8] P. UNION, Directive 2009/28/EC of the European Parliament and of the Council of 23 April 2009 on the Promotion of the Use of Energy from Renewable Sources and Amending and Subsequently Repealing Directives 2001/77/EC and 2003/30/EC, 2009.
[9] Directive 2010/31/EU of the European Parliament and of the Council, 19 May 2010 on the Energy Performance of Buildings. (recast), 2010.
[10] A.J. Marszal, P. Heiselberg, J.S. Bourrelle, E. Musall, K. Voss, I. Sartori, A. Napolitano, Zero energy building–a review of definitions and calculation methodologies, Energy Build. 43 (4) (2011) 971–979.
[11] ECOFYS, Towards Nearly Zero-Energy Buildings – Definition of Common Principles Under the EPBD (Final report), 2013.
[12] F. Mauthner, W. Weiss, M. Spörk-Dür, Markets and Contribution to the Energy Supply 2013, Solar Heat Worldwide, Solar Heating and Cooling Program IEA, 2015.
[13] R. Fedrizi, Identification of Most Promising Market and Promotion of Standardized System Configuration for the Market Entry of Small Scale Combined Solar Heating & Cooling Applications (Technical Report), 2010.
[14] A.I. Alkhamis, S.A. Sherif, Performance analysis of a solar-assisted swimming pool heating system, Energy 17 (12) (1992) 1165–1172.
[15] B. Molineaux, B. Lachal, O. Guisan, Thermal analysis of five outdoor swimming pools heated by unglazed solar collectors, Sol. Energy 53 (1) (1994) 21–26.
[16] W. Haaf, U. Luboschik, B. Tesche, Solar swimming pool heating: description of a validated model, Sol. energy 53 (1) (1994) 41–46.
[17] D.A. Katsaprakakis, Comparison of swimming pools alternative passive and active heating systems based on renewable energy sources in Southern Europe, Energy 81 (2015) 738–753.
[18] F. DeWinter, Solar Collectors, Energy Storage, and Materials, vol. 5, Mit Press, 1990.
[19] S. Kalogirou, Thermal performance, economic and environmental life cycle analysis of thermosiphon solar water heaters, Sol. Energy 83 (1) (2009) 39–48.
[20] G. Tsilingiridis, G. Martinopoulos, N. Kyriakis, Life cycle environmental impact of a thermosyphonic domestic solar hot water system in comparison with electrical and gas water heating, Renewable Energy 29 (8) (2004) 1277–1288.
[21] G. Tsilingiridis, G. Martinopoulos, N. Kyriakis, Environmental performance of thermosyphonic domestic solar hot water systems under different climatic conditions: a case study for Greece, Global Nest: The Int. J. 6 (3) (2004) 183–195.
[22] B.J. Brinkworth, Solar DHW system performance correlation revisited, Sol. energy 71 (6) (2001) 377–387.
[23] A. Argiriou, N. Klitsikas, C.A. Balaras, D.N. Asimakopoulos, Active solar space heating of residential buildings in northern Hellas—a case study, Energy Build. 26 (2) (1997) 215–221.
[24] V. Badescu, M.D. Staicovici, Renewable energy for passive house heating: model of the active solar heating system, Energy Build. 38 (2) (2006) 129–141.
[25] J.D. Marcos, M. Izquierdo, D. Parra, Solar space heating and cooling for Spanish housing: potential energy savings and emissions reduction, Sol. Energy 85 (11) (2011) 2622–2641.
[26] T.T. Chow, K.F. Fong, A.L.S. Chan, Z. Lin, Potential application of a centralized solar water-heating system for a high-rise residential building in Hong Kong, Appl. Energy 83 (1) (2006) 42–54.

[27] M. Leckner, R. Zmeureanu, Life cycle cost and energy analysis of a net zero energy house with solar combisystem, Appl. Energy 88 (1) (2011) 232–241.
[28] G. Martinopoulos, G. Tsalikis, Active solar heating systems for energy efficient buildings in Greece: a technical economic and environmental evaluation, Energy Build. 68 (2014) 130–137.
[29] CRES, Climasol. http://www.cres.gr/climasol/index_files/pdf/climasol.pdf.
[30] E. Podesser, C1: State of the Art–Survey on New Solar Cooling Developments, IEA, 2010.
[31] J.C. McVeigh, A.A.M. Sayigh (Eds.), Solar Air Conditioning and Refrigeration, Newnes, 2012.
[32] M. Ortiz, H. Barsun, H. He, P. Vorobieff, A. Mammoli, Modeling of a solar-assisted HVAC system with thermal storage, Energy Build. 42 (4) (2010) 500–509.
[33] O. Marc, F. Lucas, F. Sinama, E. Monceyron, Experimental investigation of a solar cooling absorption system operating without any backup system under tropical climate, Energy Build. 42 (6) (2010) 774–782.
[34] N. Molero-Villar, J.M. Cejudo-López, F. Domínguez-Muñoz, A. Carrillo-Andrés, A comparison of solar absorption system configurations, Sol. energy 86 (1) (2012) 242–252.
[35] C.A. Balaras, P.E.E.G. Dascalaki, P.P. Tsekouras, D. Chasapis, C. Karytsas, High Solar Combi-plus Systems–A Pilot Application in a Hellenic Office Building, 2010.
[36] Task, I.E.A. 38. Solar Air-Conditioning and Refrigeration.
[37] Solar Energy Industries Association, Solar Heating & Cooling: Energy for a Secure Future, 2013.
[38] G. Tsilingiridis, G. Martinopoulos, Thirty years of domestic solar hot water systems use in Greece–energy and environmental benefits–future perspectives, Renewable Energy 35 (2) (2010) 490–497.
[39] D. Borge-Diez, A. Colmenar-Santos, C. Pérez-Molina, M. Castro-Gil, Experimental validation of a fully solar-driven triple-state absorption system in small residential buildings, Energy Build. 55 (2012) 227–237.
[40] G. Rebitzer, Life cycle assessment: Part 1: framework, goal and scope definition, inventory analysis, and applications, Environ. Int. 30 (5) (2004) 701–720.
[41] M. Goedkoop, R. Spriensma, The Eco-indicator 99: A Damage Oriented Method for Life Cycle Impact Assessment: Methodology Report, 2001.
[42] Oko Institute, Global Emission Model for Integrated Systems Manual, 2008.
[43] PE International AG, GaBi Manual, 2013.
[44] R.H. Crawford, G.J. Treloar, Net energy analysis of solar and conventional domestic hot water systems in Melbourne, Australia, Sol. Energy 76 (1) (2004) 159–163.
[45] J. Mathur, N.K. Bansal, Energy analysis of solar water heating systems in India, Int. J. Life Cycle Assess. 4 (2) (1999) 113–116.
[46] M. Asif, J. Currie, T. Muneer, Comparison of aluminium and stainless steel built-in-storage solar water heater, Build. Serv. Eng. Res. Technol. 28 (4) (2007) 337–346.
[47] F. Ardente, G. Beccali, M. Cellura, V.L. Brano, Life cycle assessment of a solar thermal collector, Renewable Energy 30 (7) (2005) 1031–1054.
[48] G. Martinopoulos, G. Tsilingiridis, N. Kyriakis, Identification of the environmental impact from the use of different materials in domestic solar hot water systems, Appl. Energy 102 (2013) 545–555.
[49] M. Beccali (Ed.), Life Cycle Assessment of Solar Cooling Systems – Technical Report, IEA, 2010.

Part Two

Solar heating systems

Nontracking solar collection technologies for solar heating and cooling systems

4

S.A. Kalogirou
Cyprus University of Technology, Limassol, Cyprus

4.1 Introduction

This chapter deals with nontracking solar collectors, which are suitable for space heating and cooling systems. These are collectors which are permanently fixed in position and are usually nonconcentrating or have some form of concentration, but their concentration ratio, defined as the ratio of the aperture area to the absorber area, is not more than about 5.

Solar energy collectors in general are a special kind of heat exchanger that transforms solar radiation energy into internal energy of the transport medium. The major component of any solar system is the solar collector. This is a device which absorbs the incoming solar radiation, converts it into heat, and transfers this heat to a fluid flowing through the collector. The solar energy thus collected is carried from the circulating fluid either directly to the hot water or space conditioning equipment or to a thermal energy storage tank from which it can be drawn for use at night and/or on cloudy days [6].

The collectors belonging in this category are usually steady, or if some form of tracking is applied, it is intermittent and not very accurate. Fixed concentrators are very important because of the practical advantages enjoyed by fixed solar systems. By increasing the concentration ratio (C) the frequency of tracking increases. Thus a collector with $C \approx 3$ needs only biannual adjustment, whereas a collector with $C \approx 4-5$ requires seasonal adjustment. Generally speaking, the higher the concentration ratio the higher the temperature a collector can attain, but the higher the tracking requirements. Because of their low concentration ratio these collectors usually collect both direct and diffuse solar radiation as opposed to the high-concentration ones, which collect only direct solar radiation.

For a collector moved about a north–south axis with a single daily adjustment so that its surface–normal coincides with the solar beam at noon each day, the solar incidence angle θ is equal to [2,6]:

$$\cos(\theta) = \sin^2(\delta) + \cos^2(\delta)\cos(h) \qquad [4.1]$$

where the declination $\delta = 23.45 \sin\left[\frac{360}{365}(284 + \mathrm{N})\right]$ and the hour angle $h = 0.25 \times$ the number of minutes from local solar noon.

Advances in Solar Heating and Cooling. http://dx.doi.org/10.1016/B978-0-08-100301-5.00004-7

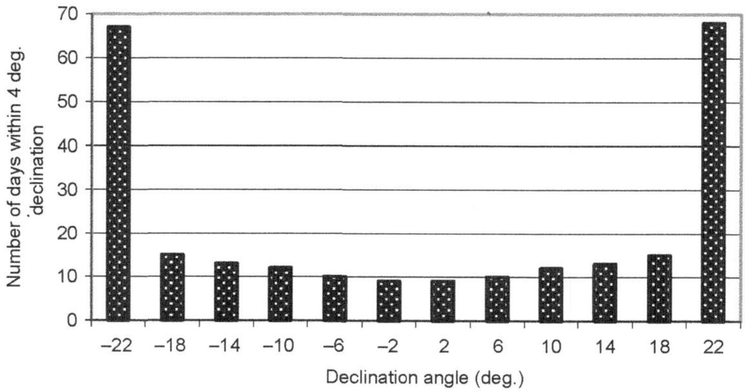

Figure 4.1 Number of consecutive days that the sun is within a 4-degree declination window.

For this mode of tracking, if we can accept that when the sun is at noon the angle of its rays and the normal to the collector can be up to 4 degrees of declination, then a seasonal adjustment is possible. This is because for small angles $\cos(4°) = 0.998 \sim 1$, which can be considered normal. Fig. 4.1 shows the number of consecutive days that the sun remains within this 4-degree "declination window" at noon. As can be seen in Fig. 4.1 the sun remains most of the time close to either the summer solstice or the winter solstice, moving rapidly between the two extremes. For nearly 70 consecutive days the sun is within 4 degrees of an extreme position, spending only 9 days in the 4-degree window at the equinox. This means that a seasonally tilted collector needs to be adjusted only occasionally and more frequently during spring (March−April) and autumn (September−October).

The types of collectors examined in this chapter are the flat plate collectors; flat plate collectors with diffuse reflectors; reverse flat plate collectors; compound parabolic collectors (CPCs), which are the representative type of collector in the steady concentrator category; and various designs applied to evacuated tube collectors with either external or internal reflectors.

4.2 Flat plate collectors

A schematic of a typical flat plate solar collector of the header and riser design is shown in Fig. 4.2(a). It consists basically of an absorbing plate covered with glazing and insulated at the back and sides. Solar radiation passing through the transparent cover is absorbed by the blackened absorber plate and then transferred to the transport medium flowing through the fluid tubes carried away for use or storage. The liquid tubes can be securely fixed or welded to the absorbing plate, or they can be an integral part of the plate. This is very important because most of the solar radiation is absorbed by the absorbing plate and conducted to the fluid contained in the tubes. The liquid tubes, called riser tubes, are connected at both ends by header tubes, which have a

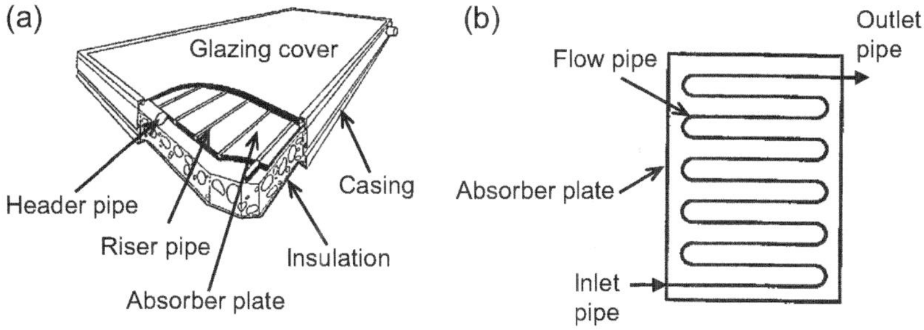

Figure 4.2 Typical flat plate collector details. (a) Header and riser collector. (b) Serpentine collector.

larger diameter than the risers. There are basically two typical designs of the flat plate collector, the header and riser and the serpentine design, shown in Fig. 4.2(b). The serpentine design does not present any problem concerning the uneven flow distribution in the various riser tubes of the header and riser design (especially for very wide collectors), but owing to the comparatively large pressure drop the serpentine collectors cannot work effectively in natural circulation mode and need a pump to circulate the heat transfer fluid. The absorber plate can be constructed as a single sheet on which all riser pipes are fixed, one next to the other, or each riser pipe can be fixed on a separate fin before being welded on the header pipe.

The transparent cover is used to reduce the wind convection losses from the absorber plate. The air between the absorber plate and the glass is stagnant so any convection losses are by natural convection, which has a much lower effect than the forced convection that would be caused by wind blowing over the absorbing plate if the glass cover were absent. The cover also helps to trap the solar radiation by the greenhouse effect as according to the temperatures involved the glass is transparent to the shortwave radiation coming from the sun but is almost opaque to long-wave thermal radiation emitted by the absorber plate [5].

The bigger advantage of flat plate collectors is that they are easy to construct and thus not very expensive to manufacture. Additionally, they are permanently fixed in position so no tracking of the sun is required and they collect both beam and diffuse radiation. The best orientation of the collectors to take advantage of the diurnal apparent motion of the sun in the sky is toward the equator, facing south in the Northern Hemisphere and north in the Southern. The optimum tilt angle of the collector depends on the application. For applications that require heat year-round this is equal to the latitude of the location [5]. If the application is solar cooling, then the optimum angle is latitude $-10°$, so that the sun would be perpendicular to the collector during summertime when the energy will be most required. If the application is space heating, then the optimal angle is latitude $+10°$, whereas for annual hot water production it is latitude $+5°$ so as to have relatively better performance during wintertime when hot water is most required [6].

The main components of a flat plate collector are the cover, which is one or more sheets of glass or other radiation-transmitting material; the absorber plate, which can

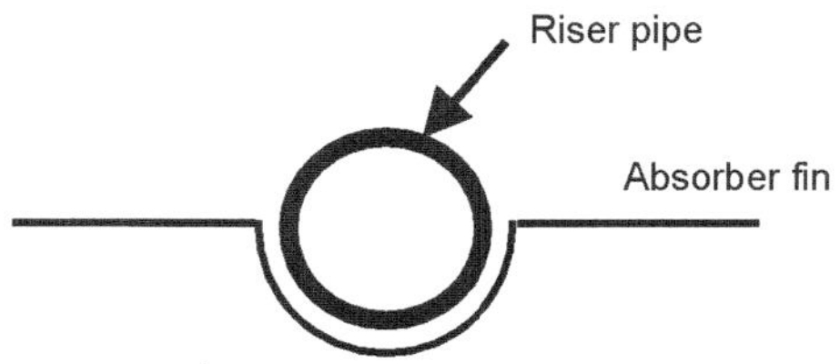

Figure 4.3 The embedded fixing of a riser on an absorber plate in a flat plate collector (no space exists between the grove and the riser pipe but is shown here for clarity).

be flat, corrugated, or grooved plates to which the tubes, fins, or passages are attached; and the insulation, which is used to minimize the heat loss from the back and sides of the collector. A typical attachment method of the pipes on the absorber plate is the embedded fixing shown in Fig. 4.3. The plate is usually coated with a high-absorptance and low-emittance layer. The back insulation of a flat plate collector can be made from fiberglass or mineral fiber mat that will not outgas at elevated temperatures. Building-grade fiberglass is not suitable as the binders evaporate at high temperature and would condense on the collector cover, blocking incoming solar radiation. Another important part of the collector is the heat removal fluid passageways. These are tubes, fins, or passages, which conduct or direct the heat transfer fluid from the inlet to the outlet. All this equipment is enclosed in a container. This is a casing which surrounds the aforementioned components and protects them from dust, moisture, and any other material.

Over the years flat plate collectors have been built in a large variety of designs and materials. They are used to heat fluids such as water, water plus antifreeze solutions, refrigerants, and air. Despite the adverse effects of the sun's ultraviolet radiation, the collector should have a long effective life and be able to withstand corrosion because of alkalinity or hardness of the heat transfer fluid. The collector must also be able to work with the deposition of dust or moisture on the glazing. An area of possible problems is the breakage of the glazing because of thermal expansion, hail, or vandalism, but these can be substantially minimized by the use of tempered glass.

A common glazing material is glass. This is widely used to glaze solar collectors because it can transmit as much as 90% of the incoming shortwave solar irradiation while transmitting virtually none of the long-wave thermal radiation emitted by the absorber plate. A glass with low iron content has a relatively high transmittance for solar radiation and its transmittance is essentially zero for the long-wave thermal radiation (5.0–50 μm) emitted by the sun-heated absorbing plate. Normal window glass usually has a high iron content and is not suitable for use in solar collectors.

Plastic films and sheets also possess good optical properties but are generally limited in the temperatures they can sustain without deterioration or undergoing dimensional changes. Additionally, only a few types of plastics can withstand the sun's ultraviolet radiation for long periods. They have, however, various advantages; for example, they are not broken by hail or stones and, in the form of thin films, they are flexible and have low weight.

The normal incidence transmittance of glass is about 0.85−0.87 for direct radiation and varies considerably with the angle of incidence from the above values, to 0 at 90 degrees incidence. Glass antireflective coatings and surface texture can also improve transmission significantly.

The glazing should admit as much solar irradiation as possible and reduce the loss of reradiating heat as much as possible. Although glass is essentially opaque to the long-wave radiation emitted by the collector plate, absorption of that radiation by the glass causes an increase in its temperature and thus there is a loss of heat to the surrounding atmosphere mainly by convection.

The effect of dirt and dust on collector glazing may be quite small and the cleansing effect of occasional rainfall is usually adequate to clean it. Dust is mostly collected during summertime when rainfall is less frequent, but during this period the dust protects the collector from overheating owing to the high magnitude of solar irradiation.

Low-cost and high-temperature transparent insulating glazing (TIG), which is a polymer-made material, has been developed to be used as a cover for solar collectors. These are called honeycomb collectors. The TIG honeycomb is transparent to solar irradiation, allowing the energy to enter the collector and heat the absorber plate. However, it creates a layer of air that cannot circulate (because of the honeycomb structure), thus dramatically reducing losses related to convection, which is the major reason for energy losses and thus the lower efficiency of flat plate collectors. In addition, the polymer blocks reradiation of the infrared thermal energy, which further reduces energy losses.

A prototype flat plate collector covered by TIG developed by [1] was tested and proved that the efficiency of the collector was comparable to that of evacuated tube collectors (see Section 4.6). Various other prototypes of transparently insulated flat plate collectors and CPCs (see Section 4.4) have been built and were tested in the 1990s [11].

Flat plate collectors (FPCs) are by far the most used type of collector. FPCs are usually employed for low-temperature applications up to 80°C, although the types of collectors employing vacuum insulation and/or transparent insulation (TI) can achieve slightly higher values [1].

The collector plate must have appropriate properties so as to absorb as much of the irradiation as possible, while losing as little heat as possible upward to the atmosphere through the glazing. Therefore, to maximize the energy collection, the absorber plate of a collector should have a coating that has high absorptance for solar radiation and a low emittance for reradiated thermal energy. Such a surface is referred to as a selective surface. Generally, the absorptance of the collector surface for the shortwave solar radiation depends on the nature and color of the coating and on the incident angle. The usual color used is black, but various other colors have been proposed by [14,18], and [9], mainly for esthetic reasons.

An energy-efficient solar collector should absorb the incident solar radiation, convert it to thermal energy, and deliver the thermal energy to a heat transfer medium with minimum losses at each step.

By suitable treatments, using mainly electrolytic or chemical methods, surfaces can be produced with low values of long-wave emittance (ε) and high values of solar radiation absorptance (α). Basically, selective surfaces consist of a thin upper layer, which is highly absorbent to shortwave solar radiation but relatively transparent to long-wave thermal radiation, deposited on a surface that has a high reflectance and a low emittance for long-wave radiation [6]. Selective surfaces are particularly important for collectors operating at high temperatures. The cheapest coating used is mat black paint but this is not selective, and the performance of a collector painted in this way would be very low and could be used only for relatively low ambient temperatures of about T_{amb}+40°C.

For good performance the pipes or passages for fluid-heating collectors must be very well bonded to the absorber plate. This bonding should be done without incurring excessive costs for labor or materials. The most frequently used materials for collector plates are copper, aluminum, and stainless steel. Copper tubes are used most often because of their superior resistance to corrosion and high thermal conductivity.

Over the years a number of absorber plate designs for solar water and air heaters have been applied with varying degrees of success. A basic design is the circular one with pipes used as risers which can be applied above, in the middle, or below the absorbing plate. Maybe the most applied and most efficient design is the middle one, with tubes soldered or brazed on sheets or strips of copper as in Fig. 4.3. It should be noted that soft solder must be avoided because of the high plate temperatures encountered at stagnation conditions, which could melt the solder. Some modern manufacturing techniques have been introduced by the industry, such as the use of ultrasonic welding, which improves both the speed and the quality of welds. This is used for the welding of fins on risers to improve heat conduction. The greatest advantage of this method is that the welding is performed at room temperature and therefore deformation of the welded parts is avoided.

A bonded sheet design, in which the fluid passages are integral with the plate to ensure good thermal conduction, has also been tried. The shape of the passage can be extruded rectangular or hexagonal tubing to obtain a larger heat transfer area between tube and plate.

The major difference between air- and water-based collectors is the need to design an absorber that overcomes the reduced heat transfer caused by the lower heat transfer coefficients between the air and the solar absorber. Therefore, air can be heated with FPCs, particularly if some type of extended surface is used. Thin corrugated metal sheets or porous absorbers may be used and when a high level of performance is required a selective coating is used. It should be noted that the thermal capacity of air is much lower than that of water; thus larger volume flow rates of air are required, which result in a higher pumping power.

The advantages of air systems are that air does not need to be protected from freezing or boiling, is noncorrosive, and does not suffer from heat transfer fluid degradation. Additionally, the system is more cost-effective as no safety values or expansion vessel are required and the air is completely free. The disadvantages are that air-handling equipment (ducts and fans) needs more space than pipes and pumps; air leaks are difficult to detect; parasitic power consumption, ie, electricity used to drive the fans, is generally higher than that of liquid systems; and the systems generally are noisier during operation. Another disadvantage is that air collectors are operated at lower fluid capacitance rates and thus with lower heat removal values (F_R) than the liquid heating collectors [6].

Another category of collectors is the uncovered or unglazed solar collectors. These are usually low-cost collectors, which can offer cost-effective solar thermal energy in low-temperature applications such as water preheating for domestic or industrial use, heating of swimming pools [8], and space heating and air heating for industrial or agricultural applications. If the entire collector area is in contact with the heat transfer fluid, the thermal conductance of the material is not important. UV-resistant plastic extrusions are used for low-temperature applications. These collectors are usually called *panel collectors* and consist of a wide absorber sheet, made of plastic and containing closed-spaced fluid passages. Materials used for plastic panel collectors include polypropylene, polyethylene, acrylic, and polycarbonate.

4.3 Flat plate collectors with diffuse reflectors

A simple way to increase the radiation input on a flat plate solar collector and thus its operating temperature is by using simple flat diffuse reflectors as shown in Fig. 4.4. This is actually a concentrator because the aperture is bigger than the absorber but the system is permanently fixed in position. This enhancement of FPCs was initially suggested by [12]. An analysis and a model of such a system are presented by [13]. The model can be used for the prediction of the total energy absorbed by the collector at any hour of the day for any latitude for random tilt angles and azimuth angles of the collector and reflectors.

The expression "diffuse reflector" denotes a material which is not a specular mirror and thus it does not form an image of the sun on the absorber, which could create uneven radiation distribution and thermal stresses. Diffuse reflectors are usually made

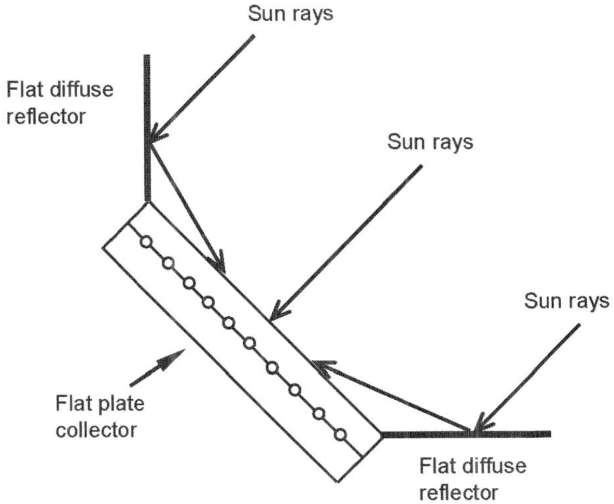

Figure 4.4 Schematic of a flat plate collector with flat diffuse reflectors.

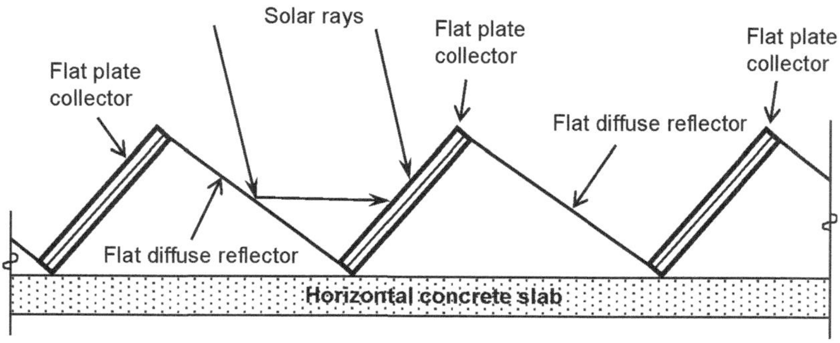

Figure 4.5 Flat plate collectors with reflectors in a sawtooth arrangement.

from stainless steel or galvanized sheets and their cost is usually very small compared to the cost of the collector.

Individual FPCs can be equipped with flat reflectors in the way shown in Fig. 4.4, and this can be on all four sides or on just two suitable sides; however, for multirow collector installations a sawtooth arrangement shown in Fig. 4.5 may be used.

Extensive, mostly experimental, studies on this type of system are presented by [14,15] as part of his studies with collectors employing color absorbers and hybrid photovoltaic/thermal systems [16,17].

4.4 Compound parabolic collectors

CPCs belong in the category of nonimaging concentrators as they do not form an image of the sun on the absorber. These are capable of reflecting to the absorber all of the incident radiation. The first designs of these collectors were developed by [19]; thus sometimes they are also called Winston-type collectors. The basic idea is that the necessity of moving the concentrator to accommodate the diurnal apparent motion of the sun can be reduced by using a trough with two sections of a parabola facing each other, as shown in Fig. 4.6.

Compound parabolic concentrators can accept incoming radiation over a relatively wide range of angles depending on the solar incidence angle by using one or multiple internal reflections. Any radiation that is entering the aperture, within the collector acceptance angle, will hit the absorber surface located at the bottom of the collector. If the reflectivity of the concentrating surface is not high, optical losses may be significant [22]. The absorber of a CPC can take a variety of shapes and as can be seen in Fig. 4.6 it can be flat, bifacial, wedge-shaped, or cylindrical. The first three are a fin type with pipes embedded on the fins (shown in details in Fig. 4.6).

Two basic types of CPCs have been developed: the symmetric, shown in Fig. 4.6, and the asymmetric, which have shapes similar to those shown in the next section.

The collector can be stationary or tracking, depending on the acceptance angle. When tracking is used this is very coarse or intermittent as the concentration ratio is

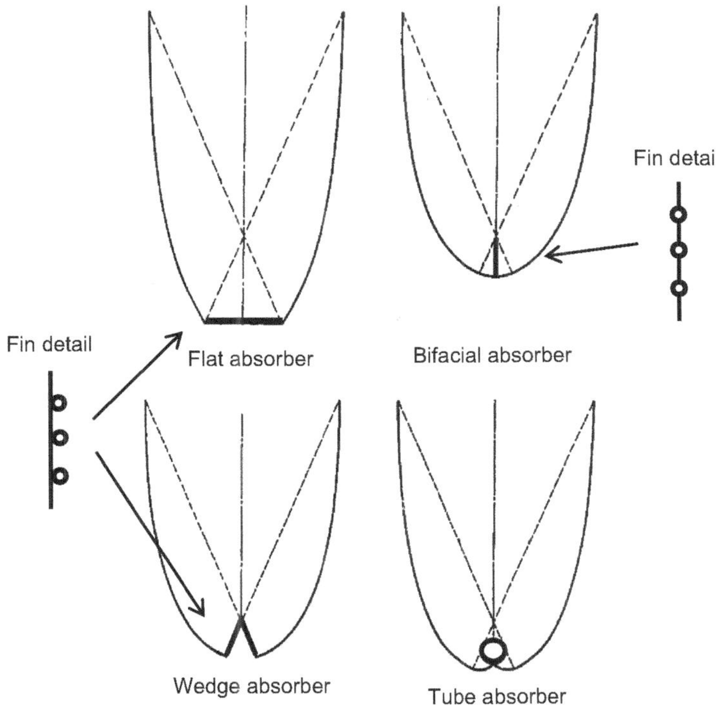

Figure 4.6 Possible absorber types for compound parabolic collectors and fin details.

small. Solar radiation is collected and concentrated by one or more reflections on the parabolic surfaces. For higher temperature applications higher concentration ratios and more refined tracking are required.

CPCs can be designed either as one large unit with one opening and one receiver as shown in Fig. 4.6 or as a panel, which looks like an FPC, as shown in Fig. 4.7.

In the following an optical and thermal analysis of CPCs is presented.

A Winston design CPC [20] is shown in Fig. 4.8. This is a nonimaging concentrator. It is a linear two-dimensional concentrator consisting of two parabolas, A and B, the axes of which are inclined at the collector half-acceptance angle ($\pm\theta_c$) on either side

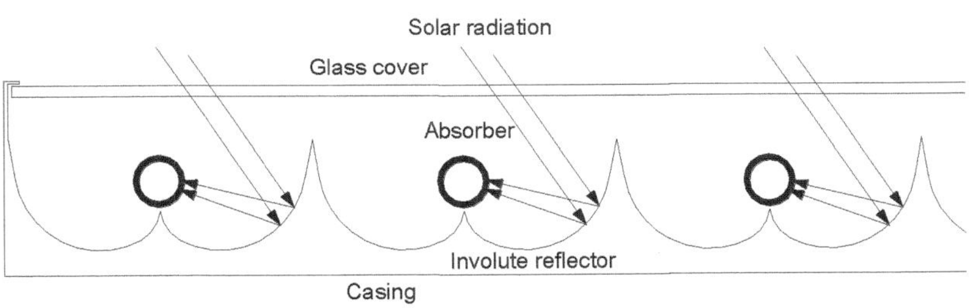

Figure 4.7 Detail of a panel compound parabolic collector with cylindrical absorbers.

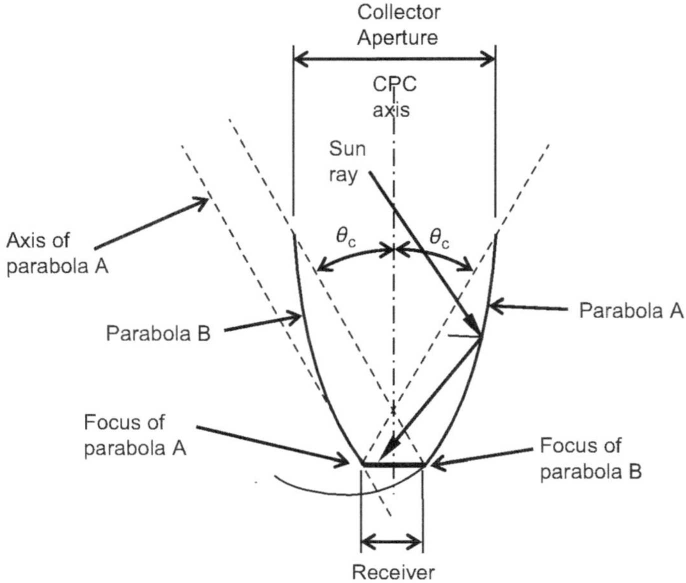

Figure 4.8 Design details of a flat receiver compound parabolic collector (CPC).

of the collector optical axis. The collector θ_c is defined as the angle through which a source of light can be moved from the normal to the collector axis and still converge at the absorber. CPCs have a constant acceptance angle over the entire aperture area [22].

A cylindrical receiver collector is shown in Fig. 4.9. As was indicated earlier, the receiver of the CPC does not have to be flat and parallel but can be bifacial,

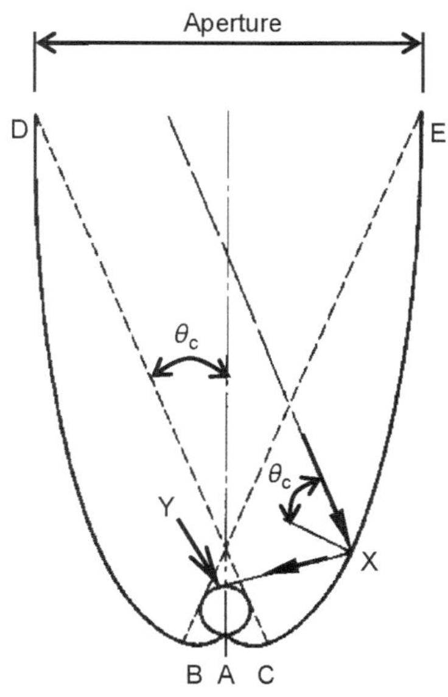

Figure 4.9 Schematic diagram of a compound parabolic collector with cylindrical receiver.

wedge-shaped, or cylindrical. In the cylindrical receiver collector, the small lower portions of the reflector (AB and AC) are of circular shape and the upper portions (BD and CE) are of parabolic shape. In this design, it is required that, for the parabolic portion of the collector at any point X, the normal to the collector must bisect the angle between the tangent line to the receiver XY and the incident ray at point X at angle θ_c with respect to the collector axis, as shown. The side wall profile of fully developed CPCs may terminate when it becomes parallel to the optical axis. Usually very little concentration is lost by truncating these devices by some fraction (about 0.6–0.9) relative to their full height [22]. Therefore, a shorter version of the CPC is obtained with less reflective material, which affects marginally the acceptance angle but changes the height-to-aperture ratio, the concentration ratio, and the average number of reflections. As the reflectivity of the mirrored surfaces is affected by dust and other material deposits, CPCs are usually covered with glass.

The orientation of a CPC is related to its acceptance angle. A two-dimensional CPC is an ideal concentrator, ie, it perfectly utilizes all rays within the acceptance angle $2\theta_c$. The collector can be stationary or tracking depending on the collector acceptance angle. Both north–south and east–west directions can be employed with respect to the orientation of its long axis. In both cases its aperture is tilted directly toward the equator at an angle equal to the latitude of the location.

When the collector axis is oriented along the north–south direction this must track the sun periodically by turning its axis so that the solar incidence is within the acceptance angle of the concentrator. Depending on the application (requirement over a certain time span) the collector can also be stationary but radiation will be received only during the hours when the sun is within the collector acceptance angle [6].

When the concentrator is oriented with its long axis along the east–west direction, the collector is able to utilize the sun's rays effectively through its acceptance angle. For stationary CPCs mounted in this mode the minimum acceptance angle should be 47 degrees. This angle covers the declination of the sun from the summer to the winter solstice ($2 \times 23.5°$). Bigger angles are used in practice with lower concentration ratio to enable the collector to also collect diffuse radiation. Small CPCs, with a concentration ratio less than 3, are of the greatest practical interest [10]. These are able to accept a large amount of diffuse radiation incident on their apertures and concentrate the beam radiation without the need of tracking the sun. As a general guideline, the required frequency of collector adjustment is related to the collector concentration ratio. For $C \leq 2$ the collector can be steady, whereas for $C = 3$ the collector needs only biannual adjustment, and for C close to 10 it requires almost daily adjustment, and these systems are also called quasi-static [6].

4.5 Reverse flat plate collectors

Another way to increase the radiation falling on FPCs to extend their operation to medium temperatures has been designed, in which the absorber plate is upside down, and this type of system is called a *reverse flat plate collector*. Initial attempts to design this

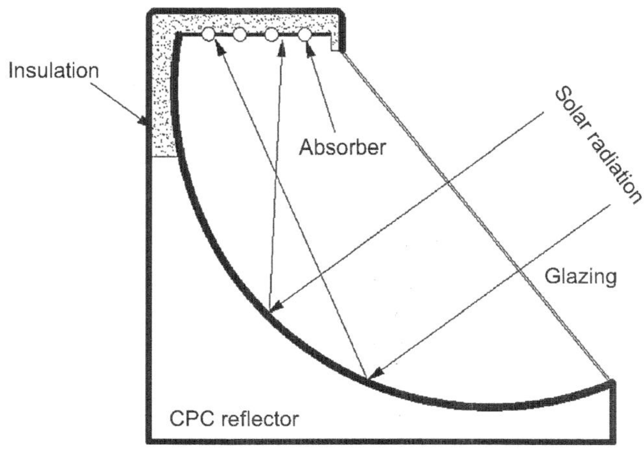

Figure 4.10 Detail of inverted flat plate collector. *CPC*, compound parabolic collector.

type of system were carried out by [7]. On these systems radiation is directed to the underside of the absorber plate by a stationary CPC-type concentrator of the shape shown in Fig. 4.10. This is in fact a CPC described in previous section. As the upper side of the plate is insulated and as the plate is upside down, the convection losses are greatly reduced; thus heat losses from the absorber are significantly reduced. A variation of this type is the inclined design shown in Fig. 4.11. Compared to a normal FPC, the reverse plate design has higher optical losses, due to the scattering losses in the reflector, but has higher incident radiation and lower thermal losses.

Another variation of the concept is the double-sided FPC developed by [3] and [15], shown in Fig. 4.12(a,b). As can be seen in this design the absorber plate is bifacially irradiated as this is illuminated at both sides with solar radiation. In the design developed by [3] the absorber is enclosed in a TI casing, whereas in the

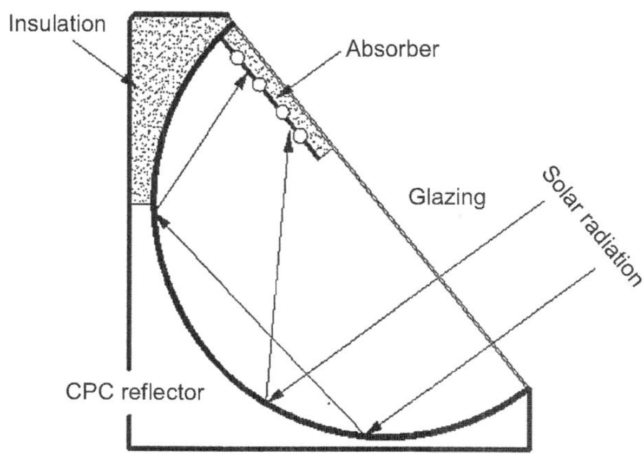

Figure 4.11 Detail of inclined flat plate collector. *CPC*, compound parabolic collector.

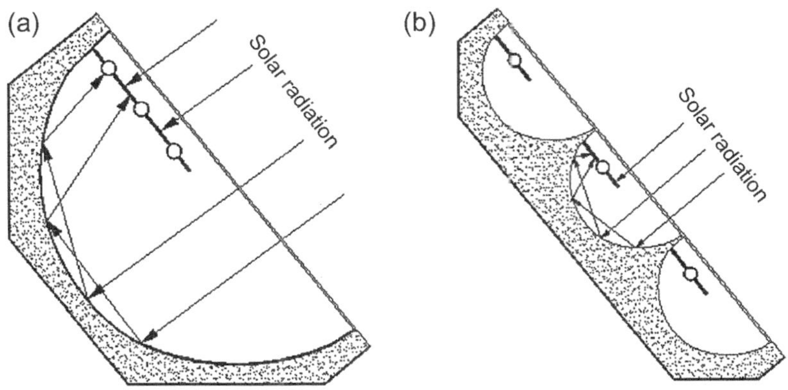

Figure 4.12 Cross-sectional detail of an inverse flat plate compound parabolic collector (CPC). (a) CPC with one mirror-absorber unit. (b) CPC with three mirror-absorber units. Modified from Y. Tripanagnostopoulos, P. Yianoulis, S. Papaefthimiou, S. Zafeiratos, CPC solar collectors with flat bifacial absorbers, Sol. Energy 69 (3) (2000) 191–203.

design developed by [15] a simple glazing is used, either in one mirror-absorber unit or in three mirror-absorber units, as shown in Fig. 4.12(a,b), respectively, adapted from [15] with many design details removed for clarity. In all these units the absorber is bifacial, ie, illuminated with solar radiation in both sides.

4.6 Evacuated tube collectors

Evacuated tube solar collectors operate different from the collectors presented so far as they usually consist of a heat pipe inside a vacuum-sealed tube. The area of one tube is small; therefore to increase the heat collection area many tubes are connected to the same manifold as shown in Fig. 4.13 (usually 10–20 tubes are used). The evacuated tube collector (ETC) usually employs a fin with a tube as in FPCs as shown in the cross-sectional detail of Fig. 4.13.

Relatively high temperatures can be obtained from ETCs owing to the combination of the selective surface used for the fin and the effective convection suppression obtained because of the vacuum insulation. Like FPCs they collect both direct and diffuse radiation; their efficiency, however, is higher at low incidence angles, which gives an advantage of the ETC over FPCs in daylong performance [6].

ETCs use liquid–vapor phase change to transfer heat at high efficiency. The pipe is a sealed copper pipe, which is attached to a black copper fin that fills the tube and forms the collector absorber plate. Each tube is terminated at a metal tip, which is attached to the sealed pipe and acts as a condenser. The heat pipe contains a small amount of volatile fluid (usually methanol or ethanol) that undergoes, as long as there is sunshine, an evaporating–condensing cycle, ie, solar heat evaporates the liquid and,

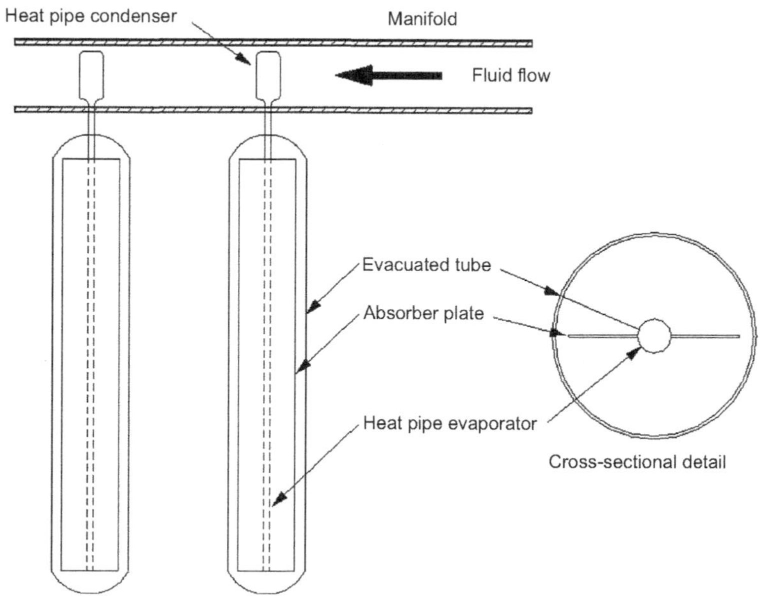

Figure 4.13 Schematic diagram of an evacuated tube collector and cross-sectional detail.

owing to lower density, the vapor rises to the heat sink region where it condenses and releases its latent heat to a flowing fluid. Then the condensed fluid returns to the solar collector owing to gravity and the process is repeated as long as there is solar heat to evaporate the fluid. Usually water, or a water–glycol mixture, flows through the manifold to pick up the heat from the condensation of the vapor. The circulated heated liquid either goes through a heat exchanger to give its heat to a process or is stored in a storage tank for later use.

One of the greatest advantages, which is a unique feature of the evacuated heat pipe collector, is that because no evaporation or condensation above the phase-change temperature is possible, the heat pipe offers inherent protection from freezing and overheating [6].

Another design of ETC that exists on the market consists of an all-glass Dewar-type ETC. This type is also called a *wet-tube ETC*. In this, two concentric glass tubes separated by a vacuum space are used. The selective coating is deposited on the outside surface of the inner glass tube, which is domed at one end. A second larger diameter domed glass tube is then inserted on the outside of the first tube and these are joined at the open end. Therefore, this design has the advantage that it is made entirely of glass and avoids the necessity of penetrating the glass envelope to extract the heat from the tube. This eliminates possible leakage losses and is cheaper than the single envelope system [6].

As the tubes cannot be very near to each other, some space is wasted between the tubes; therefore it is very much cost-effective to install a diffuse reflector at the back of the tubes to utilize any, otherwise wasted, solar radiation. This gives an opportunity to increase the distance between the tubes and allow the diffuse reflector to utilize all the available radiation as shown in Fig. 4.14. A diffuse reflector (reflectivity, $\rho = 0.6$)

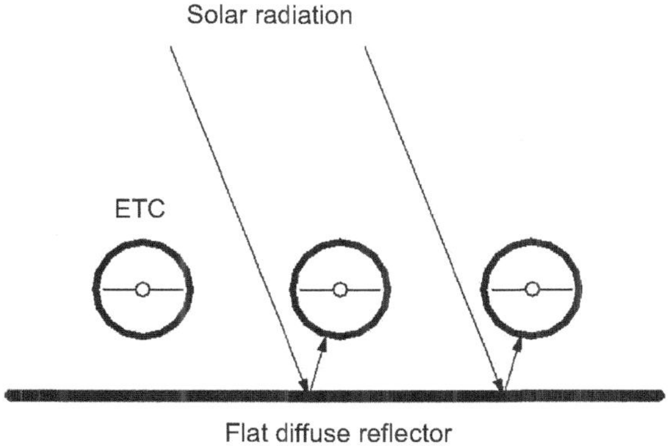

Figure 4.14 Detail of evacuated tube collectors (ETCs) with external flat diffuse reflector and tube spacing equal to one tube diameter.

mounted behind the tubes spaced one tube diameter apart increases the absorbed energy in each tube by more than 25% for normal incidence. This system presents also a 10% increase in energy collection over a full day because of incidence angle effects.

A better enhancement per tube can be achieved by using CPC-type reflectors as shown in Fig. 4.15. In this design the number of tubes is decreased even more and reflectors are used to concentrate the solar radiation onto the tubes. Evacuated tube arrays with this type of stationary concentrators may have stagnation temperatures exceeding 300°C.

Evacuated tubes with external or internal (inside the glass tube) reflectors are also commercialized by several manufacturers. When the reflector is installed inside the tube the system is called an integrated CPC (ICPC). This is an ETC in which at the bottom part of the glass tube a reflective material is used [21]. To allow solar radiation to reach the reflector more efficiently, no fin is used in this case. Depending on the operating temperature required, either a CPC reflector, shown in Fig. 4.16(a), or a cylindrical reflector, shown in Fig. 4.16(b), is used. As can be understood the cylindrical reflector does not achieve the concentration of the shaped reflector but has a lower manufacturing cost. The CPC-type collector, shown in Fig. 4.16(a), combines into a single unit the advantages of vacuum insulation and nonimaging stationary concentration. In another variation, a tracking ICPC has been developed which is suitable for higher temperature applications [4].

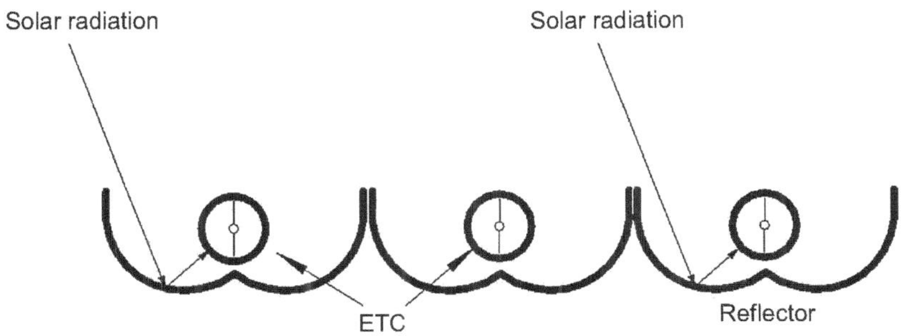

Figure 4.15 Evacuated tube collectors (ETCs) acting as receivers of CPC-type reflectors.

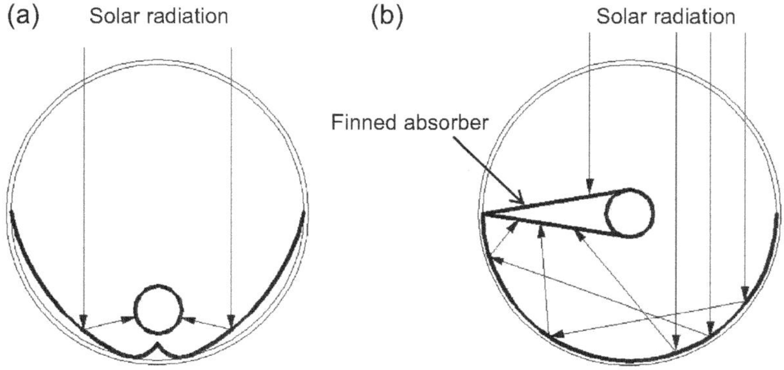

Figure 4.16 Integrated compound parabolic collector tube collectors. (a) Internal compound parabolic. (b) Circular reflector with finned absorber.

ETCs are produced in sizes with outer diameters varying from 30 mm to about 100 mm. The usual length of these collectors is about 2 m.

4.7 Conclusions

A number of nontracking solar collector types which are suitable for heating and cooling systems are presented in this chapter. These include the FPCs, FPCs with diffuse reflectors, CPCs, reverse FPCs, and ETCs.

Many more details and methods for the design of these systems are presented in the book by the author, *Solar Energy Engineering: Processes and Systems* [6].

Future trends in research and development for these types of collectors are focused on the use of improved material properties so as to increase the performance of the collectors and the use of nanofluids as the heat transfer medium, which have shown to increase the heat transfer coefficients of the collectors.

Glossary

C	Concentration ratio
CPC	Compound parabolic collector
ETC	Evacuated tube collector
ICPC	Integrated compound parabolic collector
ICS	Integrated collector storage
TI	Transparent insulation

References

[1] N. Benz, W. Hasler, J. Hetfleish, S. Tratzky, B. Klein, Flat-plate solar collector with glass TI, in: Proceedings of Eurosun'98 Conference on CD ROM, Portoroz, Slovenia, 1998.

[2] J.A. Duffie, W.A. Beckman, Solar Engineering of Thermal Processes, John Wiley & Sons, New York, 1991.

[3] A. Goetzberger, J. Dengler, M. Rommel, J. Goettsche, V. Wittwer, New transparently insulated, bifacially irradiated solar flat-plate collector, Sol. Energy 49 (5) (1992) 403–411.

[4] C. Grass, N. Benz, Z. Hacker, A. Timinger, Tube collector with integrated tracking parabolic concentrator, in: Proceedings of the Eurosun'2000 Conference on CD ROM, Copenhagen, Denmark, 2000.

[5] S. Kalogirou, Solar thermal collectors and applications, Prog. Energy Combust. Sci. 30 (3) (2004) 231–295.

[6] S.A. Kalogirou, Solar Energy Engineering: Processes and Systems, second ed., Academic Press, Elsevier Science, 2013, ISBN 978-0-12-397270-5.

[7] V. Kienzlen, J.M. Gordon, J.F. Kreider, The reverse flat-plate collector: a stationary non-evacuated, low technology, medium temperature solar collector, ASME J. Sol. Energy Eng. 110 (1988) 23–30.

[8] B. Molineaux, B. Lachal, O. Gusian, Thermal analysis of five outdoor swimming pools heated by unglazed solar collectors, Sol. Energy 53 (1) (1994) 21–26.

[9] Z.C. Orel, M.K. Gunde, M.G. Hutchins, Spectrally selective solar absorbers in different non-black colors, in: Proceedings of WREC VII, Cologne on CD ROM, 2002.

[10] M. Pereira, Design and performance of a novel non-evacuated 1.2x CPC type concentrator, in: Proceedings of Intersol Biennial Congress of ISES, Montreal, Canada, vol. 2, 1985, pp. 1199–1204.

[11] F. Spate, B. Hafner, K. Schwarzer, A system for solar process heat for decentralized applications in developing countries, in: Proceedings of ISES Solar World Congress on CD ROM, Jerusalem, Israel, 1999.

[12] H. Tabor, Mirror boosters for solar collectors, Sol. Energy 10 (3) (1966) 111–118.

[13] H.P. Garg, D.S. Hrishikesan, Enhancement of solar energy on flat-plate collector by plane booster mirrors, Sol. Energy 40 (4) (1998) 295–307.

[14] Y. Tripanagnostopoulos, M. Souliotis, Th. Nousia, Solar collectors with colored absorbers, Sol. Energy 68 (4) (2000) 343–356.

[15] Y. Tripanagnostopoulos, P. Yianoulis, S. Papaefthimiou, S. Zafeiratos, CPC solar collectors with flat bifacial absorbers, Sol. Energy 69 (3) (2000) 191–203.

[16] Y. Tripanagnostopoulos, Th. Nousia, M. Souliotis, P. Yianoulis, Hybrid photovoltaic/thermal solar systems, Sol. Energy 72 (3) (2002) 217–234.

[17] Y. Tripanagnostopoulos, Aspects and improvements of hybrid photovoltaic/thermal solar energy systems, Sol. Energy 81 (9) (2007) 1117–1131.

[18] J. Wazwaz, H. Salmi, R. Hallak, Solar thermal performance of a nickel-pigmented aluminum oxide selective absorber, Renewable Energy 27 (2) (2002) 277–292.

[19] R. Winston, Solar concentrators of novel design, Sol. Energy 16 (1974) 89–95.

[20] R. Winston, H. Hinterberger, Principles of cylindrical concentrators for solar energy, Sol. Energy 17 (4) (1975) 255–258.

[21] R. Winston, J. O'Gallagher, J. Muschaweck, A. Mahoney, V. Dudley, Comparison of predicted and measured performance of an integrated compound parabolic concentrator (ICPC), in: Proceedings of ISES Solar World Congress on CD ROM, Jerusalem, Israel, 1999.
[22] R. Winston, Solar concentrators, in: J. Gordon (Ed.), Solar Energy: The State of the Art, ISES, 2001, pp. 357–436.

Tracking solar collection technologies for solar heating and cooling systems

5

C. Chang
Key Laboratory of Solar Thermal Energy and Photovoltaic System, Institute of Electrical Engineering, Chinese Academy of Sciences, Beijing, China

5.1 Definition of solar tracking technology

The solar tracking device (also called a solar tracker) is a key component to improve the performance of solar collectors. A solar tracker can keep the collector aperture perpendicular to the incident light to maximize the solar radiation. It must be reliable and able to follow the sun with a certain degree of accuracy, return the collector to its original position at the end of the day or during the night, and track during periods of intermittent cloud cover. In addition, a solar tracker can turn the collector out of focus to protect it from hazardous environmental and working conditions, such as wind gusts, overheating, and failure of the system. A solar tracker is usually composed of three parts: the structure, the transmission device, and the control system. According to the operation mode, solar trackers can be divided into two categories: manual tracking and automatic tracking. Automatic tracking is divided into time tracking and servo tracking. According to the number of degrees of freedom of the drive device, the solar tracker can be divided into single-axis tracking and dual-axis tracking. According to the control mode of the control system, the solar tracker can be divided into photoelectric sensor tracking and sun position tracking. According to the control information feedback, the solar tracker can be divided into closed-loop, open-loop, and hybrid control.

5.2 Classification and features

5.2.1 Manual tracking

A manual solar tracking device does not need to track the sun's position in real time. It needs only to be directed in accordance with the number of hours of sun or in accordance with the size of the acceptance angle. It can keep the collector working effectively by regular or seasonal adjustment of the dip angle. The advantages of this device are its simple structure and convenient operation and control, and the disadvantage is that the collector temperature is low; a temperature range of only 100–200°C can be achieved. Because these collectors do not need to track the sun in real time, this kind of technology is often regarded as nontracking or adjustable. A typical manual

Advances in Solar Heating and Cooling. http://dx.doi.org/10.1016/B978-0-08-100301-5.00005-9

solar tracking device is the compound parabolic concentrator (CPC). CPCs are nonimaging concentrators [1]. They are adjustable concentrators and have the capability of reflecting to the absorber all of the incident radiation within a wide limit. Thus systems of this kind can be characterized as quasi-static. The required frequency of adjustment goes up rapidly with the concentration ratio C. When $C = 3$ only biannual adjustments are needed, whereas $C = 10$ requires almost daily updates.

θ_{max} is the CPC's maximum acceptance angle; it can be simply defined as half of the angle between the upper and the lower ends of the parabolas A and B, as shown in Fig. 5.1.

5.2.2 Automatic tracking

The main components and functions of the solar tracking system often contain the following parts:

Sun position calculation program: The sun position calculation program is used to confirm the direction of the sun's rays with respect to the focal plane of the collector, and the output value of the program is generally the azimuth and height of the sun.

Processor: The main functions of the processor include the implementation of the sun position calculation program, calculating the sun's azimuth and height angle; the control logic under various working conditions; running the instruction control motor; and the collection and processing of the sensor, the information of the instrument. An industrial-grade standard PLC, controllers developed by enterprise or research institutions based on the microcontroller, an ARM, or a DSP can all be used as a processor.

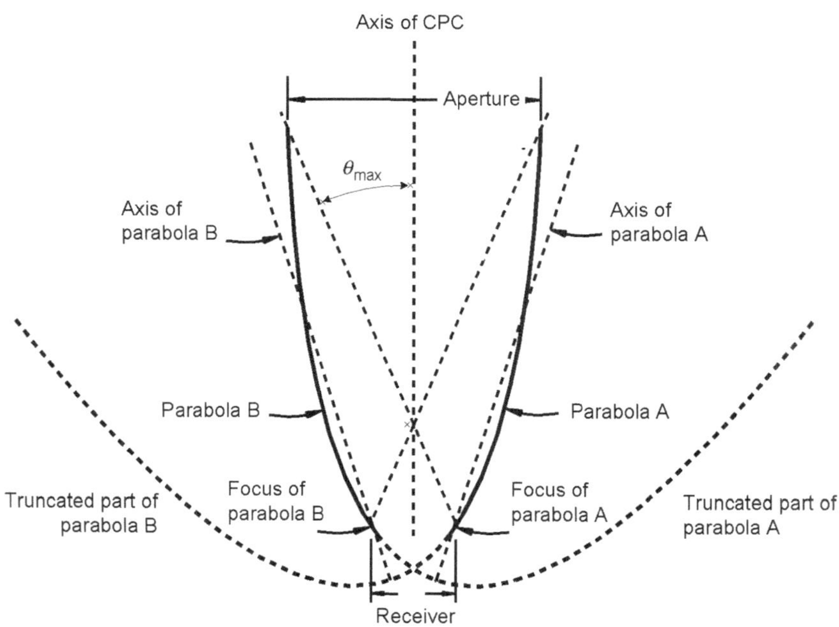

Figure 5.1 Structure of a compound parabolic concentrator (CPC).

Sensors: Sensors are used to judge the status information of the collector. According to the various control modes, the sensor can choose the rotary encoder, the angle sensor, the photoelectric sensor, and so on. A rotary encoder is generally installed in the motor shaft end. It is used to detect the angle of rotation of the motor. The processor analyzes the motor rotation angle information, which is equivalent to the reduction ratio, and then calculates the collector's actual angle. The drive shaft angle sensor is usually installed in the parabolic trough concentrator or a heliostat mirror is installed in parallel. It can directly measure the actual tilt angle of concentrator. Voltage signals of the photoelectric sensor correspond to the incident direction of the sun's rays, which can be used to determine the relative position between the sun's rays and the photoelectric sensor, and thus instruct the concentrator to follow by starting, stopping, and turning.

Transmission device: The transmission device can directly output the torque, driving the collector to track the sun. According to the various forms, the drive device may be a hydraulic drive, gear reducer drive, and so on.

Measurement and control system: The measurement and control system mainly includes the information on the solar radiation, the wind speed, the temperature, and so on. The solar radiation includes direct normal irradiation, scattered irradiation, and total irradiation.

5.3 Control system

5.3.1 *Principle of manual tracking control*

As shown in Fig. 5.2, θ_i is the angle of incidence; it is the angle between the incident light and the CPC axis. When $\theta_i < \theta_{max}$, the incident light can be reflected to the surface of the solar absorber tube. When $\theta_i > \theta_{max}$, the incident light can be reflected out

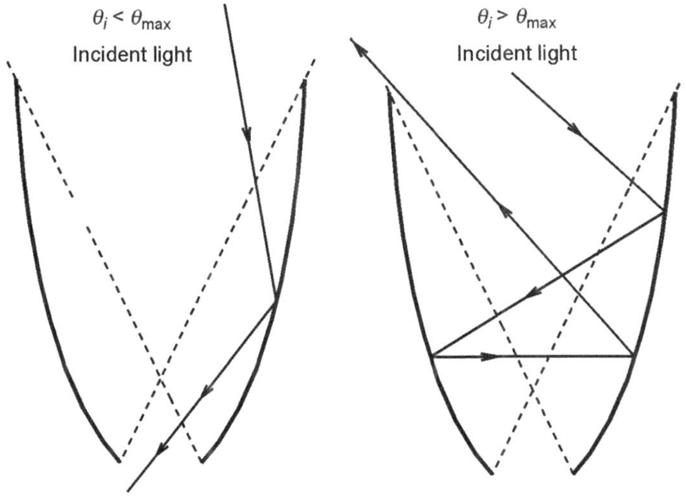

Figure 5.2 Reflection principle of a compound parabolic concentrator (CPC).

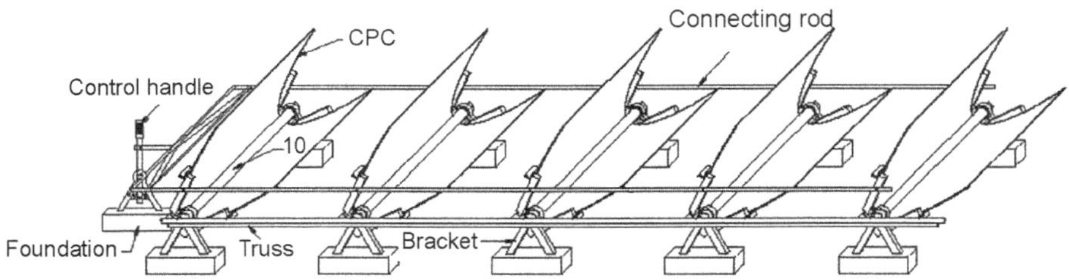

Figure 5.3 Manual tracking device of a compound parabolic concentrator (CPC).

from the collector aperture, unable to reach the surface of the solar absorber tube. The receiver can take a variety of configurations. It can be flat, wedge-shaped, bifacial, or cylindrical.

Usually, the CPC needs to be placed along the horizontal direction concentrator, the collector aperture to the sun. Using the mechanical structure shown in Fig. 5.3, it can adjust the angle of incidence θ_i to ensure $\theta_i \leq \theta_{\max}$.

5.3.2 Principle of closed-loop control

The closed-loop control method is used in sensors to measure the deviation between the incident light and the focal plane of the collector. When the deviation exceeds a threshold, the mechanical part of the machine is rotated, and the deviation is reduced, so as to realize the tracking of the altitude angle and azimuth angle of the sun. The control diagram is shown in Fig. 5.4. The commonly used sensors are solar photovoltaic cells, photosensitive resistance, photoelectric tubes, one-dimensional (1D) PSD and 2D PSD photoelectric position sensors, photoelectric angle sensors, etc.

For photoelectric sensor tracking, when the sun's position changes, the change in the intensity of the sunlight can cause a change in the electrical signal output of the photoelectric sensor. Then after analysis, judgment, and processing, the obtained results are used to control the operation of the drive motor to change the position of the solar tracking device.

Photoelectric sensors will capture the sun's position in real time. Computers can do an accounting analysis and comparison of the intensity of the sun and drive the device to track the sun. This method is not affected by the installation location of the solar tracking device or the influence and restrictions of wintertime or summertime. The device is easy to use, is flexible, and has a high tracking accuracy. The main drawback of this approach is that on cloudy days, the solar radiation is relatively weak, and it is very difficult for the photoelectric converter to respond to the changes in light. In cloudy weather, when the sun is obscured by the clouds or, because of thin clouds, appears as a relatively bright spot in the sky, the photoelectric tracking method may cause the tracking device to make mistakes and even cause serious accidents.

5.3.3 Principle of open-loop control

Open-loop control is the same as sun position tracking. According to astronomical formulas, the sun's azimuth and height angle real-time data can be calculated from sunrise to sunset every day and every year. All data are stored in the computer, according to

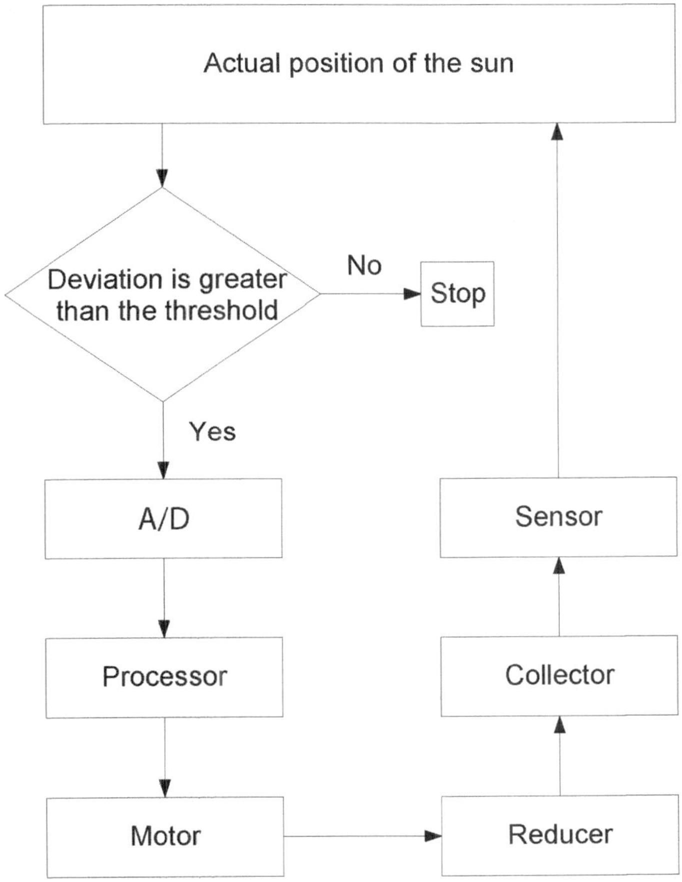

Figure 5.4 The control diagram of a closed-loop control.

the sun's designated position at the time, to control the motor rotation driven by the tracking device to track the sun.

The open-loop control method determines the location of the sun by using the location and running time information in the collector. Calculate angles of the sun through the time and latitude and longitude information, and then calculate the target angle of tracking solar concentrator, and then program control driving solar concentrator from present position to the target angle. When the actual angle of solar concentrator is equal to the target angle, open-loop control is completed. This method is not affected by weather conditions, but the sun's trajectory changes with the location of the tracking device, the latitude, and seasonal variations. The tracking position at any time is determined by the time and space functions of the tracking device. The premise of accurate tracking is to ensure the accurate positioning of the sun tracking device and the algorithm of the sun's motion. At the same time, it is also required that the limit switch is set to achieve the desired reduction of the device. In terms of hardware, the open-loop control system generally uses a limit switch, with the collocation of a stepper motor, a servomotor, or an encoder motor to achieve the positioning function of the condenser. As for software, the open-loop control is required to carry out the complex positioning control of the motor, and the complexity of the program is relatively high.

5.3.4 Principle of hybrid control

Hybrid control is the combination of open-loop and closed-loop control. In the actual operating system, the two tracking modes will coexist. According to the state of the environment, when the weather is sunny and solar irradiance is strong enough, the system will give priority to the photoelectric sensor tracking mode, and on rainy days, when solar irradiance is not strong enough, the system will automatically switch to the solar trajectory tracking mode. The two kinds of modes cooperate with each other and thus can achieve high-precision all-weather automatic tracking.

5.4 Practical examples

5.4.1 Single-axis tracking

Single-axis tracking has three kinds of layout methods:

- Inclined shaft installation, east–west direction for tracking
- South–north axis horizontal installation, east–west direction for tracking
- East–west axis horizontal installation, south–north direction for tracking

These three methods are all single-axis rotation tracking, and the work principles are similar. According to the results of the calculation of the rotation angle of the sun, the collectors rotate around the axis of rotation to track the sun. At high noon, the angle between the incident direction of the sun and the normal direction of the collector's aperture is the smallest; collectors can then obtain the maximum heat flux. In the morning or afternoon the sunlight is oblique. Single-axis tracking has the advantage of simple structure; the disadvantage is that the incident light cannot always be perpendicular to the collector's aperture, so the collection of solar energy is not enough to maintain the maximum.

Collectors that usually use single-axis tracking are CPCs, parabolic trough collectors (PTCs), and linear Fresnel collectors.

An example of a single-axis manual tracking device with a low concentration ratio (Fig. 5.5) is a CPC solar thermal utilization system located in Tibet, China. The width of the CPC is 0.4 m, and the length is 1.8 m. The total number of collectors is 96. This system needs to adjust its state only four times a year. The heat transfer medium is mineral oil and the working temperature is 160°C.

The PTC is a kind of single-axis automatic tracking collector with a high concentration ratio, usually around 40–80 times the sun. PTCs are made by bending a sheet glass mirror or reflective aluminum into a parabolic shape. PTCs can produce heat at high temperatures, between 150°C and 400°C. Parabolic trough receivers are the heat absorbers for parabolic trough concentrators. Generally, a vacuum tube with two terminal openings is composed mainly of a glass tube, a metal heat-absorbing tube, a vacuum interlayer, and a thermal stress buffer section. When the parabolic trough is

Figure 5.5 A compound parabolic concentrator solar thermal utilization system, Tibet, China.

pointed toward the sun, the solar radiation will be reflected onto the receiver tube. The concentrated radiation reaching the receiver tube heats the working fluid, which circulates through it, thus transforming the solar radiation into useful heat. The collector can be oriented in an east–west direction, tracking the sun from north to south, or in a north–south direction, tracking the sun from east to west. During the period of a whole year, a horizontal north–south trough field usually collects slightly more energy than a horizontal east–west one. However, the north–south field collects a lot of energy in summer and much less in winter. The east–west field collects more energy in winter than a north–south field and less in summer, providing a more constant annual output. Therefore, the choice of orientation usually depends on the application and whether more energy is needed during summer or winter.

Parabolic trough technology is the most advanced of the solar thermal technologies because of considerable experience with the systems and the development of a small

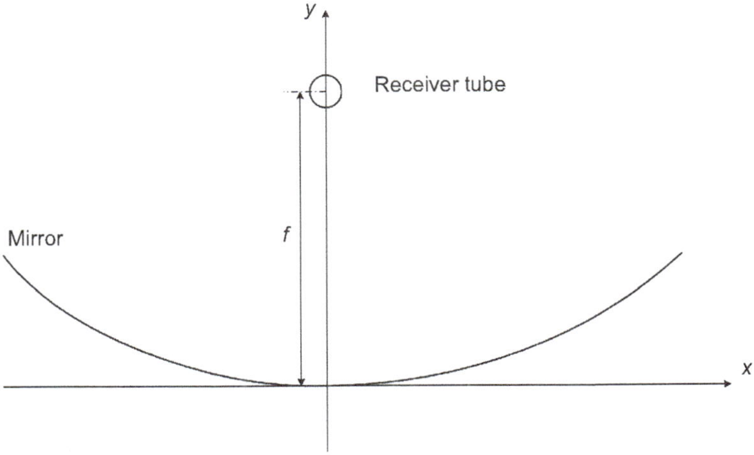

Figure 5.6 The principle of a parabolic trough concentrator.

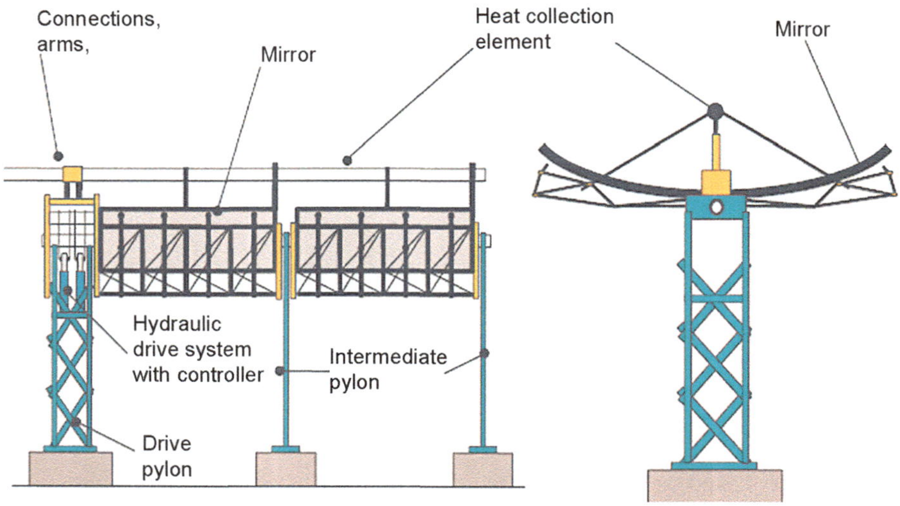

Figure 5.7 The structure of a parabolic trough concentrator.

commercial industry to produce and market this system. Fig. 5.6 shows the principle of the concentrating PTC, and its structure is shown in Fig. 5.7.

Fig. 5.8 is a parabolic trough solar heating system in Hainan Province, China. It is a solar thermal power generation and saltwater desalination system; its working temperature is 390°C.

Fresnel collectors are also single-axis tracking collectors with a high concentration ratio, usually around 50–100 times the sun. Fresnel collectors have two variations: the Fresnel lens collector and the linear Fresnel reflector. The former is made from a plastic material and is shaped in such a way as to focus the solar rays to a point receiver, as shown in Fig. 5.9.

Figure 5.8 A parabolic trough solar heating system in Hainan, China.

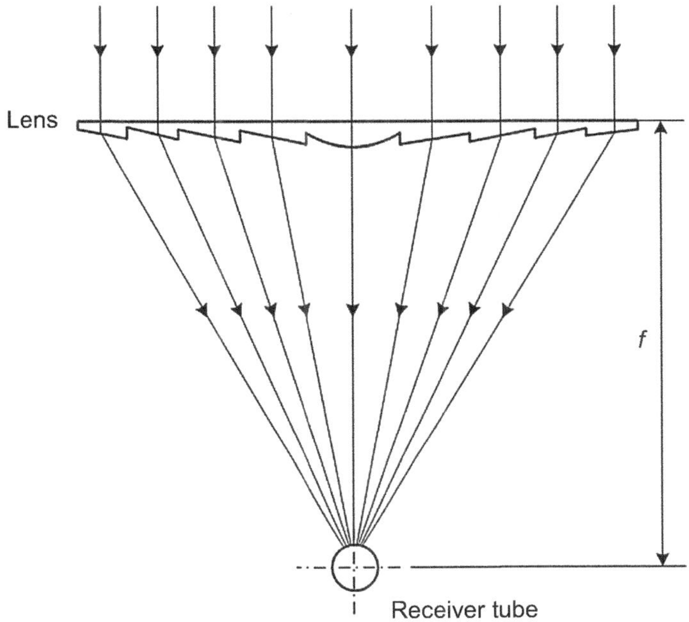

Figure 5.9 Fresnel lens collector.

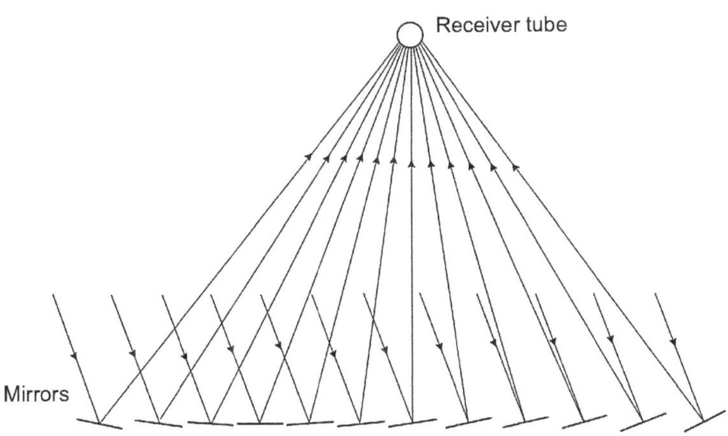

Figure 5.10 The principle of a linear Fresnel reflector.

The linear Fresnel reflector relies on an array of linear mirror strips that concentrate light onto a linear receiver. The linear Fresnel reflector can be imagined as a broken-up parabolic trough reflector; the individual strips can be parabolic or flat in shape as shown in Fig. 5.10. Large absorbers can be constructed and the absorber does not have to move. The greatest advantage of this type of system is that it uses flat or elastically curved reflectors, which are cheaper than parabolic glass reflectors. These are

Figure 5.11 A linear Fresnel reflector solar heating system in Shandong, China.

mounted close to the ground, thus minimizing structural requirements. The heat-absorbing tube may be installed with an inclined shaft and can be installed horizontally.

Fig. 5.11 is a linear Fresnel reflector solar heating system in Shandong Province, China. This is a solar thermal power generation system and the total capacity is 2.5 MW.

5.4.2 Dual-axis tracking

The altitude angle and azimuth angle of the sun are changing all the time. The dual-axis tracking device tracks the sun to collect more solar energy. According to the type of axis, the dual-axis tracking device can be divided into two types: polar-axis tracking and altitude—azimuth tracking.

Polar-axis tracking is also called spinning-elevation tracking. The principle of the polar-axis tracking device is one axis of the collector or concentrator pointing to the celestial north pole, which is parallel to the earth's rotation axis. The other axis is perpendicular to the polar axis, called the declination axis. When the device is working, the mirror or the collector rotates around the pole axis and the rotation speed is the same as that of the earth's rotation angle, but the rotation direction is opposite. Fig. 5.12 shows the principle of a polar-axis tracking collector and Fig. 5.13 shows a polar-axis tracking heliostat in Shanxi Province, China; its total aperture is 16 m^2.

The tracking equation, obtained from Chen [2–6], can be expressed as

$$\theta = \frac{\pi}{4} - \frac{1}{2}\arcsin\left\{\begin{array}{l} -\cos\delta\cos\omega(\sin\lambda\cos\Phi + \cos\lambda\cos\varphi\sin\Phi) \\ -\cos\delta\sin\omega\cos\lambda\sin\varphi + \sin\delta(\cos\lambda\cos\varphi\cos\Phi - \sin\Phi\sin\lambda) \end{array}\right\} \quad [5.1]$$

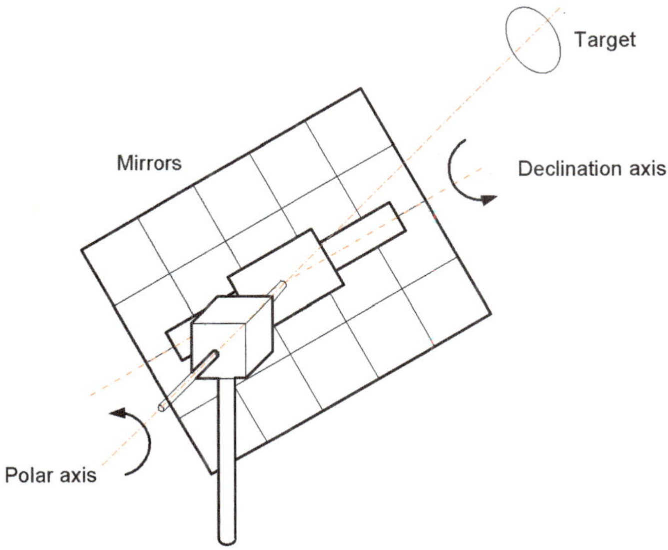

Figure 5.12 The principle of a polar-axis tracking collector.

Figure 5.13 A polar-axis tracking concentrator in Shanxi, China.

$$\rho = \arcsin\left\{\frac{-\cos\delta\cos\omega\sin\varphi\sin\Phi + \cos\delta\sin\omega\cos\varphi + \sin\delta\sin\varphi\cos\Phi}{\sin(2\theta)}\right\} \quad [5.2]$$

where θ is the tracking elevation angle; ρ is the spinning tracking angle of the collector; δ is the sun's equatorial latitude angle; ω is the solar hour angle, it is 0 when the sun reaches its highest position during the day; φ is the collector spinning axis (fixed axis) to the northeast angle, to the north is 0 and to the east is 90 degrees; λ is the angle between the spin axis and the horizontal plane; and Φ is the local latitude.

In altitude–azimuth tracking, the azimuth axis is perpendicular to the ground, and the axis perpendicular to the azimuth axis is called the altitude axis. When the system is working, the mirror or collector rotates in the direction of the sun, and the azimuth angle changes. When it rotates around the altitude axis, the altitude angle changes. Dish collectors (Fig. 5.14) and heliostats (Fig. 5.15) are both altitude–azimuth tracking collectors.

The tracking equation for altitude–azimuth tracking can usually be expressed as

$$\alpha = \arcsin(\sin\delta\sin\Phi + \cos\delta\cos\omega\cos\Phi) \quad [5.3]$$

$$A = 180^\circ + \mathrm{sign}(\sin\omega)\arccos\left(\frac{\sin\alpha\sin\Phi - \sin\delta}{\cos\alpha\cos\Phi}\right) \quad [5.4]$$

Figure 5.14 Dish collector in Shanxi, China.

Figure 5.15 Heliostats in the Dahan concentrating solar power plant in Beijing, China.

where α is the altitude angle; A is the azimuth angle; δ is the sun's equatorial latitude angle; ω is the solar hour angle, it is 0 when the sun reaches its highest position during the day; and Φ is the local latitude.

References

[1] Duffie JA, Beckman WA. Solar engineering of thermal process. 2006.
[2] Chen YT, Chong KK, Bligh TP, et al. Non-imaging, focusing heliostat. Sol Energy 2001; 71(3):155–64.
[3] Chen YT, Lim BH, Lim CS. Off-axis aberration correction surface in solar energy application. Sol Energy 2006;80(3):268–71.
[4] Chen YT, Chong KK, Lim CS, et al. Report of the first prototype of non-imaging focusing heliostat and its application in high temperature solar furnace. Sol Energy 2002;72(6): 531–44.
[5] Chen YT, Chong KK, Lim CS, et al. Report on the second prototype of non-imaging focusing heliostat and its application in food processing. Sol Energy 2005;79:280–9.
[6] Chen YT, Lim BH, Lim CS. General sun tracking formula for heliostats with arbitrarily oriented axes. ASME J Sol Energy Eng 2006;128:1–6.

Passive solar space heating

6

L. Finocchiaro, L. Georges, A.G. Hestnes
Norwegian University of Science and Technology, Trondheim, Norway

6.1 Introduction

Passive houses, which emerged in response to the need to reduce heating demands in European buildings, are generally characterized by the combination of a highly insulated envelope and a properly dimensioned passive solar heating system, together with an efficient energy system. Their architecture represents the results of an effort, initiated in the early 20th century and strengthened by the energy crisis of the 1970s, to develop environmentally friendly building concepts.

Solar houses, developed throughout the 20th century, aimed at taking maximum advantage of solar heat gains for covering most of their heating demand. Natural heat transfer phenomena such as conduction, convection, and radiation, were extensively used in these buildings to distribute the heat in accordance with the internal space's environmental requirements. Solar houses were usually characterized by an extensive use of solar systems such as greenhouses or glazed balconies, whose diffusion was made possible by industrial developments in the production of cheaper and higher performing glasses.

Experimental buildings developed throughout the 20th century played a fundamental pioneering role in the analysis and quantification of potential energy savings deriving from the use of passive solar systems. Data extracted from pilot experimental buildings often became the basis for the definition of new energy targets and national standards (Fig. 6.1).

Despite their simple aesthetics, the architectural design of a solar house is often characterized by a high degree of complexity. Passive solar heating systems adopted in these houses need to be conceived as an integral part of the whole building architectural system, taking into account construction, morphology, internal program environmental requirements, and their distribution within the building form (Finocchiaro and Hestnes, 2011). Light construction ensures that solar heat gains are quickly transferred into the space without being stored in the thermal mass of the inner envelope. An unbalanced dimensioning of solar capture and building heat capacity may still result in an inhomogeneous temperature distribution, creating discomfort.

Passive solar heating systems (PSHSs) are often nowadays optimized through the use of advanced simulation software able to predict environmental performance already during the early stages of the architectural design process. The use of small electrical devices, such as fans, automated shades, or insulated shutters is often justified by the significantly increased performance of the PSHS. The architectural design of buildings requiring the optimization of advanced PSHSs, typically powered by

Advances in Solar Heating and Cooling. http://dx.doi.org/10.1016/B978-0-08-100301-5.00006-0

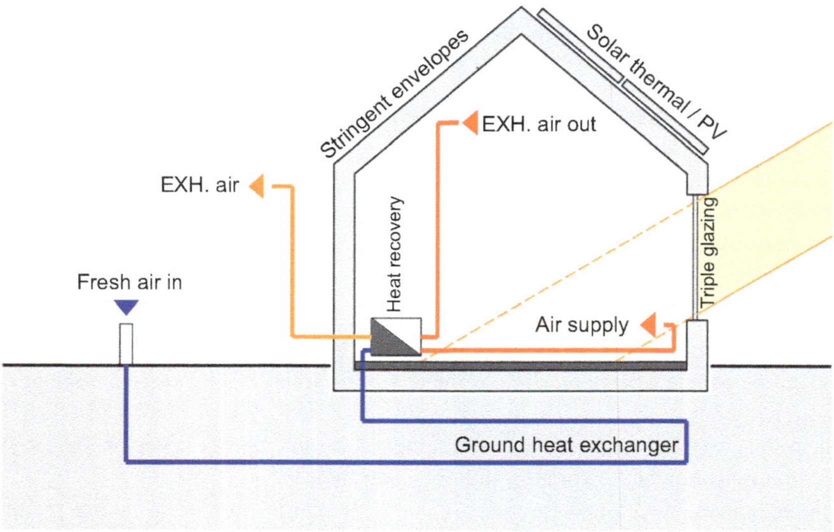

Figure 6.1 Passive house building elements and components. *EXH*, exhaust; *PV*, photovoltaic panels.

active systems based on renewable energy use, often requires a symbiotic collaboration between architects and engineers.

6.2 Sun and built forms

Despite the fact that only 35–40% of the radiation emitted by the sun is able to reach the earth's surface, in many cases solar energy striking a building is often significantly larger than that used for heating the building itself throughout the whole year. Atmospheric depletion of the sun's rays is related to the seasonal tilt of the earth's axis that determines the length of atmosphere through which the sun's rays need to pass before actually striking the globe. Solar energy hitting any surface can be quantified through numerical equations the taking sun's position in account as well as the surface orientation. Solar radiation analyses can be effectively used as a numerical basis for the optimization of the morphological characteristics of buildings toward their best environmental and energy performance. Optimal form and orientation are tightly related to the thermal stress a building is exposed to, determining the adverse or beneficial role of solar heat gains for thermal comfort. An optimal form is generally able to collect most of the available solar radiation during the heating season while rejecting most of it during the overheated season.

In temperate climatic contexts, forms elongated along the east–west axis may be the most efficient solution, being able to maximize solar heat gains from the south during the cold season while limiting the access of the sun from the east and west when it is close to the horizon during the warm season. In more extreme climatic contexts, however, such as hot arid or cold ones, thermal stress due to air temperature differences between the inside and the outside would suggest the use of compact forms able to limit heat transfer through the envelope (Olgyay, 1963).

In high-latitude cold climatic contexts, such as in Trondheim, Norway, the south side of a building generally receives nearly six times as much solar radiation in the months of April and May than in the winter (Fig. 6.2). East and west sides receive in the hottest month of July more than six times the solar radiation that they get in the winter. In the overheated season the west facade receives significantly more energy than that collected from the south. Because of the large thermal stress in winter, low temperatures therefore overrule the sun's effort to elongate the structure in the east–west direction, pressing the optimal form in this climate toward nearly squared shapes.

Technological development of highly stringent envelopes is nowadays questioning traditional assumptions regarding the interrelation between optimal forms and climatic contexts. Internal load–driven typologies such as office buildings are nowadays suffering from overheating problems for large parts of the year even in cold climatic contexts. Today the combination of elevated internal heat gains and highly stringent envelopes is opening up to more complex architectural scenarios in which

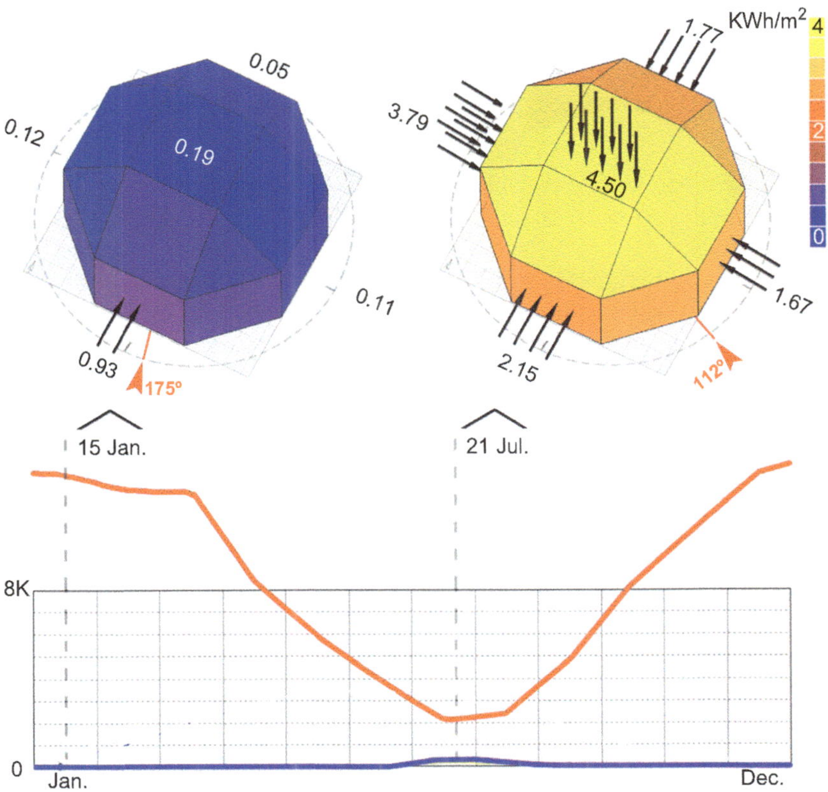

Figure 6.2 Trondheim, solar energy and thermal stress on the coldest and the warmest day.

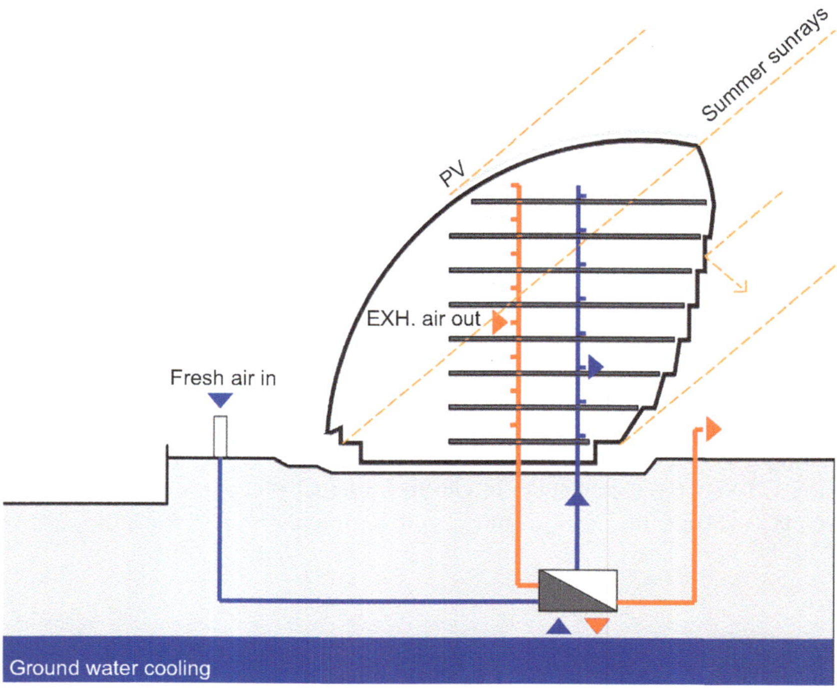

Figure 6.3 The London city hall, Norman Foster. *PV*, photovoltaic panels.

morphological and technical solutions need to be implemented to cope with the increased cooling demand of this typology (Finocchiaro, Murphy, Wigenstad, & Hestnes, 2011).

The London City Hall (Fig. 6.3), designed by Norman Foster, is a clear example of how such a challenge was solved in the cold climatic context of London by optimizing the morphological characteristics of the building toward maximum efficiency. The compact form of the building, conceived to limit energy losses in the winter, is thus tilted toward the sun angle in the summer to minimize the access of solar energy during the hot season.

6.3 Passive solar heating systems materials and components

The amount of solar radiation that is effectively absorbed within the building form depends on specific characteristics of the building envelope such as the distribution of glazed areas, overall transmittance, and eventually the use of additional components such as shading devices, insulated shutters, or solar reflectors. The heat capacity of a building determines its ability to store the absorbed solar energy before releasing it in the form of shortwave solar radiation, perceived as heat. A balanced dimensioning

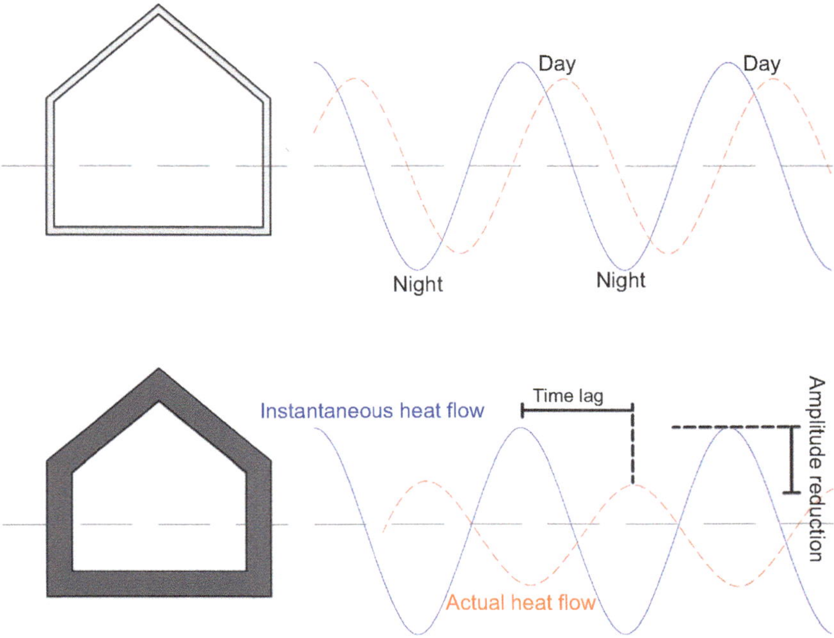

Figure 6.4 Dynamic response of buildings: light versus heavy construction.

of solar capture, heat storage, and distribution systems within the building form is fundamental to ensure that the heat gained throughout the envelope is not translated into large temperature fluctuations that might result in discomfort (Fig. 6.4).

6.3.1 Solar capture systems

Solar capture systems aim at collecting most of the solar radiation available during the underheated season. The solar heat transmittance (*g* value) of a window determines the amount of solar energy effectively absorbed through the glazed area, whereas the thermal conductivity (*u* value) determines the possible heat transfer through conduction. A state-of-the-art window is nowadays generally characterized by a *g* value of 0.5, corresponding to around 50%, and a *u* value of 0.8 W/m^2K. Advanced glazed surfaces would thus still dissipate around seven to eight times more than a highly insulated opaque wood stud wall provided with 40 cm rock wool insulation (characterized by a transmittance of 0.11 W/m^2K according to Norwegian passive house standards: PrNS3700:2010) (Standard Norge, 2010). It is thus fundamental that solar capture systems—such as windows or clerestories—are properly dimensioned and distributed throughout the building form taking into account orientation and internal layout of the building.

In the Northern Hemisphere a typical solar capture system (Fig. 6.5) is represented by a south-oriented glazed area, unshaded during the wintertime. Solar capture systems in orientations other than south are possible but should take into account a possibly

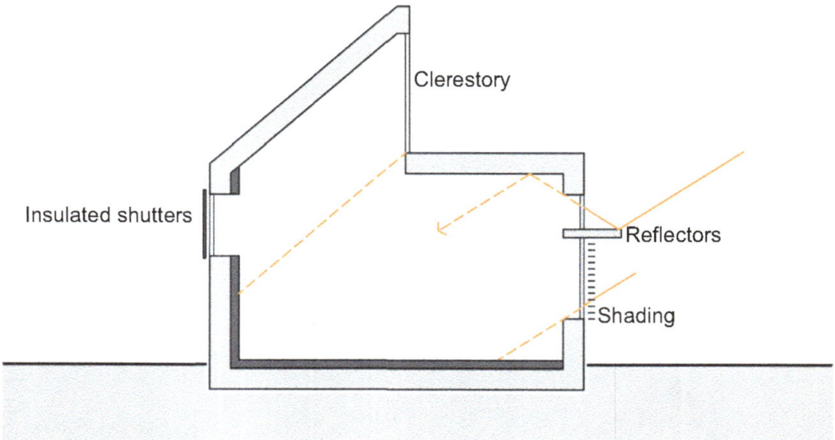

Figure 6.5 Solar capture systems and regulating devices.

negative heat balance during the cold season and the higher energy contribution of the sun when it is lower at the horizon during summer. East and west-oriented PSHSs are for this reason generally double-glazed and equipped with adjustable shading devices.

Clerestories are solar capture systems usually placed at the roof level. The advantage of such systems is that they can be more freely oriented toward optimal orientation for maximizing solar heat gains while providing optimal daylight distribution.

The environmental adaptability of solar capture systems is generally enhanced through the use of additional devices, such as shading systems, reflectors, or insulated shutters, able to better tune the access of solar radiation to the internal environmental requirements, increasing the overall performance of the PSHS throughout the whole year.

Fixed shading devices such as roof overhangs or pergolas, commonly adopted in the south facade, are optimized to be able to effectively shade the glass during the summer while still giving access to the sun's rays during the winter when the sun is lower in the sky. Because of the low angle of the sun, on facades facing east and west it is more effective to use vertical devices such as sliding shutters to more effectively block the access of the sun's rays.

Reflective devices such as mirrors or reflector profiles on the south facade can increase solar capture by 30–40%. Commercially available prismatic panels or louvers can also be optimized to limit the amount of transmitted solar energy (to 14%) while enhancing daylight distribution within the space (up to 50%) (Arnesen, Kolås, & Matusiak, 2011).

In cold climatic contexts, around 65% of the heat gained through optimally oriented windows can be lost during a winter night with clear skies. Insulating shutters can be used as effective devices for limiting heat losses that would reduce the overall PSHS efficiency. Regulating systems such as shading systems or insulated shutters can all be automated on the basis of inputs sent by timers or electronic sensing devices such as thermostats or daylight sensors.

6.3.2 Heat storage systems

The heat storage capacity of a building ensures that indoor temperature fluctuations are not affected by the solar contribution, but that energy is first stored in the building internal surfaces and then released with a convenient time lag for thermal comfort. Inside passive houses in central European climates, temperatures generally do not drop below 10°C even without heating (Woss & Musall, 2011). The question is to what point it is possible to enlarge the heat capacity of the building in such a way that the temperature would not fall below the comfort zone by simply relying on the contributions of solar energy and internal heat loads.

The heat storage system is generally composed of a dark-colored surface, working as an absorber, and a high-capacity thermal mass located in the direct path of the sun's rays. As a general principle, the higher the heat capacity of the material used, the better. Storage mediums must also be good heat conductors to easily release the absorbed heat when needed. Thermal lag, determined by heat capacity and thermal conductivity of the material used, represents a fundamental parameter for determining if the dynamic response of the adopted material would fit the climatic contexts in which we are working. Materials such as masonry or stone, characterized by a high heat capacity and high conductivity, are generally unable to quickly store the heat captured by the absorber. For this reason increasing the thickness of the storage system beyond 20 cm generally has no beneficial effect. It is, on the other hand, convenient to extend the storage surface area to large parts of the living area if we want to avoid overheating risks. If water is used as the thermal storage medium, heat is conveyed by natural convection throughout the mass, in which case it is possible to use thicknesses of the heat absorber beyond 20 cm. There is no perfect storage medium in terms of volume, heat capacity, and conductivity, but all these qualities need to be tuned to the PSHS dimensioning in relation to climatic context and building environmental requirements.

Phase-change materials (PCMs) introduced on the market in the past years, represent a light construction alternative to heavier thermal storage systems such as bricks or stones. Paraffin, encapsulated in PCM panels, is characterized by a melting-point temperature within the comfort zone (generally between 21 and 23°C). PCM panels thus act as a sort of artificial thermal mass, absorbing heat during the change of phase, whenever the melting point is reached, and then releasing it back whenever the temperature drops below the melting point. Generally, 1 m^2 of PCM for every 3 m^3 of air is necessary to ensure an effective functioning of the heat storage system using a PCM. The biggest advantage of PCM panels is lightness, a characteristic that makes them particularly advantageous for use in prefabricated construction systems. PCMs act mostly on daily temperature fluctuations. This potential is thus limited by the fact that they are not able to take advantage of larger fluctuations of the climate's environmental parameters (Woss & Musall, 2011, p. 18).

6.3.3 Heat distribution

Heat distribution in PSHSs is generally based on natural heat transfer phenomena such as conduction, convection, and irradiation. A homogeneous distribution of heat gains is first of all determined by spreading both solar energy access and heat storage within the building form. As a rule of thumb, thermal mass should cover at least 50% of the overall surface enclosing the space. Windows, clerestories, and skylights should be distributed in tight connection with the internal environmental requirements, taking care to array habitable spaces along the south facade and services and circulation on the north, acting as a thermal buffer protecting the living area.

Heat transfer through convection is regulated through operable vents or dampers driving heat from the solar space into the living area. The use of electric fans, ducts, and blowers is justified by a significantly increased performance of the whole distribution system. If water is used as the thermal storage medium, pumps may be used to assist heat circulation through convection within the distribution system.

6.4 Passive solar heating systems technologies

PSHSs are commonly classified on the basis of the disposal of the capture and storage systems components within the space to heat. Direct, indirect, and isolated PSHSs are thus defined depending on whether the heat is stored directly inside, in-between, or outside the space (Fig. 6.6). Direct systems are generally characterized by an absorbance of 60–75% of the solar energy, whereas only 30–45% of the solar radiation striking an indirect system is actually stored in the space.

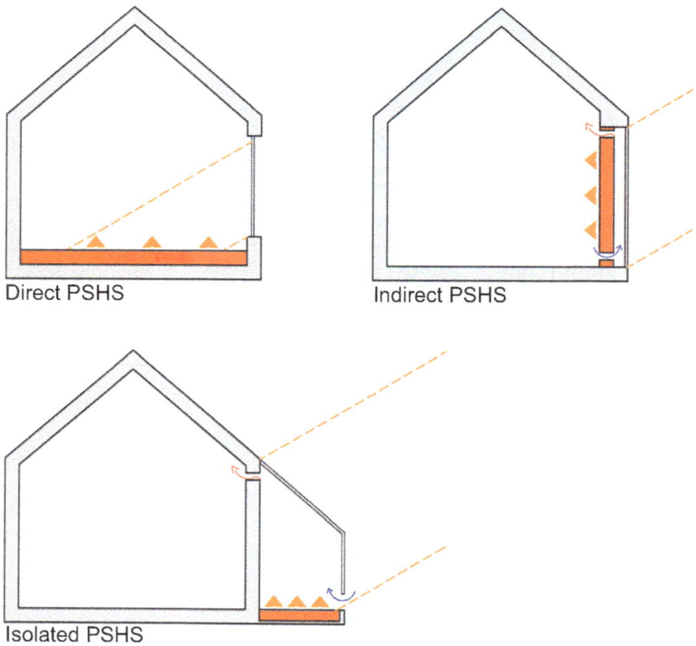

Figure 6.6 Passive solar heating system types.

Generally no one PSHS is the most advantageous in all climatic contexts, but specific climatic conditions will suggest the most effective solution to employ in a specific building typology. In vernacular architecture, materials available on site determined which system was most convenient to adopt. Nowadays advanced materials and components, commercially available on a globalized market, have made it possible to adopt passive strategies once intimately connected to different climatic contexts.

6.4.1 Direct passive solar heating systems

Direct systems represent the simplest PSHS technology, in which the sun's rays are admitted and stored directly into the space to be heated. The storage system generally detains the heat as long as the room temperature is high, to then slowly release it when the temperature drops. This usually happens during the night because of the combination of higher thermal losses and lower heat gains related to space occupancy. Heat radiation can even last several days after the heat has actually been stored. Material to be used for the heat storage medium must be defined in accordance with an accurate comparison between climatic data and environmental space requirements with the aim of stabilizing temperature fluctuations within the comfort zone.

Optimal dimensioning of the solar capture system in a direct PSHS is tightly connected to external climate conditions and building tectonics determining the amount of heat that can actually be stored within the space. A correct dimensioning of the components is fundamental to avoid large temperature fluctuations that might result in discomfort. When calculating the potential of PSHSs for optimal environmental performance, heat gains during wintertime should first at least match heat losses through the whole window (Szokolay, 2008). Around 50% of the exposed surface of a living area must be able to store heat.

A slightly increased area of the solar capture system would make it possible to let a direct PSHS work properly under cloudy sky conditions with the only contribution being diffused solar radiation. In climatic contexts in which indirect solar radiation is dominant, it is advisable to use the minimum necessary thickness of the absorbance system to let the mass respond as quickly as possible when the sun is available.

The Jacob II House—named the Hemicycle by its architect Frank Lloyd Wright—was based on the maximum use of a direct PSHS facing south and equipped with a selective shading overhang optimally dimensioned for limiting the sun access during the summer while still giving access in the wintertime (Fig. 6.7). Heat losses from the north side have been minimized by attaching the building to the terrain. The concave south facade admits significant amounts of solar radiation that is directly stored in the concrete floor slab. Thanks to the curved shape, heat losses through the window should be reduced by 8% if compared with a linear plan, according to the architect. The concrete slab accommodates a hot-water coil, transferring solar heat gains to the inner space through a siphon.

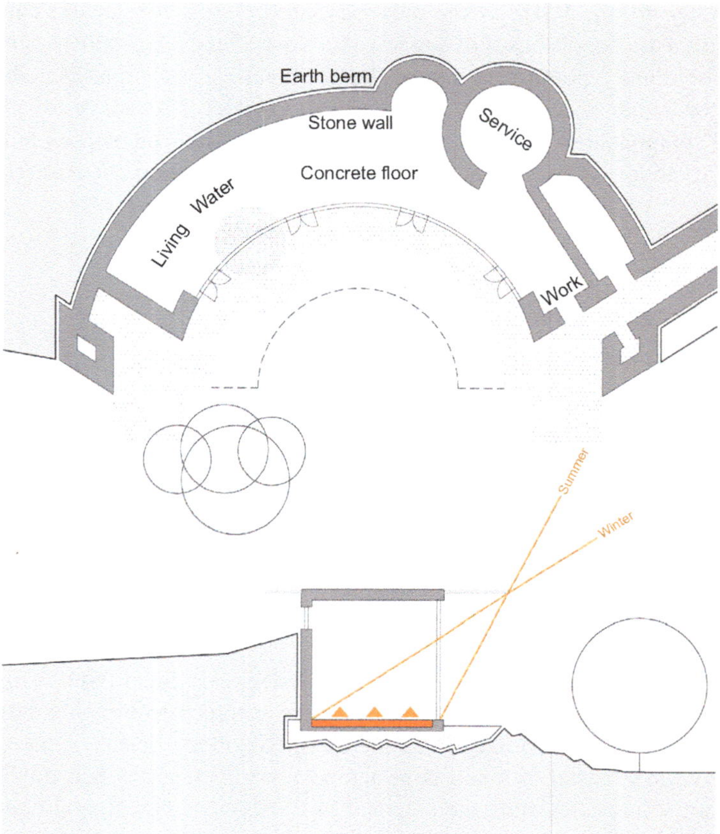

Figure 6.7 Jacob II House, plan and section.

6.4.2 Indirect systems

Indirect PSHSs are characterized by a storage mass placed in-between the solar absorber and the space to be heated. Solar energy is in this case indirectly transported through convection or irradiation into the space.

Trombe walls are the most common indirect PSHS and are characterized by an opaque wall exposed to direct solar radiation and interposed between the glass and the space to be heated. Solar radiation passing through the glazed area is converted straight into heat inside the cavity, but not all of this is quickly absorbed into the thermal mass. Because of this the air temperature in the cavity can easily pass 50°C in summer. Regulating devices should therefore be used in this period to avoid overheating risks. Vents placed at the bottom and at the top of the thermal mass spontaneously activate a natural convective loop, pouring hot air from the cavity into the space. This process will be active until there is a difference of temperature between the bottom and the top of the wall. To avoid thermal losses that could reduce the efficiency of the system or even reverse the heat transfer process, air vents should be closed in the night when heat stored into the wall can still be slowly released into the space through irradiation.

A thickness of 20–30 cm of concrete roughly corresponds to a time lag of 6–8 h, optimal for transferring heat gains on a daily basis from day to night. In the house in Pullach, Germany, designed by Thomas Herzog, a 30-cm-wide zone where installation, heat recirculation, and water flow are concentrated determines the whole layout of the building. An external envelope made of translucent insulation or glass modules covers a concrete unit painted in black working as the heat storage system (Herzog, 1996).

In the experimental building "Energieautarkes Solarhaus," built in Freiburg, Germany, in 1992 for testing the potential of hydrogen economy through electrolysis, the efficiency of the indirect system wall was drastically improved by filling the air cavity between the glazed surface and the Trombe wall with a transparent insulating material (Wigginton & Harris, 2002).

Indirect PSHSs, when properly dimensioned, can provide a high percentage of the total heat necessary in a room to ensure comfortable conditions even in cold climatic contexts. Optimal dimensioning of indirect PSHSs should be based on winter design week conditions and should aim at providing enough heat for reaching comfort on a daily basis. In comparison with direct systems, indirect ones are less problematic because the differences between captured heat and stored heat do not necessarily result in overheating risks or can simply be avoided by reducing convective transfer through the valves. Convection phenomena are generally slower and can continue for up to 3 h after sunset. Vents, generally around 0.2 m^2 in size, should be equally distributed between the top and the bottom of the thermal mass.

6.4.3 Isolated passive solar heating systems

Isolated PSHSs are heating devices detached from the living area and are typically constituted by a flat panel placed over a storage mass such as a rock bed or a water pond. Heat is transferred into the living area through convection by a water- or air-based siphon system, until an equilibrium in the temperature distribution in the medium is reached. Typical examples of isolated PSHSs are the Barra-Costantini wall or greenhouses. The Barra-Costantini is a variation of the Trombe wall, in which the only difference is the interposition of an insulation layer between the thermal mass and the space to be heated. This limits heat transfers through irradiation but makes it possible to better control the heat transfer from the PSHS into the space to be heated. This system is therefore more suitable in colder climatic contexts where extreme climatic conditions may result in significant heat losses through the system.

Attached greenhouses are the most common among isolated PSHSs, serving the double purpose of heat reservoir and functional space used to extend the living area when environmental conditions are favorable. These types of isolated PSHS are generally elongated along the south side of the building to maximize thermal exchange to the living area while minimizing exposure to the east and west. Because of their dual nature, the architectural design of buffer spaces such as greenhouses is generally slightly more difficult. In their dimensioning a correct balance between the space as a living area and as a heat reservoir should be aimed for. To optimize their environmental performance throughout the whole year, greenhouses are generally equipped with a complex system of valves that make it possible to adjust thermal exchange according to a wider set of boundary conditions.

In an experimental house built in Greve, Italy, by Aude, Lundgaard, Sorensen, and Rotne in 1985, the energy consumption of the building is reduced by dividing the building into three different zones matching the solar radiation contribution (Fig. 6.8) (Dahl, 2010). The north thermal zone is more tightly insulated, whereas the living area in the center can be extended toward the south-facing greenhouse when convenient. All the zones can work independent of the others or can be joined into one by sliding insulated shutters whenever environmental conditions are favorable.

As a rule of thumb, between 0.6 and 1.5 m^2 of south-facing greenhouse surface is needed for each square meter of floor area in the living space to cover the whole heating demand on a daily basis. Because of their light construction, temperature fluctuations inside greenhouses often depend on external climatic conditions and as a result are highly sensitive to the solar energy contribution. The use of water ponds or rock beds below the floor can effectively mitigate overheating risks, helping to create more favorable microclimatic conditions. In the Akademie Mont-Cenis, Herne, Germany, by architects Jourda and Perraudin, the greenhouse wrapping nine pavilions aims at creating a new more favorable microclimate whose conditions can be controlled from an adaptive shelter.

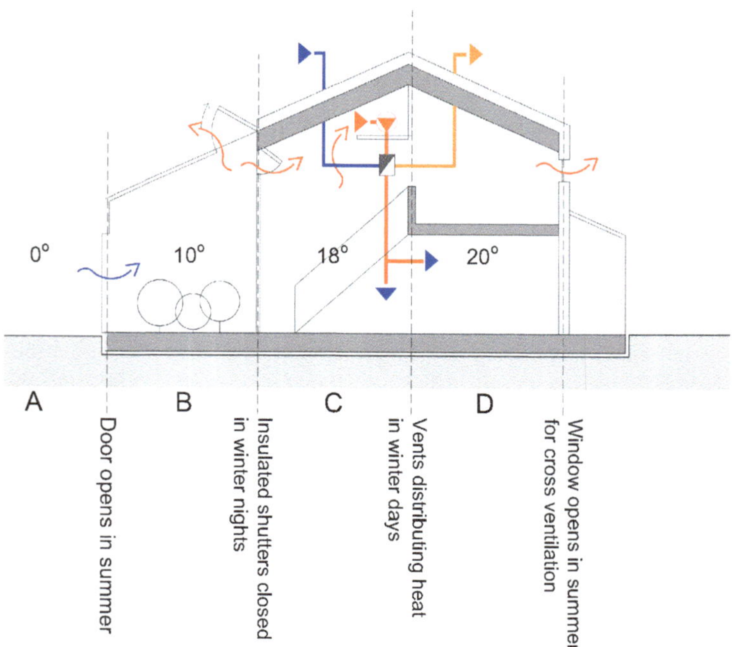

Figure 6.8 Aude, Lundgaard, Sorensen, and Rotne, experimental house, Greve, Italy, 1985.

6.4.4 Annual geo-solar systems

Annual geo-solar systems are PSHSs aiming at taking maximum advantage of seasonal differences for balancing energy demand and supply of the building on a yearly basis. In cold climatic contexts such systems thus aim at storing heat during the overheating season to then release it months later into the living area. In warm climates, on the other hand, they are generally used for cooling purposes, in which case the solar collector is exposed to cold night temperatures during wintertime.

To work properly, geo-solar systems require a considerable amount of storage mass, usually bedrock or earth placed below the building. Attics, sun spaces, ground-based radiators, and flat plates can all be used as capturing solar systems and dimensioned according to the thermal capacity of the material adopted. Even if properly dimensioned, however, it might take several years before the storage system is able to cover a large part of the whole building heating demands. An air- or water-based thermal siphon is commonly used in these systems for heat distribution. Heat collected at a low point of the structure is thus transferred into the building through convection. If an attic is used as a collecting device, the use of electric fans or pumps is required. The rock bed is generally insulated at the perimeter to minimize thermal losses or moisture transfer from the external ground. It is also important that the floor above the heat storage system is made of a high-density material able to radiate the heat into the living space. Electric fans or dampers, activated on the basis of data collected from thermostats, can regulate the access of the air from the storage zone into the living area.

Annual geo-solar systems can be employed at different scales in the architectural design of solar buildings (Fig. 6.9). The Jenny House built in Oberburg, Switzerland, in 1991 aimed at collecting enough thermal energy to serve the building throughout the whole year by using an 84-m^2 collector. A heat storage device of 118 m^3 was needed to serve the house, measuring 200 m^2 of heated surface. The Samer Mösl passive house residential complex in Salzburg, Austria, showed that a smaller heat capacity system

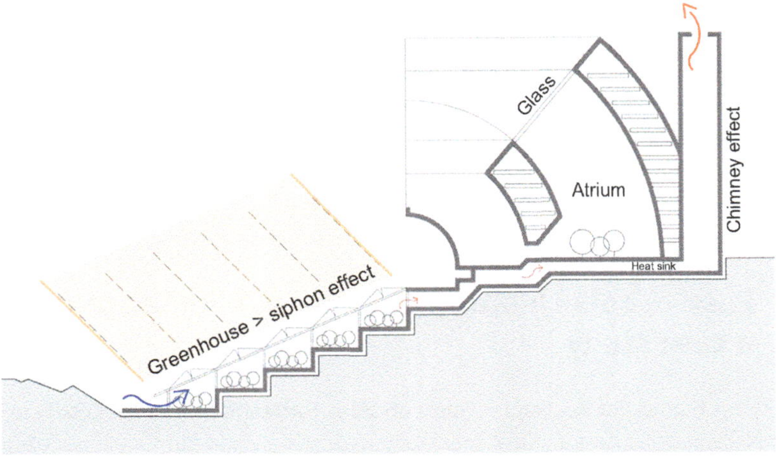

Figure 6.9 Arcosanti, Arizona, USA, section of the geo-solar system transferring heat from the greenhouse to the building through a siphon.

supplemented with furnaces or heat pumps might be a more viable solution. In this building 11-m-tall water columns, integrated in the staircases, connected to 200 m^2 of flat tube collectors and supplemented by a pellet boiler of 100 kW are able to cover 100% of the heating demand throughout the whole year (Voss and Musall, 2011).

6.5 Economics and energy efficiency of passive solar heating systems

Direct PSHSs are commonly based on a reasoned disposal of conventional architectural components that can be found in any regular building. A slightly larger window or a building construction with higher thermal capacity would make it possible to save up to 25% of the energy consumption of the building, paying for itself in a reasonable amount of time. Studies on houses in Milton Keynes, England, have shown that low-cost passive solar design features, draught proofing, and insulating measures can easily reduce heating bills by 40%, paying back initial additional costs in 2 years' time (Fuller, Doggart, & Everett, 1982).

If no significant extra costs can generally be appreciated for direct PSHSs, this is not necessarily true for indirect or isolated PSHSs, requiring the use of additional components for their correct functioning. The payback period for the investment is in those cases difficult to quantify on a general basis, because it depends on many different parameters such as climatic context, typology, and dimension of the project, in addition to changing energy prices. A well-designed conservatory acts as extra insulation for the house, preheating the ventilation air and providing direct solar heating to the intervening wall, which is drawn into the house, making it possible to save up to 20% of the annual heating demand. Architectural design of solar buildings is also generally slightly more complex. Their optimization often requires a more complex reiteration of the design process based on both quantitative and qualitative attributes of the project.

Heat conservatories such as atria or greenhouses can be expensive and cannot be justified on the grounds of energy efficiency and low operation costs only. Because of a tighter connection to the external environment, solar houses are also considered to be healthier buildings able to provide better environmental conditions for their occupants regarding both thermal and luminous comfort. Another direct advantage of solar houses is that of requiring much lighter technical equipment, reducing both initial investments and maintenance cost throughout the whole life cycle of the building.

6.6 Passive solar heating systems at high latitudes: a case study

The Living Lab at the Research Centre on Zero Emission Buildings (ZEB) in Trondheim, Norway, is a test facility representative of a single-family house of approximately 100 m^2, built in accordance with passive house standards. The building is currently used as a laboratory for action research on users' interaction with state-of-

Figure 6.10 The Zero Emission Buildings Living Lab in Trondheim.

the-art technologies for carbon neutrality. For this reason, the house has been equipped with a data acquisition system able to record any information related to environmental performance and energy consumption and generation from integrated renewable energy systems such as photovoltaic and solar thermal panels (Finocchiaro, Dokka & Gustavsen, 2012) (Fig. 6.10).

The plan of the Living Lab is organized into two main areas: a living area facing south and a studio/sleeping area oriented toward north. The entrance in the southwest corner of the house gives access to the living room including a low bench all along the south window and a kitchen on the east side. On the north side, a large studio, equipped with a long writing desk, gives access to the bathroom and two bedrooms of similar dimensions (one facing east and one west). A technical room, toilet, and kitchen have been placed along the central spine of the building to optimize the distribution of plumbing and ventilation ducts. The house is ventilated through a mixed-mode hybrid system characterized

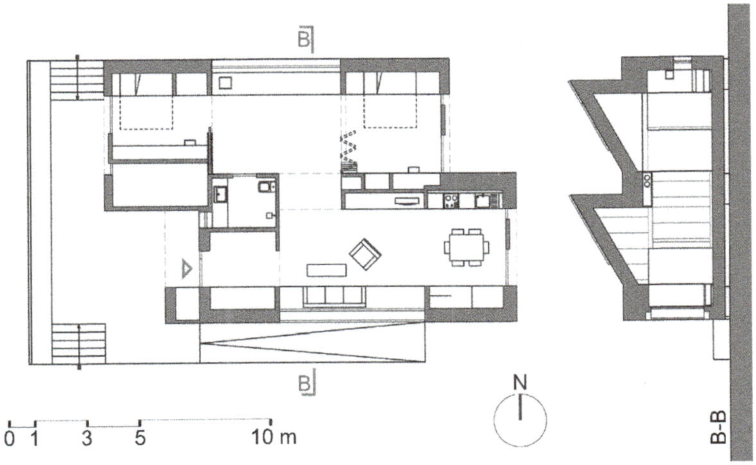

Figure 6.11 Living Lab, plan and section.

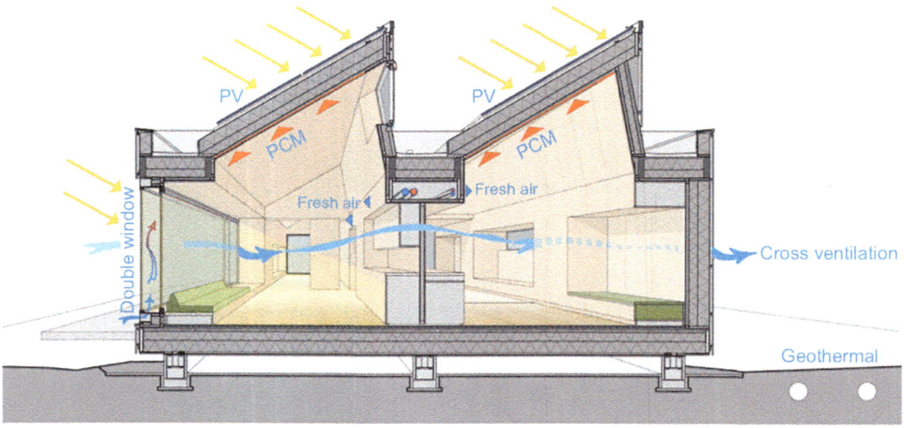

Figure 6.12 Living Lab. Bioclimatic section and passive strategies for optimizing environmental performance of the model. *PCM*, Phase-change materials; *PV*, photovoltaic panels.

by a heat wheel unit with efficiency of 85% and a hydronic heating coil capable of warming the inlet air up to 40°C. Supply air terminals are located in the living room and in the bedrooms, whereas exhausted air is extracted from the bathroom and kitchen (Finocchiaro, Goia, Grynning, & Gustavsen, 2014) (Fig. 6.11).

The building construction has been optimized through various sets of simulations resulting in a highly insulated and hermetic envelope characterized by a 20% glass ratio. Envelope components, such as walls, floors, and roofs, are characterized by a double layer of rock wool insulation for a total of 40, 40, and 45 cm, respectively (with a *u* value of 0.11, 0.10, and 0.11, respectively). Paper placed in-between the two layers of rock wool avoids convection phenomena that could reduce the effective resistance of the envelope. All windows in the house except the skylights (characterized by a *u* value of 1.0) respect the passive house standards of 0.8 W/m^{2}K. The double window facing south, in particular, is characterized by a markedly low *u* value, varying from 0.65 to 0.69 depending on the ventilation rate within the air cavity. Both south and north windows have been coupled with an automated control system able to open them 39 and 37 degrees, respectively, and activate, when required, a natural ventilation flow through the building. Skylights facing north can be opened independently to let the exhaust air out through stack ventilation. Natural ventilation and PCM panels integrated in the ceiling construction aim at avoiding overheating problems that may occur because of the combination of large heat gains from the south and the light construction of the building (Fig. 6.12).

6.6.1 Environmental performance analyses of the Living Lab

Trondheim—63°25′N, 10°27′E—is located a few kilometers east of the oceanic cost of Norway and is characterized by a climate on the border between oceanic semicontinental and subarctic (Cfb and Dfc according to the Köppen-Geiger climate classification). Annual average temperature in the city is around 5°C. Winters are generally cold with average temperatures between −2.5°C and −7°C and at least 22 days with daily minimum temperatures below −10°C. Summers are also generally cool with average temperature values between 17°C and 9.7°C and rarely exceeding 25°C. On average there are 1346 h of sunshine per year with a peak of over 20 h per day around the summer solstice and a minimum of 4 h around the winter solstice. As a consequence, direct solar radiation is abundant in the summer period with a peak in June of 432 kWh/m^2 and a minimum of 153 kWh/m^2 in December. Because of the average high cloudiness, diffuse radiation is also significant and accounts for 290 and 80 kWh/m^2 in the winter and summer months, respectively. The scarce solar radiation in the winter months is reflected in a small temperature excursion, between 4°C and 6°C, whereas temperature daily variation in summer is between 8°C and 10°C.

Simulations on a virtual model of the Living Lab were run with the specific purpose of analyzing how solar heat gains would be distributed within thermal zones characterized by different orientations, glass ratios, and activities. The passive environmental performance of the building was simulated assuming a "natural ventilation, no heating/no cooling" HVAC schedule, making it possible to analyze the building's spontaneous ability to get close to comfort conditions. The percentage of time spent within the comfort zone and Fanger's adaptive comfort model were assumed as criteria for effectiveness and optimization of the PSHS in the climatic context of Trondheim.

Simulation analyses were conducted on a virtual model of the Living Lab in two different sets: the first focusing on the heat capture and storage system and the second on the distribution. In the first set, two alternative PSHSs were applied in the south wall of the building (Fig. 6.13):

SW—A large window able to provide direct heat gains throughout the whole year, characterized by a low-e coating triple glazing of 6-6-4 and two gaps filled with argon (resulting in a u value of 0.8 W/m^{2}K in accordance with PrNS3700).

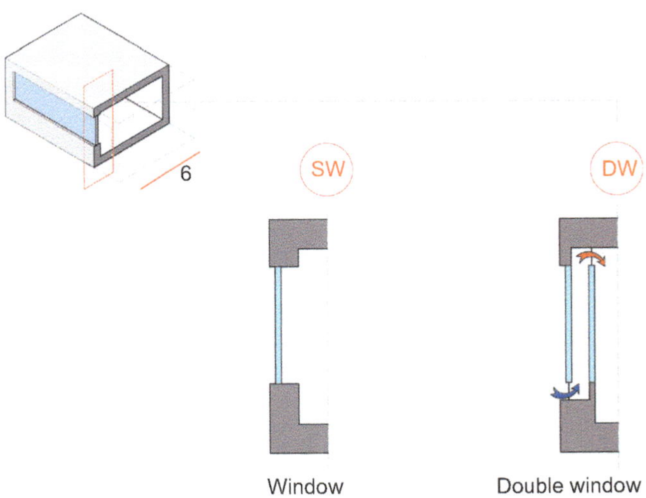

Figure 6.13 Passive solar heating systems applied to the simulated Living Lab model.

DW—A double window also providing indirect solar gains through convection. The inner layer of this window is exactly the same as SW, but on the outer facade, an extra window has been placed, creating a 30-cm ventilated air gap in-between the two. Ventilation valves have been placed on the exterior window's bottom and interior window's top.

For the direct system, SW, heat gains due to solar radiation, Qs, and conductance through the window, Qcw, were calculated. For the indirect system, ventilative heat gains from the air cavity, Qiv, were also accounted for. The air gap included between the exterior and the interior window was modeled as a cavity to let the software calculate indirect solar heat gains through the interior window, Qsi.

Simulations were run for both summer and winter design weeks. The results of this first set of analyses were synthesized into diagrams in Figs. 6.14 and 6.15. The scale relative to heat gains, reported on the right side of each diagram, has been adjusted differently for direct and indirect systems, the first being able to provide significantly higher gains than the second.

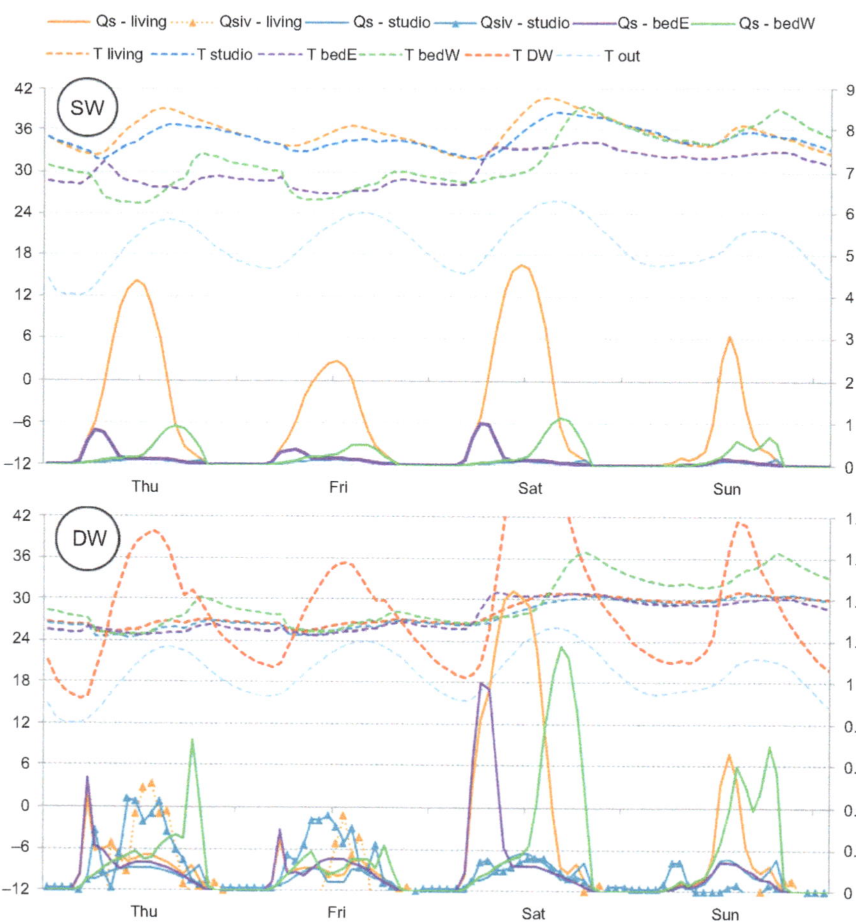

Figure 6.14 Living Lab. Extract from the simulation in the summer design week.

Figure 6.15 Living Lab. Extract from the simulation in the winter design week.

Results showed that moving from a direct PSHS (SW) into an indirect one (DW) results in a reduction of almost 50% of the total heat gains. Indoor temperatures, however, were more stable and closer to the comfort zone in the second case, in which an indirect system, DW, was employed. In the summer week in particular, indoor temperature values would not be able to reach 31°C when a DW was used, whereas they would be between 30 and 41°C if a direct system were employed. The bedroom facing west would be the one most exposed to overheating problems in both April and August, when the sun is lower in the sky, whereas the air temperature within the DW air cavity would reach temperatures of almost 40°C in summertime and 18°C in wintertime under clear sky conditions.

The second set of simulations was run with the intention of analyzing the relation between the building construction and the internal temperature fluctuations. On the basis of simulations conducted by Alonso et al. at the ZEB research center in Trondheim, a concurrent mixed-mode system, active during the day when indoor air temperatures are in the range 22–24°C, would be able to reduce the number of overheated hours recorded in the building through ventilative cooling by 99%, in comparison with a house ventilated through mechanical ventilation (Alonso, Kirkøen, & Mathisen, 2015). Simulations regarding the relation between construction and environmental performance were conducted in a week in April, assumed as representative of midseason climatic conditions, when outdoor temperature values are still below the comfort zone

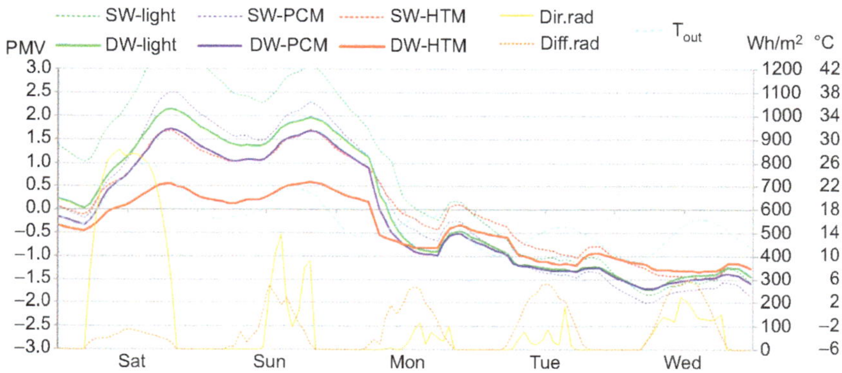

Figure 6.16 Living Lab. Environmental performance assuming three different construction systems.

(thus no NatVent is activated) and solar energy has a significantly higher potential than in the winter.

A virtual model of the Living Lab was modeled assuming three alternative construction systems characterized by increasing heat capacity. The first model was characterized by the lightest construction in wood studs and rock wool (*u* values according to passive house standards), the second model was equipped with 90 m^2 of PCMs, and the third model was characterized by a much larger heat capacity, assuming an 8-cm layer of concrete in the ground floor and parts of the inner walls. Results showed that the third model, characterized by the highest heat capacity (HTM in Fig. 6.16), performed best, maintaining Fanger's PMV parameter closer to the acceptable range of −0.5 to 0.5.

6.7 Conclusions and future trends

Passive house buildings, which emerged in response to the need to reduce space heating demands in northern European residential buildings, are generally characterized by the combination of a highly stringent envelope and a properly dimensioned PSHS. Because of the light construction often characterizing these buildings, a critical difference between heat absorbance and heat capacity of the building can potentially result in overheating problems.

In 2011, Nielsen studied the cooling demands of a Danish low-energy house, giving evidence that without any cooling control system, interior temperatures could easily go above 26°C for 30% of the time, resulting in a cooling demand up to 60 W/m^2. According to Mlecnik et al. (2012), even in cold climatic contexts such as in Scandinavia, 34% of house occupants experience overheating problems in the living room, whereas 49% complain about bedroom temperatures during the summer. This is confirmed by studies conducted by Samuelsson and Lüddeckens (2009) respectively in Sweden and Norway (Kleiven, 2007) that identified dissatisfaction in around 50% of users of passive houses due to temperatures above the comfort zone.

In high-latitude climatic contexts where temperatures are below the comfort zone for most of the year, ventilative cooling solutions would be able to reduce the number of hours when mechanical ventilation is needed by 90.4% (Oropeza-Perez & Østergaard, 2014), whereas the use of appropriate solar shading systems would make it possible to reduce the cooling demand of buildings by 50% (Nielsen, 2011).

As a matter of fact, office buildings hardly require, for most of the year, more heat than is provided by internal loads, such as occupants, appliances, and lighting, even in cold climatic contexts. The technological development of highly stringent envelopes is making it possible to imagine a new generation of buildings relying only on internal gains and the intelligent use of a ventilation system, equipped with a heat recovery system, for covering their entire heating demand. In the transition from solar houses to passive houses, compliance with existing standards for energy efficiency is making it unnecessary to further explore and exploit the potential of PSHSs.

The use of PSHSs, such as atria or greenhouses, cannot be justified on the grounds of energy efficiency and low operation costs alone. Solar spaces make it possible to replace the external environment conditions with a more favorable microclimate into which the living area may be extended. Atria or greenhouses may also become a tool for solving contradictions behind the optimization of morphological characteristics of energy-efficient buildings. Articulated forms characterized by permeable envelopes, determined on the basis of optimal performance in summer, are wrapped within a compact form minimizing heat losses in the winter (Fig. 6.17).

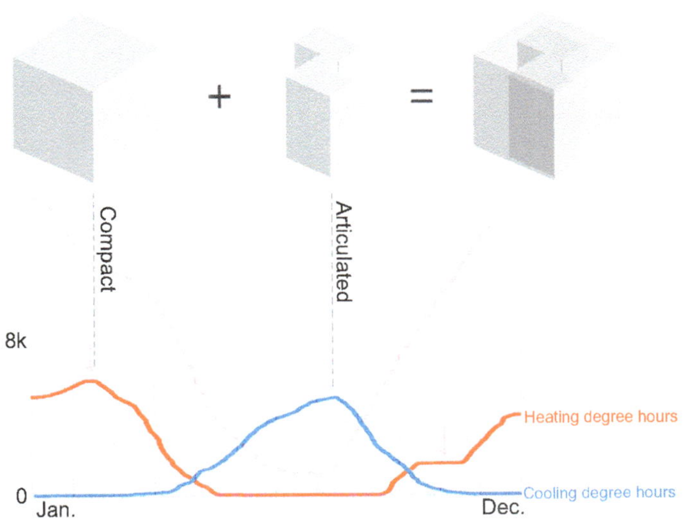

Figure 6.17 Optimizing morphological characteristics of buildings through passive solar heating systems.

References

Alonso, M. J., Kirkøen, C., & Mathisen, H. M. (2015). *Energy use consequences of ventilative cooling in a zeb residential building*.

Arnesen, H., Kolås, T., & Matusiak, B. (2011). *A guide to daylighting and solar shading systems at high latitude, ZEB project report 3 – 2011*. Norway: SINTEF Academic Press.

Dahl, T. (2010). *Climate and architecture* (p. 69). Oxon: Routledge.

Finocchiaro, L., Dokka, T. H., & Gustavsen, A. (2012). +Hytte. A versatile construction system for zero emission buildings. In *ZEMCH 2012 conference proceedings* (pp. 383–396).

Finocchiaro, L., Goia, F., Grynning, S., & Gustavsen, A. (2014). The zeb living lab: a multi-purpose experimental facility. In *IEA EBC annex 58 expert meeting, Ghent*.

Finocchiaro, L., & Hestnes, A. G. (2011). *Symbiosis and mimesis in the built environment, aesthetics of sustainable architecture*. Netherlands: 010 Publishers.

Finocchiaro, L., Murphy, M. A., Wigenstad, T., & Hestnes, A. G. (2011). The climate/comfort comparison and the basis of sustainable design: impact of climate change and technological development. In *Architecture & sustainable development. PLEA conference proceedings*.

Fuller, S., Doggart, J., & Everett, R. (1982). *Energy Projects in Milton Keynes: Energy Consultative Unit Progress Report 1976–1981*. Milton Keynes, UK: Milton Keynes Development Corporation (MKDC).

Herzog, T. (1996). *Solar energy in architecture and urban planning*. Munich: Prestel-Verlag.

Kleiven, T. (2007). *Brukerunderskelse i husby amfi*. Norway: Sintef Byggforsk.

Mlecnik, E., Scutze, T., Jansen, S., de Vries, G., Visscher, H., & van Hal, A. (June 2012). End-user experiences in nearly zero-energy houses. *Energy and Buildings, 49*, 471–478.

Nielsen, L. S. (2011). Heating and cooling challenge in nZEB experiences from built houses and detailed building simulations calls for a careful and integrated approach when designing heating and cooling systems for nearly zero energy buildings. *Rehnva Journal, 3*, 41–46.

Olgyay, V. (1963). *Design with climate: Bioclimatic approach to architectural regionalism* (p. 89). USA: Princeton University Press.

Oropeza-Perez, I., & Østergaard, P. A. (2014). Potential of natural ventilation in temperate countries – a case study of Denmark. *Energy and Buildings, 114*, 520–530.

Samuelsson, M., & Lüddeckens, T. (2009). *Passive houses from a user's perspective*.

Standard Norge. (2010). *NS3700:2010. Kriterier for passivhus og lavenergihus*. Norway: Boligbygninger. ICS 91.040.01; 91.120.01.

Szokolay, S. V. (2008). *Introduction to architectural science, the basis of sustainable design* (p. 58). USA: Elsevier.

Wigginton, M., & Harris, J. (2002). *Intelligent skins* (pp. 143–148). Oxford: Architectural press.

Woss, K. E., & Musall, E. (2011). *Net zero energy buildings*. Munich: Institute for Architectural Documentation, Detail Green Books.

Innovations in passive solar water heating systems

7

R.T.A. Prado[1], D.S. Sowmy[1,2]
[1]University of Sao Paulo, Sao Paulo, Brazil; [2]Institute of Technological Research of Sao Paulo, Sao Paulo, Brazil

7.1 Introduction

The intensive use of fossil fuels adversely affects the climate, increasing CO_2 emissions in the atmosphere and causing global warming throughout the world. Awareness of and actions toward renewable energy have grown in the past years; and relevant to this, electric and gas water heaters not only are a significant end use of energy compared to stoves, refrigerators, air conditioners, and lighting, but also are very well deployed in households, allowing the replacement of conventional with renewable energy through decentralized generation and investments. Actually, this is aligned with the distributed energy generation trend in most countries and shown by the growth of the collector market.

As aforementioned, the average growth rate of glazed collectors worldwide was about 18.7% between 2001 and 2013; the total annual installed collector area increased 5.4 times between 2002 and 2013. The market growth curve shows a fall after 2008, which can be explained by the housing economic crisis, which disrupted several countries that year, because the collector market follows civil construction and, in turn, the economy. Nevertheless, there was a 13.0% growth from 2009 to 2013 thanks to China.

Solar collectors can be classified into various categories or criteria. Regarding their functions, the most common are those for domestic or industrial use, which generally have glass covers (flat plates) or glass tubes (evacuated); for heated swimming pools, they are unglazed. There are also air collectors for thermal comfort, but only the first category is discussed in this chapter.

First, evacuated tube collectors (ETCs) and flat plate collectors (FPCs) together totaled 93% of the total installed collector area in the world, the ETCs being employed three times more than FPCs, seeing that China had 75% of the world total operating collector area (m^2) in late 2013, and in China ETCs cover 93% of its total installed area. In the European Union (EU), FPCs account for 89% of the total installed area. These differences may have several reasons, such as temperature level required, collector materials available, climate, and industrial public policies established by authorities, but certainly the low cost of sputter coaters performs an important role in China, according to Yin et al., cited by Morrison, Budihardjo, and Behnia (2004).

Advances in Solar Heating and Cooling. http://dx.doi.org/10.1016/B978-0-08-100301-5.00007-2

A marked growth in the solar collector area in operation has been observed as a response to the uncontrolled fossil fuel burning in past years. The technological advances in solar collectors that began in the mid-1900s aimed at reaching higher thermal performances for FPCs, ETCs, integrated collector storage systems (ICSs), and hybrid photovoltaic/thermal collectors (PV/Ts). This technology started from PV modules developed to passively convert solar radiation into electrical energy based on semiconductors, which received the attached thermal collectors after some decades, resulting in the PV/T. Yet the market in PV-only modules has markedly grown in past years, because electricity replaces many more end uses of fossil energy in quantity and in diversity than those relative to hot water alone.

Although solar water heating systems (SWHs) have lower costs than PV modules, a comparison between the aforementioned results suggests a quite stronger growth for the latter. However, the PV/T technology is newer and still under development. Thus, forecasting the market for PV/T and its predecessors, thermal and PV collectors, is not an easy task.

This chapter describes the operation principles of four types of passive SWHs: FPC, ETC, ICS allied to a compound parabolic collector (CPC), and the hybrid PV/T. It briefly focuses on verifying the origin of each of these collectors and on reporting on how experts have selected common issues and worked toward enhancing thermal performance. Solar water heating is an old application and has a rich history of major innovations, although many technical questions are still unsolved; in addition, better knowledge about which collector is the most suitable to a specific climate is necessary. This information is mandatory as an aid for authorities and designers to plan and decide upon various types of SWH for a specific location and climate.

A solar collector is a simple type of heat exchanger that converts solar radiation into heat, which is conveyed by an HTF (heat transfer fluid) and stored to perform various activities, such as domestic and industrial (Duffie & Beckman, 2006; Shukla, Sumathy, Erickson, & Gong, 2013; Suman, Khan, & Pathak, 2015), or transformed into power and electricity. Small collectors are used for supplying household needs; medium or large concentrating collectors are more common for electric energy generation. Currently, several types of domestic SWH are operating in the world, having as basic elements a collector, piping, and a hot-water tank.

7.2 Flat plate collector–thermosiphon

An FPC is composed of a thermal-conductive absorber plate, one or two top transparent covers to prevent heat loss by convection to the ambient air, a back heat insulator to reduce heat loss by conduction, piping to convey the HTF to the storage tank, and a case to protect the collector against dust and moisture and also to provide mechanical strength to the set. A passive collector system transfers heat from the collector to the storage tank by the natural convection principle, also referred to as the stack effect, or a thermosiphon without the support of an electric pump, as shown in Fig. 7.1. As solar radiation intensity (W/m^2) varies according to the location, season, and cloudy skies, in mild and cold climates, electric or gas heat sources are usually

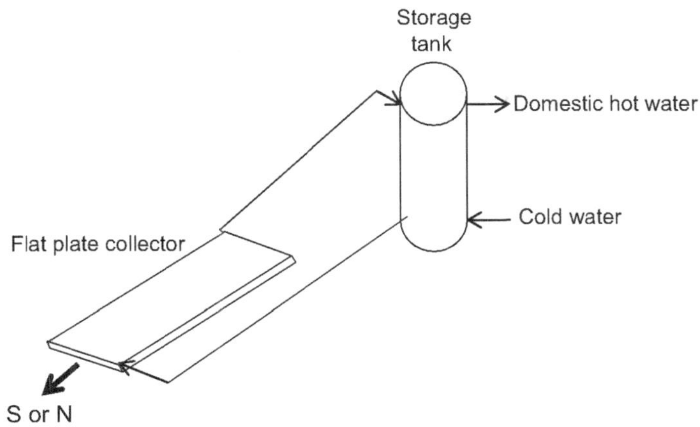

Figure 7.1 A passive solar water heater.

installed as an auxiliary system inside or outside the tank for the water to reach the desired temperature (45–50°C or even higher). It can be mixed with cold water, reaching about 38°C to provide a comfortable bath or shower.

Ideally, this simple and completely passive SWH would be enough to meet all the hot water needs of a home (Kalogirou, 2009). However, producing all the needed heat by solar energy alone rarely takes place, and the part produced by solar radiation is referred to as the solar fraction (SF), mainly in the winter, owing to either insufficient solar radiation or the location of the collector. Thus, consider SWH a preheating water technology as the best design starting point, complemented by electric energy or gas when necessary. This inevitably leads to the consumption of conventional forms of energy to produce an auxiliary fraction, which can be mandatory under adverse conditions. SWH research mostly attempts to maximize the SF. However, in very hot climates, auxiliary heating may be unnecessary because its need for hot water is much smaller or even nonexistent.

A primary design rule of solar collectors is the orientation or azimuth angle. If the collector is installed in the Northern Hemisphere, the best orientation for the flat plate is to the south; conversely, if the collector is in the south, a northward orientation is best. Thus, the designer will adopt the roof plan of a northern or southern orientation to install the SWH. If the only possible orientation still has a deflection, the adoption of a larger area may be necessary. Yet, if an independent PV panel for electric energy generation also occupies the roof, the designer must decide which, the thermal collector or the PV panel, will receive the best location considering price and energy production magnitude. In the case of the hybrid PV/T discussed later, it should be installed at the best roof location.

The difference between a more traditional passive and an active system is that the former is in charge of heat transfer from the collector to the tank by natural circulation or thermosiphon (Wang, Yin, Qi, Ma, & Liu, 2015; Wang, Yang, Qiu, Zhang, & Zhao, 2015). Passive systems are deployed round the world as domestic systems, with 77% of the total (Mauthner & Weiss, 2015). The active system (with 23%) employs an

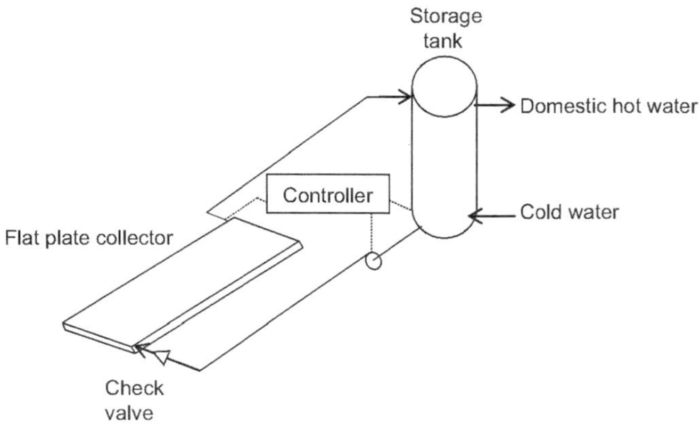

Figure 7.2 An active solar water heater.

electric pump, controlled by temperature sensors and relay or by a differential thermostat. When the temperature differential between the collector's hot water outlet at the top header and the water in the bottom of the tank is achieved, eg, 5°C, the relay turns the pump on. Active system efficiencies are higher than passive ones (Khalifa, 1998; Ogueke, Anyanwu, & Ekechukwu, 2009; Wang, Yang, et al., 2015). An active system is shown in Fig. 7.2.

The FPC may be of the direct or indirect type. In the former, the HTF is proper for domestic hot water use, whereas in the latter there is an exclusive HTF, usually anti-freezing, such as ethylene glycol, and a heat exchanger between the HTF and the domestic water (Wang & Yang, 2014). The active system is suitable for two conditions: when the storage tank position is not above the collector and when the thermosiphon effect is not possible, because the hot water inside the piping system does not go down naturally or the distance to the collector is too long or complex with many connections; that is, the head losses are high and the buoyancy is not enough to provide a satisfactory thermosiphon effect.

7.2.1 Thermal performance of solar collectors

Historically, as reported by Duffie and Beckman (2006), a simple but effective analysis of FPC efficiency was derived from studies by Willier and Hottel, beginning in the 1950s, and going up to the American Society of Heating, Refrigerating, and Air-Conditioning Engineers (ASHRAE) standard procedure 93-77 (ASHRAE, 1977). As suggested by Hossain et al. (2011), from the total solar radiation incident on the collector, 10% is reflected and 5% is absorbed by the cover, 85% being transmitted through the cover and 80% reaching the absorber plate, because it also reflects 5%. The plate loss is 10–35% by convection and radiation, thus 45% is produced as useful heat. However, the thermal performance equations that follow, inferred by the authors, are very clear. The radiation reaching the absorber plate is Q_i.

$$Q_i = G(\tau\alpha)A \quad [7.1]$$

where G (W/m^2) is the incident radiation intensity on the collector aperture, A is its area (m^2), τ is the fraction crossing the cover, and α is the fraction absorbed by the plate. The rate of heat loss is Q_o.

$$Q_o = U_L A(T_c - T_a) \quad [7.2]$$

where U_L is the thermal transmittance (W/m^2 · °C), T_c is its temperature, and T_a is the ambient temperature (°C). Then, the useful heat is Q_U.

$$Q_U = Q_i - Q_o = G\tau\alpha A - U_L A(T_c - T_a) \quad [7.3]$$

But,

$$Q_U = \dot{m}c_p(T_o - T_i) \quad [7.4]$$

where $\dot{m}$ is the mass flow rate (kg/s), c_p is the heat capacity (kJ/kg · °C), and T_o and T_i are the water outlet and inlet temperature, respectively. Thus, F_R is defined as a fractional number that relates the actual useful energy gain of a collector to the useful gain if the whole collector surface was at the fluid inlet temperature:

$$F_R = \frac{\dot{m}c_p(T_o - T_i)}{A[G\tau\alpha - U_L(T_c - T_a)]} \quad [7.5]$$

The instantaneous efficiency of the collector is the ratio between the useful heat gain and the maximum possible heat gain:

$$\eta_i = \frac{\int Q_u dt}{A\int G dt} = \frac{Q_u}{AG} \quad [7.6]$$

Thus,

$$\eta_i = \frac{F_R A[G\tau\alpha - U_L(T_c - T_a)]}{AG} \quad [7.7]$$

or

$$\eta_i = F_R\tau\alpha - F_R U_L\left(\frac{T_i - T_a}{G}\right) \quad [7.8]$$

The standard test method for liquid heater collectors was proposed by the International Organization for Standardization (ISO) standard EN 12975-2 (ISO, 2006), and the results are presented in graphic form. The instantaneous efficiency is plotted on the vertical axis, and reduced temperature $(T_m - T_a)/G_T$ on the horizontal axis, as shown in Fig. 7.3. This figure is referred to hereinafter as the "efficiency curve."

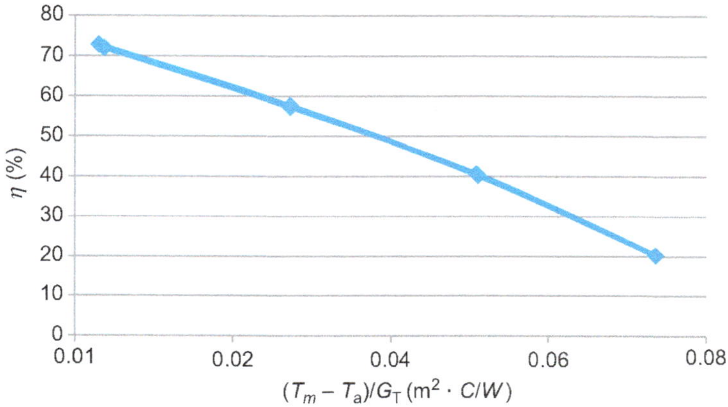

Figure 7.3 Experimental collector efficiency data measured by a solar simulator for a liquid-heating flat plate collector.
Authorized by the Institute of Technological Research of Sao Paulo, Brazil.

The curve resulting from regression is as follows.

$$\eta = 0.74 - 5.198\, T_r - 0.036 T_r^2 \qquad [7.9]$$

Considering $T_m = T_a$, at the beginning of the test, the instantaneous (or maximum) efficiency is equal to the data curve × the vertical axis intercept at instant t, when the heat transfer becomes steady state. On the other hand, considering the regression equation a straight line, then $\eta_i = F_R\,(\tau\alpha)_n$, and the slope is $F_R U_L$. Thus, the instantaneous efficiency varies along the test duration, decreasing as the inlet temperature increases, because of the collector heat losses by convection and radiation increase.

7.2.2 Absorber plate design

The main component of the SWH is the absorber plate in charge of receiving solar radiation and converting it into heat, although the radiation first has to cross the collector cover, if there is one or more. Covers are installed on the collector to reduce convection losses from the absorber plate to the ambient air. Most of the collector efficiency is due to the plate and cover according to the product $(\tau\alpha)$ presented in Eq. [7.1]. The cover transmittance to solar radiation, τ, must be as close as possible to the unit, as well as the plate absorptance, α. Thus, the collector efficiency depends on these two optical properties and on its thermal performance regarding the convection. The plate material conductivity is crucial in transferring heat to the HTF, as discussed by Sadhishkumar and Balusamy (2014). Historically, the optical issues were solved by coatings, and the thermal issue was termed convection suppression. The research into coatings began in the 1950s, gaining momentum in the late 1970s and first half of the 1980s, lasting until now.

Nonetheless, a suitable material to perform convection suppression must not overwhelm the radiation transmittance. The transparent material types for convection suppression between the plate and the cover, or between two glass covers, were thin flexible polymer foil, polymer honeycomb, bubble foams and fibers, and inorganic microporous materials, such as silica aerogels, as reported in a review paper by Granqvist (1991). The honeycomb and silica aerogels are the materials with the highest solar radiation throughput. To further boost the collector efficiency, the glass cover may have an antireflection solar coating or a texturized treatment applied to its upper surface and infrared reflecting coatings to the under surface. Simulations carried out by Hellstrom, Adsten, Nostell, Karlsson, and Wackelgard (2003) of an absorber enhancement combined with a Teflon honeycomb and of an antireflection-treated glass cover yielded a total annual performance increase of 24% at a 50°C operating temperature.

Several configurations of polycarbonate honeycombs and air gap thicknesses located below and above this device were investigated experimentally by Abdullah, Abou-Ziyan, and Ghoneim (2003) for minimizing the optical losses and for comparing the results with those of an FPC without honeycomb units. Considering Eq. [7.8], the authors reported reductions in $F_R U_L$ of up to 49% for a single honeycomb and 56% for double honeycomb units compared with the $F_R U_L$ for the solar collector without honeycomb. However, the reductions in $F_R(\tau\alpha)$ due to the single and double honeycomb units were about 15% and 32%, respectively. Furthermore, the honeycomb design requires suitable thicknesses of honeycomb and gaps.

7.2.3 Coatings and nanofluids

As regards solar selective absorber coatings, Hottel and Unger (1959) sprayed a copper oxide layer onto a heated aluminum sheet to make a selective surface. They claimed that Tabor accounted for calling attention to the high importance of selectivity in solar collector design. In fact, it was evident to the scientific community, Tabor (1958) included, that solar radiation absorbers were necessarily black, and that was recognized as inevitable because such absorbers should act substantially as blackbody radiators with high thermal emissivity. Tabor realized that this was not necessarily the only possible arrangement and suggested producing surfaces with different optical properties at different wavelengths, obtaining stable selective surfaces with absorptivities of about 0.90 and emissivities in the region of 0.08–0.012 in his laboratory.

Selective absorbers are thus expected to be able to absorb most of the incident solar radiation (high absorptance, α) and reemit the minimum infrared radiation from the heated plate (low emittance, ε). A coating having absorptance α higher than 0.9 in the solar spectrum range, or between λ 0.3 and 2.5 μm, and thermal emittance ε lower than 0.1 in the mid- to far-infrared wavelength range, with $\lambda > 2.5$ μm, is considered to perform well. Yet the market asks for materials presenting $\alpha > 0.94$ and $\varepsilon < 0.04$ in line with Amri, Jiang, Pryor, Yin, and Djordjevic (2014).

There are several techniques for preparing the selective coating. The following are the most usual: vacuum evaporation, ion exchange, vacuum sputtering, chemical vapor deposition, chemical oxidation, dipping in specific chemical baths, electroplating, electroless plating, spraying, and painting (Karthick Kumar, Suresh, Murugesan, & Raj, 2013; Wazwaz, Salmi, & Bes, 2010). Table 7.1 presents some research results and their respective techniques.

Table 7.1 **Selective coating materials and methods**

Coating	Substrate	Absorptance (α)	Emittance (ε)	Method	References
CuO	Al	0.93	0.11	Spraying a dilute solution of cupric nitrate onto an Al sheet	Hottel and Unger (1959)
CoO	Cu or Ni	0.93	0.24	Plating Cu or Ni with cobalt and anodizing the surface	Gillette (1960)
CoO or Co_3O_4 (black cobalt)	Al or galvanized iron	0.92	0.13	Spray pyrolysis	Choudhury and Sehgal (1982)
NiS–ZnS (black nickel)	Cu	0.95	0.25	Electrodeposition	Patel, Inal, Singh, and Scherer (1985)
CrO–$CrNO_3$ (cermet)	Cu	>0.93	<0.07	DC magnetron sputtering	Graf et al. (1997)
RuO_2	Ti	0.74	0.12	Spraying and dipping at room temperature	Morales-Ortiz, Avila-García, and Hugo Lara (2006)
Ni–Al_2O_3 (Ni/alumina)	Al	0.82	0.07	Ni inclusions into nanoporous anodized aluminum electrochemically prepared	Galione et al. (2010)
Ni–Al_2O_3 (Ni/alumina)	Al alloy	0.92	0.11	Anodic oxidation and electroplating	Wazwaz et al. (2010)
NiO–Al_2O_3 (Ni/alumina)	Al	0.92	0.03	Sol-gel spray–pyrolysis deposition	Ienei, Isac, Cazan, and Duta (2010)
CuO	Cu	>0.95	<0.07	Chemical oxidation	Karthick Kumar et al. (2013)
a-C:H/Cr	Cu	0.74	0.072	Magnetron co-sputter deposition	Cheng, Chiou, Ting, and Tzeng (2013)
C–TiO_2 (porous carbon–titania)	Quartz and Cu	0.95	0.12	Sol-gel spin-coating	Cheng et al. (2015)

Concerning the coating materials in Table 7.1, aluminum has many advantages over other materials, such as high thermal conductivity, low emissivity, light weight, and low cost. Its derivative, nickel alumina, has a 25-year service life and is very resistant to degradation, which are important factors in the selective coatings area (Belghith, Arurault, & Bes, 2013). In turn, black chrome is a very much used solar selective absorber component; cermets are metal particles in ceramic structures produced by electrochemical or vacuum deposition techniques. Unfortunately, black chromium is currently known to be toxic (AlShamaileh, 2010).

Copper has been researched since the 1950s, achieving a small increase in absorptivity and a more significant increase in emissivity. The decreasing size of the particles represents a great advance from micrometers to nanometers, as shown by Karthick Kumar et al. (2013), who prepared CuO thin films characterized by one-dimensional crystalline nanostructures shaped as nanofibers and nanoneedles by chemically oxidizing a copper substrate. Nanocrystalline materials have properties superior to those of conventional coarse-grained materials, both polycrystalline and amorphous, since they have the same composition (Wazwaz et al., 2010). Fig. 7.4 depicts a selective coating of nanostructured alumina over an aluminum substrate.

Fig. 7.4 shows metallic nickel embedded in the alumina porous layer by an anodic signal and its content determines the coating optical properties process to obtain the nanostructure presented in the figure.

Moreover, some nanomaterials used as coating are employed also as nanofluids, such as alumina (Al_2O_3), cupric oxide (CuO), silicon dioxide (SiO_2), and many other metallic and nonmetallic nanoparticles in FPCs. Nanofluids are mixtures of liquid (base fluid) and nanoparticles in suspension (sized 1−100 nm) (Al-Shamani et al., 2014; Moghadam, Farzane-Gord, Sajadi, & Hoseyn-Zadeh, 2014; Verma & Tiwari, 2015). Nanofluids have intensified thermophysical properties as HTFs, such as thermal conductivity, viscosity, convection heat transfer coefficient, and specific heat.

Some studies show the strong potential of nanofluids to improve FPC thermal performance, as follows. Shojaeizadeh, Veysi, and Kamandi (2015), comparatively investigating the performance of an FPC with a suspension of Al_2O_3−water nanofluid as base fluid, obtained optimum values of mass flow rate of fluid and collector inlet fluid

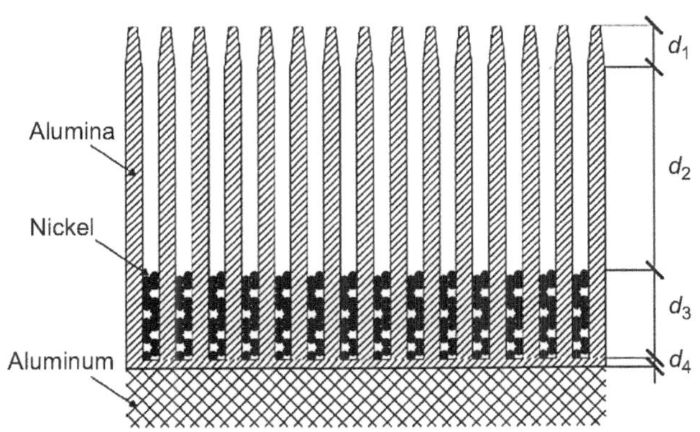

Figure 7.4 Coating sketch of alumina (Galione et al., 2010).

temperature that were 68% and 2% lower, respectively. He, Zeng, & Wang (2014) analyzed the influence of various mass fractions and sizes of $Cu-H_2O$ nanofluids on FPC thermal efficiency, finding for a (25 nm, 0.1 wt%) nanofluid with a mass flow rate of $0.019\ kg/m^2 \cdot s$ an increase of 23.83%. Salavati, Kianifar, Niazmand, Mahian, and Wongwises (2015), in an experimental study on an FPC using as HFT an SiO_2/ethylene glycol, increasing its volume fraction from 0.0% to 1.0% and mass flow rate from 0.011 to $0.028\ kg/m^2 \cdot s$, reported an efficiency enhancement between 4% and 8% despite the silicon dioxide low thermal conductivity. Thus, research results from nanofluids sound promising.

7.3 Evacuated tube collector

This collector has two glass tubes with vacuum between them, also known as the Dewar type, in which the solar radiation transmitted and reemitted by the outer tube is absorbed by the inner one and removed by the working fluid, which has some type of contact with the absorber tube.

There are three basic types of ETC: the all-glass, the heat pipe, and the U-tube. The great innovation of these collectors is their thermal isolation by means of vacuum. One of the best descriptions of the vacuum effect, convection suppression, is found in Garg, Chakravertty, Shukla, Agnihotri, and Khas (1983). They describe the ETC state of the art, the first ETC designs in 1909, and that a couple of them were commercially available in the 1980s.

Furthermore, the three basic types of ETCs announced in the 1980s have alternatively been a focus of research since then. For SunMaxx Solar (2015), the ETC technology was also developed by Quing Hua University in Beijing in that same decade, which explains the high level and current deployment of this technology in China.

7.3.1 All-glass evacuated tube collector

This collector type is the most widely used ETC. The water fills the inner tube, with all collector tubes sharing a manifold (Fig. 7.5(a)). The outer wall of the inner tube is treated with an absorbing selective coating. This type is associated with the water-in-glass concept by Morrison et al. (2004) for whom this collector is the most successful because of its simplicity and low cost. Fig. 7.5 shows two specimens of tested collectors having a storage tank as a horizontal manifold.

As can be seen in Fig. 7.5, the all-glass collector is a long single-ended thermosiphon tube, which prevents the use of a metal tube with two seals assembly, because the different linear dilatation of glass and metal materials would make this type of design fragile. The evacuated tubes are inserted directly into the tank making a type of integrated collector. This part is the water-in-glass manifold, which consists in a simple and early conception of a hot-water storage tank feasible for a wide range of system designs and operation conditions, as observed by Harding and Zhiqiang (1985). As the all-glass is a thermosiphon collector, its first design problem is heat extraction, because a region of heat stagnation is created near the tube's closed

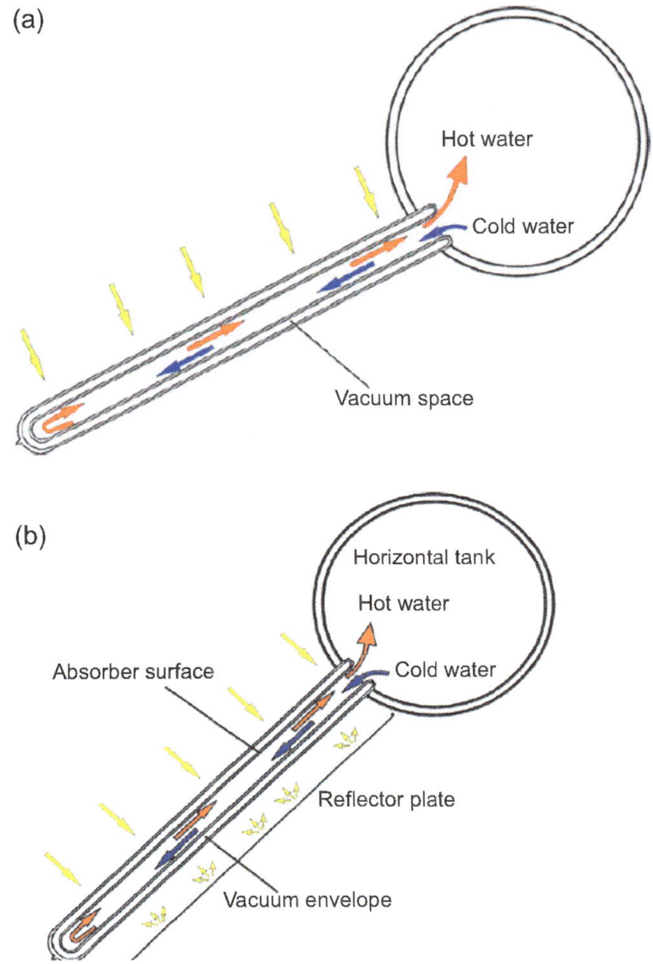

Figure 7.5 (a) Simple all-glass evacuated tube collector (Morrison et al., 2004). (b) All-glass evacuated tube collector with a back diffuse reflector (Budihardjo, Morrison, & Behnia, 2007).

side. The heat stagnation zones are very unwelcome because they not only worsen the collector thermal performance, but also increase the risk of the tubes being fractured by thermal shocks.

The longer and thinner the tube, the more difficult is the heat extraction from the tube. Regarding this relation, there is an expression, the aspect ratio, here translated to L/d, where L (m) is the tube length and d its diameter (m). The natural circulation flow rate can be expressed by the Reynolds number as a correlation of the dimensionless parameters of natural convection and L/d, as proposed by Budihardjo et al. (2007) in their numerical and experimental research on all-glass collectors. Despite using a forced circulation in a computational fluid dynamics study, Shah and Furbo (2007) concluded that the collector with the shortest tube length achieved the highest efficiency.

In other words, the thermosiphon rate through the circuit is computed based on the difference between the weight of the hot and cold fluid columns and the frictional resistance along the loop (Budihardjo & Morrison, 2009). The authors also found that, when the collector tilt was increased from the 22° of a standard roof to 45°, the annual saving during winter months was 12% higher. The new tilt corroborates older (Behnia & Morrison, 1991) research concluding that the stagnant region length near the closed end of the tube varies with the tilt angle and showed a minimum at 45°.

Regarding the optical properties of all-glass tubes and current technologies, the transmittance of the outer tube has reached 0.94 with porous SiO_2 antireflection coating by the sol-gel method, whereas the inner tube absorptance and emittance are 0.95 and 0.05, respectively (at 180°C); by dual-cathode co-sputtering, Wang, Yin, et al. (2015) achieved an instantaneous solar thermal performance of 50.2%.

7.3.2 Heat pipe collector

In the second type, a metallic heat exchanger pipe embedded in an absorber glass pipe has a small quantity of mass inside (liquid acetone or alcohol), which receives heat from solar radiation and is converted into vapor (latent heat), going up to the header at the top of the collector and transferring the heat to the water (or water–glycol) inside it. Thus, the fluid changes to liquid and goes down inside the heat pipe by gravity and capillary forces to the evaporator (Zhang, Zhao, Smith, Xu, & Yu, 2012). The vacuum between the glass pipes minimizes heat loss by convection; and a selective coating is also applied to the outer surface of the inner tube to maximize the solar radiation absorption and to minimize emittance. The transverse section of this type can be observed in Fig. 7.6.

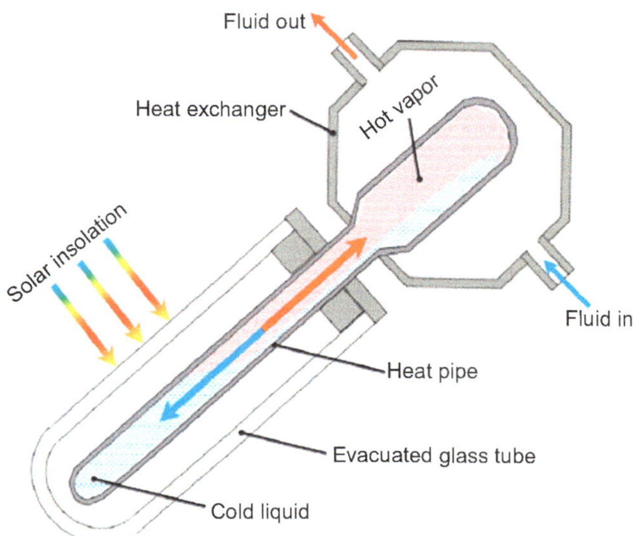

Figure 7.6 Cross section of a linear heat pipe (Suman et al., 2015).

Each pipe has a condenser and the series of them is the manifold. Thus, reverting to liquid, this solution goes down to the bottom of the heat pipe, where the evaporator is located and the cycle is repeated. The HTF (for instance, water or water–glycol) flows inside the header (large-diameter pipe) where it obtains the heat released by the condensers, transferring it to a storage tank (Asif & Muneer, 2015).

The condensate flows through a capillary structure (wick of stainless steel screen) in the inner container wall to the heating zone, completing the cycle. Heat pipes are compared with thermal diodes, because properly oriented they transfer heat only from the evaporator to the condenser, preventing reverse circulation and cutting off losses when the liquid temperature in the heat exchanger is higher than that in the absorber (Azad, 2008).

The collector's geometric relations or its aspect ratio is an important parameter for the heat pipe collector thermal performance. Comparing theoretical and experimental efficiencies in a heat pipe rig in Iran, Azad (2008) obtained 60.3% and 55.6%, respectively, with an optimum length relation between evaporator (e) and condenser (c) of 8.25, or more strictly: $L_e = 1.65$ m and $L_c = 0.2$ m. Also, in experimental research in Lebanon comparing all-glass and heat pipe evacuated collectors, Hayek, Assaf, and Lteif (2011) concluded that for tubes that are too long compared to their diameters, the effectiveness of the buoyancy forces decreases in transporting the working fluid along the tube, with the upgrowth of significant dead zones at the tube's bottom, increasing stagnant regions and lowering efficiency. The efficiency of the heat pipe was 15–20% above that of the all-glass collector, but the payback of the former was three to four times that of the latter. However, the condenser's position inside or outside the collector's manifold is also another factor, which has decreased the thermal performance from 63% to 47.4% as measured by Redpath (2012) in the northern maritime climate in Ireland. In Iran, Noie, Heris, Kahani, and Nowee (2009) examined a heat pipe and its HTF nanofluid of aqueous Al_2O_3 content, varying the volume fraction between 1% and 3% and obtaining a 14.7% rise in its thermal efficiency in relation to pure water.

7.3.3 U-tube collector

The U-tube ETC can be without fin or with fin and in detail in Fig. 7.7. A U-tube is inserted into the inner tube (absorber) and water flows in the U-tube to collect and to convey useful energy outside. The U-tube-type evacuated solar collector is a recent development, which was also an attempt to solve poor resistance to the pressure in an

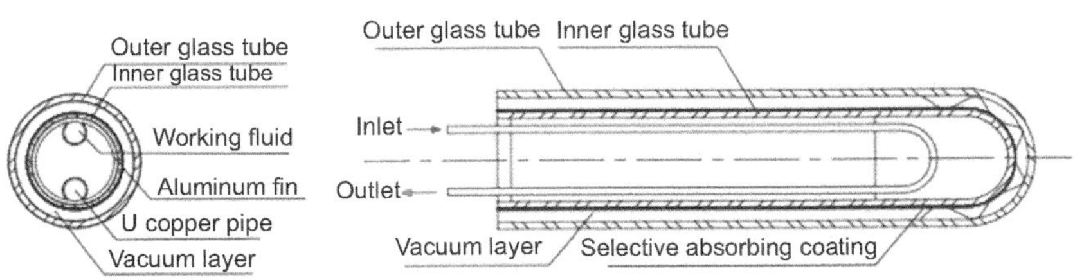

Figure 7.7 U-tube cross section and longitudinal view (Gao et al., 2014).

all-glass ETC. It has an inserted metallic tube shaped like a U with a copper fin to transfer the energy collected by the absorber tube, as described by Gao et al. (2014) and also by Liang, Ma, Zhang, and Zhao (2012).

Several types of fins have been investigated so far in the 21st century according to experimental or numeric methods to improve the U-tube thermal performance, such as by Kim and Seo (2007), who employed both methods and proposed four models combining the U-tube with copper circular or plate types of fins and considering the incidence angle, the diffuse solar radiation, and the shadow effect due to adjacent tubes, verifying their thermal performances.

More recently, the copper circular fin around the U-tube seems to be a stronger trend. In 2010, Ma, Lu, Zhang, and Liang performed a thermal performance analysis of a U-tube collector with the section shown in Fig. 7.8(a), arguing that the presence of

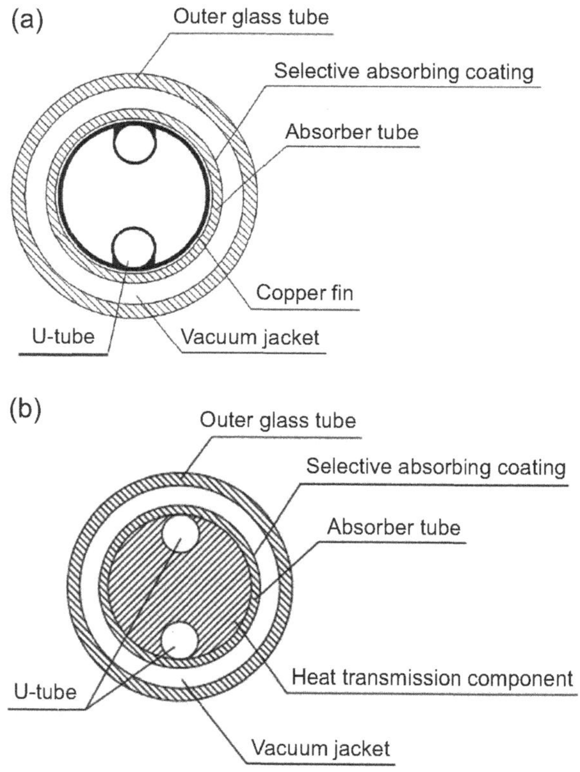

Figure 7.8 Cross section of a U-tube (a) with copper circular fin *(Ma, L., Lu, Z., Zhang, J., & Liang, R. (2010). Thermal performance analysis of the glass evacuated tube solar collector with U-tube.* Building and Environment, 45*(9), 1959–1967. http://doi.org/10.1016/j.buildenv.2010.01.015)* and (b) filled with a heat transmission component *(Liang, R., Ma, L., Zhang, J., & Zhao, D. (2012). Experimental study on thermal performance of filled-type evacuated tube with U-tube.* Heat and Mass Transfer/Waerme- Und Stoffuebertragung, 48*(6), 989–997. http://doi.org/10.1007/s00231-011-0912-7).*
Authorized by Springer.

air gaps between the absorber glass tube and the copper circular fin tube increased the thermal resistance and the difference in temperature from the selective absorbing coating and the fin, greatly damaging the heat transfer process in this part of the collector. Two years later, some of those authors proposed replacing the copper fin by filling the space with a compressed graphite expandable in the vacuum tube with thermal conductivity of 147.4 W/(m · K), obtaining a maximum collector efficiency, the intercept of the line of the η_i axis of Fig. 7.3 ("efficiency curve"), equal to 0.84, which is 12% higher than that of the copper fin.

One year later, Liang, Ma, Zhang, and Zhao (2013) proposed a new alternative with two U-tubes crossed at 90° inside the absorber tube, reaching an intercept of the line of the η_i axis 4% higher in relation to the filled-type evacuated tube with a single U-tube.

Other authors considered filling a U-tube flat absorber with a minichannel-based material and a selective coating on its outer wall, obtaining higher efficiencies compared to similar ETCs described in the literature (Sharma & Diaz, 2011). And other authors, such as Zambolin and Del Col (2012), combined ETCs with CPCs. The former attained maximum efficiencies between 80% and 85% using these components and filling, whereas the latter reached around 70% with, and 42% without, concentrators or filling, respectively. Thus, there is consistency between these values of thermal performance.

The types of ETCs described above can operate as thermosiphon systems unless the piping from the collector to the storage tank is too long or has many connections excessively increasing head losses. Through the thermosiphon, the flow goes directly to the storage tank by means of the direct systems.

7.4 Integrated collector storage systems and compound parabolic collectors

7.4.1 The solar water heating system dawn

According to Bainbridge (1983), the integrated collector or built-in storage known as the "fundamental black can" was the first solar water heater used in the United States in the 1800s and evolved to its commercial phase, being patented in 1891. Many patents were also issued in Japan until the first commercialization by Yamamoto in the 1940s (Smyth, Eames, & Norton, 2006). In the 1980s, a considerable number of ICSs were constructed and tested by researchers, beginning with simpler models, like that developed by Srivastava, Hussain, Sharma, and Tiwari (1982) having concrete under a glass cover as absorber and also as storage medium. This material was replaced by a metallic plate as an absorber over the storage tank as shown in Fig. 7.9, changing into other more sophisticated collectors, like those proposed by Gostzberger and Rommel (1987) with transparent insulation material at the front. Prakash, Garg, Hrishikesan, and Jha (1989) studied the collector performance of transparent and opaque insulations, because thermal losses from the storage tank are the main concern regarding ICSs at night and on cloudy days with low ambient temperature, which prevents perfect tank insulation for the absorber as well.

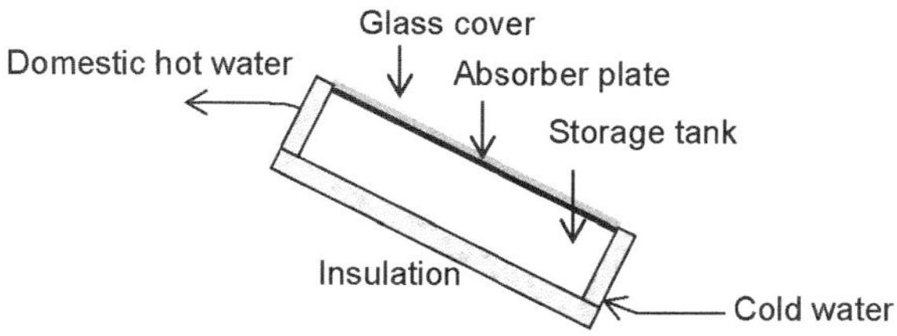

Figure 7.9 Integrated collector and storage system.

In the 1970s, ICSs with CPCs made great progress as a way to obtain higher temperatures on the vessel absorber surface by adopting reflectors, which increase the ratio of incident radiation to heat losses or attain a higher concentration ratio. The reflectors are flat or curved, circular, parabolic or compound parabolic, symmetrical or asymmetrical, and made of materials such as polished stainless steel sheet and aluminized Mylar.

7.4.2 *Facing heat losses*

The crucial question regarding ICSs is how to minimize thermal losses, which occur by convection and radiation from the absorber; in this case, the vessel. A comparative test carried out by Burton and Zweig was described by Smyth et al. (2006). Two absorbers, one painted dull black and the other receiving a selective coating of metallic nickel-chrome, with absorptance of 0.97 and emittance of 0.10, showed that the latter always produced the warmest water. To prevent losses by convection, mainly during noncollection times, the glass was placed at the aperture of the ICS, which also protects the absorber from the environment. Using a low-iron-content glass, its reflective property sends IR (Infrared) radiation back to the absorber, keeping a high value for glass solar transmittance. To further reduce losses by convection, using evacuation or an inverted absorber are possible steps. After construction of an ICS with an inverted absorber, thermal losses were appreciably reduced by using transparent baffles in various positions of the upper portion of the collector cavity, without significantly decreasing its optical efficiency (Smyth, McGarrigle, Eames, & Norton, 2005).

To minimize thermal losses from the absorber to the ambient air, (Tripanagnostopoulos & Yianoulis, 1992) designed ICSs with a cylindrical tank placed horizontally and an amplified collector consisting of a curved mirror reflecting solar radiation to the storage tank. The suitable match between the tank and the mirror curve make the basic design fundamental. Nevertheless, as stressed by Kalogirou (1997), several other design parameters should be considered, such as operation characteristics and environmental conditions of the installation site. The usual type of ICS described is 100–200 L/day of hot water consumption with an absorber area-to-storage volume ratio of 35 L/m^2. To improve the collector efficiency, a reflective concentrator was added to the ICS according to Fig. 7.10.

This collector is a CPC able to reflect most incident radiation to the absorber. Among the various types of concentrator collectors, the author recommends the linear or trough one (Fig. 7.11), in which the solar radiation enters the collector aperture,

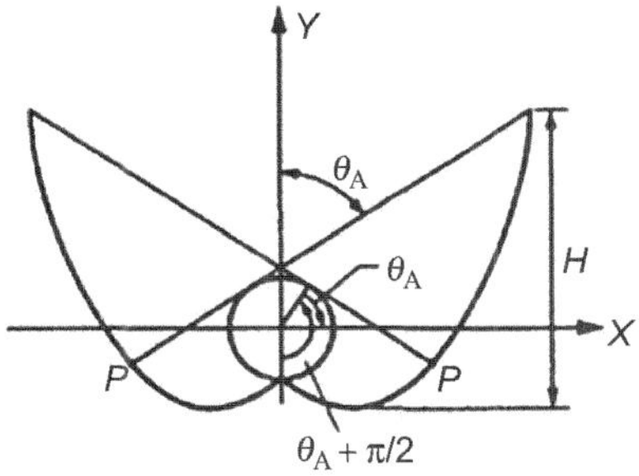

Figure 7.10 Integrated collector storage system with concentrator for fully developed cusp (Kalogirou, 1997).

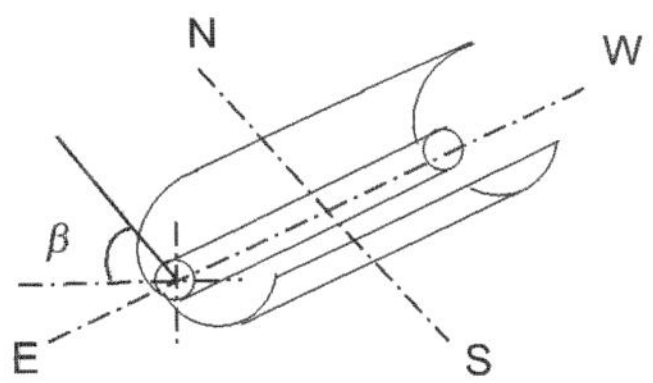

Figure 7.11 Trough concentrator collector oriented toward the south with tilt angle β equal to the local latitude.

being reflected specularly to the absorber, limited by the acceptance angle (θ in Fig. 7.10), which is that in which the radiation source may vary its direction and still reach the absorber. Concentrator collectors are known to be suitable only for places with enough clear sky direct radiation (Bojić, Marjanović, Miletić, & Bojić, 2015) unless the absorber is exposed to a diffuse component. Increasing the acceptance angle of a stationary collector allows direct absorption of more diffuse radiation and can take advantage of this, using a significant share of the solar energy. Meanwhile, ICSs suffer severe damage when exposed to very cold weather (Hossain et al., 2011).

Introducing a primary and small cold-water vessel above the large hot-water storage vessel, Kalogirou (1999) conceived the former to work as a baffle, reducing losses by convection, increasing the storage capacity by 30%, and providing a sort of barrier such that the view of the sky by the large vessel was obstructed, thus preventing losses by radiation. The advances in ICSs, not only for storage tanks, but also for their absorbers and CPC concentrator, were diverse in the beginning of the 21st century, as described by Smyth et al. (2006).

Starting from cylindrical absorbers, Devanarayanan and Kalidasa Murugavel (2014) reported a remarkable combination among concentrators and storage tanks. The former may have asymmetric and symmetric reflectors, three parabolic sections, and nonimaging cusp concentrator, as illustrated in Fig. 7.12.

Each of the ICSs in the figure has different advantages and disadvantages, according to its features. The asymmetric one in Fig. 7.12(a) is very good at reducing transfer losses because it generates a region of trapped warm air by free convection involving

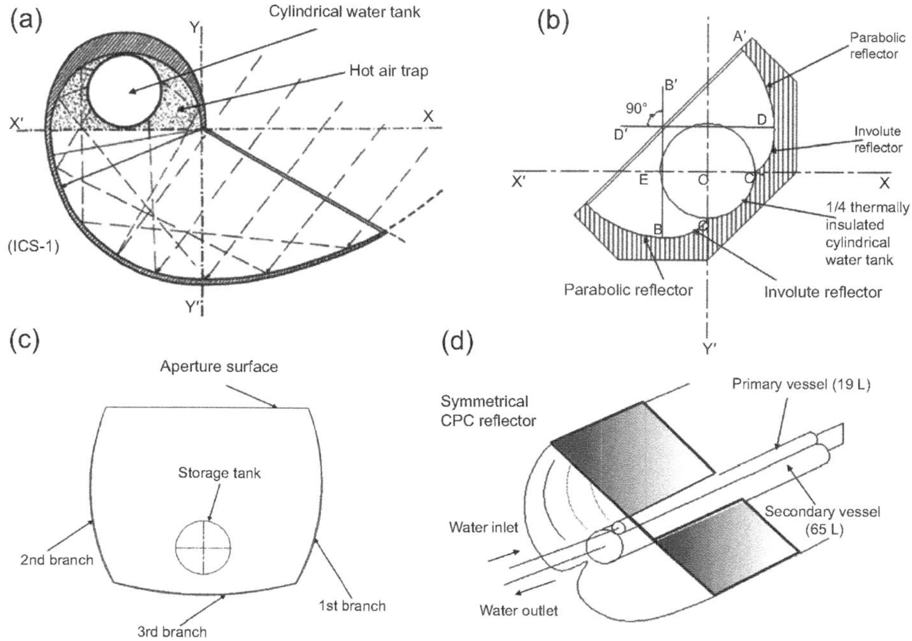

Figure 7.12 (a) Asymmetric integrated collector storage system (ICS) with one tank (Devanarayanan & Kalidasa Murugavel, 2014; Tripanagnostopoulos & Yianoulis, 1992). (b) Symmetric ICS (Tripanagnostopoulos, Souliotis, & Nousia, 2002). (c) Three-parabolic-sections ICS (Devanarayanan & Kalidasa Murugavel, 2014; Helal, Chaouachi, & Gabsi, 2011). (d) Symmetric double-tank ICS (Devanarayanan & Kalidasa Murugavel, 2014; Kalogirou, 1999).

practically the whole cylindrical tank, which is also guarded by thick insulation; although it uses expensive materials, it has low optical efficiency. The symmetric collector (Fig. 7.12(b)) performs reasonably well, raising the water temperature in daytime, despite presenting high conduction and convection losses in cold weather and null or low levels of solar radiation. The third ICS (Fig. 7.12(c)), with three parabolic branches, increases the number of reflections toward the absorber, resulting in better efficiency; however, it presents high thermal losses at night. The symmetric double-cylindrical water tank (Fig. 7.12(d)) attains water temperature stratification because the two horizontal tanks are connected in series; it not only has 30% greater storage capacity, but also reduces nighttime thermal losses (Devanarayanan & Kalidasa Murugavel, 2014).

Maximizing stratification is useful for taking hot water from the upper part of the vessel before receiving cold water, restraining its velocity for preventing mixing of heated and unheated water. Smyth et al. (2006) describe a number of vessels and their configurations in ICSs combined with CPC reflectors to maximize stratification. The factors causing the mixture are: (1) cold-water charge and hot-water discharge cycles; (2) the size and locations of inlets and outlets; (3) flow, temperatures, and velocities;

and (4) geometries. One way to enhance stratification is to mount the vessel inclined vertically, as suggested by the authors, and in a north–south orientation, although this position restricts the duration of daily collection depending on the acceptance angle of the concentrating systems. Thus, this configuration is better for high latitudes, where the strategy concerns tilting the plane of the aperture of the collector normal to the beam of solar radiation at the equinox. If the strategy is to maximize collection in winter, then the recommendation is to tilt the plane 5–10° more than the latitude angle. The horizontal vessel positioning is suitable around the equator with east–west orientation, collecting radiation all day long, although there is a risk of a low SF result.

The ICSs with double tanks may have an asymmetric reflector, or two or three reflectors, and the inverted absorber type as shown in Fig. 7.13.

The asymmetric CPC with double tanks in Fig. 7.13(a) is cost-effective, but takes a long time to attain thermal equilibrium and has high thermal loss. The second CPC (Fig. 7.13(b)) presents reflector symmetry for the lower water tank with the addition of a parabolic part next to the upper tank, increasing insulation tick in this location, and eventually attaining a lower thermal loss coefficient and 75% more mean daily efficiency. The inverter absorber ICS (Fig. 7.13(c)) is composed of (1) an inverted vessel

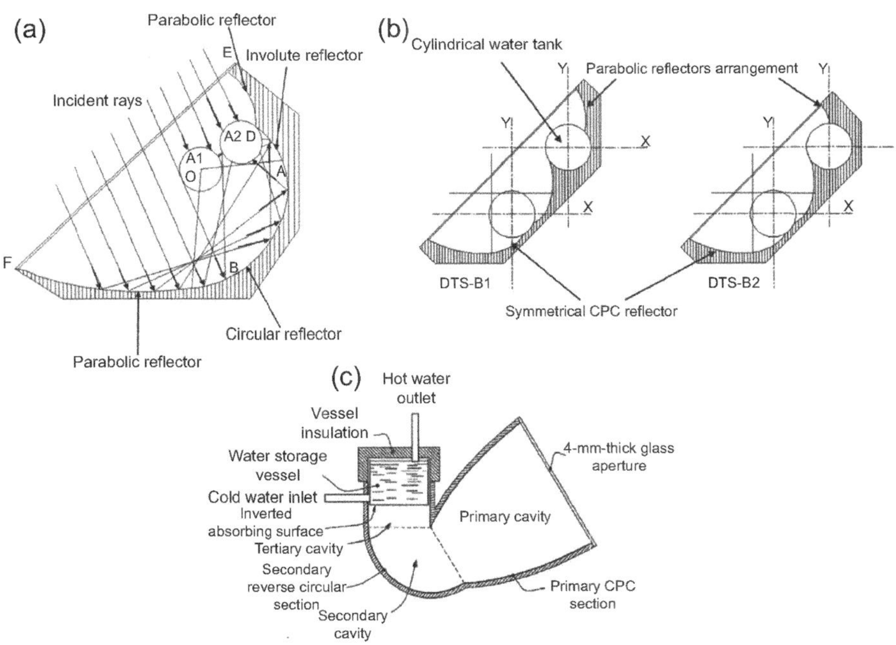

Figure 7.13 (a) Asymmetric compound parabolic collector (CPC) with double tanks and three reflectors (Devanarayanan & Kalidasa Murugavel, 2014; Tripanagnostopoulos, Souliotis, & Nousia, 1999). (b) Integrated collector storage system (ICS) with double tanks and two reflectors (Devanarayanan & Kalidasa Murugavel, 2014; Tripanagnostopoulos & Souliotis, 2006). (c) ICS of the inverted absorber type (Smyth et al., 2005, 2006; Devanarayanan & Kalidasa Murugavel, 2014).

Figure 7.14 A coupled collector.

made of an aluminum sheet coated with selective absorber film, and other surfaces guarded with polystyrene insulation, and (2) a CPC with a circular reflector and another straight reflector at its exit.

7.4.3 The coupled collector

This type of collector has its storage tank installed over the roof like an ICS, but it is not an integrated collector, because the collector and tank are linked by pipes. It has the same components as the FPC, the absorber plate and the storage tank, positioned vertically, to make good use of the stratification, or horizontally, as the easiest installation and security, as illustrated in Fig. 7.14.

The coupled collector is very common today for low-income housing in mild or hot regions where roofs have a low slope, with small room off the attic to fit a storage tank inside, and comply with the vertical distances needed for a good thermosiphon at the same time.

7.5 Hybrid photovoltaic/thermal collector

7.5.1 Combining technologies

The earlier concepts of hybrid PV/Ts were published in 1978 and a first commercial thermosiphon system of this type about 20 years later, according to Kalogirou and Tripanagnostopoulos (2006). PV cells convert solar radiation into electricity and cells are connected to form PV modules, which in turn are also grouped in series or in

Figure 7.15 A photovoltaic/thermal collector.

parallel or a combination of both series and parallel (Tiwari, Mishra, & Solanki, 2011). Fig. 7.15 shows a PV/T.

Fig. 7.15 shows a PV/T hot-water copper piping, electrical connections, and PV strings. A typical PV/T is designed to generate electrical energy through PV cells and simultaneously convey the heat absorbed by the cells, and not converted to electricity, by a medium such as air or water, to be used immediately or reserved in a hot-water tank. The heat at the PV module results from the cells not being able to convert solar radiation of all wavelengths into electricity. Removing the heat is useful not only to provide hot water, but mainly to keep the cells reasonably efficient, because the electrical power output can drop by 0.2–0.5% for every 1°C rise in the PV module temperature as stated by Makki, Omer, and Sabir (2015). The higher specific heat capacity of the water compared to the air medium, also offering less temperature fluctuation, makes the liquid more suitable as a working fluid for hybrid collectors, which presents some flexibility in design, as illustrated in Fig. 7.16.

The first type of collector shown in Fig. 7.16, the sheet-and-tube, is the most common design with an absorber plate and tubes, such as the ubiquitous FPC. It could hardly be otherwise; with this in mind Florschuetz (1979), Lalović, Kiss, and Weakliem (1986), Agarwal and Garg (1994), and Garg and Agarwal (1995) conceived their studies on PV/Ts, modeling sheet-and-tube PV/Ts by pasting PV cells over the plate. The fluid circulation can be obtained by the thermosiphon effect and gravity. Even with the high feasibility of the FPC as a support for the PV cells, the second type, the box channel design, was enhanced by Chow, He, and Ji (2006), by employing extruded aluminum-alloy box-structure modules as absorber material and the PV module over it, reaching instantaneous thermal and electric efficiencies of up to 76.3% and 12.3%, respectively. Nonetheless, packing a PV/T collector creates a higher number of layers, as illustrated in Fig. 7.17.

From top to bottom, the PV/T collector of Fig. 7.17 has a first layer of glass and a second of air, a transparent layer of adhered TPT (Tedlar–polyester–Tedlar) and EVA (ethylene vinyl acetate) encapsulating the laminated PV cell, with another layer of EVA and opaque TPT underneath. The TPT provides good electrical insulation, and

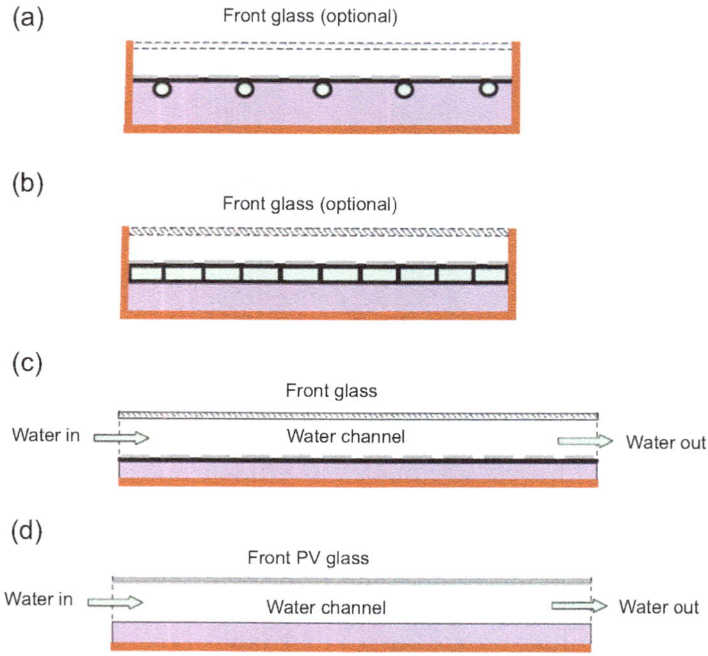

Figure 7.16 Cross sections of photovoltaic/thermal collectors with water as the working fluid (a) Sheet-and-tube design. (b) Box channel design. (c) Channel above PV design. (d) Channel below PV (transparent) design (Chow, 2010).

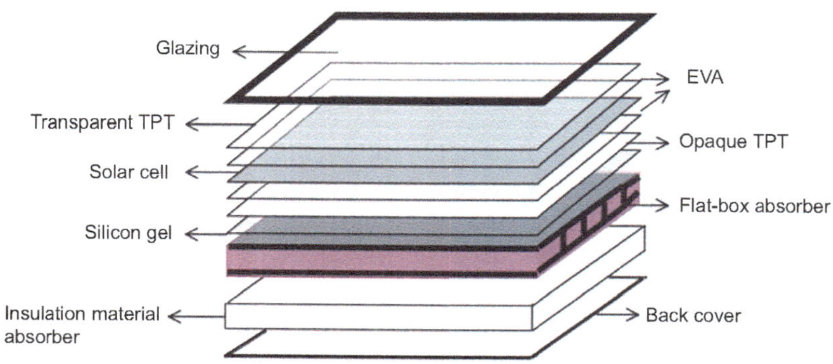

Figure 7.17 Constituent layers of a photovoltaic/thermal collector (Chow et al., 2006). *EVA*, ethylene vinyl acetate; *TPT*, Tedlar–polyester–Tedlar.

the EVA, adhesion to the absorber plate. Silicone gel is an alternative encapsulation material to EVA, used for sealing the cell in the frame, strengthening against high temperatures, UV radiation, and other aggressive agents. With the addition of the remaining components, the absorber plate, insulation material, back cover, and thermal part, the hybrid collector is complete.

Regarding the PV material, there are two manufacturing processes, one of them intensive in energy, yielding the crystalline silicon single crystal or mono (sc-Si) and polycrystalline, and the other consisting of thin-film solar cells, the most funded being the amorphous silicon, copper–indium–diselenide, copper–gallium–diselenide, copper–indium–gallium–diselenide, and cadmium–telluride, which are thin enough to be flexible. Crystalline silicon PV cells convert all the visible spectrum and part of the infrared of solar radiation into electricity as shown in Fig. 7.18, with the longer wavelengths without enough energy to excite electrons in the semiconductor of the PV device, whereas amorphous cells operate in a narrower range, not harvesting in the infrared.

The most relevant PV cell feature is efficiency (η_{el}) in converting solar radiation into electric energy, with sc-Si occupying the first position in the ranking. The second important factor for PV/T is the temperature coefficient (β). In regions where the cells operate at high temperatures, which is usual in places with high levels of solar irradiance and temperatures, a more intense degradation occurs not only in the conversion efficiency, but also in their service life. As stated by Makki et al. (2015), the PV power decreases with increasing temperature owing to the dark saturation current of the p–n junction, which causes the open-circuit voltage to decrease linearly, despite an increase in current. Moreover, as the design of the PV modules has a number of cells connected in series to obtain the necessary value of voltage, the output voltage increases as the current decreases to minimize the ohmic losses. As the resulting current in a series string is determined by the cell having the lowest current, this cell will have the highest temperature and will limit the efficiency of the entire string. Eventually, maintaining a low and homogeneous temperature all over the string will help keep its high performance. Thus, cooling the PV module by removing its unnecessary heat is a task for the thermal part of the PV/T.

The cell thickness is also an important factor not only as a problem related to costs, but also as an increase in mass to storage heat which must be removed by the HTF. The thinner the cell, the more flexible it is, and it will not require a rigid substrate

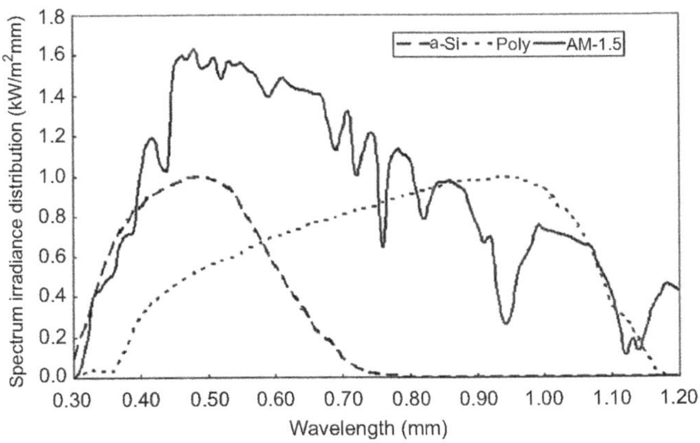

Figure 7.18 Spectrum response of amorphous (a-Si) and polycrystalline (Poly) silicon (Sirisamphanwong & Ketjoy, 2012).

like other types of semiconductors. The thermal effect on PV/T cells makes its analysis more complex than in other passive collectors, though.

7.5.2 The photovoltaic/thermal collector efficiency

The overall collector efficiency model has to include other parameters, such as cell efficiency; packing factor (*pF*), understood as the fraction of absorber plate area covered by PV cells; ratio of water mass to collector area; solar radiation; ambient temperature; and wind velocity, called operating parameters by Chow et al. (2009). If the priority is to maximize the electric energy yield, then the mass flow rate and PV cell packing factor should increase, whereas the water inlet temperature should be low. Meanwhile, other researchers consider the issues in terms of design parameters, namely mass flow rate, water inlet temperature, PV cell packing factor, number of covers, absorber to fluid thermal conductance, collector length, and all those related to the absorber plate design (Makki et al., 2015). Here, we discuss only efficiency, packing factor, and mass flow rate.

From the point of view of the PV/T thermodynamic performance, first of all one must determine the best ratio of electrical to thermal output based on the most logical renewable energy market approach, according to Coventry and Lovegrove (2003). The extremes of the ratio are 1.0 for the first law of thermodynamic analysis (energy) and 16.8 for the second law (exergy) analysis. Exergy is the available energy obtained by subtracting the unavailable energy from the total energy and is equivalent to the quantity for conversion to useful work. Once the actual demands of electricity and heat are defined, the *pF* can be set. The increase in *pF* does not always increase the annual energy gain or electrical efficiency, and the exergy can decrease. Thus, by raising the *pF* too much, the fluid exit temperature (T_o) will get higher owing to the large amount of thermal energy absorbed, increasing cell temperature and decreasing electrical efficiency. On the other hand, decreasing the *pF* too much will decrease the electrical efficiency because the radiation absorber area is smaller, as demonstrated by Moradi, Ali Ebadian, and Lin (2013). After these considerations, the efficiency analysis is as follows (Aste et al., 2014; Chow, 2010; Michael, Iniyan, & Goic, 2015; Zondag, de Vries, van Helden, van Zolingen, & van Steenhoven, 2003).

Initially, the total efficiency of a hybrid PV/T collector, η_t, would be defined by adding the electrical and thermal efficiencies:

$$\eta_t = \eta_{th} + \eta_{el} \tag{7.10}$$

The thermal efficiency (η_{th}) comes from (Duffie & Beckman, 2006) as mentioned earlier:

$$\eta_{th} = \frac{Q_u}{G \cdot A} \tag{7.11}$$

To define the electrical efficiency, we have to use the reference variables of PV cells used under standard test conditions, where the subscript mpp means the maximum power point in a PV cell curve of voltage (V) × current (I). Thus, the power (P) simply follows Ohm's law:

$$P_{\text{mpp}} = V_{\text{mpp}} \cdot I_{\text{mpp}} \quad [7.12]$$

and,

$$\eta_{\text{el},n} = \frac{V_{\text{mpp}} \cdot I_{\text{mpp}}}{G \cdot A} \quad [7.13]$$

with G (W/m^2) as the irradiance and A (m^2) as the collector area. Yet the temperature of a collector in operation will be higher than the reference (25°C) and its electrical efficiency should be adjusted by the cells' temperature coefficient, β (°C^{-1}) (Tiwari et al., 2011):

$$\eta_{\text{el}} = \eta_{\text{el},n}[1 - \beta(T_{\text{cell}} - 25)] \quad [7.14]$$

Huang, Lin, Hung, and Sun (2001) define the energy-saving efficiency (η_{f}), considering that a unit of heat is not equivalent to a unit of electric energy, although both are produced by the same PV/T collector, thus preferring to compare them using primary energy as the reference:

$$\eta_{\text{f}} = \eta_{\text{th}} + \frac{\eta_{\text{el}}}{\eta_{\text{Tpower}}} \quad [7.15]$$

where the η_{Tpower} is the electric power generation efficiency for a conventional power plant, adopted as 0.38 by the authors.

Thereby, the PV/T collector exergetic efficiency (ε) can be calculated:

$$\varepsilon = \frac{\int_{t1}^{t2} \left(A_{\text{pv}} \cdot \dot{E}_{\chi\text{el}} + A_{\text{c}} \cdot \dot{E}_{\chi\text{th}}\right) dt}{A \int_{t1}^{t2} \dot{E}_{\chi\text{sun}} \cdot dt} \quad [7.16]$$

where $\dot{E}_{\chi\text{el}}$ is the PV exergy output per unit of PV cell area, $\dot{E}_{\chi\text{th}}$ is the thermal exergy output per unit of collector area, and $\dot{E}_{\chi\text{sun}}$ is the exergy input of solar radiation.

The exergies are estimated as follows (Carnot cycle):

$$\dot{E}_{\chi\text{th}} = \dot{E}_{\text{th}}\left(1 - \frac{T_{\text{a}}}{T_{\text{o}}}\right) \quad [7.17]$$

$$\dot{E}_{\chi\text{el}} = \dot{E}_{\text{el}} \quad [7.18]$$

$$\dot{E}_{\chi\text{sun}} = \left(1 - \frac{T_{\text{a}}}{T_{\text{s}}}\right) \quad [7.19]$$

where $\dot{E}$ is the energy rate output per unit area (W/m^2), with the subscripts already defined. T_a, T_o, and T_s are the ambient, fluid outlet, and sun temperatures (K), respectively. In an experimental rig, comparing a PV/T to a PV module and a thermal collector, Saitoh et al. (2003) concluded that the former had better exergy than the latter and an expected area reduction of 27%, although they found some drop in the collector efficiency.

The mass flow rate is also an important parameter related to electrical and thermal efficiencies of the PV/T collector, because a high flow rate implies a high convection heat transfer coefficient, which improves the overall collector efficiency with a lower water exit temperature. However, the latter might not be so low, requiring a very high water velocity to the point of causing a nonlinear reduction of the absorber plate temperature as expected (Moradi et al., 2013), and might not be as useful as it could be to the user, who would have to add on conventional energy to the water heating process. Kalogirou (2001) found an optimum value for mass flow rate of 0.014 kg/m^2 · s in his TRNSYS simulations on PV/Ts, while Moradi et al. (2013) collected results from several researchers, stating a minimum average of 0.014 up to a maximum average of 0.088 kg/m^2 · s. However, Charalambous, Maidment, Kalogirou, and Yiakoumetti (2007) highlighted the influence of the geometric parameter W/D (W is the distance between tubes, and D is the tube diameter) in the thermal and electrical efficiencies, which is related to the mass flow rate.

Finally, there is a prolific field for the application of nanofluids in hybrid PV/T collectors. According to Verma and Tiwari (2015), further research should find those nanomaterials that can increase the surface absorption capability of the absorbing plates in FPCs to improve their thermal efficiency.

7.6 Conclusion and future trends

This chapter focused on the four main types of passive SWHs, namely, the FPC, the ETC, the ICS, and the hybrid PV/T. Two relevant innovative trends were identified in this research field. First, the combination of technological details from various types of collectors has occurred in recent decades to propose a more efficient appliance. The second is the application of nanofluids on the more established technologies of collectors.

FPCs are the simplest collectors among the most installed types of collectors in the European Union and in the developing world, having an acceptable efficiency in mild and hot climates with competitive prices. Research endeavors have focused on the combination of FPCs with PV modules, resulting in the shell-and-tube PV/T, and on the use of nanofluids in FPCs, setting the next innovations in passive solar collectors for low-temperature water.

ETCs are flexible enough to yield low and average water temperatures, mainly when coupled to concentrators, even reaching feasibility for power generation. Concerning their domestic use, these collectors are more suitable for cold climates because of the huge heat loss reduction obtained by convection suppression. New strategies of heat withdrawal from the absorber tube, such as heat pipe and U-tube development, may be as significant as the selective coating was for ETCs in this century.

ICSs are a good option when the lack of space is a problem for the installation of the hot water tank. It is placed over the roof occupying a small area, avoiding household attics, which are very complex places. These collectors have been modified to obtain convection suppression and higher efficiency through radiation reflectors. Until recently, they were suitable only for regions with enough clear-sky beam radiation. However, the developing world foments the installation of thousands of square meters of FPCs coupled with hot water tanks on the roofs, which is not exactly an ICS, but can benefit from its last innovations.

The PV/T electric and thermal parts have each their own efficiency, the former being similar to that of a PV module and the latter to that of the FPC, which is the thermal collector type most combined to PV/T. Considering the concept of distributed electric energy generation, the installation of a PV/T alone, instead of a split PV module and an FPC, reduces the cost of installation and maintenance, as the services can be accomplished by a single practitioner. Furthermore, the PV modules ensure their efficiency from the free cooling of the thermal collector possibly by circulating nanofluid, with the assembly seen today from an aesthetic or architectural point of view as that of the building-integrated photovoltaics.

Nomenclature

A	area (m^2)
a-Si	Amorphous silicon
c_p	Specific heat (kJ/kg · °C)
c-Si	Monocrystalline silicon
CPC	Compound parabolic collectors
CFD	Computer fluid dynamics
d	Diameter (m)
E	Energy
$\dot{E}$	Exergy
ETC	Evacuated tube collector
F_R	Collector heat removal factor
FPC	Flat plate collector
G	Incident radiation intensity (W/m^2)
HTF	Heat transfer fluid
I	Current (A)
ICS	Integrated collector storage
L	Length (m)
$\dot{m}$	Mass flow rate (kg/s)

Continued

p-si	Polycrystalline silicon
PV	Photovoltaic
PV/T	Photovoltaic/thermal collector
Q_i	Radiation reaching the absorber plate (W)
Q_o	Rate of heat loss (kJ)
Q_u	Useful heat (kJ)
SF	Solar fraction
SWH	Solar water heating systems
T_a	Ambient temperature (°C)
T_c	Collector temperature (°C)
T_i	Water inlet temperature (°C)
T_m	Average water temperature (°C)
T_o	Water outlet temperature (°C)
T_r	Reduced temperature (°C)
U_L	Overall heat transfer coefficient ($W/m^2 \cdot °C$)
V	Voltage (V)

Subscripts

c	Condenser
e	Electric, evaporator
mpp	Point of maximum power
th	Thermal

Greek letters

α	Absorptivity
β	Cells temperature coefficient, slope
ε	Emissivity, exergetic efficiency
τ	Transmissivity
η_{el}	Electrical efficiency
η_{th}	Thermal efficiency
η_{Tpower}	Electric power generation efficiency

References

Abdullah, A. H., Abou-Ziyan, H. Z., & Ghoneim, A. A. (2003). Thermal performance of flat plate solar collector using various arrangements of compound honeycomb. *Energy Conversion and Management, 44*(19), 3093–3112. http://dx.doi.org/10.1016/S0196-8904(03)00013-X.

Agarwal, R. K., & Garg, H. P. (1994). Study of a photovoltaic-thermal system – thermosyphonic solar water heater combined with solar cells. *Energy Conversion and Management, 35*(7), 605–620.

Al-Shamani, A. N., Yazdi, M. H., Alghoul, M. A., Abed, A. M., Ruslan, M. H., Mat, S., & Sopian, K. (2014). Nanofluids for improved efficiency in cooling solar collectors – a review. *Renewable and Sustainable Energy Reviews, 38*, 348–367. http://dx.doi.org/10.1016/j.rser.2014.05.041.

AlShamaileh, E. (2010). Testing of a new solar coating for solar water heating applications. *Solar Energy, 84*(9), 1637–1643. http://dx.doi.org/10.1016/j.solener.2010.06.003.

Amri, A., Jiang, Z. T., Pryor, T., Yin, C. Y., & Djordjevic, S. (2014). Developments in the synthesis of flat plate solar selective absorber materials via sol-gel methods: a review. *Renewable and Sustainable Energy Reviews, 36*, 316–328. http://dx.doi.org/10.1016/j.rser.2014.04.062.

ASHRAE – American Society of Heating, Refrigerating and Air-Conditioning Engineers. (1977). *Methods of testing to determining the thermal performance of solar collectors*. New York: ASHRAE.

Asif, M., & Muneer, T. (2015). Solar water Heating: domestic and industrial applications. In *Encyclopedia of energy engineering and technology* (2nd ed., pp. 1331–1339). New York: Taylor and Francis. http://dx.doi.org/10.1081/E-EEE2-120042349.

Aste, N., del Pero, C., & Leonforte, F. (2014). Water flat plate PV-thermal collectors: a review. *Solar Energy, 102*, 98–115. http://dx.doi.org/10.1016/j.solener.2014.01.025.

Azad, E. (2008). Theoretical and experimental investigation of heat pipe solar collector. *Experimental Thermal and Fluid Science, 32*(8), 1666–1672. http://dx.doi.org/10.1016/j.expthermflusci.2008.05.011.

Bainbridge, D. A. (1983). The integral passive solar water heater book. *Solar Energy, 30*. http://dx.doi.org/10.1016/0038-092X(83)90167-6.

Behnia, M., & Morrison, G. L. (1991). An experimental investigation of inclined open thermosyphons. *Solar Energy, 47*(4), 313–326. http://dx.doi.org/10.1016/0038-092X(91)90124-F.

Belghith, M., Arurault, L., & Bes, R. S. (2013). Selective absorber obtained by nickel-pigmented anodized 6060 aluminium surface. *Arabian Journal for Science and Engineering, 38*(4), 751–757. http://dx.doi.org/10.1007/s13369-012-0367-2.

Bojić, M., Marjanović, N., Miletić, I., & Bojić, L. (2015). Comparison of optical performances of sea-shell trough solar concentrators. *Energy and Buildings, 98*, 144–150. http://dx.doi.org/10.1016/j.enbuild.2014.08.037.

Budihardjo, I., & Morrison, G. L. (2009). Performance of water-in-glass evacuated tube solar water heaters. *Solar Energy, 83*(1), 49–56. http://dx.doi.org/10.1016/j.solener.2008.06.010.

Budihardjo, I., Morrison, G. L., & Behnia, M. (2007). Natural circulation flow through water-in-glass evacuated tube solar collectors. *Solar Energy, 81*(12), 1460–1472. http://dx.doi.org/10.1016/j.solener.2007.03.002.

Charalambous, P. G., Maidment, G. G., Kalogirou, S. A., & Yiakoumetti, K. (2007). Photovoltaic thermal (PV/T) collectors: a review. *Applied Thermal Engineering, 27*(2–3), 275–286. http://dx.doi.org/10.1016/j.applthermaleng.2006.06.007.

Cheng, H. Y., Chiou, J. W., Ting, J. M., & Tzeng, Y. (2013). Reactively co-sputter deposited a-C:H/Cr thin films: material characteristics and optical properties. *Thin Solid Films, 529*, 164–168. http://dx.doi.org/10.1016/j.tsf.2012.10.025.

Cheng, B., Wang, K. K., Wang, K. P., Jiang, W., Cong, B. J., Song, C. L. … Liu, Y. (2015). Porous carbon–titania nanocomposite films for spectrally solar selective absorbers. *Solar Energy Materials and Solar Cells, 133*, 126–132. http://dx.doi.org/10.1016/j.solmat.2014.10.032.

Choudhury, C., & Sehgal, H. K. (1982). Black cobalt selective coatings by spray pyrolysis for photothermal conversion of solar energy. *Solar Energy, 28*(1), 25–31. http://dx.doi.org/10.1016/0038-092X(82)90220-1.

Chow, T. T. (2010). A review on photovoltaic/thermal hybrid solar technology. *Applied Energy, 87*(2), 365–379. http://dx.doi.org/10.1016/j.apenergy.2009.06.037.

Chow, T. T., He, W., & Ji, J. (2006). Hybrid photovoltaic-thermosyphon water heating system for residential application. *Solar Energy, 80*(3), 298–306. http://dx.doi.org/10.1016/j.solener.2005.02.003.

Chow, T. T., Pei, G., Fong, K. F., Lin, Z., Chan, A. L. S., & Ji, J. (2009). Energy and exergy analysis of photovoltaic–thermal collector with and without glass cover. *Applied Energy, 86*(3), 310–316. http://dx.doi.org/10.1016/j.apenergy.2008.04.016.

Coventry, J. S., & Lovegrove, K. (2003). Development of an approach to compare the "value" of electrical and thermal output from a domestic PV/thermal system. *Solar Energy, 75*, 63–72. http://dx.doi.org/10.1016/S0038-092X(03)00231-7.

Devanarayanan, K., & Kalidasa Murugavel, K. (2014). Integrated collector storage solar water heater with compound parabolic concentrator – development and progress. *Renewable and Sustainable Energy Reviews, 39*, 51–64. http://dx.doi.org/10.1016/j.rser.2014.07.076.

Duffie, J., & Beckman, W. (2006). *Solar engineering of thermal processes* (3rd ed.). Hoboken: Wiley.

Florschuetz, L. W. (1979). Extension of the Hottel-Whillier model to the analysis of combined photovoltaic/thermal flat plate collectors. *Solar Energy, 22*(4), 361–366. http://dx.doi.org/10.1016/0038-092X(79)90190-7.

Galione, P. A., Baroni, A. L., Ramos-Barrado, J. R., Leinen, D., Martín, F., Marotti, R. E., & Dalchiele, E. A. (2010). Origin of solar thermal selectivity and interference effects in nickel-alumina nanostructured films. *Surface and Coatings Technology, 204*(14), 2197–2201. http://dx.doi.org/10.1016/j.surfcoat.2009.12.008.

Gao, Y., Fan, R., Zhang, X. Y., An, Y. J., Wang, M. X., Gao, Y. K., & Yu, Y. (2014). Thermal performance and parameter analysis of a U-pipe evacuated solar tube collector. *Solar Energy, 107*, 714–727. http://dx.doi.org/10.1016/j.solener.2014.05.023.

Garg, H. P., & Agarwal, R. K. (1995). Some aspects of a PV/T collector/forced circulation flat plate solar water heater with solar cells. *Energy Conversion and Management, 36*(2), 87–99. http://dx.doi.org/10.1016/0196-8904(94)00046-3.

Garg, H. P., Chakravertty, S., Shukla, A. R., Agnihotri, R. C., & Khas, H. (1983). Advanced tubular solar energy collector – a state of the art. *Energy Conversion and Management, 23*(3), 157–169.

Gillette, R. B. (1960). Selectively emissive materials for solar heat absorbers. *Solar Energy, 4*(4), 24–32. http://dx.doi.org/10.1016/0038-092X(60)90004-9.

Gostzberger, A., & Rommel, M. (1987). Prospects for integrated storage collector systems in central Europe. *Solar Energy, 39*(3), 211–219.

Graf, W., Brucker, F., Köhl, M., Tröscher, T., Wittwer, V., & Herlitze, L. (1997). Development of large area sputtered solar absorber coatings. *Journal of Non-Crystalline Solids, 218*, 380–387. http://dx.doi.org/10.1016/S0022-3093(97)00283-4.

Granqvist, C. G. (1991). Solar energy materials. *Applied Physics A Solids and Surfaces, 52*(2), 83–93. http://dx.doi.org/10.1007/BF00323721.

Harding, G. L., & Zhiqiang, Y. I. N. (1985). Thermosiphon circulation in solar water heaters incorporating evacuated tubular collectors and a novel water-in-glass manifold. *Solar Energy, 34*(I), 13–18.

Hayek, M., Assaf, J., & Lteif, W. (2011). Experimental investigation of the performance of evacuated-tube solar collectors under Eastern Mediterranean climatic conditions. *Energy Procedia, 6*, 618–626. http://dx.doi.org/10.1016/j.egypro.2011.05.071.

Helal, O., Chaouachi, B., & Gabsi, S. (2011). Design and thermal performance of an ICS solar water heater based on three parabolic sections. *Solar Energy, 85*(10), 2421–2432. http://dx.doi.org/10.1016/j.solener.2011.06.021.

Hellstrom, B., Adsten, M., Nostell, P., Karlsson, B., & Wackelgard, E. (2003). The impact of optical and thermal properties on the performance of flat plate solar collectors. *Renewable Energy, 28*(3), 331–344. http://dx.doi.org/10.1016/S0960-1481(02)00040-X.

He, Q., Zeng, S., & Wang, S. (2014). Experimental investigation on the efficiency of flat-plate solar collectors with nanofluids. *Applied Thermal Engineering, 88*, 1–7. http://dx.doi.org/10.1016/j.applthermaleng.2014.09.053.

Hossain, M. S., Saidur, R., Fayaz, H., Rahim, N. A., Islam, M. R., Ahamed, J. U., & Rahman, M. M. (2011). Review on solar water heater collector and thermal energy performance of circulating pipe. *Renewable and Sustainable Energy Reviews, 15*(8), 3801–3812. http://dx.doi.org/10.1016/j.rser.2011.06.008.

Hottel, H. C., & Unger, T. A. (1959). 2 The properties of a copper oxide-aluminum selective black surface absorber of solar energy. *Solar Energy, 3*(3), 10–15. http://dx.doi.org/10.1016/0038-092X(59)90134-3.

Huang, B. J., Lin, T. H., Hung, W. C., & Sun, F. S. (2001). Performance evaluation of solar photovoltaic/thermal systems. *Solar Energy, 70*(5), 443–448. http://dx.doi.org/10.1016/S0038-092X(00)00153-5.

Ienei, E., Isac, L., Cazan, C., & Duta, A. (2010). Characterization of $Al/Al_2O_3/NiO_x$ solar absorber obtained by spray pyrolysis. *Solid State Sciences, 12*(11), 1894–1897. http://dx.doi.org/10.1016/j.solidstatesciences.2010.05.028.

ISO – International Organization for Standardization. (2006). *EN 12975–2: Thermal solar systems and components – solar collectors e Part 2: Test methods.*

Kalogirou, S. (1997). Design, construction, performance evaluation and economic analysis of an integrated collector storage system. *Renewable Energy, 12*(2), 179–192. http://dx.doi.org/10.1016/S0960-1481(97)00029-3.

Kalogirou, S. A. (1999). Performance enhancement of an integrated collector storage hot water system. *Renewable Energy, 16*(1–4), 652–655. http://dx.doi.org/10.1016/S0960-1481(98)00245-6.

Kalogirou, S. A. (2001). Use a TRNSYS for modelling and simulation of a hybrid pv-thermal solar system for Cyprus. *Renewable Energy, 23*(2), 247–260. http://dx.doi.org/10.1016/S0960-1481(00)00176-2.

Kalogirou, S. (2009). Thermal performance, economic and environmental life cycle analysis of thermosiphon solar water heaters. *Solar Energy, 83*(1), 39–48. http://dx.doi.org/10.1016/j.solener.2008.06.005.

Kalogirou, S. A., & Tripanagnostopoulos, Y. (2006). Hybrid PV/T solar systems for domestic hot water and electricity production. *Energy Conversion and Management, 47*(18–19), 3368–3382. http://dx.doi.org/10.1016/j.enconman.2006.01.012.

Karthick Kumar, S., Suresh, S., Murugesan, S., & Raj, S. P. (2013). CuO thin films made of nanofibers for solar selective absorber applications. *Solar Energy, 94*, 299–304. http://dx.doi.org/10.1016/j.solener.2013.05.018.

Khalifa, N. (1998). Forced versus natural circulation solar water heaters: a comparative performance study. *Renewable Energy, 14*(98), 77–82.

Kim, Y., & Seo, T. (2007). Thermal performances comparisons of the glass evacuated tube solar collectors with shapes of absorber tube. *Renewable Energy, 32*(5), 772–795. http://dx.doi.org/10.1016/j.renene.2006.03.016.

Lalović, B., Kiss, Z., & Weakliem, H. (1986). A hybrid amorphous silicon photovoltaic and thermal solar collector. *Solar Cells, 19*(2), 131–138. http://dx.doi.org/10.1016/0379-6787(86)90038-4.

Liang, R., Ma, L., Zhang, J., & Zhao, D. (2012). Experimental study on thermal performance of filled-type evacuated tube with U-tube. *Heat and Mass Transfer/Waerme- Und Stoffuebertragung, 48*(6), 989–997. http://dx.doi.org/10.1007/s00231-011-0912-7.

Liang, R., Ma, L., Zhang, J., & Zhao, L. (2013). Performance analysis of a new-design filled-type solar collector with double U-tubes. *Energy and Buildings, 57*, 220–226. http://dx.doi.org/10.1016/j.enbuild.2012.11.004.

Makki, A., Omer, S., & Sabir, H. (2015). Advancements in hybrid photovoltaic systems for enhanced solar cells performance. *Renewable and Sustainable Energy Reviews, 41*, 658–684. http://dx.doi.org/10.1016/j.rser.2014.08.069.

Ma, L., Lu, Z., Zhang, J., & Liang, R. (2010). Thermal performance analysis of the glass evacuated tube solar collector with U-tube. *Building and Environment, 45*(9), 1959–1967. http://dx.doi.org/10.1016/j.buildenv.2010.01.015.

Mauthner, F., & Weiss, W. (2015). *Solar-Heat-Worldwide-2015. Gleisdof – Austria*. International Energy Agency. Retrieved from http://www.iea-shc.org/solar-heat-worldwide.

Michael, J. J., Iniyan, S., & Goic, R. (2015). Flat plate solar photovoltaic-thermal (PV/T) systems: a reference guide. *Renewable and Sustainable Energy Reviews, 51*, 62–88. http://dx.doi.org/10.1016/j.rser.2015.06.022.

Moghadam, A. J., Farzane-Gord, M., Sajadi, M., & Hoseyn-Zadeh, M. (2014). Effects of CuO/water nanofluid on the efficiency of a flat-plate solar collector. *Experimental Thermal and Fluid Science, 58*, 9–14. http://dx.doi.org/10.1016/j.expthermflusci.2014.06.014.

Moradi, K., Ali Ebadian, M., & Lin, C. X. (2013). A review of PV/T technologies: effects of control parameters. *International Journal of Heat and Mass Transfer, 64*, 483–500. http://dx.doi.org/10.1016/j.ijheatmasstransfer.2013.04.044.

Morales-Ortiz, U., Avila-García, A., & Hugo Lara, C. V. (2006). Ruthenium oxide films for selective coatings. *Solar Energy Materials and Solar Cells, 90*(6 Special Issue), 832–840. http://dx.doi.org/10.1016/j.solmat.2005.04.019.

Morrison, G. L., Budihardjo, I., & Behnia, M. (2004). 04/01964 Water-in-glass evacuated tube solar water heaters. *Fuel and Energy Abstracts, 45*(4), 272. http://dx.doi.org/10.1016/S0140-6701(04)94551-1.

Noie, S. H., Heris, S. Z., Kahani, M., & Nowee, S. M. (2009). Heat transfer enhancement using Al_2O_3/water nanofluid in a two-phase closed thermosyphon. *International Journal of Heat and Fluid Flow, 30*(4), 700–705. http://dx.doi.org/10.1016/j.ijheatfluidflow.2009.03.001.

Ogueke, N. V., Anyanwu, E. E., & Ekechukwu, O. V. (2009). A review of solar water heating systems. *Journal of Renewable and Sustainable Energy, 1*(4), 043106. http://dx.doi.org/10.1063/1.3167285.

Patel, S. N., Inal, O. T., Singh, A. J., & Scherer, A. (1985). *Solar Energy Materials, 11*(1985), 381–399.

Prakash, J., Garg, H. P., Hrishikesan, D. S., & Jha, R. (1989). Performance studies of an integrated solar collector-cum-storage water heating system with transparent insulation cover. *Solar and Wind Technology, 6*(2), 171–176.

Redpath, D. A. G. (2012). Thermosyphon heat-pipe evacuated tube solar water heaters for northern maritime climates. *Solar Energy, 86*(2), 705–715. http://dx.doi.org/10.1016/j.solener.2011.11.015.

Sadhishkumar, S., & Balusamy, T. (2014). Performance improvement in solar water heating systems—a review. *Renewable and Sustainable Energy Reviews, 37*, 191–198. http://dx.doi.org/10.1016/j.rser.2014.04.072.

Saitoh, H., Hamada, Y., Kubota, H., Nakamura, M., Ochifuji, K., Yokoyama, S., & Nagano, K. (2003). Field experiments and analyses on a hybrid solar collector. *Applied Thermal Engineering, 23*(16), 2089–2105. http://dx.doi.org/10.1016/S1359-4311(03)00166-2.

Salavati, S., Kianifar, A., Niazmand, H., Mahian, O., & Wongwises, S. (2015). Experimental investigation on the thermal efficiency and performance characteristics of a flat plate solar collector using SiO_2/EG–water nanofluids. *International Communications in Heat and Mass Transfer*, 71–75. http://dx.doi.org/10.1016/j.icheatmasstransfer.2015.02.011.

Shah, L. J., & Furbo, S. (2007). Theoretical flow investigations of an all glass evacuated tubular collector. *Solar Energy, 81*(6), 822–828. http://dx.doi.org/10.1016/j.solener.2006.09.007.

Sharma, N., & Diaz, G. (2011). Performance model of a novel evacuated-tube solar collector based on minichannels. *Solar Energy, 85*(5), 881–890. http://dx.doi.org/10.1016/j.solener.2011.02.001.

Shojaeizadeh, E., Veysi, F., & Kamandi, A. (2015). Exergy efficiency investigation and optimization of an Al_2O_3–water nanofluid based flat-plate solar collector. *Energy and Buildings, 101*, 12–23. http://dx.doi.org/10.1016/j.enbuild.2015.04.048.

Shukla, R., Sumathy, K., Erickson, P., & Gong, J. (2013). Recent advances in the solar water heating systems: a review. *Renewable and Sustainable Energy Reviews, 19*, 173–190. http://dx.doi.org/10.1016/j.rser.2012.10.048.

Sirisamphanwong, C., & Ketjoy, N. (2012). Impact of spectral irradiance distribution on the outdoor performance of photovoltaic system under Thai climatic conditions. *Renewable Energy, 38*(1), 69–74. http://dx.doi.org/10.1016/j.renene.2011.07.008.

Smyth, M., Eames, P. C., & Norton, B. (2006). Integrated collector storage solar water heaters. *Renewable and Sustainable Energy Reviews, 10*(6), 503–538. http://dx.doi.org/10.1016/j.rser.2004.11.001.

Smyth, M., McGarrigle, P., Eames, P. C., & Norton, B. (2005). Experimental comparison of alternative convection suppression arrangements for concentrating integral collector storage solar water heaters. *Solar Energy, 78*(2), 223–233. http://dx.doi.org/10.1016/j.solener.2004.06.004.

Srivastava, A., Hussain, S. Q., Sharma, B. N., & Tiwari, G. N. (1982). Thermal performance of ground as an inexpensive solar collector and storage system. *Energy Conversion and Management, 22*(3), 197–204. http://dx.doi.org/10.1016/0196-8904(82)90042-5.

Suman, S., Khan, M. K., & Pathak, M. (2015). Performance enhancement of solar collectors—a review. *Renewable and Sustainable Energy Reviews, 49*, 192–210. http://dx.doi.org/10.1016/j.rser.2015.04.087.

SunMaxx Solar. (n.d.). Information guide – evacuated tube solar collectors. Retrieved August 26, 2015, from http://www.sunmaxxsolar.com/info-guide-evacuated-tube-collectors.

Tabor, H. (1958). Solar energy research: program in the new desert research institute in Beersheba. *Solar Energy, 2*(1), 3–6. http://dx.doi.org/10.1016/0038-092X(58)90001-X.

Tiwari, G. N., Mishra, R. K., & Solanki, S. C. (2011). Photovoltaic modules and their applications: a review on thermal modelling. *Applied Energy, 88*(7), 2287–2304. http://dx.doi.org/10.1016/j.apenergy.2011.01.005.

Tripanagnostopoulos, Y., & Souliotis, M. (2006). ICS solar systems with two water tanks. *Renewable Energy, 31*(11), 1698–1717. http://dx.doi.org/10.1016/j.renene.2005.08.028.

Tripanagnostopoulos, Y., Souliotis, M., & Nousia, T. (1999). Solar ICS systems with two cylindrical storage tanks. *Renewable Energy, 16*(1–4), 665–668. http://dx.doi.org/10.1016/S0960-1481(98)00248-1.

Tripanagnostopoulos, Y., Souliotis, M., & Nousia, T. (2002). CPC type integrated collector storage systems. *Solar Energy, 72*(4), 327–350. http://dx.doi.org/10.1016/S0038-092X(02)00005-1.

Tripanagnostopoulos, Y., & Yianoulis, P. (1992). Integrated collector-storage systems with suppressed thermal losses. *Solar Energy, 48*(1), 31–43. http://dx.doi.org/10.1016/0038-092X(92)90174-9.

Verma, S. K., & Tiwari, A. K. (2015). Progress of nanofluid application in solar collectors: a review. *Energy Conversion and Management, 100*, 324–346. http://dx.doi.org/10.1016/j.enconman.2015.04.071.

Wang, Z., & Yang, W. (2014). A review on loop heat pipe for use in solar water heating. *Energy and Buildings, 79*, 143–154. http://dx.doi.org/10.1016/j.enbuild.2014.04.051.

Wang, Z., Yang, W., Qiu, F., Zhang, X., & Zhao, X. (2015). Solar water heating : from theory, application, marketing and research. *Renewable and Sustainable Energy Reviews, 41*, 68–84. http://dx.doi.org/10.1016/j.rser.2014.08.026.

Wang, J., Yin, Z., Qi, J., Ma, G., & Liu, X. (2015). Medium-temperature solar collectors with all-glass solar evacuated tubes. *Energy Procedia, 70*, 126–129. http://dx.doi.org/10.1016/j.egypro.2015.02.107.

Wazwaz, A., Salmi, J., & Bes, R. (2010). The effects of nickel-pigmented aluminium oxide selective coating over aluminium alloy on the optical properties and thermal efficiency of the selective absorber prepared by alternate and reverse periodic plating technique. *Energy Conversion and Management, 51*(8), 1679–1683. http://dx.doi.org/10.1016/j.enconman.2009.11.047.

Zambolin, E., & Del Col, D. (2012). An improved procedure for the experimental characterization of optical efficiency in evacuated tube solar collectors. *Renewable Energy, 43*, 37–46. http://dx.doi.org/10.1016/j.renene.2011.11.011.

Zhang, X., Zhao, X., Smith, S., Xu, J., & Yu, X. (2012). Review of R&D progress and practical application of the solar photovoltaic/thermal (PV/T) technologies. *Renewable and Sustainable Energy Reviews, 16*(1), 599–617. http://dx.doi.org/10.1016/j.rser.2011.08.026.

Zondag, H. A., de Vries, D. W., van Helden, W. G. J., van Zolingen, R. J. C., & van Steenhoven, A. A. (2003). The yield of different combined PV-thermal collector designs. *Solar Energy, 74*(3), 253–269. http://dx.doi.org/10.1016/S0038-092X(03)00121-X.

Active solar space heating

8

D.A. Chwieduk
Institute of Heat Engineering, Faculty of Power and Aeronautical Engineering, Warsaw University of Technology, Warsaw, Poland

8.1 Background on active space heating

8.1.1 Improvement of energy efficiency in buildings

Energy consumption in buildings used to be very high and still corresponds to about 40% of the final energy demand in most developed countries [1–3]. Residential buildings are on average responsible for 22% of energy use and commercial ones for 18%. A large part of the energy consumption in residential buildings is for space heating, which accounts for more than 50% in International Energy Agency (IEA) countries [1]. A trend toward a reduction in energy use has been seen since about 1995.

Surprisingly the high energy demand of buildings does not really depend on climatic conditions. In Europe, for example, countries of high latitudes (northern countries) and countries of low latitudes (Mediterranean countries) are characterized by the same energy consumption in buildings per capita. The consumption in buildings depends mainly on energy and fuel prices, local traditions, and national regulations for building construction. There are many different building codes in the world. In countries with more severe climatic conditions usually more strict regulations are put into force to reduce heat losses through the building's envelope and ventilation. The big effort to reduce energy consumption for space heating is a recent phenomenon.

In Europe major improvements in the energy efficiency of buildings have been made since the beginning of the new millennium when the European Union (EU) Directive of Energy Performance of Buildings was implemented [2]. To enhance the role and necessity of energy efficiency in buildings, in 2010 the European Commission prepared a new version of the Directive [3]. The Directive requires implementation of new energy standards not only for new buildings, but for existing ones as well, during a thermal refurbishment process. According to the Directive, all new buildings in the EU after 2021 will be nearly zero energy buildings. However, national regulations, being a consequence of the Directive, show different interpretations of the meaning of "nearly zero." Nevertheless, the improvement in the energy efficiency in buildings is evident.

It is interesting to consider historical national changes in building codes and energy consumption regulations. Such historical changes in energy standards for buildings in Poland are presented in Table 8.1.

It should be emphasized that regulations for final and primary energy consumption are given in the form of limits that cannot be exceeded and depend on the type of building (residential, public, industrial, etc.). As has been mentioned, in Europe the energy reduction requirements have resulted from the revision of the EU Directive of Energy

Advances in Solar Heating and Cooling. http://dx.doi.org/10.1016/B978-0-08-100301-5.00008-4

Table 8.1 Historical changes in energy standards for buildings in Poland

Years of building construction	Heat transfer coefficient for external walls (W/($m^2 \cdot K$))	Annual average energy demand for space heating	
		Final energy (kWh/m^2)	Primary energy (kWh/m^2)
Up to 1966	1.16 1.40	240–280 300–350	364–447 489–569
1967–1985	1.16	240–280	364–447
1986–1992	0.75	160–200	244–325
1993–2002	0.55	120–160	203–244
2002–2008	0.30	90–120	69–183
2009–2014	0.3	90–120	69–183
Since 2014	0.25	120	80–122

Performance in Buildings [3]. New regulations (limits) for energy consumption in buildings, being a consequence of this revision, take into account climate and tradition, as well as the national fuel (energy) mix. In Poland energy production is based on hard and brown coal. As a result, the limits are not as close to "nearly zero energy" consumption. The Polish national energy limits for the annual average primary energy demand for space heating plus hot water per square meter of heated floor area for various types of buildings are presented in Table 8.2.

Table 8.3 shows present and future building codes, ie, heat transfer coefficients, for various elements of residential building envelopes in Poland. Comparing data from Tables 8.1 and 8.3, it is evident how much energy standards for a building envelope have changed since 1966. As a result the heat losses from buildings have been reduced a lot, especially through the walls and windows.

Fig. 8.1 presents an example of changes in building construction technology connected to changes in the appearance of buildings from various decades, as a result of different space heating demands. The building on the left was constructed at the end of the 1980s, the building on the right was constructed about 20 years later.

Nowadays the energy standards for windows require the heat transfer coefficient to be equal to 1.6 W/($m^2 \cdot K$), when about 15 years ago it was at 3.6 W/($m^2 \cdot K$). As a consequence the demand for space heating, which used to be responsible for 60–70% of total energy demand in buildings (depending on the type), decreased significantly to 50% of total energy demand. The final energy consumption index (per square meter of heated floor area, expressed in kWh/m^2; see Table 8.1) has been reduced by half and correspondingly the primary energy consumption index by about four times.

Table 8.2 Energy limits for annual average primary energy demand for space heating plus hot water for various types of buildings in Poland [4]

Type of building	Annual average primary energy demand for space heating + hot water (kWh/m^2)		
	After 2014	After 2017	After 2021
Residential buildings			
Single-family houses	120	95	70
Multifamily blocks of buildings	105	85	65
Public buildings[a]			
Health care	390	290	190
Others	65	60	45
Enterprise, production, and warehouse buildings[a]	110	90	70

[a]There are extra energy demand limits for electricity for lighting (not included in the table) and for cooling energy, if cooling is available.

Table 8.3 Present and future limits for heat transfer coefficients, for various elements of a residential building envelope in Poland

Construction element of a building envelope	Heat transfer coefficient U (W/(m^2 · K))		
	After 1 January 2014	After 1 January 2017	After 1 January 2021
External walls	0.25	0.23	0.20
Roofs, flat roofs, ceilings	0.20	0.18	0.15
Floors on ground	0.30	0.30	0.30

In IEA countries, worldwide generally, space heating and cooling account for over one-third of all energy consumed in buildings, but in cold climates space heating accounts for about 50%, and in the residential sector for 60% [5–7]. It is evident that energy efficiency in buildings is a priority of national energy policies. There has been a significant reduction in energy use in buildings for space heating. According to present building codes space heating requirements are a few times lower than they used to be. In residential buildings space heating energy needs have become equivalent to domestic hot water energy demand. They are no more a dominant component of total energy use in buildings.

(a) (b)

Figure 8.1 An example of changes in building construction technology connected to changes in the look of buildings through recent decades.

A considerable decrease in space heating demand leads to more likely use of heating systems (HVAC—heating, ventilation, air conditioning) that are based on renewables and that are not of such high performance in winter, such as solar energy. It is a characteristic that space heating needs in buildings are not coincident with solar energy availability. When solar irradiance is low and the daytime (from sunrise to sunset) is short, then the ambient temperature is low and as a consequence space heating demands are high. However, nowadays, as was described above, buildings are more independent of climatic conditions and resistant even to cold winters. When heating demand is moderate or even small, then it is much more efficient to use solar energy for space heating than it used to be when space heating demand was high. As a result nowadays the share of solar energy supplying heat to space heating systems can be 20–30% of the total space heating demand in cold climates. It should be emphasized that at present, when buildings are well insulated and have high thermal capacity, it is not the ambient air temperature that is crucial for the heat balance of the building, but the availability of solar radiation that directly reaches the interior of a building through its glazing [8].

8.1.2 Changes in space heating concepts and technology applied

The road map to heating and cooling technologies for buildings prepared by the IEA [6] presents active solar thermal as a key technology for the future. Solar energy can be used in an effective way especially for low-temperature applications. Therefore, to use solar energy for space heating it is important to apply low-temperature heating systems. However, traditional space heating systems that have been applied for many years, and are still very common in most countries, are not low-temperature systems.

For many years space heating systems have operated at high temperature, ie, the temperature of the working heating fluid has been at a high level. It can be especially evident at the large scale of central district heating systems that supply high-temperature heat to buildings. In such a case, usually at the heat exchanger station in a building, the inlet temperature of the working heating fluid, ie, the heating water, has been 90°C and the return temperature of the water 70°C. When heat was produced in a cogeneration plant, then very often superheated steam was used as the working fluid, distributing heat through the heating network of a city, region, etc. When a local boiler plant was used, then the heat supplied to a building could have been at a lower temperature, ie, about 70°C.

A temperature of 70–90°C for the working fluid is quite high compared to the ambient air temperature in winter. Because of the high temperature difference between the working fluid of a heat distribution system and the ambient surroundings, thermal losses through the piping system can be high. To reduce the losses high-quality and thick thermal insulation can be used. However, it is much more efficient to distribute heat at a lower temperature, to make this temperature difference smaller. Nowadays, this tendency in reducing the temperature of the working heating fluid is evident. For example, it is typical for distributed energy plants of small-scale, local thermal plants and boiler houses to use the heating working fluid at a temperature of 50°C. It should be emphasized that the temperature of a space-heating working fluid can be especially low in microscale systems, ie, in individual buildings. Single-family houses can be equipped with their own heat sources and then the heat is produced just on site. The temperature of the heating fluid can be close to the required indoor temperature of rooms and equal to 25–30°C, but it requires the application of special room heating technology, like floor or wall heating systems. Nowadays, such technologies are commonly available on the market. In new buildings they are in use more often than traditional radiators (eg, convectors, radiant heaters). This is because of their high efficiency and good aesthetic features—they cannot be seen in a room, being hidden in a floor or in a wall.

The new energy-efficient concept of space heating requires a low-temperature heating medium, operating close to the required indoor air temperature of heated spaces. This means that space heating systems should be based on a suitable low-temperature technology, ie, a heat source can produce the low-temperature heat and the heat distribution system can operate in the low-temperature mode. All this makes solar energy attractive for use in active space heating systems.

8.1.3 Development of solar combi systems

Solar heating systems are usually distributed-energy systems. They are located close to the end-user site or on site. For many years solar energy has been utilized for water heating for low-temperature applications in agriculture and buildings.

In the case of buildings solar energy has been used mainly for domestic hot water (DHW) systems. In such a system the hot water temperature should be about 50°C. These systems have been mainly applied in residential buildings, with the majority of applications in single-family houses. Most solar DHW systems are factory made. They have been constructed, sold, and installed as fully independent of space heating systems. However, nowadays, when space heating needs are reduced a lot, because of the high thermal standards of buildings, and with the development of new low-temperature (floor or wall) space heating technology, solar thermal systems are designed, dimensioned, and constructed for all heating needs of the end users, ie, DHW and space heating. Such systems are called solar combi systems [9,10], because they combine two main heating functions: hot water heating and space heating.

Solar combi systems are different from solar DHW systems in their philosophy of operation, construction, strategy of control and regulation, and energy management. During the modernization of a water heating system the solar DHW system can be added in a relatively simple way, as a separate module of the system, working with its own rules of operation. When a solar combi system is to be applied, then usually it is a completely new space heating system, or its application needs strong and very complex intervention into the structure and modes of operation of the existing system.

Implementation of a solar combi system requires a very comprehensive design of the system structure, its elements, and their internal integration. From the beginning it is important to develop the concept of a multiple energy source system, in which the energy sources are complementary to one another, but the priority of using different energy sources under given conditions must be determined. Priority for operation usually is based on energy efficiency. In a solar combi system the use of different energy sources depends on their thermal energy performance under given conditions. These conditions are connected to the weather and climate, type of end-user requirements, and quantity of energy needed and its main parameters; usually the temperature of the working fluid is crucial. At a given time the energy performance of different sources of energy can be different. The main idea is to select the most logical energy source for the energy needs in a specific situation and time.

Solar combi systems are complex from the point of view of their structure, their configuration and connections of many elements and subsystems, and the control and regulation of all those elements and their mutual cooperation. Their design and concept of operation are interdisciplinary, combining various energy technologies and utilizing various energy sources. Their management requires smart automatic control. It is very important that the requirements of energy efficiency and reduction of fossil fuel consumption are at a high level in such a system. This is mainly because heating energy is produced on site and is distributed and used there. The energy producer is also the energy distributor and consumer, and this is the best way to ensure energy efficiency for all three processes.

8.2 Operation of active solar space heating systems

8.2.1 Basic classification of the systems

Progress in solar technologies and materials requires that solar energy heating systems cover various heating demands, including space heating, with increasing efficiency and reliability [6–17]. It is characteristic that many technologies recently considered as inefficient in high-latitude countries nowadays have become quite popular. This happened not only because of development of solar technologies, but also owing to progress in the building construction sector, mainly improvements in thermal standards of buildings, as a result of which the heat losses from buildings decreased (as was described in Section 8.1). Reduced heating demand may now be covered by increasingly popular low-temperature heating systems. Such low-temperature space heating systems allow solar thermal heating systems to be applied in an efficient way.

In active solar space heating systems the working fluid in the solar collector loop and the working fluid in the room space heating circuit are at a low temperature. In the case of indoor space heating systems (heating circuit) with water as the heating medium, floor and wall heating systems are used. In the case of indoor space heating systems with air as the heating medium, traditional air space heating can be used, usually coupled with a ventilation system.

An active solar space heating system can be defined as a heating system with a flow of working fluid, in a solar collector loop, being forced by mechanical devices. In the case of systems with water or an antifreeze mixture working as the heat transfer fluid in a collector loop, the flow is accomplished by a circulating pump. In the case of systems with air as the working medium in a collector loop, the flow is accomplished by fans. In indoor space heating systems water or air is used as the heat transfer medium.

Taking into account the type of working fluid of the solar collector loop and the heating medium of the indoor space heating loop, the following types of active space heating systems can be considered:

- Water or antifreeze mixture–water
- Water or antifreeze mixture–air
- Air–air
- Air–water

Rapid development of active solar space heating technology can be seen mainly in the water or antifreeze mixture–water heating systems, so-called solar combi systems. Solar combi systems can be recognized as an upgraded technical solution to DHW solar heating systems, which are the most popular and mature solar thermal technology. Water is a much better heat transfer medium than air, because of its higher thermal capacity and heat transfer coefficients. There are also systems of the water or antifreeze mixture–air type. In these systems technology improvements are mainly connected to the solar heating system, heat exchanger station, boiler house, etc. Air distribution through the air heating system (eg, ducts) in a room is organized as in a traditional air space heating system. In high-latitude countries and in cold climates, water space heating is mainly used. In low-latitude countries and in warm climates, air space heating is commonly used.

Air as the heat transfer medium has some advantages, the main one being the problem with freezing and boiling in solar collectors. In air–air active space heating systems the solar collector loop and the air duct loop in a heated building are closed [8]. Nowadays, in modern buildings solar air collectors are integrated with the building envelope. Solar collectors can be integrated with a roof or they can constitute solar walls. In solar walls there is natural heat flow in the envelope elements with air ducts. The natural heat flow takes place owing to the solar irradiation influence on the air in the ducts. The density of the air in the ducts decreases with the height of the duct, because the temperature of the air increases. As a result buoyancy forces develop to force the air flow through the ducts. In cases of low insolation the flow of collected heat can be enhanced by mechanical equipment (fans). This short description shows that air–air space heating systems nowadays belong to the passive or semipassive solar systems, rather than to the active. There is a strong interest in developing and implementing such solar technologies in buildings. Solar thermal systems that are mainly semipassive solutions integrated with the envelope of the building are known as BIST (building-integrated solar thermal) systems [8]. Being not fully active heating systems, they are not discussed in this chapter.

It is possible to use air as the heat transfer medium in a solar collector loop and water as the heat transfer medium in indoor space heating systems. However, the thermal efficiency of such a combination of heat transfer media is low. A large heat-exchange surface area is required for air to exchange heat to water and a high flow of air is required; as a result the energy needed for fans is high and noise problems can arise.

As has been already mentioned active solar heating systems are usually used for at least two functions, ie, DHW and space heating. Sometimes other functions can be supplied by the system, eg, swimming pool water heating. The extra heating function for summer is very important. Problems with too much solar gain resulting in too high a temperature of the working fluid and storage medium may occur in summer, because the solar collector surface area of solar combi systems must be much larger than for standard DHW systems (heat load is much bigger in times of less solar insolation).

Thus active solar space heating systems can be classified with regard to the number of heating functions that are to be ensured. This classification is as follows:

- Single-function systems, only for space heating
- Double-function systems, typically for space heating and DHW heating
- Multifunction systems, for space heating, DHW heating, and swimming pool water heating or other low-temperature applications that are used seasonally, eg, in agriculture (water heating for animals, greenery, etc.)

Because of differences in structure the double- or multifunction systems can be classified as:

- A common system for all functions—no subsystems (usually it is a double-function system: for space heating and DHW); this means that the main part of the system, including heat storage, is common for all functions

- A system with separate subsystems; every subsystem is to ensure the individual function of that system; this means that, eg, DHW and space heating systems have separate storage tanks and auxiliary heaters and their parameters are different in operation; only the solar collector loop is common to both functions
- A combined-function system, in which the main elements of the system, eg, storage, are common to all functions and some elements are dedicated to specific functions, eg, additional storage tanks, auxiliary heaters, etc.

In active solar space heating systems we can consider different types of system configuration with regards to:

- the number of circuits in a system and
- the form of the circuits (loops): open or closed.

The solar collector loop is usually closed. In a simple air space heating system with direct flow of heated air through the solar collector and to openings in heated rooms, the loop can be open. In such a case this is the open direct active solar air space heating system.

In a liquid solar space heating system the solar collector loop and the space heating loop (eg, floor heating, wall heating, radiator systems) are always closed. The space heating circuit in an air space heating system can be open or closed. In open circuit the heated air is used as the heat transfer medium, as well as fresh air supplied to the building through the ventilation system. In closed-loop systems the heated air has only the space heating function. DHW is always open circuit; there is an inlet of cold water from the water grid and an outlet of hot water for use: washing, bathing, showering, etc. However, in a DHW system the recirculation of water can be applied to ensure there is always hot tap water, and in this case the recirculation loop is closed.

Examples of standard active solar space heating systems are shown in Figs. 8.2–8.5. Figs. 8.2 and 8.3 present the basic idea of the configuration of an active solar system only for space heating, in which liquid and air are used as the working media,

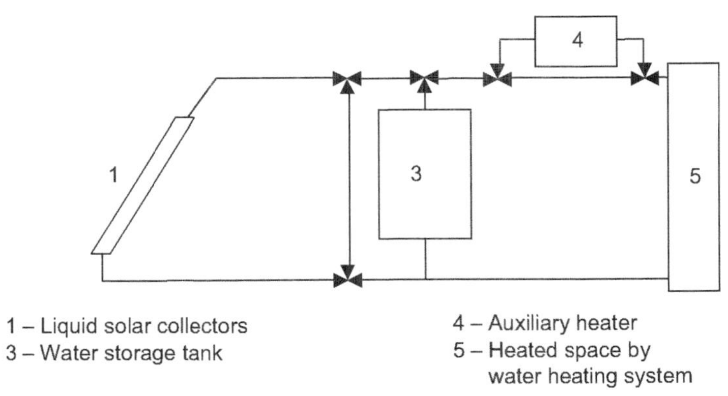

Figure 8.2 Simplified schematic of configuration of standard active solar liquid system for space heating.

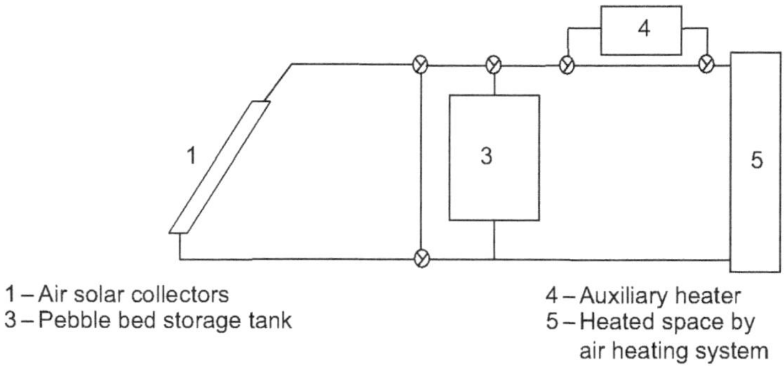

Figure 8.3 Simplified schematic of configuration of standard active solar air system for space heating.

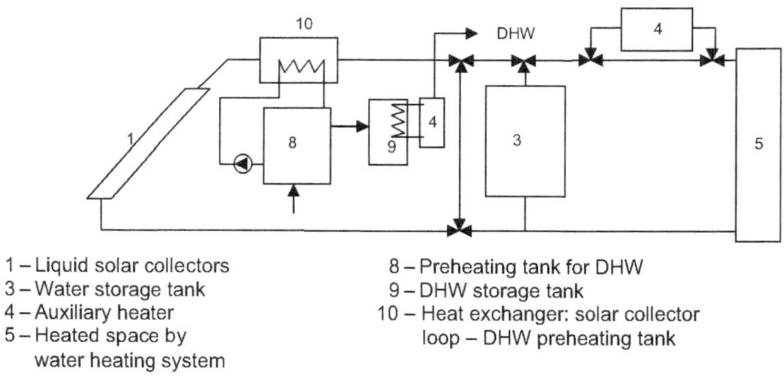

Figure 8.4 Simplified schematic of standard active solar liquid system for space heating and domestic hot water (DHW).

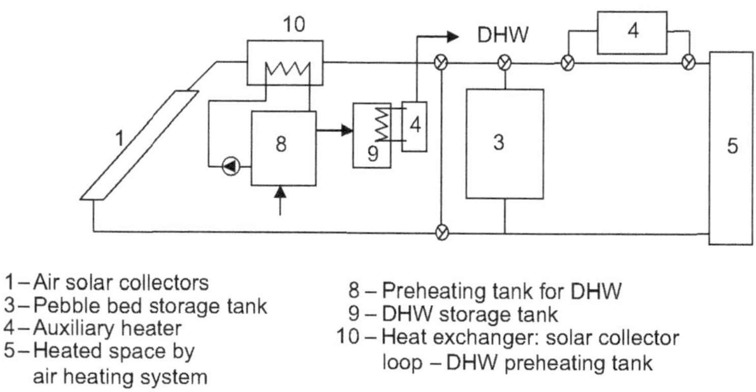

Figure 8.5 Simplified schematic of standard active solar air system for space heating and domestic hot water (DHW).

respectively. Figs. 8.4 and 8.5 show the system for both functions, space heating and DHW heating, for both working fluids, respectively. The figures show exactly which loop is closed and which is open. It can be noticed that the configurations of both systems with two types of working media do not differ significantly.

It can be noticed that the configuration of the standard active solar system for space heating and DHW presented in Figs. 8.4 and 8.5 is one possible idea. The location of the DHW heat exchanger (No. 10) gives priority to DHW heating and ensures the highest temperature of the working fluid of the solar collector loop in this heat exchanger. The DHW loop can be integrated with a space heating system through the main storage tank or after this tank (in both Figs. 8.4 and 8.5 the DHW system is integrated before the tank).

Other classifications of active solar space heating systems can be based on the method of integration of the solar heating system with other devices and systems. This method results from the number of heat sources used within the heating system. It should be emphasized that an active solar space heating system cannot be a pure (100%) solar space heating system. It should always have at least one extra energy source to supplement the space heating demand. Taking into account a number of energy sources used in space heating systems, the following classifications can be made:

- Double energy source heating system, which is based on solar energy and one extra auxiliary heat source, usually a conventional one, eg, an electric heater
- Multisource systems; these are solar systems with a few auxiliary/peak heat sources; typically solar systems are coupled with a heat pump and an auxiliary conventional heater for peak loads, eg, electric heater or a gas/oil boiler

From the point of view of the manufacturing method applied to construct a solar heating system, the following classifications can be made:

- Factory-made solar heating systems. These are products made under one trade name, ready-to-install kits with a fixed configuration; they are usually not offered as space heating systems, but rather as DHW systems.
- Custom-built solar heating systems. These are based on a standard assortment of components; they can be constructed for large-scale solar combi systems, designed by HVAC engineers, and for small-scale applications, offered by a supplier company and marketed by it.

Depending on the scale of the active solar space heating system, ie, micro, small, or large, the complexity of the system may be very different, and a system may include different subsystems and elements.

8.2.2 Main components of the systems

In active solar heating systems solar energy is converted into useful heat in an active way. Collection of solar energy and its conversion into heat take place in solar collectors. Heat gained by solar collectors can be used directly or stored in a storage unit to be supplied later to the consumer in a planned way. The operation of active systems is possible thanks to mechanical devices, which enforce circulation of the working medium, ie, the heat transfer medium. Those devices are circulation pumps in water heating systems and fans in air heating systems.

Active solar space heating systems are equipped with flat plate or evacuated tube solar collectors. In Europe the most popular collector type is a flat plate solar collector with a selective absorber coating. It can be mentioned that at the global scale evacuated tube collectors dominate, mainly owing to applications in Asia (but they are mainly used for DHW heating). Active solar space heating systems are always equipped with auxiliary heaters (devices or systems), working the whole year for the individual heating purposes of the user in the residential sector, public buildings, and others in the tertiary sector [9,10]. It should be emphasized that auxiliary heaters could be in the form of traditional electrical heaters (eg, electrical immersion heaters in storage tanks); traditional boilers fired by coal, gas, or oil; or nontraditional devices and systems based on renewables. There is always a heat store or stores in a space heating system. They are very often recognized as the key elements of the solar heating system. Thus the main elements of active space heating systems are as follows:

- Solar collector
- Storage tank(s)
- Auxiliary conventional or unconventional heater (device or system)
- Room space heating installation in the form of a floor heating, wall heating, radiators, or air-duct heating system.

There are also many other elements and devices, which are necessary for the operation of a heating system. They are as follows:

- Pipelines for supply and return of the working fluid (antifreeze mixture/water) or ducts for the supply and return of air, in a collector loop
- Pipelines for supply and return of heating water or ducts for supply and return of heating air, in rooms
- Pipelines or ducts in any closed loop, for supply and return, and in an open loop
- Circulation pumps and fans for forced circulation of working fluid in every loop (one pump or fan in one closed loop)
- Heat exchangers separating fluids circulating in closed or open loops in a system
- Safety devices such as expansion vessels, safety valves, and vents
- Instrumentation and control systems for individual loops and devices such as temperature sensors, thermostats, control valves, etc.

Some elements of the indoor part of a solar combi system, based on a liquid heat transfer medium, located in a "boiler" room of a single-family house, are shown in Figs. 8.6–8.9.

The elements presented in Figs. 8.6–8.9, apart from the solar collectors, are typical devices of a solar combi heating system used in single-family buildings.

8.2.3 Heat storage as a crucial element of the system

Solar radiation reaching the earth's surface has an intermittent nature and is available periodically. This means that storage of energy gained through conversion of solar energy into heat is necessary, if we want to use this gained energy for heating purposes.

Figure 8.6 Main storage tank (volume 700 L) and expansion vessel (over the tank) for the tank.

When solar energy is used for space heating, the distribution in time of the heating load and its quantity is quite opposite the availability of solar radiation in time and its peak values. This is especially true in high-latitude countries and in cold climates. Winter is characterized there by low ambient air temperature and low solar irradiation and its duration is short per day. Consequently, there are large space heating demands. Therefore efficient storage of solar energy is the key issue for the effective operation of active solar space heating systems. Heat storage can help meet the heating needs in the short run, from a few hours to a few days, or in the long run, seasonally, ie, from summer to winter.

For solar combi systems the most common type of storage is a short-term store, with capacity for one to a few days. Solar combi systems usually have at least two loops for two different loads: DHW and space heating. They usually use at least two separate heat sources, solar collectors and auxiliary heaters. They could have more than one storage tank, eg, a solar buffer tank and a main tank. Any modern storage tank requires heat exchangers. The number of heat exchangers increases with the complexity of the solar system. A storage tank is the central element that integrates all the components of the system. The design of the stores in solar combi systems greatly affects the overall system performance. The store is the "heart" of the active solar space heating system.

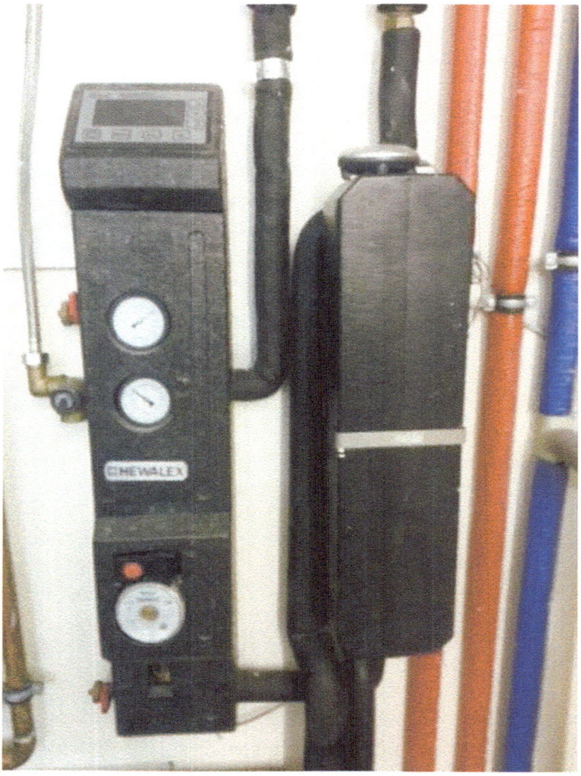

Figure 8.7 Circulation pump group (left) of the solar collectors loop and heat exchanger between the solar collectors loop and the storage tank (right).

In active space heating systems short-term heat storage is usually accomplished through the use of sensible heat from the storage medium, mainly water in water heating systems or a solid medium in a pebble bed store for air heating systems. The thermal capacity of a given medium C (kJ/K) is described by its density, ρ (kg/m^3); specific heat, c (kJ/(kg · K)), and volume (m^3). It is important to use a storage medium of high C (kJ/K) and high storage volumetric density Q_V (kJ/m^3). A store should have high C per unit volume C_V (kJ/(K · m^3)) and so-called volumetric heat capacity and, of course, heat losses should be small.

To describe the operation of a solar thermal heating system it is necessary to write down the energy balance of the storage tank. If it is only one storage tank, then it is relatively simple to determine all energy fluxes going into the storage and going out of the storage. Changes in these fluxes cause changes in the thermal state of the storage medium in a tank, described by changes in storage medium (water) temperature T_w in time. The storage medium, usually water, is described by its volume V, specific heat c_p, and density ρ. The heat gained by the solar collectors is the heat rate going into the store. It is described by the useful energy of solar collectors $\dot{Q}_u(t)$ transferred by working fluid to a water storage tank. Heat taken out of the store is used to meet heating

Figure 8.8 Expansion vessel for the solar collectors loop (right) and circulating pump group (left) of an under-floor low-temperature heating system.

loads, when it is required. When stored heat cannot meet heating requirements fully, ie, the accumulated amount of heat and its temperature are below the required level, then an auxiliary heater (electric heater, peak boiler) or other heating device/system is turned on. In such a situation the accumulated heat taken out of the store $\dot{Q}_{\mathrm{h}}(t)$ can provide only part of the heating demand $\dot{Q}_{\mathrm{hd}}(t)$, and an additional rate of heat $\dot{Q}_{\mathrm{aux}}(t)$ is supplied by the auxiliary heater. These two situations can be mathematically described in the following way:

$$\dot{Q}_{\mathrm{hd}}(t) = \dot{Q}_{\mathrm{h}}(t) \quad \text{or} \quad \dot{Q}_{\mathrm{hd}}(t) = \dot{Q}_{\mathrm{h}}(t) + \dot{Q}_{\mathrm{aux}}(t) \qquad [8.1]$$

When heating losses out of the store $\dot{Q}_{\mathrm{L}}(t)$ are taken into account, then the energy balance of the store, assuming that the storage medium is fully mixed, can be written in a general form as follows:

$$Vc_{\mathrm{p}}\rho\frac{dT_{\mathrm{w}}}{dt} = \dot{Q}_{\mathrm{u}}(t) - \dot{Q}_{\mathrm{L}}(t) - \dot{Q}_{\mathrm{h}}(t) \qquad [8.2]$$

Figure 8.9 Domestic hot water storage tank with auxiliary electric heater.

The useful energy rate of solar collectors $\dot{Q}_u(t)$ can be expressed by the Hottel–Whillier–Bliss equation [18], which gives the energy balance of a solar collector (the heat capacity of the collector is not taken into account):

$$\dot{Q}_u(t) = A_c F_R \left[(\tau\alpha)G(t) - U_L\left(T_{f,i}(t) - T_a(t)\right)\right] \quad [8.3]$$

where G_s is the solar irradiance incident on the tilted collector surface (W/m^2), Q_u is the useful energy rate (W), $\tau\alpha$ is the transmission–absorption factor, U_L is the collector heat loss coefficient (W/(m^2 K)), A_c is the collector surface (aperture) area (m^2), F_R is the collector heat removal factor, $T_{f,i}$ is the fluid inlet temperature (K), and T_a is the ambient air temperature (K).

The heat loss from a storage tank $\dot{Q}_L(t)$ can be expressed in a general way as follows:

$$\dot{Q}_L(t) = U_s A_s (T_w(t) - T_v(t)) \quad [8.4]$$

where U_s is the storage tank heat loss coefficient (W/(m^2 · K)), A_s is the storage tank surface (aperture) area (m^2), T_w is the water temperature in the storage tank (K), and T_v is the storage tank surrounding temperature (K).

Sometimes Eq. [8.4] is presented in a more specific way [18] as:

$$\dot{Q}_L(t) = U_s V^{\frac{2}{3}} C(T_w(t) - T_v(t)) \qquad [8.4']$$

where C is the shape coefficient of the store.

The heat rate taken out of the store $\dot{Q}_h(t)$ is the heat rate removed from the store. This heat rate can be equal to the heating demand $\dot{Q}_{hd}(t)$ at the given time or less than the heating load when the auxiliary heater is applied, as expressed by Eq. [8.1].

Eq. [8.2] can be integrated over time to determine the performance of the store. Assuming time of integration Δt (eg, equal to 1 h) and using the Euler integration, the storage medium temperature for water, at the next time step (at the end of the time increment, ie, after 1 h) can be expressed in the following way:

$$T_w^{t+1} = T_w^t + \frac{\Delta t}{V c_p \rho}\left[\dot{Q}_u^t - \dot{Q}_L^t + \dot{Q}_h^t\right] \qquad [8.5]$$

In Eq. [8.5] sometimes the mass of the storage medium m (kg) is used ($V\rho = m$). The heat rate extracted from the store $\dot{Q}_h(t)$ for space heating depends on the storage temperature T_w^t at a given time, the temperature of the return water $T_{sp,ret}^t$ from the space heating system, and the mass of the space heating medium m_{sph}^t that flows through the store in the given time. Thus, taking into account the notation as given in Eq. [8.5], the heat taken out of the store can be written as:

$$Q_{h,}^t = \dot{m}_{sph}^t c_p\left(T_w^t - T_{sp,ret}^t\right) \qquad [8.6]$$

The heat taken out of storage flows through the closed loop of the space heating system with flow m_{sph}^t, which could be constant for a whole year of operation of the system or could be changed depending on the heating load. However, the maximum level of the flow of a space heating fluid $m_{sph,max}$ is determined by the construction and safety features of the system and it cannot be exceeded. The temperature of the working medium at the inlet (storage temperature) and the outlet (return temperature) of the space heating system also differs in time depending on the storage tank's thermal state and the space heating load. It must be emphasized that the temperature of return water $T_{sp,ret}^t$ from a space heating system has a lower limit $T_{sp,ret,min}$. This is because the temperature of return water (from heating a room) cannot be lower than the indoor (room air) temperature T_{in} plus a few degrees, ΔT_{in} ($T_{sp,ret,min} = T_{in} + \Delta T_{in}$). There is one more obvious thing: heat taken out of the storage for the space heating $Q_h^t(t)$ must be available in the storage tank $Q_s^t(t)$ at least at the same quantity, which can be written as:

$$Q_h^t \leq Q_s^t = m_W c_p\left(T_w^t - T_{sp,ret}^t\right) \qquad [8.7]$$

Thus, taking into account all the above conditions, it can be written that:

(1) if $m^t_{\text{sph}} \leq m_{\text{sph,max}}$ and $T^t_{\text{sp,ret}} \geq T_{\text{sp,ret,min}}$ and $Q^t_{\text{h}} \leq Q^t_{\text{s}}$, then Q^t_{h} is expressed by Eq. [8.6] and $Q^t_{\text{h}} = Q^t_{\text{hd}}$, so the space heating demand is covered by the heat stored in a tank, or
(2) if $m^t_{\text{sph}} \geq m_{\text{sph,max}}$ or $T^t_{\text{sp,ret}} \leq T_{\text{sp,ret,min}}$ or $Q^t_{\text{h}} \geq Q^t_{\text{s}}$, then $Q^t_{\text{h}} < Q^t_{\text{hd}}$ and the space heating demand is covered partly by the heat stored in the tank and the rest by the auxiliary heater, as can be written as the right side of Eq. [8.1], ie,

$$Q^t_{\text{hd}} = Q^t_{\text{h}} + Q^t_{\text{aux}}$$

To calculate the heat demand for space heating in a building usually a simple formula may be used:

$$\dot{Q}_{\text{hd}}(t) = U_{\text{b}}A_{\text{b}}(T_{\text{r}}(t) - T_{\text{a}}(t))$$

where U_{b} is the building heat loss coefficient (W/(m^2 · K)), A_{b} is the building surface area (m^2), T_{r} is the air temperature in the room (K), and T_{a} is the ambient air temperature (K).

This formula, however, gives results that are too simplified. It does not take into account the heat capacity of the building, which influences the energy balance of the building significantly. Very often in such models ventilation is neglected. The formula does not take into account internal heat gains and solar energy gains through glazing of windows (or glazed facades). As a detailed study has shown [8], nowadays in modern buildings solar energy gains dominate the energy balance of a building, because thermal losses through the building envelope are reduced significantly (high thermal resistance of insulation) and losses through ventilation are also reduced thanks to the application of heat recovery in ventilation systems (recuperative heat exchangers). Therefore, it is very important to calculate space heating demands using more complicated models that take into account the dynamics of processes in the building and its surroundings. It is necessary to solve the regular energy balance equation (see similarity to Eq. [8.2]) of the air in the building, which takes the following form:

$$Vc_{\text{p}}\rho\frac{dT_{\text{in}}}{dt} = \dot{Q}_{\text{in}}(t) - \dot{Q}_{\text{out}}(t) + \dot{Q}_{\text{qv}}(t) \quad [8.8]$$

This is an energy balance equation with lumped parameters. Energy rates flowing into $\dot{Q}_{\text{in}}(t)$ and out of $\dot{Q}_{\text{out}}(t)$ a building can be calculated using separate models of different complexity. Energy flow into a room represents the solar energy passing through glazing $\dot{Q}_{\text{sol}}(t)$ during the daytime and under specific thermal conditions also the heat transferred because of ventilation $\dot{Q}_{\text{ven}}(t)$ and through the building envelope (under the specific temperature gradient and thermal capacity of the building). The energy flow out of a room is represented by the heat transferred through the building envelope, windows $\dot{Q}_{\text{win}}(t)$, walls $\dot{Q}_{\text{wall}}(t)$, roof $\dot{Q}_{\text{roof}}(t)$, and floor on the ground (or in the basement) $\dot{Q}_{\text{floor}}(t)$ and through ventilation $\dot{Q}_{\text{ven}}(t)$, under specific thermal conditions. Assuming that the air temperature inside the building is the same in all rooms (in space) and constant in time, the left part of Eq. [8.8] is equal to 0. Thus, the energy balance of air in the building, Eq. [8.8], can be written in a form that allows one directly to determine the flux supplied for space heating into a building:

$$\sum \dot{Q}_{\text{hd}}(t) = \sum \dot{Q}_{\text{wall}}(t) + \sum \dot{Q}_{\text{win}}(t) + \sum \dot{Q}_{\text{roof}}(t) + \sum \dot{Q}_{\text{floor}}(t) + \sum \dot{Q}_{\text{ven}}(t) - \sum \dot{Q}_{\text{sol}}(t) \quad [8.9]$$

Using energy balance Eq. [8.2] and knowing all the heat rates flowing into and out of the store, and the parameters of the store and storage medium, the distribution in time of the water temperature in the storage tank can be determined. Fig. 8.10 gives an example of the results of calculations of water temperature changes in a storage tank during an averaged day in winter in Warsaw. Hourly data of solar radiation, ambient air temperature, DHW consumption, and space heating loads for a single-family house in Warsaw have been used for the calculations. Fig. 8.10 is presented here to show how an active solar heating system can operate in winter conditions (December 1) in a low-latitude country, like Poland. Calculations were performed for a single-family house of 100 m^2 of floor surface, with an energy demand of 100 kWh/m^2 per year. An under-floor heating system was used. The temperature of the heat-transferring fluid (water) circulating in the floor heating system was 35°C. Hot water for the DHW system was required at an amount of 50 L per person for

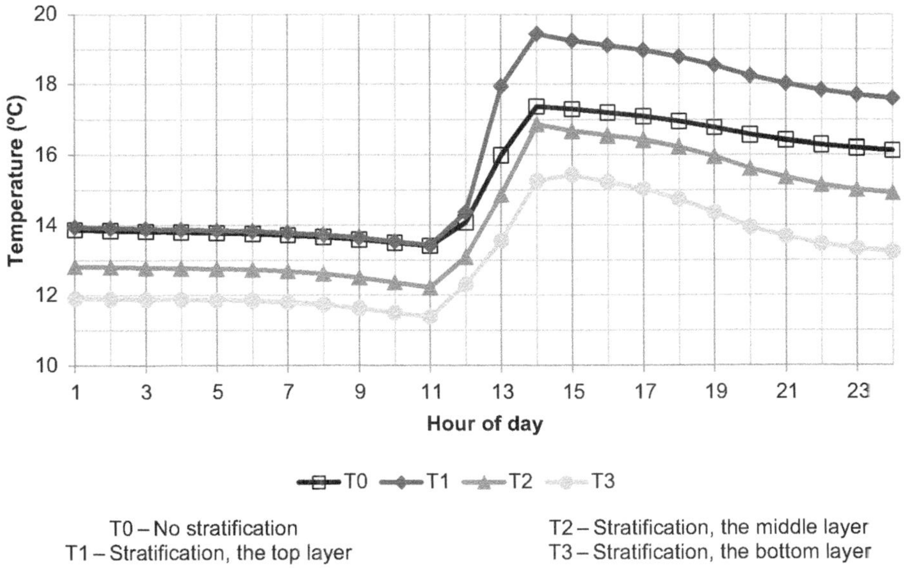

Figure 8.10 Averaged daily distribution of water storage temperature in a storage tank in a solar heating system in winter in Warsaw, Poland.

four people per day, at a temperature of 45°C (heating from 10 to 45°C). Flat plate solar collectors were used with a surface area of 8 m^2 and a heat storage tank with volume of 560 L. By analyzing Fig. 8.10 it is evident that the storage temperature is low and it is necessary to apply an extra energy source for space heating of the building and providing heat for the DHW. As a result the share of solar energy in the total heat supplied to the building is low.

Consideration of the modeling phenomena that take place for a storage tank and the calculations of storage temperature, with results shown in Fig. 8.10, refers to a tank with a fully mixed storage medium and with stratification. Nowadays, all storage tanks in solar space heating systems are more complicated in their construction. They are designed to utilize the effects of stratification in a planned effective way. Stratification is one of the important issues of energy storage in solar water tanks. Water in the physical store is naturally stratified in zones of different temperatures, required for the loads for DHW and space heating. At the top of the store the water temperature is highest (density is lowest) and it is used for DHW. With the decrease of the height of the tank the water temperature decreases, too, and is lowest at the bottom. In the middle and the lower part of the tank an outlet for supplying water for low-temperature heating applications can be located, eg, for a floor heating system. There are several methods for maintaining stratification in stores: increased number of internal heat exchangers, stratifying tubes, stratifying units, etc.

For modeling stratification phenomena in tanks, two main categories can be applied [18]:

- In the multimode approach, the store under consideration is divided into sections (n nodes) and the energy balance (see Eq. [8.2]) is written for each section of the store (for each node); in the result there is a set of differential equations (n equations) that are solved for the temperature of the section (n) of time (see Eq. [8.5]).
- In the plug-flow approach, the store considered is divided into segments characterized by the different temperatures of the storage fluid, and segments are assumed to move through the store in a plugged flow.

The idea of modeling stratifications in a storage tank, given by Duffie and Beckmann [18], is shown in Fig. 8.11, for the case of a three-node tank.

Liquid flows to the collector from the bottom—node 3—the lowest temperature, and returns to node 1, 2, or 3 depending on its temperature. Node 1 is at the top. Here is the highest temperature and water always flows to the space heating system from the top (if it is only a one-function store). Duffie and Beckmann introduced a collector control function $F_{\mathrm{i}}^{\mathrm{c}}$, which shows which node receives working liquid from the solar collectors, and a load return control function $F_{\mathrm{i}}^{\mathrm{l}}$, which shows which node receives return water from the space heating system. On the basis of this model, presented in detail in the literature [18], calculations of a space heating system have been performed. Space heating load has been calculated for many buildings of different types under Polish climatic conditions for the whole heating season [19]. To show the idea of stratification in a water storage tank and compare it with a tank with fully mixed water, one result of the numerical calculations of the operation of space heating systems has been selected for presentation here. Fig. 8.12 presents the

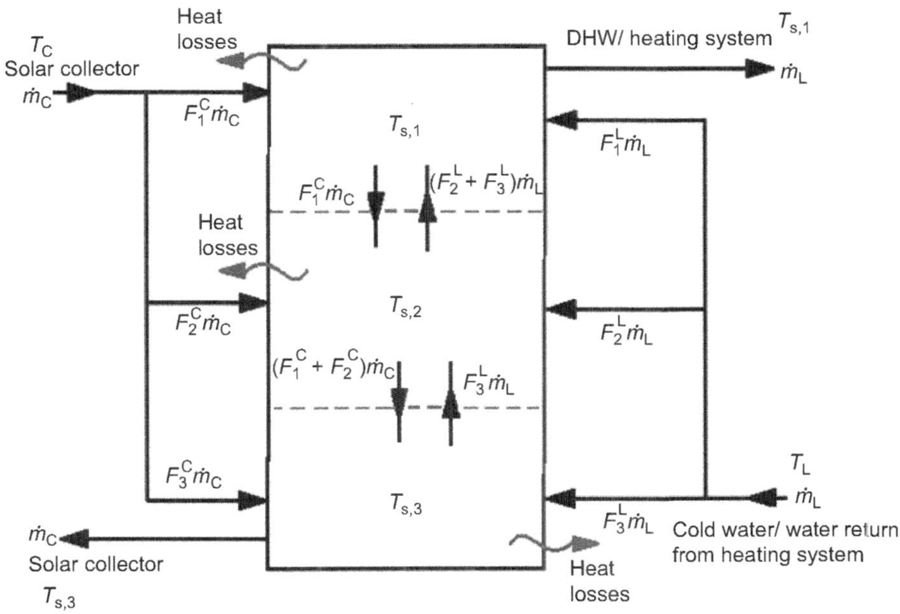

Figure 8.11 Three-node stratified liquid storage tank model in a space heating system. From D. Chwieduk, Solar Energy in Buildings. Thermal Balance for Efficient Heating and Cooling, Elsevier. Academic Press, 2014.

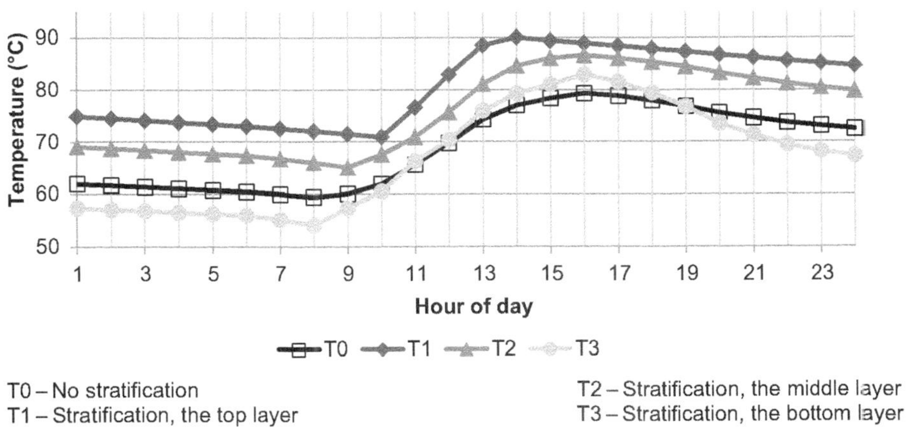

Figure 8.12 Daily distribution of water temperature in a storage tank on April 22.

distribution of water temperature in a three-node stratified liquid storage tank, in all three nodes, and in a tank without stratification, on one day (April 22). Input data for the calculations (weather data, building, heat demand) were the same as for Fig. 8.10.

Fig. 8.12 presents the benefits of a storage tank with stratification in comparison with a tank with a fully mixed storage medium. It is evident that the difference between the highest and the lowest temperature in a tank can be about 15K. The difference depends on the quantity of heat supplied to the tank and heat taken out of the tank for space heating needs in a given time. Fig. 8.13 presents a schematic of a liquid space heating system with a heat store, in which appropriate location of the heat exchangers ensures a stratification effect. This refers to a solar collector loop, with two different locations for heat supply piping and return piping from the storage to the solar collectors, and to the location of the heat exchangers for supplying heat from storage to the heated space with return piping (as is shown in Fig. 8.13). If the storage tank is connected to the low-temperature source of heat (eg, through a heat pump), then the location of the heat exchanger of the low-temperature auxiliary loop is also determined. A heat exchanger should be in the middle of the tank. Fig. 8.14 presents the same solar heating system, with a store as the center of the heating system, plus a DHW heating system.

Considering water storage in a solar combi system, it is important that the daily turnover in the store is not much more than the volume of the store and the heat sources should be able to generate a significant temperature difference, which is capable of generating stratification in the store. As was already mentioned, for the connection of solar collectors to the store the pipe connecting the heat exchangers to the collector should be mounted at the bottom of the store to keep the efficiency high. For high-flow systems (approximately 50 L/h of flow in the collector circuit per square meter of collector surface area) with one heat exchanger in the collector loop, the input into the store should be at the bottom. For low-flow systems (approximately 10–15 L/ $(h \cdot m^2)$ of flow in the collector circuit) the input to the store should be higher [20]. For the connection of an auxiliary heater to the storage it is important that the inlet piping from the auxiliary heater is at the top of the tank and the outlet to the auxiliary heater should be lower, but as high as possible. It is necessary to mention that DHW usually needs the highest temperature of the combi system (50–60°C) and space

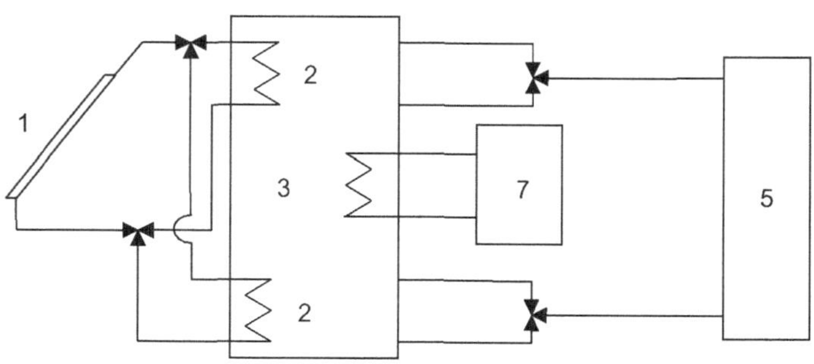

1 – Liquid solar collectors
2 – Water to water heat exchanger
3 – Water storage tank
5 – Space heated by water heating system
7 – Low-temperature auxiliary heating, eg, via a heat pump

Figure 8.13 Configuration of standard active solar liquid system for space heating with utilization of the stratification effect in the storage tank.

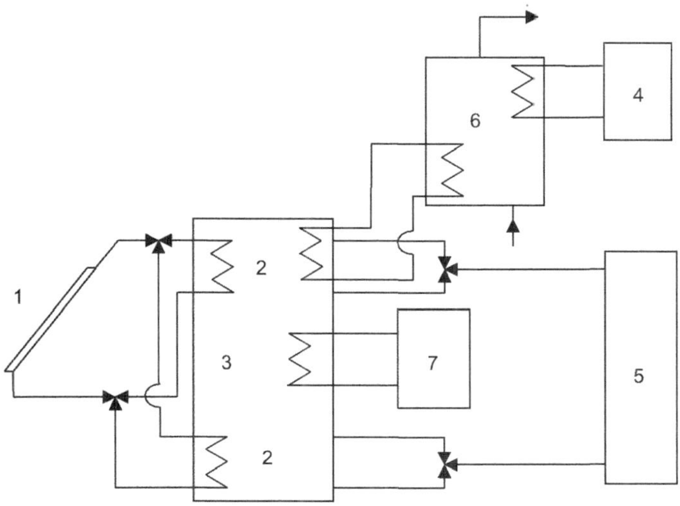

1 – Liquid solar collectors
2 – Water to water heat exchanger
3 – Water storage tank
4 – Auxiliary DHW heater
5 – Space heated by water heating system
6 – DHW tank
7 – Low-temperature auxiliary heating, eg, via a heat pump

Figure 8.14 Configuration of standard active solar liquid system for space heating and DHW with utilization of the stratification effect in the storage tank.

heating is at the lower temperature, eg, floor heating. Therefore the DHW outlet is located at the top of the tank, and the freshwater inlet is at the bottom. The connection to the space heating, being a low-temperature heating system, is positioned in the middle of the tank.

To make storage efficient it is crucial to reduce heat losses as much as possible. There are losses due to convection of the air inside the insulation, air leakages, and thermal bridges and due to the natural convection in external circuits (this can be avoided by nonreturn valves). The use of polyurethane (PU) foam is the most economical solution. Further improvement can be achieved by vacuum insulation and PU foam, which should be used to fill in irregularly shaped gaps.

Storage is a very important component of the solar heating system and it integrates all the other components containing intermediate heat exchanger loops. A solar system can be integrated with a heat pump. There are various types of solar-assisted heat pump (SAHP) systems and differences are also evident in their integration with a storage unit [21]. Storage tanks can contain heat exchanger loops, including loops with direct evaporation or condensation, if the heat pump evaporator or condenser is located directly in the store.

To reduce the volume of storage tanks, ie, to increase storage density, the storage medium is crucial. Improvement of the thermal performance of storage and reduction in storage volume can be achieved through utilization of the latent heat gained during a phase-change process (melting/solidifying phenomenon) of the medium. The phase-change material (PCM) absorbs energy as it goes from a solid to a liquid (melting process) and releases energy as it goes back to a solid (solidifying process) and this phenomenon makes PCMs attractive for heat storage. There is a great interest in the application of PCMs in buildings, especially for PCMs integrated with the buildings' structure [22–27]. The concept of the thermal behavior of PCMs under the transition process is presented in Fig. 8.15.

A storage tank can be filled with PCMs that melt at a given temperature. An example of a triple-medium (water, PCM1, and PCM2) storage tank integrated into a solar collector loop is shown in Fig. 8.16.

Around 1990 at the Solar Laboratory of the Institute of Heat Engineering of Warsaw University of Technology, experimental tests were performed with two different, commercially available PCMs, ie, stearine in the upper part and *n*-octadecane in the bottom part of the storage. Stearine showed its phase-change transition, spread over the temperature range from 33 to 50°C with maximum at 46°C, accompanied by a heat storage effect of 160 kJ/kg. For *n*-octadecane, its phase change was spread over

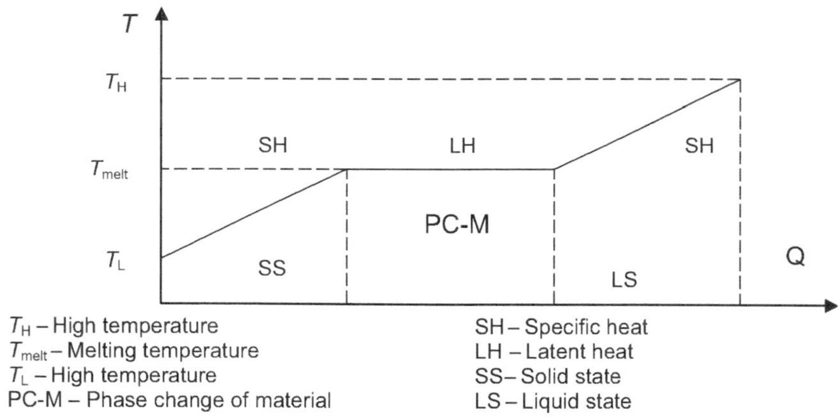

Figure 8.15 Thermal behavior of a phase-change material under the transition process.

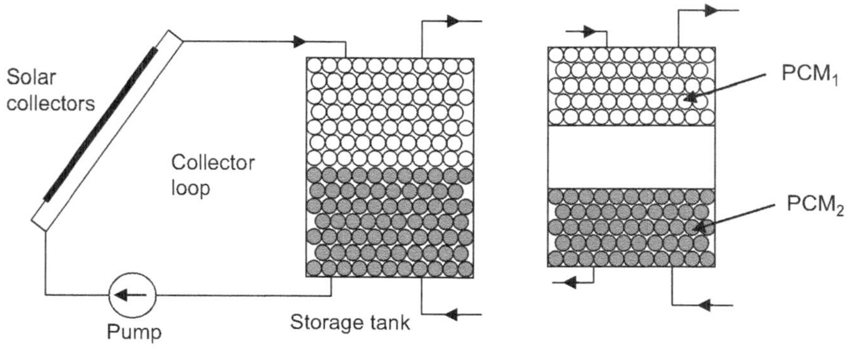

Figure 8.16 Concept of a triple-medium (water, phase-change material (PCM) 1, and PCM2) storage tank integrated into a solar collector loop.

27–29°C, with a heat storage effect of 243 kJ/kg. It turned out that application of two different PCMs at different places inside the storage tank was favorable for keeping the stratification at the required level of temperature at both the upper and the lower part of the tank (filled respectively with both PCMs) [22]. It should be emphasized that even though research on the application of PCMs in storage tanks started many years ago, up to now this storage technology has not been in common use.

For long-term heat storage, mainly underground thermal energy storage (UTES) can be applied [28–31]. A large amount of energy is needed to meet the heating load in winter and therefore the seasonal storage capacity must be large to fulfill the heating requirements. Owing to this fact the ground is most often taken into consideration as the storage medium. The seasonal storage facility can be designed in many different ways. Heat can be stored in the ground (clay, sand), in unfractured rocks, and in water.

To summarize, in active solar space heating systems energy can be accumulated through the use of:

- Sensible heat of:
  - Liquid, usually water in water storage tanks
  - A solid medium (stones, rocks) in a pebble bed heat store, with air flow through the bed
  - A solid and/or liquid medium in ground, rock, underground water, or aquifer in UTES systems
- Latent heat of PCMs located in water storage tanks.

However, in some modern space heating systems it is also possible to use thermo-chemical processes for energy storage.

8.2.4 Modes of operation, configuration, and functions of the systems

Modes of operation, configuration, and functions of the systems are strictly connected to one another. First are the functions of the active solar heating system. Depending on the number of functions of the system (see Section 8.2.1) the configuration of the system differs, more functions means more complex configuration. Then, as a result the modes of operation differ. With more functions, a more complex configuration is needed, which means application of more advanced technology and solutions. As a consequence the number of modes of operation increases, they are more complex, and they need a sophisticated control strategy.

The monovalent mode of operation means that a heating system uses only one source of energy, in this case solar energy. Typically solar space heating systems use more than one energy source. However, solar air space heating systems can be based on solar energy only. It is possible in low-latitude countries, where heating load is very low. Usually in such situations, active space heating is coupled with passive solar heating. In the monovalent mode of operation, it is possible to use one of the following options:

- Direct solar heating (directly from solar collectors)
- Heating through a solar storage tank

Standard active solar space heating systems work in a bivalent mode, using two energy sources; usually it is solar energy and electric energy. In such a case solar energy can be used directly for space heating. However, if the collected and stored energy cannot provide enough heat for space heating, then the electric heater is turned on, as an auxiliary heater. From the point of view of reduction of fossil fuel consumption, it would be beneficial if this electric energy were gained from renewables, eg, photovoltaic plants, wind farms, biomass plants, etc. If not an electric heater then another auxiliary heat source can be used, eg, a boiler fired by gas, oil, coal, or biomass. In all these cases solar energy can be used as the primary energy and heat from an electric heater or a boiler as supplemented energy. This is a series mode of operation of the system. If available solar energy is at a very low level and it is not efficient to utilize this energy, then the auxiliary heater or boiler can operate alone as the peak heat source. In bivalent modes of operation it is possible to use the following options:

- Direct solar heating (directly from solar collectors)
- Heating through a solar storage tank
- Series heating: solar as primary (directly or from a storage) and then auxiliary heating
- Auxiliary heating by electric heater or boiler, or other heat source, in the form of peak heating

The multivalent mode of operation means that a heating system is designed to operate using more than two energy sources. This mode is applied in most new technologies of space heating systems to make them not only a very reliable solution, but also more energy efficient. Here the priority of utilization of energy source is given to that which is the most energy and economically efficient at the moment when heat is used. Apart from solar energy, usually electric energy and heat from boilers fired by fossil fuel or biomass are applied. However, SAHPs are also one of the solutions. They have been used for many years, but recently they have become especially popular. They are considered in more detail in a later section.

In the multivalent mode of operation the following options are possible:

- Direct solar heating (directly from solar collectors)
- Heating through a solar storage tank
- Series heating: solar as primary (directly or from a storage) and then auxiliary heating using one auxiliary source for each function (eg, space heating plus DHW, two different auxiliary heaters can be used)
- Triple source heating with solar energy, series, and parallel:
  - Series heating for DHW and space heating: solar as primary plus electric heater for DHW, and solar as primary plus a peak boiler for space heating
  - Series heating for DHW and parallel for space heating: two functions are ensured by two different modes, series, eg, solar as primary plus auxiliary electric heater for DHW, and parallel third source, eg, gas boiler for space heating
- Auxiliary heating by one auxiliary energy source: electric heater or boiler, or other heat source, can be in the form of peak heating and the same auxiliary energy source for DHW and space heating
- Auxiliary dual-source parallel heating by auxiliary—peak energy sources—separate for DHW and for space heating, usually an electric heater for DHW and a boiler for space heating

Every solar heating system is equipped with automatic control systems. It should be emphasized that the multivalent mode of operation of active solar space heating systems requires especially well-planned effective control of system operation and that all elements are integrated into one heating system cooperating with interior microclimate sensors and controllers for the heated spaces (rooms).

To consider various configuration and functions of the solar heating system Figs. 8.17–8.20 are presented. The focus is on space heating, as this is a main topic of consideration in this chapter. A DHW function is also included in some of the schemes of the system configurations. When a system is a two-function system then of course its configuration is more complicated than that of a one-function system. The type of working fluid is taken into account.

Figs. 8.17–8.20 present systems with liquid, usually an antifreeze mixture, as the solar collector working fluid. There is a heat exchanger between the solar collector loop and the storage vessel. The main reason is that water is used as a storage medium (its thermal parameters are much better than those of an antifreeze mixture). There is heat stratification in the storage tank. This phenomenon was described earlier, when the schematic idea of a storage tank integrated with other solar system elements was drawn. Various positions (heights) of inlets from the solar collector can be seen. It depends on the temperature rise of the working medium gained in the solar collector. Also various positions of outlets from the storage to the user have been shown. The height of the outlets depends on the level of required temperature of heated working fluid, dependent on the function and type of heating system (floor temperature, radiators). Fig. 8.19 presents a standard active liquid space heating system, with liquid as the working medium in the solar collector loop and water as the heating medium in the space heating system.

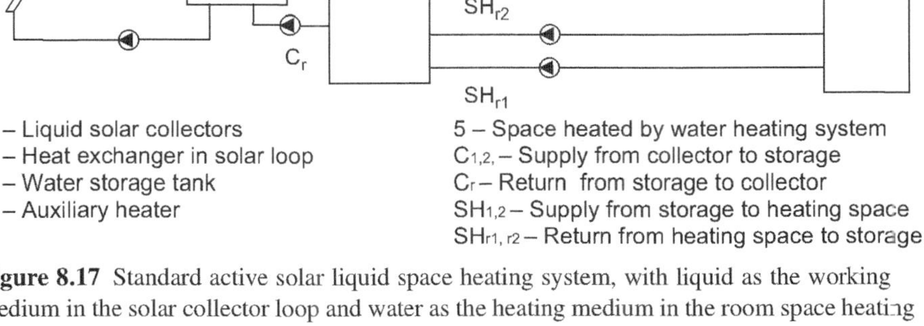

1 – Liquid solar collectors
2 – Heat exchanger in solar loop
3 – Water storage tank
4 – Auxiliary heater
5 – Space heated by water heating system
$C_{1,2,}$ – Supply from collector to storage
C_r – Return from storage to collector
$SH_{1,2}$ – Supply from storage to heating space
$SH_{r1,\,r2}$ – Return from heating space to storage

Figure 8.17 Standard active solar liquid space heating system, with liquid as the working medium in the solar collector loop and water as the heating medium in the room space heating system.

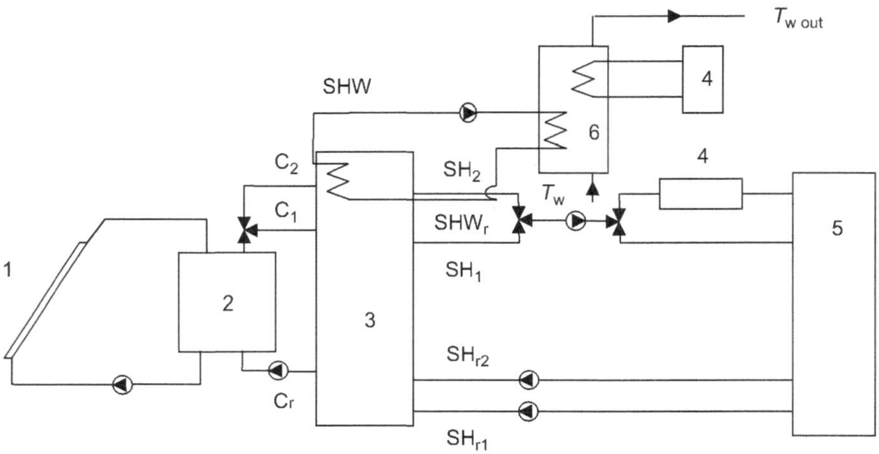

1 – Liquid solar collectors
2 – Heat exchanger in solar loop
3 – Water storage tank
4 – Auxiliary heater
5 – Space heated by water heating system
6 – DHW tank

$C_{1,2,}$ – Supply from collector to storage
C_r – Return from storage to collector
$SH_{1,2}$ – Supply from storage to heating space
$SH_{r1,\ r2}$ – Return from heating space to storage
SHW – Supply DHW
SHW_r – Return from DHW

Figure 8.18 Standard active solar combi space heating system, with liquid as the working medium in the solar collector loop and water as the heating medium in the room space heating system and DHW system.

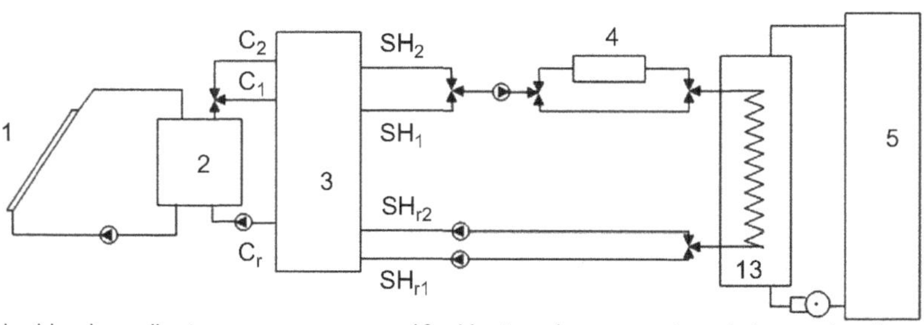

1 – Liquid solar collectors
2 – Heat exchanger in solar loop
3 – Water storage tank
4 – Auxiliary heater
5 – Space heated by water–air heating system

13 – Heat exchanger water–air (space heating system)
$C_{1,2,}$ – Supply from collector to storage
C_r – Return from storage to collector
$SH_{1,2}$ – Supply from storage to heating space
$SH_{r1,\ r2}$ – Return from heating space to storage

Figure 8.19 Standard active solar liquid space heating system, with liquid as the working medium in the solar collector loop, water as the storage medium, and air as the heating medium in the room space heating system.

In Fig. 8.18, apart from the space heating function, the DHW heating is also included, which makes the system and its control more complicated. Fig. 8.19 presents an active space heating system, in which liquid is the working medium in the solar collector loop and water is the storage medium, but the space heating system uses air.

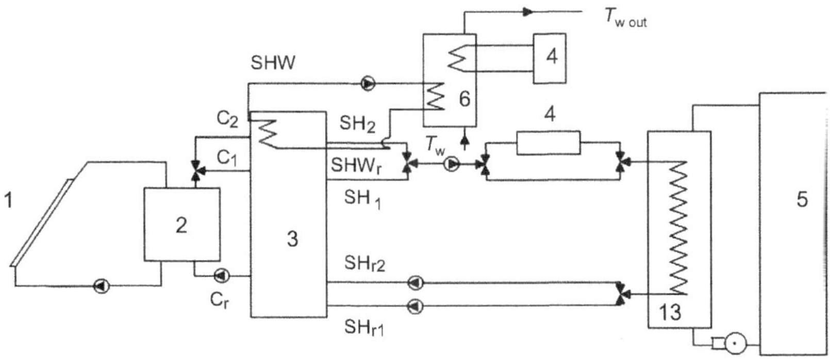

1 – Liquid solar collectors
2 – Heat exchanger in solar loop
3 – Water storage tank
4 – Auxiliary heater
5 – Space heated by water heating system
6 – DHW tank

$C_{1,2,}$ – Supply from collector to storage
C_r – Return from storage to collector
$SH_{1,2}$ – Supply from storage to heating space
$SH_{r1,\ r2}$ – Return from heating space to storage
SHW – Supply DHW
SHW_r – Return from DHW

Figure 8.20 Standard active solar liquid space heating system, with liquid as the working medium in the solar collector loop, water as the storage medium and working fluid in the DHW system, and air as the heating medium in the room space heating system.

Because of that an extra heat exchanger must be added between the air space heating loop and the water storage loop, and as a consequence the thermal performance of the system must be lower (than that of the system in Fig. 8.17). Fig. 8.20 shows a system similar to that presented in Fig. 8.19, but the DHW function is added.

Figs. 8.21 and 8.22 present an air space heating system in which air is used as the working medium in the solar collector loop and space heating loop. Of course, for DHW heating, water is needed as the medium to be heated, and therefore an air-to-water heat exchanger is applied, as is shown in Fig. 8.22.

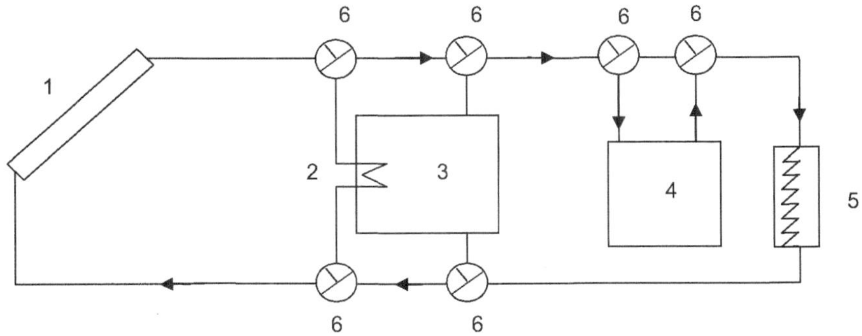

1 – Air solar collectors
2 – Heat exchanger in solar loop
3 – Pebble bed storage tank

4 – Auxiliary heater
5 – Space heated by air heating system
6 – Three-way dampers

Figure 8.21 Standard active solar air space heating system, with air as the working medium in the solar collector and space heating loop.

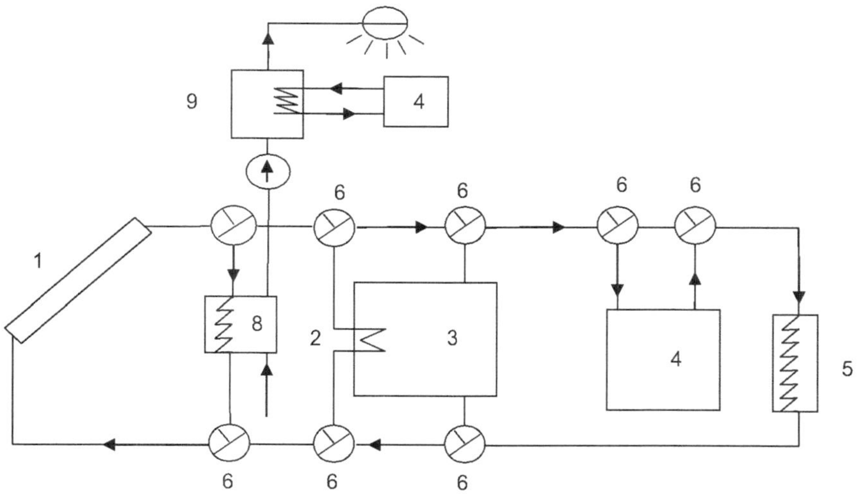

1 – Air solar collectors
2 – Heat exchanger in solar loop
3 – Pebble bed storage tank
4 – Auxiliary heater
5 – Space heated by air heating system
6 – Three-way dampers
8 – Preheating tank for DHW
9 – DHW storage tank

Figure 8.22 Standard active solar air space heating system, with air as the working medium in the solar collector loop and space heating loop and water as the storage medium and working fluid in the DHW system.

As was mentioned, in the case of solar space heating too much solar energy could be a problem, and as a result too much heat is gained by the system. To solve this problem extra heating loads are required. One such load can be swimming pool heating.

In the case of a single-family house too much solar gain in summer should not be a problem. When flat plate solar collectors are used, reverse flow can be used at night. In such a situation the working medium transports heat accumulated in the storage tank back to the solar collectors. At night the ambient temperature is much lower than the temperature of water in the store. The solar collector, being a normal heat exchanger, releases heat to the ambient surroundings, and as a result the temperature of the storage medium in the storage tank decreases. In the case of vacuum tube solar collectors, which are so well thermally constructed that they cannot lose the heat, the solar collector loop can be connected to a roof-mounted chiller. The chiller releases excess heat directly to the surroundings, and liquid flowing to the storage tank is cooled down.

In the case of large solar space heating systems the best idea is to find extra heating loads in summer. The situation is especially difficult in the cases of buildings that are not used in summer, eg, schools. In this case perhaps a good idea is to build a sports center with an open-air water park close to the school and use the heat produced by the active solar space heating to supply heat to this sports center. Another idea of using the heat could be some agricultural applications.

Previously modes of active solar heating systems were classified by taking into account various sources of energy that could be used by the system. Now the classification is made with regard to the loops or subsystems of the whole heating system that are in use, where the flow of the working medium takes place. For every heating system an exact procedure for operation must be determined. Detailed limits for the operation of a solar space heating system in different modes must be defined and utilized by an automatic control system. Usually the working medium of the solar collector loop can circulate if the temperature difference between the working medium at the outlet of the solar collector and at the outlet of the heat exchanger or storage tank is above a determined limit equal to a few degrees. If the temperature of the solar collector working medium is at the required level to supply heat, directly or through a heat exchanger in a collector loop (for heating purposes), then there is

- a direct solar heating (directly from the solar collectors) mode.

If the temperature of the solar collector working medium and the gained heat are high enough to be used directly for heating needs and to be accumulated at the same time, then the mode of system operation that takes place is

- a direct solar heating and heat storage (in a tank) mode (at the same time).

If the temperature difference between the working medium at the outlet of the solar collector and the outlet of the heat exchanger, or a storage tank, is above a determined limit, and there is no heating load, then the following mode of operation that takes place is

- a heat storage mode.

If the temperature difference between the working medium at the outlet of a solar collector and the outlet of the heat exchanger or storage tank is above a determined limit, but not at the required level to supply heat directly, and this temperature is higher than the temperature of the storage medium in the storage tank, then there is primary solar heating by the solar collector loop and then auxiliary heating by the auxiliary heater. This mode of operation can be classified as

- a primary solar heating (from the solar collector loop) and auxiliary heating mode.

If the temperature difference between the working medium at the outlet of the solar collector and the outlet of the heat exchanger or storage tank is above a determined limit, but not high enough for direct solar heating, and this temperature is lower than the temperature of the storage medium in the upper part of the storage tank, then primary solar heating is used based on the heat stored in the tank and then auxiliary heating by the auxiliary heater is used. It is good to note that this mode of operation gives the opportunity for the solar collectors to operate even under suboptimal solar irradiation and to store the heat in the lower part of the tank, when the heat stratification effect is used. This mode of operation can be classified as

- a primary solar heating by stored energy (from the storage tank) and auxiliary heating with loading of the storage tank (from the solar collectors) mode.

If the temperature difference between the working medium at the outlet of the solar collector and the outlet of the heat exchanger or storage tank is below a determined limit, and the temperature of the storage medium is high enough to be used directly for heating, then the mode of solar system operation that takes place is

- a solar heating by stored energy (from a storage tank) mode.

If the temperature difference between the working medium at the outlet of the solar collector and the outlet of the heat exchanger or storage tank is below a determined limit, and the temperature of the storage medium is such that it is possible to use the heat accumulated in the tank, but this temperature is not as high as the required temperature for the space (or water) heating, then primary solar heating is used based on the heat stored in the tank and auxiliary heating by the auxiliary heater is used. This mode of operation can be classified as

- a primary solar heating by stored energy (from the storage tank) and auxiliary heating mode.

Of course, there is one more mode of operation possible, when the solar collector loop is not working, the storage tank is not loaded, and the temperature in storage is too low to be used for heating purposes.

- an auxiliary heating mode.

The examples described above show how many different options of operation of active solar space heating loops and subsystems can be realized. Thus a role for the monitoring and control systems is to manage the operation of the solar space heating system in an effective way, realizing the appropriate mode of operation for existing weather conditions, the thermal state of the system and its subsystems, and the heating load.

8.3 Solar hybrid systems

8.3.1 Integration and complementarity of various energy sources

Because of the way solar combi systems operate, solar energy cannot be the only energy source for such systems. The case of active solar liquid space heating systems is similar, as they are mainly used in middle- and high-latitude countries, where the level of solar irradiation in winter is low and heat demand is high or moderate. In the case of active air space heating systems applied in low-latitude countries (with weak winters), where heating systems operate at low-temperature of the heating medium (air), it is possible to utilize only solar energy. However, in such a case the energy storage, ie, the pebble bed storage tank, must be well designed (big enough in volume and with a big stratification effect). Nowadays, very often active air space heating systems cooperate with passive solutions.

A solar hybrid system can be defined as a system that uses more than one energy source, and there are specially designed systems based on these energy sources, which supply heat to a building cooperating together to ensure efficient and effective energy production, distribution, and use. Systems based on different energy sources are integrated into one system, constituting multiple-energy-source subsystems. All subsystems are complementary to one another operating in control mode. An automatic control system manages energy production, distribution, and use, giving priority to this subsystem—this energy source, which is the most effective at a given time. Having such a definition, it is evident that a solar heating system with an electrical immersion heater in a storage tank, or electric flow heater in one of the system loops, although based on two energy sources, cannot be called a solar hybrid system. It is just a solar heating system with an auxiliary heater.

When considering different energy sources that are to be integrated with a solar energy source in one active space heating system, it is necessary to take into account the nature of these sources, having in mind that they should be complementary to the solar energy source. Because of the stochastic and intermittent character of solar energy the complementary or auxiliary source of energy should be very reliable and have predictable thermal performance, with the ability to be steady for long periods and achieve higher performance when necessary. These requirements caused electric energy to be used for many years as an auxiliary as well as peak energy. However, the consumption of electricity is not cheap. Apart from that if electricity is produced in a power plant fired by fossil fuel, then consumption of primary energy is high. First combustion of fossil fuel takes place to produce high-temperature heat to produce steam, then the steam is used to drive turbines to produce electricity. If now this electricity is used to produce low-temperature heat for a floor heating system, for example, then of course it is a very inefficient method. It is much better to use low-temperature heat directly.

Heat can be produced by any type of boiler that can be fired by fossil fuel, eg, coal, gas, or oil. Heat can be produced at different levels of temperature. However, from the point of view of reducing consumption of fossil fuels, this is not a good solution. Therefore nowadays, biomass boilers are used. For a single-family house it could be a boiler fired by biomass pellets. In bigger systems, it can be a boiler fed with wood residue, straw, or biogas.

The best idea for production of low-temperature heat is application of a heat pump, which is a well-known low-temperature heat source. Apart from that, the consumption of electricity by a mechanically driven (compressor) heat pump is a few times lower than that used by an electric heater. As a result it is easier to utilize renewable energy, for example, through application of photovoltaic panels, to produce electricity. A well-designed heat pump can operate in a reliable, predictable, and efficient way. These benefits of coupling a solar energy system with a heat pump have meant that SAHPs are very popular for applications in active space heating systems. There is another important fact, that the only traditional energy source needed for such a system is electricity, which is the most accessible form of energy for the end user.

8.3.2 Solar-assisted heat pump systems

As was mentioned above, nowadays a solar heating system is very often coupled with a heat pump to supply heat to a building for space heating and DHW. There are many different possibilities for integration of a solar system with a heat pump. One of the oldest and best known classifications of SAHP systems refers to the heat source of the heat pump and its connection to the solar heating system [32–36]. Thus, there are the following types of SAHP systems, which are presented in Figs. 8.23–8.25:

- A series SAHP system. Solar collectors and a heat pump are connected in series, directly or through a storage tank. Solar energy can be used directly from a solar collector loop or a storage tank, and the heat pump uses solar energy as a heat source, from a solar collector loop directly or through a storage tank.
- A parallel SAHP system. Solar collectors and a heat pump are connected in parallel; they are an alternative solution for heating. Solar energy can be used directly from a solar collector loop or from a storage tank. The heat pump is separated from the solar subsystem. It uses

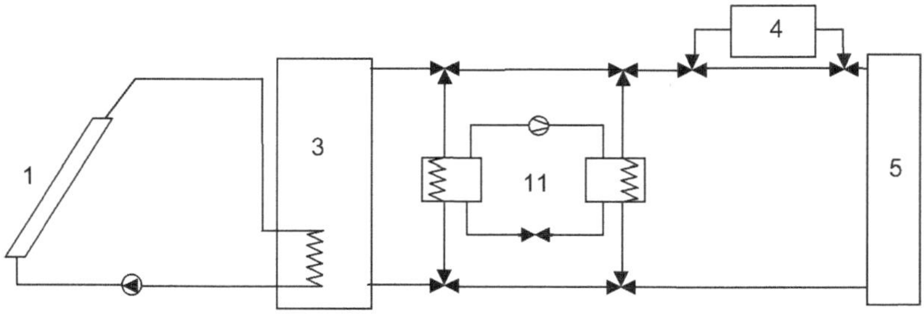

1 – Liquid solar collectors
3 – Water storage tank
4 – Auxiliary heater
5 – Space heated by water heating system
11 – Heat pump

Figure 8.23 Schematic of a traditional series solar-assisted heat pump system.

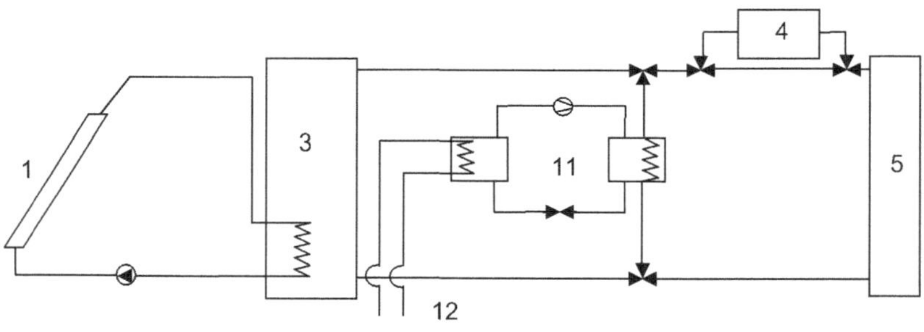

1 – Liquid solar collectors
3 – Water storage tank
4 – Auxiliary heater
5 – Space heated by water heating system
11 – Heat pump
12 – Alternative heat source for a heat pump

Figure 8.24 Schematic of a traditional parallel solar-assisted heat pump system.

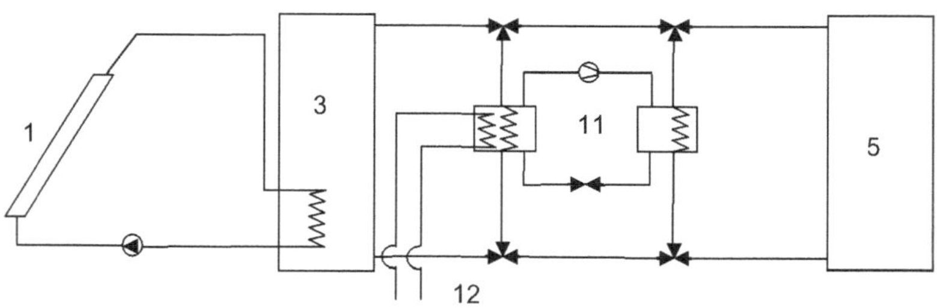

1 – Liquid solar collectors
3 – Water storage tank
5 – Space heated by water heating system
11 – Heat pump
12 – Alternative heat source for a heat pump

Figure 8.25 Schematic of a traditional dual-source solar-assisted heat pump system.

a heat source other than solar energy (eg, air, ground, underground water, etc.). The solar subsystem is independent of the heat pump subsystem. An automatic control decides which subsystem should be used at a given time, ie, which subsystem has better thermal performance at that time.

- A dual-source SAHP system. This system is a combination of a series and a parallel system, and it could be called an alternative dual-source system. Solar energy can be used directly from the solar collector loop or from a storage tank. The heat pump can use solar energy as a heat source, from a solar collector loop directly or through a storage tank, or it can use another heat source, but not both together. Another heat source is independent of the solar subsystem, it is the alternative solution for heating. Although both heat sources are completely independent of each other (eg, solar and ground) the operation of a heat pump depends on the thermal state of both of these heat sources. Thus the automatic control system can select the heat source whose thermal performance at a given time is higher. It should be mentioned that when a heat pump uses another source of energy (air, ground, groundwater), then at the same time solar energy can be collected by the solar collectors and stored in a tank, so the sources are complementary to one another.

SAHP systems can use liquids and air as working media. Liquid (water or antifreeze mixture) and air can be used in solar collectors. The heat pump can use different fluids circulating in the loop, extracting heat from a heat source, eg, water, brine, or air, and in the loop of the heating space, where water or air is used as the heat transfer medium. Figs. 8.23–8.25 present liquid SAHP systems; however, the idea and schematic of these traditional air space heating systems are nearly the same. The difference is that in an air SAHP system a storage tank is not always used, air can transfer heat directly to a heat pump evaporator. In a liquid SAHP system solar collectors are always coupled to a storage tank.

Apart from traditional types of SAHP systems (series, parallel, and dual source), there is one more, which has been known for many years [37]. It is a direct SAHP system. In such a system solar energy is used to heat the working fluid—a refrigerant—directly in the evaporator of the heat pump. A solar collector and an evaporator constitute one integrated unit. Two types of solar collectors can be used for such solar evaporator: unglazed solar collectors and regular glazed solar collectors with one cover. A refrigerant flows through the integrated unit: solar collector–evaporator. Under the influence of solar radiation the refrigerant starts evaporating (going from liquid to vapor). The refrigerant is directly evaporated in the solar collector–evaporator and then expanded, and because of this it is called a direct expansion (DX) SAHP system [38–40]. In a DX SAHP system the intermediate heat exchanger between the solar collector loop and the heat pump is eliminated, which makes the construction of the system simpler compared to the standard one. For effective operation of a DX SAHP system very careful selection of a suitable refrigerant is needed, because it must operate at very different temperatures [38]. There is always a problem of uncertainty with DX SAHP system operation in unstable weather conditions. Perhaps because of that, up to now this technology has not become popular, even if the idea of application of direct SAHP systems was presented many years ago.

As has been already mentioned, nowadays storage becomes the center of the heating system. A heat pump condenser usually is connected to a storage tank (see Fig. 8.26), not as it was before, when the evaporator was connected to a store (see Figs. 8.23–8.25).

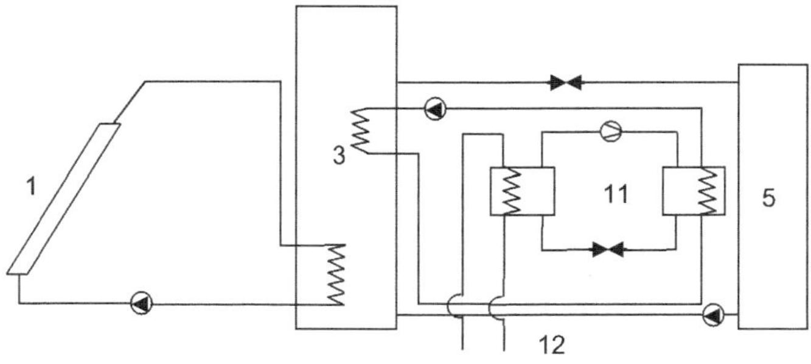

Figure 8.26 Schematic of a typical modern configuration of an SAHP system for space heating.

A storage tank usually contains intermediate heat-exchanger loops. It can also contain direct evaporation or condensation loops, but this is rather rare.

Fig. 8.26 presents a typical modern configuration of an SAHP system, in which an active solar heating system is coupled to a heat pump through the storage tank. What is going in the storage tank is crucial for the operation of the whole heating system. This system could be considered as a parallel one, but it is not exactly like that. In parallel systems two energy sources are fully independent. Here it is different, because two sources supply heat to the same store, so loading the store by one heat source can influence the operation of the other one. It is possible that at the same time both sources are utilized and both load the store. Therefore, an automatic control system that manages the utilization of energy from both sources in an effective way is very important. The SAHP system presented in Fig. 8.26 is a typical modern hybrid system. It is designed to provide all heating demands using different energy sources based on different subsystems, which are integrated in one system in the center of this system, ie, in the storage tank, to be complementary to each other and to ensure effective space heating of the whole system.

The SAHP system presented in Fig. 8.26 is a space heating system. In the past heat pumps were used only for space heating. Hot water preparation was provided by a separate system, eg, a solar DHW system with an auxiliary heater. Nowadays, hot water preparation is one of the functions of an SAHP system. Application of the stratification effect in the storage tank is standard and it allows for the temperature requirements of the heating medium in positioning the outlet of the heat exchanger from the storage tank. When a low-temperature space heating system is used, then an outlet for the space heating is in the middle of the tank and an outlet for DHW is located at the upper part of the storage. Usually there is another tank only for DHW. The inlet for cold water is at the bottom of this tank, above is an inlet for heated water supplied from the main store, and higher above there is a heat exchanger connected to the auxiliary heater (or an electric immersion heater is placed directly in the tank). At the top of this store there is an outlet for DHW. A schematic of such a system is shown in Fig. 8.27.

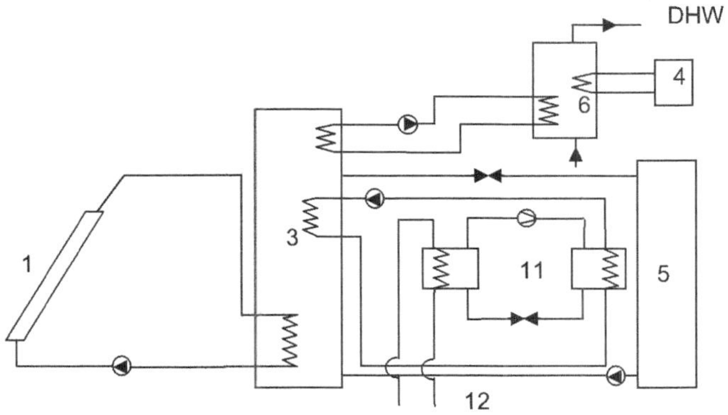

1 – Liquid solar collectors
3 – Water storage tank
4 – Auxiliary heater
5 – Space heated by water heating system
6 – DHW tank
11 – Heat pump
12 – Alternative heat source for a heat pump

Figure 8.27 Schematic of a typical modern configuration of an SAHP system for space heating and DHW heating.

Nowadays, there are many different modifications of SAHP systems [41–47]. Modern SAHP systems are no longer a typical series, parallel, or dual-source SAHP system. SAHP systems have received interest for applications in micro-, small-, and middle-scale buildings. In low-latitude countries SAHP systems are usually coupled with air to air heat pumps. In high-latitude countries solar heating systems are usually coupled to ground source heat pumps. This is mainly because of the very good complementarity of solar energy and ground energy. In autumn, at the beginning of the heating season solar radiation and ambient temperature are at moderate levels. The thermal state of the ground is undisturbed. In winter when solar irradiation and ambient temperature are low, the thermal state of the ground is still characterized by relatively high thermal performance. Deeper than 10 m the temperature of the ground is nearly the level of annually averaged ambient air temperature (in Poland it is about 10–11°C). In spring, when the ground is cooled down, because of heat extraction, then solar irradiation is relatively high and ambient air temperature moderate. All this makes solar and ground energy very complementary to each other, and utilization of the ground at bigger depths, through vertical heat exchangers, can be a very effective solution, especially with cooperation with a solar heating system. In central European countries the freezing zone in the soil is about 1 m. More to the north (high-latitude countries) the freezing depth is bigger. This makes it preferable to use vertical ground heat exchangers rather than horizontal ones [48]. When a ground source heat pump is used, heat is extracted from the ground by vertical ground heat exchangers that are mainly in the form of U-tubes (so-called ground ducts or probes) or double U-tubes (so-called duplex probes).

The view of a modern SAHP system, in which a solar heating system is coupled to a ground heat pump, is presented in Fig. 8.28. This is a picture of a so-called boiler

Figure 8.28 A room with a solar-assisted heat pump system—a solar heating system coupled to a ground heat pump.

room; however, there is no boiler there, but there is no other name for a place with such a heating system. A solar collector loop (flat plate solar collectors with antifreeze mixture as a working fluid) is closed and connected through a heat exchanger to a buffer storage tank, with water as the storage medium. There is a ground source heat pump with vertical ground heat exchangers (U-type, antifreeze mixture as a working fluid) operating in a closed loop and connected through a heat exchanger with the same buffer storage tank. There is another storage tank only for DHW, which can be supplied from the buffer storage. At the top of this DHW tank is an electric heater. The scheme of this system is nearly the same as the one shown in Fig. 8.27. There is a low-temperature floor heating circuit in the building. Heating of the building is accomplished by the heat stored in the big tank of combined buffer storage.

Various operation modes of the system presented in Fig. 8.28 are possible thanks to well-developed automatic control systems. The print screen of a display of configuration and operation of an SAHP system is presented in Fig. 8.29. The solar combi heating system is coupled to a ground heat pump through the storage tank.

Some of the SAHP systems using the ground as a heat source for the heat pump have a function of storing extra solar energy that is not used for heating purposes. This function can be also called recharging (regeneration) the ground (duct, borehole, etc.) [49]. Usually, this type of storage is called UTES. An example of such a system is presented in Fig. 8.30.

The SAHP system presented in Fig. 8.30 has more modes of operations than the systems presented up to now. Solar collectors of the SAHP system can constitute different closed loops. Solar collectors can supply heat directly to the storage tank or to the heat pump evaporator. A heat pump also supplies heat to the storage tank from its condenser. Depending on the temperature of the working fluid supplying heat to the storage the upper or the lower heat exchanger in a storage tank can be

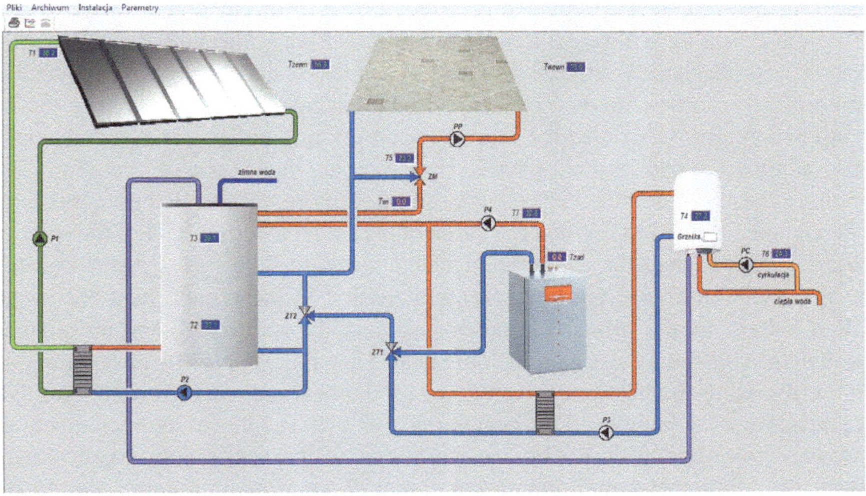

Figure 8.29 The print screen of a display of configuration and operation of a solar-assisted ground heat pump system.

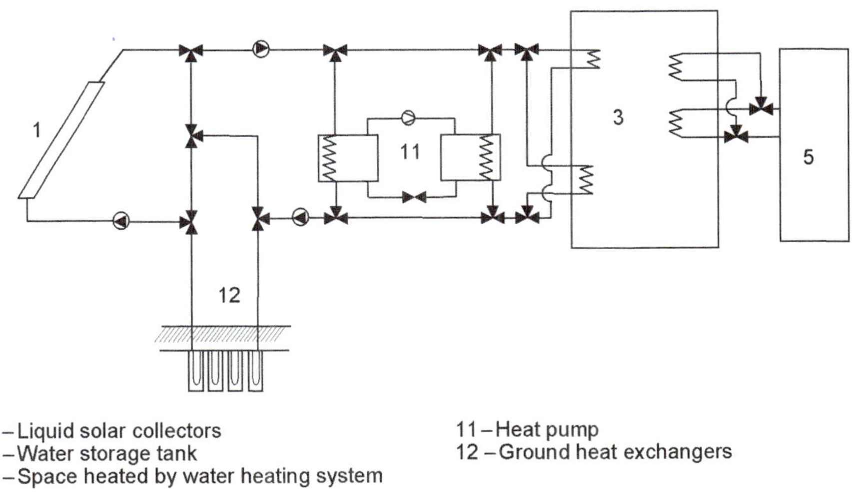

Figure 8.30 Schematic of a solar-assisted heat pump system—a solar space heating system coupled to a ground heat pump with a seasonal storage, so-called underground thermal energy storage.

used, as is shown in the figure. In addition solar collectors can be connected directly to the piping of vertical ground heat exchangers loops. In summer, when too much solar energy is gained, heat gained by the solar collectors is transferred by the working fluid of the solar collector loop, ie, the antifreeze mixture (based on glycol), to the ground heat exchanger loops. In such a way solar energy is accumulated in a ground body causing an increase in the temperature of this storage medium. Solar collectors and ground heat exchangers can be linked together when the same type of working fluid is used (the same antifreeze mixture). In winter solar energy stored in the underground store can be extracted by vertical heat exchangers and directed to the heat pump evaporator. Of course the loop between the ground heat exchangers and the heat pump evaporator is then closed. However, another mode of operation of the heating system is possible, thanks to the same antifreeze mixture in the solar collector loop and the ground loop. When the temperature of the working fluid of the solar collectors is higher than the temperature of the working fluid in the ground heat exchangers, the antifreeze mixture can circulate first in the ground heat exchangers, which behave as the primary heat source. Next the antifreeze mixture circulates through the solar collectors gaining more heat. Then it flows into the heat pump evaporator, releasing the heat gained. In this way the heat extracted from the ground is upgraded by solar energy and the heat pump uses a heat source of higher temperature, and as a consequence the thermal performance of the heat pump is better, which means the coefficient of performance is higher. What is more, the heat accumulated in the ground is not extracted so quickly. Utilization of solar energy helps the ground heat source to be of high thermal performance for a longer time. This example of an SAHP system shows how multifunctional such heating systems can be nowadays and how important is the complementarity of the energy sources selected for operation in one integrated system. Fig. 8.31 presents the same heating system as in Fig. 8.30 but upgraded to supply also DHW, as is standard in modern SAHP systems.

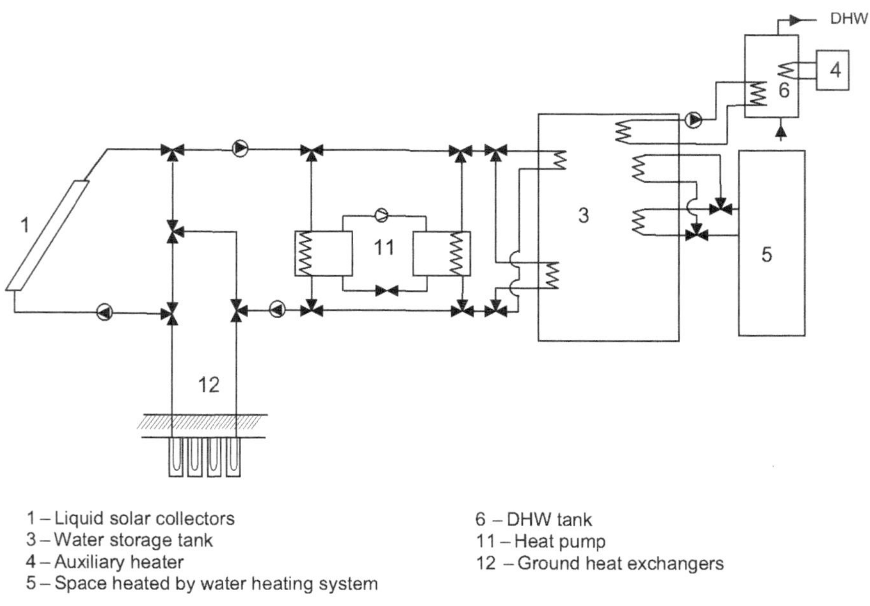

Figure 8.31 Schematic of a solar-assisted heat pump system—a solar space heating and DHW heating system coupled to a ground heat pump with seasonal storage.

8.3.3 Small- and large-scale systems: autonomous, distributed, and centralized systems

Active solar space heating systems are applied on small, medium, and even large scales. Small- or microscale systems are used in small buildings, eg, in single-family houses. Medium systems are used in multifamily buildings, public buildings, etc. Large scale means supplying heat to a city, a district, or a region. When the scale is small, then it is easier to apply autonomous systems; when the scale is bigger it is more complicated; however, it does not mean that it is impossible.

As was already mentioned, space heating systems usually are not 100% solar systems, they use other auxiliary energy. Solar energy is available at the place—at the building, where it is converted into heat, distributed, and used. If it is the same with the auxiliary energy, then it is an autonomous system. However, electricity is always needed for operation of any heating system, to drive fans, pumps, compressors of the heat pumps, the automatic control system, etc. Electric energy is usually produced in large power plants and distributed by the grid to end users, to provide electricity to all these systems and devices. So of course in such a situation, the active solar heating system is not autonomous. It will be fully autonomous only when not only heat but also electricity is produced on site.

Zero energy buildings are equipped with such autonomous energy systems. They are designed to produce energy by themselves. A zero energy (from the grid) building must of course be designed in a planned way to have small energy requirements and to use energy available in the vicinity (surroundings), like solar energy. However, the use of solar energy must be ensured not only through application of advanced active heating technology, but by passive as well. It is difficult for any building to be a zero energy building if its architecture, construction materials, and structure are not focused on the reduction of energy consumption and passive energy solutions.

Zero energy buildings on small and middle scales must have a well-developed automatic control system. The complexity of energy systems can be very different depending on the functions of these systems, required temperature of the heating medium, and expected heating load. The control system must be adapted to the type of consumer and heating requirements. When small heat loads, low heating temperatures, and two heating functions are required, and the heat load distribution can be predicted, then uncomplicated technical solutions and control strategies for the heating system may be used. The complexity of a system rises along with heating requirements and demanded reliability in all conditions. When a system is more complex it requires more elements, devices, loops, controls, and safety protection. Auxiliary heating devices or systems may support the solar system nearly continuously or provide a backup.

Solar energy heating systems are very good examples of distributed energy systems. Solar energy, like other renewables, can be named a local energy source, from the point of view of its utilization, even if it is available globally. Its extraction, conversion into another form of energy, distribution, and use are very effective, because all these processes take place at one location under one management and control. However, this does not mean that the utilization of solar energy and operation of solar heating systems cannot be done in a centralized way. They can, but then the complexity of the system is bigger.

Centralized solar heating systems are used for the large-scale heat demand. When a solar heating system is designed to supply heat to a village, a city, or a region, then it becomes a centralized system of very large size. A centralized solar district heating system is equipped usually with flat plate solar collectors, with an area of a few hundred to several thousand square meters [50–53]. An example of a solar collector field for a large centralized heating system is shown in Fig. 8.32.

Solar collectors are divided into many loops, which are integrated together to supply heat to a storage system. In the case of a centralized solar heating plant the storage of heat gained can be accomplished through one big store, usually with water as the storage medium, or through many smaller tanks. Such tanks can be connected in series and in parallel. Fig. 8.33 presents a schematic of such a system.

Figure 8.32 An example of a solar collector field for a large centralized heating system in Graz, Austria.

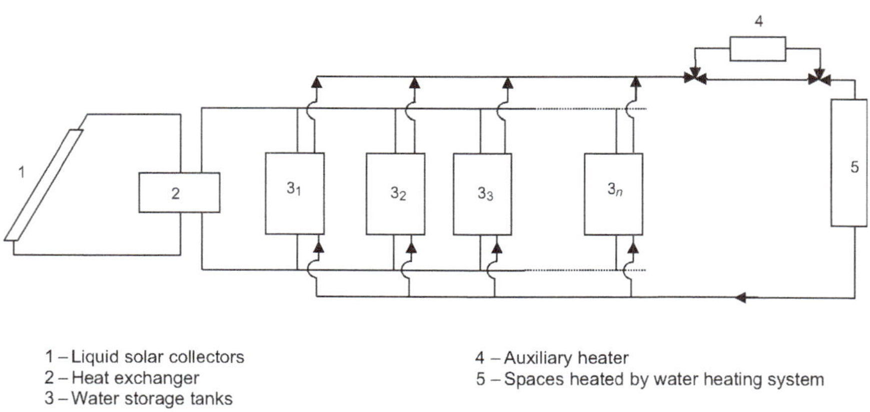

Figure 8.33 A schematic of a centralized solar heating system with multiple storage tanks.

A centralized solar heating system is always coupled to a conventional peak heat source (plant). Such centralized solar district heating systems can be used for many heating purposes including space heating. It is of course an active solar heating system, but its idea, structure, configuration of elements, and modes of operation are different from those of standard small- and middle-scale active solar space heating systems. In some ways, the organization of the control and operation of such a system is similar to that of central heating plants fired by biomass or conventional fuels (gas, oil, coal). This is required to join the knowledge of the utilization of solar energy for heating purposes with the operation of centralized standard heating systems.

8.4 Energy efficiency of active solar space heating

8.4.1 Conditions influencing energy efficiency of the systems

The efficiency of active solar space heating systems depends on solar energy availability, which is very important, but it is not the only factor that affects the operation of the system. Other conditions influencing energy efficiency of the space heating system are as follows:

- Climatic conditions: Solar irradiance and its distribution in time throughout the heating season, duration of the day, ambient air temperature and its distribution in a day and throughout the heating season, wind velocity and direction
- Space heating demands of a building resulting from the building's location, architectural concept, construction, materials, civil engineering design, and surroundings
- Number of heating functions to be fulfilled by the system and their operating parameters
- Distribution in time of space heating demand in a day, in a heating season, and in a year for solar combi systems
- Type of end user: Type of building (residential, public, school, hospital, etc.)
- Type of space heating system in the building: Temperature of heating medium (low, middle, high) and physical state of the medium, ie, liquid (water) or gas (air)
- Thermal characteristic (performance) of the system as a whole and its components, influenced by size of elements, materials used, construction, location, etc.
- Type of operation and parameters of storage:
  - Time of storage (eg, a day, a few days, season)
  - Size (volume), shape, materials used
  - Type of storage medium: solid (pebble bed, ground), liquid (water), air
  - Temperature of storage: low, medium, high
  - Stratification effect used, number of heat exchangers and their position in the storage tank, or storage with a fully mixed medium
  - Surroundings of the store and heat losses per day, month, year
- Automatic control system operation, priority procedure for selection of different energy sources and their complementarity in a given time
- Awareness of the end use of the modes of operation of the system and methods of energy conservation

As can be seen, there are many factors influencing the efficient operation of a solar heating system. The energy efficiency of the system is very often understood as the fraction of solar energy supplying heat to a building to cover all heating demands. However, it is not the same, as can be seen according to the bullets just given above. The share of solar energy can show partly how energy efficient the solar heating system is.

Figs. 8.34–8.36 present the share of solar energy supplied by an active space heating system to a single-family house of defined size (floor heating area), in defined climatic conditions, ie, for Warsaw, Poland. When climate, location of a building, and its size do not change, some other parameters differ. These are:

- Thermal quality of the building, resulting from the thermal resistance of the building envelope, which influences the heat demand of the building
- Size of the solar collectors area, which influences the solar energy gained by the system

Fig. 8.34 presents the share of solar energy supplied by an active solar heating system, with flat plate solar collectors of surface area 8 m^2 and a storage tank of 560 m^3, to the heating system of a standard single-family house. The energy demand for space heating is 100 kWh/m^2 per year. A low-temperature under-floor heating system is used, with the hot water temperature at 35°C. Every day there is a need of 200 L of hot water to be supplied at a temperature of 45°C (heating from 10 to 45°C) by the DHW. The annual DHW demand is 2974 kWh. The total solar energy used for heating purposes is 2456 kWh per year. In this case the solar energy fraction is 19%. It can be mentioned that this example of a house and its heating load has already been taken into account, when the distribution of the temperature of the storage medium during selected days was presented (see Section 8.2.3).

The same active solar combi system was considered for a single-family low-energy house of the same size as before, but with highly reduced consumption for space heating, being 20 kWh/m^2 per year. The results are presented in Fig. 8.35. Now the total solar energy used for heating purposes is 2344 kWh per year, and total solar energy

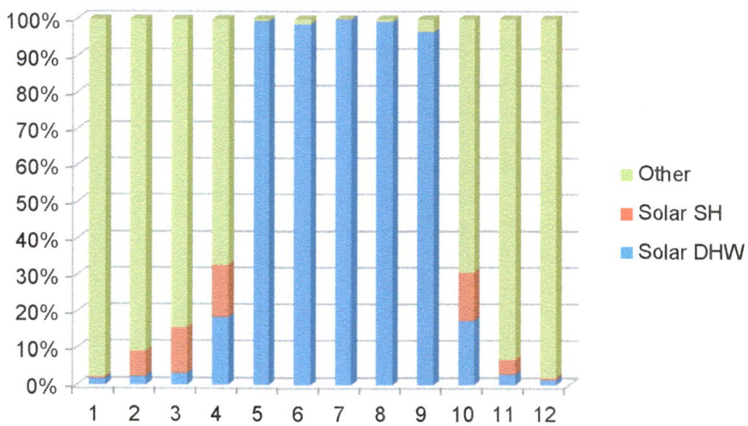

Figure 8.34 Share of solar energy supplied by an active space heating system (solar collectors of 8 m^2, storage tank of 560 m^3) for space heating (SH; 10,000 kWh per year) and domestic hot water (DHW; 2974 kWh per year) to a single-family house in Warsaw.

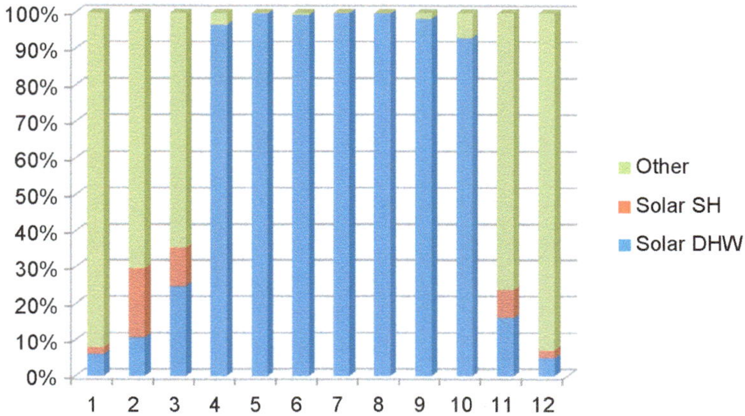

Figure 8.35 Share of solar energy supplied by an active space heating system (solar collectors of 8 m^2, storage tank of 560 m^3) for space heating (SH; 2000 kWh per year) and domestic hot water (DHW; 2974 kWh per year) to a single-family house in Warsaw.

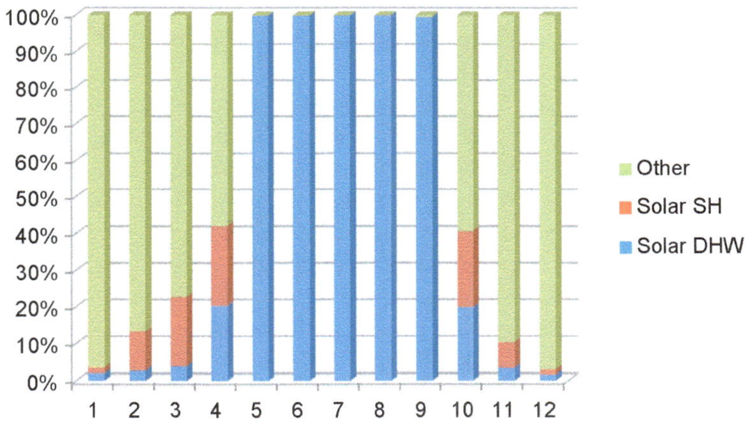

Figure 8.36 Share of solar energy supplied by an active space heating system (solar collectors of 12 m^2, storage tank of 560 m^3) for space heating (SH; 100,000 kWh per year) and domestic hot water (DHW; 2974 kWh per year) to a single-family house in Warsaw.

share is 47%. It can be noticed that even if the total amount of solar energy supplied by the solar heating system is a little bit lower, thanks to significantly reduced space heating demand, the solar heating system can provide nearly 50% of the total heating load. It can be stated that good thermal characteristics of the building is crucial for obtaining a high solar energy fraction.

Figs. 8.36 and 8.37 present the results of calculations of the share of solar energy supplied by the active solar combi system for space heating and DHW for the first of the considered buildings, the standard one with a space heating load of 100 kWh/m^2 (DHW load is not changed). In this case the surface area of the solar collectors

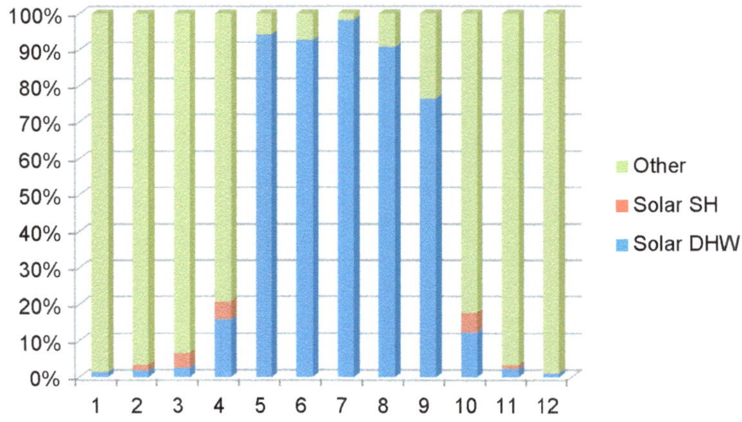

Figure 8.37 Share of solar energy supplied by an active space heating system (solar collectors of 4 m^2, storage tank of 560 m^3) for space heating (SH; 100,000 kWh per year) and domestic hot water (DHW; 2974 kWh per year) to a single-family house in Warsaw.

differs. In Fig. 8.36 the results are for a bigger surface area, of 12 m^2, and in Fig. 8.37 they are for a smaller area, of 4 m^2.

Analyzing Figs. 8.36 and 8.37, it is evident how the surface area of solar collectors influences the share of solar energy in providing heat to a building. When the solar collector area is equal to 4 m^2, then the heat gained by the solar heating system goes mainly for DHW in summer (there is 1 m^2 of solar collector per person and the heated house is of 100 m^2 floor area). In this case the total solar energy used for heating purposes is 1751 kWh per year, thus the total solar energy share is only about 13%. In the case of a large area of solar collectors, ie, 12 m^2 (now there are 3 m^2 of solar collectors per person, the heated house remaining the same), the total solar energy used for heating purposes is now 2954 kWh per year, and the total solar energy share is 23%. Comparing Fig. 8.36 with Fig. 8.37 it can be noticed how with the increase in solar collector area the share of solar energy in providing heat for the space heating also increases. Of course, to decide what is the recommended surface area of solar collectors for space heating a detailed optimization analysis should be performed for different criteria, taking into account the different conditions of operation and climate.

Of course the solar energy fraction, just analyzed, shows the input of solar energy in providing heating energy to a building, but it does not express the total efficiency of the system, and not even to its thermal part. Solar collector loops and other loops are equipped with pumps or fans and automatic control systems, which need energy, usually electricity, to operate. An auxiliary source can work with different efficiency and use different energy sources or fuels. Energy efficiency means a global approach to energy conservation. Thus the energy consumed by all subsystems, devices and equipment should be small. So the solar fraction is not the only factor important for total energy efficiency.

8.4.2 Evaluation of the seasonal performance factor

The seasonal performance factor (SPF) is an indicator used for heat pumps to evaluate their efficiency expressed as a ratio of the total heat supplied to a building (by the heating system) to the electricity used by the heat pump (to drive the compressor of a heat pump) and other devices of the heating system over the year [53–57]. This factor is used to calculate the renewable portion of the heat pump's heat output.

The SPF can also be used for active solar heating systems, being a measure of energy efficiency and utilization of renewables. In active solar space heating systems the SPF can be defined in a similar way as the ratio of the total heat $Q_{suptotal}$ supplied to a building by the heating system over the year to the electricity $E_{el,total}$ used by all electrically driven devices and subsystems, such as an electrical auxiliary heater E_{aux} and other auxiliary heating systems or devices Q_{fossil}, based on fossil fuels, used to make the heating system operate over a year. Thus the SPF can be described by the following:

$$\mathrm{SPF} = \frac{Q_{suptotal}}{E_{el,total} + E_{aux} + Q_{fossil}} \qquad [8.10]$$

Of course, depending on the type and configuration, the SPF can have a simpler or more complicated form. Thus in the case of a solar space heating system providing heat for space heating and DHW partly by solar energy Q_{sol} and partly by the auxiliary electric heater $Q_{aux,el}$, Eq. [8.10] will take the following form:

$$\mathrm{SPF} = \frac{Q_{suptotal}}{E_{el,total} + E_{aux}} = \frac{Q_{sol} + Q_{aux,el}}{E_{el,total} + E_{aux,el}} \qquad [8.11]$$

In the case of an SAHP system the SPF is determined in a standard way, ie, as for every system with a heat pump, although different approaches are sometimes taken by different groups of engineers, because not all situations are clear. For a standard SAHP, series, parallel, or dual-source system, heat is supplied by solar collectors directly or through a store Q_{sol}, by a heat pump Q_{hp} using solar or another heat source. A heat pump uses electric energy $Q_{comp,hp}$ to drive a compressor. The system is usually equipped with an auxiliary heater, eg, for DHW (and usually $Q_{aux,el} = E_{aux}$), and the SPF can be written as:

$$\mathrm{SPF} = \frac{Q_{suptotal}}{E_{el,total} + E_{aux} + E_{comp,hp}} = \frac{Q_{sol} + Q_{hp} + Q_{aux,el}}{E_{el,total} + E_{aux} + E_{comp,hp}} \qquad [8.12]$$

When a solar heating system cooperates with different energy sources and a storage tank is the center of the system, and the system does not have enough sensors to monitor and measure the energy coming from the different energy sources, then heat supplied to a building is written in a general form as $Q_{suptotal}$ (a sum of all forms of heat supplied). Electrical energy used by all subsystems and devices E_{el}, auxiliary heaters E_{aux}, and the heat pump compressor $Q_{comp,hp}$ (if a heat pump is applied) is included in the denominator. If the heating system uses also energy supplied by systems fired by fossil fuels (gas, oil, coal) Q_{fossil}, it is also put in the denominator. Heat supplied by a biomass (biogas) Q_{bio} boiler/plant or electricity produced based on renewables (wind, hydro, solar—photovoltaics) shall not be included in the dominator. Thus the SPF for a multiple-energy-source system is:

$$\mathrm{SPF} = \frac{\sum_i Q_{\mathrm{sup},i}}{E_{\mathrm{el,total}} + E_{\mathrm{aux}} + E_{\mathrm{comp,hp}} + Q_{\mathrm{fossil}}} = \frac{Q_{\mathrm{suptotal}}}{E_{\mathrm{el,fossil}} + Q_{\mathrm{fossil}}} \qquad [8.13]$$

It can be stated that if the SPF is to be the real measure of energy efficiency and use of renewables in the heating system, including space heating systems, then also some correction factors should be assigned to the various energy sources/fuels and efficiency of energy conversion from primary to final.

8.4.3 Future improvements

It can be expected that future active solar space heating technologies will be integrated with the building envelope, constituting partly active and partly passive systems. Perhaps most of such systems would use air as the heat transfer medium. BIST systems could be developed very quickly, if the construction sector cooperated with engineers and specialists in solar energy. Of course, the involvement of architects in the process of shaping the building structure, not only from the artistic and civil engineering points of view, but also owing to the necessity of treating a building as a living energy system, is very important. Hybrid solar air heating systems can, for example, operate in a passive way for southern rooms and active for northern. Experiments with hybrid solar air heating systems performed in the Solar Demonstration Building at the University of Science and Technology in China have given very good results for the thermal performance of such a system and its share of solar energy [58]. A lot can be changed in the technology of solar walls, which usually operate in a passive or semipassive way [59]. A prototype of an active solar thermoelectric radiant wall system has been evaluated and is described in the literature [60].

Liquid solar thermal systems would mainly be used in high-latitude countries. Simple examples presented in this chapter have shown how important it is to install solar collectors of large surface area to supply heat for space heating a building in a cold climate. Therefore BIST systems in the form of flat plate liquid collectors integrated with roofs seem to be a good solution for high-efficiency space heating systems. When a liquid is used as the working fluid of solar collector loops, then usually water is the heat storage and heat transfer medium in a space heating system in a building. To improve the efficiency of the system it is good to increase the density of the heat stored. Perhaps storage tanks filled with a different PCM (different melting temperature), to keep the stratification at the required level, are a good solution. The space heating system in a building should be of low temperature, in the form of under-floor or wall heating systems. Perhaps putting floor heating piping in a PCM cover would help the space heating system to be more effective in meeting space heating requirements.

To make solar space heating technology work effectively in buildings, however, first the space heating demand should be reduced as much as possible. With a low-energy-consumption building the active solar space heating system operation is more efficient and the share of solar energy is higher. Thanks to improvements in the thermal characteristics of buildings, now solar thermal technologies can be successfully applied in buildings, even in high-latitude countries, where because of very high energy demand in winter, it used to be not possible.

References

[1] Energy Efficiency Requirements in Building Codes, Energy efficiency policies for new buildings IEA information paper, in: Support of the G8 Plan of Action, Jens LAUSTSEN, IEA, 2008.

[2] Directive 2002/91/CE of the European Parliament and the Council of 16 December 2002 on Energy Performance of Buildings, 2002.

[3] Directive 2010/31/EU of the European Parliament and of the Council of 19 May 2010 on the Energy Performance of Building, 2010.

[4] Rozporządzenie Ministra Infrastruktury W Sprawie Warunków Technicznych, Jakim Powinny Odpowiadać Budynki I Ich Usytuowanie Warszawa, Dziennik Ustaw Dnia 13 Sierpnia 2013 r. Poz. 926.

[5] Technology Roadmap, Energy Efficient Building Envelopes. Energy Technology Perspective, IEA, 2014.

[6] Technology Roadmap, Energy-efficient Buildings: Heating and Cooling Equipment, IEA, 2011.

[7] Renewables for Heating and Cooling, Untapped Potential. RETD Renewable Energy Technology Deployment, OECD/IEA, 2007.

[8] D. Chwieduk, Solar Energy in Buildings. Thermal Balance for Efficient Heating and Cooling, Elsevier. Academic Press, 2014.

[9] W. Weiss, Solar Heating Systems for Houses: A Design Handbook for Solar Combisystems. Solar Cooling and Heating Programme, IEA, 2003.

[10] W. Weiss, The challenge to exploit the solar thermal potential, in: EUROSUN2010 Proceedings (Electronic Version), Graz, 2010.

[11] J.O. Dalenback, Take off for Solar District Heating in Europe. Polska Energetyka Słoneczna, No 1-4/2009, 1/2010, PTES—ISES, 2010, pp. 9—13.

[12] M. Thirugnanasambandam, S. Iniyan, R. Goic, A review of solar thermal technologies, Renewable Sustainable Energy Rev. 14 (2010) 312—322.

[13] S. Mekhilef, R. Saidur, A. Safari, A review of solar energy use in industries, Renewable Sustainable Energy Rev. 15 (2011) 1777—1790.

[14] V.V. Tyagi, N.L. Panwar, N.A. Rahim, R. Kothari, Review on solar air heating system with and without thermal energy storage systems, Renewable Sustainable Energy Rev. 16 (2012) 2289—2303.

[15] G. Faninger, Solar hot water heating systems, in: Comprehensive Renewable Energy, Solar Thermal Systems: Components and Application, vol. 3, Elsevier, 2012, pp. 419—448.

[16] S.A. Kalogirou, G.A. Florides, Solar space heating and cooling systems, in: Comprehensive Renewable Energy, vol. 3, Elsevier, 2012, pp. 449—480.

[17] R. Shukla, K. Sumathy, P. Erickson, J. Gong, Recent advances in the solar water heating systems. A review, Renewable Sustainable Energy Rev. 19 (2013) 173–190.
[18] J.A. Duffie, W.A. Beckman, Solar Engineering of Thermal Processes, John Wiley & Sons, Inc., New York, 1991.
[19] D. Chwieduk, M. Chwieduk, J. Bigorajski, Modelowanie stratyfikacji ciepła w zasobnikach ciepła słonecznych instalacji grzewczych. Modeling of heat stratification in storage tanks of solar heating systems (in Polish), Build. Phys. Theory Pract. 7 (3) (2015) 17–22.
[20] D. Chwieduk, M. Zawidzki, Energy Storage, Internal report. CAMELIA Project, VI FR, TREN/04/FP6EN/S07.31777/506486, 2006.
[21] D. Chwieduk, Solar assisted heat pumps, in: Comprehensive Renewable Energy, Solar Thermal Systems: Components and Application, vol. 3, Elsevier, 2012, pp. 495–528.
[22] R. Wnuk, M. Jaworski, Badania charakterystyk cieplnych elementów budowlanych akumulujących ciepło, zawierających materiały PCM (Experimental study of thermal characteristics of thermal storage building materials with incorporated PCM). Polska Energetyka Słoneczna 1 (2011) 5–11, (in Polish).
[23] N. Zhu, Z. Ma, S. Wang, Dynamic characteristics and energy performance of buildings using phase change materials, Energy Convers. Manage. 50 (12) (2009) 3169–3181.
[24] F. Kuznik, J. Virgone, K. Johannes, J.-J. Roux, A review on phase change materials integrated in building walls, Renewable Sustainable Energy Rev. 15 (1) (2011) 379–391.
[25] Y. Zhang, G. Zhou, K. Lin, Q. Zhang, H. Di, Application of latent heat thermal energy storage in buildings. State-of-the-art and outlook, Build. Environ. 42 (2007) 2197–2209.
[26] Y. Tian, C.Y. Zhao, A numerical investigation of heat transfer in phase change materials (PCMs) embedded in porous metals, Energy 36 (9) (2011) 5539–5546.
[27] D. Chwieduk, Dynamics of external wall structures with a PCM (phase change materials) in high latitude countries, Energy 59 (2013) 301–313.
[28] G. Bakema, Snijders, B. Nordell, Underground Thermal Energy Storage, State of the Art 1994, IEA Energy Storage Programme Report, IF Technology, 1995.
[29] B. Sanner, A different approach to shallow geothermal energy – underground thermal energy storage (UTES), in: International Summer School on Direct Application on Geothermal Energy, 2002.
[30] H. Schulz, D. Chwieduk, Wärme aus Sonne und Erde Energiesparende Heizungssysteme mit Erdwärmespeicher, Solarabsorber und Wärmepumpe, Okobuch Verlage, Staufen bei Freiburg, 1995.
[31] P. Pinel, C.A. Cruickshank, I. Beausoleil-Morrison, A. Wills, A review of available methods for seasonal storage of solar thermal energy in residential applications, Renewable Sustainable Energy Rev. 15 (2011) 3341–3359.
[32] B.W. Tleimat, E.D. Howe, A solar assisted heat pump system for heating and cooling residence, Sol. Energy 21 (1) (1978) 45–51.
[33] T.L. Freeman, J.W. Mitchell, T.E. Audit, Performance of combined solar heat pump systems, Sol. Energy 22 (2) (1979) 125–135.
[34] J.V. Anderson, J.W. Mitchell, W.A. Beckmann, A design method for parallel solar – heat pump system, Sol. Energy 25 (2) (1980).
[35] R.E. Terrell, Performance and analysis of a series heat pump – assisted solar heat residence in Maison, Sol. Energy 23 (5) (1979).
[36] T. Sakoi, H. Takagi, K. Terakowa, J. Ohue, Solar space heating and cooling with bi- heat source heat pump and hot water supply system, Sol. Energy 18 (6) (1976) 525–532.
[37] P. Sporn, E.R. Ambrose, The heat pump and solar energy, in: Proceedings of the World Symposium on Applied Solar Energy, Phoenix, Arizona, 1955, pp. 1–5.

[38] F.B. Gorozabel Chata, S.K. Chaturvedi, A. Almogbel, Analysis of a direct expansion solar assisted heat pump using different refrigerants, Energy Convers. Manage. 46 (2005) 2614–2624.

[39] Y.H. Kuang, R.Z. Wang, Performance of multi-functional direct expansion solar assisted heat pump system, Sol. Energy 80 (2006) 795–803.

[40] M.N.A. Hawlader, S.M.A. Rahman, K.A. Jahangeer, Performance of evaporator - collector and air collector in solar assisted heat pump dryer, Energy Convers. Manage. 49 (2008) 1612–1619.

[41] E. Frank, M. Haller, S. Herkel, J. Ruschenburg, Systematic classification of combined solar thermal and heat pump systems, in: Proceedings of the EuroSun 2010 Conference, September 29–October 1, Graz, Austria, 2010.

[42] Z. Han, M. Zheng, F. Kong, F. Wang, Z. Li, T. Bai, Numerical simulation of solar assisted ground – source heat pump heating system with latent heat energy storage in severely cold area, Appl. Therm. Eng. 28 (2008) 1427–1436.

[43] M. Koragiorgas, K. Galatis, M. Tsagouri, T. Tsoutsos, A. Botzios-Valaskakis, Solar assisted heat pump on air collectors: a simulation tool, Sol. Energy 84 (2010) 66–78.

[44] O. Ozgener, Use of solar assisted geothermal heat pump and small wind turbine systems for heating agricultural and residential buildings, Energy 35 (2010) 262–268.

[45] X. Ch, L. Lin, Y. Hongxing, Long term operation of a solar assisted ground coupled heat pump system for space heating and domestic hot water, Energy Build. 43 (2011) 1835–1844.

[46] F.M. Rad, A.S. Fung, W.H. Leong, Feasibility of combined solar thermal and source heat pumps in cold climate, Canada, Energy Build. 61 (2013) 224–232.

[47] F. Busato, R. Lazzarin, M. Noro, Ground or solar heat pump systems for space heating: which is better? Energetic assessment based on a case history, Energy Build. 102 (2015) 347–356.

[48] D. Chwieduk, Analysis of utilisation of renewable energies as heat sources for heat pumps in building sector, Renewable Energy, International Journal (1–4) (September–December 1996). Pergamon, Elsevier Sciences, Oxford.

[49] E. Kjelsson, G. Hellstrom, B. Perers, Optimization of systems with combination of ground source heat pump and solar collectors in dwellings, Energy 35 (2009) 2667–2673.

[50] T. Schmidt, D. Mangold, H. Müller-Steinhagen, Central solar heating plants with seasonal storage in Germany, Sol. Energy 76 (2004) 165–174.

[51] B. Sibbitt, D. McClenahan, R. Djebbar, J. Thornton, B. Wong, J. Carriere, J. Kokko, The performance of a high solar fraction seasonal storage district heating system – five years of operation, Energy Procedia 30 (2012) 856–865.

[52] W. Weiss, F. Mauthner, Solar Heat Worldwide Markets and Contribution to the Energy Supply 2010, IEA Solar Heating & Cooling Programme, 2012.

[53] VDI-Richtlinie: VDI 4650 Blatt 1 Berechnung der Jahresarbeitszahl von Wärmepump enanlagen – Elektrowärmepumpen zur Raumheizung und Trinkwassererwärmung, 2009.

[54] https://www.vdi.de/technik/fachthemen/energie-und-umwelt/fachbereiche/energiewand lung-und-anwendung/richtlinien/vdi-4650/.

[55] https://www.ofgem.gov.uk/key-term-explained/seasonal-performance-factor-spf.

[56] http://www.icax.co.uk/Seasonal_Performance_Factor.html.

[57] http://www.engineeringtoolbox.com/heat-pump-efficiency-ratings-d_1117.html.

[58] Z. Yu, J. Ji, W. Sun, W. Wang, G. Li, J. Cai, H. Chen, Experiment and prediction of hybrid solar air heating system applied in solar demonstration building, Energy Build. 78 (2014) 59–65.

[59] D. Kolokotsa, A. Santamouris, T. Karlessi, Passive solar architecture, in: Comprehensive Renewable Energy, vol. 3, Elsevier, 2012, pp. 637–665.
[60] Z. Liu, L. Zhang, G. Gong, Y. Luo, F. Meng, Evaluation of a prototype active solar thermoelectric radiant wall system in winter conditions, Appl. Therm. Eng. 89 (2015) 36–43.

Active solar water heating systems

9

J. Gong, K. Sumathy
North Dakota State University, Fargo, United States

9.1 History

The usage of solar water heaters dates back to 200 BC when Romans used this concept to heat their public baths. The transformation of solar water heating (SWH) from hypothesis to a prototype took place in the year 1767 when Swiss naturalist De Saussure introduced the concept of SWH by trapping the heat within two glass panes. However, the first commercially available SWH, called Climax, was introduced by Clarence Kemp in 1891 [2]. Climax was an integrated system in which the metal tank served the purpose of collector and storage tank. Inspired by Kemp's initial work, by the end of 19th century several researchers started to focus their studies on improving the efficiency and durability of SWH. William Bailey (1909) introduced a new form of an SWH system in which the solar collector and the storage tank are separated from each other. This model utilized the thermosyphon principle for the first time in the history of SWH to circulate the water between collector and the storage tank [3]. Although such passive SWH systems are less expensive because of the simplicity, active SWHs are widely used at present and proven to be most cost-effective, especially in cold climates. Until the 1930s hot water for domestic purposes was produced using coal-fired boilers. In the second half of the 20th century, many countries such as Israel, Japan, China, Australia, South Africa, India, and Greece started experimenting with SWH and developed their own indigenous SWH systems. Between 1960 and 1980 Japan had approximately 1.5 million active SWH system installations whereas Israel had approximately 2.5 million SWH system installations. In the next few decades, the growing awareness among people regarding the pollution and health hazards posed by the use of conventional energy resources supplemented by the change in government's policy in response to the Kyoto Protocol on climate change helped the active SWH market to attain remarkable growth.

Currently active SWH has been spread all over the world. In recent years, the prominence for SWH has increased, mainly because of the rising costs of fossil fuels. By the end of 2012, a total of 384.7 million square meters of collector area was in operation across the world, corresponding to an installed capacity of 269.3 GWth [4] reported by the International Energy Agency (IEA). The vast majority of the total capacity in operation has been installed in China (180.4 GWth) and Europe (42.8 GWth), which together account for 83% of the total installed capacity. The remaining installed capacity was shared between the United States and Canada (17.2 GWth); Asia excluding China (10.3 GWth); Latin America (7.4 GWth); Australia and New Zealand

Advances in Solar Heating and Cooling. http://dx.doi.org/10.1016/B978-0-08-100301-5.00009-6

(5.4 GWth); the Middle Eastern/North African countries Israel, Jordan, Lebanon, Morocco, and Tunisia (4.9 GWth); and the sub-Sahara African countries Mozambique, Namibia, South Africa, and Zimbabwe (1.0 GWth). The share of the total installed capacity is detailed in Fig. 9.1. The data show that the total number of water-based active solar thermal systems in operation was approximately 78 million in the year 2012. Among these 78 million SWH systems, 8% are accounted for by swimming pool heating, 78% for domestic hot water heating, and 9% exclusively for large-scale applications.

9.2 Overview of technologies for active solar water heating systems

Although solar water heating systems (SWHS) have been in use for centuries, with today's technological advances, SWH technologies can be operated efficiently and affordably in any climate. Depending on the heating needs the systems are specifically designed for various climatic and geographical areas of the country. Several types of basic components are listed here:

1. *Collector:* Flat-plate collector (FPC), evacuated tube, parabolic dish/concentrating collectors
2. *Heat transfer fluid (HTF):* Air, water, glycol/water mixtures (ethylene and propylene glycol are antifreezes), hydrocarbon oils, silicones, refrigerant/phase change fluids
3. *HX design:* Coil-in-tank, shell-and-tube, tube-in-tube
4. *Controls:* Differential controller, photovoltaic (PV) powered controller
5. *Auxiliary heating:* Gas, electric
6. *Storage tank:* One-tank system with back-up heating element, two-tank system (in which solar hot water storage is connected to a conventional gas or electric water heater)

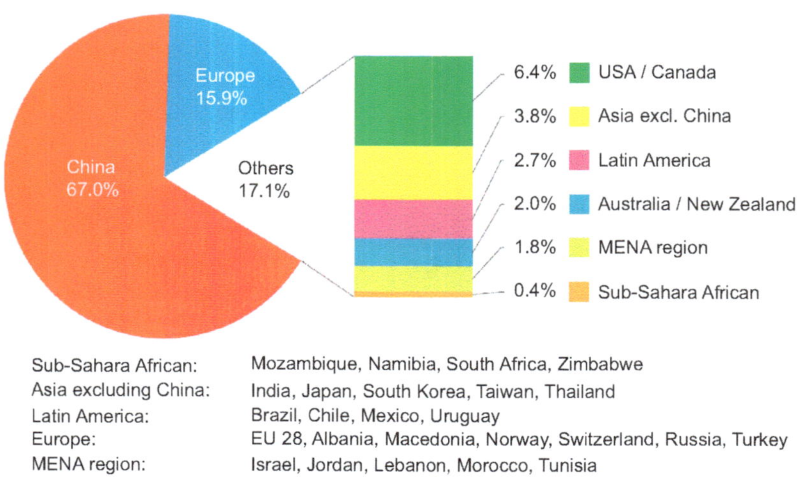

Figure 9.1 Share of the total installed capacity in operation (glazed and unglazed and air collectors) by economic region at the end of 2012 [5].

SWHS can be broadly categorized into five major types [6]: (1) direct circulation systems, (2) indirect water heating systems, (3) drain-back systems, (4) air systems, and (5) pool heaters.

9.2.1 Direct (open-loop) solar water heating

Open-loop systems use pumps to circulate water through collectors and operate at standard line pressure. These systems are suitable for single-application domestic hot water supply and are appropriate in areas that do not freeze for long periods and do not have hard or acidic water. They are simple to design and operate and the least expensive to install among all active systems. No HX is required, which allows efficient heat transfer directly to the water. Water used as the working fluid is heated within the collector to the temperature range of 50–60°C. These systems are also simple to add capacity and integrate with existing systems if there is an increase in the hot water demand. In general, direct SWHS produces the highest operating performance because there are no nighttime heat losses from hot water stored on the roof as in a passive system nor is any efficiency lost through a heat exchange process as in a closed-loop system. However, some stored heat can be lost when the system recirculates.

Direct systems are not feasible to operate in freezing conditions because it leads to pipe damage. The primary freeze protection needs electricity or battery backup. For instance, a variable capacity direct expansion solar-assisted heat pump system can be used for domestic water heating purposes. As shown in Fig. 9.2, a bare solar collector can be used as the source (as an evaporator) for the heat pump system. The coefficient of performance of such systems can be enhanced extensively by lowering the speed of the compressor when the ambient temperatures were higher [7]. Hence, such systems perform better in summer compared with winter.

A drain-down system is one modification of the direct circulation. When there is no danger of freezing, this open loop is used where collectors are filled with domestic water under supply pressure. Once the system is filled, a differential controller operates

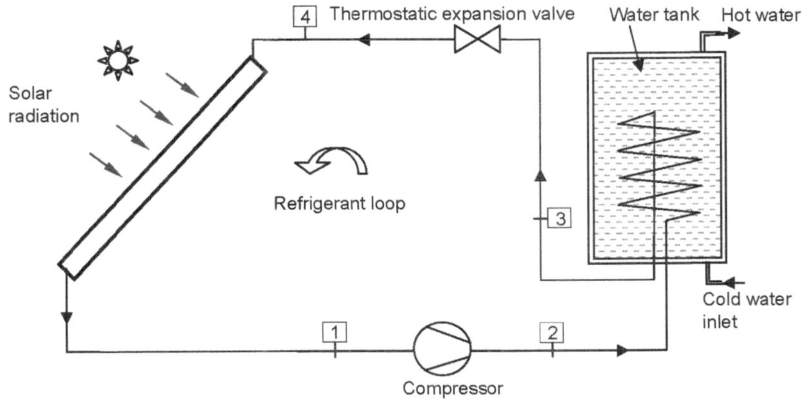

Figure 9.2 Direct expansion solar-assisted heat pump system.

a pump to move water from the tank through collectors. A drain-down valve provides the freeze protection function. While the drain-down valve is activated by the controller, it will isolate the collector inlet and outlet from the tank. It simultaneously opens a valve that allows water in the collector to drain away. A vacuum breaker is always installed at the top of the collectors to allow air to enter the collectors at the top so that water can drain out the bottom.

9.2.2 Indirect (closed-loop) solar water heating

Closed-loop active systems pump HTFs such as a mixture of glycol and water antifreeze through collectors. HXs are utilized to transfer the heat from the fluid to the water stored in the tanks. Pumps circulate a nontoxic, nonfreezing HTF through the collectors and HX. These systems are popular in cold climates because of excellent freeze protection. However, the antifreeze must be recharged on a 3- to 5-year basis depending on antifreeze quality and system temperatures. The system is generally more complicated than an open-loop system because either a tank with a heat exchange coil or an external HX is required. Because an HX is required the collector loop will run at slightly higher temperature than an open-loop system. The collector loop also needs to be pressurized (8−12 psi).

These close-looped systems use two circulation loops to aid heating: (1) the closed-collector loop and (2) the open storage tank loop. The working fluid is circulated within the closed collector loop to gain the heat and pass the heat through an HX to the potable water that flows in an open loop to the storage tank. In the closed loop several different types of working fluids can be used, such as air, water, glycol/water mixtures, hydrocarbon oils, and refrigerants/phase change fluids. In addition, the HX can be either an internal system or external system. In an internal system (Fig. 9.3) an

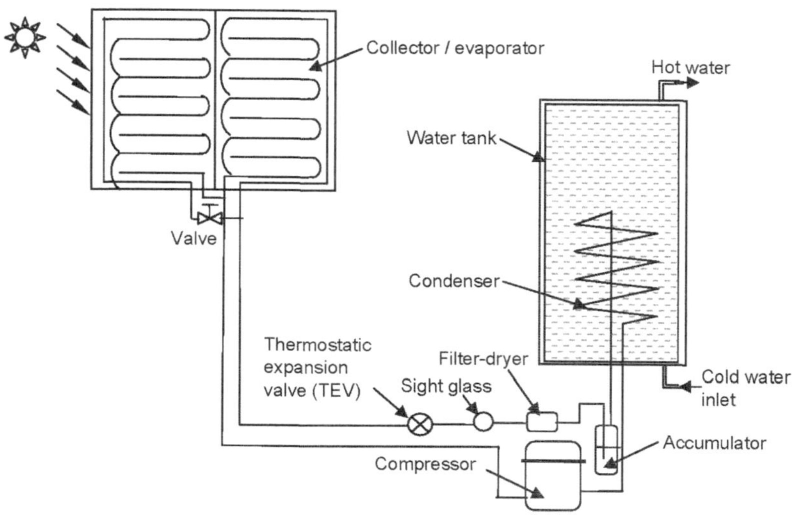

Figure 9.3 Schematic diagram of indirect water heating presented by Kuang et al. [8].

HX is placed inside of the water storage tank whereas in an external system the HX is placed outside of the storage tank. These systems are suitable for single and multiple SWH application systems.

9.2.3 Drain-back systems

Drain-back systems are modified indirect systems that are completely nonpressurized and use pumps to circulate water through the collectors. Because the water in the collector loop drains into a reservoir tank when the pumps are off, this is a good system for colder climates. In addition, it allows the systems to turn off when the water in the storage tank becomes too hot.

The drain-back tank has to be appropriately sized so that all of the water in the collector and lines can drain down into it when the pumps turn off. The water in the system is separate from the domestic water; therefore it needs an HX to transfer the heat from the collected water to the domestic water. Because water is used as HTF, it never needs to be changed like pressurized antifreeze systems. Most plumbing codes do not require double-wall HXs for drain-back systems using distilled water. The system is simple and has no check valves, no air vents, no pressure gauges, and no expansion tanks. However, these drain-back systems use larger piping (3/4-in. copper pipe) and insulation, leading to a higher components cost compared with an active open direct loop SWHS for residential water heating.

9.2.4 Air systems

In an indirect water heating system, air can be used as working fluid to avoid extreme temperatures such as overheating in summer or freezing in winter. A fan is used to circulate air through the tubes and the concentric air-to-water HX that delivers this heat to water in the horizontal storage tank. The advantages of this system are that it is noncorrosive and requires less maintenance. In addition, it can be used in very low temperature conditions. However, the SWH system that uses air as a working fluid generally requires a large area because of the air handling unit. A typical arrangement of a solar air heating system along with a pebble bed storage unit is illustrated in Fig. 9.4. Air is passed through the collector duct, where it gets heated up and rejects the heat to the water with the help of an HX. Fans and dampers are incorporated to aid the system operation. This system can supply hot water up to 80°C. One of the main drawbacks for this system is low heating capacity, and the formation of leaks over years could lead to a gradual reduction in performance.

The solar-assisted heat pump/heat pipe technology for residence water heating has been developed for more than half a century. In terms of high costs, inflexible applications, and unreliable operation, the air-to-water heat pumps and other heat pumps for water heating are not yet competitive. Another way to affect residential water heating is to use a photovoltaic thermal (PVT) system. Most PVT systems are unglazed and use air or water to cool down the cells and use the heat for water heating. Most studies show significant increased performance for these combined systems [9,10].

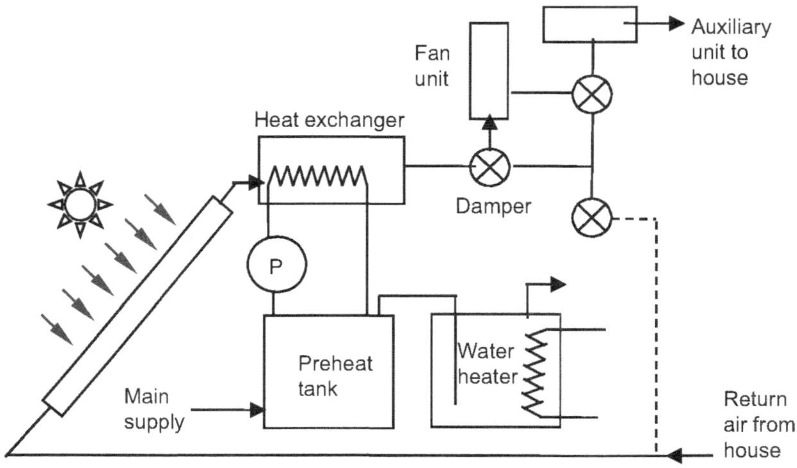

Figure 9.4 Schematic diagram of standard air system configuration [11].

9.2.5 *Pool heaters*

These systems are not used for heating domestic water. Pool collectors are usually made up of many parallel poly pipes sitting on a roof near the pool. The pool water is first pumped through the filter before being circulated through a circulation pump. The controller operates a valve to divert and pump the pool water through the collector. When not operating, the collector water drains back into the pool. These systems are exclusively used for swimming pool heating and more commonly used in warm climate zones.

9.3 Economics and energy efficiency of active solar water heating systems

The technoeconomic evaluations play an important role in establishing a strong market strategy for SWHS and persuading necessary information for energy policy decisions. A comprehensive nomograph has been developed to determine the potential number of households who can use water heating systems in any given country [12]. Factors such as the availability of shadow-free sunshine hours, space availability, practical awareness of system operation, cost of the SWHS, financial constraints, and motivation play a role in the establishment of water heating systems in the residential buildings. A simple framework for financial evaluation of SWHS has also been developed [12] that takes into account the net present value of the system, the benefit-to-cost ratio, the internal rate of return, and the payback period. On the basis of a thorough investigation, the study reported that the operation and maintenance cost of the system as well as capacity utilization play as key factors in the penetration of this solar water technology in the domestic market. A generic framework can be used to provide accurate information about the SWHS required for any specific area, and policy-makers could henceforth use this as a tool to track and promote SWHS nationwide [13].

Another simple method to evaluate the effectiveness of SWHS is to assess the energy savings through SWH integrated buildings. A study has reported that the current technical potential of SWH in the United States is estimated to be approximately 1 quad (1.055 × 1018 J) of primary energy savings per year [14]. To more effectively implement the cost-benefit analysis, the concept of number of effective solar days and effective solar radiation (ESR) is used instead of using the total annual solar radiation parameter because it may overestimate the amount of energy benefits [15]. For instance, in different geographical regions of Taiwan this model estimated the ratios of ESR to total annual solar radiation to be in the range of 82–89% and the payback periods from 6 to 15 years.

The popular f-chart method [16] and the demographic data for a target region can also be used to assess the estimated power savings due to the installation of domestic SWH systems. Financial profits can be estimated using economic indices, and the profit potential of the manufacturing may be estimated by comparing all of the economic costs and benefits regarding SWH. These costs include material and labor costs as well as taxes. The market prices of an SWH product generally depend on the size of the system, its brand name, and the place of purchase. A comparison of cost and typical characteristics of SWH systems used by the US and China markets is presented in Table 9.1.

Table 9.1 Comparison of costs and other general characteristics of solar water heating market between United States and China [17]

Characteristics	US market	China market
Typical installed cost (domestic, 2–4 people)	\$5000–10,000	\$300–1000
Most common technology	Indirect (with pump)	Thermosyphon (no pump)
Tank capacity	80 gal	30 ~ 50 gal
Collector sizes	~50 ft^2	~20 ft^2
Back-up system	Conventional electric/gas	Electric heating element
Quality	Highest, SRCC certified	Low, many not certified and shorter system life
Typical installation	Collectors on pitched roof, indoor tank, complex design, building not designed for SWH, limited SWH experience, high labor costs	Collectors and tank on the roof (flat or pitched type roof), simple system, experienced installers, low labor costs
Market volume	30,000 installations/year	6,000,000 installations/year

SRCC, Solar Rating & Certification Corporation; *SWH*, solar water heating.

To obtain technoeconomically feasible standardized water heating systems, it is also important to evaluate the viability of the chosen SWH system design. Such studies will serve as one of the promotional measures in deploying them further in the market. The performance of a typical SWH system can be evaluated by various performance parameters such as heat gain, temperature increase, and collector/overall efficiency (Table 9.2). SWH performance can also be evaluated by the use of energy factors—similar to the efficiency rating used for conventional electrical/gas-operated water heaters (Table 9.2). The solar water heater's energy efficiency can also be measured based on the solar energy factor (SEF) and solar fraction (SF). The SEF is defined as the energy delivered by the system divided by the electrical or gas energy put into the system. The higher the number, the more energy efficient. SEFs generally range from 1.0 to 11. The SF refers to the portion of the total conventional hot water heating load (delivered energy and tank standby losses). The higher the SF, the greater the solar contribution to water heating, which reduces the energy required by the back-up water heater. The SF varies from 0 to 1.0. Typical solar factors are 0.5–0.75. Hence, it is desirable to have higher SEF and SF values for a given SWH system. However, it is essential to choose an SWH system not just based on these factors, but it is also important to consider the size and overall cost of the system. As such, the SWH systems are cost-competitive with electric water heaters, they can be cost-competitive with natural gas-fired water heaters, and they have a design life of approximately 20–25 years (Table 9.3) [18].

9.3.1 Performance improvement of basic solar water heating components

Although SWH is considered to be a matured technology, to make SWH more competitive and effective, research is currently focused on improving the main components such as the collector, HX, and storage tank. In addition, different working fluids have been attempted to improve the system efficiency and make the system functional under different operating conditions. Significant studies on the design modifications are reported in the respective subsections.

9.3.1.1 Solar thermal collectors

The key component of an SWHS is the solar thermal collector, which can be viewed as an HX that collects heat by absorbing sunlight and transferring it to a fluid flowing through the collector. The efficiency of an SWH system mainly depends on the effectiveness of the collector, which in turn is influenced by the collector type and its size. Thus most work has been concentrated in improving the performance of the collector—more specifically the absorber plate design and glazing material. The most common solar collectors used for the purpose of water heating are FPCs, evacuated-tube collectors (ETCs), and concentrating collectors. The collector for a particular type of SWHS can be selected based on the heating requirements and the location.

Table 9.2 Solar water heater performance based on collector type

Collector	Working fluid	Water temperature (°C)	Collector efficiency (%)	Remarks
FPC direct circulation	Water	40–60	50–80	• High-performance collector • Limited to low-temperature applications
FPC indirect circulation	R-12/R-22	50–60	40–60	• Suitable to cold regions • Possible for drain-back system (protects from freezing/overheating)
ETC direct circulation	Water	30–40	60–80	• Can be used in different climatic zone • Difficult to maintain vacuum environment
ETC indirect circulation	R-134a	37	76	• Noncorrosive • No freezing issues
Heat pipe indirect circulation	R-134a	70–85	40–52	• Can be used year-round in the cold climate • Can be easily combined with existing pipeline
CPC direct circulation (V-trough absorber)	Water	50–80	65–75	• High thermal stratification and stability • High labor and material cost
Fresnel lens direct circulation	Water	80–250	55–65	• Concentration is affected through reflection • Lens can be prepared by biopolymer materials • Relatively inexpensive
Parabolic trough direct circulation	Water	150–300	25–45	• 220% more efficient than FPC • Single axis tracking • Large-scale industrial application
Paraboloidal dish indirect circulation	Molten salt	200–500	60	• Requires dual-axis tracking; not recommended for domestic water heating • Suitable for power generation • Manufacturing/maintenance costs are high

FPC, flat-plate collector; *ETC*, evacuated-tube collector; *CPC*, compound parabolic concentrator.

Table 9.3 **Comparison of water heaters [18]**

High-efficiency water heater type	Energy saving vs. minimum standards (%)	Best climates	Expected energy savings over equipment lifetime	Expected lifetime	Major advantages
High-efficiency storage tank (oil, gas, electric)	10–20	Any	Up to $500	8–10 years	Lowest initial cost
Demand (tankless) using gas or electricity	45–60	Any	Up to $1800	20 years	Unlimited supply of hot water
Heat pump	65 (compared with electric resistance)	Mild-hot	Up to $900	10 years	Most efficient electric fuel option
Solar with electric backup	70–90	Mild-hot	Up to $2200	25 years	Largest energy saving using a renewable energy source

Flat-plate collector

The FPC is most widely used in SWHS because of its low initial cost and capability of absorbing direct and diffuse radiation. One of the commonly used designs in FPCs is the parallel-tube collector layout. Although this design is widely used, some of the disadvantages are nonuniform temperature distribution over the absorbed plate surface, unequal distribution of the working fluid through the collector risers, and the increase in the collector's heat loss caused by the higher temperature of the absorber plate in low flow rate conditions. To overcome these issues a serpentine-tube collector was introduced. This design was typically used to compensate for low flow rate conditions; the design enables the total mass flow rate to pass through the tube, increasing the heat transfer coefficient [19]. It was found that the serpentine geometry showed better thermal performance than the parallel one. In an attempt to further improve the thermal performance of a parallel-tube collector, two parallel-tube collectors were connected in parallel and this system was compared to a serpentine-tube collector having an equivalent exposed area [19]. It was found that the efficiency was approximately 6% more than a serpentine-tube collector and 10% more than a single parallel-tube collector under similar ambient conditions. Other geometric designs [20,21] include a fin-and-tube collector, in which fluid is circulated through a corrugated channel that is in direct contact with the absorbing surface.

In an FPC the absorber plays an important role in absorbing the solar radiation and conducting heat to the HTF. The absorber is available in different profiles, among which the parabolic profile is the most efficient [22]. Other profiles include the rectangular profile [23] and recto-trapezoidal profile [24]. These absorber plates are usually made of highly conductive metals such as copper or aluminum. However, corrosion is somewhat of a concern. In recent years there has been an upswing in the usage of polymer-based FPCs because of their noncorrosive nature, light weight, and ease of manufacturability. Polymer-based collectors are flexible to accommodate the expansion of tubes/passages under freezing conditions; hence water can be used as an HTF in polymer-based collectors with any antifreeze. Moreover, the fabrication cost for producing polymer components is less compared with metallic components. For example, the cost of a nylon-based absorber is approximately 0.8 times the cost of a copper-based absorber [25]. Although polymer collectors are resistant to corrosion, they do have several limitations, such as low thermal conductivity and they can be used only for moderate temperature applications. To overcome these limitations, to a certain extent, an extruded parallel-plate absorber design has been proposed [26]. It is one of the most promising designs, which comprises a pair of parallel plates with water flowing between the two plates. The gap between the plates can be varied from a few millimeters to a few centimeters according to the design parameters.

Evacuated-tube collector

Compared with FPCs, ETCs have better performance in producing high temperatures. ETCs minimize the heat losses due to convection and radiation [27]. Hence ETCs have proven to be a natural fit for colder regions although they are expensive. One of the important design factors for the glass ETC is the shape of the absorber tube [28]. Different absorber tube designs exist, such as finned tube, a U-tube welded inside of a circular fin, and a U-tube welded on a copper plate or placed within a rectangular duct. Among these four designs a U-tube welded inside of a circular fin is shown to have the best performance. Apart from the previously listed designs, a double glass tubular absorber is found to be more efficient, which could absorb radiation from all directions. The inner tube is selectively coated to further maximize the solar absorption. Compared with conventional FPCs, the efficiency of ETCs ranges between 70% and 80%. As such, the difference in efficiency could become more pronounced when operated in colder regions. For domestic water heating, heat pipe ETCs and U-tube glass ETCs are the two most widely used because of their simple structure and tolerance to high-pressure conditions.

Concentrating collector

FPCs and ETCs are widely used in solar thermal applications, mainly to provide low to intermediate temperatures (20–120°C). However, to attain temperatures above this range, concentrators such as parabolic collectors, paraboloidal collectors, or Fresnel reflectors have to be used to maximize the incident radiation to result in high temperatures. For instance, compound parabolic concentrators are nonimaging concentrators, having the ability to reflect most of the incident radiation to the absorber, attaining temperatures on the order of 150–200°C with a reasonable efficiency. The concentrators are often used to provide industrial process heating or cooling such as air conditioning.

9.3.1.2 Heat transfer fluid

HTF plays a very significant role in indirect (closed-loop) SWH. HTF acts as a medium for transporting heat that is collected by the solar collectors to the actual water that is required to be heated up with the help of an HX. The selection of an HTF for an SWHS depends on several factors, which include thermodynamic and heat transfer properties of HTF as well as the location. For the successful operation of such a solar water heater, careful selection of the working fluid is essential. The selected fluid must have most of the desirable properties from the viewpoint of thermodynamics and heat transfer, such as coefficient of expansion, viscosity, specific heat, freezing point, boiling point, and flash point. Air and water have been commonly used as HTFs in SWHS. Air has certain advantages compared with water, such as its noncorrosive nature and it is not prone to boiling/freezing. However, because of its very low heat capacity, it could only be utilized for low-temperature applications and not for domestic water heating purposes. On the other hand, water's high specific heat, low viscosity, nontoxicity, and less expensive features have positioned water to be the most popular working fluid in SWHS. However, its corrosive nature (especially at high temperatures) as well as freezing and scaling issues pose a challenge in collector tubing and plumbing. To overcome the relatively high freezing point of water, a glycol additive is used along with water to act as an antifreeze [29].

In indirect (closed-loop) SWH systems, chlorofluorocarbon refrigerants are more commonly used as HTFs because of their stability, nonflammability, low toxicity, noncorrosiveness, and low freezing point. Specific examples include R-11, R-12, R-13, R-113, R-114, and R-115 Natural fluids have been considered as long-term HTFs because they are halogen-free, environmentally benign, and have very low or near-zero ozone depletion potential (ODP) and global warming potential (GWP) [30]. Typical natural HTFs include propane (R-290), butane (R-600), isobutane (R-600a), propylene (R-600), ammonia (R-717), and carbon dioxide (CO_2; R-744). Research on CO_2 has gained momentum with an objective to investigate the possibility of using CO_2 in a heat pump SWHS and particularly evaluate the performance when operated under transcritical conditions [31–33]. CO_2 is a promising natural fluid because it is nonflammable, noncorrosive, and nontoxic, and it has a low critical point (31.1°C at 73.7 bar). A CO_2 heat pump water heater may produce hot water temperature up to 90°C without any operational problems, and the primary energy consumption can be reduced by more than 75% compared with electrical systems.

9.3.1.3 Storage tank

The storage tank is yet another essential component of the SWH system. It also plays a major role in dictating the system performance. Storage tanks are generally constructed using steel, concrete, plastic, or fiberglass to store hot water [33]. Steel tanks are most commonly used because they are easy to install in the required sizes and are generally free of corrosion [27]. One major problem experienced with the storage tank is the heat losses due to the mixing of hot and cold water. To minimize the mixing and aid thermal stratification, several designs of storage tanks have been proposed. Thermal stratification can be attained by using inlet stratifiers combined with low-flow operation in the solar collector loop. Flexible fabric piping is generally used as an inlet pipe to affect stratification because they can expand or contract, equalizing the pressure in the pipe. There are several different fabric pipes, including Nylon, filament polyester, spun polyester, and acrylic [34]. Compared with the conventional nonflexible inlet stratifiers, the two fabric layered stratifier with a spacing of 10 mm enhances the thermal stratification.

To further enhance stratification, different inlet designs of the storage tank exist. The effect of 12 different obstacles on the thermal stratification in a cylindrical storage tank has been tested by predicting the temperature distribution within the tank [35]. Fig. 9.5 shows the 12 different obstacle geometries analyzed in the study. Results had shown that placing an obstacle in the tank provided better thermal stratification compared with the case with no obstacle. In addition, the obstacles 7, 8, and 11, which had a gap in the center, resulted in better thermal stratification than those that had a gap near the tank wall. The study had concluded that, in terms of hot water supply, obstacle 11 provided the best thermal stratification in the tank.

Thermal stratification can also be improved by incorporating baffle plates within the tank. Baffle plates help in guiding the fluid flow and effectively decrease the impingement of the fluid. Diverse baffle plate designs include hemispherical baffle plates, large flat baffle plates, and raw pipes, and they were tested at different discharge flow rates. It was found that the conical baffle plates followed by the flat baffle plates resulted in

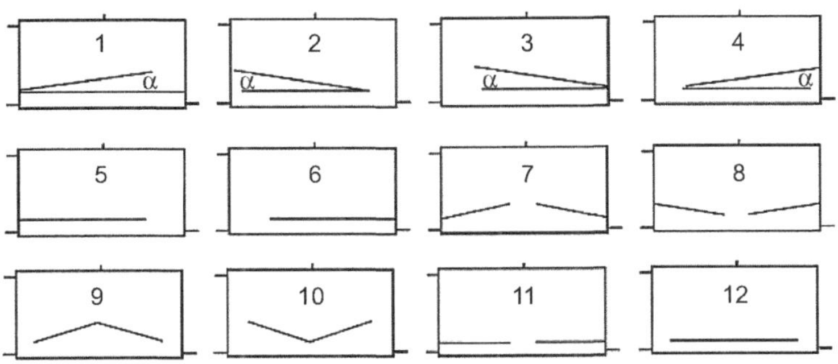

Figure 9.5 Obstacles geometries and their assembly in the tank [35].

the best performance because they progressively increased the flow rate in the tank within the space between the tank walls.

9.3.1.4 Heat exchangers

HXs are widely used in indirect SWH systems because they could use a freeze-resistant fluid in the collector to facilitate the freeze protection technique and to separate potable water from working fluid in the collector. HXs are basically classified into internal HXs and external HXs. Internal HXs generally include immersed coil tube HXs [36], finned tube HXs, and plain tube HXs [37], out of which the plain tube HX is easier to design and install. However, the main disadvantage of an internal HX is the formation of scales, which greatly deteriorates its efficiency. The most common type of external HX is the shell and tube HX [38]. In this type, the hot and cold fluids transfer heat based on a cross-flow method. External HXs have higher heat transfer capacity and are well insulated when compared with internal HXs. In addition, it is possible to heat multiple storage tanks at a given time. However, the external HXs are expensive compared with internal HXs.

A few other HXs exist that are not that popular; however, research has been performed on load-side immersed HXs [39] and mantle HXs [40–43] in which hot working fluid flows through the annulus around the cylindrical storage tank, where it rejects the heat. It helps in providing a large heat transfer area for a given volume of collector fluid flow in the mantle. This makes the mantle HX one of the most efficient and simplest ways of facilitating high HX effectiveness in promoting thermal stratification. One of the prominent advantages of this design is that the hot water tank and the HX are combined into one unit [44].

9.4 Applications of active solar water heating systems: case study

Active SWH technology encompasses a wide range of applications such as water heating, space-heating/cooling, and air conditioning in the residential (low temperature), commercial (medium temperature), and industrial (high temperature) sectors. Some of the basic components such as solar collectors and storage tanks remain in principle the same for most types of solar thermal applications.

9.4.1 Domestic hot water

Solar domestic hot water heating technology is mature and commercially available in many countries. Domestic uses of hot water include cooking, cleaning, bathing, and space-heating. Typical systems used for single-family homes in North America have hot water storage with volumes of approximately 300 L and a collector area between 4 and 6 m^2, and they can supply 60–90% of the annual hot water demand depending on the type of collector and local climatic conditions [45]. An energy-efficient household might use less than half of that amount.

9.4.2 Space-heating

In contrast to SWH, solar space-heating requires significantly larger solar collector areas. They require a 20- to 40-m^2 collector area, and the volume of the hot water store is in the range of 2–4 m^3. Solar space-heating systems are usually combined with water heating, and they are sized to accommodate both uses. For example, radiant floor heating is an ideal home heating system. Solar-heated liquid circulates through pipes embedded in a thin concrete slab floor, which then radiates heat to the room. The slab is typically furnished with tile. Radiant slab systems take longer to heat the home than other types of heat distribution systems. However, once they are operating, they provide a consistent level of heat. This technology shows great promise for further market success.

9.4.3 Space cooling

Solar cooling/air conditioning of buildings is an attractive idea because the cooling loads and availability of solar radiation are in phase. In addition, the combination of solar cooling and heating (Fig. 9.6) greatly improves the use factors of collectors compared with heating alone [46]. Solar air conditioning can be accomplished by three types of systems: absorption cycles, adsorption (desiccant) cycles, and solar mechanical processes. Solar thermal cooling is an important market in countries and regions with high cooling demands.

9.4.4 Pool heating

Solar pool heating is the most widely used application of an active SWHS. These systems are commonly configured as drain-back systems where the water drains into the pool when the water pump is switched off. Compared with conventional natural gas

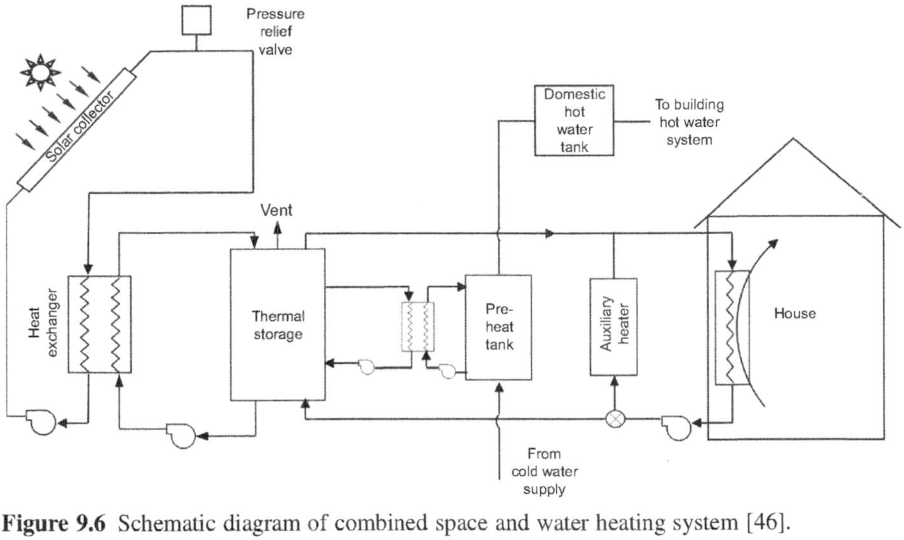

Figure 9.6 Schematic diagram of combined space and water heating system [46].

and electric heaters, SWHS for swimming pools are cost-competitive and cost-effective because swimming pools require a low-temperature heat source, which a relatively small solar collector can easily provide. A simple rule of thumb for the required panel area needed is 50% of the pool's surface area [47].

9.4.5 Commercial applications

Typical commercial applications include space-heating/cooling and water heating. Building types that are particularly well suited include apartment buildings, government facilities, hotels, resorts, etc. SWHS are ideal because the load demand usually complements the peak operation of the solar system. In addition, they can dramatically reduce water heating expenses while also promoting business sustainability. SWH systems can offset most heating costs for up to 25 years or longer with little maintenance. The more hot water a system has to produce, the more cost-effective it becomes. Large systems must be properly designed and installed.

9.4.6 Industrial applications

Although popularly used to meet the domestic hot water supply, SWHS are also utilized to serve the hot water demand in the industrial sector. The relatively modest temperatures that can be achieved using solar heating technologies (ie, temperatures below 180°C) are potentially useful for many processes in the agriculture and industry sectors (including in the food, beverage, tobacco, textile, and chemical industries). In particular, cleaning processes (eg, for containers and equipment in the food and chemical industries and for laundering in the textile industry) and subsequent drying processes often require temperatures in the range attainable with solar heating. Furthermore, it is often possible and cost-effective to retrofit solar heating technology because SWH equipment can be installed on existing water storage tanks. Other processes that can make use of solar heating include evaporation, sterilization, pasteurization (which needs temperatures below 105°C), and preheating of boilers. The cost of the industrial SWHS varies depending on the thermal capacity, storage convenience, and pressure requirement for the system. These systems require comprehensive design to be installed for optimal performance.

9.4.6.1 Case study

If properly designed, an SWHS could be utilized not only to serve the domestic hot water supply but could also provide space-heating in winter and provide space-cooling during summer by energizing a liquid desiccant air-conditioning system or other type of cooling system. A case study on such multipurposed SWHS was implemented in a kindergarten building in Beijing (cold climate) [48]. An SWHS was designed to serve three applications—space-heating, space-cooling, and domestic hot water—throughout the year. Evacuated tubular collectors were utilized to heat the water in the storage tank. On the basis of the law of diminishing marginal utility, the collector area and storage volume were optimized. The optimal values of the

Table 9.4 Information about the solar water heating system

Location	Beijing (East 112°23′, North 39°54′)
Hot water load	Load profiles of space-heating, space cooling and domestic hot water are shown in Fig. 9.7
Collector	Evacuated tubular collectors, the area to be optimized $F_R(\tau\alpha)$: 0.80 and $F_R U_L$: 0.99 W/m^2 °C
Storage tank	Cylindrical with (h/d) = 1, well mixed, the volume to be optimized Initial water temperature: 15°C Minimal working temperature: 65°C Maximal working temperature: 95°C Tank loss coefficient: 0.30 W/m^2 °C

collector area and storage tank volume were determined to be 194 m^2 and 89 m^3, respectively. Salient details about the system design are listed in Table 9.4, and Fig. 9.7 shows the hourly load profiles of these three heating applications.

The performance of this multipurpose system is summarized in Table 9.5. The system could effectively harvest solar energy and could attain an SF of approximately 73%. However, the effectiveness of solar heat gain was only marginal (59.3%), which indicates that only approximately 60% of the harvested solar heat gain is effectively utilized and the rest is wasted as heat loss or heat discharge caused by the energy mismatch between the heating load demands and solar energy supply. The energy consumption of the additional solar circulation pump is also taken into account while evaluating the actual energy saving of the system. The energy payback time of this multipurpose SWH system was predicted to be approximately 7 years. On the basis

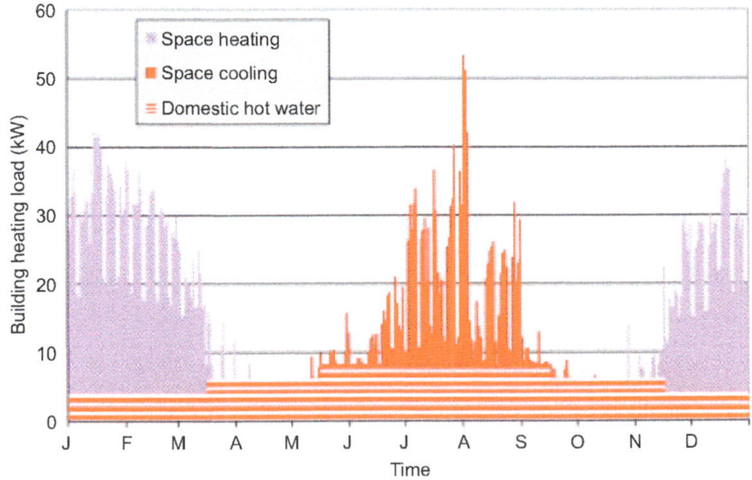

Figure 9.7 Hourly load profile of three heating applications [48].

Table 9.5 **Energy performance of the optimized system**

	Harvested solar heat gain (Q_{solar})	Building heating load (Q_{load})	System energy saving (Q_{saving})	Annualized embodied energy of the system (Q_{EE})
Annual performance data	210,933 kWh	170,529 kWh	114,622 kWh	37,153 kWh
Solar fraction	$F = Q_{tank}/Q_{load} = 73.4\%$			
Effectiveness of solar system	$E_{solar} = Q_{tank}/Q_{solar} = 59.3\%$			
Energy payback time (EPT)	$EPT = EE_{system}/Q_{saving} = 6.5$ years			
Annualized net energy saving	$Q_{net} = Q_{saving} - Q_{EE} = 77,469$ kWh			
Life-cycle net energy saving	$Q_{life\text{-}cycle} = 20 \times Q_{net} = 1,549,378$ kWh			

of the 20-year life span of the system, it is predicted that the annualized net energy saving and life cycle net energy saving would be approximately 77,469 kWh and 1.55 million kWh, respectively.

9.5 Conclusions and future trends

Active SWH is a mature technology and has been successfully deployed in several countries for more than 30 years. Governments in these countries have been instrumental in mandating their deployment, especially into new construction sites to eliminate the entry barriers. In addition, considerable advances in the absorber coating, HXs, and absorber plate design have increased the reliability of SWHS. However, key challenges exist, including high upfront installation costs, complex integration process, and competition with heat pumps and PV panels.

To address the challenges, innovations aim to make SWHS thinner, cheaper, and more durable and to better integrate them to rooftops. One possibility for enhanced penetration of active solar heating systems is through district heating. In these systems the heat gathered by the solar collectors is fed into a district heating network either directly (without heat storage) or via large heat storage facilities, which are charged with solar heat during the summer season and discharged in late autumn and winter. Likewise, these large-scale systems could be used to support district cooling. At present, only approximately 1% of the market consists of large systems connected to district heating (>350 kW_{th}), and the rest are small systems (3–10 kW_{th}) such as domestic hot water preparation in single and small multifamily homes [1]. As a complement to solar heating, solar cooling can provide an effective solution to reduce the peak electricity consumption as it operates when the cooling demand is highest. Such systems have to be integrated into the construction process at the earliest stages of building planning. The walls can function as a component of the active heating and cooling systems, supporting thermal energy storage through the application of advanced materials (eg, phase-change materials). Although solar cooling has been growing rapidly in recent years, the market is still in a nascent stage.

Future residential heating demand is highly dependent on technical factors and demographic development. In general, space-heating demand is expected to remain stable and decline in developed countries because of improved home insulation and energy efficiency whereas the water/space-heating and cooling demand is expected to increase in developing countries because SWHS are often the most economical option. Solar combi-systems that combine water and space-heating have been developed to provide hot water and space-heating and consequently require significantly larger solar collector areas. Such systems for residential applications operate at temperature ranging between 20°C and 90°C and reduce fuel consumption by 50–70% for hot water and by 30–60% for space-heating. For industrial applications, solar heating is mainly used for low-temperature processes, ranging from 20°C to 100°C. Although most combi-systems are experimental and relatively small scale, great potential exists for market and technological developments.

Tapping into this potential would provide a significant solar contribution to future industrial energy requirements. The use of active SWHS can certainly have a great effect on the economic, environmental, and energy conservation perspectives.

References

[1] Solar heating and cooling for residential applications. January 2015.
[2] Perlin J, Butti K. A golden thread. London: Marion Boyars Publishers Ltd; 1981.
[3] Perlin J, Butti K. Early solar water heaters. A golden tread. New York: Van Nostrand Reinhold Company; 1979. p. 117–27.
[4] http://cleantechnica.com/2015/05/04/solar-thermal-panels-heating-cooling/.
[5] Weiss W, Franz M. Solar heat worldwide, markets and contributions to the energy supply 2012. 2014.
[6] Kalogirou SA. Solar thermal collectors and applications. Prog Energy Combust Sci 2004;30(3):231–95.
[7] Chaturvedi SK, Chen DT, Kheireddine A. Thermal performance of a variable capacity direct expansion solar-assisted heat pump. Energy Convers Manage 1998;39(3–4):181–91.
[8] Kuang YH, Sumathy K, Wang RZ. Study on a direct-expansion solar-assisted heat pump water heating system. Int J Energy Res 2003;27(5):531–48.
[9] Carbonell D, et al. Simulations of combined solar thermal and heat pump systems for domestic hot water and space heating. Energy Procedia 2014;48:524–34.
[10] Mauthner F, Weiss W. Solar heat worldwide: markets and contribution to the energy supply 2011. Gleisdorf, Austria: IEA Solar Heating and Cooling Programme; 2013.
[11] Duffie JA, Beckman WA. Solar engineering of thermal processes. New York: John Wiley; 2006.
[12] Chandrasekar B, Kandpal TC. Techno-economic evaluation of domestic solar water heating systems in India. Renew Energy 2004;29(3):319–32.
[13] Pillai IR, Banerjee R. Methodology for estimation of potential for solar water heating in a target area. Sol Energy 2007;81(2):162–72.
[14] Denholm P. The technical potential of solar water heating to reduce fossil fuel use and greenhouse gas emissions in the United States. National Renewable Energy Laboratory; 2007.
[15] Pan T-C, Kao J-J, Wong C-P. Effective solar radiation based benefit and cost analyses for solar water heater development in Taiwan. Renew Sustain Energy Rev 2012;16(4):1874–82.
[16] Beckman WA, Klein SA, Duffie JA. Solar heating design, by the f-chart method. NASA STI/Recon Tech Rep A 1977;78:31071.
[17] Cassard H, Denholm P, Ong S. Break-even cost for residential solar water heating in the United States: key drivers and sensitivities. Contract 2011;303:275–300.
[18] http://www.energystar.gov/ia/new_homes/features/WaterHtrs_062906.pdf.
[19] Matrawy KK, Farkas I. Comparison study for three types of solar collectors for water heating. Energy Convers Manage 1997;38(9):861–9.
[20] Rommel M, Moock W. Collector efficiency factor F' for absorbers with rectangular fluid ducts contacting the entire surface. Sol Energy 1997;60(3–4):199–207.
[21] Alvarez A, et al. Experimental and numerical investigation of a flat-plate solar collector. Energy 2010;35(9):3707–16.

[22] Kovarik M. Optimal distribution of heat conducting material in the finned pipe solar energy collector. Sol Energy 1978;21(6):477–84.

[23] Hollands KGT, Stedman BA. Optimization of an absorber plate fin having a step-change in local thickness. Sol Energy 1992;49(6):493–5.

[24] Kundu B. Performance analysis and optimization of absorber plates of different geometry for a flat-plate solar collector: a comparative study. Appl Therm Eng 2002;22(9): 999–1012.

[25] Liu W, Davidson J, Mantell S. Thermal analysis of polymer heat exchangers for solar water heating: a case study. J Sol Energy Eng 2000;122(2):84–91.

[26] Tsilingiris PT. Back absorbing parallel plate polymer absorbers in solar collector design. Energy Convers Manage 2002;43(1):135–50.

[27] Hossain MS, et al. Review on solar water heater collector and thermal energy performance of circulating pipe. Renew Sustain Energy Rev 2011;15(8):3801–12.

[28] Kim Y, Seo T. Thermal performances comparisons of the glass evacuated tube solar collectors with shapes of absorber tube. Renew Energy 2007;32(5):772–95.

[29] O'Dell MP, Mitchell JW, Beckman WA. Design method and performance of heat pumps with refrigerant-filled solar collectors. J Sol Energy Eng 1984;106(2):159–64.

[30] Chaichana C, Aye L, Charters WWS. Natural working fluids for solar-boosted heat pumps. Int J Refrig 2003;26(6):637–43.

[31] Kim M-H, Pettersen J, Bullard CW. Fundamental process and system design issues in CO_2 vapor compression systems. Prog Energy Combust Sci 2004;30(2):119–74.

[32] Nekså P, et al. CO_2-heat pump water heater: characteristics, system design and experimental results. Int J Refrig 1998;21(3):172–9.

[33] Shukla R, et al. Recent advances in the solar water heating systems: a review. Renew Sustain Energy Rev 2013;19(0):173–90.

[34] Andersen E, Furbo S, Fan J. Multilayer fabric stratification pipes for solar tanks. Sol Energy 2007;81(10):1219–26.

[35] Altuntop N, et al. Effect of obstacles on thermal stratification in hot water storage tanks. Appl Therm Eng 2005;25(14–15):2285–98.

[36] Spur R, et al. Performances of modern domestic hot-water stores. Appl Energy 2006;83(8): 893–910.

[37] Sonnenenergie DGf. Planning and installing solar thermal systems: a guide for installers, architects, and engineers. Earthscan; 2005.

[38] Wang X, Wang R, Wu J. Experimental investigation of a new-style double-tube heat exchanger for heating crude oil using solar hot water. Appl Therm Eng 2005;25(11–12): 1753–63.

[39] Farrington RB, Bingham CE. Testing and analysis of load-side immersed heat exchangers for solar domestic hot water systems. Golden (CO, USA): Solar Energy Research Institute; 1987.

[40] Morrison GL, et al. Analysis of horizontal mantle heat exchangers in solar water heating systems. Sol Energy 1998;64(1–3):19–31.

[41] Baur JM, Klein SA, Beckman WA. Simulation of water tanks with mantle heat exchangers. In: Proceedings of the… Annual Conference. USA: American Solar Energy Society, Inc; 1993.

[42] Shah LJ, Furbo S. Optimisation of mantle tanks for low flow solar heating systems. In: EuroSun'96; 1996. p. 369–75.

[43] Knudsen S, et al. Analysis of the flow structure and heat transfer in a vertical mantle heat exchanger. Sol Energy 2005;78(2):281–9.

[44] Han YM, Wang RZ, Dai YJ. Thermal stratification within the water tank. Renew Sustain Energy Rev 2009;13(5):1014−26.
[45] Müller-Steinhagen H. Applications of solar heat for temperatures ranging from 50−2000 C. In: 5th European thermal-sciences conference, Netherlands; 2008.
[46] http://www2.cde.state.co.us/artemis/ucsumonos/ucsu6382p881978internet/ucsu6382p88197806internet.pdf.
[47] Lane D. Solar pool heating basics, Part 2. HomePower; 2003. p. 95.
[48] Yan C, et al. A simplified method for optimal design of solar water heating systems based on life-cycle energy analysis. Renew Energy 2015;74:271−8.

Part Three

Solar cooling technologies

Photovoltaic-powered solar cooling systems

10

Y. Li, R.Z. Wang
Institute of Refrigeration and Cryogenics, Shanghai Jiao Tong University, Shanghai, China

10.1 Introduction

The concern for the increasing effects of climate change and environmental pollution due to the use of fossil fuels has promoted great interest in the application of renewable energy. Operating silently and without any moving parts or environmental emissions, solar photovoltaic (PV) systems have developed from being advanced science and technology applications into a mature technology used for electricity generation. A PV system consists of several components, including PV panels to absorb the solar radiation and generate electricity; an inverter to convert the electric current from direct current (DC) to alternating current (AC); and mounting, cabling, and other electrical accessories to set up a working system.

With the development of manufacturing technology and the maturity of the market, the price of PV panels and other components has steadily decreased. Fig. 10.1 shows the installed price of residential and commercial PV systems in the United States and the Global Module Price Index from 1998 through 2012 [1]. Strong PV technology price decreases and electricity prices on the rise lead to rapid progress in the application of PV systems. The installation capacity of the world has increased steadily in recent years. Solar PV capacity has grown at the most phenomenal rate—up 48-fold from 2004 (3.7 GW) to 2014 (177 GW). The annual growth rate is 50% from 2009 to 2013. In 2014 the PV installation capacity added was approximately 40 GW. The leading countries include China, Japan, and the United States. In 2014 Inner Mongolia, Qinghai, and the coastal province of Jiangsu in China were at the forefront of additions, each installing more than 1 GW. China generated approximately 25 billion kWh of electricity with solar PV in 2014, an increase of more than 200% over 2013 [2].

The energy used in buildings accounts for 30% of the total primary energy used. With the use of heating, ventilating, and air-conditioning (HVAC) systems for indoor environments, air conditioning has become a very common practice around the world. Approximately 70% of the energy used in buildings is for heating and cooling. With the economic development in developing countries, more energy is used for buildings, especially use of cooling for air conditioning and refrigeration. The carbon dioxide emissions related to buildings has also steadily increased in recent decades. To meet the challenge of an energy resource shortage and to solve the environmental problems, solar-assisted cooling technologies have been extensively studied in recent decades. Solar cooling offers a wide variety of cooling

Advances in Solar Heating and Cooling. http://dx.doi.org/10.1016/B978-0-08-100301-5.00010-2

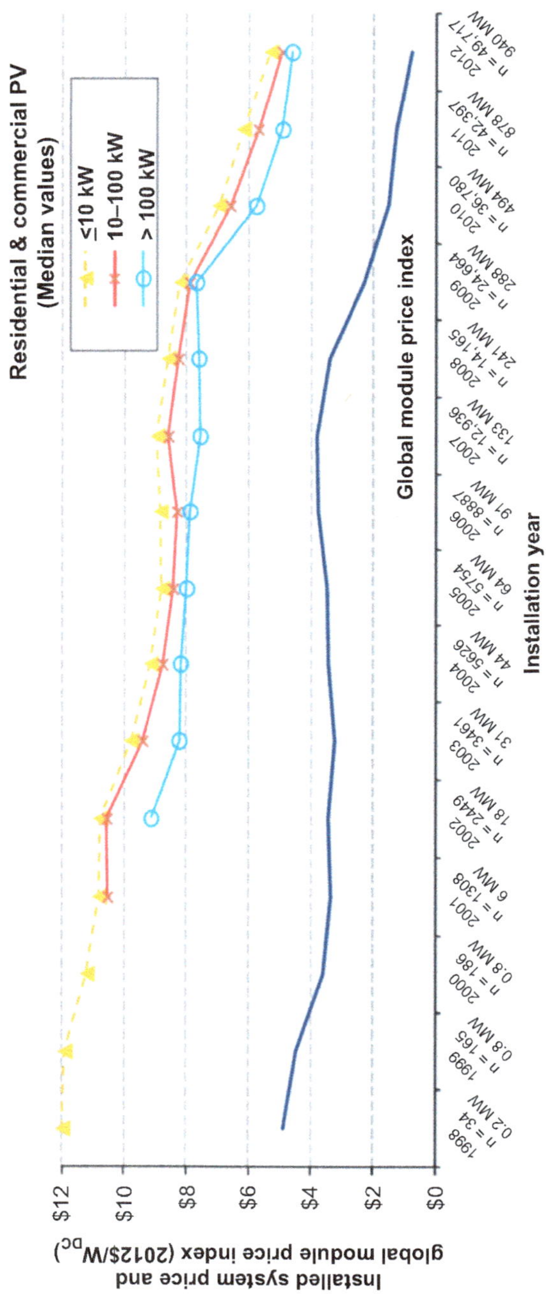

Figure 10.1 The installed price of residential and commercial photovoltaic (PV) systems in the United States and Global Module Price Index from 1998 through 2012.

techniques powered by solar collector–based thermally driven cycles and PV-based electrical cooling systems [3–5]. Because solar energy is time-dependent, the successful utilization of all of these systems is to a very large degree dependent on the thermal storage units used. Many demonstration projects have been built around the world. Most of them are solar thermal-driven cooling systems. The PV-powered cooling system has attracted increasing research and development in recent years with the wide application of PV systems.

There is not a large difference between PV cooling systems and other PV systems. A PV cooling system usually consists of a PV array, a vapor compression refrigeration system, and other necessary equipment. Compared with a thermal-driven cooling system, the advantages of a vapor compression refrigeration system include being compact, easy control, mature technology, and easy maintenance. The vapor compression refrigeration system is usually driven by an electrical motor.

The PV cooling system can be categorized based on different criteria, including according to the different electrical current type used to power the electrical motor of the compressor. The PV cooling system can be classified into a DC and AC cooling system. Because the PV array produces DC current, the DC system can be directly connected to the PV array without the inverters. While an inverter have to be used for an AC system which to be powered by the electricity produced by PV, the AC cooling system is widely available on the market and can be selected from commercial products.

There are two main types of solar PV application technologies according to the connection to the public grid: the stand-alone PV system and the grid-connected PV system. Grid-connected systems are relative simple and consist of PV panels, inverters, power conditioning units, and grid connection equipment. These systems seldom have batteries. The grid-connected PV system supplies the excess power, beyond consumption by the connected load, to the power grid. In stand-alone PV systems, because the demand from the load does not always equal the solar panel capacity, battery banks are generally used. On the basis of the electrical current used, the stand-alone system can be subdivided into the AC system and DC system.

According to the different working temperature level, the PV cooling system can be categorized into refrigeration, air-conditioning systems, heat pump systems, and combined cooling, heating, and power (CCHP) systems. The working temperature level of a refrigeration system is around or lower than 0°C. It is mainly used for cold storage in daily life or some process applications in industry. The air-conditioning system is typically used for thermal comfort in buildings and some process applications such as temperature control for electrical devices. The temperature level is approximately 15–20°C. Because the applications and research on PV-powered heat pump systems and CCHP systems are few at the present time, this chapter only deals with PV refrigeration and PV air-conditioning systems.

There are four parts in a PV cooling system: the PV array, control unit, storage unit, and compressive refrigeration unit. For the stand-alone system, the control unit regulates the electricity from the PV array to the appropriate voltage to charge the DC electrical energy into the storage unit or drive the compressor of the refrigeration unit.

For the grid-connected system, an inverter is included in the control unit, which converts the DC electricity from the PV array to AC with appropriate voltage to drive the compressor of the refrigeration unit or charge into the grid. In the DC side, different types of batteries have been used as the storage unit, with a lead-acid battery being the most used type. By now almost all conventional compressive refrigeration machines can be used for PV cooling. However, to increase the energy efficiency and reduce the system initial and operation cost, research and development are still needed on refrigeration systems dedicated for PV cooling.

The PV cooling system has become more and more attractive in recent years for the following reasons:

1. *Simple structure*: The PV cooling system consists of several important components that are mature and widely available in the market. Compared with a thermal-driven system for which the power system must be connected through a metal pipeline, the connections between the components of a PV system are mainly electrical wire and can be easily installed.
2. *Easy control and short response time*: The electrified system can be easily controlled and incorporated with information technology, which means optimal and remote control and remote monitoring can be easily achieved. The use of a vapor compression refrigeration system can achieve a quick change of output cooling power and quickly respond to the variation of cooling load.
3. *Green energy and lower environmental pollution*: The system can consume less electricity from the grid and make full use of local produced power. This feature will be more meaningful in the hot summer, when the peak power puts great strain on the power network.
4. *High energy efficiency and potential reducing cost*: The cost of a PV system will continue to decrease in the following years. The price of PV electricity might be even lower than the price of electricity generated from fossil fuel. Furthermore, the cost of components such as the inverter, battery, vapor compression refrigeration system, and grid connection will continue to decrease.
5. *Working as a distributed energy system with more flexibility for energy transition and conversion*: The system can be grid connected and reduce the influence of the intermittence of solar radiation. The off-grid system can be used to provide heating, cooling, and electrical power, which is of great significance in improving the living standard of a remote area where grid connection is not economical.

Although the PV cooling system can just be looked as a load, there are still some special features for this system: (1) The load varies seasonally; it is an energy-intensive user in some seasons whereas in other seasons it consumes almost no energy. In designing a PV system, this must be considered. (2) The load is coincident with PV-generated power in summer whereas it is not coincident in winter in daytime. In summer, usually around noon, the cooling load is high and the radiation is also high, which makes it possible that the electricity produced from photovoltaics matches the consumption of an air conditioner. (3) The control is more complicated and very sensitive to weather conditions. The power supply is influenced by the radiation whereas the cooling load is determined by the environmental temperature. (4) In many cases in which the load might not match the supply, grid connection or energy storage must be considered. Different from other electrical systems that just depend on the electricity from the grid or from the battery, it can also back up its energy in the form of cold energy and use it when it is needed. (5) If the system design is not mature, then a simulation might be needed for different climate zones. In some climate zones, it might be a good solution whereas in other climate zones, the PV cooling system might not be cost-effective or not even be feasible. For example, in the area where solar radiation is weak while the environmental temperature is high, the PV cooling might not be a feasible solution to meet the cooling loads.

The objective of this chapter is to survey the progress of current researches on solar cooling system. The main achievements have been summarized and important issues of this technology have been discussed. The review is divided into two sections: the first is the solar refrigeration system and the second is the solar air-conditioning system. The studied results about the economic aspects of solar PV cooling system are also reported.

10.2 Performance index

Because a PV cooling system is actually an energy conversion system, the energy performance is always important in evaluating the system. Various indexes have been developed to access the performance and applicability of the PV cooling system, including (1) the efficiency of important subsystems such as a PV array, inverter, and battery; (2) the coefficient of performance (COP) of a cooling system; and (3) the solar fraction (SF).

The average PV conversion efficiency is defined as the ratio of the total energy delivered from the PV array to the energy of the solar radiation on the PV:

$$\eta_{pv} = \frac{E_{pv}}{E_{irr}} \quad \eta_{PV} = \frac{E_{PV}}{E_{Solar}} = \frac{9.092\ \text{kWh}}{96.50\ \text{kWh}} = 9.42\% \qquad [10.1]$$

where E_{pv} is the electricity energy generated by the PV array, and E_{irr} is the energy of solar radiation.

For any solar energy system, it is intended to achieve a higher output under a certain collection area of a solar device. Hence, the system performance is evaluated by the solar COP (COP_{solar}), which is given as the following:

$$COP_{solar} = \frac{Q_{h/c}}{E_{irr}} \qquad [10.2]$$

where $Q_{h/c}$ is heating/cooling power.

The SF is commonly used to evaluate the performance of a solar cooling system. It shows the portion of the solar energy contribution as compared with the total energy required to drive the refrigeration part of a solar cooling system. This total amount of energy input is commonly supplemented by the auxiliary provision. The SF is calculated as the ratio of the electrical energy provided by the solar energy to the total electrical energy (E_{total}) used to drive the air conditioner:

$$SF = E_{pv}/E_{total} \quad [10.3]$$

A seasonal energy efficiency ratio can be calculated with the following expressions [28]:

$$EER = \frac{31 \cdot EER_{JU} + 31 \cdot EER_A + 2 \cdot 30 \cdot EER_S + 2 \cdot 15 \cdot EER_O}{152} \quad [10.4]$$

10.3 Photovoltaic-powered refrigeration system

Cooling is always needed in daily life. In remote and undeveloped areas where electricity is not available, the storage of perishable food, vaccines, and medicine have always been big problems [6] because of the lack of appropriate reliable and cost-effective technologies. It is expected that solar PV refrigeration technology would provide a more reliable, safer, and cleaner form of refrigeration for perishable foods and improve the cold chain for vaccines. Various solar PV-powered refrigeration systems in small capacity have been proposed and investigated from the beginning of this century [15,18,19]. Studies on solar PV refrigeration have been focused on the performance of different components, energy storage, and the overall performance of the system. The study methods include theoretical analysis, modeling, and experiments.

10.3.1 Studies on photovoltaic-powered refrigerators

To maintain the properties of vaccines, they need to be kept in a lower temperature container ranging between −3°C and 8°C [6]. The earliest PV-powered refrigerator was mainly used for vaccine storage in remote, underdeveloped villages. In the earlier investigation, the commercial refrigerators were modified to be powered by a PV system to solve the problem of shortage in electricity [7−10]. Different components of a PV refrigerator have been investigated. Kattakayam et al. [7] used commercial inverters that produce a quasisquare wave output and proved that conventional domestic refrigerators can be operated on nonsinusoidal AC inputs without any degradation of the thermal performance. However, a slight additional heating of the hermetic compressor was observed. Thermal mapping of the temperatures at various points on the refrigerator is provided for steady state, cool down, warm up, periodic opening of the door, and ice-making. An energy flow diagram was given for a steady-state sunny day operation. Major sources of losses are identified. The experimental system was the most widely used domestic refrigerator in India in the 1990s. The refrigerator had an internal volume of 165 L and 100-W cooling capacity using R-12 as refrigerant. It had a single door hinged on the right, a top freezer compartment of approximately 40-L volume, and it used a capillary expansion device. A 25-mm-thick polyurethane foam insulation is provided all around. Characterization of a lead-acid battery bank consists of a set of four 6-V, 180-Ah deep discharge tubular lead-acid batteries were studied later by the same research team [8]. Each battery has three cells of 2 V each. All four batteries were connected in series to provide a nominal output of 24 V. The charging and discharging characteristics were obtained for the cases when the battery bank was charged from conventional mains; a generator set; and PV panels with two, three, and four pairs of panels charging it. It was shown that through a judicious choice of capacity and load matching, the maximum power point tracker can be eliminated.

The fabrication, experimentation, and simulation of converting a 165-L domestic electric refrigerator to a solar-powered one were reported by Modi et al. [9]. A conventional domestic refrigerator was chosen and was redesigned by adding a battery bank, inverter, and transformer, and it was powered by solar PV panels. Various performance tests were carried out to study the performance of the system. The COP was observed to decrease with time from morning to afternoon and a maximum COP of 2.102 was observed at 7:00 am. Simulations regarding the economic feasibility of the system for the climatic conditions of Jaipur city (India) were also carried out using RETScreen 4. It was observed that the system can only be economically viable with a carbon trading option taken into account and an initial subsidy or a reduction in the component costs, mainly PV panels and a battery bank.

A conventional refrigerator was chosen, and some changes were introduced to reduce the cooling load and consequently the power required, even at low solar insolation. The modifications included the following:

1. An increase of the polyurethane insulation by 25 mm results a reduction in the internal volume of the refrigerator by 30%.
2. Replacement of the door's double glass by insulation to reduce the heat losses.
3. A three-phase brushless motor with a permanent magnet was used in the compressor. This is a DC compressor with variable speed running from 2000 to 3500 rpm. Because of this change no DC/AC inverter was required.
4. The battery system could be switched on/off during the operation through the control panel. The development of such a system was effected and the results of this work to provide a cost-effective version were analyzed. Tests were carried out to study the performance of the refrigerator components, especially the compressor's, as well as the refrigerator as a whole. It was found that the energy consumption decreased to 1.53 kWh/day for refrigeration purposes with a 15-h operation of the compressor whereas for conservation the load was 1.7 kWh with the system operating for 24 h. The alteration from an AC motor to a DC one implies that no inverter was required for the PV-powered operation. Hence, less power loss and better economics were achieved. The use of a variable-speed motor of the compressor increased the power transfer or solar insolation utilizability to the level of 92% for a speed value ratio of 2.5.

A PV-powered refrigerator was constructed and a study of the performance in Shanghai and other climate conditions was carried out [11]. The tested refrigerator consists of a PV panel, an electric charge controller, a lead-acid battery specially designed for deep cycling, and a refrigerator. The electricity generated by the PV panel is regulated by the charge controller and then is stored in a lead-acid battery in the form of electrochemical energy. The nominal charging voltage of the battery is 12 V with the capacity of 200 Ah. The refrigerator was modified from a commercially available AC unit, in which the AC vapor compressor was removed and replaced by a DC vapor compressor and some essential controlling parts were accordingly modified. The refrigerant used in the refrigerator is R-134a, which is ozone friendly. The total volume of the refrigeration is 86 L, including a 12-L freezer and a 74-L chilled storage cabinet. The field performance test results showed that the refrigerator can work well in the Shanghai area, the lowest temperature in the freezer can reach near −15°C, and the power consumption is also reasonable. Furthermore, the energy flow of the system and applicability of it to different climatic regions has been analyzed and discussed. It was revealed that the solar insolation and ambient temperature are the two main factors that affect the performance of the PV-powered refrigerator. The former decides mainly the amount of the electricity generated and the latter is mainly responsible for the load of the refrigerator. It is essential to reduce the energy consumption by the controller and connecting cables, which account for approximately 15% of the total electricity generated by the PV panel, and optimization of the controller and proper sizing of the cable should be performed.

A parametric study was performed to determine the best combination of the PV panel wattage, battery capacity, and insulation level to operate the refrigerator only on solar power [12]. The Transient Simulation Software (TRNSYS) is used to simulate the PV-based refrigeration system. The system consists of a refrigerator with 50-L capacity, a 24-V battery bank, a 1-kVA inverter, and PV panels with a charge controller. The simulation model has been validated with experimental data. The results concluded that it is feasible to operate a refrigerator (25-mm insulation thickness) on solar photovoltaics without grid power with 320-W panel arrays with 50-Ah battery capacity. As the thickness of insulation is increased to 50 mm, then 200-W panel capacity is sufficient to drive the refrigerator.

10.3.2 Energy storage in a photovoltaic-powered refrigerator

An important factor of refrigerators in remote areas is their cold autonomy, which depends on several factors, including the capacity of the energy storage, the ambient temperature, insulation, and airtightness. The autonomy of the PV refrigerator can be extended by increasing the battery capacity. However, this solution is rather expensive. Another way of increasing the autonomy is to create a cold storage area in the refrigerator. This solution might be cheaper than the first one. A solar PV-powered icemaker that operates without the use of batteries and is therefore environmentally friendly may be used in truly autonomous applications in remote areas. Theoretical modeling and experiments have been conducted to study the solar PV refrigerator with three compartments [13], which include a compartment for vaccine and medicine storage with a volume of 28 L, one for the personal use of the medical staff with a volume of 24 L, and a third compartment of 4.5 L for ice-making. A cold storage area was made around the evaporator to give the system several days of autonomy. The experimental study found out that the power of the PV generator is 249 Wp, and the value of the thermal loss is 11 W. At a steady rate, the mean temperatures of the vaccine compartment and medical staff compartment are 4°C and 5°C, respectively. The autonomy of the system due to the cold storage is approximately 3 days. The capacity of the battery used is 150 Ah. In addition to the primary role as energy storage devices, they also provide a power buffer between the PV array and load. They do add to system complexity and overall cost. However, an optimal choice could result in an excellent energy inventory and management of the overall power system, with high reliability, low maintenance costs, long lifetime, high charging efficiency, and high voltage stability. The parameters affecting optimization are the storage capacity, the operating voltage, and matching with the load it is expected to meet.

A new storage strategy of stand-alone PV plants that substitutes the battery storage with thermal, eutectic, latent, or a hydraulic storage has been studied by Cherif and Dhouib [14]. The performances, the simulation responses, and the dynamic behavior of a PV refrigeration plant using latent storage were reported. The measurements and the evaluation of these less battery storage systems at several climatic conditions and under load disturbances were used to evaluate the PV system reliability and to compare its performances with classic battery storage systems.

The successful operation of the refrigeration compressors by the PV panels was ensured by the use of a novel concept dedicated controller, which provided easy startup, maximum power tracking, and power management for the four compressors of the system. The prototype results showed very good ice-making capability and reliable operation as well as a great improvement in the start-up characteristics of the compressors, which remain operational even during days with low solar irradiation and operate with improved utilization of the available PV power [16].

An experimental investigation on the performance of a solar-driven, DC motor, vapor compression refrigerator through indoor and outdoor tests with/without thermal storage and with/without loading were performed in a hot, arid area [17]. The main components of the experimental setup were a multicrystalline PV module, battery as a buffer for constant 12-V DC, and a 50-L portable refrigerator with/without phase change material (PCM) thermal energy storage. In addition, a theoretical model was established to evaluate the refrigerator performance when operated under different environmental conditions of the design point. The model was used to size a larger cooling capacity solar-driven refrigerator. The results show that this system could be used in remote, hot, arid areas for refrigeration of postharvest crop transportation activities. From the outdoor results at the PCM—full load condition, a COP of 1.22 was achieved, and a storage temperature of 5°C was achieved on the third day and 0°C on the sixth day.

10.3.3 Innovative application in photovoltaic-powered refrigerators

To enhance the autonomy and expand the application of PV refrigerators, some innovative strategies have been proposed. A novel PV DC refrigerator that uses hydrogen as the energy storage material has been proposed. The system consisted of a PV panel, electrolyzer, hydrogen storage tank, fuel cell, and DC refrigerator. When the PV-produced electricity is more than that needed from the refrigerator, the electricity is used to produce hydrogen by an electrolyzer and stored in the hydrogen storage tank. When the solar radiation is not available, such as in the evening or on rainy days, the hydrogen in the storage tank is discharged into the fuel cell to produce the hydrogen and drive the refrigerator.

A monitoring architecture for stand-alone PV systems was proposed with a web application being adopted, thus allowing for online monitoring and control of remote installations [20]. To check the effectiveness of the proposed remote monitoring system, the performance of a PV-powered outdoor refrigerator was evaluated. In the case of temperature-sensitive applications, such as vaccine or perishable good storage, critical operating conditions have been detected. Experimental measurements executed on a prototype, adapted for outdoor operation with a horizontal PV module, show that the refrigerator works adequately in the given place (Catania, Italy), where a typical Mediterranean climate is present, if an additional vertical PV module is adopted.

An Ethernet-based real-time remote monitoring system, specifically developed for stand-alone PV-powered appliances, was capable of monitoring the main electrical parameters of the device under test, which were stored on a database together with ambient parameters (Fig. 10.2).

A PV-powered container system that can suitably be used in many rural regions where electricity is unreliable or nonexistent but refrigeration is continuously critical has been proposed and studied [21]. It is composed of four parts: (1) the cooling unit (container), (2) the energy production unit (PV panels), (3) the energy control unit, and (4) the energy storage unit (the solar lead-acid battery bank, 550 Ah). The PV-container system worked with an input voltage of 220 V and AC. R-134a refrigerant was used in the container. The energy demand of the PV-container system was provided by two-unit PV panels (DC 24 V, 150 W), each one with seven panels connected in parallel. PV panels were selected by polycrystalline type because of their high efficiency. A solar control unit regulates the

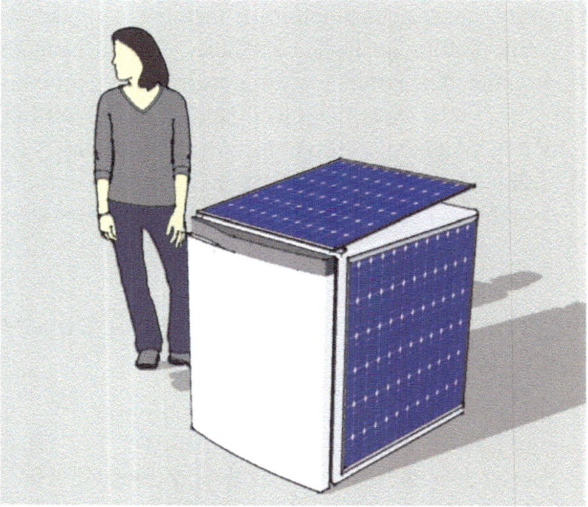

Figure 10.2 Conceptual model of the refrigerator with horizontal and vertical integrated photovoltaic modules [20].

DC output of the PV panel and supplies energy to the battery bank. It prevents battery overcharge and full discharge. The battery bank consists of two units of 550-Ah/24-V dry-type batteries connected in parallel. The power of the compressor was 380 W with consumption of approximately 4460 Wh/day. The evaporator was regulated to keep the temperature inside of the container between 3°C and 9°C. The inverter converted the DC 24 V to AC 230 V and provides a perfect sinusoidal current with a capacity of 2 kVA. The controller was designed to regulate the load and incorporated three functions to charge the batteries with high reliability and to regulate the load or the bypass. It operated according to a choice of three modes of operation. An experimental study for the operation state of the container under different loads and without load was performed to determine the daily and long-term behaviors of the PV-container systems during some typical days in summer and winter and in two different climate types: mild (along the Mediterranean coast) and arid (in the Sahara desert). In all of the experiments the inside temperature of the refrigerator was set between 4°C and 8°C.

10.3.4 Photovoltaic-powered refrigerators available in the market

Some commercial products for refrigeration have been developed based on the PV energy source. One of the most famous was SolarChill [22], which was developed by the Danish Technological Institute and funded by the Danish Energy Agency. SolarChill technology improved on the existing solar vaccine-cooling technology by bypassing the use of conventional lead batteries, which have proven to be a major obstacle to the uptake of solar technology in developing countries. Energy is stored in ice-packs through the use of a DC compressor; no lead battery is required. It uses R-600 hydrocarbon refrigerant and cyclopentane blown insulation foam, which is environmentally friendly with no contribution to ozone depletion or global warming. It is powered by three 60-W PV panels and used as a grid-independent vaccine cooler or as a household or small commercial refrigerator. The temperature control is achieved through natural convection between the ice storage compartment and vaccine compartment. It was declared that one SolarChill unit can serve a population of 50,000 people for preserving vaccines with an expected commercialization price between $1500–2000 (US; including solar panels)—50–60% cheaper than the currently available solar refrigerators meeting World Health Organization specifications. It was estimated that the product has the potential to enter the domestic market in remote, off-the-grid areas of developed and developing countries. The estimated need in developing countries for vaccine storage purposes is between 10,000 and 20,000 units per year.

Several refrigerators, freezers labeled as "solar-powered" or "solar-ready," are currently available on the market today. A survey of 32 refrigerator and freezer models across six different manufacturers [23–28] showed that, although they may use the same refrigeration cycle as any grid-powered refrigerator, solar-powered refrigerators and freezers make use of a DC compressor. This eliminates the need for an inverter when running on PV energy. The compressor can typically accept either 12 or 24 VDC.

This distinction comes from the common use of 12 VDC lead-acid batteries in modern PV systems. Using two such batteries in series will increase the voltage to 24 V and consequently halve the current at the same power input, which is useful for reducing heat generated by the system. More than half of the solar refrigerators and freezers surveyed also accept grid power as an energy source. This allows for more versatility in suboptimal weather conditions.

SunDanzer [23] was the only manufacturer to advertise a variable voltage input allowance, either 10–31 VDC or 10.4–17 VDC, depending on the specific model number. This is due to the variable speed compressor used in SunDanzer cooling systems. The variable speed compressor also allows for the unit to run on lower levels of light than a single speed compressor as well as generate greater cooling in optimal conditions. This combination means that they are typically far less reliant on batteries and grid connections.

Of the units surveyed, many quoted a single energy usage but had storage capacity for refrigeration and for freezing; similarly, some quoted a single energy usage but stated that the unit could be used as either a refrigerator or a freezer. Looking only at the models that paired a single energy use to a single specified capacity of refrigerator or freezer, Fig. 10.3 was constructed.

As expected, the energy requirement for the freezer units is greater than that of the refrigeration units, size independent. There also seems to be a negative trend as the capacity increases, suggesting that a larger capacity requires less energy per unit capacity. This trend is more apparent with the freezer units than with the refrigeration units, but with such a small number of samples, the analysis is inconclusive.

All of the refrigerator and freezer units surveyed rely on batteries. The batteries are charged during the day when there is excess light and discharged at night and in other low-light conditions. Because batteries need to be replaced and because their production and disposal is harmful to the environment, research is being conducted to create refrigerators that store thermal energy rather than electrical energy. Such

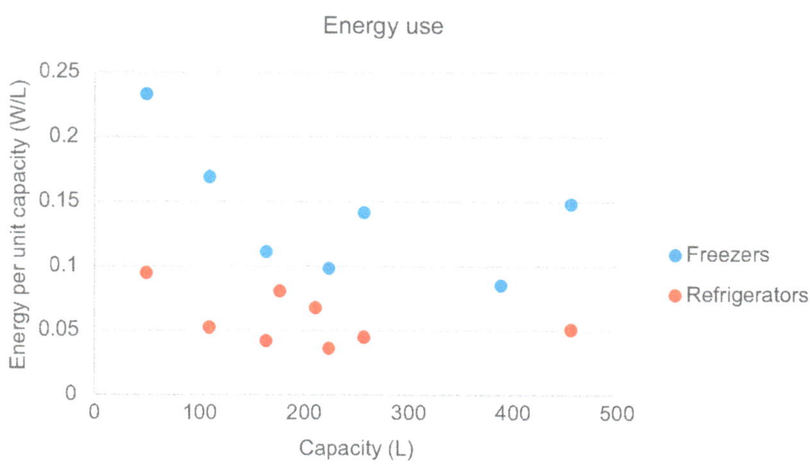

Figure 10.3 Normal power versus storage capacity of a commercial photovoltaic refrigerator.

units have significantly improved insulation, and rather than simply cooling the air, they produce ice. Rather than increasing the temperature of the refrigerator contents, environmental heat that enters the refrigerator melts the ice at a constant temperature. Such a system has been shown to be effective for several days without power input.

10.4 Photovoltaic-powered air-conditioning system

HVAC is a necessary equipment in modern buildings, be they residential or public building environments. Normally one-third to half of the annual total electricity consumption is used for air conditioning and refrigeration in buildings worldwide. Solar cooling is considered as the sustainable technology to provide air conditioning and refrigeration because solar energy is the primary energy source and widely available, especially in hot areas where air conditioning is needed. Actually, solar PV air conditioning is a feasible way to replace the conventional electric refrigeration machines for buildings because of the much higher energy efficiency of electric chillers compared with sorption cooling systems.

When the PV price is high, solar PV air conditioning is not economically feasible; hence, study on PV air conditioners has only been conducted in recent years. The initial researches were mainly based on modeling and theoretical analysis [29 ~ 32].

10.4.1 Energy performance of photovoltaic-powered air conditioners

Simulation models were developed to compare five types of solar cooling systems and their performances throughout a year in a subtropical city that commonly features long, hot, and humid summers [29]. The solar cooling systems included solar electric compression refrigeration, solar mechanical compression refrigeration, solar absorption refrigeration, solar adsorption refrigeration, and solar solid desiccant cooling. The solar PV cooling system, the air side system, and the corresponding control provisions were built using TRNSYS and TESS. The schematic diagram of the PV cooling system is shown in Fig. 10.4.

The key performance indicators used in the study included SF, COP, solar thermal gain, and primary energy consumption. In addition, different installation strategies and types of solar collectors were compared for each kind of solar cooling system. It was found that solar PV cooling and solar absorption cooling had the highest energy saving potential in subtropical Hong Kong. These two solar cooling systems would have even better performances through the continual advancement of the solar collectors. It will provide a promising application potential of solar cooling for buildings in subtropical regions.

In another modeling analysis, a comparison of solar thermal and solar electric cooling for a typical small office building exposed to two different European climates (Freiburg, Germany and Madrid, Spain) was conducted by Hartmanna et al. [30]. The investigation was based on load series for heating and cooling obtained from annual building simulations in TRNSYS. A conventional compression chiller was

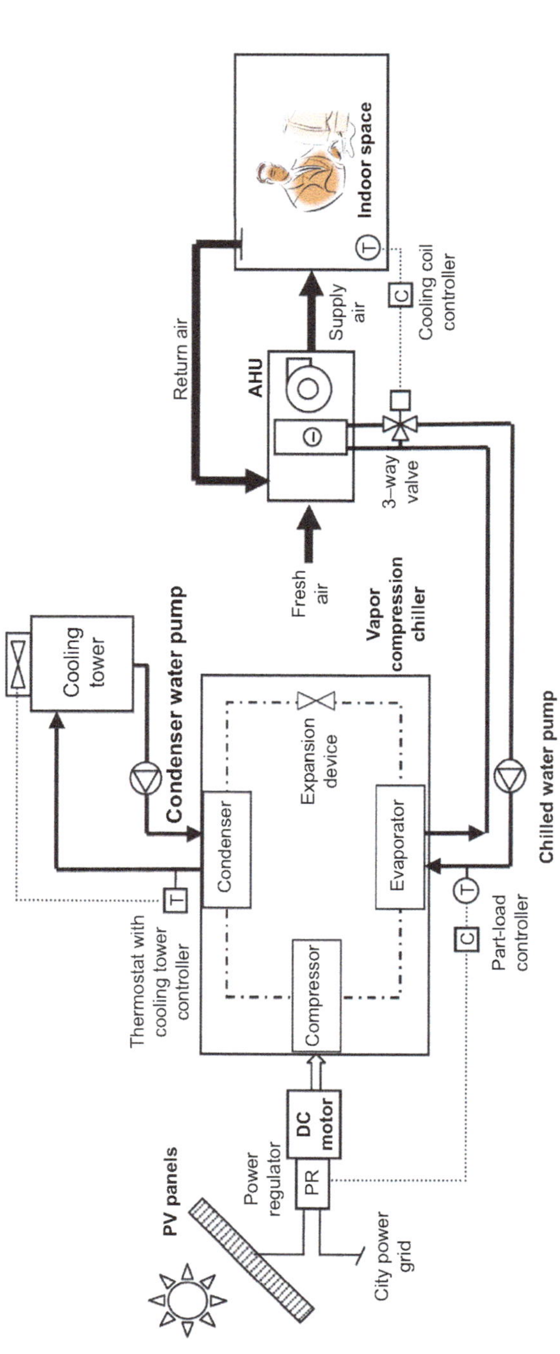

Figure 10.4 Schematic diagram of heating, ventilating, and air-conditioning system designed with PV cooling [29]. *PV*, photovoltaic; *PR* power regulator; *DC*, direct current; *AHU*, air handling unit; *C*, controller; *T*, temperature sensor.

used as the reference system against which the solar options were evaluated with respect to primary energy savings and additional cost. A parametric study on collector and storage size was performed for the solar thermal system to achieve the minimal cost per unit of primary energy saved. The simulated solar electric system consists of the reference system equipped with a grid-connected PV module that can be varied in size. For cost comparison of the two systems, the electric grid is assumed to function as a cost-free storage. A method to include macroeconomic effects in the comparison was presented and discussed. It was found that within the system parameters and assumptions used in the study, the grid-coupled PV system leads to lower costs of primary energy savings than the solar thermal system at both locations. The presumed macroeconomic advantages of the solar thermal system, due to the nonusage of energy during peak demand, were confirmed for Madrid.

A solar PV air-conditioning system was proposed and investigated for different evaporating temperatures and months in the city of Adana, located in the southern region of Turkey [31]. The hourly cooling load capacities (heat gain) of a sample building during the 23rd days of May, June, July, August, and September were determined by using meteorological data such as hourly average solar radiations and atmospheric temperatures. The hourly total heat gain of the sample building comprising a wall, a window, humans, illumination, and devices were determined by using the Cooling Load Hourly Analysis Program (HAP) 4.4. The hourly variations of various parameters such as the COP, condenser capacity, and compressor power consumption were calculated. In addition, the minimum PV panel surface area was determined to meet the compressor power demand according to the hourly average solar radiation data. For evaporating temperature $T_e = 0°C$, the maximum compressor power consumption was obtained as 2.53 kW at 3:00 pm on August 23. The required PV panel surface area was found to be approximately 31.26 m^2. It was concluded that the PV air-conditioning system could be used for home/office-cooling purposes during the day in the southern region of Turkey.

A systematic simulation study was performed to evaluate the overall performance of PV compression cooling systems in office buildings for six different climatic conditions worldwide [32]: Cologne, Germany; Stuttgart, Germany; Barcelona, Spain; Madrid, Spain; Jakarta, Indonesia; and Riyadh, Saudi Arabia (Table 10.1). For each climate three different building load scenarios were considered and simulated: Case A with high internal loads for a well-insulated building, Case B with low internal loads for a well-insulated building, and Case C with average internal loads in a poorly insulated building.

The performance of PV-powered compression chiller systems was systematically analyzed. The office building heating and cooling demand was obtained from dynamic building simulations using EnergyPlus. A detailed analysis of the Solar Fractions, energy efficiency ratios (EERs), full load hours, and the system efficiency were conducted for office buildings in various climatic, load, and insulation conditions. The results were compared with the performance of solar thermal single-effect absorption systems. For each technology investigated, an energy optimized control strategy was developed that maximizes the primary energy efficiency. This control strategy was implemented in the modular simulation environment INSEL, and system models were developed for a range of cooling systems. As it has detailed validated models

Table 10.1 Annual mean air temperature, global horizontal irradiance, and optimum tilt angles/photovoltaics for all locations investigated in reference [32]

Location	Latitude–longitude	Global horizontal irradiance (kWh/m^2a)	Mean ambient temperature (°C)	Optimum tilt angle PV
Cologne, Germany	N50°52′–E7°10′	965	9.9	30°
Stuttgart, Germany	N48°40–E9°13	1092	9.1	30°
Barcelona, Spain	N41°16′–E2°4′	1454	15.7	30°
Madrid, Spain	N40°24′–W3°40	1610	14.3	29°
Jakarta, Indonesia	S6°7′–E106°45′	1639	27.2	12°
Riyadh, Saudi Arabia	N24°42′–E46°47′	2222	26.2	20°

PV, photovoltaic.

both for absorption chillers, solar thermal and photovoltaic modules, inverters and others. This modular simulation environment allows to program individual control strategies, which is extremely important to minimize the auxiliary energy consumption.

It was found that the primary energy savings for solar electric cooling and heating are comparable to solar thermal systems. Because solar thermal systems include hot storage, they mostly provide higher SFs and in some cases higher primary energy savings.

In recent years some demonstration projects have been built and experimental results were reported. Experimental study was conducted on a PV-powered inverter air-conditioning unit without the use of batteries or any inverter regulators [33]. The aim of the project was to study an "inverter" heat pump with a nominal cooling capacity of 3.52 kW and a nominal heating capacity of 3.81 kW. Three 235-Wp PV panels were connected directly to the machine at 24 Vcc. The "inverter module" simultaneously takes the electricity from the grid and solar panels. It was found that during summer months the average EER of the installation could be close to 14 and the solar contribution of PV panels is approximately 65%. When the machine works in heating mode the results show that the average COP is approximately 9 and the solar contribution of PV panels is approximately 50%.

A solar PV air-conditioner system was studied on the heating and cooling performance of the system in a hot summer and cold winter zone such as Shanghai, China [34]. Four working modes have been investigated: cooling in summer and heating in winter, both for daytime and nighttime. Results are reported in terms of COP_{solar} based on solar radiation, SF, and the solar direct consumption ratio. It was found that the COP_{solar} for cooling of the system was approximately 0.32, which is higher than a solar thermal-driven cooling machine, whereas the heating COP_{solar} for heating was approximately 0.37, which is lower than the thermal efficiency of the conventional solar thermal collector. The energy losses in different devices of the system have been evaluated. This work has demonstrated that consistent and reliable air conditioning can be achieved by the present system in winter and in summer and thereby can be used as a good solution to reduce the peak load of the electrical grid in the hot summer and cold winter zone of China or the area with similar weather conditions around the world.

Solar photovoltaics are hampered by the lack of solar radiation during peak energy demand hours of the day and variation of solar radiation. The ability to shift the PV power curve and make the energy accessible during peak hours can be accomplished through pairing solar photovoltaics with energy storage technologies. A system of combining solar photovoltaics and ice thermal storage to operate conventional air-conditioning units has been developed [35]. As shown in Fig. 10.5, the prototype system consists of PV panels, a DC compressor, glycol thermal storage (75-gal freezer filled with Cryogel balls, 1-qt oil containers, 12-oz bottles of water immersed in a weak glycol solution), two air handlers (a conventional 1-T HVAC air handler and a glycol air handler), and a power selector box. The Cryogel balls consist of a plastic spherical honeycomb casing filled with water. These balls are used to increase the capacity of the thermal storage. The system combines PV panels and a DC compressor cycling R-134a refrigerant through two separate evaporators directed by two solenoid valves. The use of a DC compressor minimizes the losses of the energy harnessed from the PV panels because no inverter is needed. The first evaporator loop cycles the refrigerant from the compressor through the conventional HVAC air handler. The second loop consists of cycling the refrigerant through the ice thermal storage evaporator contained within a commercial chest freezer.

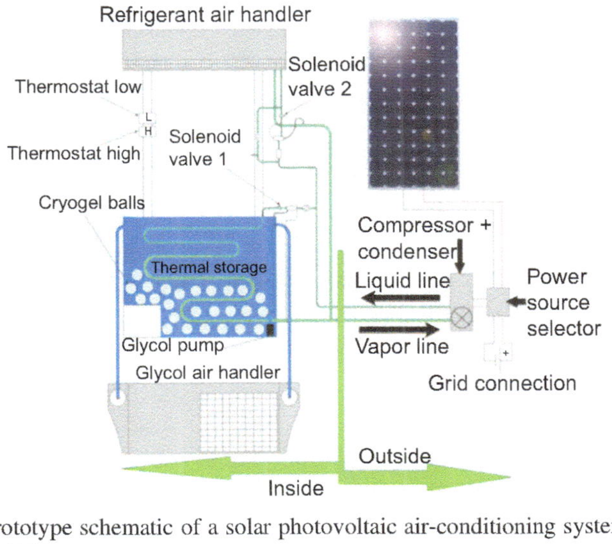

Figure 10.5 Prototype schematic of a solar photovoltaic air-conditioning system and ice thermal storage [35].

10.4.2 Economic evaluation of photovoltaic-powered air conditioner

To achieve a large-scale application PV air conditioner, the system must be reliable and cost-effective. A modeling comparison was made on a solar thermal-powered absorption chiller and a solar PV-driven cooling system that uses a vapor compression machine [36]. Both solar cooling systems are compared against a standard air-cooled cooling system that uses electricity from the grid. Results are presented in two figures. Each figure has one curve for the solar thermal cooling system and one for the solar PV cooling system. One figure allows estimation of savings calculated based the present value of the discounted energy consumption cost. The other figure allows estimating primary energy consumption reduction and emissions reduction. Both figures present the result per ton of refrigeration and as a function of area of solar collectors and/or the area of the PV modules. It was declared that the approach to present the result of the simulations of the systems makes these figures general. This means that the results can be used to compare both solar cooling systems independent of the cooling demand (capacity of the system) and allow for the analysis for different sizes of the solar system used to harvest the solar energy (collectors or PV modules).

Also an economic analysis was performed by Eicker et al. [32] based on an annual full-cost accounting and compared with the results of a reference system. It was found that if electricity export to the grid or for appliance use is included in the primary energy analysis, then PV cooling systems always have higher primary energy savings. The total cooling costs for solar electric cooling are comparable to solar thermal cooling systems if there is no feed in tariff for the excess PV electricity. If a grid export is possible and paid for, then solar electric cooling systems are always more advantageous. If no remuneration is available for excess PV electricity, then the total annual costs of both systems are comparable depending on the assumptions above. If a feed in tariff is considered (eg, with equal costs for grid injection and extraction), then PV AC systems are always advantageous. In these cases total costs and payback times are lower for a PV air conditioner than for solar thermal systems. Therefore the lowest payback times of 2 years are obtained in situations with high grid export of PV electricity. If there is no feed in tariff, then this situation drastically changes and the payback times are best for locations with high cooling operating hours.

10.4.3 Commercial products of photovoltaic-powered air conditioners

A PV direct-driven inverter centrifugal chiller as a central HVAC system is available on the market in China [37,38] The system adopts innovative ternary current commutation technology and integrates the function of the DC to AC converter. The system cannot only provide the PV-generated electricity to power the air conditioner, but it can also supply surplus electricity to the power grid. Because of the investment on the PV system, the initial cost will be 100−150% more than a conventional centrifugal chiller. The parameters for a central PV centrifugal chiller are shown in Table 10.2.

Table 10.2 Typical parameters for a central photovoltaic centrifugal chiller

Parameter	Value
AC power supply	380 V to ~3–50 Hz
DC power supply	370–800 V DC
Integrated part load value	11.68
Capacity range	1582–3516 kW
Switch time between operation mode	<10 ms
PV power utilization ratio	99%

AC, alternating current; *DC*, direct current; *PV*, photovoltaic.

As shown in Table 10.2, it takes less than 10 ms to switch among the following five working modes:

1. *Pure grid power mode*: The chiller is driven by electricity from the power grid instead of the PV system. The PV system does no work, and the centrifugal chiller works the same as a common inverter centrifugal chiller.
2. *Grid charge mode*: When the air conditioner does not work and the PV system charges all of its power to the grid line. The air-conditioner unit will function as an inverter, and the overall system works as a PV power plant.
3. *PV air-conditioning mode*: When the load and the PV-generated electricity is matched. Power generated by the PV system provides all of the power consumed by the chiller.
4. *PV air-conditioning and power generation mode*: When the PV-generated power is more than the air-conditioning consumption demand, the PV power will give priority to the air-conditioning operation and then the excess power will be sent to the grid line.
5. *PV air-conditioning and power consumption mode*: When PV-generated power is less than the air-conditioning consumption demand, air-conditioning equipment is driven by the electricity from the grid and from the photovoltaics.

The advantage of the central centrifugal chiller is that it can realize real-time switchover among these five mentioned working modes according to the condition of PV power generation and cooling load demands, which ensures the full utilization of PV power and the stable operation of the air-conditioning system. The variable refrigerant flow/volume system is also available on the market in China. It has similar advantages as the central centrifugal chiller mentioned before. The specification parameters are shown in Table 10.3.

A survey was also conducted for air-conditioning products available on the international market labeled as being "solar-powered." The survey included 25 models across four manufacturers [39–42]. Unlike the refrigerators and freezers, the input voltage requirements were not as uniform in the case of air conditioners. For products that only accepted DC power, 48 V DC is the uniform requirement. Two of the four manufacturers and more than half of the total models require AC power, typically allowing either 208 or 230 V AC. This is likely because air conditioners generally have much larger

Table 10.3 Typical parameters for central photovoltaic centrifugal chiller

Parameter	Value
AC power supply	380 V ~3–50 Hz
DC power supply	370–800 V DC
Operation temperature	−25–60°C
Integrated part load value	7.6
COP	4.17
Capacity range	25.2–1340 kW
Unit power	8/10/12 hp
Switch time between operation mode	<10 ms
Combination range	18–48

AC, alternating current; *DC*, direct current; *COP*, coefficient of performance.

power requirements than refrigerators and freezers; therefore running entirely on solar power is not as feasible. These units are expected to maintain a grid connection at all times, only to be supplemented by solar power as a way to reduce energy costs rather than eliminate it all together. Only one of the four manufacturers uses R-134a as the refrigerant whereas the remainder use R-410a in all of their solar units. In addition, only 3 of the 25 total units are advertised having a heating function as well as cooling.

The air conditioners in this survey all operate on a standard refrigeration cycle and can be effective in all environments. However, PV energy is most effective in hot, dry climates. Likewise, evaporative cooling is at its greatest efficiency in the same climates. Evaporative cooling also runs on much lower power levels, generally meaning that a grid connection may not be as essential as with a standard refrigeration cycle. Some researchers believe that these commonalities make PV and evaporative cooling technologies ideal when combined in a hot, desert environment [43].

10.5 Conclusions

In the process of designing a PV cooling system, some important aspects should be considered:

1. The system should achieve maximum power generation and high electricity generation efficiency η.
2. The electrical voltage, current, and power capacity should be matched among the subsystems.
3. Fit for the climate condition with the smallest storage capacity and the lowest investment to provide the needed cooling loads.

4. If the PV cooling is the best choice in the specified climate zone.
5. It is important to find out the most appropriate climate area for applying PV cooling with the shortest payback year.

The refrigeration system is usually in small capacity and can be used as a cold chain in remote areas. Energy storage can be easily incorporated. The storage method can include cold storage or electricity storage. The control is simple. The system can be grid connected or grid independent.

There are different power capacities for PV-powered air-conditioning systems, from several kilowatts to several thousand kilowatts. For a large-capacity air-conditioning system, grid connection is preferred because of the fluctuation of PV electricity. Commercial products are already available on the market from the window type and split type to the variable refrigerant flow system in small and medium capacity and a central unit in large capacity. The design guides of the PV-powered air-conditioning system in different climate conditions are not available at the time of this writing. Reports on the operation experiences in different climate conditions are rare. More researches and demonstration projects are needed to improve its application and make it more cost-effective.

In daily operation, the cold storage system can be incorporated as an alternative storage technology instead of using a battery. This might be cost-effective because the initial investment on the thermal energy storage is much lower than that for a battery. For the grid-connection system, the cold storage system can also make full use of the valley electricity in the nighttime, which makes it possible to reduce the initial cost of the PV array as well as the operation cost of the overall air-conditioning system. However, in the transition seasons when cooling or heating is not needed, the electricity produced by PV array must be transferred to the grid or used for another purpose. Use of the electricity to produce biofuel as a seasonal storage might be a promising solution to make full use of the photovoltaics. For the large-scale application of a PV-powered AC system, the effective use of the electricity produced by a PV system must be considered for the low working load in the transition season.

For PV refrigeration application, previous research work has found the following:

1. The performance is related to the insulation.
2. A DC compressor might be a good choice when inverter use is canceled.
3. Ice storage or PCM storage might be an alternative solution to battery energy storage. The economic advantage might be justified.
4. Grid connection is not so important as compared with PV air-conditioning application. Because the variation of load is not so big in all seasons, the PV-generated electricity is still needed in spring and autumn.
5. The load pattern of refrigeration is different from that of air conditioning. The electricity produced from the PV system might not be used by the refrigerator in the winter season. The control strategy will be very different.

For PV air-conditioning application, previous research work reveals the following:

1. The performance is related to the insulation, but the load is much higher than refrigeration application.
2. A DC compressor might be a good choice when inverter use is canceled.

3. The grid connection is very important. Because the variation of load in different seasons is very significant, the PV-generated electricity might need to feed into the grid in spring and autumn. The electricity might be charged or discharged from the grid whenever the heating/cooling load does not match the PV-generated electricity.
4. The ice storage or PCM storage might not be justified. In addition, the electricity storage should be carefully analyzed.
5. The load pattern is significantly influenced by the occupants behavior, such as for office application and domestic application. Further study is needed on the control strategy of solar PV air conditioners.
6. Different from the PV refrigerator, the PV air conditioner can work in heat pump mode when heating is needed.

The advantages for PV-powered cooling system are significant. There are many applications and great market potential for a solar PV-powered cooling system. There are already some commercial products available on the market. However, there is not wide application of PV-powered cooling systems, and they are now in the initial stages. The main obstacles for application might include the following:

1. The electricity cost of photovoltaics is still higher than the electricity price from conventional fossil energy.
2. The incentive policy is still not available.
3. The lack of demonstration projects to provide operation experiences under different climate conditions.
4. The lack of a design guide for the system in different climate conditions.
5. Grid connection is not convenient in some areas, and the energy-storage section is still expensive. It can be expected that the application of photovoltaics will expand steadily with the further decrease of PV price, the maturation of related technologies, and as more incentive policies are available.

References

[1] Barbose G, Darghouth NR, Weaver S, et al. Tracking US photovoltaic system prices 1998–2012: a rapidly changing market. Prog Photovoltaics Res Appl 2015;23(6): 692–704.

[2] REN21's renewables 2015 global Status report. www.ren21.net/gsr.

[3] Henning HM. Solar assisted air conditioning of buildings – an overview. Appl Therm Eng 2007;27:1734–49.

[4] Chidambaram LA, Ramana AS, et al. Review of solar cooling methods and thermal storage options. Renew Sustain Energy Rev 2011;15(6):3220–8.

[5] Bansal PK, Martin A. Comparative study of vapour compression, thermoelectric and absorption refrigerators. Int J Energy Res 2000;24:93–107.

[6] Tom O, Omer SA, Taha AZ, Sayigh A. Performance of a photovoltaic solar refrigerator in tropical climate conditions. Renew Energy 1991;1(2):199–205.

[7] Kattakayam TA, Srinivasan K. Thermal performance characterization of a photovoltaic driven domestic refrigerator. Int J Refrig 2000;23:190–6.

[8] Kattakayam TA, Srinivasan K. Lead acid batteries in solar refrigeration systems. Renew Energy 2004;29:1243–50.

[9] Modi A, Chaudhuri A, Vijay B, Mathur J. Performance analysis of a solar photovoltaic operated domestic refrigerator. Appl Energy 2009;86:2583–91.

[10] Kaplanis S, Papanastasiou N. The study and performance of a modified conventional refrigerator to serve as a PV powered one. Renew Energy 2006;31:771–80.

[11] Zhang P, Liu QS, Wang RZ. Performance and applicability of a dc refrigerator powered by the photovoltaics. J Renew Sustain Energy 2010:013101. 1–;1–8.

[12] Gupta BL, Bhatnagar M, Mathur J. Optimum sizing of PV panel, battery capacity and insulation thickness for a photovoltaic operated domestic refrigerator. Sustain Energy Technol Assessments 2014;7:55–67.

[13] Toure S, Fassinou WF. Cold storage and autonomy in a three compartments photovoltaic solar refrigerator: experimental and thermodynamic study. Renew Energy 1999;17: 587–602.

[14] Cherif A, Dhouib A. Dynamic modeling and simulation of a photovoltaic refrigeration plant. Renew Energy 2002;26:143–53.

[15] Yang S, Liu ZB, Ma QB, Liu T. Feasibility analysis of solar photovoltaic DC refrigerator system with cold storage. Refrig Air Conditioning 2010;10 [in Chinese].

[16] Axaopoulos PJ, Theodoridis MP. Design and experimental performance of a PV ice-maker without battery. Sol Energy 2009;83:1360–9.

[17] El-Bahloul A, Ali A, Ookawara S. Performance and sizing of solar driven dc motor vapor compression refrigerator with thermal storage in hot arid remote areas. Energy Procedia 2015;70:634–43.

[18] Xi GN, Cai CS, LI J DY. Experimental study on a solar photovoltaic DC refrigerator. Appl Mech Mater 2013;317:109–12.

[19] Su Q, Liu YF. Solar photovoltaic refrigerator with hydrogen energy and fuel cell. In: Proceeding of Chinese Refrigeration Association; 2009 [in Chinese)].

[20] Tina GM, Grasso AD. Remote monitoring system for stand-alone photovoltaic power plants: the case study of a PV-powered outdoor refrigerator. Energy Convers Manage 2014;78:862–71.

[21] Laidi M, Abbad B, Berdja M, Chikh M. Performance of a photovoltaic solar container under mediterranean and arid climate conditions in Algeria. Energy Procedia 2012;18: 1452–7.

[22] Pedersen PH, Poulsen S, Katic I. SolarChill – a solar PV refrigerator without batteries. Paper Presented at EuroSun 2004 (Freiburg, June 2004).

[23] http://www.sundanzer.com/product/dcf390/ [accessed May 2015].

[24] http://www.sunfrost.com/efficient_refrigerator_specs.html [accessed May 2015].

[25] http://www.thetford.com/product/1210-ultraline/ [accessed May 2015].

[26] http://www.warehouseappliance.com/ [accessed May 2015].

[27] http://www.grapesolar.com/portable-products/refrigerators.html [accessed May 2015].

[28] http://www.whynter.com/productdetail/refrigeration/portable_freezers/309 [accessed May 2015].

[29] Fong KF, Chow TT, et al. Comparative study of different solar cooling systems for buildings in subtropical city. Sol Energy 2010;84(2):227–44.

[30] Hartmann N, Glueckb C, Schmidt FP. Solar cooling for small office buildings: comparison of solar thermal and photovoltaic options for two different European climates. Renew Energy 2011;36(5):1329–38.

[31] Bilgili M. Hourly simulation and performance of solar electric vapor compression refrigeration system. Sol Energy 2011;85:2720–31.

[32] Eicker U, Pietruschka, et al. Comparison of photovoltaic and solar thermal cooling systems for office buildings in different climates. Sol Energy 2015;118:243–55.

[33] Aguilar FJ, Quiles PV, Aledo S. Operation and energy efficiency of a hybrid air conditioner simultaneously connected to the grid and to photovoltaic panels. Energy Procedia 2014;48: 768–77.
[34] Li Y, Zhang G, Lv GZ, Zhang AN, Wang RZ. Performance study of a solar photovoltaic air conditioner in the hot summer and cold winter zone. Sol Energy 2015;117:167–79.
[35] Tobin PL, Sherbeck J, Magerman, B., & Phelan, PE. Solar cooling with ice storage. Proceedings of the ASME 2012 International Mechanical Engineering Congress & Exposition, November 9–15, 2012, Houston, Texas, USA.
[36] Fumo N, Bortone V, Zambrano JC. Comparative analysis of solar thermal cooling and solar photovoltaic cooling systems. J Sol Energy Eng 2013;135. 021002–021011.
[37] http://www.alibaba.com/showroom/gree-air-conditioner.html [accessed July 2015].
[38] http://www.alibaba.com/product-detail/Gree-photovoltaic-direct-driven-inverter-multi_60250536089.html?spm=a2700.7724857.35.1.nAKeRR [accessed July 2015].
[39] http://tech.lennoxintl.com/C03e7o14l/R4j4FSuqYL/ehb_xc17_1504.pdf [accessed July 2015].
[40] http://kingtecsolar.com/products/air-conditioner-solar-k25z-4/ [accessed July 2015].
[41] http://www.hotspotenergy.com/ [accessed July 2015].
[42] http://www.chiltrix.com/solar-chiller-applications.html [accessed July 2015].
[43] http://www.nasa.gov/centers/johnson/techtransfer/technology/MSC-22970-1_Solar-Refrigerator-TOP.html#.Vag24Hj9Cpo [accessed July 2015].

Solar-powered absorption cooling systems

11

Z.Y. Xu, R.Z. Wang
Institute of Refrigeration and Cryogenics, Shanghai Jiao Tong University, Shanghai, China

11.1 Overview

Solar-powered absorption cooling is one of the most promising solar cooling technologies. Owing to the technological maturity and commercialization of the absorption chiller, the solar-powered absorption cooling systems are easier to operate than other solar-powered cooling systems. In this chapter, the concept and the working pair of absorption refrigeration are introduced first. Then, various absorption cooling systems are introduced based on their couplings between the solar collectors.

11.1.1 Absorption refrigeration

The absorption cooling system is a heat-driven cooling system. It is able to utilize low-grade heat, including solar heating power and industry waste heat. Similar to the compression cooling system, the cooling effect of the absorption cooling system also comes from the evaporation of a liquid refrigerant. The difference between the two systems lies in the way in which the refrigerant vapor is pressurized. In the compression cooling system, the refrigerant vapor is pressurized by mechanical compression. In the absorption cooling system, the refrigerant vapor is pressurized by the solution circulating between two pressure levels. Different from the compression cooling system, the absorption cooling system operates with a binary working fluid. The working fluid contains the refrigerant and the absorbent, and it is usually called the working pair. The refrigerant has a lower boiling temperature than the absorbent. In this case, the equilibrium concentration of the binary working fluid varies with temperature and pressure.

Fig. 11.1 shows a schematic diagram of an absorption cooling system. There are two circulation loops in the system. One is the refrigerant loop and the other is the solution loop. In the refrigerant loop, refrigerant flows through the generation, condensation, throttling, evaporation, and absorption processes. In the solution loop, the solution flows through the generation, throttling, absorption, and pumping processes. Two circulation loops are connected by the absorption and generation processes. The absorption and evaporation processes have the same pressure. The generation and condensation processes also have the same pressure. In the absorption process, the refrigerant vapor is absorbed by the working pair under low pressure. This process releases heat to the environment. In the generation process, the refrigerant vapor is boiled off from the working pair under high pressure. This process needs heat input.

Advances in Solar Heating and Cooling. http://dx.doi.org/10.1016/B978-0-08-100301-5.00011-4

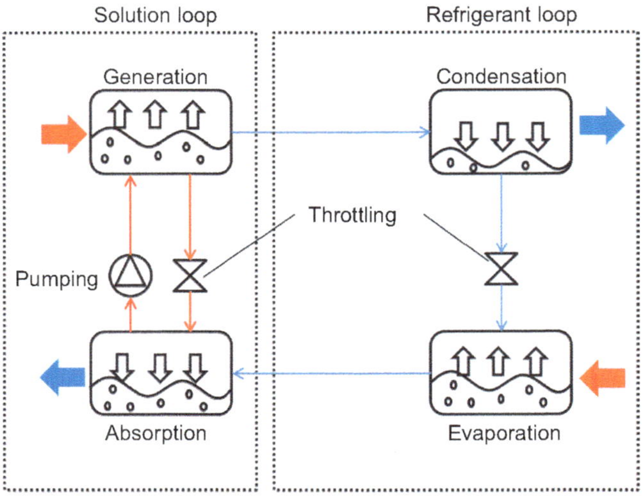

Figure 11.1 Schematic diagram of an absorption cooling system.

The generation and absorption processes act as a heat-driven compressor for the refrigerant, and the condensation and evaporation processes are the same, with a compression cooling system.

11.1.2 Working pair for absorption cooling

The performance of an absorption chiller is determined by the properties of its working pair to a large extent. The refrigerant of the absorption chiller and that of the compression chiller should have properties including high latent heat, suitable pressure, high heat/mass transfer coefficient, and low viscosity. The absorbent should have low specific heat, high affinity with the refrigerant, large boiling temperature difference with the refrigerant, and a high heat/mass transfer coefficient [1]. Both refrigerant and absorbent should be nontoxic, nonflammable, and nonexplosive. In addition, the working pair should be nonazeotropic. Some of the desired properties are mutually exclusive. Compromises should be made for the choice of working pair. Conventional working pairs for absorption cooling include water–LiBr and ammonia–water. They have been widely adopted in real operation.

In the water–LiBr solution, water is the refrigerant and the water–LiBr solution is the absorbent. Water is normal in daily life. It has excellent latent heat, high heat/mass transfer efficiencies, and low viscosity. Water is also cheap and easy to get. Lithium bromide has a relative molecular mass of 86.844. The density of LiBr is 3464 kg/m^3 at 25°C. Its melting point and boiling point are 549 and 1265°C, respectively [2]. LiBr is stable, is nontoxic, and does not volatilize together with water. It is prone to dissolving in water. The concentration and saturation pressure relation of water–LiBr solution deviates strongly from Raoult's law, which indicates a high affinity with the refrigerant. The water–LiBr solution is suitable for absorption cooling. However, the water–LiBr solution will freeze under 0°C and crystalize when the concentration is high. The crystallization concentration varies with temperature. To avoid crystallization in an absorption chiller, the concentration should be less than 65% (referring to the mass concentration of LiBr). In addition, the water–LiBr has a corrosion problem under high temperature and the solution equilibrium pressure is low, which requires vacuum working conditions. Fig. 11.2 shows the P-T-x (pressure–temperature–concentration) diagram of water–LiBr. The concentration refers to the concentration of LiBr.

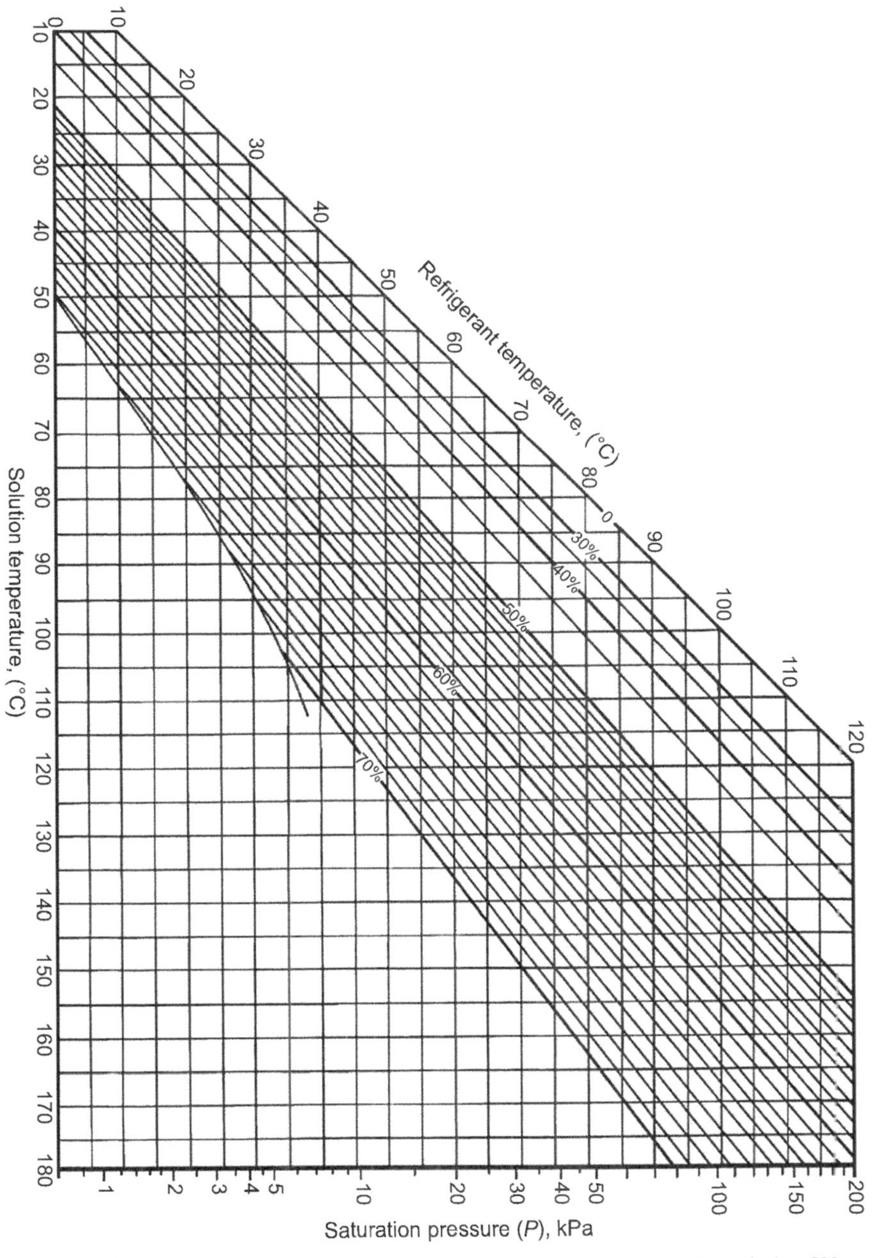

Figure 11.2 Pressure–temperature–concentration diagram of a water–LiBr solution [3].

The equilibrium property of the solution can be calculated from two independent parameters including pressure, temperature, and concentration. Once two independent parameters are confirmed, the other parameters, including enthalpy and entropy, can be calculated.

In an ammonia—water solution, ammonia is the refrigerant and the ammonia—water solution is the absorbent. Ammonia is a colorless gas of low density at room temperature with a pungent smell. It has a relative molecular mass of 17.03 and is lighter than air at atmospheric conditions. It can be stored and transported as a liquid under a pressure of 1 MPa at 25°C. The critical point of ammonia is at 132.3°C and 11.3 MPa. The critical density is 235 kg/m^3 [1]. Ammonia also has high latent heat. Ammonia—water has a low freezing point and does not crystallize. It has an advantage over water—LiBr at low-temperature refrigeration. The vital shortcomings of the ammonia—water working pair are toxicity, flammability, and explosivity when mixed with air. The potential safety problems limit its application. The boiling temperature difference between ammonia and water is small. In this case, a rectifier is essential in the ammonia—water absorption chiller. In addition, the solution equilibrium pressure is high under normal cooling conditions. Fig. 11.3 shows the enthalpy—concentration diagram of ammonia—water. The concentration refers to the concentration of ammonia.

Other than water—LiBr and ammonia—water, many other working pairs can possibly be used for absorption cooling. These working pairs are being researched to improve the efficiency or adaptability of absorption cooling systems. There are four main kinds of working pairs. The water-based working pairs include water—NaOH, water—H_2SO_4, water—LiCl, water—LiI, water—LiBr—LiCl, and water—LiBr—$ZnBr_2$. The alcohol-based working pairs include methyl alcohol—LiBr and methyl alcohol—$ZnBr_2$. The ammonia-based working pairs include methylamine—water, ethylamine—water, and ammonia—NaSCN. The chlorofluorocarbon (CFC)-based working pairs use CFC as the refrigerant and E181, dimethylformamide, or dimethylacetamide as absorbent. These working pairs may have advantages over water—LiBr and ammonia—water in one or two aspects, but in general, they are not as appropriate under normal refrigeration conditions.

11.1.3 Solar-powered absorption cooling system

Solar-powered absorption cooling systems utilize solar heat power to drive an absorption chiller and produce a cooling effect. This is an efficient method for solar-driven refrigeration. Fig. 11.4 shows the systematic diagram of a typical solar-powered absorption cooling system. The system is made up of the following components: the solar collector, the storage tank, the absorption chiller, the auxiliary heat source, and some other accessories. The solar collector offers heat input to the system. Options for solar collectors include flat plat collector, evacuated tube collector, concentrating parabolic collector (CPC), parabolic trough collector (PTC), and so on. The storage tank stores the heat and stabilizes the heat source temperature. The storage tank can be replaced by other thermal storage devices. The absorption chiller utilizes the heat

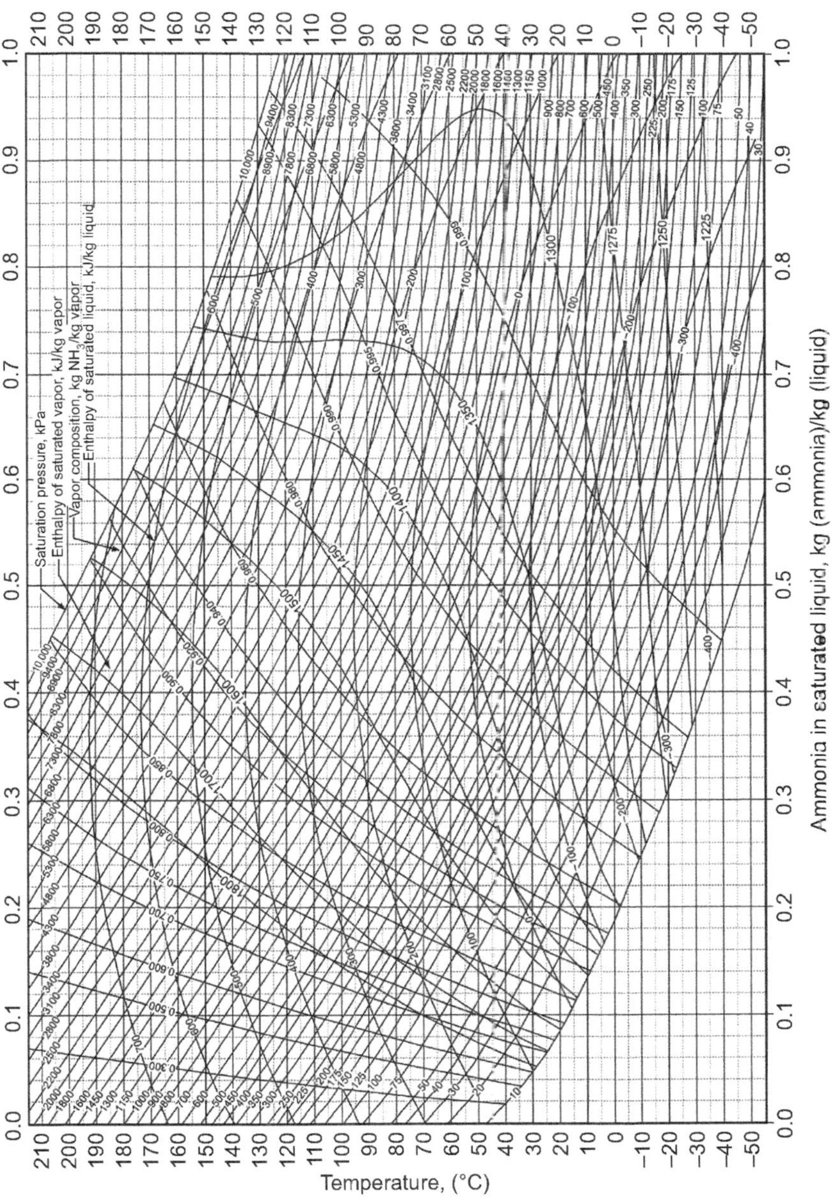

Figure 11.3 Enthalpy–concentration diagram of an ammonia–water solution [3].

from the water tank and delivers cooling output. Options for absorption chillers include half-effect, single-effect, and double-effect absorption chillers with a working pair of water–LiBr or ammonia–water. The auxiliary heat source offers heat input when the solar collector cannot ensure the system's operation. It increases the continuous operation ability of the system.

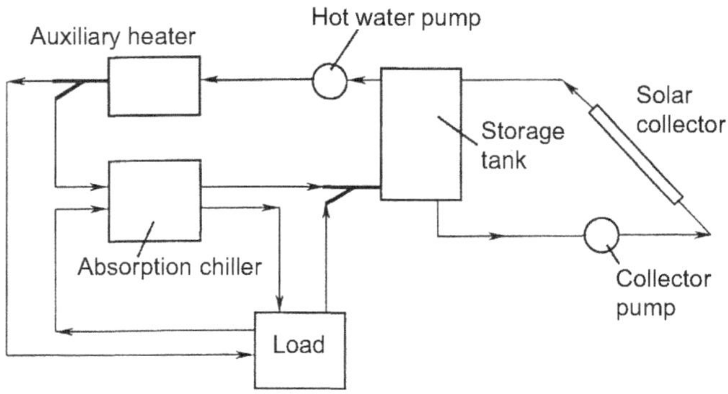

Figure 11.4 Solar-powered absorption cooling system.

Different solar collectors offer heat sources with different temperatures. Absorption chillers require different driving temperatures for various configurations. For example, the double-lift, single-effect, and double-effect absorption chillers require heat input higher than 50, 80, and 140°C for air conditioning (chilled water temperature about 7–10°C), respectively. The combination of solar collector and absorption chiller mainly depends on the temperature coupling. In the following sections, solar absorption cooling technologies are introduced based on the driving temperature of the solar collector.

11.2 Low-temperature solar power-driven systems

Low-temperature solar collectors, including the flat plat collector and evacuated tube collector, offer a heat source lower than 100°C. The low-temperature collector is enough to activate the half-effect and single-effect absorption chillers.

11.2.1 Single-effect water–LiBr absorption cooling system

11.2.1.1 Working principle

The single-effect absorption chiller is the simplest continuously operating absorption chiller, which has been marketed for decades. It is the most commonly used absorption chiller in solar-powered absorption cooling systems. From the real operational perspective, it is also the state of the art. The single-effect absorption chillers are marketed products. Companies including Broad, Carrier, Colibri, Mitsubishi, Robur, Sanyo, Trane, York, and some others all do business in single-effect water–LiBr chillers or single-effect ammonia–water chillers (introduced later). With the continual improvements in single-effect absorption chillers, the chiller coefficient of performance (COP) has reached 0.8.

Fig. 11.5(a) shows the systematic diagram of a single-effect water–LiBr absorption cycle. Fig. 11.5(b) shows a hot-water-powered water–LiBr absorption chiller

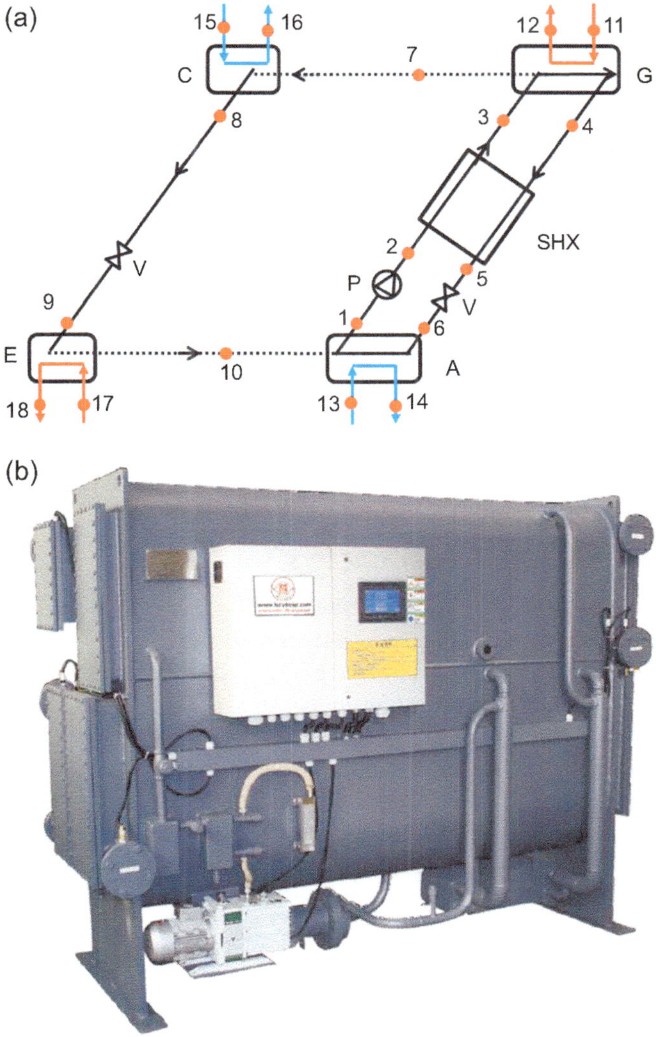

Figure 11.5 Single-effect water–LiBr absorption cycle. (a) Cycle schematic. (b) Chiller product [4]. *A*, absorber; *C*, condenser; *E*, evaporator; *G*, generator; *P*, pump; *SHX*, solution heat exchanger; *V*, valve.

manufactured by Lucy New Energy Technology Co., Ltd. In Fig. 11.5(a), dashed lines represent the flow of vapor and solid lines represent the flow of liquid. The components and connecting lines are placed according to the P-T-x diagram. The generation, absorption, evaporation, and condensation take place in the generator (G), absorber (A), evaporator (E), and condenser (C), respectively. The throttling and pumping processes are conducted by the valve (V) and the pump (P). The generator is heated to boiling by the heat source. The absorber and condenser are cooled by the cooling water continuously or at the same time. The evaporator produces the cooling power and cools the chilled water. To enhance the chiller performance, a solution heat exchanger (SHX) is used. The strong solution from the generator is cooled by the weak solution from the absorber. In this way, the sensitive heat of the strong solution is recovered and the heat input to the generator is reduced.

The detailed circulation of the solution is as follows: the weak solution from A is pumped into the SHX and heated by the strong solution; the preheated weak solution from the SHX flows into G and is heated by the heat source; superheated vapor is generated and the weak solution becomes a strong solution; the strong solution from G flows into the SHX and is cooled by the weak solution; the cooled strong solution flows through V1 and back to A; the strong solution in A absorbs vapor from E, becomes a weak solution, and releases the absorption heat. The detailed circulation of the refrigerant is as follows: the superheated vapor from G is cooled and condensed to a liquid refrigerant in C; the liquid refrigerant flows through V2 to E; the refrigerant in E evaporates into vapor and delivers the cooling output; refrigerant vapor in E flows to A.

11.2.1.2 Modeling and parameters

The major state points in the chiller are labeled in Fig. 11.5. The state point parameters and chiller performance can be calculated based on a physical–mathematical model. In general, the model should contain three kinds of equations: mass balance equations, energy balance equations, and heat transfer equations, as shown in Eqs. [11.1]–[11.3]. The COP of the absorption chiller is defined as the ratio between the cooling output and the heat input as shown in Eq. [11.4]. The balance equations can be applied to each component. Usually, the following assumptions will be made to simplify the model: (1) generator and condenser have the same pressure, (2) absorber and evaporator have the same pressure, (3) refrigerant leaving the condenser or evaporator is saturated, (4) solution leaving the generator or absorber is saturated, (5) throttling processes are adiabatic, (6) pumping process is isentropic, and (7) the temperature difference can be presented by the countercurrent log mean temperature difference method. Based on the balance equations, assumptions, and property equations of the working pair, the state point parameters can be calculated:

$$\sum_i \dot{m}_{\text{ref},i} + \sum_j \dot{m}_{\text{sol},j} = 0 \qquad [11.1]$$

$$\sum_i \dot{m}_{\text{ref},i} \cdot h_{\text{ref},i} + \sum_j \dot{m}_{\text{sol},j} \cdot h_{\text{sol},j} + \dot{Q} = 0 \qquad [11.2]$$

$$UA \cdot \Delta t = \dot{Q} \qquad [11.3]$$

$$\text{COP} = \dot{Q}_{\text{evaporator}} \Big/ \dot{Q}_{\text{generator}} \qquad [11.4]$$

Table 11.1 Parameters of the single-effect water–LiBr absorption chiller—1 [1]

No.	t (°C)	p (kPa)	x (% LiBr)	m (kg/s)	h (kJ/kg)
1	32.9	0.679	56.7	0.0500	85.8225
2	32.9	7.347	56.7	0.0500	85.8266
3	63.2	7.347	56.7	0.0500	147.0
4	89.4	7.347	62.4	0.0455	221.2
5	53.3	7.347	62.4	0.0455	153.9
6	44.7	0.679	62.4	0.0455	153.9
7	76.8	7.347	0.0	0.0045	2644.6
8	39.9	7.347	0.0	0.0045	167.2
9	1.5	0.679	0.0	0.0045	167.2
10	1.5	0.679	0.0	0.0045	2503.4

There are two kinds of physical–mathematical models for an absorption chiller. If only the mass and energy balance equations are imposed, the model calculates the cycle parameters under design conditions. As this model does not consider the heat transfer of each heat exchanger, internal parameters of the cycle including the evaporation temperature and condensation temperature are necessary for the model calculation. Parameters of each state point and heat transfer amount can be obtained. This model mainly gives a parametric description of the cycle under the given conditions. Table 11.1 shows a group of parameters derived from this model. Table 11.2 shows the corresponding heat transfer amount, COP, and SHX effectiveness.

If the heat transfer equations are also imposed, the heat exchange processes between the cycle and the surroundings are counted. This model is also called the heat transfer model. The chiller performance can be calculated from the external conditions including the heat transfer area of the SHX, cooling water temperature, and chilled water temperature. The model can calculate the chiller performance under various operating conditions. It mainly gives a prediction of the chiller performance under off-design conditions. The prediction proceeds in the following order: (1) calculate the UA (heat transfer coefficient multiplied by heat transfer area; it is the same later on in this chapter) parameters with the external and internal cycle parameters;

Table 11.2 Parameters of the single-effect water–LiBr absorption chiller—2 [1]

$Q_A = 14.039$ kW	$Q_C = 11.213$ kW	$Q_E = 10.574$ kW	$Q_G = 14.678$ kW
$Q_{SHX} = 3.06$ kW	$W_P = 0.000206$ kW	$\varepsilon_{SHX} = 0.640$	COP $= 0.720$

Table 11.3 External parameters of the absorption chiller [1]

No.	11	12	13	14	15	16	17	18
t (°C)	100.0	96.5	25.0	37.0	25.0	34.6	10.0	3.7
m (kg/s)	1.00	1.00	0.28	0.28	0.28	0.28	0.40	0.40
h (kJ/kg)	418.9	404.2	104.8	154.9	104.8	144.8	42.0	15.6
UA (kW/K)	$UA_A = 1.8$		$UA_C = 1.2$		$UA_E = 2.25$		$UA_G = 1$	

(2) change one of the external parameters, internal parameters, or UA parameters; (3) solve the equation group to obtain the cycle parameters. Table 11.3 gives the external data and UA parameters based on the same conditions as in Table 11.2. The state point numbers referred to are labeled in Fig. 11.5.

According to the simulation results based on the heat transfer model, when the hot water inlet temperature increases from 50 to 120°C, the COP varies in a small range between 0.625 and 0.75, and the cooling power increases from 2 to 13.5 kW. When the chilled water inlet temperature increases from 8.0 to 20.0°C, the COP increases a little around 0.75, and the cooling output increases from 10.0 to 12°C. When the cooling water inlet temperature increases from 15 to 35°C, the COP decreases from 0.75 to 0.67, and the cooling power decreases from 15 to 6.5 kW [1]. These results indicate that high heat-source temperature, high chilled-water temperature, and low cooling-water temperature are beneficial for good chiller performance. In addition, a large concentration glide and high SHX effectiveness increase the COP. A large solution flow rate increases the cooling power but decreases the COP.

Real absorption chillers vary in many aspects. Hot water, steam, smoke, or a burner can be used as the heat source. Absorption chillers driven by different heat sources need different generators. The high-pressure chamber (generator and condenser) and the low-pressure chamber (absorber and evaporator) could be put in one shell or two shells. A single-shell chiller is compact, is short, and does not need to connect tubes between the chambers. A double-shell chiller has low heat loss, low thermal stress, and small installation area.

11.2.1.3 Solar-powered case

Table 11.1 indicates that the single-effect water–LiBr absorption chiller needs a driving temperature of less than 100°C. The low-temperature solar collector is enough to drive this chiller. The low-temperature solar collector-driven single-effect water–LiBr absorption cooling system is also the most popular solar-powered cooling system owing to its simplicity. In this section, an experimental study of solar cooling systems will be introduced. The system was built according to Fig. 11.4 without the auxiliary heater. The experiment was carried out in Madrid during the summer of 2003. Fig. 11.6 shows the solar collector array installed on the roof and the absorption chiller.

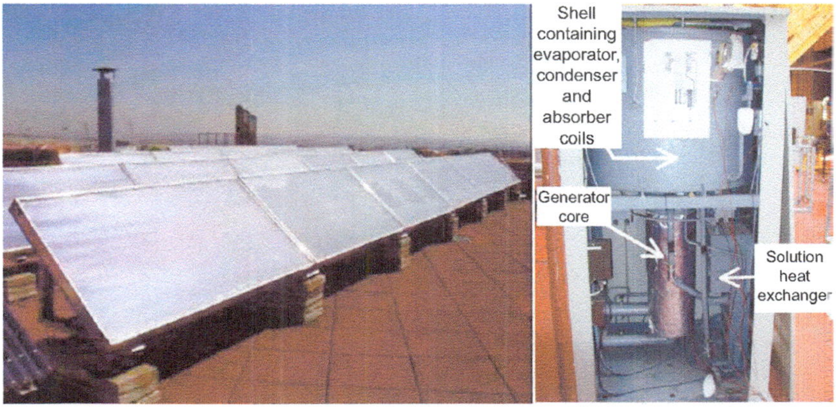

Figure 11.6 Solar collector and absorption chiller in the system [5].

A 35-kW single-effect water–LiBr absorption chiller was integrated with flat plate collectors. Twenty flat plate collector modules with an absorber area of 2.5 m^2 were used. The collector supplied hot water to a 2-m^3 storage tank through a plate heat exchanger when the collector outlet temperature exceeded the average tank temperature by 2–3°C. The storage tank then offered hot water to the absorption chiller. The chiller was cooled by cooling water with inlet/outlet temperature about 20/25°C. Chilled water from the absorption chiller delivered a cooling effect to the fan coil at about 8°C. Fig. 11.7 shows the parameters variation of the solar cooling system during the daytime on August 8, 2003. In Fig. 11.7(a) the ambient conditions including dry bulb temperature, wet bulb temperature, and relative humidity are shown. In Fig. 11.7(b), the temperature changes in the system are shown. Taking the time of 1150 as an example, the collector outlet temperature (87°C) was the highest, followed by the generator inlet temperature (77°C), collector inlet temperature (75°C), generator outlet temperature (67°C), cooling water outlet temperature (25°C), cooling water inlet temperature (21°C), chilled water inlet temperature (13°C), and chilled water outlet temperature (10°C). The chiller operated steadily during the day.

11.2.2 Single-effect ammonia–water absorption cooling system

11.2.2.1 Working principle

The performance of the ammonia–water single-effect absorption chiller is decided by its working fluid. It has a lower efficiency than the water–LiBr single-effect absorption chiller under the same conditions owing to the following reasons. (1) An ammonia–water system requires a rectifier, which is not needed in a water–LiBr system. (2) Ammonia's latent heat is about half that of water. For the same cooling output, the circulation rate of ammonia–water is double that of water–LiBr. A high circulation rate increases the preheating consumption in the generator and decreases the cycle

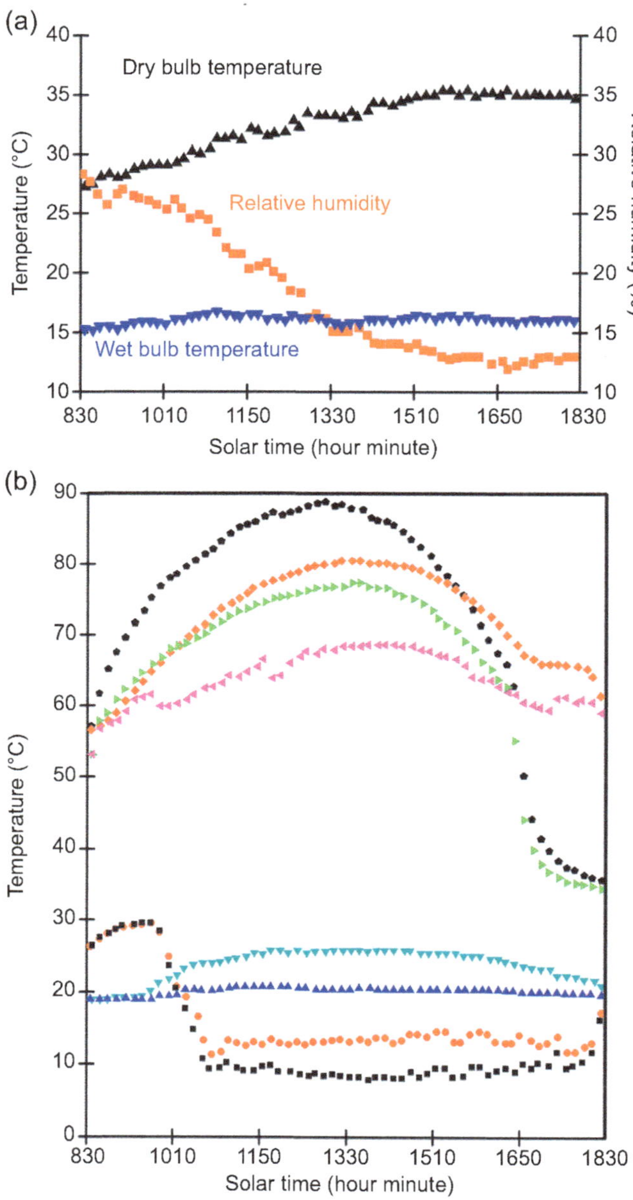

Figure 11.7 System parameter variation during a day [5].

efficiencies. (3) The specific heat of ammonia is half that of water. For the same heat exchange amount in an SHX, an ammonia–water solution has a larger temperature change than a water–LiBr solution. In another words, higher exchanger effectiveness is essential for the ammonia–water system. In addition, ammonia–water has some drawbacks, including toxicity and a high working pressure. However, the ammonia–water absorption chiller is still an efficient way to obtain refrigeration under 0°C. Manufacturers of ammonia–water absorption chillers include Robur, Energy Concepts, and some other companies.

There are several ways to improve the efficiency of ammonia–water single-effect absorption chillers, including the rectifier heat recovery and condensate precooling. (1) As the absorbent of the ammonia–water working pair is water, the absorbent volatilizes with the refrigerant. About 5~10% water exists in the vapor generated from the generator. Rectification is needed to improve the purity of the superheated refrigerant vapor. This part of the rectification heat can be used to heat the rich solution. (2) The refrigerant vapor from the evaporator still has a low temperature, which is much lower than that of the refrigerant liquid from the condenser. A precooler can be integrated to precool the refrigerant out of the condenser using the refrigerant out of the evaporator. The precooler is also beneficial for the water–LiBr absorption cycle, but the low pressure of the water–LiBr system limits its application.

Fig. 11.8 shows the schematic diagram for the single-effect ammonia–water absorption cycle with a precooler (PC): the vapor out of the evaporator (E) flows through the PC to the absorber. The lines through the PC do not represent the real parameter change in the P-T-x diagram. To be precise, only solid lines are used here to represent the flow direction. The components are still placed in the order of the P-T-x diagram. The state points are labeled in the diagram and the corresponding parameters of each state point are shown in Table 11.4. The calculation is based on the following assumptions: (1) steady state; (2) no pressure changes except flow through the throttling valve and pump; (3) states at points 1, 4, 8, 10, and 13 are saturated liquid; (4) states at points 7 and 14 are saturated vapor; (5) throttling is adiabatic and pumping is isentropic; (6) vapor leaving the generator is at equilibrium temperature of the entering solution stream; (7) the effectiveness values of SHX and PC are 0.692 and 0.629, respectively [3].

Attention to several parameters should be paid. (1) The highest system pressure is 1461 kPa, whereas it is 7.347 kPa for the water–LiBr system. The ammonia–water

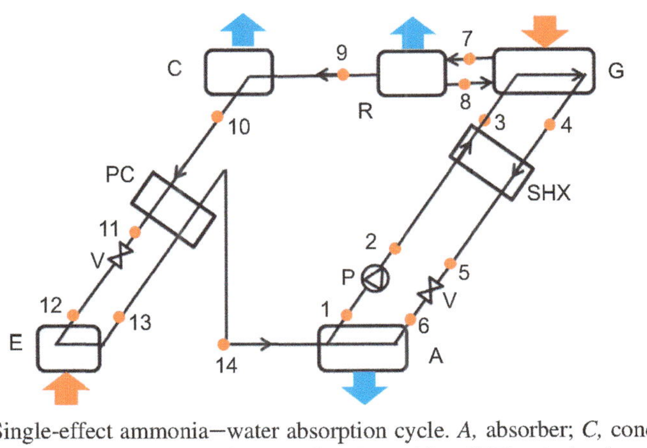

Figure 11.8 Single-effect ammonia–water absorption cycle. *A,* absorber; *C,* condenser; *E,* evaporator; *G,* generator; *P,* pump; *PC,* precooler; *R,* rectifier; *SHX,* solution heat exchanger; *V,* valve.

Table 11.4 Parameters of the single-effect ammonia–water absorption chiller [3]

No.	t (°C)	p (kPa)	x (% ammonia)	m (kg/s)	h (kJ/kg)
1	40.56	515.0	50.094	10.65	−57.2
2	40.84	1461	50.094	10.65	−56.0
3	78.21	1461	50.094	10.65	89.6
4	95.00	1461	41.612	9.09	195.1
5	57.52	1461	41.612	9.09	24.6
6	55.55	515.0	41.612	9.09	24.6
7	79.15	1461	99.809	1.59	1429
8	79.15	1461	50.094	0.04	120.4
9	55.00	1461	99.809	1.55	1349
10	37.82	1461	99.809	1.55	178.3
11	17.80	1461	99.809	1.55	82.1
12	5.06	515.0	99.809	1.55	82.1
13	6.00	515.0	99.809	1.55	1216
14	30.57	515.0	99.809	1.55	1313
$Q_G = 3083$ kW		$Q_A = 2869$ kW		$Q_C = 1862.2$ kW	
$Q_E = 1760$ kW		$Q_{SHX} = 1550$ kW		$Q_{PC} = 149$ kW	
$Q_R = 170$ kW		$W_P = 12.4$ kW		COP = 0.571	

system has much higher pressure. (2) The pump consumption is 12.4 kW for 1760 kW cooling output, whereas it is 0.000206 kW for 10.574 kW in the water–LiBr system. The pump consumption for unit cooling output of the ammonia–water system is 361 times that of the water–LiBr system. (3) The PC increases 149 kW heat recovery, which constitutes 8.5% of the total cooling output. (4) The COP of 0.571 is obtained under a generator temperature and evaporator temperature of 95.00 and 6.00°C, respectively. The system is less efficient than the water–LiBr system.

11.2.2.2 Solar-powered case

In this section, an experimental study of a solar-powered single-effect ammonia–water cooling system will be introduced. Fig. 11.9(a) shows the schematic diagram of the system. An ammonia–water absorption chiller was used. This system integrated a parabolic collector to offer a heat source to the chiller. The collector was controlled by an electric control system through north to south axes. The measured temperatures of each component are shown in Fig. 11.9(b). In the experiments, generator temperatures of 80–100°C were obtained from the solar collector. Condenser temperatures of

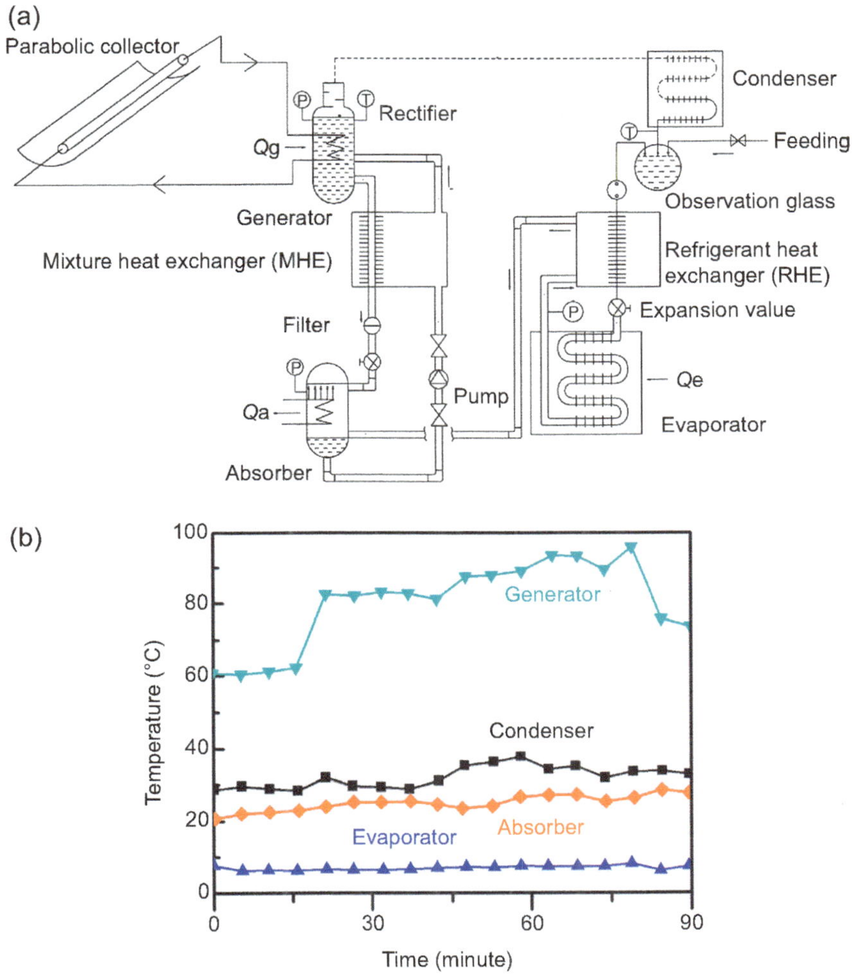

Figure 11.9 The solar-powered single-effect ammonia–water system [6].

25–30°C were obtained. Absorber temperatures of 20–23°C were obtained. Evaporation temperature decreased from 7.5 to 2°C [6].

11.2.3 Double-lift absorption cooling system

11.2.3.1 Working principle

Double-lift absorption cooling systems have lower driving temperature than single-effect absorption cooling systems. A double-lift water–LiBr absorption chiller, which is also called a half-effect water–LiBr absorption chiller, is able to use a heat source of 70–80°C. It is important for the utilization of low-grade heat. However, the double-lift chiller has a COP of only 0.3–0.4, which is about 55% that of the single-effect chiller. In addition, the double-lift chiller needs more components, and this increases the cost.

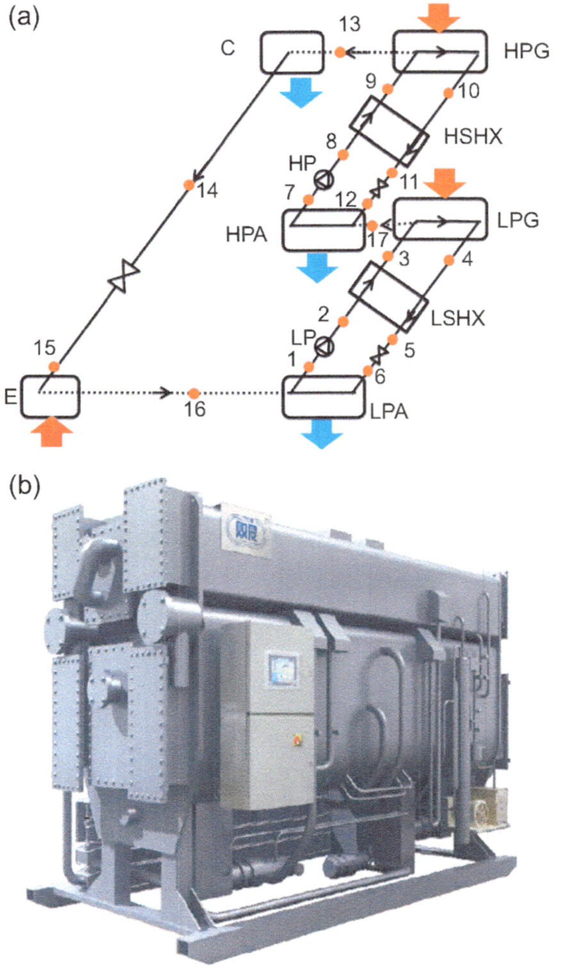

Figure 11.10 Double-lift absorption cycle. (a) Cycle schematic. (b) Chiller product [7]. *C*, condenser; *E*, evaporator; *HP*, high-pressure pump; *HPA*, high-pressure absorber; *HPG*, high-pressure generator; *HSHX*, high-temperature solution heat exchanger; *LPA*, low-pressure absorber; *LPG*, low-pressure generator; *LSHX*, low-temperature solution heat exchanger.

Fig. 11.10(a) shows the schematic diagram of a double-lift absorption chiller. Fig. 11.10(b) shows a hot-water-driven double-lift absorption chiller manufactured by Shuangliang Eco-Energy Systems Co., Ltd. Compared with the single-effect absorption chiller, this chiller has one more group comprising a generator, absorber, valve, and solution pump. This chiller has two solution loops and one refrigerant loop. The refrigerant and solution loops of this chiller are the same as those of the single-effect absorption chiller, but the connections are different. The detailed circulation of the double-lift cycle is as follows. (1) In the low-pressure solution loop, the solution flows through an LPA (low-pressure absorber), low-pressure pump (LP), LSHX (low-temperature SHX), LPG (low-pressure generator), LSHX, valve 1, and back to the LPA. The LPG is heated to boiling by the heat source and the LPA is cooled by the ambient air. This solution loop is the same as the solution loop in a single-effect absorption chiller. (2) In the high-pressure solution loop, the solution flows through an HPA (high-pressure absorber), high-pressure pump (HP), HSHX (high-temperature SHX), HPG (high-pressure generator), HSHX, valve 2, and back to the HPA. The HPG is heated to boiling by the heat source and the LPA is cooled by the ambient air. (3) The refrigerant that is boiled in the LPG is absorbed by the solution in the HPA. (4) The refrigerant boiled off from the HPG is condensed in the C, throttled in valve 3, evaporated in the E, and then absorbed by the LPA.

The parameters of each key state point are shown in Table 11.5. The calculation is based on a water–LiBr solution. The effectiveness of the SHXs is set as 0.3. As is

Table 11.5 Parameters of the double-lift water–LiBr absorption chiller [1]

No.	t (°C)	p (kPa)	x (% LiBr)	m (kg/s)	h (kJ/kg)
1	30.0	0.657	55.4	3.000	74.7
2	30.0	1.932	55.4	3.000	74.7
3	38.1	1.932	55.4	3.000	91.4
4	65.0	1.932	63.6	2.614	182.7
5	54.5	1.932	63.6	2.614	163.5
6	46.6	0.657	63.6	2.614	163.5
7	30.0	1.932	42.7	2.500	57.1
8	30.0	7.368	42.7	2.500	57.1
9	38.2	7.368	42.7	2.500	76.5
10	65.0	7.368	50.5	2.114	137.3
11	54.5	7.368	50.5	2.114	114.4
12	40.3	1.932	50.5	2.114	114.4
13	54.8	7.368	0.0	0.386	2602.0
14	40.0	7.368	0.0	0.386	167.4
15	1.0	0.657	0.0	0.386	167.4
16	1.0	0.657	0.0	0.386	2502.3
17	48.1	1.932	0.0	0.386	2591.7
Q_{HPG} = 1090.1 kW		Q_{HPA} = 1099.2 kW		Q_{LPG} = 1203.7 kW	
Q_{LPA} = 1103.5 kW		Q_C = 939.9 kW		Q_E = 901.1 kW	
Q_{HSHX} = 48.5 kW		Q_{LSHX} = 50.1 kW		W_{HP} = 0.0097 kW	
W_{LP} = 0.0024 kW			COP = 0.391		

shown in Table 11.5, the generation temperature, condensation temperature, absorption temperature, and evaporation temperature are set as 65.0, 40.0, 30.0, and 1.0°C, respectively. The heat consumption in the HPG, LPG, and E is 1090.1, 1203.7 and 901.1 kW, respectively. A COP of 0.391 is obtained. The HPG needs less heat input than the LPG because the solution concentration in the HPG is smaller than that of the LPG.

11.2.3.2 Solar-powered case

The double-lift water–LiBr absorption chiller has a lower driving temperature than the single-effect chiller. This makes the solar absorption cooling system with a double-lift chiller work longer. In addition, the low driving temperature also increases the efficiency of the solar collector. To better describe the performance of a solar-powered double-lift cooling system, some experimental research is introduced in this chapter. Fig. 11.11 shows the schematic diagram of a solar-driven double-lift cooling system for real operation. This 100-kW system was built on a 24-floor building in Jiangmen, China (longitude 113°E, latitude 22.40°N). The system was built to supply hot water and air conditioning for part of the building [8]. The corresponding components in the system for Fig. 11.11 are as follows: (1) solar collector, (2) hot water tank for cooling, (3) double-lift absorption chiller, (4) cooling water tank, (5) cooling tower, (6) chilled water tank, (7) hot water for daily lift, (8) oil burner, (9) air conditioning user, and (10) daily lift for hot water use.

The detailed parameters of the main components are as follows: (1) A modified flat plate solar collector with area of 500 m^2 was used, the hot water supply amount was 30 m^3/day, the supply temperatures were 60°C for daily use and 75°C for driving the

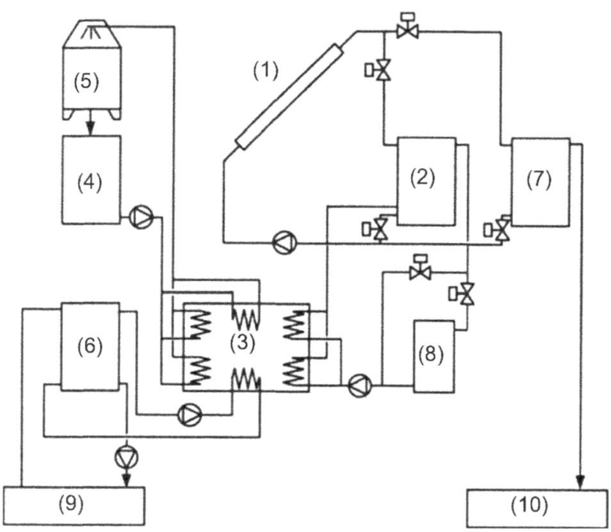

Figure 11.11 Schematic of a solar-driven double-lift absorption cooling system [8]. (1) Solar collector, (2) hot water tank for cooling, (3) double-lift absorption chiller, (4) cooling water tank, (5) cooling tower, (6) chilled water tank, (7) hot water for daily lift, (8) oil burner, (9) air conditioning user, and (10) daily lift for hot water use.

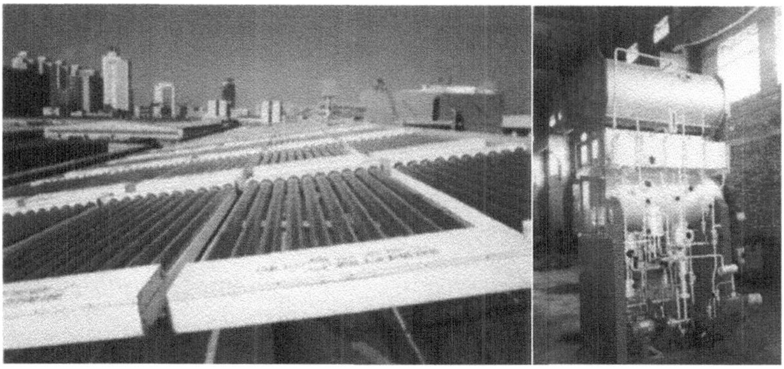

Figure 11.12 Pictures of the solar collector and the double-lift absorption chiller [8].

chiller. (2) A double-lift water–LiBr absorption chiller with a nominal cooling capacity of 100 kW was used; the heat input temperature and chilled water temperature were 75 and 9°C, respectively. (3) The air-conditioning area was 600 m^2 and additional heat from an oil burner was used to ensure continuous operation [8]. As is shown in Fig. 11.11, this system was similar to that of Fig. 11.4. In this system, the hot water from the solar collector flowed into a hot water tank (2) for absorption cooling and flowed into a hot water tank (7) for daily hot water supply. During the summer, priority was given to providing hot water for absorption cooling. Fig. 11.12 shows the solar collector and the 20-RT double-lift absorption chiller used in this system.

A group of chiller operating data is shown in Table 11.6. It can be seen that the chiller was able to work even when the heat source temperature was 60.8°C, which

Table 11.6 Performance of the double-lift absorption chiller under solar-driven conditions [8]

Date and time (24-h clock)	Hot water temp (°C)		Chilled water temp (°C)		Cooling water temp (°C)		Cooling output (kW)	COP
	Inlet	Outlet	Inlet	Outlet	Inlet	Outlet		
April 29 1130	60.8	50.4	14.7	10.1	28.4	33.0	71.3	0.338
April 29 1300	62.0	51.9	11.1	6.8	28.2	32.1	66.7	0.373
May 4 1200	69.1	55.7	15.5	10.1	29.6	35.7	102.8	0.434
May 4 1530	68.6	56.0	13.5	8.7	30.1	35.8	91.3	0.410
May 5 1300	69.1	56.9	12.0	7.5	29.6	35.1	86.6	0.397
May 6 1000	62.6	52.2	14.4	10.1	29.1	33.9	81.8	0.440
May 7 1130	70.8	56.8	15.1	9.5	29.2	35.7	106.5	0.426
May 8 1100	66.4	54.1	17.3	12.0	29.9	35.5	100.8	0.458
May 8 1530	72.6	59.7	14.5	8.8	30.6	36.9	108.4	0.437

COP, coefficient of performance.

was much lower than the driving temperature of a single-effect absorption chiller. The COP varied a little with the driving temperature. The low driving temperature only decreased the cooling power. This indicated that the integration of a double-lift water–LiBr absorption chiller into the solar cooling system expanded the serviceable range of the system.

11.2.4 Other configurations

11.2.4.1 Double-lift ammonia–water absorption chiller

The solar absorption cooling technology is energy saving, but the system is large in both size and power. To promote this green technology, the domestic size system without a cooling tower should be developed. However, the double-lift water–LiBr absorption chiller shown in Fig. 11.10 is not suitable for air cooling because the high cooling temperature increases the crystallization risk. In this case, a double-lift absorption chiller with a noncrystallizing working pair, for example, ammonia–water, is reasonable.

Fig. 11.13 shows a simulation and experimental study of an air-cooled double-lift ammonia–water absorption chiller. The calculation was based on the cooling output of 5 kW. Fig. 11.13(a) shows the calculation results on a P-T-x diagram. The cycle obtained a calculated COP of 0.34 when the hot water temperature was 85°C, cooling air temperature was 35°C, and evaporation temperature was 10°C [9]. An experimental prototype was designed and tested based on this cycle as shown in Fig. 11.13(b), and the COP of the prototype stabilized between 0.18 and 0.25 [10].

11.2.4.2 Double-lift water–LiBr absorption chiller for air-cooling conditions

To make an absorption chiller that operates well under air-cooling conditions, the double-lift absorption chiller with an ammonia–water solution is not the only way. Fig. 11.14(a) shows another configuration for a double-lift absorption chiller that is suitable for water–LiBr with a low crystallization risk.

In this cycle, the strong solution from the G flows into the medium-pressure absorber (MPA) and LPA. The refrigerant liquid from C flows into a medium-pressure evaporator (MPE) and low-pressure evaporator (LPE). The refrigerant liquid in the MPE absorbs the heat from the LPA, evaporates into vapor, and is then absorbed by the strong solution in the MPA, which is cooled by the ambient air. The refrigerant liquid in the LPE delivers cooling output, evaporates into vapor, and is then absorbed by the strong solution in the LPA, which is cooled by the MPE. The weak solution from the LPA and MPA then flows together into the G. In this cycle, the cooling of the LPA consumes part of the cooling output and decreases the cooling output temperature. Serial and parallel configurations can be adopted for the solution flow. In serial flow, the solution from the G flows into the MPA and LPA successively. In parallel flow, the solution from the G flows into the MPA and LPA at the same time. To obtain better performance from the chiller, SHXs can be integrated. As the cycle can be

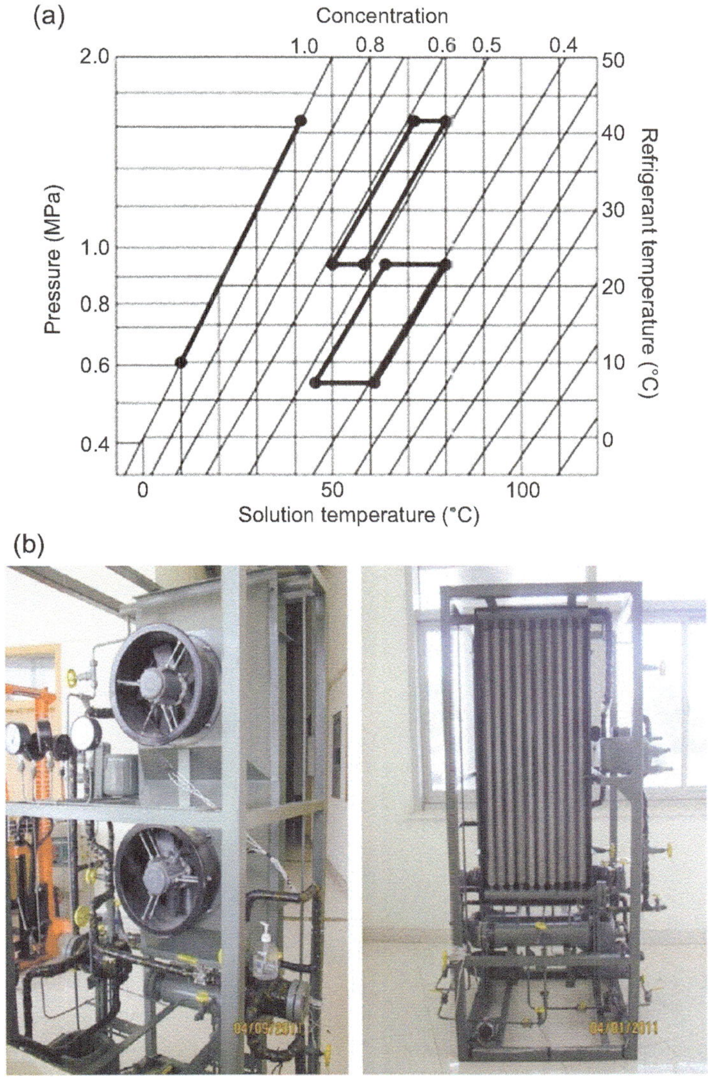

Figure 11.13 Two-stage air-cooled ammonia–water absorption refrigeration system. (a) Cycle parameters [9]. (b) Prototype [10].

regarded as an integration of two single-effect absorption cycles (G-MPA-MPE-C and G-LPA-LPE-C) through the coupling between the MPE and the LPA, this cycle is also called a heat-coupled double-lift cycle.

If the working conditions are set the same as in Table 11.5, the concentrations at the generator outlets of two double-lift water–LiBr chillers are the same, while it is different for the LPA inlet. In the double-lift absorption chiller in Fig. 11.10, the generator outlet concentration is 50.5% and the LPA inlet concentration is 63.6%.

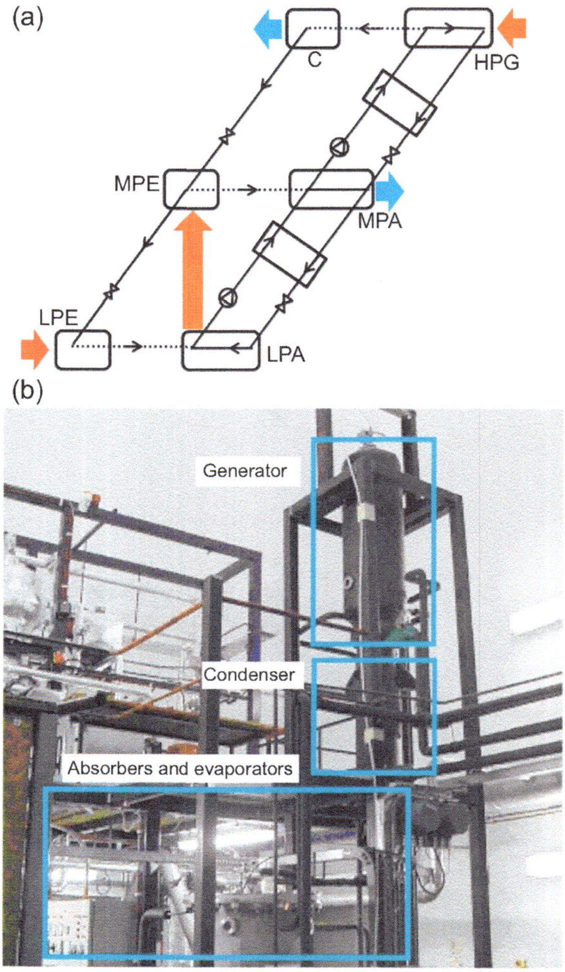

Figure 11.14 Double-lift water–LiBr absorption cycle for air-cooling conditions. (a) Cycle schematic. (b) Experimental setup [11]. *C,* condenser; *HPG,* high-pressure generator; *LPA,* low-pressure absorber; *LPE,* low-pressure evaporator; *MPA,* medium-pressure absorber; *MPE,* medium pressure evaporator.

In the double-lift absorption chiller in Fig. 11.14, the inlet solution concentration of the LPA is less than 50.5%. The huge decrease in LPA inlet concentration weakens the crystallization risk greatly. This makes the double-lift absorption chiller in Fig. 11.14 suitable for air-cooling conditions, which is important in the development of a domestic size chiller. Fig. 11.14(b) shows an experimental setup of this chiller. According to the test, the experimental chiller obtained a COP of 0.25–0.3 under a cooling water inlet temperature of 28–44°C [11]. The COP was stable with the cooling water temperature variation.

11.3 Medium-temperature solar power-driven systems

Medium-temperature solar collectors, including CPCs and PTCs, offer a heat source with temperature between 100 and 200°C. The medium-temperature collector is enough to activate a double-effect water−LiBr absorption chiller, a generator absorber heat exchange (GAX) absorption chiller, a single-effect ammonia−water absorption ice-maker, and a diffusion−absorption chiller.

11.3.1 Double-effect water−LiBr absorption cooling system

11.3.1.1 Working principle and parameters

The single-effect water−LiBr absorption chiller has a COP about 0.7 under a driving temperature less than 100°C. However, the chiller COP does not increase when the driving temperature increases. To better utilize the high-temperature heat source, a double-effect water−LiBr absorption chiller is needed. A double-effect water−LiBr absorption chiller obtains a COP of 1.0−1.2 with a driving temperature of about 140°C. The gas-fired chiller is a mature technology, which is competitive among the gas-fired cooling technologies. It is produced by many manufacturers mentioned earlier. Fig. 11.15(b) shows the double-effect absorption chiller product (direct-fired type) manufactured by Shuangliang Eco-Energy Systems Co., Ltd. In addition to the direct-fired type, flue gas type, steam type, and hot-water type double-effect chillers are also available. For solar cooling systems, the double-effect water−LiBr chiller is a highly efficient option aimed at air conditioning. Steam type or hot-water type chillers can also be used.

Fig. 11.15(a) shows the schematic diagram of the double-effect cycle. The HPG is boiled by the heat source and the E delivers cooling output. This cycle could be interpreted as the coupling between two single-effect subcycles. The first subcycle is composed of the HPG, A, E, and high-pressure condenser (HPC). The second subcycle is composed of the LPG, A, E, and low-pressure condenser (LPC). To recover the solution-sensitive heat, two SHXs are integrated in each subcycle. The two subcycles are integrated together through the heat coupling between the HPC and the LPG. In other words, the condensation heat of the high-pressure vapor is recovered to heat the LPG to boiling. In this case, one unit of heat input to the HPG generates two units of refrigerant vapor in the HPG and LPG, and the system is called the double-effect absorption cycle. The key point ensuring the operation of the double-effect absorption chiller is that the HPC has a higher temperature than the LPG.

Table 11.7 shows the calculation results for the double-effect water−LiBr absorption cycle corresponding to Fig. 11.15. Considering the importance of the double-effect water−LiBr absorption cycle, a heat transfer model is used for the performance calculation. The calculation is based on the following settings: (1) the heat source temperature, cooling water temperature, and chilled water temperature are 150, 25, and 12°C, respectively; (2) the outlet of each component is at saturated conditions except for the SHX and the super-heated vapor from the generator; (3) the effectiveness of the SHX is 0.5; (4) the UA (referred to in Section 11.2.1.2) values of HPG, HPC−LPG,

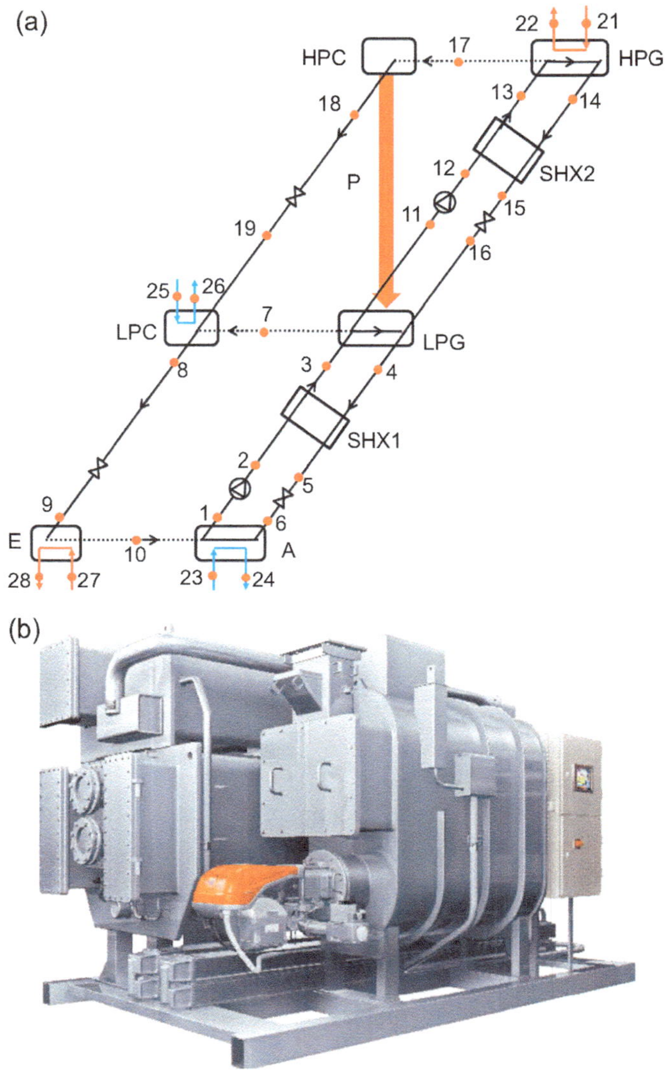

Figure 11.15 Double-effect absorption cycle. (a) Cycle schematic. (b) Chiller product [7]. *A,* absorber; *E,* evaporator; *HPC,* high-pressure condenser; *HPG,* high-pressure generator; *LPC,* low-pressure condenser; *LPG,* low-pressure generator; *SHX,* solution heat exchanger.

LPC, A, and E are 25, 10, 65, 50, and 40 kW/K, respectively. According to the calculation, the generation temperature, condensation temperature, absorption temperature, and evaporation temperature are 144.84, 29.72, 29.85, and 5.13°C, respectively. The driving temperature is complementary to the medium solar collector. The outlet temperatures of HPC and LPG are 87.73 and 76.39°C, respectively [1]. Compared with the single-effect system, the LPG outlet temperature is lower and the concentrations are similar. This is caused by the low condensation temperature here.

Table 11.7 Parameters of double-effect water−LiBr cycle in Fig. 11.15 [1]

No.	t (°C)	p (kPa)	x (% LiBr)	m (kg/s)	h (kJ/kg)
1	29.85	0.880	52.765	1.000	65.5911
2	29.85	4.171	52.765	1.000	65.5932
3	47.33	4.171	52.765	1.000	102.7
4	76.39	4.171	61.967	0.852	195.0
5	53.12	4.171	61.967	0.852	151.4
6	47.91	0.880	61.967	0.852	151.4
7	57.47	4.171	0.000	0.067	2608.7
8	29.72	4.171	0.000	0.148	124.5
9	5.13	0.880	0.000	0.148	124.5
10	5.13	0.880	0.000	0.148	2511.0
11	57.47	4.171	52.765	0.550	124.284
12	57.49	64.231	52.765	0.550	124.322
13	90.18	64.231	52.765	0.550	194.1
14	144.84	64.231	61.967	0.469	323.3
15	101.16	64.231	61.967	0.469	241.4
16	78.60	4.171	61.967	0.469	241.4
17	122.80	64.231	0.000	0.082	2726.2
18	87.73	64.231	0.000	0.082	367.2
19	29.72	4.171	0.000	0.082	367.2
20	150.00			8.000	
21	142.04				
22	25.00			12.000	
23	33.65				
24	25.00			14.000	
25	28.16				
26	12.00			20.000	
27	7.78				

$Q_{HPG} = 267.492$ kW	$Q_{LPG} = 192.776$ kW	$Q_A = 436.179$ kW
$Q_{HPC} = 192.776$ kW	$Q_{LPC} = 185.702$ kW	$Q_E = 354.366$ kW
$Q_{SHX1} = 37.116$ kW	$Q_{SHX2} = 38.387$ kW	$W_1 = 0.002$ kW
$W_2 = 0.022$ kW	COP = 1.325	

Table 11.8 Flow configuration comparison of double-effect water–LiBr absorption cycle [1]

Configuration	COP	Capacity (kW)
Serial	1.244	371.1
Parallel	1.325	354.4
Reverse	1.238	370.2

COP, coefficient of performance.

The solution in A flows into both Gs in this cycle. Parallel configuration, serial configuration, and reverse configuration can be adopted for the solution circulation. In the serial flow configuration, the solution from A flows into HPG and LPG successively. The serial configuration is easy to control. In the parallel flow configuration, the solution from A flows into HPG and LPG at the same time. The parallel configuration has one more solution flow rate to be adjusted. The control strategy is complex, but the coupling between HPC and LPG can be optimized. In the reverse flow configuration, the solution from A flows into LPG and HPG successively. The cycle in Fig. 11.15 employs the serial flow configuration. The heat coupling between HPC and LPG is affected by the flow configuration and the COP varies with it. Table 11.8 shows the COP and cooling capacity calculation of various flow configurations under the same conditions. The serial and reverse configurations have higher cooling capacity but lower COP. A better temperature match between HPC and LPG can be easily obtained in parallel by adjusting the solution flow rate, thus increasing the COP. However, the temperature match also decreases the temperature difference, which weakens the cooling capacity.

11.3.1.2 Solar-powered case

In this section, experimental research on the solar-powered double-effect water–LiBr cooling system is introduced. The system was installed at Carnegie Mellon University, Pittsburgh, Pennsylvania, USA. Fig. 11.16(a) shows the schematic diagram of the system. Fig. 11.16(b) shows the picture of the linear PTC integrated in the experiment. It had an area of 52 m^2. A 16-kW double-effect absorption chiller manufactured by Broad Air Conditioning Co. is used in the experiment. In addition, a heat-recovery heat exchanger with circulation pumps and control valves was integrated in this system. The absorption chiller installed was a dual-fired, double-effect water–LiBr chiller integrated with a cooling tower. A natural gas burner was integrated to provide heat when solar energy was inadequate. This system was successfully operated for more than 1 year [12].

According to the experiments under typical weather conditions in Pittsburgh in summer, the overall solar efficiency of the parabolic trough solar collectors was

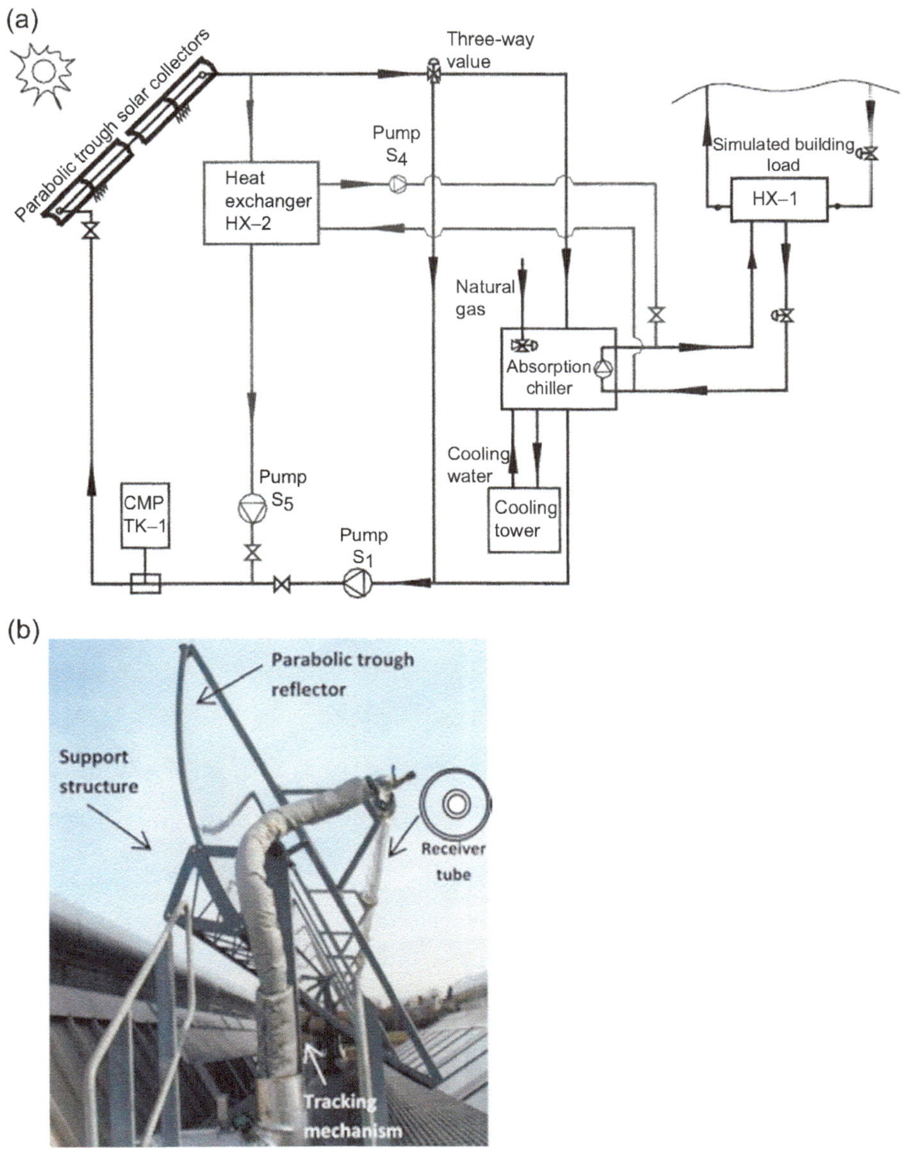

Figure 11.16 The solar-powered double-effect water–LiBr absorption cooling system at Carnegie Mellon University [12].

approximately 33 ~ 40% when the heat transfer fluid was operated at 150 ~ 160°C. The COP of the installed absorption chiller ranged between 1.0 and 1.1. The overall solar COP of the system, which is the product of the absorption chiller COP and the solar collector efficiency, was about 0.33–0.44. The maximum output of the absorption chiller was 12 kW. The reason for this capacity, lower than the chiller's design capacity of 16 kW, was mainly related to the intensity of direct solar radiation. Owing to the relatively high humidity of Pittsburgh in summer, the direct solar radiation was relatively low, with typical values of 600 ~ 850 W/m^2.

Fig. 11.17(a) shows the measured temperatures of the heat transfer fluid at the inlet/exit of the parabolic trough solar collectors (T_SC-in/out) and at the inlet/outlet of the chiller (T_HX-2_cs_in/out) and the chilled water temperature at the inlet/outlet of the chiller (T_HX-1_cs_in/out) throughout a day. Fig. 11.17(b) shows the corresponding

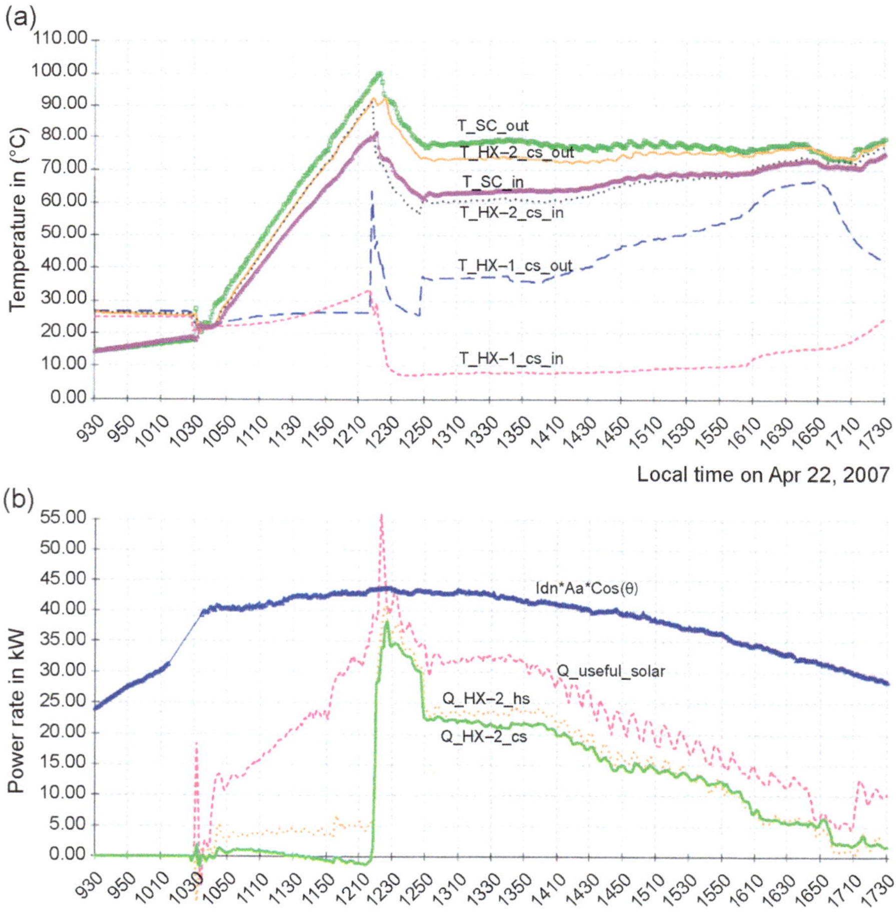

Figure 11.17 Parameters of the solar double-effect absorption cooling system. (a) Temperature. (b) Energy flow [12].

energy flow diagram. The lines represent the solar energy that could be collected, useful solar energy gained, heat input of the absorption chiller, and cooling output of the absorption chiller.

11.3.2 Generator absorber heat exchange absorption cooling system

11.3.2.1 Generator absorber heat exchange absorption refrigeration cycle

Because the heat couplings in multiple-effect cycles are constrained by the tight matching of temperatures between the heat-rejecting components and the heat-receiving components, the heat source temperature decides the refrigeration temperature when the heat sink temperature is confirmed. The cycle has poor flexibility against the temperature variation. The GAX absorption cycle improves the flexibility through absorption heat recovery. Fig. 11.18 shows the schematic diagram of a GAX absorption cycle. Solution flows through the A, GAX-generator (GAX-G), G, GAX-absorber (GAX-A), and back to A. Temperature overlap exists between GAX-A and GAX-G. G is heated to boiling by the heat source. GAX-G is heated to boiling by the absorption heat from GAX-A. Both G and GAX-G generate refrigerant vapor to C. The vapor is then condensed in C, flows to E, and delivers cooling output. In this case, one unit of heat input to G generates more than one unit of cooling output.

Table 11.9 shows the parameters of the ammonia–water GAX cycle. The state points correspond to Fig. 11.18. The calculation is carried out with the following assumptions: (1) Saturated vapor leaves the generator and the rectifier. (2) The rectifier is analyzed as a reversible device that produces a vapor mass fraction of 0.995. (3) The pump efficiency is 50% and the condensate precooler has an effectiveness of 0.8. Before entering the generator, the rich solution is preheated at the low temperature of the absorber with an approach temperature of 0 K. The results show that the generation temperature and evaporation temperature are 163.3 and 5°C, respectively. Owing to the GAX heat recovery, only 422 kW of the generator heat consumption 822 kW

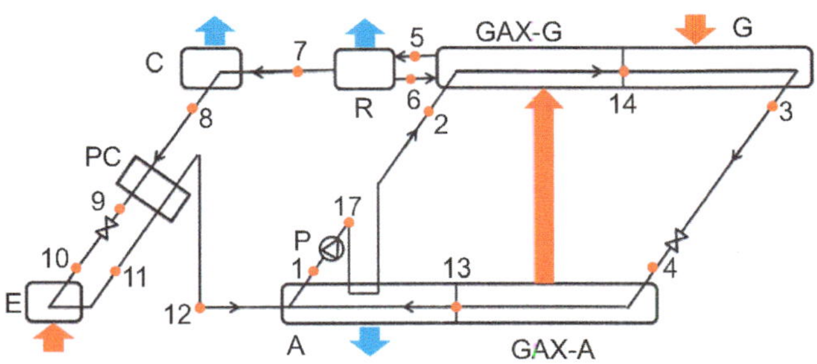

Figure 11.18 Schematic diagram of a generator absorber heat exchange (GAX) absorption cycle. *A*, absorber; *C*, condenser; *E*, evaporator; *G*, generator; *P*, pump; *PC*, precooler; *R*, rectifier.

Table 11.9 Parameters of the ammonia–water generator absorber heat exchange cycle [1]

No.	t (°C)	p (kPa)	x (% ammonia)	m (kg/s)	h (kJ/kg)
1	40.0	478.4	48.9	1.0	−60.6
2	83.7	1548.0	48.9	1.0	139.8
3	163.3	1548.0	13.9	0.591	610.1
4	123.7	478.4	13.9	0.591	610.1
5	83.7	1548.0	98.4	0.418	1440.4
6	83.7	1548.0	48.9	0.010	140.0
7	67.1	1548.0	99.5	0.409	1365.4
8	40.0	1548.0	99.5	0.409	187.6
9	12.2	1548.0	99.5	0.409	53.6
10	3.1	478.4	99.5	0.409	53.6
11	5.0	478.4	99.5	0.409	1198.0
12	30.1	478.4	99.5	0.409	1332.1
13	83.7	478.4			
14	123.7	1548.0			
15	40.4	1548.0	0.489		−58.0
$Q_G = 422$ kW		$Q_A = 369$ kW		$Q_C = 482$ kW	
$Q_E = 468$ kW		$Q_{GAX} = 400$ kW		$Q_{PC} = 55$ kW	
$Q_R = 198$ kW		$W_P = 2.6$ kW		COP = 1.110	

needs outside input. The cycle COP is 1.110. Without the GAX heat recovery, the cycle COP will be only 0.570.

The temperature overlap and COP of the GAX cycle vary with the driving temperature. The temperature overlap between G and A ensures the operation, and a large concentration field is necessary. In this case, the water–LiBr solution is not suitable for the GAX cycle owing to the crystallization risk. Usually, an ammonia–water solution is used as the working pair. Compared with the single-effect cycle, the GAX cycle has a higher driving temperature and COP; the solution flow rate is also reduced for the same cooling output.

11.3.2.2 Branched generator absorber heat exchange absorption cooling system

In real operation of a GAX cycle, the concentration of the A overlap is always higher than that of the G overlap. For the same amount of heat, the GAX-A has a larger

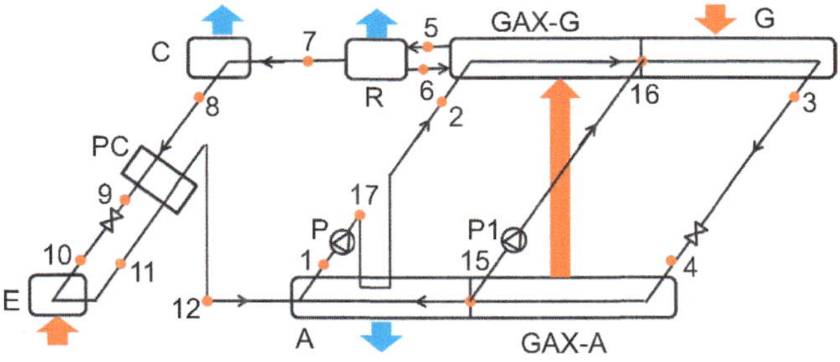

Figure 11.19 Schematic diagram of a branched generator absorber heat exchange (GAX) absorption cycle. *A,* absorber; *C,* condenser; *E,* evaporator; *G,* generator; *P,* pump; *PC,* precooler; *R,* rectifier.

temperature glide. In this case, the heat-release capability of GAX-A is weaker than the heat-accepting capability of the GAX-G. This mismatch between GAX-A and GAX-G decreases the COP. To improve the coupling between A and G, the branched GAX cycle is proposed. Fig. 11.19 shows the schematic diagram of it. In the branched GAX cycle, the solution in G is separated into two branches with different concentrations. The flow rates of the two branches are controlled by independent pumps. When the flow rate of pump-1 is increased, the heat load of GAX-A is increased and the heat load of GAX-G is decreased. The GAX heat coupling can be optimized by adjusting the flow rate of the two branches. In this case, this cycle has a higher COP than the standard GAX cycle in real operation.

Table 11.10 shows the parameters of the branched GAX absorption cycle under the same conditions as Table 11.9. The ammonia–water is still used as the working fluid. The state points are the same as in Fig. 11.18 except for points 13, 14, 15, and 16. The results show that the load of GAX is increased from 400 to 472 kW and the COP is increased from 1.110 to 1.174 with an improvement of 5.77%. This indicates that the branches in this cycle enhance the temperature match of GAX and increase the cycle performance as analyzed.

11.3.2.3 Solar-powered case

The ammonia–water GAX chiller is suitable for utilization in a medium-temperature solar collector. Also, the ammonia–water chiller has no crystallization risk, which makes the solar GAX system able to work under air-cooling conditions. A simulation study of a linear Fresnel reflector concentrator (LFRC) driving an air-cooled GAX chiller will be introduced in this section.

An ammonia–water GAX chiller with cooling capacity of 10.6 kW was used. The LFRC was designed for 12.23 kW of heat output. The LFRC was used as a direct generator for the ammonia–water GAX chiller in this study. Heat storage was neglected to obtain higher input temperature in the generator of the chiller and to reduce the thermal losses. In addition, the flexibility of the GAX chiller also allowed temperature variations of the heat source. Fig. 11.20 shows the schematic diagram and the corresponding parameters of the system. The inlet ammonia–water solution of the

Table 11.10 Parameters of the ammonia–water branched generator absorber heat exchange cycle [1]

No.	t (°C)	p (kPa)	x (% ammonia)	m (kg/s)	h (kJ/kg)
1	40.0	478.4	48.9	1.0	−60.6
2	83.6	1548.0	48.9	1.0	139.8
3	163.3	1548.0	13.9	0.591	610.1
4	123.7	478.4	13.9	0.591	610.1
5	83.7	1548.0	98.4	0.418	1440.4
6	83.7	1548.0	48.9	0.010	140.0
7	67.1	1548.0	99.5	0.409	1365.4
8	40.0	1548.0	99.5	0.409	187.6
9	12.2	1548.0	99.5	0.409	53.6
10	3.1	478.4	99.5	0.409	53.6
11	5.0	478.4	99.5	0.409	1198.0
12	30.1	478.4	99.5	0.409	1332.1
13	83.7	478.4			
14	84.1	1548.0			
15	40.4	1548.0	0.489		−58.0
$Q_G = 417$ kW		$Q_A = 363$ kW		$Q_C = 503$ kW	
$Q_E = 489$ kW		$Q_{GAX} = 472$ kW		$Q_{PC} = 58$ kW	
$Q_R = 198$ kW		$W_P = 2.6$ kW	$W_{P1} = 0.3$ kW	COP = 1.174	

LFRC generator had 6.4% composition at 187.21°C, 18 bar, and 0.0197 kg/s. Under the designed conditions, the LFRC efficiency was 0.63 and the solar–GAX cycle efficiency was 0.85, which made the global efficiency 0.54 [13].

11.3.3 Other configurations

11.3.3.1 Single-effect ammonia–water absorption ice-making system

Fig. 11.21 shows the schematic diagram and picture of the single-effect ammonia–water absorption chiller with rectification heat recovery: the rich solution from A is pumped through the rectifier (R) and SHX to G. The state points are labeled in the

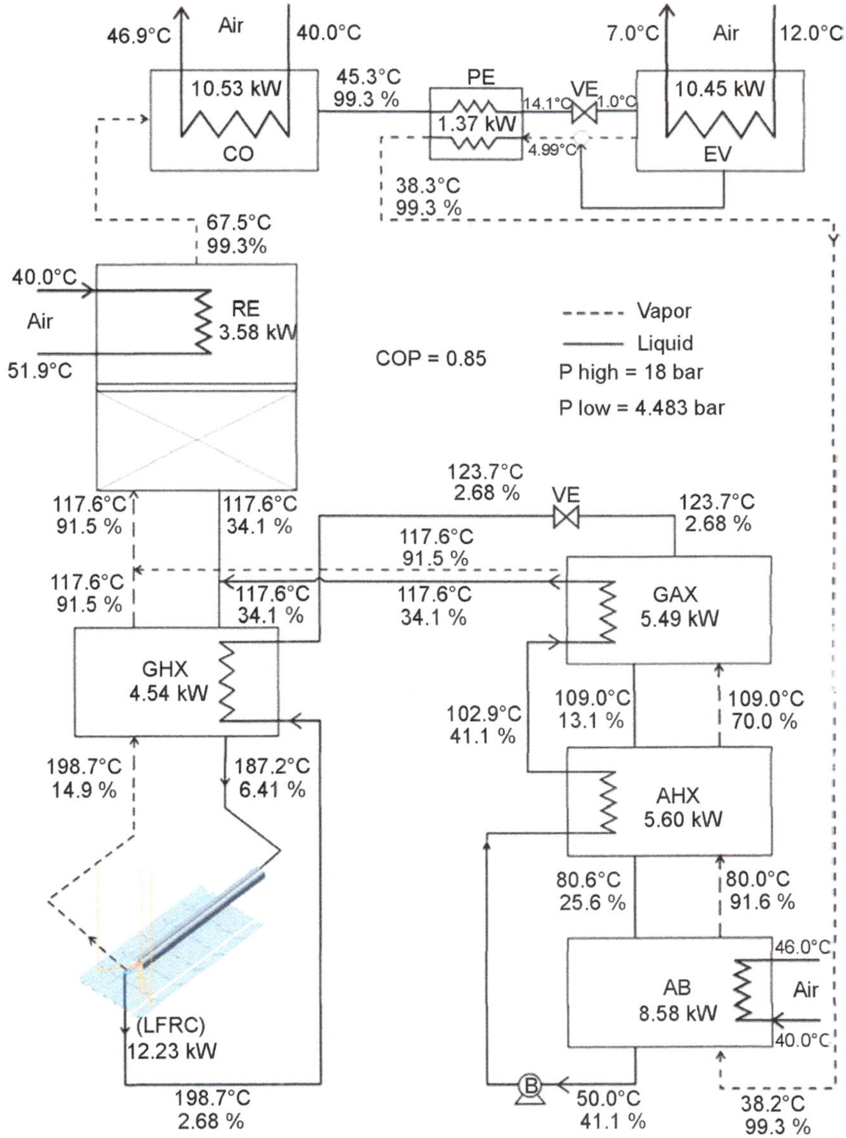

Figure 11.20 Parameters of a solar-powered generator absorber heat exchange (GAX) absorption chiller [13].

diagram and the corresponding parameters of each state point are shown in Table 11.11. The calculation is based on the following assumptions: (1) The evaporator saturation temperature at the outlet is assumed to be −10°C with a vapor quality of 0.975. (2) The mass flow rate of solution through the solution pump is 1 kg/s.

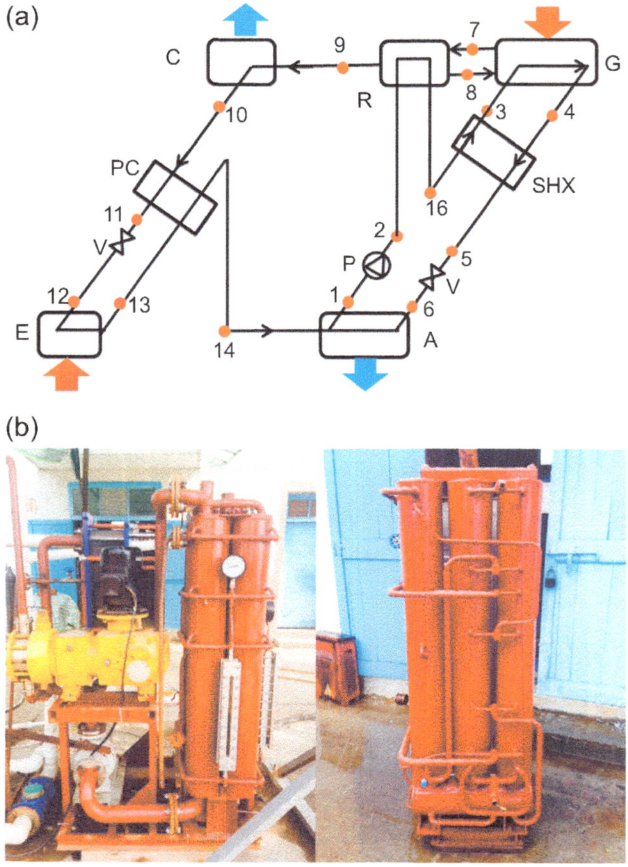

Figure 11.21 Single-effect ammonia–water absorption chiller. (a) Schematic diagram. (b) Chiller. *A*, absorber; *C*, condenser; *E*, evaporator; *G*, generator; *P*, pump; *PC*, precooler; *R*, rectifier, *SHX*, solution heat exchanger; *V*, valve.

(3) The temperature of the saturated liquid leaving the absorber and the condenser is 40°C. (4) The concentration change is assumed to be 10%. (5) The R produces a vapor with mass faction of 0.999634. (6) The pump efficiency is 100% and the effectiveness of the SHX is 80% [1].

The rectification heat recovery increased the COP by 5.4% according to the calculation. A COP of 0.409 is obtained with a G temperature and E temperature of 152.4 and −27.5°C, respectively. The driving temperature is complementary to the medium-temperature solar collector. The single-effect ammonia—water chiller is able to work with a low-temperature solar collector as shown in Section 11.2.2. The chiller has a higher COP and higher evaporation temperature when it is activated by a heat source under 100°C. However, it is still less efficient than the water–LiBr system. It is better to run the water–LiBr system with a low-temperature collector for air conditioning and run the ammonia–water system with a medium-temperature collector for ice-making.

Table 11.11 Parameters of the single-effect ammonia–water absorption chiller (ice-making) [1]

No.	*t* (°C)	*p* (kPa)	*x* (% ammonia)	*m* (kg/s)	*h* (kJ/kg)
1	40.0	132.8	0.2811	1.000	−9.2
2	40.0	1540	0.2811	1.000	−7.6
3	122.5	1540	0.2811	1.000	356.2
4	152.4	1540	0.1811	0.876	538.6
5	75.1	1540	0.1811	0.876	196.5
6	64.3	132.8	0.1811	0.876	196.5
7	127.4	1540	0.8736	0.148	1674
8	127.4	1540	0.2811	0.024	378.6
9	76.540.0	1540	0.99	0.124	1410
10	13.5	1540	0.99	0.124	184.1
11	−27.5	1540	0.99	0.124	55.9
12	−10.0	132.8	0.99	0.124	55.9
13	25.4	132.8	0.99	0.124	1234
14	25.4	132.8	0.99	0.124	1362
15	54.9	1540	0.99	1.000	56.4
Q_G = 354.1 kW		Q_A = 349.8 kW		Q_C = 151.6 kW	
Q_E = 145.6 kW		Q_{SHX} = 299.8 kW		Q_{PC} = 15.8 kW	
Q_R = 64.0 kW		W_P = 1.6 kW		COP = 0.409	

11.3.3.2 Diffusion–absorption cooling system

The absorption refrigeration cycles introduced earlier all need mechanical work input to ensure circulation of the solution. The diffusion–absorption refrigeration cycle is a special technology that does not need mechanical work input; instead, a noise-free bubble pump driven by heat is used for solution circulation. It is also a marketed product for residential, hotel, and recreational vehicle use. In a diffusion–absorption cooling system, the working pair is mixed with the pressure compensation gas. The system's total pressure stays constant, while the partial pressure of the refrigerant varies. For example, ammonia–water can be used as the working pair. Helium or hydrogen can be used for pressure compensation. In addition, organic working pairs, for instance, C_4H_{10}–C_9H_{20} and R124–DMAC, are popular in diffusion–absorption cooling systems. However, the diffusion–absorption cooling system is not favorable in solar cooling because of its low efficiency.

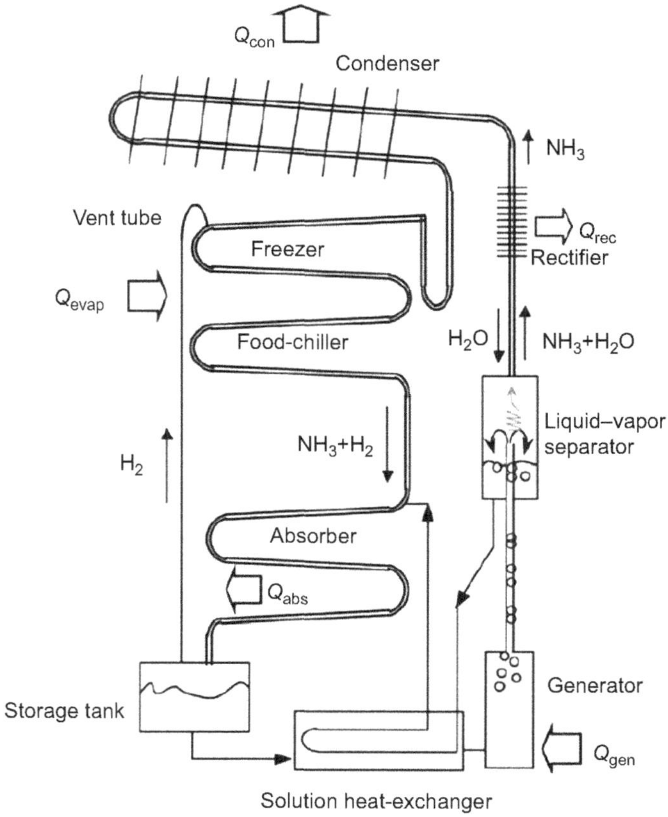

Figure 11.22 Schematic diagram of a diffusion–absorption chiller [14].

The schematic diagram of a diffusion–absorption chiller is shown in Fig. 11.22. The ammonia–water working pair and compensation gas hydrogen are used. The chiller is composed of a generator, a condenser, an absorber, an evaporator, a gas heat exchanger, and an SHX. Specifically, the generator is integrated with a bubble pump and a rectifier. The flow of this cycle is as follows. (1) The rich solution is preheated by the SHX and flows into the generator. (2) The rich solution is then boiled by the heat source in G and the liquid–vapor mixture is pumped to a higher level by the bubble flow. This pumping process by bubble flow is called the bubble pump. The mixture is separated into vapor and weak solution at the outlet of G. The vapor is rectified and flows into C. The weak solution flows back to A. (3) The vapor (mostly ammonia) in C is cooled and condensed by the ambient air. The condensed liquid flows into E. (4) The ammonia/hydrogen vapor from A also flows into E. Considering the whole system has nearly constant pressure, the existence of hydrogen decreases the partial pressure of the condensed ammonia. The ammonia liquid evaporates and delivers cooling output. The ammonia/hydrogen vapor flows out of E with a higher fraction of ammonia. (5) The ammonia/hydrogen vapor flows into A and mixes with the weak solution from G. The solution is cooled by the ambient air and absorbs the ammonia. A strong solution and ammonia/hydrogen vapor with a lower fraction of ammonia are obtained in A.

The performance of the diffusion–absorption chiller is affected by both its configuration and the performance of the bubble pump. Generally, the chiller works under a driving temperature between 150 and 200°C. As the diffusion–absorption chiller is usually used for air conditioning, the cooling output temperature is about 0°C. The performance is about 0.1 ~ 0.2, which is actually much smaller than that of the single-effect chiller. It has an advantage over the single-effect chiller in small size and residential conditions.

11.4 Drawbacks of solar absorption cooling systems and improvement

The solar collector and absorption chiller are integrated because of their complementary working temperatures. However, they have different operating properties that will cause problems between their couplings. In this section, the problems are analyzed and corresponding solutions are given.

11.4.1 Drawbacks of solar absorption cooling systems

11.4.1.1 Continuous working ability

The solar collector works only during the daytime and it cannot ensure 24 h of continuous operation of the solar absorption cooling system. In addition, the solar heating power temperature varies with the season or weather. To make the absorption cooling system more reliable, an auxiliary heat source is necessary. For instance, direct firing of fossil fuels, including oil or natural gas, may be used.

The coupling between a single-effect absorption chiller and a low-temperature solar collector is the most commonly used solar absorption cooling system. The direct firing of fossil fuels offers temperatures far higher than necessary for a single-effect absorption chiller. Considering that the absorption chiller does obtain higher efficiency under higher driving temperatures, there will be a waste of energy grade in the gas-fired single-effect absorption cooling system.

11.4.1.2 Variable driving temperature

According to the former discussions, the double-lift, single-effect, and double-effect water–LiBr absorption chillers obtain a COP of about 0.4, 0.72, and 1.2 under driving temperatures of around 70, 90, and 150°C, respectively. The chillers are not flexible under variable driving temperatures: their COP varies little with the driving temperature; their driving temperatures are constrained in a small range. However, the solar heat power has variable temperature during operation. This makes the coupling between the solar collector and the absorption chiller work in an unstable condition. When the heat source temperature is too low, it is unable to activate the absorption chiller. When the heat source is too high, there will be a waste of energy grade.

11.4.2 Single-effect/double-effect water–LiBr absorption cooling system

11.4.2.1 Single-effect/double-effect water–LiBr absorption chiller

To make better use of the auxiliary heat source, a plan of working in the single-effect refrigeration mode with solar power and double-effect refrigeration mode with fossil fuel is reasonable. Actually, some researchers have proposed the idea of operating single-effect absorption refrigeration and double-effect absorption refrigeration in one double-effect absorption chiller. However, this plan is not practical owing to the following reason: the LPG of a double-effect absorption chiller is a gas-driven generator, the generator of a single-effect absorption chiller for solar cooling is a hot-water-driven generator, and the heat exchange form and heat source flowing channel of the two generators are different. It is better to integrate one more hot-water-driven generator in a double-effect absorption chiller. This is the single-effect/double-effect (SE/DE) absorption chiller. Fig. 11.23 shows the schematic diagram and product picture of an SE/DE water–LiBr absorption chiller.

As is shown in Fig. 11.23, the absorption chiller has three pressure levels and six major components, ie, the HPG (1), two LPGs (2) and (3), the condenser (4), the left

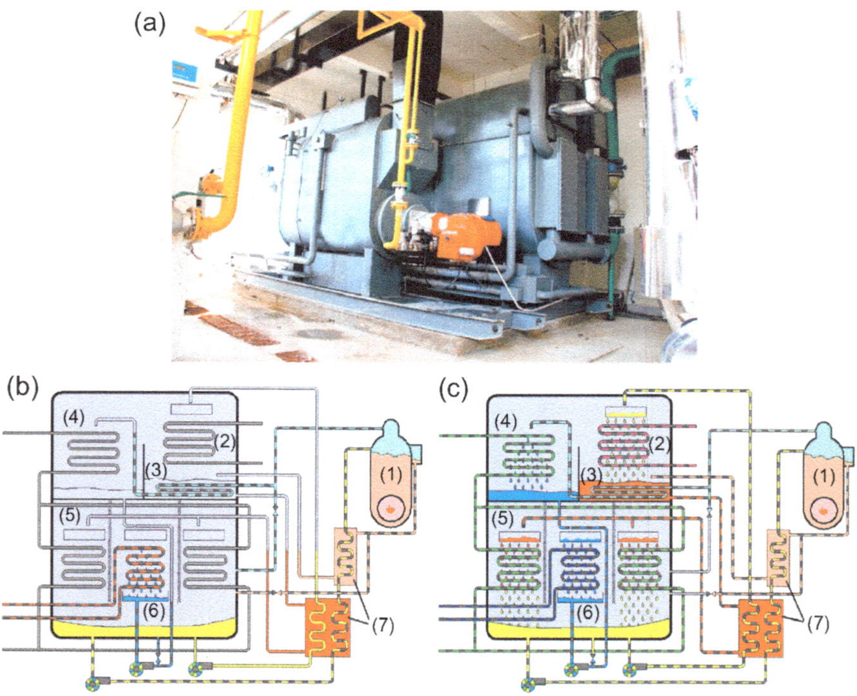

Figure 11.23 (a) Single-effect/double-effect absorption chiller. (b) Cooling mode. (c) Heating mode [15]. (1) Gas boiler (high-pressure generator), (2) low-pressure generator for single-effect circulation, (3) low-pressure generator for double-effect circulation, (4) condenser, (5) absorber, (6) evaporator, (7) SHX.

and right absorbers (5), the evaporator (6), and the SHX (7). The absorption chiller has an extra LPG compared with a conventional double-effect absorption chiller. This extra generator works for solar cooling, whereas the evaporator and absorbers are in common. In this case, the solar-powered single-effect refrigeration and gas-fired double-effect refrigeration work both independently and simultaneously. Among the various working modes, the solar-powered single-effect refrigeration mode has the highest priority. When solar power temperature is not high enough or the solar cooling cannot afford the cooling consumption, the gas-fired double-effect refrigeration is activated.

11.4.2.2 Solar/fossil fuel-driven single-effect/double-effect absorption cooling system

In this section, an experimental study of the SE/DE absorption chiller shown in Fig. 11.23 is introduced. Fig. 11.24 shows the schematic and pictures of the hybrid

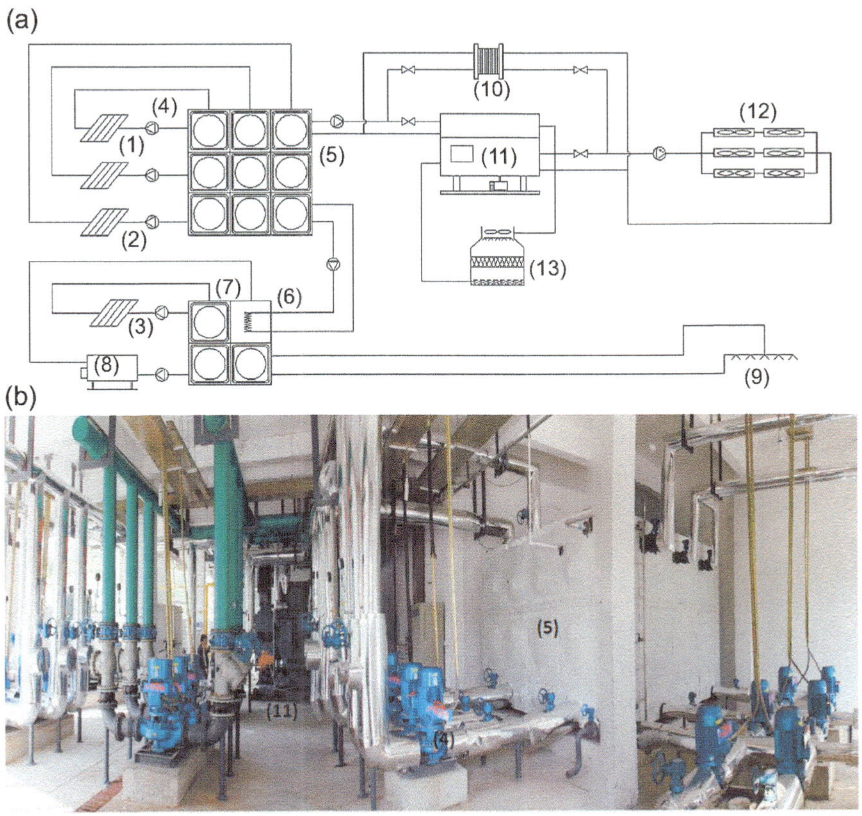

Figure 11.24 Hybrid solar/gas firing-powered single-effect/double-effect absorption chiller. (a) Schematics of the hybrid energy system. (b) Real plant room [15]. (1) Evacuated tube gravity heat pipe solar collector, (2) evacuated tube horizontal heat pipe solar collector, (3) evacuated glass tube solar collector, (4) pump, (5) water tank for air conditioning, (6) copper pipe heat exchanger, (7) hotel hot water tank, (8) auxiliary boiler, (9) hot water use, (10) plate heat exchanger, (11) absorption chiller, (12) fan coil, (13) cooling tower.

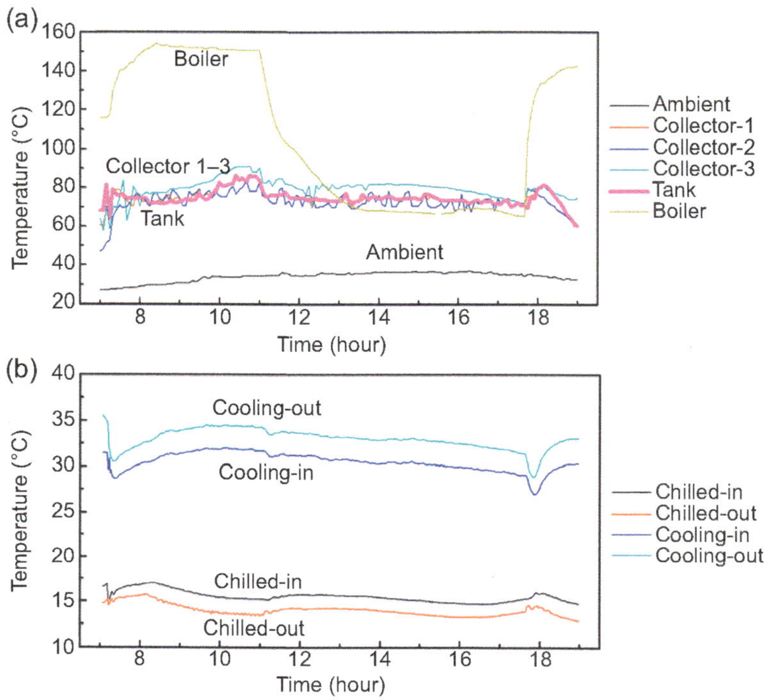

Figure 11.25 Temperature variations in the hybrid system [15].

cooling system. The system was composed of (1) evacuated tube gravity heat pipe solar collector, (2) evacuated tube horizontal heat pipe solar collector, (3) evacuated glass tube solar collector, (4) pump, (5) water tank for air conditioning, (6) copper pipe heat exchanger, (7) hotel hot water tank, (8) auxiliary boiler, (9) hot water use, (10) plate heat exchanger, (11) absorption chiller, (12) fan coil, and (13) cooling tower. The three major groups of solar collectors included horizontal heat pipe evacuated tube solar collectors (1) and gravity-assisted heat pipe evacuated tube solar collectors (2). There were 4950 pipes and 1425 pipes for two types, respectively. The whole area for the three major groups of solar collectors was 1020 m^2. The system was adopted in a hotel located in Changle, Shandong, in China. Changle is located at 36.69°N and 118.83°E.

Fig. 11.25 shows the parameter variations of the system during the daytime. The data were recorded once every minute. In Fig. 11.25(a), the ambient air, collector, tank, and boiler temperatures are shown. In Fig. 11.25(b) the chiller temperatures are shown. According to Fig. 11.25, the system worked in solar-driven mode from 1100 to 1745 hours. In this period, the average COP calculated from total energy flow achieved 0.57. For the rest of the day, the system worked in double-effect refrigeration mode.

11.4.3 Variable-effect water–LiBr absorption cooling system

To obtain a better match between solar collector and absorption chiller, the variable-effect absorption chiller is a possible solution. The variable-effect absorption chiller works under a large driving temperature range and obtains different COPs under different driving temperatures. It could make better use of a heat source with variable temperature than the traditional absorption chiller. In this section, two variable-effect absorption chillers with water–LiBr are introduced. The first is the single-effect/double-lift absorption chiller working between double-lift refrigeration and single-effect refrigeration. The second is the 1.n-effect absorption chiller, which works between single-effect refrigeration and double-effect refrigeration.

11.4.3.1 Single-effect/double-lift absorption cooling system

As shown in Table 11.3, the heat source outlet temperature of a single-effect water–LiBr chiller is 96.5°C. In real operation, this outlet temperature may be lower, but this temperature is still far more than enough to drive a double-lift water–LiBr absorption chiller. To make better use of the heat source, the single-effect/double-lift (SE/DL) absorption chiller can be used. Fig. 11.26 shows the schematic of the SE/DL absorption cycle. This cycle is proposed for the utilization of a heat source with large temperature glide. The cycle contains two solution circuits. In the first solution circuit, solution flows through HPG-1, LPG, and LPA. In the second solution circuit, solution flows through HPG-2 and high-pressure absorber (HPA). The three generators (HPG-1, HPG-2, and LPG) are heated to boiling by the heat source successively. The two absorbers (HPA and LPA) are cooled by the ambient air. This can also be interpreted as the coupling of a single-effect subcycle and a double-lift subcycle.

When the cycle works under the driving temperature of the single-effect cycle, it has lower COP, larger cooling capacity, and lower heat source outlet temperature. More energy input is gained by the SE/DL cycle from the same heat source. Flow rates of the two solution circuits can be adjusted according to the heat source temperature. The performance calculation of the SE/DL cycle is shown in Table 11.12. As is shown,

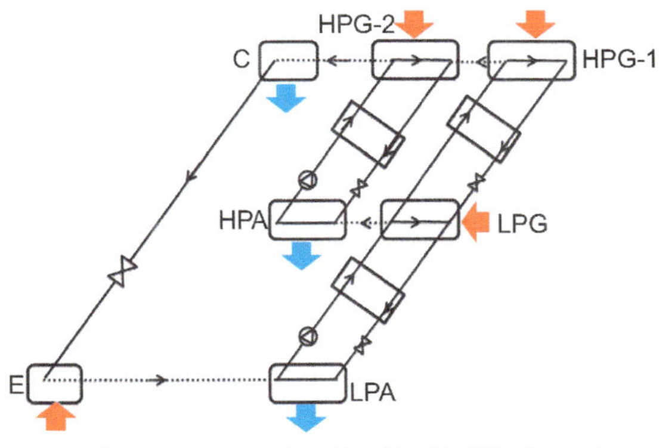

Figure 11.26 Schematic diagram of a single-effect/double-lift absorption cycle. *C*, condenser; *E*, evaporator; *HPA*, high-pressure absorber; *HPG*, high-pressure generator; *LPA*, low-pressure absorber; *LPG*, low-pressure generator.

Table 11.12 Performance calculation of the single-effect/ double-lift cycle [16]

No.	Heat source inlet/ outlet temp (°C)	Cooling water inlet/outlet temp (°C)	Chilled water inlet/outlet temp (°C)	COP	Heat source consumption [kg/(kW·h)]
1	100/60	28/33	15/10	0.604	35.61
2	95/60	28/33	15/10	0.582	42.30
3	90/60	28/33	15/10	0.556	51.61
4	85/60	28/33	15/10	0.522	66.05

the COP of the SE/DL cycle varies from 0.522 to 0.604 under different driving temperatures. The heat source temperature change in the SE/DL cycle is large. This indicates a better utilization of the heat source; hence the heat source consumption for unit cooling output of an SE/DL cycle is low. For the single-effect water–LiBr absorption cycle shown in Section 11.2.1, the heat source consumption for unit cooling output under a 100°C heat source temperature is 340.46 kg/(kW·h), whereas this value is 35.61 kg/(kW·h) for an SE/DL cycle. This is caused not only by the small heat source temperature glide of the single-effect cycle, but also by the superior heat utilization property of the SE/DL cycle.

Except for the SE/DL combined mode shown in Table 11.12, the cycle can work in full single-effect mode, which is the same as the cycle shown in Fig. 11.5 when the solution circuit of HPG-2/MPA does not flow. The cycle can also work in full double-lift mode, which is the same as the cycle shown in Fig. 11.10 when the two solution circuits both circulate but HPG-1 has not boiled. The flexibility of this cycle makes it suitable for solar cooling.

11.4.3.2 The 1.n-effect absorption refrigeration cycle

The 1.n-effect absorption cycle is shown in Fig. 11.27(a). Dotted lines stand for the flow of vapor and solid lines stand for that of liquid. This cycle has three pressure levels and seven primary components, ie, high-pressure generator (HG), high-pressure absorber (HA)–second low pressure generator (LG2), high-pressure condenser (HC)–first low-pressure generator (LG1), C, E, low-pressure absorber (LA), and SHX.

In this cycle, high-pressure refrigerant vapor generated from the HG is separated into two parts. One part flows into the LG1 and its condensation heat is utilized. This part of the vapor has double-effect refrigeration. The other part of the vapor is first absorbed in the HA and produces a large amount of absorption heat. The absorption heat is then utilized to generate refrigerant vapor in the LG. This part of the high-pressure vapor has only single-effect refrigeration. In this way, the cycle

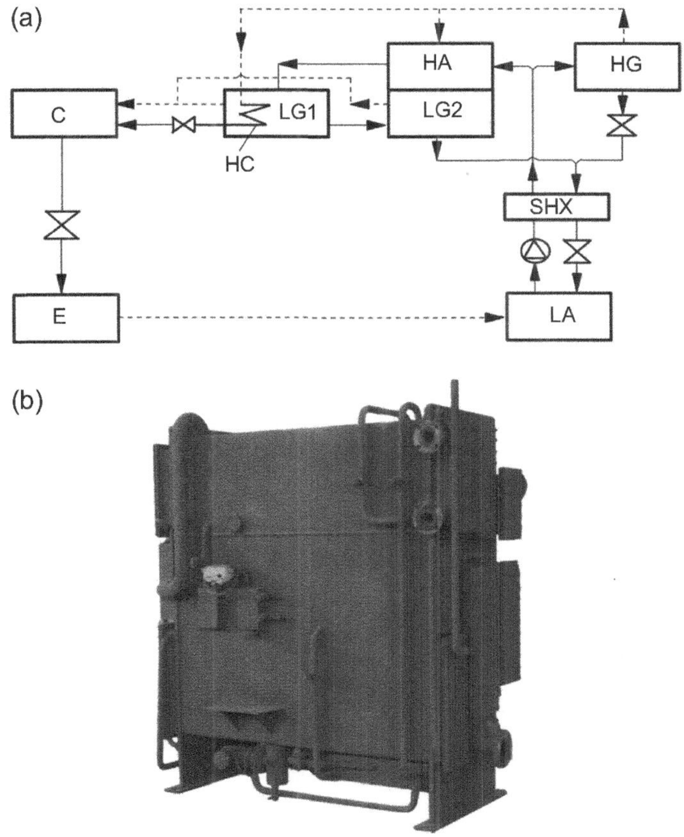

Figure 11.27 The 1.n-effect absorption cycle. (a) Schematic diagram [17]. (b) Prototype [18]. *C*, condenser; *E*, evaporator; *HA*, high-pressure absorber; *HC*, high-pressure condenser; *HG*, high-pressure generator; *LA*, low-pressure absorber; *LG*, low-pressure generator; *SHX*, solution heat exchanger.

can obtain refrigeration between single-effect and double-effect. The COP curve of the cycle with different generation temperatures was obtained from calculation and is shown in Fig. 11.28. The calculation is based on the evaporation temperature, absorption temperature, and condensation temperature of 5, 35, and 40°C, respectively.

A variable-effect water–LiBr absorption chiller prototype was designed and manufactured based on this cycle by Shanghai Jiao Tong University. The prototype picture is shown in Fig. 11.27(b). The experimental results show that the chiller obtained a COP from 0.69 to 1.08 under generation temperatures from 95.0 to 120.0°C [18]. The COP increased as is predicted by the calculation. Owing to a lower cooling temperature and higher evaporation temperature in the experiment, the experimental COP reached the highest level under a lower generation temperature than predicted by the calculation.

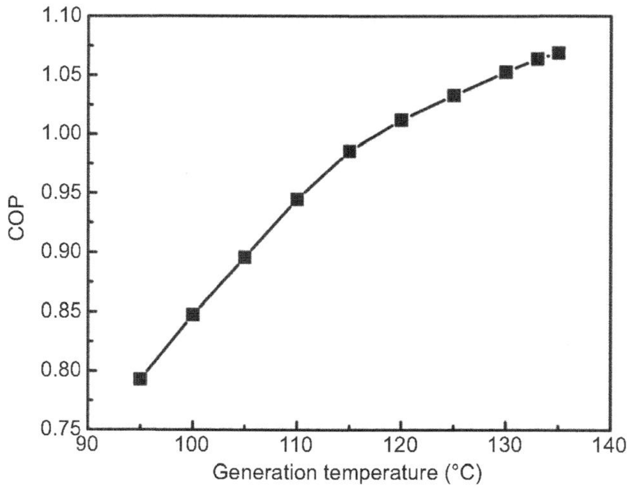

Figure 11.28 Coefficient of performance (COP) of a 1.n-effect absorption cycle under various generation temperatures [17].

11.5 Economic performance and adaptability analysis

To analyze the practicability of a solar absorption cooling system, an economic performance evaluation is essential. According to a market investigation, the prices of absorption chillers vary with type, power, efficiency, and manufacturer. The chiller is more expensive for unit cooling output in the same size. The collector prices vary mainly with the type and manufacturer, which are not constant as of this writing. Governments may also have different ratios of allowance for solar collector installation or solar cooling system building. The allowance should also be considered in the specific case analysis.

Among the absorption chillers, single-effect absorption chillers and double-effect absorption chillers are widely commercialized. The evaluation focuses mainly on the systems of single-effect and double-effect water–LiBr absorption chillers. In addition, the economic evaluation can be used for the different aims. Two cases using different systems with different aims are introduced in this chapter.

11.5.1 Case 1: comparison between solar-powered single-effect absorption cooling systems with different solar collectors

In this case, a single-effect water–LiBr absorption chiller was coupled with three different collectors, including a flat plate collector (FPC), a stationary CPC (StCPC), and an evacuated tube collector (ETC). The system was evaluated in 2007 for office building (room area 930 m^2) air conditioning in Madrid. The initial costs of FPC, StCPC, and ETC were €280, €400, and €620/m^2, respectively (€280, €400, and

Table 11.13 Economic performance of various systems [19]

Collector	Area for unit cooling output (m^2/kW)	Heat storage (m^3)	Total annual cost (k€)	Primary energy saving (%)	Value of saved primary energy (¢/kWh)
FPC	2.99	12.6	11.7	36	6.8
StCPC	2.13	9	11.8	30	8.2
ETC	2.13	12.6	12.9	45	9.5

ETC, evacuated tube collector; *FPC*, flat plate collector; *StCPC*, stationary concentrating parabolic collector.

€350/m^2 in another study in Madrid in 2014). The initial costs of an absorption chiller and fossil fuel-driven backup heater were €400 and €120/kW. Considering the annual costs, the economic performances are shown in Table 11.13. The primary energy saving was compared to a conventional reference system including a compression chiller with a COP of 3.0 and a gas burner. The value of primary energy saved was a combined cost–energy performance, which is defined by Eq. [11.5], where $C_{\mathrm{PE,saved}}$, $C_{\mathrm{annual,sol}}$, $C_{\mathrm{annual,ref}}$, and $E_{\mathrm{PE,saved}}$ represent the cost of primary energy saving, total annual cost of the solar system operation, total annual cost of the reference system, and primary energy saving of the solar operation, respectively. The FPC system obtained the lowest value of $C_{\mathrm{PE,saved}}$, namely, the best energy–cost performance [19]:

$$C_{\mathrm{PE,saved}} = \frac{C_{\mathrm{annual,sol}} - C_{\mathrm{annual,ref}}}{E_{\mathrm{PE,saved}}} \quad [11.5]$$

11.5.2 Case 2: life-cycle assessment of solar-powered double-effect absorption cooling system

In this case, a double-effect water absorption chiller was coupled with an external compound parabolic concentrator with a U-tube and reflectors. This collector was able to deliver 180~200°C fluid temperature with efficiency of 40~50%. It has a targeted market price of US$150180/$m^2$. Two configurations of the solar absorption cooling system were studied. In the first configuration, the area of the solar collectors and absorption chiller met the peak cooling demand of the tested place and natural gas was used as the backup heat source. In the second configuration, only half of the peak demand was covered by the solar system. The rest of the cooling demand was fulfilled by a vapor compression chiller. The life-cycle economic performances of the two solar-powered absorption cooling systems were compared with a conventional heating/ventilation/air-conditioning system. The study was carried out in the University of California at Merced, California, USA. The costs of the main components are listed in Table 11.14.

Table 11.14 Costs of main components [20]

Item	Cost (US$)
Natural gas boiler	$-0.0139 \times \text{capacity}^2 + 42.756 \times \text{capacity} + 1990.1$
Natural gas furnace	$-0.0139 \times \text{capacity}^2 + 42.756 \times \text{capacity} + 640.13$
Double-effect absorption chiller (small size)	500 × capacity
Double-effect absorption chiller (large size)	147.3 × capacity + 100,680
Packaged air-conditioning unit	128.45 × capacity + 6622.2
Centrifugal compression water−cool chiller	110.79 × capacity + 2454.7
XCPC solar collector	165 × area
Electricity rate	0.12834/kWh
Natural gas rate	0.03024/kWh

XCPC, external compound parabolic concentrator.

For the economic evaluation, the installation cost was assumed to be 30% of the initial cost, and the maintenance cost was assumed to be 2% of the initial cost. The annualized operation cost (OC_{annual}) was calculated based on Eq. [11.6], where FIR is the fuel inflation rate. The total present worth cost (*PWC*) of the system was calculated from the installation cost (*IC*), operating cost (*OC*), and maintenance cost (*MC*) as presented in Eq. [11.7]. According to the calculation, the second configuration had a lower *PWC* during the entire life span than the conventional system. The first configuration had a higher *PWC* mainly due to the high initial cost [20].

$$OC_{annual} = OC \times \left[\frac{1-(1+FIR)^N \times (1+d)^{-N}}{d-FIR}\right]\left[\frac{d(1+d)^N}{(1+d)^N-1}\right] \quad [11.6]$$

$$PWC = IC + \frac{(1+d)^N-1}{d(1+d)^N} \times (OC_{annual} + MC_{annual}) \quad [11.7]$$

11.6 Summary

In this chapter, solar-powered absorption cooling technology was introduced, including the fundamentals of absorption refrigeration, the various options for absorption chillers, case studies of solar-powered absorption cooling systems, the drawbacks of the technology, the corresponding solutions, and an economic evaluation.

The various options for absorption chillers were introduced based on their complementary solar collectors. For low-temperature solar collectors, single-effect absorption chillers and double-lift absorption chillers are available. For medium-temperature solar collectors, single-effect ammonia–water absorption ice-makers, double-effect water–LiBr absorption chillers, GAX ammonia–water absorption chillers, and diffusion–absorption chillers are available. In addition, to make better utilization of the backup heat source, single-effect/double-effect absorption chillers can be used. To make better use of solar power with variable temperatures, single-effect/double-lift absorption chillers and 1.n-effect absorption chillers can be used.

References

[1] Herold KE, Radermacher R, Klein SA. Absorption chillers and heat pumps. CRC press; 1996.

[2] Dai Y. Technology and application of LiBr absorption refrigeration. Chinese mechanical engineering industry publication; 1997.

[3] Handbook-fundamentals ASHRAE. NE Atlanta, GA: American society of heating, refrigerating and air-conditioning engineers. Inc; 2009.

[4] Lucy New Energy Technology Co, Ltd. Product manual of solar air conditioner. 2013.

[5] Syed A, Izquierdo M, Rodriguez P, Maidment G, Missenden J, Lecuona A, et al. A novel experimental investigation of a solar cooling system in Madrid. Int J Refrig 2005;28: 859–71.

[6] Sözen A, Altıparmak D, Usta H. Development and testing of a prototype of absorption heat pump system operated by solar energy. Appl Therm Eng 2002;22:1847–59.

[7] Shuangliang Eco-Energy Systems Co, Ltd. Absorption chiller product catalogue. 2011.

[8] Sumathy K, Huang Z, Li Z. Solar absorption cooling with low grade heat source—a strategy of development in South China. Sol Energy 2002;72:155–65.

[9] Lin P, Wang R, Xia Z. Numerical investigation of a two-stage air-cooled absorption refrigeration system for solar cooling: cycle analysis and absorption cooling performances. Renew Energy 2011;36:1401–12.

[10] Du S, Wang R, Lin P, Xu Z, Pan Q, Xu S. Experimental studies on an air-cooled two-stage NH_3-H_2O solar absorption air-conditioning prototype. Energy 2012;45:581–7.

[11] Kim D-S. Solar absorption cooling: TU Delft. Delft University of Technology; 2007.

[12] Qu M, Yin H, Archer DH. A solar thermal cooling and heating system for a building: experimental and model based performance analysis and design. Sol Energy 2010;84:166–82.

[13] Velázquez N, García-Valladares O, Sauceda D, Beltrán R. Numerical simulation of a Linear Fresnel Reflector Concentrator used as direct generator in a Solar-GAX cycle. Energy Convers Manage 2010;51:434–45.

[14] Srikhirin P, Aphornratana S. Investigation of a diffusion absorption refrigerator. Appl Therm Eng 2002;22:1181–93.

[15] Sun H, Xu Z, Wang H, Wang R. A Solar/gas fired absorption system for cooling and heating in a commercial building. Energy Procedia 2015;70:518–28.

[16] Ying C, Yuqun Z, Wei G, Jun Z. SE/DL absorption refrigeration cycle driven by low temperature heat resources. Acta Energiae Solaris Sin 2002;23:102–7.

[17] Xu Z, Wang R, Xia Z. A novel variable effect LiBr-water absorption refrigeration cycle. Energy 2013;60:457–63.

[18] Xu Z, Wang R. Experimental verification of the variable effect absorption refrigeration cycle. Energy 2014;77:703−9.
[19] Balaras CA, Grossman G, Henning H-M, Ferreira CAI, Podesser E, Wang L, et al. Solar air conditioning in Europe—an overview. Renew Sustain Energy Rev 2007;11:299−314.
[20] Hang Y, Qu M, Winston R, Jiang L, Widyolar B, Poiry H. Experimental based energy performance analysis and life cycle assessment for solar absorption cooling system at University of Californian, Merced. Energy Build 2014;82:746−57.

Solar-powered adsorption cooling systems

12

R.Z. Wang, Q.W. Pan, Z.Y. Xu
Institute of Refrigeration and Cryogenics, Shanghai Jiao Tong University, Shanghai, China

12.1 Introduction

12.1.1 Fundamental principle of adsorption refrigeration

Similar to absorption refrigeration, adsorption refrigeration is also a heat-driven refrigeration technology. It is also able to use low-grade heat, including solar heat and industrial waste heat. In absorption refrigeration, the phenomenon of refrigerant absorption by a liquid solution is utilized, whereas the phenomenon of refrigerant adsorption by a solid adsorbent is extended to adsorption refrigeration. In this case, the topology constructions of the absorption refrigeration cycle and the adsorption refrigeration cycle are similar. The differences between the two technologies come from the working pairs.

A schematic diagram of a solar-powered adsorption ice-maker is shown in Fig. 12.1, and the thermodynamic cycle is shown in Fig. 12.2. As is shown in Fig. 12.1, the solar-powered adsorption refrigerator is composed of an adsorber, condenser, evaporator, valve, and refrigerant tank. When the adsorber is cooled at night, the pressure inside the adsorber decreases, and the refrigerant inside the evaporator, which evaporates under the pressure difference between adsorber and evaporator, is adsorbed by the adsorbent inside the adsorber. The evaporation process of the refrigerant generates the refrigeration power. The refrigeration will stop when the adsorbent is saturated. In the daytime, the adsorber is heated by solar energy, and the pressure inside increases. The refrigerant inside the adsorber will be desorbed from the adsorber by the pressure difference between the adsorber and the condenser and then will be condensed inside the condenser, which was cooled by the environmental air around.

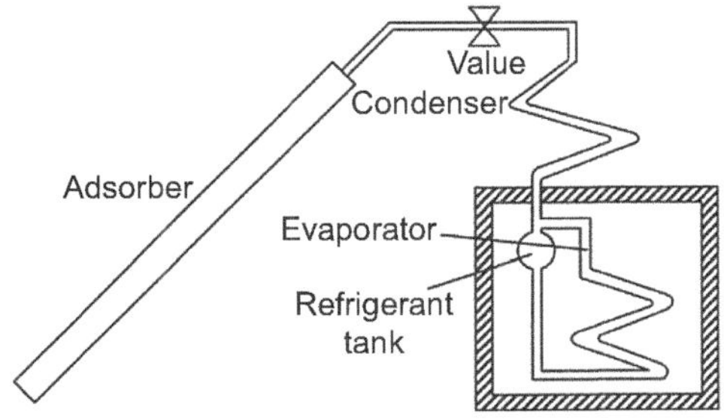

Figure 12.1 A solar-powered adsorption refrigeration system.

Advances in Solar Heating and Cooling. http://dx.doi.org/10.1016/B978-0-08-100301-5.00012-6

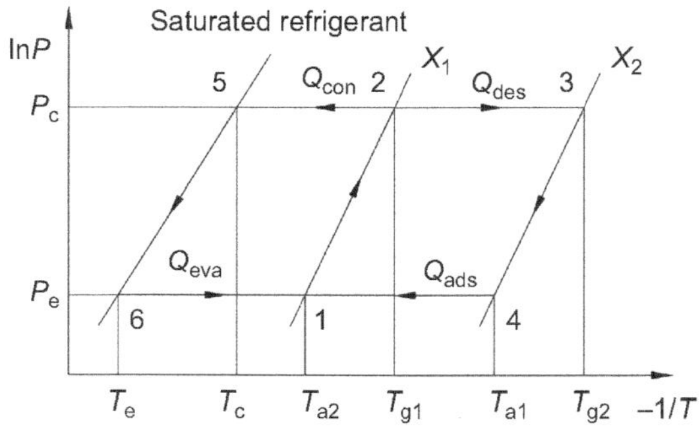

Figure 12.2 A P–T diagram of an adsorption refrigeration cycle. T_e, evaporating temperature; T_c, condensing temperature; T_{a1}, initial adsorption temperature; T_{a2}, adsorbing temperature; T_{g1}, initial desorbing temperature; T_{g2}, desorbing temperature; P_e, evaporating pressure; P_c, condensing pressure; Q_{ads}, adsorption heat; Q_{des}, desorption heat; Q_{eva}, evaporation heat; Q_{con}, condensation heat; X_1, maximum adsorption quantity; X_2, minimum adsorption quantity.

The whole process can be summarized in detail as follows (Fig. 12.2):

1. The valve is closed in the morning, assuming an environmental temperature T_{a2} of 30°C. As time passes the adsorber will be heated by solar energy, and the pressure of the adsorber will increase. Finally, the pressure of the refrigerant will be the saturated pressure for the condensing temperature of the refrigerant, which is 30°C. The temperature of the adsorber is T_{g1} in Fig. 12.2.
2. When the valve is opened, the refrigerant desorbed from the adsorber will be condensed inside the condenser, which is cooled by the natural convection heat-transfer method. After that the refrigerant will flow to the evaporator and refrigerant tank and accumulate there. In this phase the final temperature of the adsorber can be as high as T_{g2} (desorbing temperature).
3. The valve is closed in the evening. The temperature of the adsorber begins to decrease because of little or no solar energy outside. The pressure of the adsorber decreases as well, and it will decrease to the saturated pressure for the evaporating temperature; the corresponding temperature of the adsorber is T_{a1} (initial adsorption temperature).
4. The valve is opened and the refrigerant inside the evaporator will evaporate and be adsorbed by the adsorbent inside the adsorber because of the pressure difference between the adsorber and the evaporator. The evaporation process of the refrigerant provides the refrigeration power, and the adsorption heat of the adsorber will be released into the environment. This phase will proceed until the next morning, and after that a new cycle will begin.

Adsorption refrigeration has two processes, which are the heating–desorbing process and the cooling–adsorbing process. The simple traditional cycle is a type of intermittent refrigeration cycle, which is suitable for the utilization of solar energy, because solar energy is also a type of intermittent energy. If the heat source can be provided continually and the continuous refrigeration effect is required, two adsorbers or multiple adsorbers need to be designed for an adsorption refrigeration system, for which the heating and cooling processes of multiple adsorbers will be complementarily arranged. A common continuously working adsorption refrigeration system is the

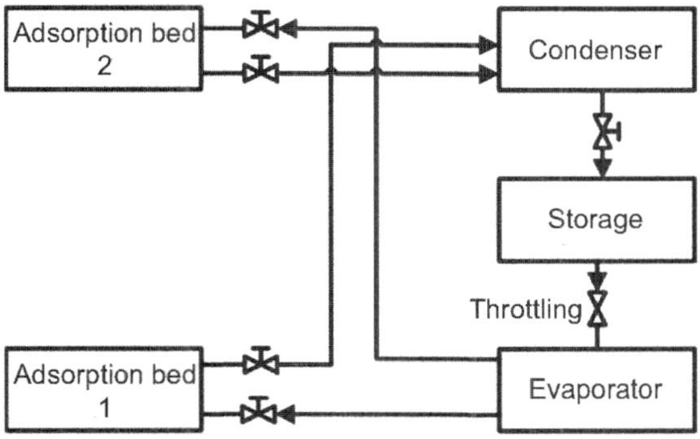

Figure 12.3 Continuously working dual–bed adsorption refrigeration system.

dual-bed system. The schematic of this system is shown in Fig. 12.3. When adsorption bed 1 starts to be heated, adsorption bed 2 starts to be cooled, and vice versa. In this case, two beds ensure that the desorption and adsorption processes work at the same time and the cooling output can be delivered continuously.

12.1.2 Adsorption working pairs

The adsorption working pair is a key element in the adsorption refrigeration and heat pump system. Thermal properties of working pairs have a great influence on the performance coefficient of the system, the temperature increment velocity of the adsorber, and the initial investment. For efficient refrigeration output, a suitable adsorption working pair needs to be selected according to the heat source temperatures, and a suitable adsorption refrigeration cycle needs to be selected according to the actual requirements. The application scope and properties are different for different adsorption refrigeration working pairs. The common adsorption refrigeration working pairs mainly include activated carbon–methanol, activated carbon fiber–methanol, activated carbon–ammonia, zeolite–water, silica gel–water, metal hydrides–hydrogen, calcium chloride–ammonia, strontium chloride–ammonia, and so on (physical and chemical adsorption) [1]. Studies also show that composite adsorption, which is a type of effective heat- and mass-transfer intensification technology for a chemical adsorbent, is a prospective technology for refrigeration [2–4]. According to the mechanism of the adsorption process, the adsorption working pairs can be classified into physical adsorption working pairs, chemical adsorption working pairs, and composite adsorption working pairs.

12.1.2.1 Physical adsorption working pairs

In the physical adsorption working pairs, the adsorbent and refrigerant are bonded mainly by the dispersion force, Debye force, and orientation force. (1) Among the various adsorption working pairs, the carbon–methanol working pair has large adsorption and desorption concentrations. Its desorption temperature is around 100°C, which is not high, and it also has the advantage of low adsorption heat, which

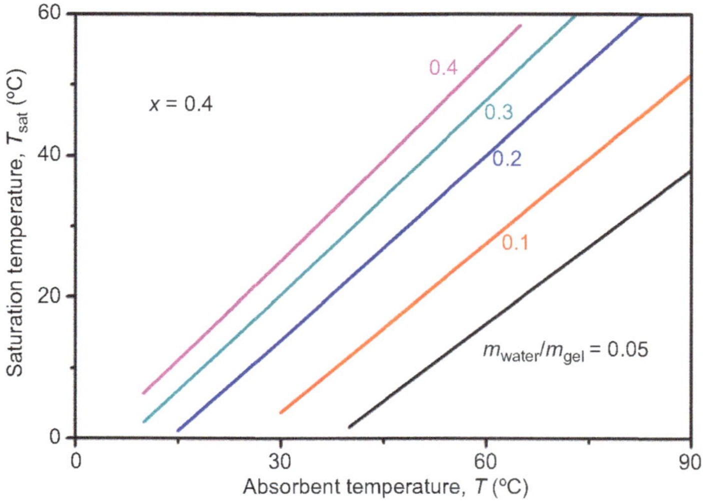

Figure 12.4 Properties of silica gel–water. x, adsorption quantity; m_{water}, adsorbed water mass; m_{gel}, silica gel mass.

is around 1800–2000 kJ/kg. The methanol refrigerant can be applied to make ice because its freezing point is below 0°C. For activated carbon–methanol working pairs, the highest desorption temperature cannot exceed 120°C, otherwise the methanol will decompose. The advantage of the activated carbon–ammonia system, which is commonly used for making ice, is the low evaporation temperature of the refrigerant. Characterized by being less sensitive to temperature changes for adsorption capacity, it is generally used for a higher heat source temperature. (2) For the working pair of silica gel–water, the desorption temperature cannot be too high. If it is higher than 120°C, the silica gel will be destroyed. Thus it is a common adsorbent for low-temperature heat sources. Fig. 12.4 shows an absorbent temperature versus saturation temperature diagram of type 3A silica gel–water. (3) The zeolite–water working pair has a wide range of desorption temperature of 70–250°C. Its adsorption heat is about 3200–4200 kJ/kg, and the evaporation latent heat of water is 2400–2600 kJ/kg. Zeolite–water is quite stable and will not be destroyed at a high temperature as happens to silica gel. However, it has the disadvantage of a higher adsorption heat, which will lead to a low coefficient of performance (COP), as well as an evaporation temperature that needs to be higher than 0°C, which cannot be utilized for making ice. In addition, the system is a vacuum system, which leads to a high requirement of vacuum sealing; meanwhile the low evaporation pressure also makes the adsorption process slower.

12.1.2.2 Chemical adsorption working pairs

In the chemical adsorption working pairs, the adsorbent and the refrigerant are bonded mainly by the functions of complexation, coordination, hydrogenation, and oxidization. Chemical adsorption working pairs mainly include hydrides–hydrogen, metal chlorides (salt)–ammonia, metal oxides–water, metal oxides–carbon dioxide, and so on. The metal hydrides–hydrogen system utilizes the adsorption process as well as the desorption process between metals or alloys and hydrogen for refrigeration, which is characterized by large adsorption and desorption heat, especially for advanced porous metal hydrides or misch metal alloy matrices including Ni, Fe, La, and Al. Such types of working pairs are generally utilized for adsorption heat pumps because they have high adsorption heat as well as high adsorption concentration. Metal chloride–ammonia working pairs are featured as having a large adsorption capacity. For example, for the calcium chloride–ammonia working pair 1 mol of calcium chloride can adsorb 8 mol of ammonia. Simultaneously, the boiling point of ammonia is lower than −34°C so that it can be used for making ice; meanwhile the refrigerator works under conditions of positive pressure, which is a feature of the simpler manufacturing techniques required for the system. Metal oxides–water and metal oxides–carbon dioxide have the advantages of being able to store high levels of energy in the hydration and carbonation processes [5,6]. With calcium oxide, for example, the storage energy in the hydration and carbonation process is 800–900 kJ/kg, which makes it possible to develop efficient heat pump systems with the application of such types of working pairs.

Fig. 12.5 shows the equilibrium lines of metal chloride–ammonia working pairs in a Clausius–Clapeyron diagram. In this diagram, the numbers correspond to different working pairs: 1, NH_3; 2, $PbCl_2$ (8/3.25); 3, $BaCl_2$ (8/0); 4, LiCl (4/3);

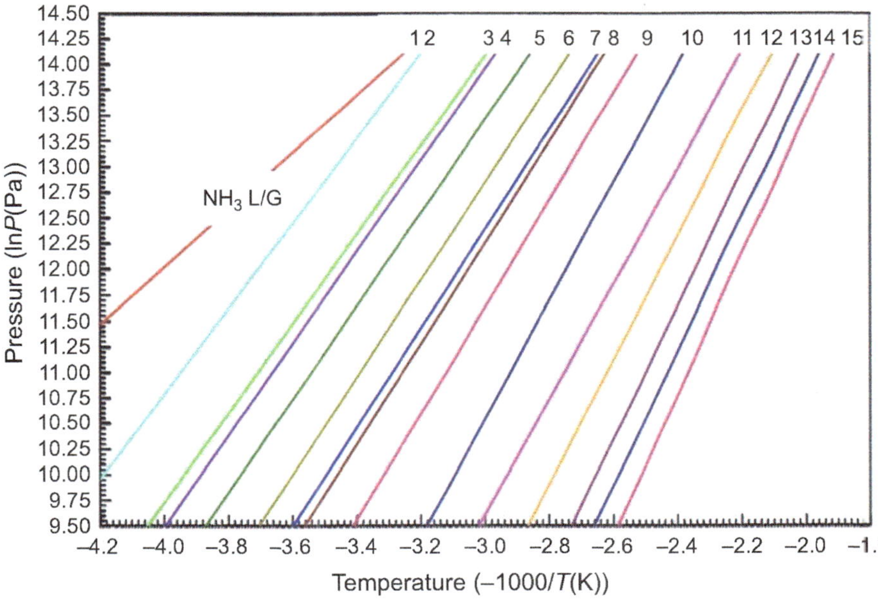

Figure 12.5 Reaction equilibrium lines of working pairs using metal chloride–ammonia [45].

5, AgCl (3/1.5); 6, $CaCl_2$ (8/4); 7, $SrCl_2$ (8/1); 8, $CaCl_2$ (4/2); 9, $CdCl_2$ (6/2); 10, $ZnCl_2$ (6/2); 11, $MnCl_2$ (6/2); 12, $FeCl_2$ (6/2); 13, $CoCl_2$ (6/2); 14, $MgCl_2$ (6/2); 15, $NiCl_2$ (6/2).

12.1.2.3 Composite adsorption working pairs

Chemical adsorption has the disadvantages of agglomeration and swelling phenomena, which will lead to problems of low permeability and poor mass-transfer performance of adsorbents. To overcome this problem, porous heat-transfer matrices were put forward for the improvement of mass transfer as well as heat transfer (by solidified adsorbents) of chemical adsorbents. Studies on such types of adsorbents mainly focus on composite adsorbents with matrices of expanded natural graphite, activated carbon fiber, and activated carbon. Research shows that such types of composite adsorbents could improve the volume filling quantity and volumetric cooling capacity [2–4] of the adsorbent. The main composite adsorption working pairs include silica gel/chlorides–water and chlorides/porous media–ammonia.

12.2 Low-temperature solar-powered adsorption systems

For the low-temperature solar heat source, under 100°C, adsorption refrigeration technology is a good solution for refrigeration. According to the cooling demands, it can supply air-conditioning and freezing. The low-temperature solar-powered adsorption refrigeration systems are classified into air-conditioning systems and ice-making systems.

12.2.1 Air-conditioning systems

Silica gel–water is the most commonly used working pair for adsorption air conditioning. Water is a nontoxic natural refrigerant with large latent heat of vaporization, leading to a bright application prospect. In general, silica gel–water adsorption chillers can be driven by 55–90°C hot water and provide 5–15°C chilled water. The operating temperature range of silica gel–water adsorption chillers is quite suitable for the common solar plate collector and solar evacuated tube collector, which can produce hot water below 100°C. They are a good choice for a low-temperature solar power-driven system. To improve the efficiency of silica gel–water adsorption chillers, various types of chillers have been developed.

12.2.1.1 Single-stage silica gel–water adsorption chiller with heat recovery

In the absorption refrigeration cycle, the heat recovery between the rich solution and the weak solution is very important to increase the cycle efficiency. Similarly, heat recovery is also important in the adsorption refrigeration system. However, owing to the differences between absorbents, the heat exchanger cannot be used for heat recovery in an

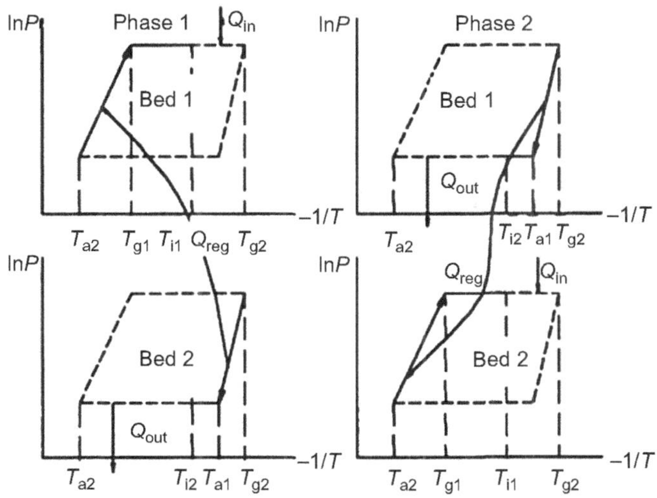

Figure 12.6 Two-bed adsorption system with heat recovery [44]. T_{a1} and T_{a2} are initial and final adsorption temperature. T_{g1} and T_{g2} are initial and final desorption temperature. T_{i1} and T_{i2} are the bed temperature after heat recovery. Q_{in}, Q_{out} and Q_{reg} are the input heat, output heat and recovered heat.

adsorption refrigeration system. Heat can be recovered between two adsorption beds. Fig. 12.6 shows a schematic diagram of the heat recovery between two adsorption beds. When the desorption and adsorption processes are close to being finished, the adsorption heat and sensible heat of the high-temperature bed can be recovered by the low-temperature bed. In this case, the system efficiency can be improved a lot.

Fig. 12.7 shows the COP comparison between the systems with and without heat recovery. The working conditions are as follows: evaporation temperature 5°C, adsorption and condensation temperature 35°C. Line 1 represents the basic cycle and line 2 represents the heat recovery cycle. Line 2 has two branches, which indicate

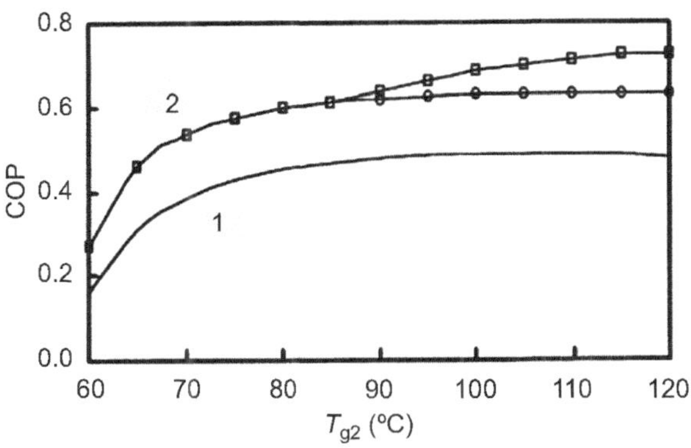

Figure 12.7 Coefficient of performance (COP) with and without heat recovery [44].

the sensible plus adsorption heat recovery and the sensible heat-only recovery systems. The COP of the basic adsorption refrigeration cycle can be increased by 25% through the heat recovery.

12.2.1.2 Single-stage silica gel–water adsorption chiller with heat and mass recovery

To improve the system efficiency further, mass recovery can be adopted. When the adsorption and desorption processes are finished in a two-bed system, the low-temperature bed and the evaporator have similar pressures and the high-temperature bed and the condenser have similar pressures. If the two beds are connected after the adsorption/desorption processes, the pressure difference between the two beds will enhance the desorption rate a lot, which increases the amount of refrigerant adsorbed in the next period. Fig. 12.8 shows the schematic and the Clapeyron diagram of the heat/mass recovery adsorption cycle. It can be seen that the refrigerant concentration glide is enlarged by the mass recovery system. Considering that mass recovery happens after the heat input period, the COP is increased by the mass recovery system without heat input.

A silica gel—water adsorption chiller with heat and mass recovery was commercialized successfully in the 1980s [7]. Nishiyodo Kuchouki Co. Ltd. developed its silica gel—water adsorption chiller 30 years ago and this product is still sold by HIJC USA, Inc. [8]. The photo and schematic of Nishiyodo's silica gel—water adsorption chiller is shown in Fig. 12.9. The main components of this chiller include two adsorbers, one condenser, one evaporator, one bypass valve, and four vacuum check valves. The connection between adsorber and condenser/evaporator is operated by the vacuum check valves, and the switch for the adsorption and desorption modes is implemented at the same time. For high system performance, this chiller uses a mass and heat recovery cycle. The bypass valve opens to equalize the pressure between adsorbers 1 and 2 at the end of the adsorption/desorption mode, which is actually a

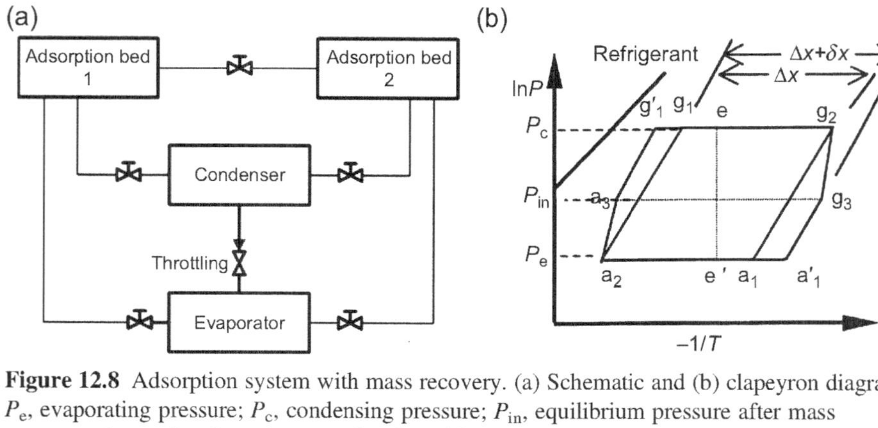

Figure 12.8 Adsorption system with mass recovery. (a) Schematic and (b) clapeyron diagram. P_e, evaporating pressure; P_c, condensing pressure; P_{in}, equilibrium pressure after mass recovery; Δx, cycle adsorption quantity; δx, additional adsorption quantity by mass recovery.

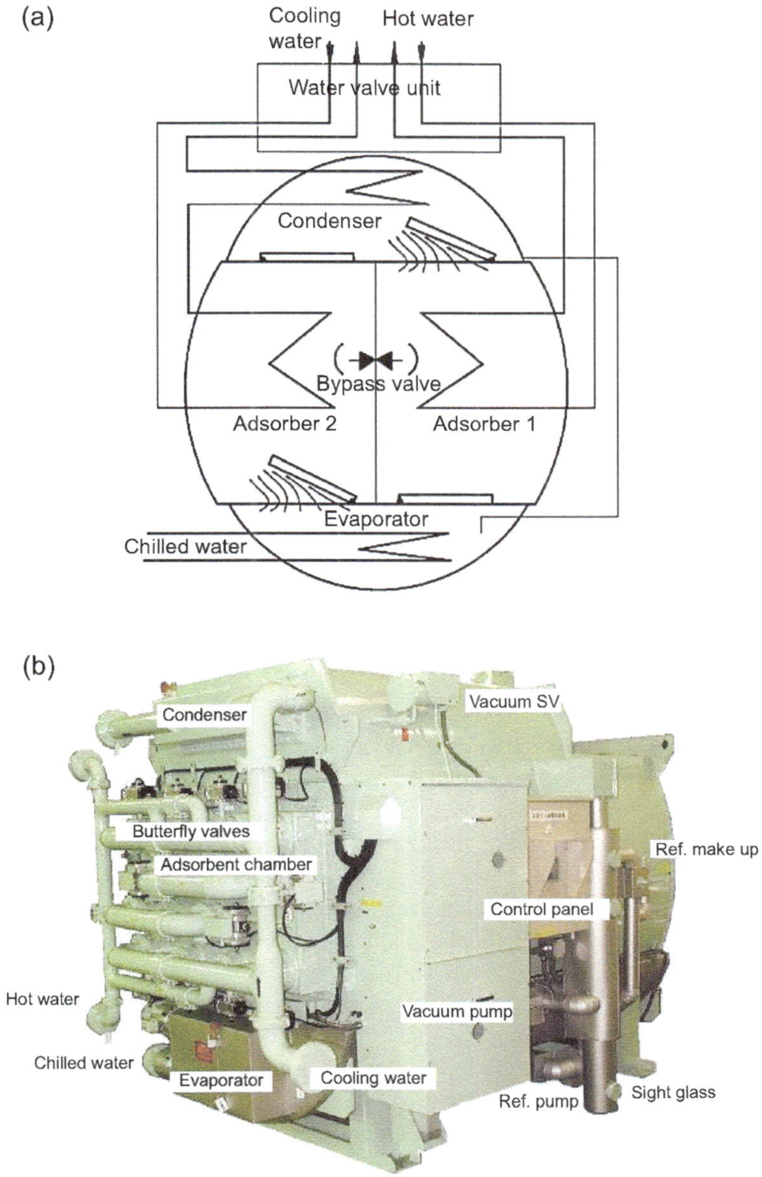

Figure 12.9 Silica gel—water adsorption chiller with mass recovery. (a) Schematic [23] of Nishiyodo's product [8] and (b) photo [45].

mass recovery process. In addition, the heat recovery process is executed before the switch to adsorption and desorption modes by using hot water in the hot adsorber to preheat the cool adsorber. According to the data documents on HIJC USA's website, this chiller has four kinds of specifications (cooling capacity is larger than 220 kW) and its COP can be 0.59 under conditions of 90°C hot water, 29.4°C cooling water, and 7°C chilled water.

SorTech AG [9], GBU [10], and PPI [11] also developed their silica gel—water adsorption chiller products using vacuum check valves. SorTech's adsorption chillers are much smaller and their cooling capacity is 7.5 or 15 kW, which is suitable for residential use. For simple and compact system constructions, SorTech's adsorption chillers use only the basic adsorption refrigeration cycle. Some residential solar cooling projects using SorTech's adsorption chillers have been successfully implemented [12]. The cooling capacity of GBU's adsorption chillers has a wide range from 71 to 1061 kW to meet different cooling demands. In addition, GBU's adsorption products use a heat recovery cycle to improve the efficiency of heat use. PPI's adsorption chillers use a system construction similar to that of Nishiyodo's adsorption chillers. All of the mentioned silica gel—water adsorption chiller products face the same question of unreliable vacuum check valves. These vacuum check valves have to operate frequently (about every 7 min) and they are rarely replaced because they are generally installed inside the vacuum chamber of the chiller. So the quality requirements and cost of these check valves are very high, which contributes to the high cost of the whole chiller.

12.2.1.3 Single-stage silica gel—water adsorption chiller with heat pipe combined evaporator

To reduce the vacuum check valve, Shanghai Jiao Tong University (SJTU) developed their silica gel—water adsorption chillers with the two-evaporator system construction [13—17]. To make the cooling output more convenient, a heat pipe evaporator is used to transfer the cooling output in both evaporators into chilled water. The schematic and prototype photo of one typical SJTU chiller are shown in Fig. 12.10.

This chiller is composed of two adsorption chambers and each adsorption chamber comprises one adsorber, one condenser, and one evaporator. So vacuum check valves are unnecessary in this chiller and the adsorption/desorption switch is realized only by adjusting the flow direction of hot and cooling water through a hot/cooling water switch system. To improve the performance and increase the adaptability to a low-temperature heat source, a vacuum valve is installed between two adsorption chambers to help the chiller fulfill the mass recovery process. A heat recovery process is also adopted in this chiller and it is controlled by the hot/cooling water system. It is noteworthy that the evaporator in this chiller, practically, is a gravitation heat pipe. The heat medium of the heat pipe is methanol. Inside the tubes in the cold end (upside) of the heat pipe is a methanol condensing surface. The water evaporation surface, outside of the tubes in the upside, is divided into two parts by the clapboard in the middle, which lies in the two adsorption chambers. The hot end (lower side) of the heat pipe is the methanol evaporating side and is also the heat exchange side between the chilled water and the methanol. The structure of the heat pipe evaporator is more complicated and more metal is consumed, compared with the evaporator in the conventional system. However, the water evaporators can automatically switch to evaporating work, and fewer vacuum valves are used.

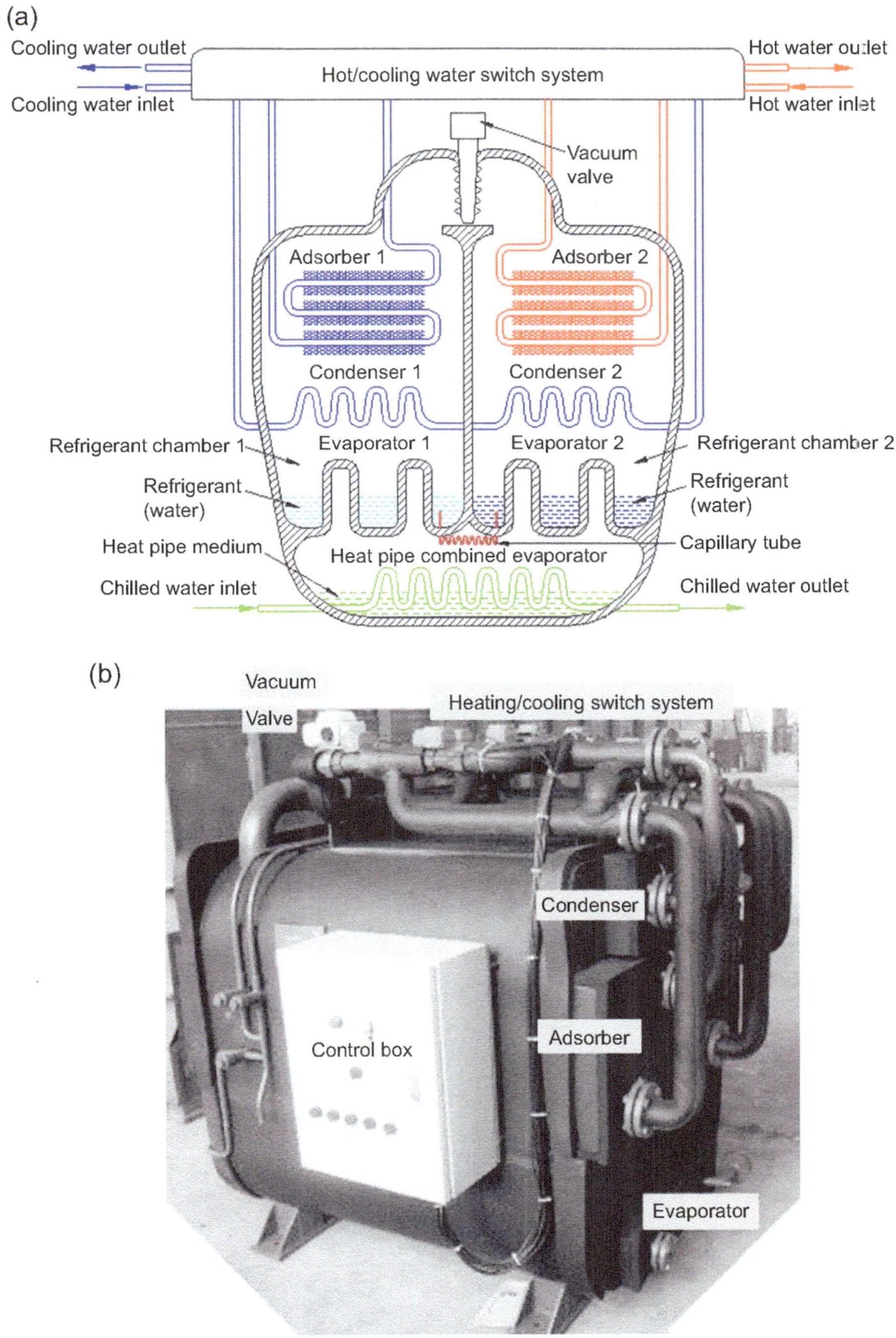

Figure 12.10 Typical Shanghai Jiao Tong University silica gel—water adsorption chiller [14,15]. (a) Schematic and (b) prototype.

In the performance test, this chiller is effectively driven by a low-grade heat source such as 65–85°C. The cooling capacity and COP of this chiller are 7.15 kW and 0.38, respectively, when the hot water temperature is 84.8°C, the cooling water temperature is 30.6°C, and the chilled water outlet temperature is 11.7°C.

Because of its good performance, this chiller has been adopted in several projects in China [18–20], such as the solar energy system in the green building of Shanghai Research Institute of Building Science, as shown in Fig. 12.11. Two chillers were powered by 150-m^2 solar collectors. This system continuously ran from 9:00 am to 5:00 pm in the summer of 2005 with an average cooling output, average system COP, and average solar COP of 10.76 kW, 0.32, and 0.12, respectively. But the

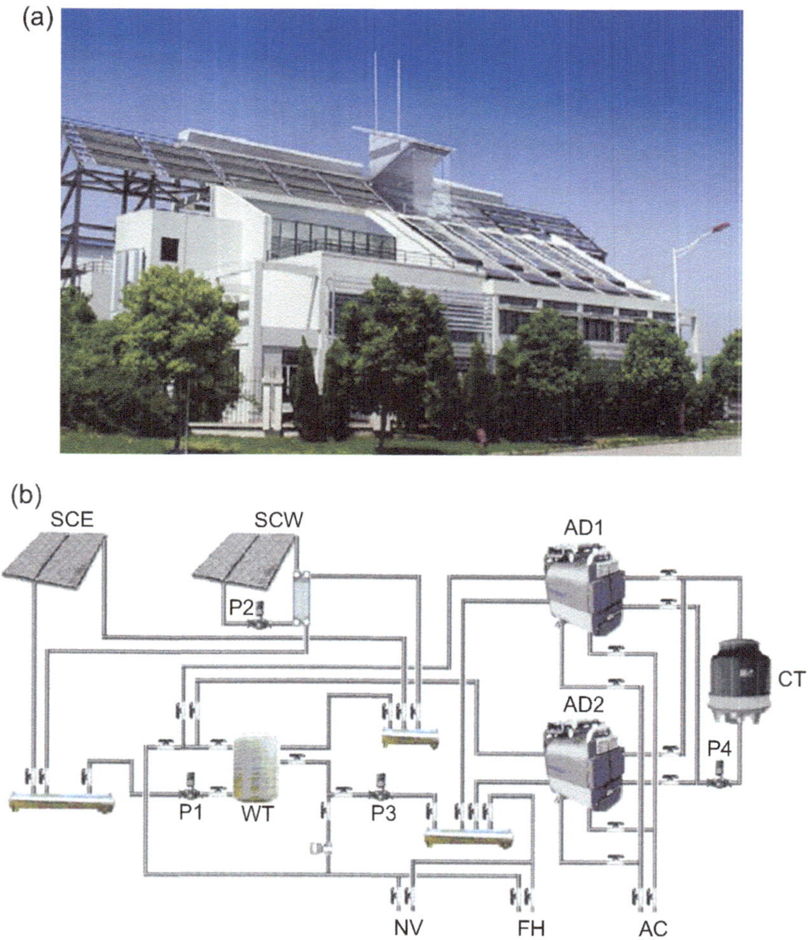

Figure 12.11 Solar energy system in the green building of Shanghai Research Institute of Building Science [18]. (a) Photo and (b) schematic. *P*, pump; *SCE*, east solar collector; *SCW*, west solar collector; *AD*, adsorption chiller; *WT*, water tank; *CT*, cooling tower; *NV*, natural ventilation; *FH*, floor radiation heating; *AC*, air conditioning.

two-evaporator system construction of the adsorption chiller has shortcomings of large system size and extra loss in cooling capacity. As a result, the heat and mass recovery process is almost required in this kind of adsorption chiller.

12.2.1.4 Single-stage silica gel—water adsorption chiller with modular design

The silica gel—water adsorption chiller has been researched a lot. However, it has not been widely used because of its shortcomings including large size, low efficiency, high cost, etc. A 50-kW silica gel—water adsorption chiller was developed by SJTU, using a modular design concept to make improvements in system size, efficiency, and cost. The heating source temperatures for the chiller are between 55°C and 90°C.

The 50-kW silica gel—water adsorption chiller with rated COP over 0.4 is shown in Fig. 12.12 and has a size of 2.88 m length, 1.92 m width, and 2.52 m height. This chiller includes two adsorbers, two condensers, two evaporators, one chilled water pump, one chilled water tank, and eight valves. The two adsorbers can alternate between adsorption and desorption phases to realize continuous cooling by switching the hot/cooling water valves (V1—V4). After the desorption phase, the adsorber is still very hot and heat can be recovered to reduce energy input. A partial heat recovery process can be fulfilled after adsorption/desorption switch via valves V1—V4. In addition, owing to the adsorption/desorption switch, the temperature of the evaporator will show a periodic change, contributing to the dissipation of the cooling capacity. To reduce this dissipation, a mass recovery-like process is adopted using chilled water valves with no vacuum valve. So the system reliability is promoted by the nonuse of a vacuum valve. As a remarkable feature of the adsorption refrigeration system,

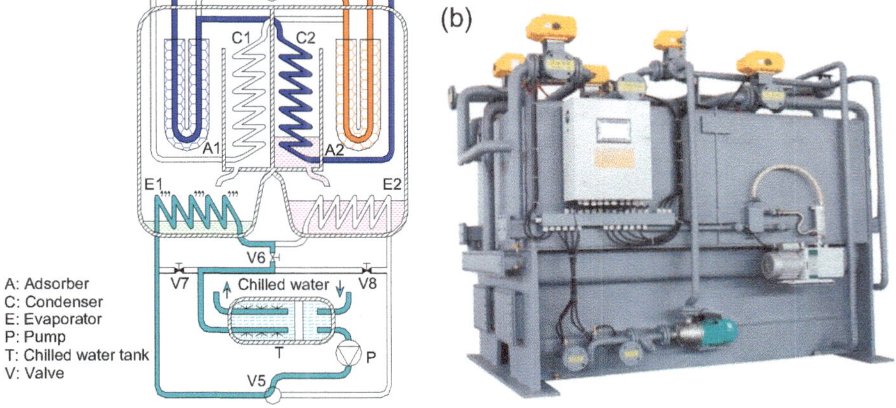

Figure 12.12 The modular silica gel—water adsorption chiller. (a) Schematic and (b) protctype.

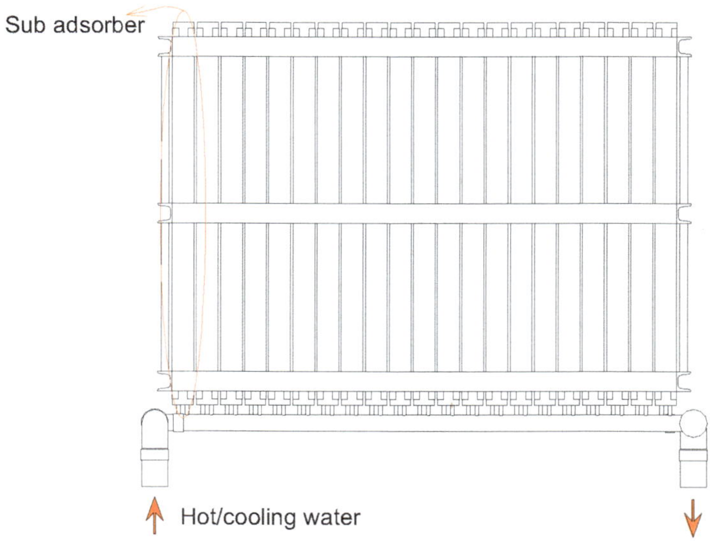

Figure 12.13 Modular adsorber.

the cooling capacity will change periodically. For better cooling effect, a chilled water pump is installed to guarantee the stable outlet temperature of the chilled water. In addition, a chilled water pump is integrated into the chiller, and fan coil units can be directly connected to the chiller, contributing to a more compact air-conditioning system.

The adsorber is the most significant and expensive component of an adsorption chiller. The adsorber used in this chiller is optimized with a modular design. In Fig. 12.13, the adsorber consists of 19 independent subadsorbers. The subadsorber is a compact fin-tube heat exchanger, which expands the heat transfer area to effectively enhance heat transfer performance. The total heat transfer area of the adsorbers is 22 m^2 (adsorbent side). Eighteen-kilogram silica gels are packed into the space between fins for each subadsorber. If the subadsorber is defined as a standard component, the different scales of adsorption chiller will depend on the quantity of subadsorbers. As a consequence, the modular adsorber can avoid redundant work in design. In this case, both cost and size of the adsorption chiller can be reduced, which makes mass production of adsorption chillers possible.

Fig. 12.14 shows the instantaneous temperatures and cooling power of the modular-designed chiller. At the beginning of cooling mode, the outlet temperatures of hot and cooling water are very low and high, respectively, because the adsorbers in the desorption and adsorption phase are still very cool and hot, separately. Then, both profiles are quickly changed to the inlet temperature profile of hot and cooling water, respectively. The inlet and outlet temperature profiles of hot/cooling water are finally very close at the end of the periodic time. The outlet temperature of chilled water is high at first because the evaporator in the adsorption chamber is hot in the last

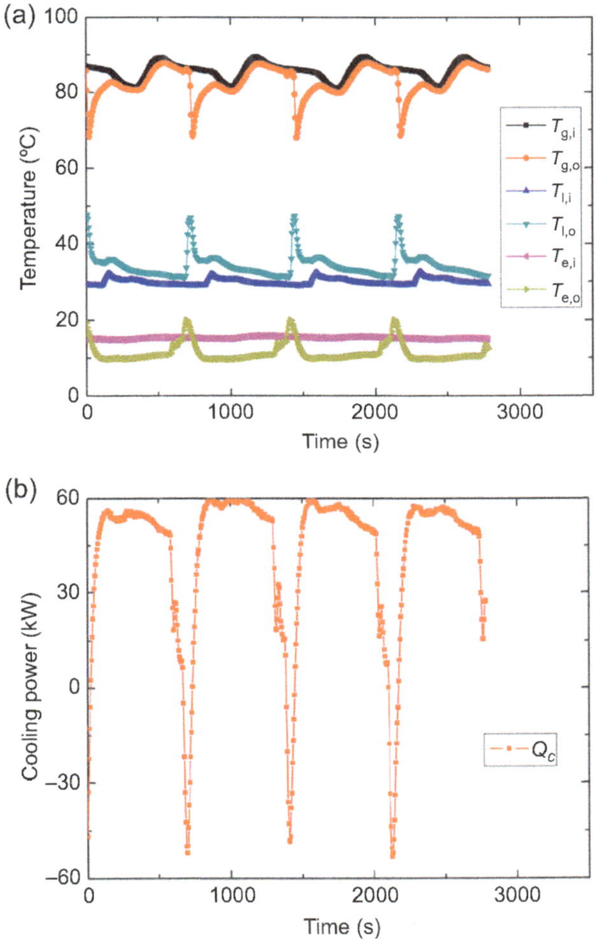

Figure 12.14 Typical conditions of the modular designed chiller. (a) Temperature profiles and (b) instantaneous cooling power profile.

mode. After that, the outlet temperature of the chilled water drops rapidly and finally reaches 10°C. The outlet temperature of the chilled water is almost constant most of the periodic time until the mass recovery mode. In mass recovery mode, the outlet temperature of the chilled water slightly rises and then becomes constant. The outlet temperature of the chilled water at this time is still lower than the inlet temperature. In other words, the cooling power in mass recovery mode is positive but very small.

The value of the instantaneous cooling power remains within the range of 50–60 kW in most of the periodic time. Once the chiller runs in mass recovery mode, it drops down to 20–30 kW, then remains within that range for tens of seconds and finally drops directly to the minimum value. A group of experimental data is shown in Table 12.1.

Table 12.1 Experimental results of the modular silica gel–water adsorption chiller

No.	$T_{hot,1}$ (°C)	$T_{c,1}$ (°C)	$T_{ch,1}$ (°C)	$T_{ch,2}$ (°C)	Q_e (kW)	COP	SCP (W/kg)	t_d (s)	t_m (s)	t_h (s)
1	85.9	29.8	15.0	11.2	41.1	0.515	120.1	800	40	40
2	85.4	30.3	11.8	9.1	29.6	0.344	86.4	600	40	24
3	82.2	32.1	19.5	14.7	50.1	0.627	146.4	600	40	24
4	61.8	29.4	14.7	13.2	17.1	0.358	50.1	600	40	24
5	85.8	30.0	15.3	11.5	40.9	0.431	119.5	600	80	40
6	86.0	30.1	15.0	11.3	40.1	0.439	117.2	600	40	40

$T_{hot,1}$, inlet temperature of hot water; $T_{c,1}$, inlet temperature of cooling water; $T_{ch,1}$, inlet temperature of chilled water; $T_{ch,2}$, outlet temperature of chilled water; Q_e, cooling power; *COP*, coefficient of performance; *SCP*, specific cooling power; t_d, cooling time; t_m, mass recovery time; t_h, heat recovery time.

12.2.1.5 Multistage silica gel–water adsorption chiller

If the temperature of the solar heat is very low, a multistage silica gel–water adsorption chiller would be a good choice. A two-stage adsorption chiller prototype, as shown in Fig. 12.15, was reported by Saha [21] in 2001. Its experimental results show the COP to be about 0.36 with a 55°C driving source in combination with a heat sink at 30°C.

Later in 2006, Saha [22] investigated another multistage (three-stage) silica gel–water adsorption refrigeration system, which is shown in Fig. 12.16. The aim of this system was to utilize low-temperature solar or waste heat sources of between 40°C and 95°C effectively with two different operation modes. A prototype was built, but there were not any experimental data or practical projects using this prototype reported.

12.2.2 Ice-making systems

Water is not a suitable refrigerant for freezing conditions so methanol and ammonia are chosen. Activated carbon can be a good choice of adsorbent when methanol is the refrigerant. Among the previous solar-activated carbon–methanol adsorption ice-makers, there are two heating methods: direct and indirect. The direct heating method requires a solar collector integrated with an adsorber and the activated carbon can be heated directly by the solar energy. In the indirect heating method activated carbon is heated by a heating medium, which is previously heated in the solar collector.

12.2.2.1 Single-stage activated carbon–methanol adsorption ice-maker

Pons [24,25], Headley [26], Li [27], and Boubakri [28] successively developed their direct solar-powered ice-maker using activated carbon–methanol as the

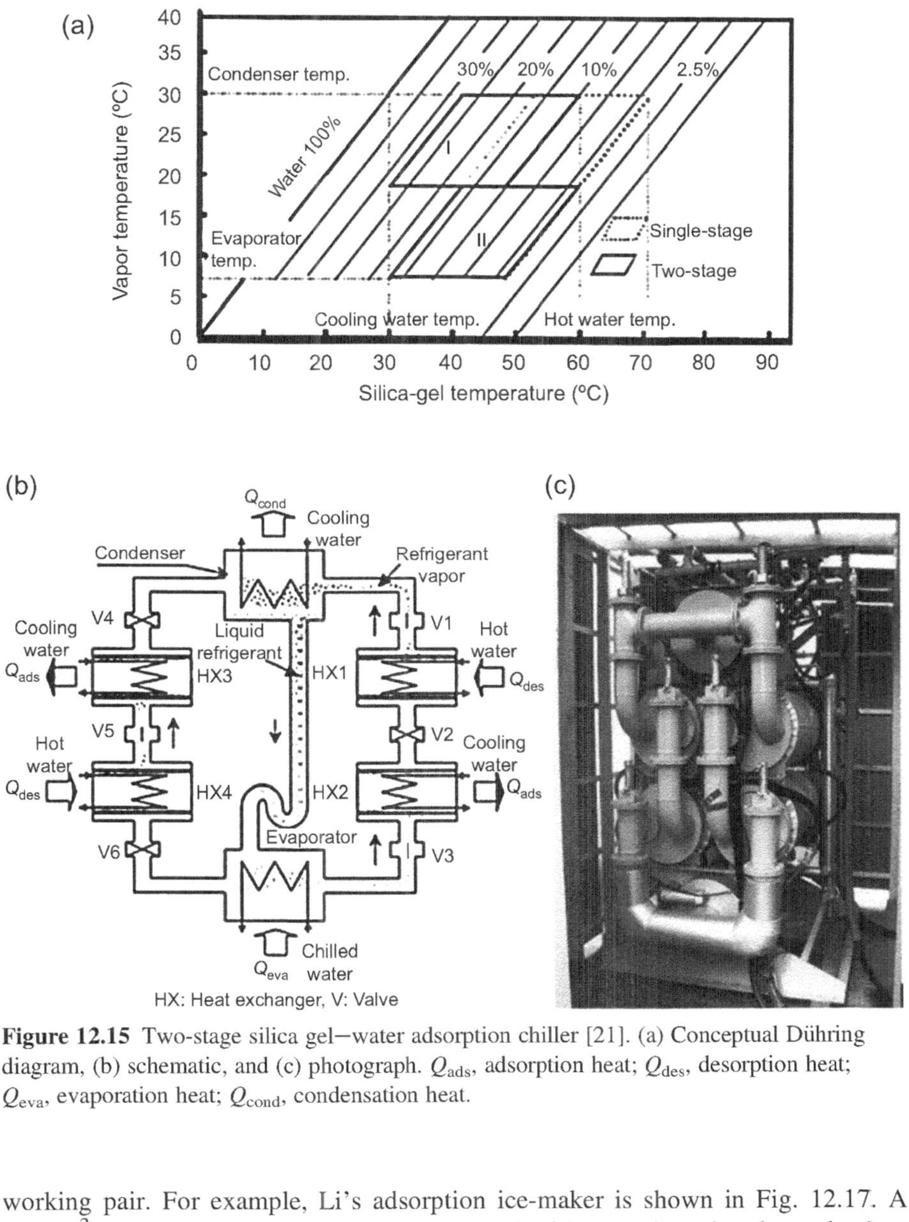

Figure 12.15 Two-stage silica gel—water adsorption chiller [21]. (a) Conceptual Dühring diagram, (b) schematic, and (c) photograph. Q_{ads}, adsorption heat; Q_{des}, desorption heat; Q_{eva}, evaporation heat; Q_{cond}, condensation heat.

working pair. For example, Li's adsorption ice-maker is shown in Fig. 12.17. A 0.92-m^2 flat plate solar collector was integrated with an activated carbon adsorber. During the daytime, solar energy absorbed by the collector directly made the adsorber temperature rise to a maximum value of 70—78°C when the solar insolation was good. Desorbed methanol would condense in the condenser and condensate flowed into the evaporator. Once night came, the adsorber temperature would decrease by air convection and the adsorption process could last the whole night, causing ice production in the evaporator. This ice-maker would produce about

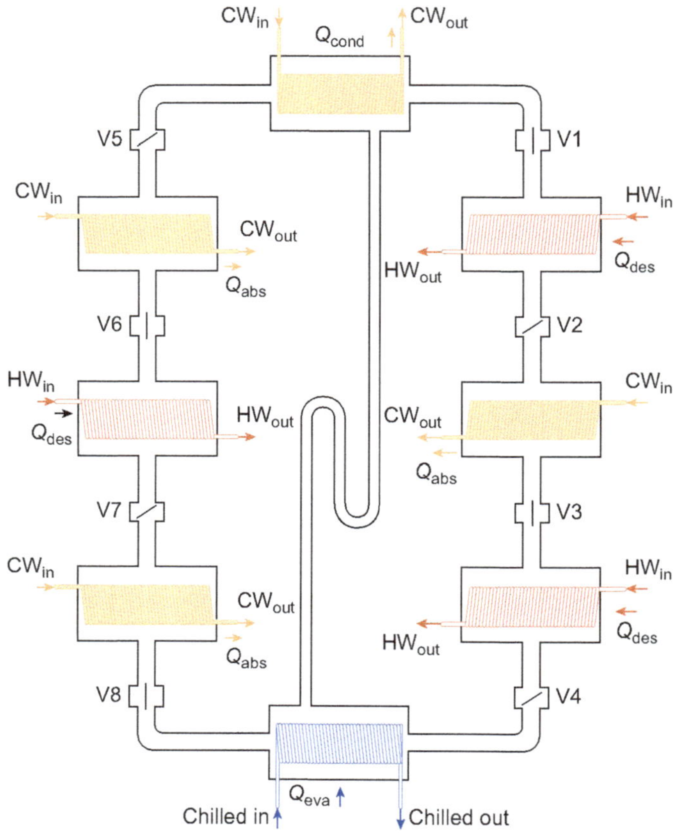

Figure 12.16 Three-stage silica gel–water adsorption chiller [23]. Q_{abs}, adsorption heat; Q_{des}, desorption heat; Q_{eva}, evaporation heat; Q_{cond}, condensation heat.

4–5 kg of ice per day with a system COP at around 0.1–0.2 when the insolation was about 17–19 MJ/m^2.

Unlike Li's system, Wang [29] developed an indirect solar-powered adsorption ice-making system, which is actually a hybrid system of a solar-powered water heater and an adsorption ice-maker using activated carbon–methanol. As is shown in Fig. 12.18, the system contained: (1) solar collector, (2) water pipe, (3) adsorber, (4) valve, (5) condenser, (6) evaporator, (7) refrigerator (with cold storage), (8) receiver, and (9) hot water container. The working principle is just a combination of a solar water heater and adsorption refrigeration. Heating of the water tank begins in the morning in the vacuum tube-type solar collector. With the increase in water temperature, the temperature in the adsorbent bed rises. When the temperature in the adsorbent rises up to a temperature that brings the vapor pressure of the desorbed refrigerant up to the condensing pressure, desorption at constant pressure is initiated, and the desorbed vapor is condensed in the condenser and collected in the receiver. This liquid flows to the evaporator via a flow-rate-regulating valve. With the refilling of the water tank with

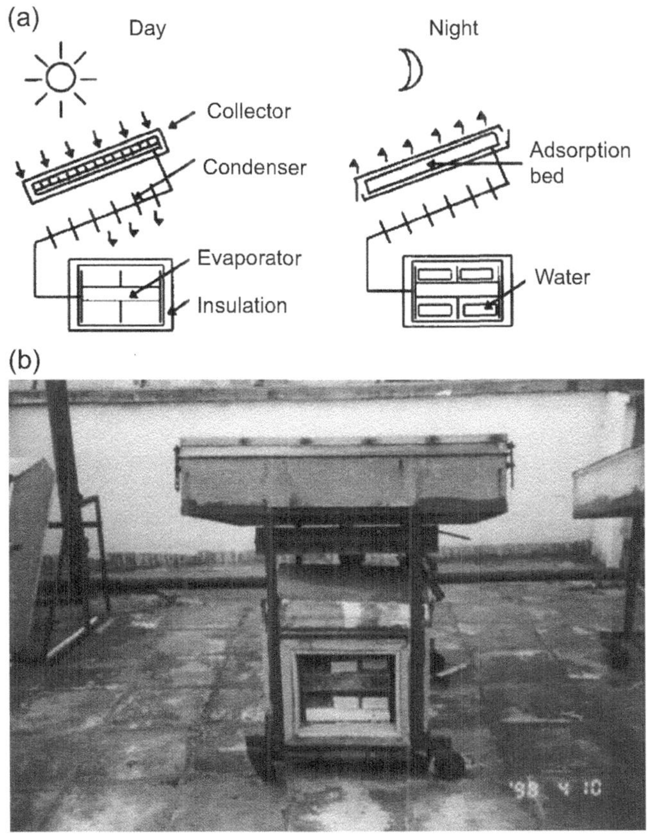

Figure 12.17 Solar-powered ice-maker with activated carbon–methanol [27]. (a) Schematic and (b) photo.

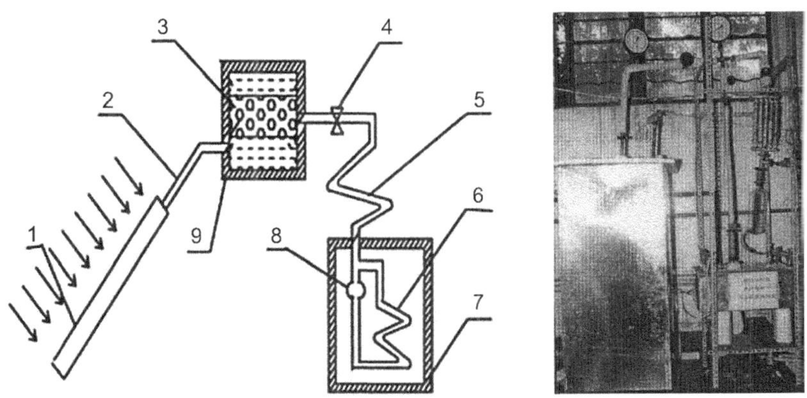

Figure 12.18 Hybrid system of solar-powered water heater and adsorption ice-maker [29].

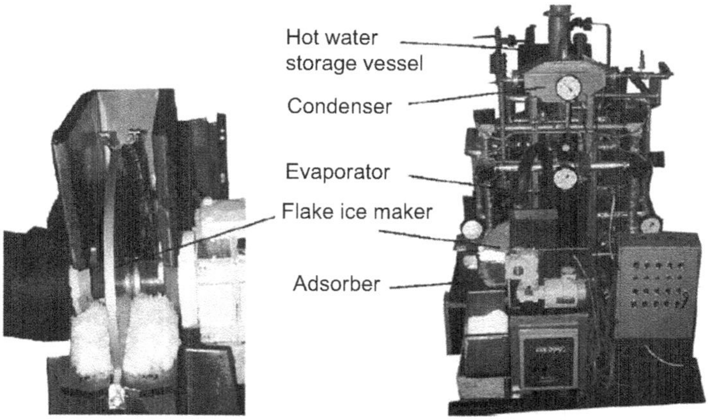

Figure 12.19 Solidified activated carbon–methanol adsorption ice-maker [30].

cold water at night, the temperature of the adsorbent bed is reduced rapidly, and the pressure in the adsorber drops to a value below evaporation pressure. Evaporation could happen if the connecting valve is open, and ice will be made in the refrigeration box. With a 2-m^2 solar collector, it is capable of making 60°C hot water and producing 10 kg of ice per day.

Furthermore, Wang's research group [30] built a solidified activated carbon–methanol adsorption ice-maker with heat and mass recovery, as shown in Fig. 12.19. Through experimental investigation, COP and ice productivity were 0.239 and 21 kg/h, respectively, when the water temperature was 25°C and evaporation temperature −7°C. The experiments also showed that cooling power could be enhanced by the mass recovery process up to 11%, and the heating power could be lowered by the heat recovery process up to 30%.

12.2.2.2 Two-stage $CaCl_2$–ammonia/$BaCl_2$–ammonia adsorption ice-maker

For ammonia refrigerant, a single-stage adsorption freezing system is rarely driven by a heat source under 100°C. But this problem can be solved by using a multistage system. A multistage ammonia adsorption freezing system needs more than one kind of working pair. Wang [31] developed a two-stage adsorption freezing machine (Fig. 12.20). This machine can be used for ice-making because its evaporation temperature is below 0°C.

$CaCl_2$–ammonia and $BaCl_2$–ammonia are the two working pairs in this machine. The $CaCl_2$ adsorber is the middle-temperature adsorber and the $BaCl_2$ adsorber is the low-temperature adsorber. In the first phase, ammonia desorbed from the $CaCl_2$ adsorber is directly adsorbed by the $BaCl_2$ adsorber. In the second phase, ammonia desorbed from the $BaCl_2$ adsorber condenses in the condenser and the $CaCl_2$ adsorber adsorbs the ammonia from the evaporator, causing a cooling effect in the evaporator. The performance test showed this machine could work effectively

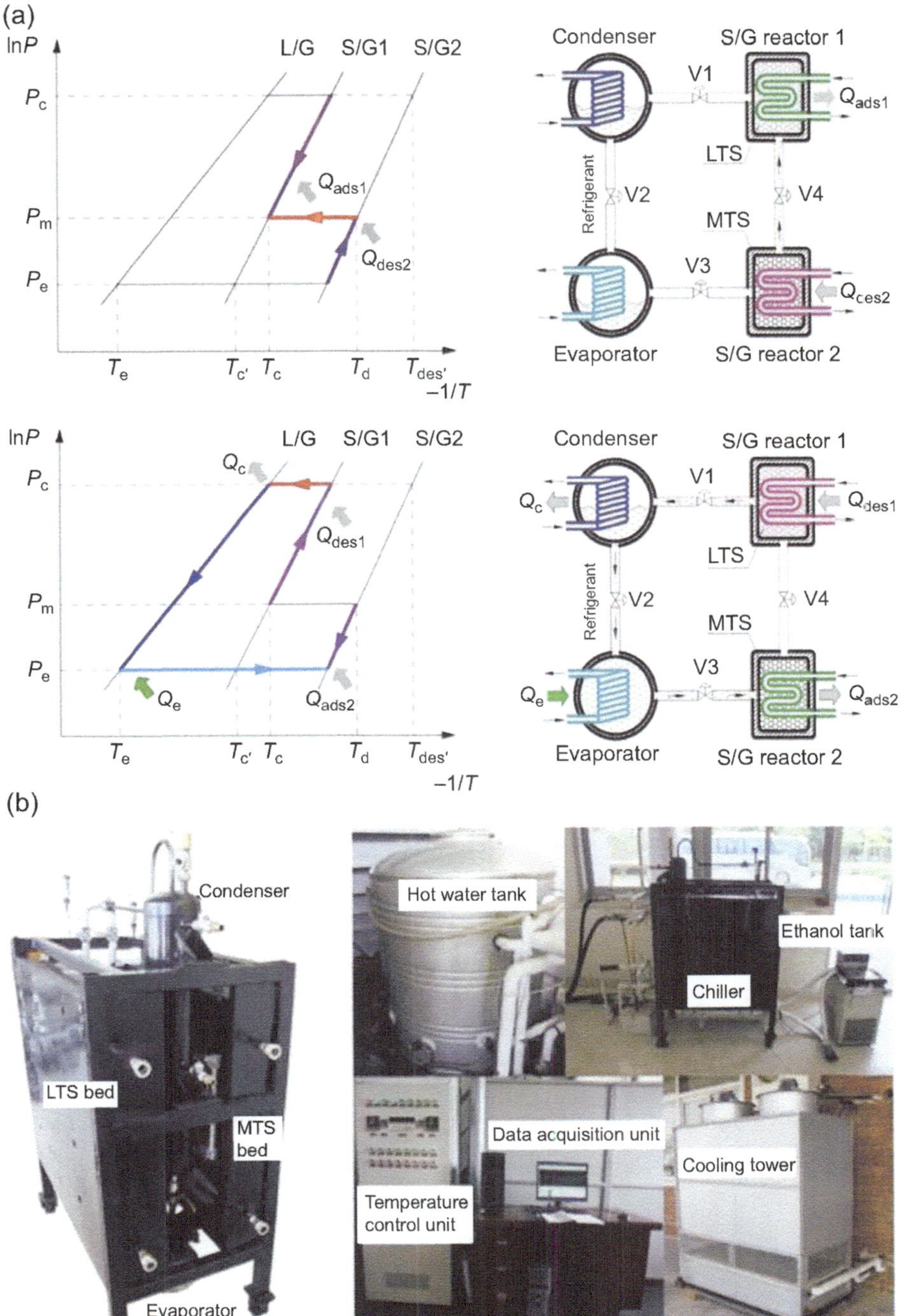

Figure 12.20 Two-stage adsorption ice-maker [31]. (a) Schematic and (b) machine and testing system. *LTS*, low-temperature salt; *MTS*, middle-temperature salt; Q_{ads}, adsorption heat; Q_{des}, desorption heat; Q_e, evaporation heat; Q_c, condensation heat; P_e, evaporating pressure; P_c, condensing pressure; P_m, middle pressure; T_e, evaporating temperature; T_c, condensing temperature; T_d, desorbing temperature; $T_{c'}$, condensing temperature for LTS; $T_{des'}$, desorbing temperature for MTS.

with a heat source of 75–85°C. For a heat source of 85°C and coolant temperature of 25°C, the system COP at 15°C refrigeration was 0.127.

12.3 Medium-temperature solar-powered adsorption systems

12.3.1 Air-conditioning systems

For the middle-temperature solar heat source at a temperature range of 100–250°C, adsorption refrigeration technology is still a very good solution.

12.3.1.1 Single-stage zeolite–water adsorption chiller

The zeolite–water working pair is a good fit for the working conditions of middle-temperature solar air conditioning. Zhang [32] built a zeolite–water adsorption cooling system for automobile air-conditioning use. The system COP is 0.38 when engine exhaust gas is the heat source at about 250°C. Lu [33] developed another zeolite–water adsorption chiller, as shown in Fig. 12.21. This chiller was powered by a

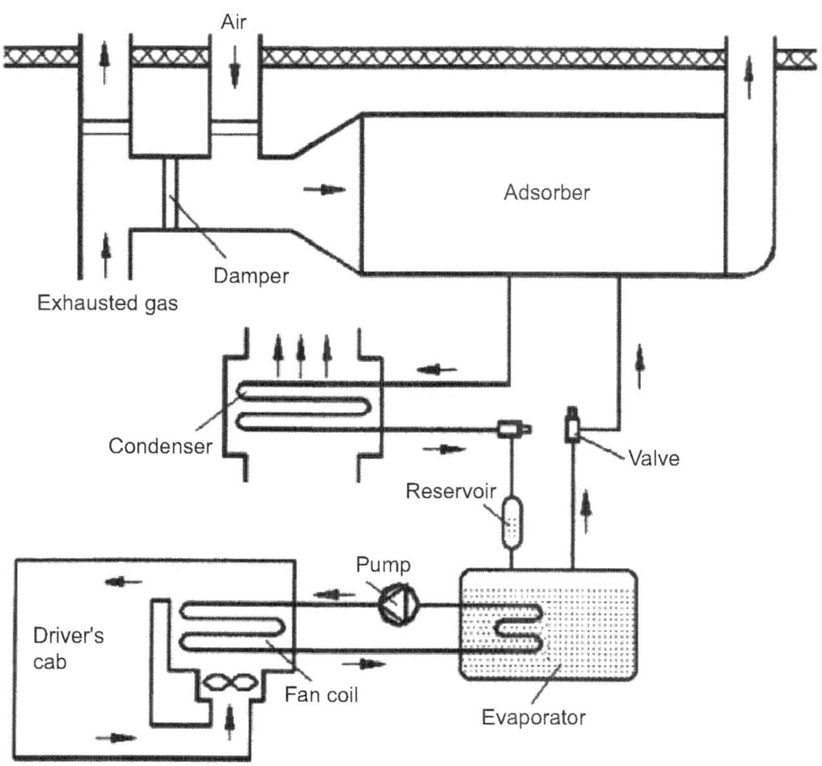

Figure 12.21 Schematic of a zeolite–water adsorption cooling system [33].

middle-temperature heat source, such as more than 200°C exhausted gas, and provided chilled water for the air conditioner of the driver's cab. This chiller was an intermittent-type adsorption chiller and included only one adsorber, which was heated by exhausted gas in the desorption phase and cooled by ambient air in the adsorption phase. In a performance test, the average cooling capacity and COP were 4.1 kW and 0.25, respectively.

A zeolite—water system might be a good match with a middle-temperature solar collector and has been also considered as a good working pair for solar thermal storage.

12.3.1.2 Cascaded adsorption refrigeration chiller with zeolite—water and activated carbon—methanol

To make better utilization of a high-temperature source, cascaded adsorption refrigeration cycles can be used. In this kind of adsorption cycle, two or more working pairs with different working temperatures are needed. For example, zeolite—water and activated carbon (AC)—methanol can be adopted. Zeolite—water has a higher working temperature, whereas AC—methanol has a lower working temperature. Different adsorption beds can be cascaded through adsorption heat recovery or condensation heat recovery. Fig. 12.22 shows one kind of cascaded sorption refrigeration cycle with zeolite—water and AC—methanol. In this cycle, the adsorption heat of the high-temperature subcycle is recovered by the low-temperature subcycle.

During the first phase, the working pair of zeolite—water was heated by an external heat source to perform the desorption process, and the other working pair, AC—methanol, was cooled by an external heat sink to perform the adsorption process.

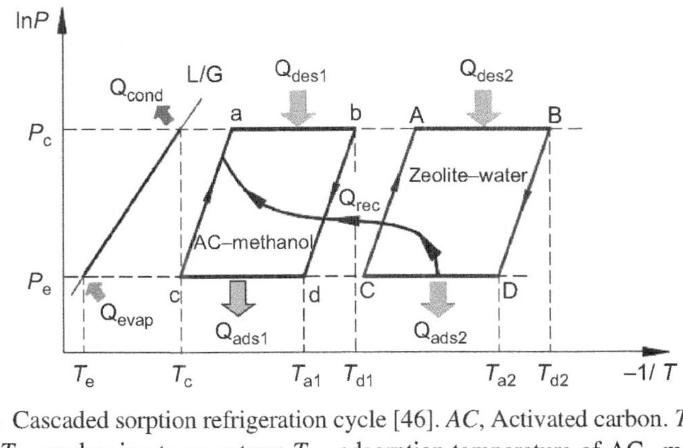

Figure 12.22 Cascaded sorption refrigeration cycle [46]. *AC*, Activated carbon. T_e, evaporating temperature; T_c, condensing temperature; T_{a1}, adsorption temperature of AC—methanol; T_{a2}, adsorption temperature of zeolite—water; T_{d1}, desorption temperature of AC—methanol; T_{d2}, desorption temperature of of zeolite—water; P_e, evaporating pressure; P_c, condensing pressure; Q_{ads1}, adsorption heat of AC—methanol; Q_{ads2}, adsorption heat of zeolite—water; Q_{des1}, desorption heat of AC—methanol; Q_{des2}, desorption heat of zeolite—water; Q_{evap}, evaporation heat; Q_{cond}, condensation heat; Q_{reg}, recovered heat.

During the second phase, the working modes of the two working pairs were interchanged, in which zeolite—water performed the adsorption process and AC—methanol underwent the desorption process. The adsorption heat released by the former sorption working pair during the adsorption phase was recovered to supply the desorption heat of the latter sorption working pair during the heating desorption phase [46].

Douss and Meunier [47] studied the cycle shown in Fig. 12.22 through experiments. The results showed that the cooling COP was as high as 1.06. The desorption temperature and adsorption temperature of the high-temperature bed were 220°C and 105°C. They were 90°C and 35°C for the low-temperature bed. The cascaded sorption refrigeration cycle has a high COP; however, the complex system and operation limited its further improvement.

12.3.2 Ice-making systems

12.3.2.1 Single-stage activated carbon—ammonia ice-maker

For the freezing condition, AC—ammonia and $CaCl_2$—ammonia are the most suitable working pairs. To further the efforts to utilize the solar energy, Critoph [34] built a laboratory-scale AC—ammonia adsorption ice-maker. With 1.4 m^2 of collector area, the peak collector temperature for the simulated day tests was 115°C. The evaporator temperature attained was up to −1°C, about 3 kg of ice was manufactured, and the solar COP was 0.04. And then, to develop a compact system, Critoph [35,36] studied a rapid-cycling ice-maker (Fig. 12.23). In this version, two adsorbers were used so that the adsorption and desorption phases could proceed simultaneously in different

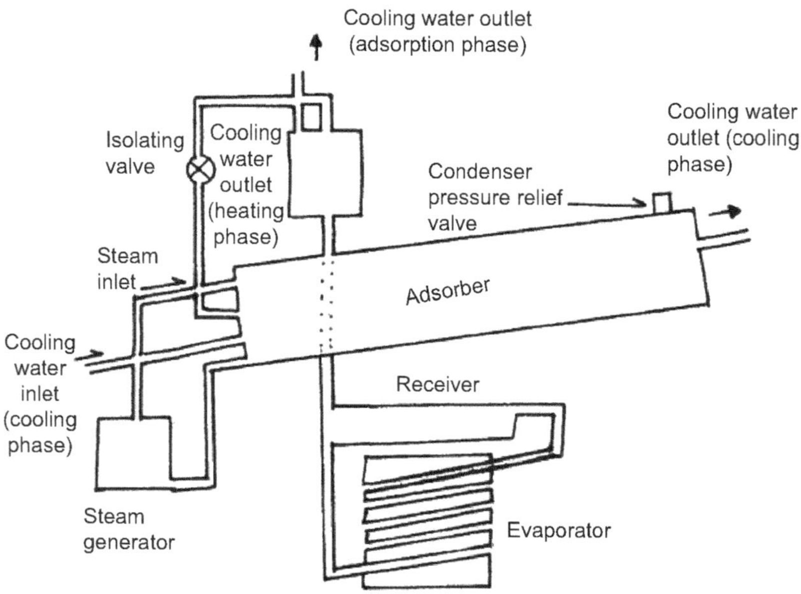

Figure 12.23 Schematic of an activated carbon—ammonia adsorption ice-maker [36].

adsorbers. The results of the experiments conducted with this unit showed that the half-cycle times for optimum ice production varied from 16 min with steam at 150°C to 26 min with steam at 100°C.

12.3.2.2 Single-stage $CaCl_2$–ammonia adsorption ice-maker

Unlike AC–ammonia, $CaCl_2$–ammonia are a chemical working pair with larger adsorption quantity. The first demonstration of a $CaCl_2$–ammonia adsorption unit was built at the Refrigeration Laboratory, Technical University of Denmark, in 1977 [37], as shown in Fig. 12.24. The project was carried out at the Institute of Solar Energy and Related Environmental Research in Khartoum, Sudan. The performance measurement of the unit on a clear day showed that net cold production was

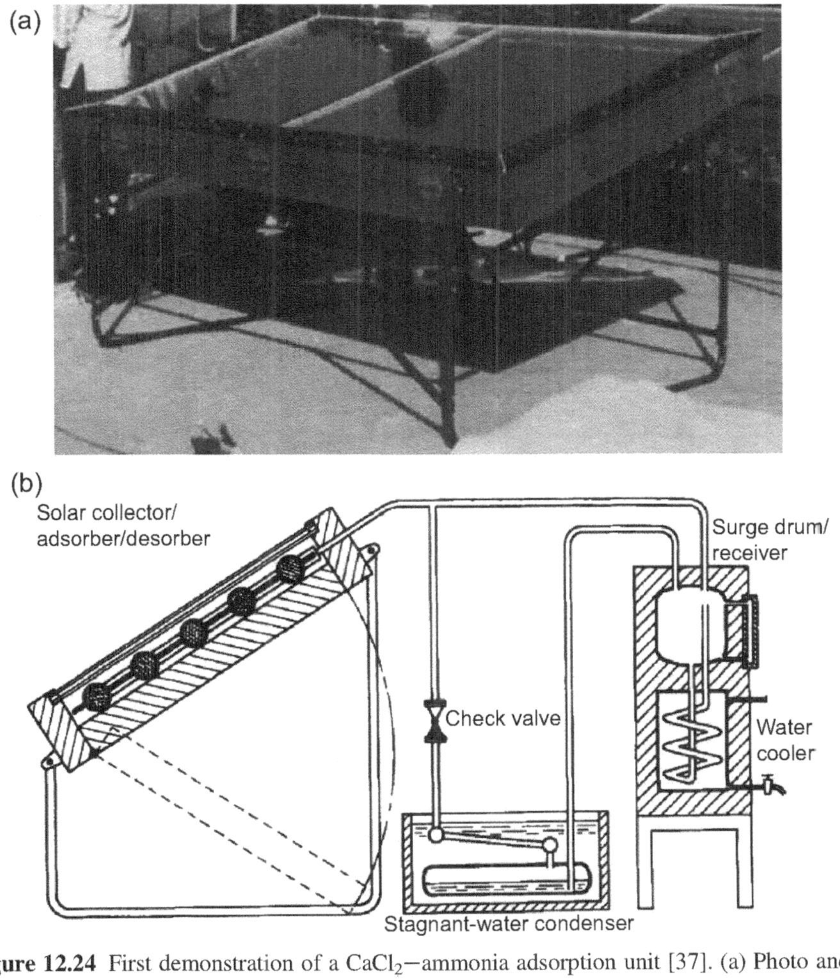

Figure 12.24 First demonstration of a $CaCl_2$–ammonia adsorption unit [37]. (a) Photo and (b) schematic.

2000 kJ/m^2 of collector surface, corresponding to the freezing of 6 kg of ice when the initial temperature of the water was 0°C. The COP of the refrigerating cycle was 0.34, and the overall COP was 0.096.

12.3.2.3 Single-stage $CaCl_2$/activated carbon–ammonia adsorption ice-maker

As the development of adsorption refrigeration technology proceeded, the agglomeration and expansion phenomena of $CaCl_2$ were discovered when it adsorbed ammonia, and they are the main reasons for performance degeneration of the $CaCl_2$–ammonia adsorption refrigeration system. The agglomeration and expansion phenomena can be solved by a composite adsorbent composed of $CaCl_2$ and a porous medium (like AC, expanded natural graphite, and so on). The efforts [38–43] made at SJTU to develop a composite adsorbent–ammonia adsorption chiller are outstanding. Lu [41] built a multifunction heat pipe adsorption ice-maker using a $CaCl_2$/AC–ammonia working pair, which is shown in Fig. 12.25. The adsorbers were heated and cooled by heat pipe heat exchangers and the heat-transfer performance was improved. For further improvement of the system performance, heat and mass recovery processes were adopted in this machine. In the test, the low-grade heat source (solar heat or waste heat) was replaced by electricity. With a desorption temperature of 126°C, cooling water temperature of 22°C, and ice-producing temperature of −7.5°C, the ice productivity and COP of this ice-maker were 17.6 kg/h and 0.2, respectively.

A hybrid system of a solar heating system and adsorption ice-maker were built at SJTU [43], as shown in Fig. 12.26. The ice-maker was a $CaCl_2$/AC–ammonia adsorption refrigeration machine and its driving source was middle-temperature solar heat, which was collected by a parabolic trough solar collector (PTC). Experimental results of this system showed that the highest COP reached 0.15 and the ice-making capacity was 50 kg per day with a 20-m^2 PTC and 30 kg $CaCl_2$/AC composite

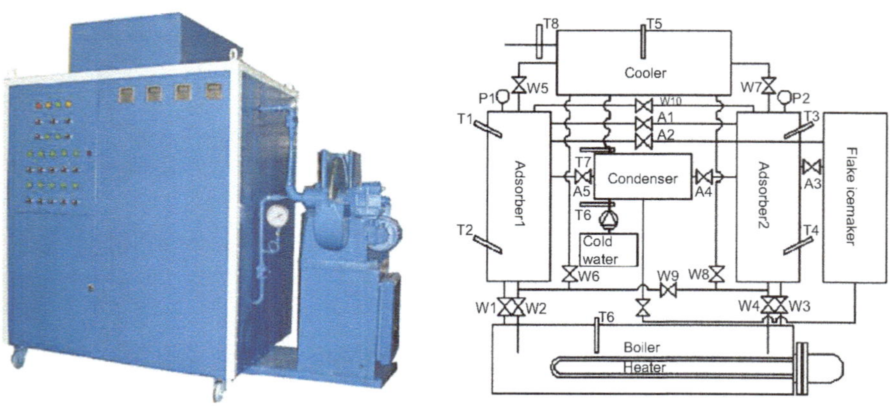

Figure 12.25 A multifunction heat pipe adsorption ice-maker using $CaCl_2$/activated carbon–ammonia [41].

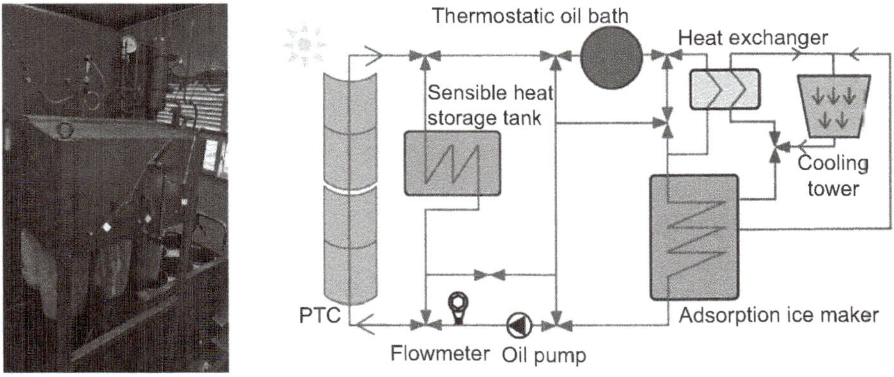

Figure 12.26 An adsorption ice-maker driven by a parabolic trough solar collector [43].

adsorbent when the desorption temperature, condensing temperature, and direct normal solar radiation were 105°C, 30°C, and 3 kWh/day m^2, respectively.

12.3.2.4 Single-stage $CaCl_2$/expanded natural graphite–ammonia adsorption ice-maker

Another adsorption ice-maker (Fig. 12.27), developed by Li [42], used $CaCl_2$/expanded natural graphite–ammonia as the working pair. The adsorbers were also heated by a heat pipe heat exchanger but cooled by a normal single-phase water heat exchanger. The difference in this machine was the installation of a mass recovery pipe. It was installed between the two evaporators and the mass recovery process would occur there, helping in the mass equilibrium between the two evaporators. The experimental results showed that the cooling capacity and COP of the refrigerator

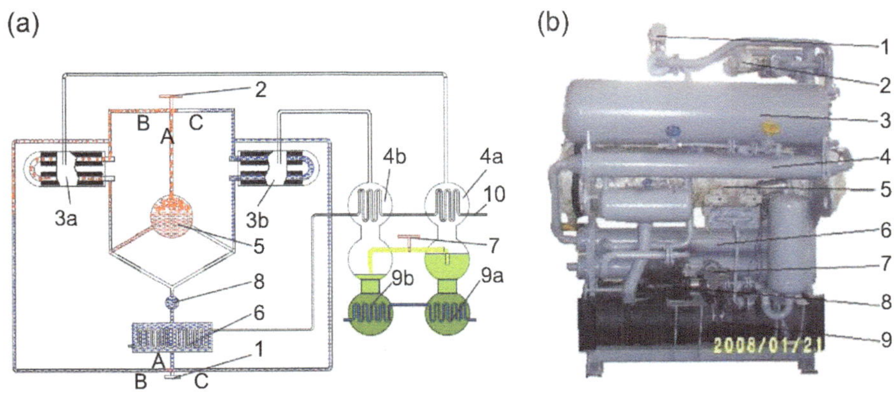

Figure 12.27 A $CaCl_2$/expanded natural graphite–ammonia adsorption ice-maker [42]. (1) Water valve, (2) vapor valve, (3) adsorbers, (4) condensers, (5) vapor generator, (6) cooler, (7) ammonia valve, (8) pump, (9) evaporators, and (10) cooling water. A, B and C are the directions of the valves.

were 11.4 kW and 0.27, respectively, under the following conditions: the temperature of the heating vapor was 120–140.3°C, the cooling water temperature was 25°C, the evaporation temperature was −15°C, and the heating/cooling time was 25 min including 45 s mass recovery time.

12.4 Summary

In this chapter, the solar-powered adsorption cooling technology was introduced, including the fundamentals of adsorption refrigeration, various adsorption working pairs, and options for adsorption refrigeration systems for both air-conditioning and ice-making. For some systems, solar-powered adsorption refrigeration cases were also introduced.

The various options introduced were based on their complementary solar collector and cooling output temperature. The classification of adsorption refrigeration systems is more complex than that of the absorption refrigeration systems owing to the variation in working pairs. In absorption refrigeration systems, water–LiBr and ammonia–water are the most commonly used working pairs. In adsorption refrigeration systems, silica gel–water, AC–methanol, $CaCl_2$–ammonia, zeolite–water, AC–ammonia, $CaCl_2$/AC–ammonia, and $CaCl_2$/expanded natural graphite–ammonia are all potential working pairs.

With low-temperature solar collectors, single-stage and multistage silica gel–water adsorption chillers are available for air conditioning; single-stage AC–methanol and two-stage $CaCl_2$–ammonia adsorption ice-makers are available for freezing conditions. With medium-temperature solar collectors, single-stage zeolite–water adsorption chillers, cascaded adsorption chillers with zeolite–water and AC–methanol adsorption chillers, and double-way thermochemical sorption chillers are available for air conditioning. Single-stage AC–ammonia, $CaCl_2$–ammonia, $CaCl_2$/AC–ammonia, and $CaCl_2$/expanded natural graphite–ammonia ice-makers are available for freezing conditions.

References

[1] Srivastava NC, Eames IW. A review of adsorbents and adsorbates in solid–vapour adsorption heat pump systems. Appl Therm Eng 1998;18(9–10):707–14.
[2] Wang LW, Wang RZ, Wu JY, Wang K. Compound adsorbent for adsorption ice maker on fishing boats. Int J Refrig 2004;27(4):401–8.
[3] Aristov YI, Restuccia G, Cacciola G, Parmon VN. A family of new working materials for solid sorption air conditioning systems. Appl Therm Eng 2002;22(2):191–204.
[4] Mauran S, Prades P, L'Haridon F. Heat and mass transfer in consolidated reacting beds for thermochemical systems. Heat Recovery Syst CHP 1993;13(4):315–9.
[5] Kato Y, Takahashi F-u, Watanabe A, Yoshizawa Y. Thermal analysis of a magnesium oxide/water chemical heat pump for cogeneration. Appl Therm Eng 2001;21(10): 1067–81.

[6] Kato Y, Yamada M, Kanie T, Yoshizawa Y. Calcium oxide/carbon dioxide reactivity in a packed bed reactor of a chemical heat pump for high-temperature gas reactors. Nucl Eng Des 2001;210(1−3):1−8.
[7] Wang RZ, Ge TS, Chen CJ, Ma Q, Xiong ZQ. Solar sorption cooling systems for residential applications: options and guidelines. Int J Refrig 2009;32(4):638−60.
[8] HIJC USA Inc. Green adsorption chiller. http://www.greenchiller.biz/[accessed 01.09.15].
[9] SorTech AG. Adsorption chiller aggregate eCoo 2.0. http://www.sortech.de/en/products/adsorption-chillers/[accessed 01.09.15].
[10] GBU. Adsorptions chiller Typ NAK-C. http://www.gbunet.de/frame-set/Frame-Set%20AK-NAK-C-e.html [accessed 01.09.15].
[11] PPI. ECO-MAX adsorption chiller. http://www.ppiway.com/brands-solutions/eco-max/products-services/adsorption-chillers [accessed 01.09.15].
[12] SolarNext AG. Cooling projects. http://www.solarnext.eu/eng/ref/coolingkitprojects.shtml [accessed 01.09.15].
[13] Liu YL, Wang RZ, Xia ZZ. Experimental performance of a silica gel−water adsorption chiller. Appl Therm Eng 2005;25(2−3):359−75.
[14] Wang DC, Xia ZZ, Wu JY, Wang RZ, Zhai H, Dou WD. Study of a novel silica gel−water adsorption chiller. Part I. Design and performance prediction. Int J Refrig 2005;28(7):1073−83.
[15] Wang DC, Wu JY, Xia ZZ, Zhai H, Wang RZ, Dou WD. Study of a novel silica gel−water adsorption chiller. Part II. Experimental study. Int J Refrig 2005;28(7):1084−91.
[16] Chen CJ, Wang RZ, Xia ZZ, Kiplagat JK, Lu ZS. Study on a compact silica gel−water adsorption chiller without vacuum valves: design and experimental study. Appl Energy 2010;87(8):2673−81.
[17] Lu ZS, Wang RZ, Xia ZZ, Wu QB, Sun YM, Chen ZY. An analysis of the performance of a novel solar silica gel−water adsorption air conditioning. Appl Therm Eng 2011;31(17−18):3636−42.
[18] Zhai XQ, Wang RZ, Dai YJ, Wu JY, Xu YX, Ma Q. Solar integrated energy system for a green building. Energy Build 2007;39(8):985−93.
[19] Luo HL, Wang RZ, Dai YJ, Wu JY, Shen JM, Zhang BB. An efficient solar-powered adsorption chiller and its application in low-temperature grain storage. Sol Energy 2007;81(5):607−13.
[20] Huangfu Y, Wu JY, Wang RZ, Xia ZZ. Experimental investigation of adsorption chiller for micro-scale Bchp system application. Energy Build 2007;39(2):120−7.
[21] Saha BB, Akisawa A, Kashiwagi T. Solar/waste heat driven two-stage adsorption chiller: the prototype. Renewable Energy 2001;23(1):93−101.
[22] Saha BB, Koyama S, Ng KC, Hamamoto Y, Akisawa A, Kashiwagi T. Study on a dual-mode, multi-stage, multi-bed regenerative adsorption chiller. Renewable Energy 2006;31(13):2076−90.
[23] Wang D, Zhang J, Tian X, Liu D, Sumathy K. Progress in silica gel−water adsorption refrigeration technology. Renewable Sustainable Energy Rev 2014;30(0):85−104.
[24] Pons M, Guilleminot JJ. Design of an experimental solar-powered, solid-adsorption ice maker. J Sol Energy Eng 1986;108(4):332−7.
[25] Pons M, Grenier P. Experimental data on a solar-powered ice maker using activated carbon and methanol adsorption pair. J Sol Energy Eng 1987;109(4):303−10.
[26] Headley OS, Kothdiwala AF, McDoom IA. Charcoal-methanol adsorption refrigerator powered by a compound parabolic concentrating solar collector. Sol Energy 1994;53(2):191−7.

[27] Li ZF, Sumathy K. A solar-powered ice-maker with the solid adsorption pair of activated carbon and methanol. Int J Energy Res 1999;23(6):517–27.
[28] Boubakri A. A new conception of an adsorptive solar-powered ice maker. Renewable Energy 2003;28(5):831–42.
[29] Wang RZ, Li M, Xu YX, Wu JY. An energy efficient hybrid system of solar powered water heater and adsorption ice maker. Sol Energy 2000;68(2):189–95.
[30] Wang LW, Wu JY, Wang RZ, Xu YX, Wang SG. Experimental study of a solidified activated carbon-methanol adsorption ice maker. Appl Therm Eng 2003;23(12):1453–62.
[31] Wang J, Wang LW, Luo WL, Wang RZ. Experimental study of a two-stage adsorption freezing machine driven by low temperature heat source. Int J Refrig 2013;36(3):1029–36.
[32] Zhang LZ. Design and testing of an automobile waste heat adsorption cooling system. Appl Therm Eng 2000;20(1):103–14.
[33] Lu YZ, Wang RZ, Zhang M, Jiangzhou S. Adsorption cold storage system with zeolite–water working pair used for locomotive air conditioning. Energy Convers Manag 2003;44(10):1733–43.
[34] Critoph RE. Laboratory testing of an ammonia carbon solar refrigerator. In: Proceedings of the solar world congress. Hungary; 1993.
[35] Critoph RE, Gong F. A rapid cycling ice maker for use in developing countries. In: Renewable energy technology and the environment. New York: Pergamon Press; 1992. p. 708–13.
[36] Anyanwu EE. Review of solid adsorption solar refrigerator I: an overview of the refrigeration cycle. Energy Convers Manag 2003;44(2):301–12.
[37] Worsøe-Schmidt P. Solar refrigeration for developing countries using a solid-absorption cycle. Int J Ambient Energy 1983;4(3):115–24.
[38] Wang K, Wu JY, Wang RZ, Wang LW. Composite adsorbent of $CaCl_2$ and expanded graphite for adsorption ice maker on fishing boats. Int J Refrig 2006;29(2):199–210.
[39] Wang LW, Wang RZ, Lu ZS, Xu YX, Wu JY. Split heat pipe type compound adsorption ice making test unit for fishing boats. Int J Refrig 2006;29(3):456–68.
[40] Wang K, Wu JY, Xia ZZ, Li SL, Wang RZ. Design and performance prediction of a novel double heat pipes type adsorption chiller for fishing boats. Renewable Energy 2008;33(4): 780–90.
[41] Lu ZS, Wang RZ, Li TX, Wang LW, Chen CJ. Experimental investigation of a novel multifunction heat pipe solid sorption icemaker for fishing boats using $CaCl_2$/activated carbon compound–ammonia. Int J Refrig 2007;30(1):76–85.
[42] Li SL, Xia ZZ, Wu JY, Li J, Wang RZ, Wang LW. Experimental study of a novel $CaCl_2$/expanded graphite-NH_3 adsorption refrigerator. Int J Refrig 2010;33(1):61–9.
[43] Li C, Wang RZ, Wang LW, Li TX, Chen Y. Experimental study on an adsorption icemaker driven by parabolic trough solar collector. Renewable Energy 2013;57:223–33.
[44] Wang RZ. Performance improvement of adsorption cooling by heat and mass recovery operation. Int J Refrig 2001;24(7):602–11.
[45] Li TX. Study on an innovative muti-effect and double-way thermochemical sorption refrigeration system. Shanghai: Shanghai JiaoTong University; 2009.
[46] Li TX, Wang RZ, Li H. Progress in the development of solid–gas sorption refrigeration thermodynamic cycle driven by low-grade thermal energy. Prog Energy Combust Sci 2014;40:1–58.
[47] Douss N, Meunier F. Experimental study of cascading adsorption cycles. Chem Eng Sci 1989;44:225–35.

Review of solar-powered desiccant cooling systems

13

T.S. Ge, J.C. Xu
Institute of Refrigeration and Cryogenics, Shanghai Jiao Tong University, Shanghai, China

13.1 Solar-powered rotary desiccant wheel cooling system

13.1.1 Introduction

Nowadays, there is a great need for air-conditioning systems, with environmental change and improvement of living standards. However, air-conditioning systems already account for a large part of the energy consumption by society, and how to effectively increase the energy utilization ratio of air-conditioning systems is crucial to realize a sustainable development. Traditional air-conditioning systems operate on a vapor compression (VC) cycle, and although the theoretical coefficient of performance (COP) of the system can reach to about 30, owing to the existence of heat transfer difference and other irreversible loss, the actual COP can reach to only 3.5. In addition, such systems rely on electricity, and the refrigerant used still makes a contribution to greenhouse gas emissions. Rotary desiccant wheel air-conditioning systems, which operate on the principle of adsorption dehumidification and evaporative cooling, have become a good alternative to conventional VC systems. In the rotary wheel desiccant cooling system, a desiccant material (solid adsorbent) is impregnated into the rotary desiccant wheel, processed air is pumped into the desiccant wheel to come into contact with the desiccant material, and owing to the vapor concentration difference between the processed air and the desiccant material, water vapor within the processed air can be adsorbed by the solid desiccant material; thus the latent load of processed air is removed and the humidity ratio of the air is always extremely low. After that, evaporative cooling is used to handle the sensible load. With the use of a dehumidification component and evaporative cooler, comfortable air with a decreased humidity ratio as well as temperature can be supplied to the conditioned space in the cooling season. To make the desiccant material reusable, a regeneration process in which air with high temperature is used to desorb water vapor from the desiccant material operates in parallel with the previous dehumidification process. In contrast, when the regeneration air is used as the supply air to the conditioned space, the system operates in a heating and humidification mode in the heating season. Compared with conventional VC systems in heat pump mode, the rotary desiccant wheel cooling system not only can control the temperature but also can perform a humidification function. It can be seen that thermal energy is the main source used to drive the operation of a solid desiccant cooling system and the refrigerant used is a natural working fluid; in other words,

Advances in Solar Heating and Cooling. http://dx.doi.org/10.1016/B978-0-08-100301-5.00013-8

the solid desiccant system is an energy-saving and environmentally friendly air-conditioning method. For these reasons, numerous studies have been conducted in this field in terms of desiccant material, cycle mode, and so on, as described in previous works [1–4].

A brief review on previous work shows that the studies mainly focus on how to increase the energy utilization ratio or the air quality, and electricity or burning of gas is used to simulate the low-grade heat source, mainly owing to the merits of easy operation and maintenance. Solar energy, with the advantages of cleanness and sustainability, is widely acknowledged as a promising low-grade thermal energy. Meanwhile, the solar energy obtained coincides with the required cooling load in the cooling season, which means that when solar radiation is abundant, there is always an increasing need in cooling power. Also, solar technology has achieved remarkable progress with increasing attention on the energy crisis in past decades. Based on these facts, solar-powered solid desiccant air-conditioning systems, especially solar-powered solid desiccant cooling systems, are attracting more and more attention, and several actual systems have been built around the world. To the authors' knowledge, there is not yet, as of this writing, a work that summarizes comprehensively the work done on solar-powered rotary desiccant wheel air-conditioning systems. Hence, the objective of this chapter is to examine the progress in this area and provide guidelines for designers. Emphasis should be placed on the fact that most solid desiccant air-conditioning systems are developed to provide cooling power and thus this chapter focuses on summarizing the development of solar-powered rotary desiccant wheel cooling systems. However, available results for the heating season are also included, although it accounts for only a small part. The chapter is categorized by whether an auxiliary refrigeration system is installed. The first category is ordered according to type of solar collector. For the second category, works are grouped together by type of auxiliary refrigeration system. For each system reviewed, an attempt has been made to gather up the specifications of the solar system, desiccant cooling system, and overall performance of the whole system.

13.1.2 Solar-powered separate rotary desiccant wheel cooling systems

13.1.2.1 Basic mode of the solar-powered rotary desiccant wheel cooling system

A solar-powered desiccant wheel evaporative cooling system (SDECS), as shown in Fig. 13.1, is the widely adopted basic type of solar rotary desiccant wheel cooling system. This system mainly consists of two subsystems: a solar subsystem and a desiccant cooling system. In the solar subsystem, the solar collector, auxiliary heater, and water tank are the main components, similar to a conventional solar hot water system. The solar collector (see Fig. 13.1, (1)) is used to absorb solar radiation and heat up the water, which is utilized in the air heater (3) to heat up the regeneration air. The water tank (2) acts like a heat storage unit, which also helps to adjust the flux of water. The desiccant cooling subsystem operates on a basic ventilation

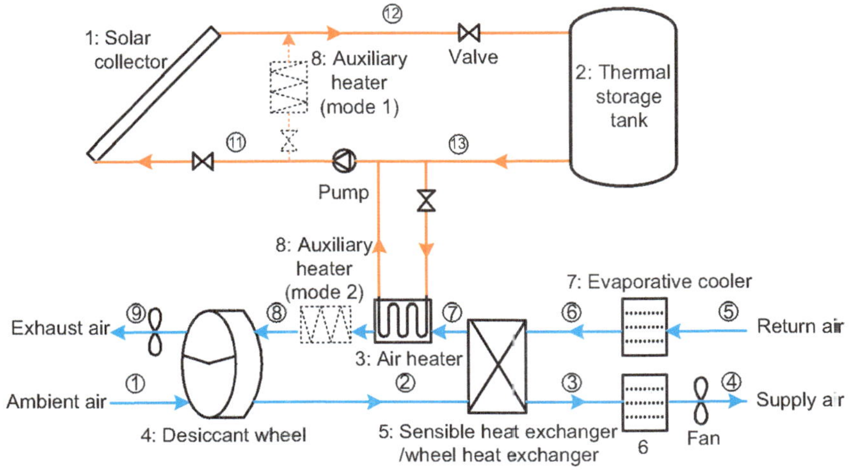

Figure 13.1 Schematic of a solar-powered desiccant wheel evaporative cooling system.

mode: processed air from the ambient condition is dehumidified and heated in the desiccant wheel (4), then it is cooled in the sensible heat exchanger (5) (heat exchanger wheel or plate-type heat exchanger) and is further cooled in the evaporative cooler (6). At last the air is supplied to the conditioned space. Regeneration air from the indoor condition flows through the evaporative cooler (7), sensible heat exchanger (5), air heater (3), and desiccant wheel (4) in series. It can be seen that these two subsystems are connected by the air heater (3), in which hot water from the solar subsystem is utilized to heat up the regeneration air in the desiccant cooling subsystem.

Emphasis should be placed on the fact that an auxiliary heat source (8), such as a gas heater or electricity, is used in some systems to ensure continuous operation when solar energy is lacking. The auxiliary heater can be installed in the solar subsystem (mode 1) or installed directly on the regeneration air side (mode 2).

13.1.2.2 Performance indicators

There exist some indices in the various literature to evaluate the performance of a solar solid desiccant cooling system. These indices are summarized and categorized in this part to make the following review much more clear.

Coefficient of performance

The COP is a common and crucial performance index for an air-conditioning system and the following parameters are used to show the energy utilization ratio of a solar rotary desiccant cooling system.

Thermal COP (COP_{th}), as shown in the following equation, is the widely adopted parameter to evaluate the performance of a solid desiccant cooling system. It is defined as the enthalpy change of processed air divided by the thermal energy for regeneration:

$$\mathrm{COP_{th}} = \frac{Q_{pc}}{Q_{reg,tot}} = \frac{m_{pa}\left(h_{pa,in} - h_{pa,out}\right)}{Q_{reg,sol} + Q_{reg,aux}} = \frac{m_1(h_1 - h_4)}{m_7(h_8 - h_7)} \qquad [13.1]$$

in which Q_{pc} is the handled cooling load from processed air, $Q_{reg,tot}$ is the total regeneration heat required to drive the system, $Q_{reg,sol}$ and $Q_{reg,aux}$ are the thermal energy obtained in the solar collector and the auxiliary heater, m_{pa} is the quality of the processed air, h is the enthalpy of the processed air, and subscripts 1, 4, 7, and 8 correspond to the state in Fig. 13.1.

In Eq. [13.1] it can be seen that when the solid desiccant cooling system operates in fresh air mode, the system has to handle a cooling load from both fresh air and indoor air, and then the enthalpy change of the processed air is used to calculate the cooling power. However, some researchers still adopt the conventional method based on the enthalpy change of indoor air to calculate the COP as follows:

$$\mathrm{COP^*_{th}} = \frac{Q_{ic}}{Q_{reg}} = \frac{m_{pa}\left(h_{ind} - h_{pa,out}\right)}{Q_{reg,sol} + Q_{reg,aux}} = \frac{m_1(h_5 - h_4)}{m_7(h_8 - h_7)} \qquad [13.2]$$

in which the enthalpy change between the supply air and the indoor air is calculated as the cooling power. Subscripts ic, ind, and pa represent cooling power based on indoor air, indoor condition, and processed air, respectively.

The electrical COP ($\mathrm{COP_{el}}$) of the system can be calculated by:

$$\mathrm{COP_{el}} = \frac{Q_{pc}}{E_{tot}} = \frac{m_{pa}\left(h_{pa,in} - h_{pa,out}\right)}{E_{fan} + E_{pump} + E_{aux}} \qquad [13.3]$$

$$\mathrm{COP^*_{el}} = \frac{Q_{c}}{E_{tot}} = \frac{m_{pa}\left(h_{indoor} - h_{pa,out}\right)}{E_{fan} + E_{pump} + E_{aux}} \qquad [13.4]$$

in which E_{tot} is the total electrical consumption, including fans E_{fan}, pumps E_{pump}, and auxiliary consumption E_{aux} of the system, and E represents electrical energy consumption.

If all energy consumption is considered to evaluate the energy utilization ratio, the following equations are considered:

$$\mathrm{COP_{tot}} = \frac{Q_{pc}}{Q_{reg,tot} + E_{tot}} = \frac{m_{pa}\left(h_{pa,in} - h_{pa,out}\right)}{Q_{reg,sol} + Q_{reg,aux} + E_{fan} + E_{pump} + E_{aux}} \qquad [13.5]$$

$$\mathrm{COP^*_{tot}} = \frac{Q_{ic}}{Q_{reg,tot} + E_{tot}} = \frac{m_{pa}\left(h_{indoor} - h_{pa,out}\right)}{Q_{reg,sol} + Q_{reg,aux} + E_{fan} + E_{pump} + E_{aux}} \qquad [13.6]$$

Also, some researchers consider solar energy as a free thermal source, and then the following equation is used to describe the energy utilization ratio:

$$COP_{tot,aux} = \frac{Q_{pc}}{Q_{reg,aux} + E_{tot}} = \frac{m_{pa}\left(h_{pa,in} - h_{pa,out}\right)}{Q_{reg,aux} + E_{fan} + E_{pump} + E_{aux}} \quad [13.7]$$

Solar thermal COP ($COP_{th,sol}$) is calculated by:

$$COP_{th,sol} = \frac{Q_{pc}}{Q_{sol}} = \frac{m_{pa}\left(h_{pa,in} - h_{pa,out}\right)}{G \cdot A} \quad [13.8]$$

Solar fraction

For a solar-powered system, the solar fraction (SF) is an important parameter to estimate the contribution of solar radiation. It is defined as the ratio of energy provided by solar collectors for regeneration to the total energy required for regeneration including the auxiliary:

$$SF = \frac{Q_{reg,solar}}{Q_{reg,tot}} = \frac{Q_{reg,solar}}{Q_{reg,solar} + Q_{reg,auxiliary}} \quad [13.9]$$

13.1.2.3 Solar water collector–powered systems

Flat plate solar collector–powered systems

The flat plate solar collector is the most common solar collector for use in solar water heating systems. It mainly consists of an insulated metal box with a transparent cover and dark-colored plate absorber. Solar radiation is absorbed by the plate absorber and transferred to water. The provided temperature of a flat plate solar collector is about 30–80°C. Also, the manufacture of flat plate solar collectors is relatively mature and the cost is relatively low. Therefore, a lot of researchers adopt flat plat solar collectors to drive the operation of solid desiccant cooling systems.

Some researchers integrate a flat plate solar collector with a conventional desiccant wheel evaporative cooling system (DECS): Preisler et al. [5] investigated the annual performance of an actual SDECS by experiment. The experimental demonstration was built in an office building in Vienna, Austria. Actual monitoring data of the system were compared with simulated data from a conventional compression chiller using TRNSYS 17 software. Results showed that the average COP^{*}_{el} of an SDECS in winter could reach 7.0, and in 2010 the SDECS achieved 60.5% primary energy savings compared to the reference system. The highest energy savings were obtained in winter for heating and humidity recovery purposes of fresh air; moderate energy savings were obtained in summer for cooling and humidity control, and there were no obvious energy savings in the transition time. Also, the performance of a solar-assisted heating and desiccant cooling system was investigated in Baghdad, Iraq, by Khalid et al. [6]. In winter, heat collected from the solar collector was supplied to a conditioned room for heating purposes, and a rock bed storage unit and auxiliary heat source were adopted. In summer, a ventilation mode was utilized to provide cooling power.

A model hybrid system was developed to study both heating and cooling performance of the system and to analyze the effects of the main operation parameters on system performance. It was found that a regeneration temperature of 62°C could be reached by the solar collector alone, and an auxiliary heat source was required to achieve a higher regeneration temperature. Also, the solar desiccant cooling system could provide satisfactory supply air to the conditioned space under local weather conditions, and the performance of the desiccant cooling system relied mainly on the effectiveness of the sensible heat exchanger and evaporative cooler, whereas the effect of the dehumidifier was relatively small.

In addition, several modified systems are investigated to obtain improved performance. Guidara et al. [7] developed precooling and final-cooling modes to satisfy the required cooling capacity. A schematic of the proposed system is shown in Fig. 13.2: two indirect evaporative coolers are installed before the ambient air enters the desiccant wheel and before processed air enters the conditioned space. The performance of the system was simulated in three representative cities with different climatic conditions in Tunisia. Meanwhile, three different operation functions, including conventional DECS mode, with precooling mode and with precooling and final-cooling mode, were compared and analyzed. Simulation results showed that the basic DECS mode was feasible under relatively cold and humid conditions in Bizerte. However, for dry conditions in Remada and moderate conditions in Djerba, systems with precooling mode and with precooling and final-cooling mode had to be applied to cover the corresponding demands. Also, it was found that the conditioned space with respect to each mode of functioning contained a comfortable environment for occupants.

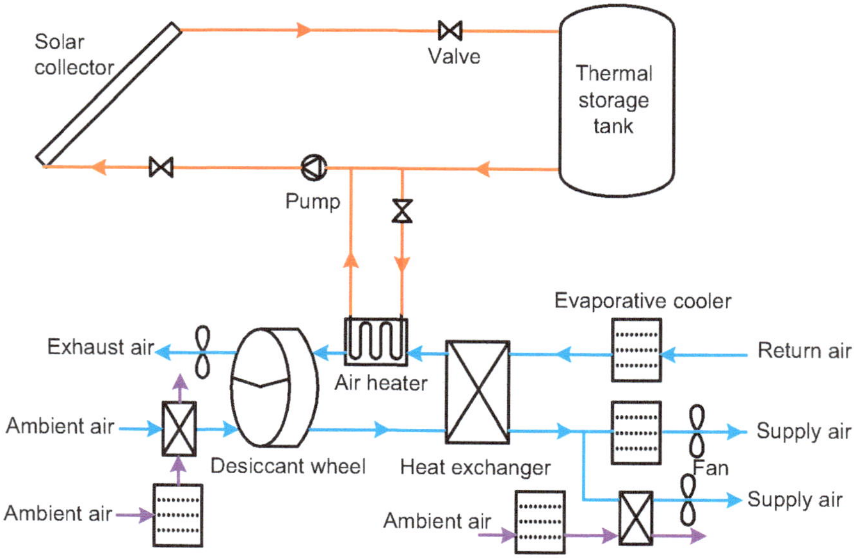

Figure 13.2 Schematic of system with precooling and final cooling [7].

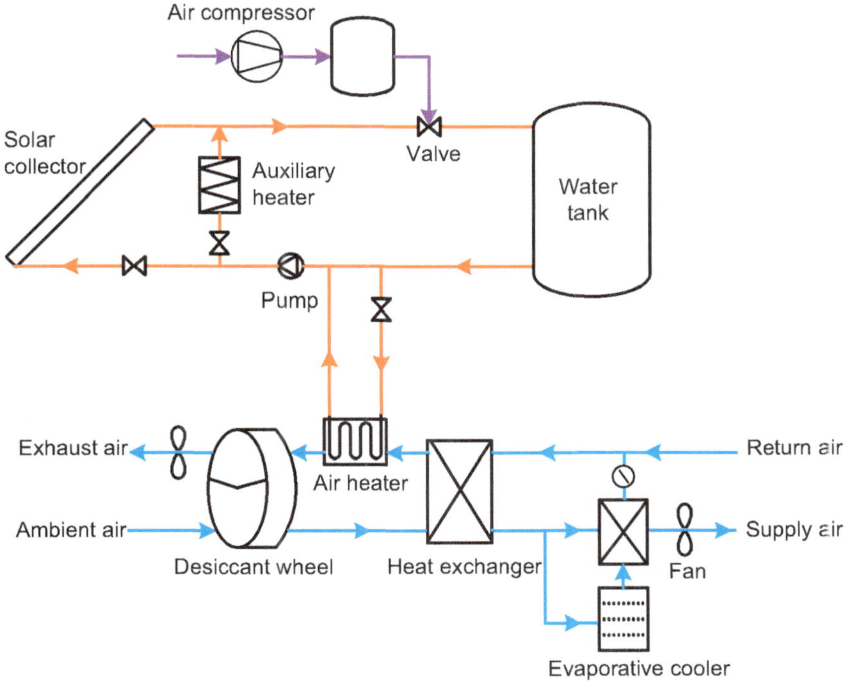

Figure 13.3 Schematic of a modified solar-powered desiccant wheel evaporative cooling system [9].

Enteria et al. [8–11] proposed a combined solar thermal and electric desiccant cooling system as shown in Figs. 13.3 and 13.4. In the system, the solar-powered desiccant cooling system operated in the daytime to produce the required cooling power, an electric heater was connected with a flat plate solar collector in parallel as the heat source for nighttime operation for thermal energy storage, and an air compressor was adopted to remove water from the tank in case of freezing. Also, compared with a conventional SDECS (Fig. 13.1), a small heat exchanger was adopted in this system to recover cooling capacity from the return air. An experimental system was constructed and tested during nighttime and daytime of typical days [9,11]. Measured results showed that the system could provide satisfactory cooling capacity to a conditioned room, and almost three-quarters of the thermal energy was from the solar collector. Also, Enteria et al. [8] extended this system to different climatic conditions in east Asia and evaluated the potential performance with different desiccant materials including silica gel and titanium dioxide. A complete system model was established in TRNSYS and validated by experimental results [12]. With the use of the model, simulation work was conducted in six representative cities from temperate, subtropical, and tropical climates. The performance of the solar subsystem, desiccant subsystem, building subsystem, and total system was examined. Results showed that SDECS with a TiO_2 wheel could produce a lower supply air humidity ratio as well as temperature

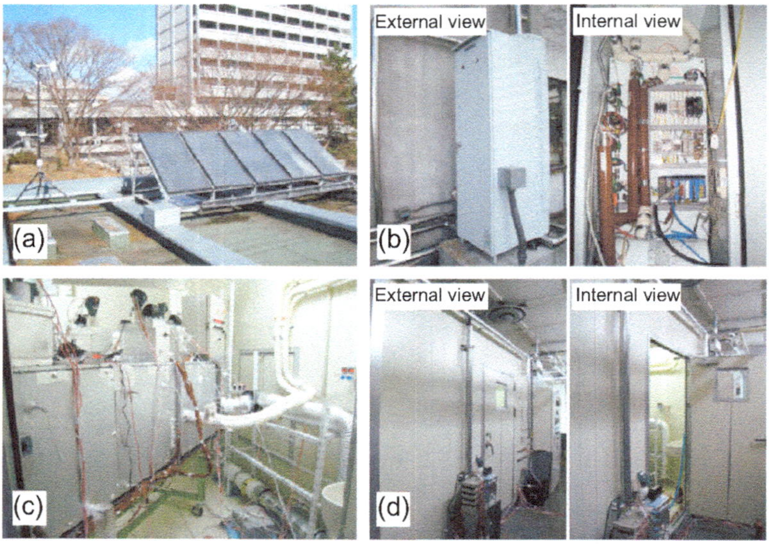

Figure 13.4 Total system actual view. (a) Solar collector; (b) thermal storage and auxiliary heater; (c) desiccant cooling; (d) controlled chambers [9].

compared with the system using SiO_2, and the specifications of the system varied depending on the climatic conditions: a bigger collector surface, larger thermal storage tank, higher airflow rate, and higher electrical consumption were required in the tropical climate. Also, the COP_{th} of the system was between 1.5 and 3. Further, the researchers investigated the performance of this system based on energy and exergy analysis methods to improve system operation on summer days [11]. The analyzed results demonstrated that the main contributors to energy loss and exergy destruction were the solar collector, water pipe, electric heater, and storage tank, and thus improvements on these components was crucial for increasing system performance.

White et al. [13] proposed a plate-type water collector-driven SDECS with indirect evaporative cooler and direct evaporative cooler installed in series and without a backup heat source as shown in Fig. 13.5. A simulation model was developed in TRNSYS to study the performance of the system in three cities in Australia: the warm temperate climates of Melbourne and Sydney and the tropical climate of Darwin. It was found that this type of two-stage evaporative cooling process could enhance the sensible cooling capacity. Especially at low primary air humidity, the temperature of the supply air could be significantly reduced. Also, the effects of the desiccant cooled air-flow rate and the area of the solar collector were discussed. Results showed that increasing collector area and airflow rate reduced the frequency of high-temperature events in the occupied space. In the warm temperate climate (Melbourne and Sydney), high ventilation rates enabled comfortable conditions to be maintained at or near acceptable levels in the occupied space, without the use of a backup thermal source. In extreme weather events, evaporative cooling appeared to be the dominant mechanism for cooling the occupied space.

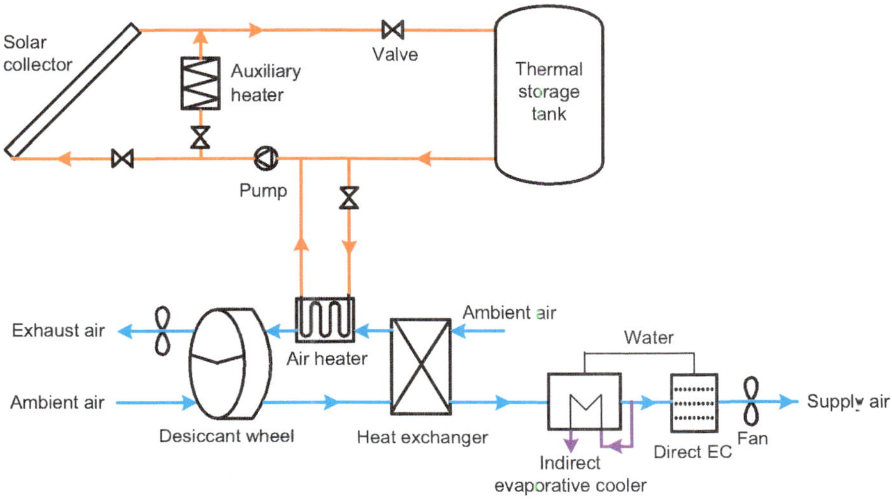

Figure 13.5 Schematic of a solar-powered desiccant wheel evaporative cooling system with indirect and direct evaporative cooling [13]. *EC*, Evaporative cooler.

Vacuum tube solar collector–powered system

Bourdoukan et al. [14] proposed that the low efficiency of flat plate collectors or air collectors results in low potential of solar use in solid desiccant cooling, and thus heat pipe vacuum tube (HPVT) collectors were adopted in the study to drive conventional DECS. Mathematical models of HPVT collectors and water tank were established and validated in the study, and then these models were combined with existing desiccant wheel models [15,16] to develop an integrated SDECS mathematical model. Based on this system model, performance of an HPVT-driven DECS was simulated to provide cooling capacity to a building with area of 1350 m^2 in three locations, including La Rochelle, France; Bolzano, Italy; and Berlin, Germany, to study the potential of such applications. It was assumed that: (1) when ambient temperature exceeded 26°C or ambient humidity ratio exceeded 10.8 g/kg, HPVT-driven DECS operated to handle both sensible and latent heat, otherwise, only an indirect evaporative cooling was utilized to realize air conditioning, and (2) an energy backup was installed to ensure enough cooling power was delivered. Simulation results over a summer season demonstrated that the system could maintain the building temperature and humidity at a comfortable level (indoor temperature lower than 26.5°C and indoor humidity ratio less than 11.8 g/kg) for different locations from relatively dry to hot and moderately humid. Meanwhile, the daily efficiency of HPVT varied from 50% to 64% and the daily SF varied from 87% to 97%, and the corresponding seasonal overall η and SF were higher than 51% and 90%, respectively. Based on this research, Bourdoukan et al. [17] further developed an experimental setup of HPVT-powered DECS. Actual operation data not only validated the feasibility of such system but also demonstrated that an HPVT collector could obtain a high temperature level and stable duration of regeneration temperature compared

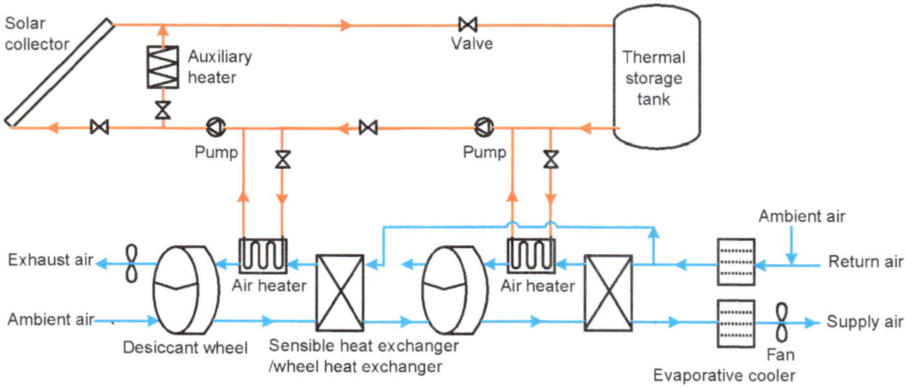

Figure 13.6 Schematic of two-stage rotary desiccant wheel cooling system [20].

with the flat plat type. Under experimental conditions of a moderately humid day (T_{amb} = 23−30°C, d_{amb} = 12−14°C; subscript amb is ambient condition), the indoor environment could be maintained at comfortable levels of 26.5°C and 12−14 g/kg, the COP^{*}_{th} of the system was 0.4, and the performance indicator defined based on electrical energy consumption, COP^{*}_{el}, was evaluated at 4.3 over the day. In addition, the effects of outside and regeneration conditions were analyzed, and the outside humidity ratio was found to be the most sensitive parameter.

Ge et al. [18,19] pointed out that the released adsorption heat in a dynamic dehumidification process was the main reason causing decreased performance of rotary desiccant wheel cooling systems. And thus an intercooling two-stage rotary desiccant wheel cooling system (TSDCS) was developed to obtain improved performance and it was found that the TSDCS had the merit of low regeneration temperature as well as high COP_{th}. Further study on solar-powered TSDCS as shown in Fig. 13.6 was conducted in Ref. [20]. In the research, the performance of a solar-driven TSDCS and a VC system (VCS) was simulated to provide cooling for one floor in a commercial office building in two cities with different climates: Berlin and Shanghai. Both thermodynamic and economic performance of the two systems was evaluated and compared. Results showed that the desiccant cooling system was able to meet the cooling demand and provided comfortable supply air in both regions. The required regeneration temperatures were 55°C in Berlin and 85°C in Shanghai. Compared to the VCS, the desiccant cooling system had better supply air quality and consumed less electricity. The results of the economic analysis demonstrate that the dynamic investment payback periods were 4.7 years in Berlin and 7.2 years in Shanghai.

13.1.2.4 Solar air collectors

Solar air collector

The previously mentioned systems use a solar water collector to heat the regeneration air. Solar water collectors have the merits of high heat flux, but a water−air heat exchanger is required in the system to realize the heat transfer between hot water and regeneration air. When a solar air collector is utilized, there is no need to install a heat exchanger component, and the volume and construction of solar solid desiccant cooling systems are much more compact and simple.

Fong et al. [21] conducted simulation optimization of SDECS in a subtropical climate in Hong Kong. A simulation model SDECS was established using TRNSYS 16.01, in which the performance of a silica gel desiccant wheel was predicted by a fitted efficiency relationship. A solar air collector instead of conventional solar water collector was adopted as the regeneration heat source in the system, and a robust evolutionary algorithm was developed to achieve a maximal annual average SF. Simulation results showed that the monthly average SF was in the range 8% to 33%, and the yearly average was 17%. The monthly average COP_{th} was from 1.08 to 1.60, with a mean of 1.38. The overall performance of the SDECS showed that it was a technically feasible option in Hong Kong for air-conditioning purposes.

Hatami et al. [22] adopted a solar air collector to power a DECS in an investigation that was carried out to obtain the optimized surface area of a solar air collector based on a mathematical model in TRNSYS. Results showed that under the simulation conditions ($T_{amb} = 30°C$, $d_{amb} = 25$ g/kg, $T_{reg} = 90°C$), the necessary area of the solar collector surface could be decreased from 33.1 to 15.1 m^2 by using all the parameters of the desiccant wheel in optimum quantities. The necessary area of the solar collector surface decreased with increased outdoor air dry bulb and wet bulb temperature.

Kabeelet al. [23] adopted a porous-type solar air heater to drive a solid desiccant cooling system. A schematic of the solar collector is shown in Fig. 13.7. It can be seen that ambient air is pumped into the channels within the solar collector and heated by solar radiation directly. The experimental setup was built to investigate adsorption as well as regeneration performance of the system with respect to various ambient air conditions and radiation intensity. Results showed that the system was highly effective in the regeneration process, and empirical equations to calculate the performance of the desiccant wheel under various operation conditions were obtained.

There are also some studies focusing on solar air collector–powered TSDCS. Li et al. [24,25] combined a solar vacuum tube air collector with the two-stage rotary desiccant wheel cooling/heating system to provide cooling/heating power for a conditioned space in Dezhou, China. First, the performance of the system was tested and then a simulation model was established in Matlab Simulink to obtain an optimized design for the solar air collector. Both experimental and simulation results showed the system could provide satisfactory supply air, especially in hot and humid climate conditions, and solar heating with desiccant humidification can improve the indoor comfort significantly in winter. Also, there was an optimal solar collector design to obtain higher cooling/heating power and SF.

Based on the research on the two-stage system, Ali et al. [26] further developed a one-rotor six-section solar-powered desiccant cooling system, in which one cross section of the desiccant wheel was divided into six parts, two precooling parts were added to cool the desiccant wheel after the regeneration process as shown in Fig. 13.8, and then the dehumidification capacity of the system was improved. A simulation model

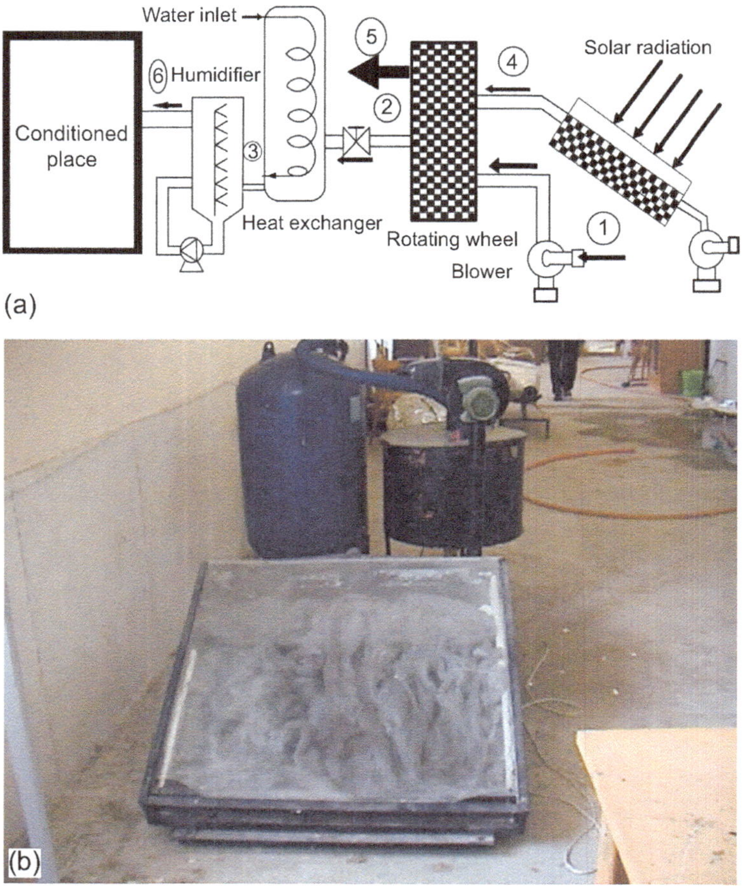

Figure 13.7 Schematic of the system investigated in Ref. [23].

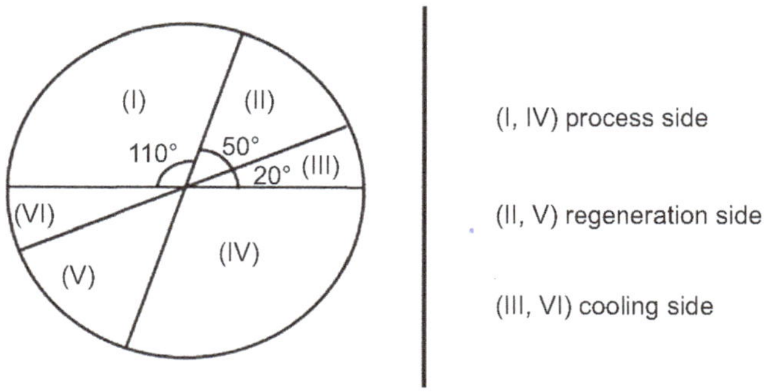

Figure 13.8 Cross section of one-rotor six-section desiccant wheel [26].

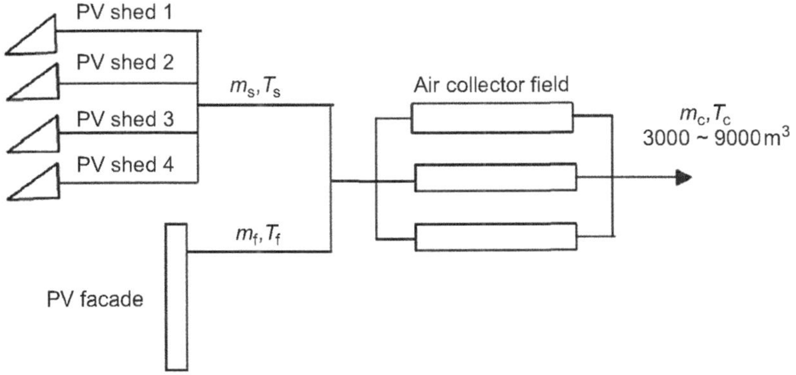

Figure 13.9 Photovoltaic/thermal solar air system. *PV*, photovoltaic.
Adopted in Mei L, Infield D, Eicker U, Loveday D, Fux V. Cooling potential of ventilated PV façade and solar air heaters combined with a desiccant cooling machine. Renewable Energy 2006;31:1265–78.

was established to investigate the effects of the main operation parameters on system performance. It was found that the payback time of this system was 4.89 years compared with conventional VCS.

Solar air collector and photovoltaic cells

Photovoltaic (PV) cells are an important application of solar energy; in most cases, there is a layer of air circulating behind the PV panel to decrease the temperature of the PV panel and thus improve the electrical efficiency, and if the heated air is utilized as a thermal source, this component is named photovoltaic/thermal (PVT). PVT can be combined with SDECS to produce cooling/heating power.

In a library in Mataró, Spain, PVT and a solar air collector heating system were integrated to regenerate a DECS [27]. The hybrid solar system included a 525-m^2 PV panel (PV facade and PV shed) and 105-m^2 solar air collectors, and the collector and PVT were connected in series to provide regeneration heat for DECS as shown in Fig. 13.9. Various system operation modes were realized by adjusting the volume flow rate of the processed air and regeneration air, and operation modes were determined based on the temperature of ambient and indoor air conditions. A system model was established and validated by actual operation data on a typical day, and then the model was adopted to analyze the performance of the system in summer. It was found that a cooling power of 43,580 kWh was obtained, which could achieve an SF of 0.75 at least, and an average COP_{th} of 0.518 was achieved.

13.1.3 Solar-powered hybrid rotary desiccant wheel cooling systems

The rotary desiccant wheel system has a good capacity for handling latent load, and conventional VCS can effectively remove sensible load compared with the evaporative cooling process. Therefore, there exist some investigations on hybrid solid desiccant cooling systems. In this part, a brief review of solar-powered hybrid rotary desiccant wheel cooling systems is conducted.

13.1.3.1 Basic hybrid SDECS and vapor compression systems

VCS is the most mature and common refrigeration technology, and thus most researchers combine SDECS with conventional VCS. In the integrated system, SDECS and VCS are utilized to remove latent and sensible load, respectively. A schematic of a hybrid SDECS/VCS is shown in Fig. 13.10.

13.1.3.2 Performance indicators

The performance parameters introduced in Section 13.1.2.2 can still be adopted to describe the hybrid system, and the subscripts dec, vc, and hs stand for solar desiccant cooling system, VCS, and the whole hybrid system, respectively. For example, E is electrical energy consumption; Q stands for thermal power; $COP_{tot,dec}$ means the total COP of the desiccant cooling system, which can be calculated by

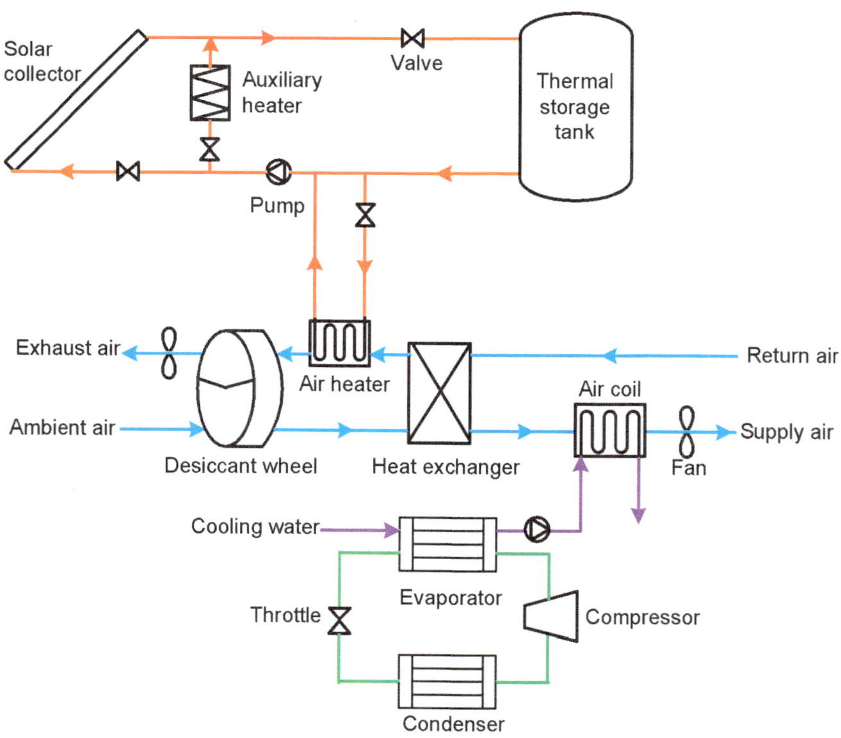

Figure 13.10 Schematic of a basic hybrid solar-powered desiccant wheel evaporative cooling system and vapor compression system.

Eq. [13.5]; and $COP_{tot,hs}$ is the total COP of the whole hybrid system and is defined as:

$$COP_{tot,hs} = \frac{Q_{pc,hy}}{Q_{reg,tot} + E_{tot,dec} + E_{tot,vc}} = \frac{m_{pa}\left(h_{pa,in} - h_{sup}\right)}{Q_{reg,sol} + Q_{reg,aux} + E_{fan} + E_{pump} + E_{compressor} + E_{aux}} \quad [13\ 10]$$

There are two kinds of energy sources, electricity and thermal energy, in the hybrid system, and thus primary energy consumption (PE) is a better performance indicator:

$$PE = \frac{Q_{reg,aux}}{\eta_{th}} + \frac{E_{tot}}{\eta_e} \quad [13.11]$$

and COP based on primary energy is calculated by:

$$COP_{tot,pe,hs} = \frac{Q_{pc,hy}}{PE} = \frac{m_{pa}\left(h_{pa,in} - h_{sup}\right)}{\frac{Q_{reg,aux}}{\eta_{th}} + \frac{E_{fan} + E_{pump} + E_{compressor} + E_{aux}}{\eta_e}} \quad [13.12]$$

in which η_{th} and η_e are the energy transfer efficiency to produce auxiliary thermal energy and electricity.

When total energy consumption is considered, the SF is:

$$SF_{tot} = \frac{Q_{reg,solar}}{Q_{reg,tot} + E_{tot}} = \frac{Q_{reg,solar}}{Q_{reg,solar} + Q_{reg,auxiliary} + E_{fan} + E_{pump} + E_{aux}} \quad [13.13]$$

A hybrid system is proposed to decrease the electrical consumption of conventional VCS, and thus the electricity savings ratio E_{save} and the primary energy savings ratio PE_{save} compared with conventional VCS with the same cooling power are utilized in some research:

$$E_{save} = \frac{E_{tot,vc} - E_{tot,hs}}{E_{tot,vc}} \quad [13.14]$$

$$PE_{save} = \frac{PE_{tot,vc} - PE_{tot,hs}}{PE_{tot,vc}} \quad [13.15]$$

13.1.3.3 Hybrid SDECS and vapor compression system

Researchers from the United Kingdom established a mathematical model to predict the performance of hybrid SDECS/VCS [28]. Then, a feasibility study in Europe was conducted with the use of real summertime meteorological data [29,30]. Several representative cities in Europe from north to south were selected as the sample locations, in which hot days occurring in July 1996 were considered. It was assumed that supply air was maintained at 18°C in the simulation process. Simulation results outlined that the solar-powered single-stage rotary desiccant wheel cooling system was feasible in most of central Europe, the Atlantic region, and the inland regions of southern Europe. Latent load rather than available solar energy was the prime constraint on the utilization of desiccant cooling. The performance of the system was evaluated by energy saving ratio compared with a conventional gas-powered desiccant cooling system. It was found that, except for Oslo, a gas energy savings of 25.1−46.5% was obtained. A gas energy savings of 93% was achieved in Oslo, because the latent load in Oslo was much lower than those in other locations.

Finocchiaro et al. [31] pointed out that because the cooling capacity of return air was limited in DECS, a relatively high outlet temperature, which was rarely lower than 26°C, was obtained. To overcome this problem, a packaged wet heat exchanger was proposed to replace the sensible heat exchanger to realize a much more sensible load removal. A detailed configuration and pictures of the wet heat exchanger are shown in Fig. 13.11. Then, an experimental setup of a hybrid SDECS/VCS was built to investigate the detailed performance. Experimental results showed that owing to the optimization of the indirect evaporative cooling process, a supply temperature in the range of 21−22°C can be achieved without the use of an auxiliary cooling coil.

Baniyounes et al. [32,33] investigated the performance of a hybrid SDECS/VCS both in theory and through experiments at Central Queensland University (North Rockhampton, Australia). In the first step, a simulation model predicting the performance of the system was developed based on TRNSYS software, and the model was adopted to quantify annual technical performance and energy savings, etc., in reference to a conventional air-conditioning system. Simulation results showed that the total annual cooling load was 6428 kWh. It reached its peak in the month of

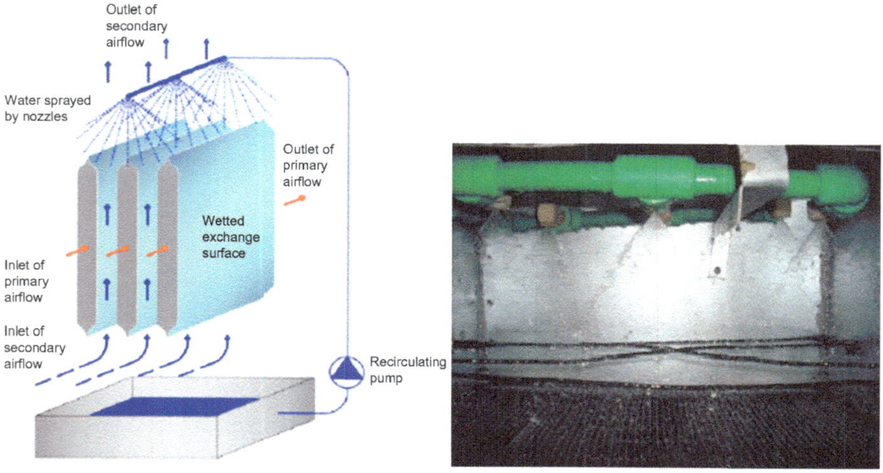

Figure 13.11 Packaged wet heat exchanger used for indirect cooling process [31].

Figure 13.12 Photo of hybrid solar desiccant cooling system [32].

December and the minimum was in the month of July. Satisfactory supply air could be provided by the hybrid SDECS/VCS, with a temperature of 25–27°C and relative humidity ratio near 60%. Also, it was found that the area of the solar collector affected the COP_{th} significantly: when the area was less than 20 m^2, COP_{th} kept increasing with the increase in solar collector area; however, the COP_{th} barely changed after more than 20 m^2 of solar collector was installed, and the highest COP_{th} was about 1.2. Based on these theoretical analyses, an experimental setup as shown in Fig. 13.12 was constructed, which consisted of 10 m^2 of collector area. An annual operation experiment, from November 2011 to October 2012, was conducted to analyze the performance of the system. The experimental results agreed well with the simulation; the actual desiccant cooling system was able to deliver treated air with conditions close enough to comfort standards. Furthermore, the system-obtained annual $COP_{tot,hs}$ was 0.5 measured and 0.52 simulated, and the annual SF was 0.25 measured and 0.21 simulated. Therefore, both experimental and simulated results of the Central Queensland University hybrid solar cooling system showed that the technology is promising and is effective even in the tropics.

Fong et al. [34] applied hybrid SDECS/VCS to a conditioned space with high latent cooling load in Hong Kong, such as in a Chinese restaurant and a wet market. A simulation model in TRNSYS was built to research system performance in terms of COP and energy-savings efficiency. The designed latent and sensible loads for the Chinese restaurant and wet market were 13 and 19 kW and 13 and 27 kW, respectively. Results demonstrated that the hybrid SDECS/VCS could maintain indoor temperature as well as humidity ratio much more steadily compared with a conventional VCS throughout different loading and climatic conditions within a year. Also, when the designed zone humidity was constant at 60% RH, the hybrid system could obtain an annual savings of PE of 49.5% for the Chinese restaurant and of 13.3% for the wet market. Meanwhile, this energy savings ratio decreased with increased designed indoor humidity ratio; however, a humidity ratio that was too high caused an uncomfortable thermal environment in such places. Consequently, such hybrid system was recommended for subtropical climates like Hong Kong in which a high latent cooling load is demanded.

Khalid et al. [35] built a hybrid SDECS/VCS in Karachi, Pakistan. This system used two indirect evaporative coolers instead of a direct evaporative cooler on the processed air side: one was installed before the desiccant wheel to precool the ambient air, and another was used between the sensible heat exchanger and the cooling coil of VCS. It was found that the desiccant wheel cooling system alone could not cool the supply air to comfort conditions because of the high latent and sensible loads in Karachi, and an auxiliary VCS was required under such conditions. Also, the indirect evaporative cooler resulted in decreased regeneration temperature by 15% and decreased dehumidification by only 6%, which saved thermal energy. Then, a TRNSYS simulation model was established to further analyze the performance of this system during the Karachi and Lahore cooling season. Four operation modes, ie, with or without precooling, with or without an auxiliary cooling coil, were proposed and discussed. Simulation results for Lahore showed that the system was able to cool the air to comfortable levels without the aid of an auxiliary cooling unit for 3 of the 7 months of cooling season.

Hurdogan et al. [36] also conducted research on a modified hybrid SDECS/VCS, in which three groups of air worked in parallel, as shown in Fig. 13.13. The investigated system used 100% fresh air, regeneration air came from the ambient air, and return air was used to recover the sensible heat of the processed air before being exhausted to the outdoors. Four heat exchangers were used in the system to obtain processed air with low temperature and to recover heat from exhausted regeneration air. Both a simulation model and an experimental setup were established by the authors, and the main aim of the research was to evaluate the utilization efficiency of solar energy. It was found that the solar heating system could increase the regeneration temperature 5−17°C depending on the ambient dry bulb temperature and solar radiation in Adana,

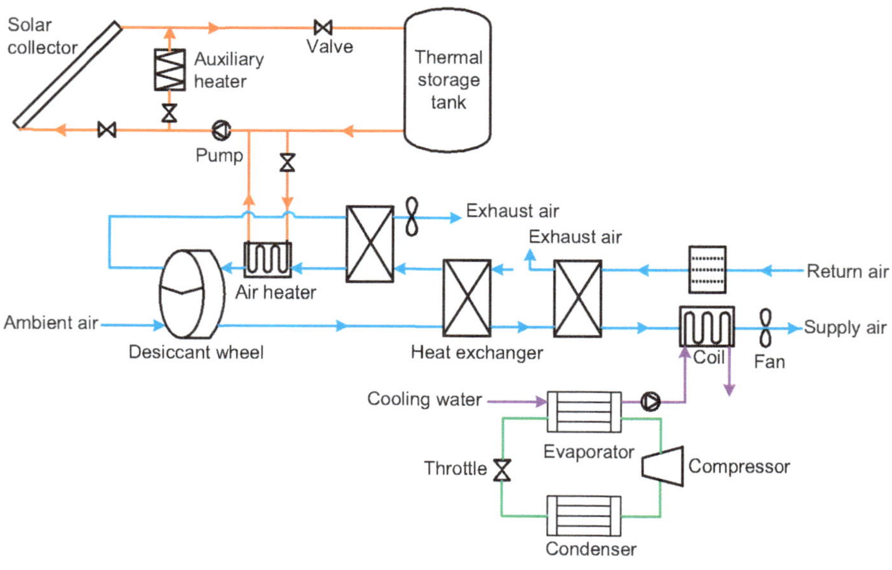

Figure 13.13 Schematic of the modified hybrid solar-powered desiccant wheel evaporative cooling system and vapor compression system in Ref. [36].

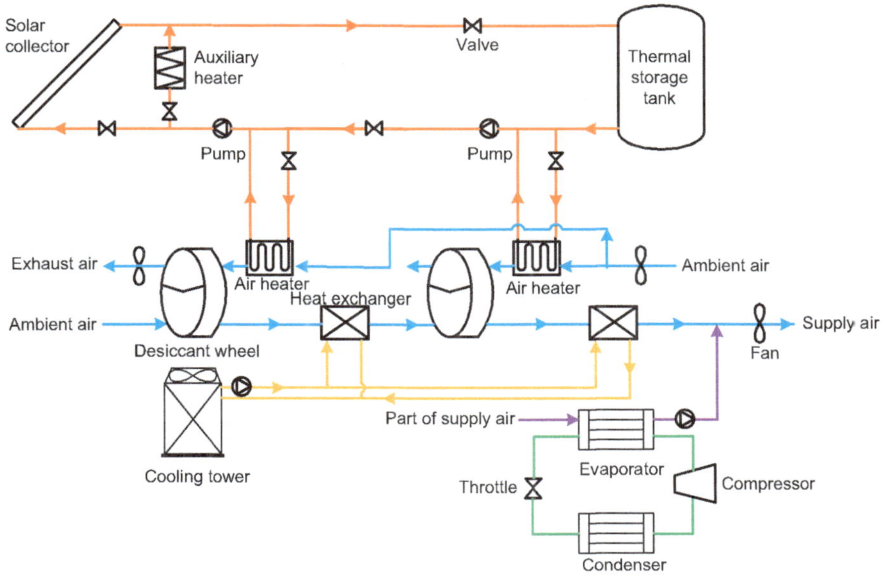

Figure 13.14 Schematic of a solar hybrid air-conditioning system.
From La D, Dai Y, Li Y, Ge T, Wang R. Case study and theoretical analysis of a solar driven two-stage rotary desiccant cooling system assisted by vapor compression air-conditioning. Sol Energy 2011;85:2997–3009.

Turkey. Meanwhile, the SF of the regeneration heat was 45–140% and utilization of solar energy in the system increased the $COP_{tot,hs}$ between 50% and 120%.

La et al. [37] investigated a flat plate solar collector–powered hybrid air-conditioning system integrating a TSDCS and air-source VCS in parallel as shown in Fig. 13.14. In the system, the supply air was constituted by two groups: one group from the ambient air was dehumidified by the TSDCS, and another one was cooled by the VCS. In other words, the solar-powered TSDCS is used to remove latent load, and the conventional VCS is utilized to remove sensible load. The experimental performance of such system powered by a 72-m^2 solar collector was examined under Chinese local summer conditions. When the ambient air possessed an average temperature of 34.6°C and humidity ratio of 21.54 g/kg, the solar radiant intensity fluctuated between 122.2 and 539.9 W/m^2, and the average efficiency of the solar collector was 0.32, experimental results showed that for the solar TSDCS, it could achieve an average cooling capacity 10.9 kW, with corresponding average COP_{th} and COP_{el} reaching 1.24 and 11.48, respectively; for the integrated hybrid system, 35.7% of the cooling output was contributed by the solar TSDCS unit, and the COP_{el} of the whole system was 4.41, which was about 35% higher than the conventional VCS. Also, a simulation model based on TRNSYS was developed to predict the performance of this hybrid system under different climate conditions. It was found that the system was feasible in humid (Shanghai), temperate (Beijing), and extremely humid (Hong Kong) weather conditions, and seasonal electrical savings rates were from 22% to 34%.

13.1.3.4 Other hybrid SDECS

In addition to conventional VCS, SDECS can be integrated with other types of refrigeration systems such as radiant cooling, absorption chiller, and PVT powered.

Radiant cooling technology can be driven by a higher temperature of chilled water, which results in high energy efficiency. However, such systems cannot realize the purpose of removing latent load. Desiccant cooling technology provides a good alternative to this problem owing to the dehumidification capacity. Therefore, several researchers have combined a solar desiccant cooling system with radiant cooling to produce the required cooling power.

Jose et al. [38] analyzed the performance of a hybrid SDECS/VCS and radiant floor system as shown in Fig. 13.15. The performance of the system was modeled and analyzed in TRNSYS 16. It was assumed that the radiant floor within the room kept the surface temperature of the floor at 2°C higher than the wet bulb temperature of the conditioned space, and hybrid SDECS/VCS was adopted to handle the odd sensible load and latent load. The performance and control strategy of the system under two climatic conditions in Spain (both with hot climate, but different water content of the outside air: humid and dry) were investigated and discussed. It was found that the desiccant system was useful in hot and humid climates, and the combined system consumed 35% less electricity than the conventional one.

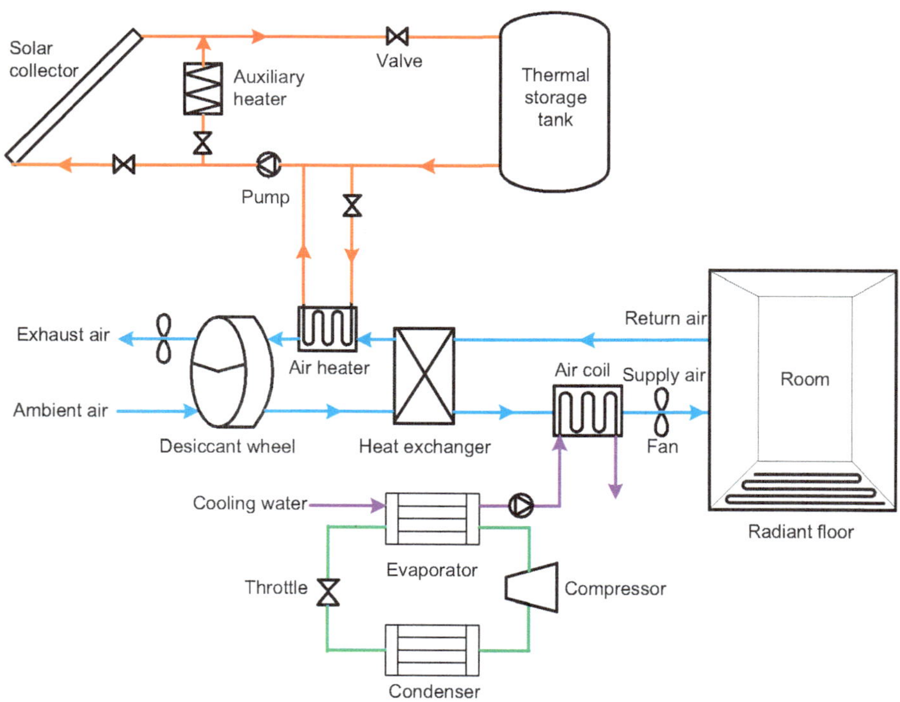

Figure 13.15 Hybrid solar-powered desiccant wheel evaporative cooling system/vapor compression system and radiant floor system [38].

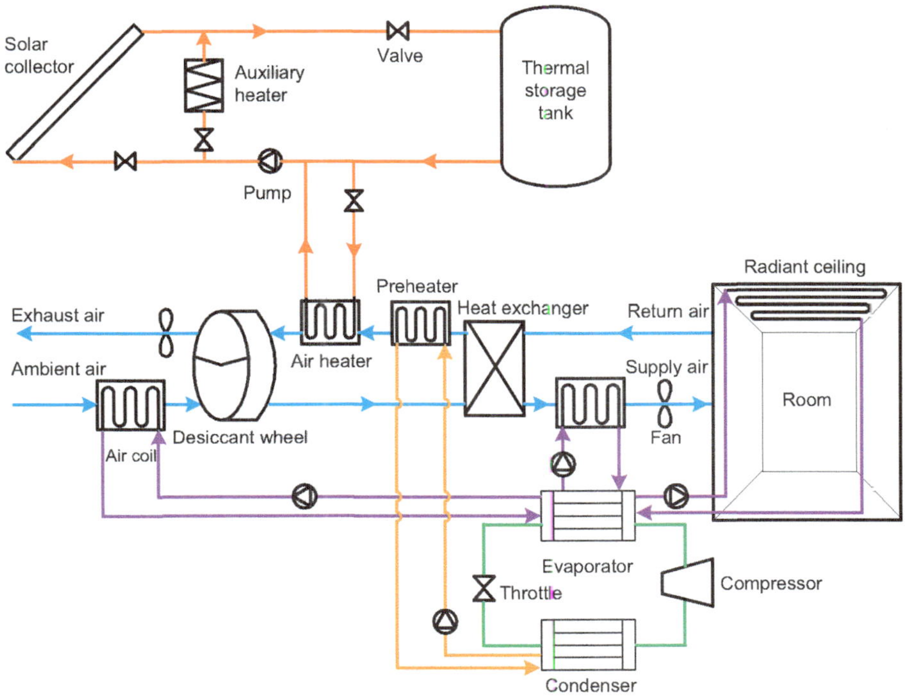

Figure 13.16 Hybrid solar-powered desiccant wheel evaporative cooling system/vapor compression system and radiant ceiling system [39].

A modified hybrid system was built and investigated experimentally by Beccali et al. [39] as shown in Fig. 13.16. In the proposed system, a radiant ceiling instead of radiant floor was adopted. Meanwhile, chilled water from the VCS was utilized in three ways: the first group of cooling coils was installed before the desiccant wheel to precool the supply air, the second was installed after the sensible heat exchanger as usual to remove sensible heat, and the last was used to provide cooling water in the radiant ceiling. Also, the ejected condensation heat from the chiller was used to preheat regeneration air before it entered the solar heater. An experimental setup was established in southern Italy to test the actual operation performance of such system. Five control modes were proposed for system operation based on ambient temperature and humidity ratio, and a control strategy was programmed in the Labview platform. Also, it was found that the use of heat ejected by the chiller to preheat regeneration airflow allowed a reduction in the solar collector area by about 30%.

Fong et al. [40] integrated an SDECS with solar absorption chiller-powered ceiling radiant cooling and investigated its performance in an office space. The zone sensible and latent loads were handled by the radiant cooling and desiccant dehumidification, respectively, and an absorption chiller was used to provide chilled water to the radiant ceiling directly and to the desiccant cooling system by a supply air coil. Regeneration

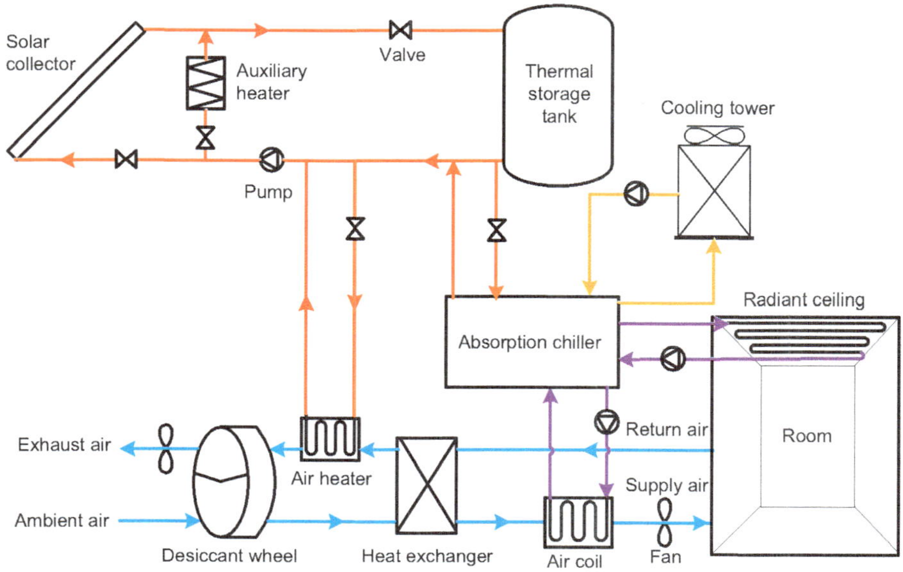

Figure 13.17 Hybrid solar-powered desiccant wheel evaporative cooling system and absorption chiller system [40].

processes of both the desiccant wheel cooling system and the absorption chiller were driven by a plate-type solar collector (Fig. 13.17). The performance of the system was evaluated based on the year-round dynamic simulation model in TRNSYS for the Hong Kong climatic conditions. Also, the effects of different radiant types including passive chilled beams and active chilled beams were analyzed. The main results obtained were as follows: performance of the solar hybrid cooling system could meet the needs of air conditioning, the primary energy savings potential of the system could be up to 36.5% compared to a conventional centralized air-conditioning system under the simulation conditions, and the system using passive chilled beams had better overall performance than that using active ones.

A PVT and solar air collector–powered hybrid SDECS/VCS study was conducted in Ref. [41]. Three different types of PVT and solar air collector-driven desiccant wheel cooling systems were proposed: the first was a PVT and solar air collector-powered conventional DECS/VCS (Fig. 13.18(a)), the second differed from the first by use of an additional heat exchanger before the desiccant wheel to recover heat from the return air as shown in Fig. 13.18(b), and in the third configuration a heat pump was utilized both for processed air for cooling and for regeneration air for heating as shown in Fig. 13.18(c). A comprehensive model with respect to different systems was developed in TRNSYS, and both the energy and the economic performance of these systems was analyzed for an office room as well as a lecture room with more persons under hot and humid conditions in Italy. Emphasis has to be placed on the fact that different area ratios between solar air collectors and PVT

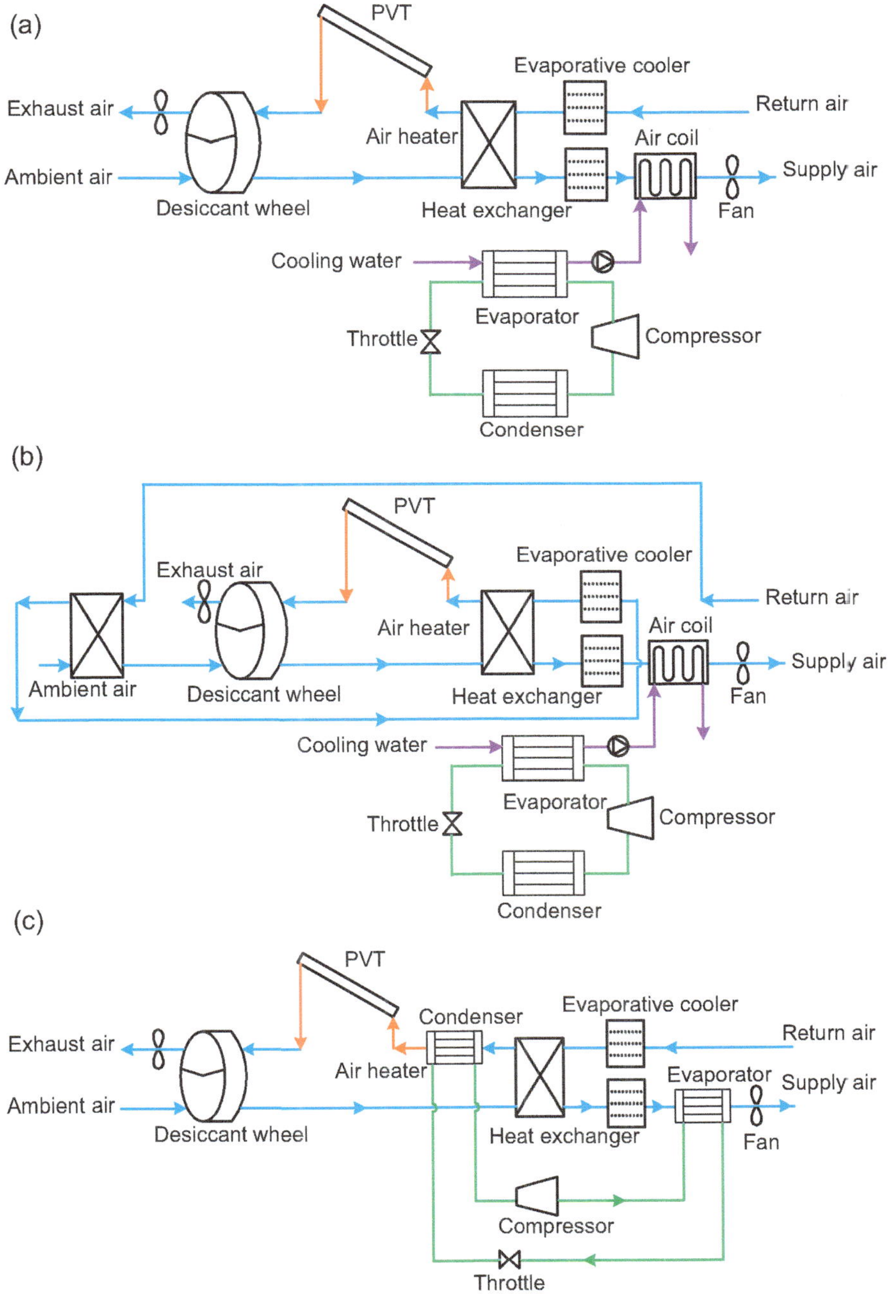

Figure 13.18 (a) Basic photovoltaic/thermal panel (PVT)-powered desiccant wheel evaporative cooling system and vapor compression system, (b) with enthalpy recovery from return air, and (c) with integrated heat pump [41].

collectors were also discussed in the study. It was shown that the energy savings ratio E_{save} of these systems compared to conventional VCS was in the range of 50–90%, and in most of the considered cases, the third configuration demonstrated the best results in terms of PE_{save}. Configurations with large PVT collector area had better primary energy savings, largely due to the electricity production of the PV cells. The economics of these systems at current equipment costs and energy prices were acceptable.

To improve the performance of SDECS in subtropical Hong Kong, Fong et al. [42] proposed six hybrid systems and evaluated their annual performance for typical office application in TRNSYS: three of them were full fresh air systems combined with conventional VCS (as shown in Fig. 13.10), PV-powered VCS (electricity produced by PV was used to drive the VCS, and an auxiliary heater was used to power desiccant cooling), and PVT-powered hybrid SDECS/VCS (electricity and thermal energy produced by PVT was used to drive the VCS and SDECS, respectively). The others were return air systems combined with PV-powered VCS (electricity produced by PV used to drive the VCS, an auxiliary heater used to power desiccant cooling), PVT-powered hybrid SDEC/VCS (electricity and thermal energy produced by PVT used to drive the VCS and SDECS, respectively), and an absorption chiller (similar to what is shown in Fig. 13.17). All the systems were feasible and could obtain up to 35.2% savings in year-round PE compared with conventional VCS. Among these systems, the SDECS hybridized with the VCS for full fresh air design and with absorption refrigeration for return air design were recommended owing to the savings potentials of both primary energy and initial cost.

13.2 Solar-powered liquid desiccant systems

13.2.1 Introduction

In addition to the solid desiccant cooling systems, there has also been great development in solar-powered liquid desiccant cooling systems. The earliest liquid desiccant system was suggested and experimentally tested by Lof. He used triethylene glycol as the absorbent and solar-heated air for regeneration purposes. Liquid desiccant systems essentially consist of an absorber for dehumidifying the air, a regenerator for regenerating the solution, and a series of heat exchangers and evaporative coolers for sensible cooling and heating of air and solution. Dehumidification and regeneration are the key processes.

Desiccant cooling systems are now being successfully incorporated into commercial air-conditioning (HVAC) systems. They are being used either as stand-alone units or as hybrid units in conjunction with VCSs or vapor absorption systems (VASs), the hybrid ones being more in use. All stand-alone desiccant cooling systems (DCSs) include at least three major components in various configurations: an absorber, an indirect/direct evaporative cooler, and an air–air heat exchanger. In a hybrid system, the desiccant dehumidifier handles the latent cooling load and the VCS/VAS handles the sensible cooling load [43].

13.2.2 Solar-powered stand-alone desiccant cooling systems

13.2.2.1 Basic mode

Fig. 13.19 presents a schematic diagram of a basic liquid desiccant dehumidification and air-conditioning system [44], which generally comprises four major units, namely, a dehumidification unit, a regeneration unit, a liquid desiccant storage unit, and a sensible heat handling unit. The task of the dehumidification unit is to remove the moisture of the inlet air by bringing it into contact with a sprinkled liquid desiccant. The regeneration unit is used to regenerate the diluted solution flowing from the dehumidification unit to an acceptable concentration (near the desiccant's initial concentration); thus, the continual operation of the cycle can be maintained. The liquid storage unit is aiming at realizing energy storage by storing strong solution. The sensible heat handling unit removes the sensible heat load of the processed air flowing from the dehumidification unit according to indoor air comfort standards. Other such systems can be made through different arrangements, replacements, and additions of components based on this system.

It is obvious that the configuration of the dehumidification unit is similar to that of a refrigeration unit, except that the former has an additional insulating layer and a filter layer, which are used separately to prevent environmental heat effects on dehumidification and remove the droplets of the liquid desiccant brought by the processed air. Both units use the same packing materials, but the processes are totally opposite. In the refrigeration unit, the surface vapor of a diluted hot liquid desiccant is higher than that of the inlet ambient air, so the direction of mass (water) transfer is from the sprayed liquid film to the processed air; the weak solution is concentrated near to the initial concentration of the desiccant solution pumped into the dehumidifier. The surface vapor pressure difference between the liquid desiccant film and the air acts as the driving force for mass transfer; therefore, the diluted solution flowing

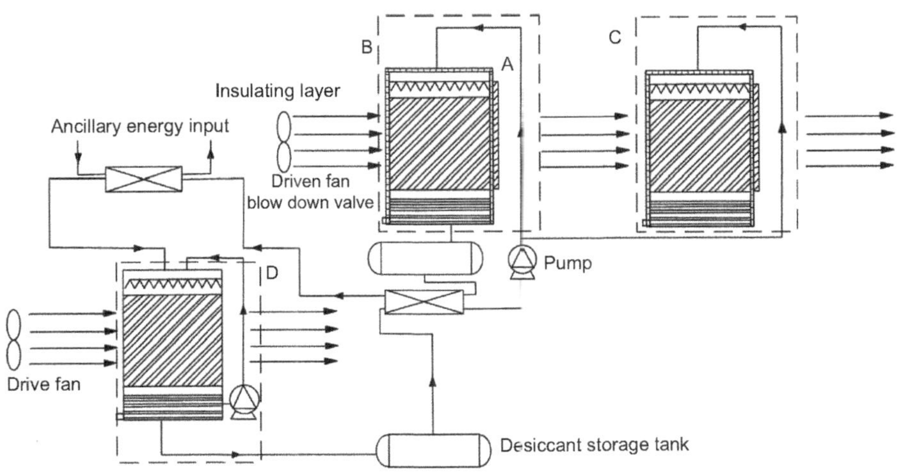

Figure 13.19 Schematic of a liquid desiccant dehumidification air-conditioning system [44].

from the dehumidification unit needs preheating to increase its surface vapor pressure to improve refrigeration efficiency. Low-grade energy hereby serves as the preheating source for the weak desiccant as well as the driving force for the whole system. The ancillary heat exchanger here is necessary lest the low-grade energy is weak or unavailable. Similarly, the inlet liquid desiccant needs to be precooled to lower its surface vapor pressure, and thus better dehumidification efficiency can be achieved. The heat exchanger between the weak solution flowing out of the dehumidification unit and the strong solution flowing into it can effectively preheat the former and precool the latter. In addition, to maintain the low surface pressure of the desiccant solution, actions such as adding an insulating layer to prevent the increase in solution temperature caused by heat transfer are taken to maintain the ideal mass transfer driving force. The desiccant storage tank is used for energy storage by storing the regenerated strong solution. It should be mentioned here that the storage tank is usually equipped with an ancillary heater to avoid crystallization, which will affect the normal operation of the system by clogging the pipe. It is self-evident that the main energy consumption occurs in the process of regeneration, which can be completed using low-grade energy; thus the application of attractive but not readily available low-temperature heat resources is greatly expanded. Additionally, their potential is enhanced by the energy storage.

13.2.2.2 Performance index

The performance index of the system is calculated as follows:

$$\mathrm{COP} = \frac{Mr_0 + m_{\mathrm{air}} C_{\mathrm{P}} \left(T_{\mathrm{air,in}} - T_{\mathrm{air,out}}\right)}{Q_{\mathrm{reg}}} \quad [13.16]$$

where $M = m_{\mathrm{air}}(d_{\mathrm{in}} - d_{\mathrm{out}})$; d is the humidity ratio of the air; r_0 is the latent heat of the air; C_{p} is the constant-pressure specific heat; Q_{reg} is the thermal energy consumption of regeneration; m_{air} is the quality of the air; and T_{air} is the temperature of the in, inlet air, or out, outlet air.

13.2.2.3 Research progress

Solar-regenerated liquid desiccant ventilation preconditioning system

Thosapon et al. [45] described the performance of a solar-regenerated liquid desiccant ventilation preconditioning system. A heat exchanger is used to cool the dehumidified air instead of typical evaporative cooling to maintain the dryness of the air. Sensible cooling using a heat exchanger was selected to cool the dehumidified air before it enters the conditioned space instead of evaporative cooling to maintain the dryness of the air.

Fig. 13.20 shows the schematic of a solar-regenerated liquid desiccant ventilation preconditioning system. It comprises two air loops and two liquid loops. The two air loops are the dehumidified air and the regenerating air loop. The two liquid loops are the liquid desiccant and the cooling water loop. Lithium chloride was selected as the desiccant because of its stability and high performance.

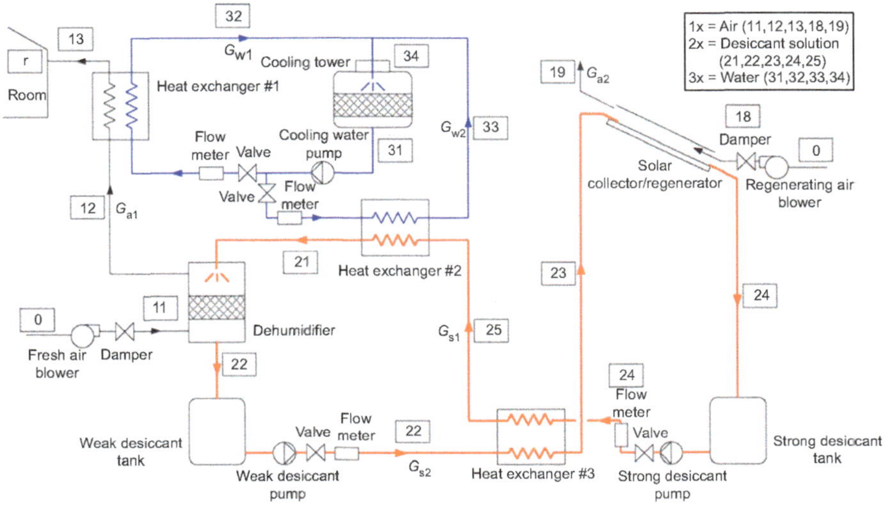

Figure 13.20 Schematic of a solar-regenerated liquid desiccant system [45].

According to the positions designated in Fig. 13.20, ambient air (11) is brought into the dehumidifier and directly contacts the strong desiccant solution (21) in a counter-flow fashion. A random packed bed type is selected for the dehumidifier because of its compactness in terms of large contact surface area per volume. The dehumidified ventilation air (12) is then sensibly cooled by cooling water (31) in a compact heat exchanger before entering the room (13). The heat exchanger is chosen instead of an evaporative cooler or a mechanical cooling system to maintain the dryness of the air and reduce the energy use. The desiccant solution (22), after it absorbs moisture from the ventilation air, becomes weaker and needs to be reconcentrated. This is accomplished by the solar collector/regenerator (C/R) through which ambient air flows as regenerating air (18, 19). Solar energy is selected as the heat source because of its many advantages. The solar C/R also helps reduce the amount of installed equipment. The strong desiccant leaving the solar C/R (24) is cooled by cooling water (31) in a compact heat exchanger before entering the dehumidifier (21) and completing the loop. Another compact heat exchanger exchanges heat between the strong (24, 25) and the weak (22, 23) desiccant solutions. This helps precool the strong desiccant entering the dehumidifier and preheat the weak desiccant entering the solar C/R at the same time. A cooling tower provides cooling water needed for the entire system.

The overall purpose of the liquid desiccant ventilation preconditioning system is to provide ventilation air that is as dry and cold as possible so that it would introduce only a minimal cooling load to the conditioned space. It is an alternative configuration of liquid desiccant systems that can hybrid with a separate air-conditioning system (which would handle mainly the sensible load) inside the zone. The results suggest that the most influential parameters are solar radiation, ventilation rate, and desiccant solution concentration. The balance between the water removed at the dehumidifier and that evaporated at the regenerator needs to be considered to maintain uniform performance during continuous operation.

Solar-powered membrane liquid desiccant air-conditioning system

Ahmed et al. [46] shows the schematic diagram for the membrane liquid desiccant air-conditioning (M-LDAC) systems studied. The same M-LDAC system is used in eight systems configurations in Fig. 13.21, while the differences between the configurations are in the liquid desiccant heating system and the energy recovery ventilator (ERV).

When an ERV is installed in an all-air HVAC system, the outdoor hot and humid air is preconditioned using the air leaving the conditioned space. The cool and concentrated desiccant solution, which is used to meet the entire latent load of the conditioned space, leaves the dehumidifier both warm and dilute. The dilute desiccant solution has to be regenerated to be reused to dehumidify the hot and humid ventilation air. The desiccant solution regeneration is performed by increasing its temperature to a level that makes its surface vapor pressure higher than that of the regeneration air. Thus, the desiccant solution is preheated in the solution-to-solution sensible heat exchanger before being heated to a specific set-point temperature in the heating coil. The hot and dilute desiccant solution leaving the heating coil enters the regenerator, where it is concentrated using the regeneration air (ie, outdoor air in this study). Although the desiccant solution coming out from the regenerator is concentrated, its temperature still

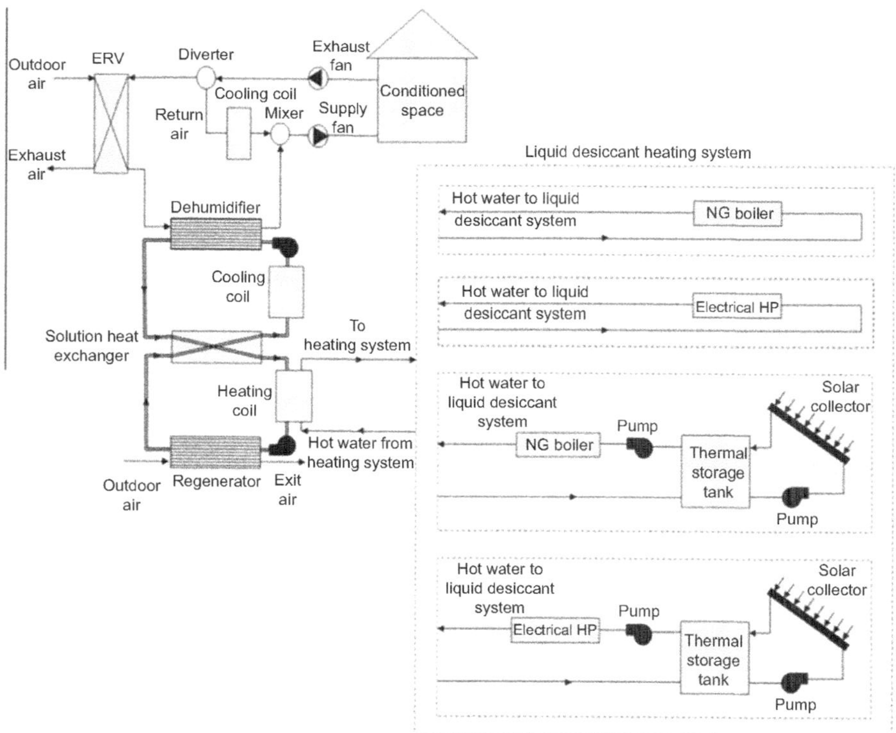

Figure 13.21 Schematic diagram for the membrane liquid desiccant air-conditioning systems studied [46]. *ERV*, Energy recovery ventilator; *HP*, heat pump; *NG*, natural gas.

high. The concentrated desiccant solution has to be cooled before entering the dehumidifier, to decrease its surface vapor pressure and be able to absorb moisture from the humid ventilation air stream. Therefore, it is precooled in the solution-to-solution sensible heat exchanger and then cooled by a dry cooling coil to a specific set point. In this study, the thermal, economic, and environmental performance of the proposed solar M-LDAC (S-M-LDAC) system, which uses liquid-to-air membrane energy exchangers as a dehumidifier and a regenerator, was investigated. It was found that the use of a solar thermal system (STS) to provide the thermal energy required for the regeneration of the dilute desiccant solution increased the initial cost but decreased the annual operating costs compared to nonsolar systems. The life-cycle cost is always lower with the STS than with no solar; however, the payback period is not always shorter when an STS is used. The payback period is shorter for the STS with a natural gas boiler backup but longer for the STS with an electrical heat pump backup, compared to the same systems without solar. The annual CO_2 emissions are always lower when an STS is used. All the S-M-LDAC configurations will produce lower annual CO_2 emissions than the conventional air-conditioning system either with or without an ERV.

The multistage dehumidification system

In single-stage dehumidification, an adverse temperature increase in the desiccant solution, which will lower the driving force for mass transfer between desiccant solution and processed air, will occur. The concept of multistage dehumidification was studied by Jiang et al. Fig. 13.22 demonstrates the multistage dehumidifier, which introduces an original dehumidifier by installing several single dehumidifier modules in series, in which the liquid desiccant flows from module to module and is separately cooled in every single dehumidifier module to better dehumidification. This new mode of dehumidification overcomes the potential disadvantage of conventional single-stage dehumidification [44,47]. The inlet air is first treated by the weakest desiccant solution, because weak desiccant solution has the highest surface vapor pressure, and the inlet air also owns the highest vapor pressure. Then, the process air is dried, and its vapor pressure is decreased. In the last module of the dehumidifier, the strongest solution with the lowest surface vapor treats the air with the lowest vapor pressure. As a result,

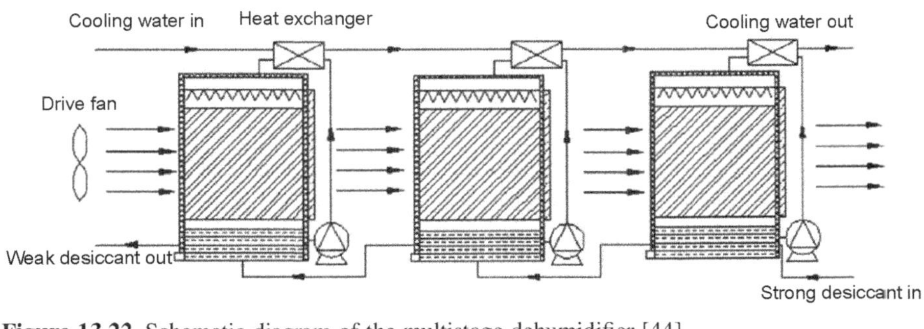

Figure 13.22 Schematic diagram of the multistage dehumidifier [44].

the vapor pressure difference between desiccant solution and processed air is kept at a moderate level without drastic change, whereas for the conventional dehumidifier, the vapor pressure difference is decreased with the proceeding of the moisture removal process. The irreversible loss in the multistage dehumidifier is greatly lowered compared with that of a conventional dehumidifier. The demand for mass flux of the desiccant solution for a conventional dehumidifier is great to realize the dehumidification requirements, and as a result, the inlet and outlet concentration difference of the desiccant is small; thus the vapor pressure difference between the desiccant solution and the processed air is increased, and consequently, the irreversible loss is enhanced, which eventually contributes to the aggravation of the system performance.

Z.Q. Xiong et al. [48] presented a novel two-stage liquid desiccant dehumidification system assisted by $CaCl_2$ solution using exergy analysis and found that it was more efficient to run the system in a high desiccant variance mode. Exergy analysis based on the second thermodynamic law has been recognized as an effective way to improve system performance. A two-stage liquid desiccant dehumidification system assisted by $CaCl_2$ solution is proposed, as shown in Fig. 13.23(a). Processed air is first dehumidified by $CaCl_2$ solution and then further dehumidified by LiCl solution to the desired air humidity. In contrast, in the regeneration process, air is first used to regenerate LiCl solution and then to regenerate $CaCl_2$ solution. This arrangement is made according to the properties of LiCl and $CaCl_2$ solutions. LiCl solution shows better dehumidification ability, but is harder to regenerate. On the other hand, $CaCl_2$ solution is easier to regenerate, but has poor dehumidification ability. Thus, two sets of liquid desiccant cycles are combined in this proposed system. One is the LiCl cycle, the other is the $CaCl_2$ cycle. They are similar to each other.

To better explain the working principle of the system, the desiccant states of $CaCl_2$ solution are shown in Fig 13.23(b), ie, the 6-7-8-9-10-11-6 cycle. Strong $CaCl_2$ solution at state point 11 is cooled in the cooling water–$CaCl_2$ heat exchanger (HE) to state point 6. Then it flows to the first-stage dehumidifier where the processed air is in direct contact with the $CaCl_2$ solution. Moisture in the processed air is absorbed by the $CaCl_2$ solution accompanied by heat ejection. After the dehumidification process, the $CaCl_2$ solution is at state point 7′, at which state its concentration decreases a little bit and its temperature increases compared with state point 6. However, the $CaCl_2$ solution still has a strong ability to absorb water if it is cooled. So it is pumped through the “f” channel of the three-way valve at the outlet of the first-stage dehumidifier back to the cooling water–$CaCl_2$ HE. Then it goes into the dehumidifier again. In this way, the $CaCl_2$ solution is used to dehumidify processed air again and again until its concentration is lower than a certain value, which is designed according to the requirement. After the dehumidification processes, the $CaCl_2$ solution is at state point 7. Then it is pumped to the $CaCl_2$–$CaCl_2$ HE to recover heat from the $CaCl_2$ solution after regeneration. Thereafter, the $CaCl_2$ solution at state point 8 is further heated by the hot water–$CaCl_2$ HE to state point 9 at the regeneration temperature and then goes into the second-stage regenerator. In the regenerator, water in the $CaCl_2$ solution evaporates into the air and the $CaCl_2$ solution reaches state point 10′. Because the concentration of the $CaCl_2$ solution out of the regenerator is still

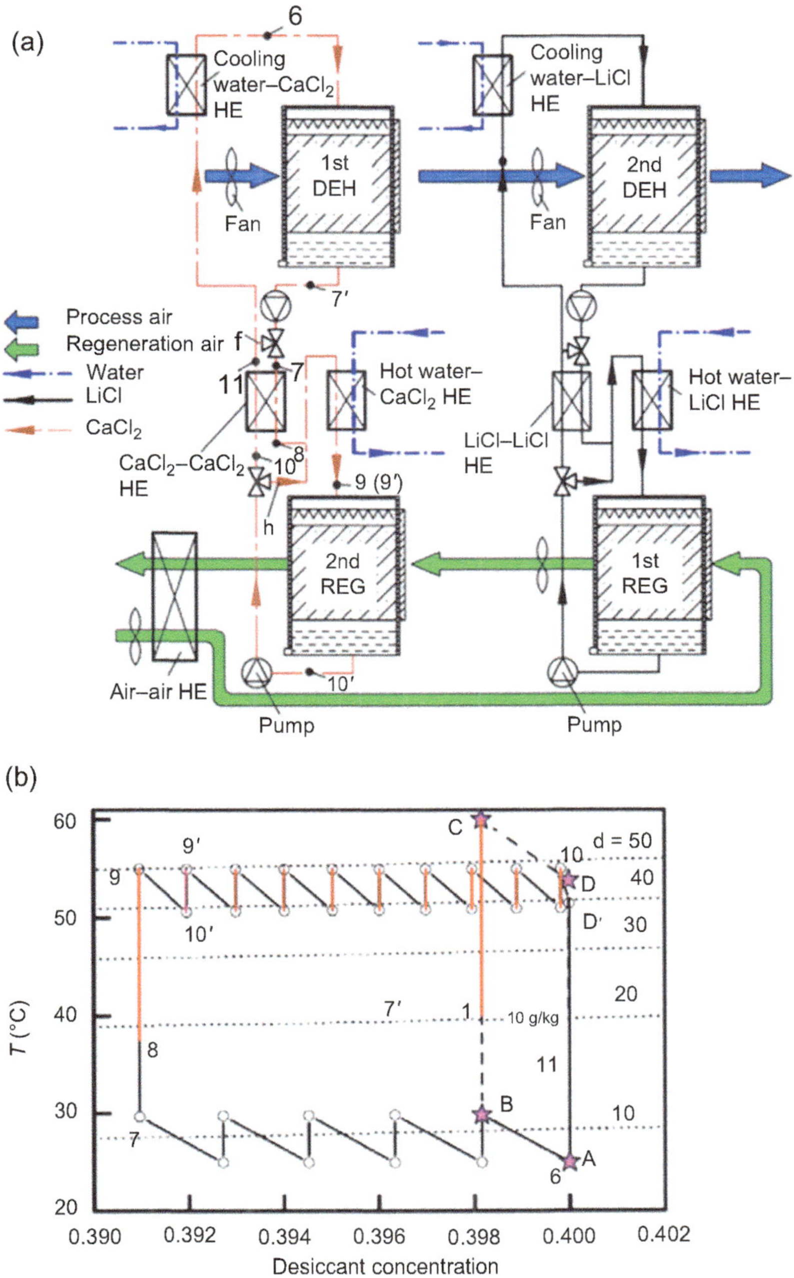

Figure 13.23 Two-stage liquid desiccant dehumidification system with big desiccant concentration variance [48]. (a) Schematic chart, (b) the states of desiccant solution *DEH*, Dehumidifier; *HE*, heat exchanger; *REG*, regenerator.

lower than its original value (40%), it needs to be regenerated again. So the $CaCl_2$ solution is pumped through channel "h" of the three-way valve located at the outlet of the second-stage regenerator. Then it goes through the cooling water–$CaCl_2$ HE to the regenerator again. In this way, it reaches state point 10 at a high concentration. Then it goes to the $CaCl_2$–$CaCl_2$ HE and is cooled to state point 11. So the $CaCl_2$ cycle is formed. The three-way valves are coarsely controlled by a time signal. For comparison, the states of the desiccant in the basic liquid desiccant dehumidification system are also given, ie, the A-B-C-D-D′-A cycle. Compared with the basic system, the proposed system has a larger concentration variance between the strong desiccant solution after regeneration and the weak desiccant solution after dehumidification. The regenerators/dehumidifiers considered are cross-flow type and filled with CELdek packing material.

In this proposed system, the corrosion problem is relieved for two reasons. First, most of the components in this system (including desiccant solution pipes, wind pipes, dehumidifiers, and regenerators) are made of polyvinyl chloride plastic material, which is anticorrosion. Second, the regeneration temperature of the desiccant solution, which is closely related to the corrosion problem, is reduced by using the two-stage design.

It is reported that the hybrid system's COP_{el} [49] (the ratio of refrigerating output gained by processed air to electrical power consumed by the hybrid system) and COP_{th} (the ratio of refrigerating output gained by processed air to heat for regeneration) reach 6.7 and 6.8, respectively, in summer mode by adding a three-stage total heat recovery device using a liquid desiccant. In winter mode, the average COP_{th} is 1.34. The good potential of the hybrid desiccant system has been proven.

Other developments

Li Yu Tong et al. [50] presented the numerical simulation results of an open-cycle liquid desiccant dehumidification system. The transient performance of the open-cycle liquid desiccant dehumidification system was simulated for the weather conditions of Hong Kong. In the system, the exhaust air of an air-conditioning system can be cycled to the solar C/R to regenerate the weak desiccant solution, instead of to the environment. An open-cycle solar C/R utilizes both solar radiation and unsaturated air, which can significantly improve the performance of the solar C/R. Thus the open-cycle liquid desiccant system performs well in humid weather conditions. By incorporating a solar desiccant dehumidification system in an HVAC system, 25–50% energy savings can be expected for various sensible load to latent load ratios. The saving percentage increases with the latent load fraction.

Xiaofeng Niu et al. [51] studied a liquid desiccant-based air-conditioning system by analyzing the system performance under various fresh air ratios using simulation tests. The simulations were conducted based on the climatic conditions of Hong Kong with fresh air ratios varying from 20% to 100% at intervals of 10%. With the fresh air ratio increasing from 20% to 100%, the efficiency of the dehumidifier gradually increases from 0.792 to 0.929, and the efficiency of the regenerator gradually drops from

0.234 to 0.157 gradually. The COP of the liquid desiccant-based system varies slightly at first and then decreases sharply as the fresh air ratio moves from 60% to 100%. The highest COP appears when the fresh air ratio is in the range of 40–50%. Hence, a larger fresh air ratio does not improve the COP of a liquid desiccant-based system that uses a heat pump as regeneration heat source. Considering both indoor air quality (IAQ) and energy consumption, the optimal fresh air ratio would be between 40% and 50%. Compared to a conventional air-conditioning system with primary return air, the liquid desiccant-based system consumes notably less power. The maximum power savings ratio is 58.9% when the fresh air ratio is 20%, and the minimum is 4.6% when the fresh air ratio is 100%.

X.H. Liu et al. [52] experimentally studied the mass transfer process between the air and the desiccant when they cross flow through a structured dehumidifier. Moisture removal rate and dehumidifier effectiveness were used as the performance indices of the mass transfer process. It was found that the moisture removal rate increased with increasing air and desiccant flow rate, air inlet humidity ratio, and desiccant inlet concentration; decreased with desiccant inlet temperature; and changed very little with air inlet temperature. The dehumidifier effectiveness increased with increasing desiccant flow rate and inlet temperature and decreased with airflow rate but was affected little by the other three inlet parameters, including the air inlet temperature and humidity ratio and desiccant inlet concentration.

Y.J. Dai et al. [53] proposed a hybrid liquid desiccant air-conditioning system in which a cross-flow liquid desiccant air dehumidifier packed with honeycomb paper is used to improve the performance of a VC air-conditioning system. In this paper, a mathematical model was developed and validated to evaluate the performance of a liquid desiccant air dehumidifier packed with such kind of honeycomb papers. They also presented liquid desiccant systems that have been proposed as alternatives to conventional air-conditioning systems to control air humidity, especially in hot and humid areas.

Esam Elsarrag [54] tested a novel regeneration system modified from a solar tilted still that was tested experimentally for calcium chloride regeneration. A corrugated blackened surface was used to heat the desiccant and an airflow was used to regenerate the solution. The effects of the liquid-to-air flow rate ratio, the desiccant temperature, the desiccant concentration, and the inlet air humidity ratio on the evaporation rate were assessed. A wide range of liquid-to-air flow rate ratios were used. The optimum value of the liquid-to-air flow rate ratio was reported.

S. Alizadeh [55] described that the regenerator used in the application is a forced parallel flow-type solar C/R. The regenerator was designed and optimized and the prototype of the solar C/R was built and tested under summer conditions of Adelaide, Australia.

A comparison between the rate of evaporation in the tests with water and liquid desiccant revealed that the water desorption rate in the case of liquid desiccant is generally higher owing to higher initial air and solution temperatures. The results of the experiments indicate that the rate of water evaporation from the weak solution increases as the air mass flow rate is increased.

13.2.3 The hybrid system

13.2.3.1 The vapor compression system combined with a liquid desiccant dehumidification hybrid system

Fig. 13.24 shows the schematic diagram of a hybrid system [56]. There exist three fluid cycles, namely, the refrigerant cycle in the compressor, condenser, expansion valve, and evaporator; the liquid desiccant cycle in the dehumidifier, HEs, and regenerator; and the cooling water cycle in the evaporative cooler. The first cycle is primarily used to remove sensible heat. The second cycle is used to deal with latent heat. The third cycle, on one hand, is used to cool the regenerated liquid desiccant, and on the other hand, is used to evaporatively cool the processed air. Three parts of airflow are involved in the operation of the system, namely, the processed air sent to the conditioned space, the cooling air at the condensation side ejected to the outside, and the regeneration air to take away the water vapor released by the dilute liquid desiccant.

Hot and humid air flows first into the liquid dehumidifier, which is primarily configured by packing honeycomb paper, sprinkler, and solution pump. The concentrated liquid desiccant is sprayed over the packing honeycomb paper by the sprinkler and a liquid film is thus formed along the honeycomb wall. Mixing heat, a little more than the condensation heat of the water vapor, is produced at the same time. In Fig. 13.24, a component like an evaporative cooler is set before the HE. It plays a role to capture the droplets of liquid desiccant leaving the dehumidifier or sometime acts as a direct evaporative cooler if the humidity ratio of the air is low enough. The air after dehumidification flows into the cross-flow HE where the air is cooled by another layer of air preliminarily cooled in the evaporative cooler. This is the so-called process of indirect evaporative cooling. The processed air can be cooled without humidification in this way. Thereafter the processed air is sent into the VC cooling system and is conditioned to delivery conditions. The air in the condensation side is ejected to the outside. The dilute liquid desiccant is pumped to the regeneration sector and is concentrated again. If the air temperature at point 8 in Fig. 13.24 is sufficiently high, the air is able to be used to regenerate the liquid desiccant and then is ejected to the outside (point 9), or the regeneration heat has to be input specially by means of a specially designed regenerator. The energy resources to regenerate the desiccant can be low-grade energy such as solar energy, waste heat, etc., because the regeneration temperature of the liquid desiccant is just 50–65°C. There are two HEs in the liquid desiccant cycle. The concentrated desiccant solution is cooled by dilute liquid desiccant from the dehumidifier and the cooling water cycling in the evaporative cooler.

Performance indices

Performance indices of the VCS combined with liquid desiccant dehumidification hybrid system are as follows:

1. COP_{el}:

$$COP_{el} = Q_C/W \qquad [13.17]$$

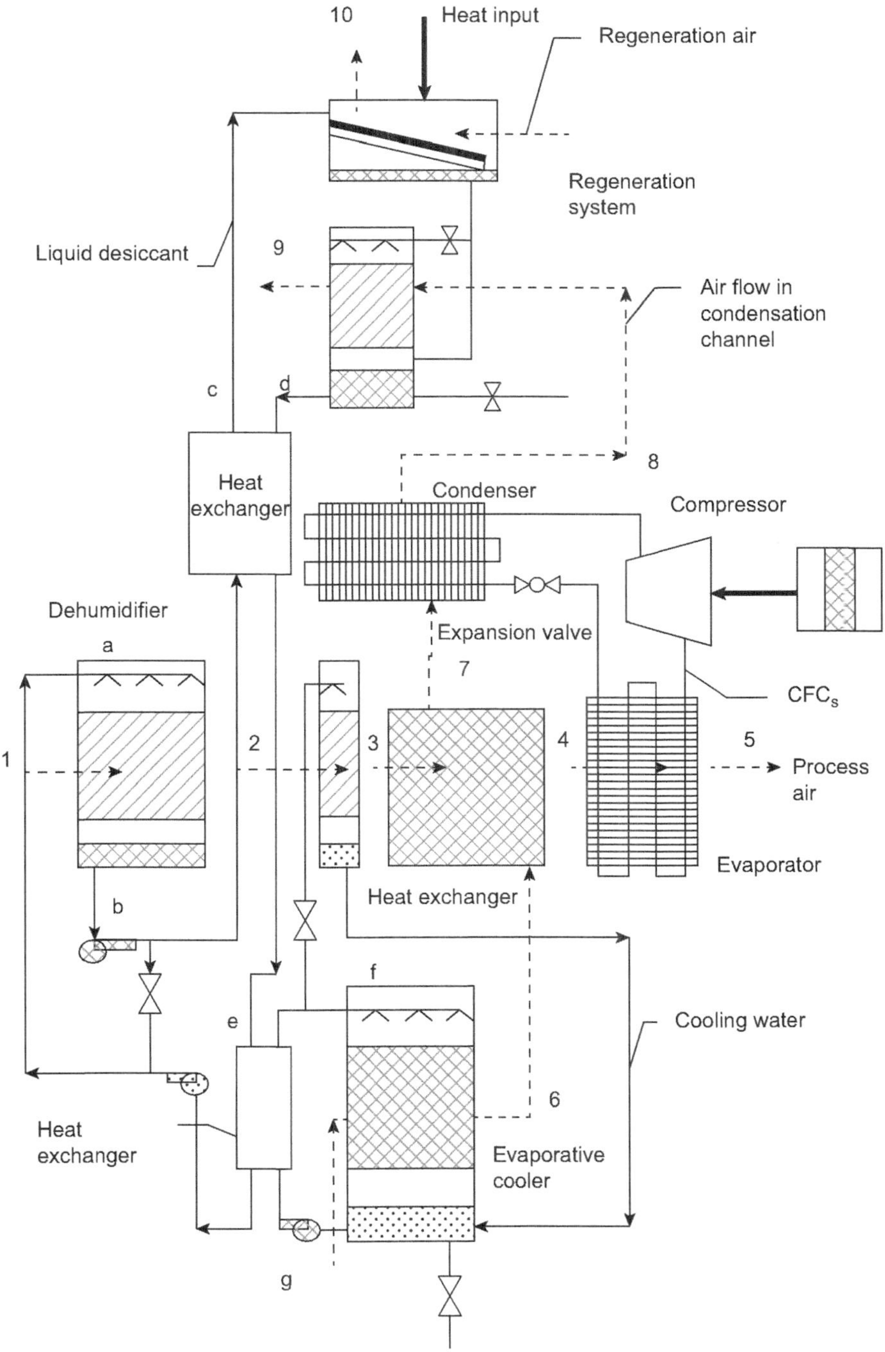

Figure 13.24 Schematic of the hybrid cooling system [56].

Here, Q_C is the cooling production of the system, in kW, and W is the total electric power consumption including compressor, fan, and pumps, in kW.

2. COP_{th}:

$$COP_{th} = \frac{Q_C}{Q_T} \quad [13.18]$$

Q_T is thermal energy consumption of the system. In this system, it is also the heat input to regenerate the dilute liquid desiccant.

3. COP of the system:

$$COP = \frac{Q_C}{Q_T + W/0.3} \quad [13.19]$$

The index is the ratio of cooling production to total energy input. Here, the equivalent coefficient of electric power and thermal energy is taken as 0.3.

4. COP of VCS (VCOP):

$$VCOP = Q_{ev}/W_c \quad [13.20]$$

Here, W_c is the power consumption of the compressor, in kW and Q_{ev} is the cooling production of the evaporator, in kW.

Generally speaking, the introduction of desiccant cooling changes the inlet air states of the VCS toward conditions helpful for improving the COP. The desiccant cooling subsystem undertakes all or part of the latent load, and lowers the condensation temperature and lifts the evaporation temperature in the meantime. The performance of the VCS is intimately related to the inlet air states and the practical latent load. The lower the latent load, the higher the COP would be. Because the cooling load for the evaporator and the load for the condenser are decreased, the size of the VCS may be further reduced [56].

The reason the hybrid system is superior to the VCS lies in its dehumidification section and evaporative cooling section. The performance of dehumidification can be described by the effectiveness of dehumidification and the performance of evaporative cooling can be described by the effectiveness of evaporative cooling instead of the programs mentioned above. Here, it is assumed that the moist air has no enthalpy change after evaporative cooling and has no temperature change after dehumidification (isotherm dehumidification) owing to the cooling effect of the cooling water cycled in the evaporative cooler to the liquid desiccant. It is acceptable according to our test results.

The hybrid cooling system, which combines desiccant cooling and VC air conditioning together, may have more cooling production than the VCS alone by 20–30%. The COP_{el} is increased to a great extent. It is also found that the electric power consumption and size of the VCS can be further reduced.

Howell et al. [57] have studied a hybrid system combining liquid desiccant dehumidification with VCS. It was found that the hybrid system reduces the area of evaporation and condensation by 34%, and power consumption by 25%, compared with VCS alone. After thorough study of a gas-fired air-conditioning system combining a VC machine with a solid desiccant dehumidifier, Parsons et al. [58] found that the cooling capacity of the hybrid system increased by 50% and the COP based on gas consumption increased by 40%; however, the initial cost had an increase of $140/kW cooling capacity.

13.2.3.2 The vapor absorption system combined liquid desiccant dehumidification hybrid system

Khalid Ahmed et al. [59] carried out a simulation on a hybrid system using LiBr as the working fluid for absorption and dehumidification. The proposed hybrid cycle is shown in Fig. 13.25. The cycle consists of a liquid desiccant system using LiBr as the liquid desiccant for both absorption and dehumidification processes and water as the

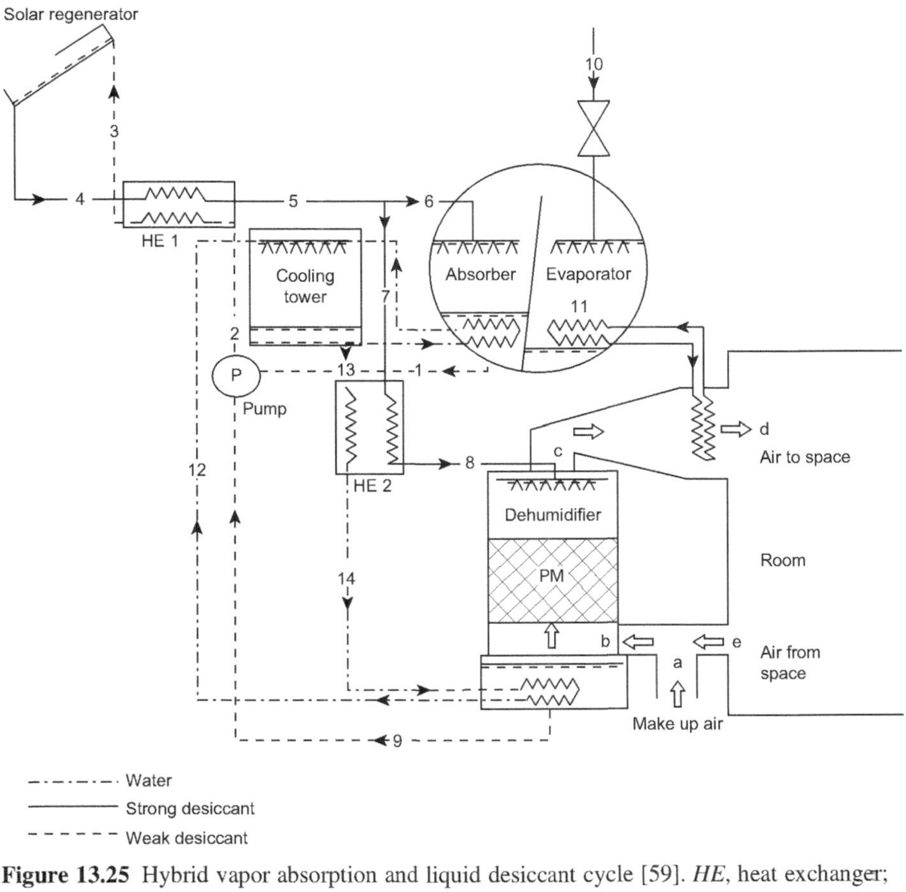

Figure 13.25 Hybrid vapor absorption and liquid desiccant cycle [59]. *HE*, heat exchanger; *PM*, packing material.

refrigerant. Cooling takes place by evaporating the refrigerant from an external source in the evaporator rather than obtaining refrigerant from the condenser. However, an external source of refrigerant (water) must always be available. The weak absorbent solution (state 1) from the absorber is pumped (state 2) to the solar regenerator, which is partly open to the atmosphere, through the regenerative HE 1. It is then heated and subsequently concentrated in the solar regenerator from state 3 to state 4. The strong solution passes through the regenerative HE 1 on its way (states 5 and 6) to the absorber. In the absorber, the strong desiccant absorbs water vapor as soon as the incoming refrigerant evaporates in the evaporator (state 11) by absorbing the heat from the cold space, maintaining the reduced pressure required by the evaporator. The heat of absorption for the refrigerant–absorbent solution is removed by the cooling water loop. In the evaporator, water from an external source (state 10) is evaporated at reduced pressure by the energy supplied by the heat from the cooled space. The resultant weak desiccant is then pumped from the absorber back to atmospheric pressure through the regenerative HE and to the solar regenerator, completing the absorption–machine cycle. Part of the strong desiccant (state 5) from the solar regenerator is passed through the regenerative HE 2 (state 7) to the dehumidifier, where the air to be conditioned is dehumidified in the countercurrent direction. The actual dehumidification takes place owing to the vapor pressure difference between the vapor in the air and the liquid desiccant. As the regenerated liquid desiccant is cooled and concentrated (state 8), its vapor pressure is less than the vapor pressure of the air, therefore the vapor present in the air tends to escape into the liquid desiccant, thereby diluting the liquid desiccant. The weak desiccant (state 9) from the dehumidifier is then pumped back for regeneration, along with the weak desiccant from the absorber (state 1). After regeneration the strong solution is supplied to the absorber and dehumidifier in the required proportions, as shown by states 6 and 7. The absorbent used in the hybrid cycle for absorption in the absorption machine, as well as the desiccant in the dehumidifier, is lithium bromide. The cooling water (states 12, 13, and 14) is circulated to the dehumidifier and completes the cooling tower cycle.

The hybrid vapor absorption and liquid desiccant system would produce a cooling effect of 700 tons of refrigeration at an evaporator temperature of 22°C per unit mass flow rate of the refrigerant. The COP obtained for the hybrid system is about 50% higher than the COP for the conventional absorption machine for the same conditions [59].

From the discussion, it is obvious that the COP of the hybrid cycle, owing to the variations in the different parameters, ranges between 0.96 and 1.25. Hence this hybrid cycle has a tendency to accommodate the various effects of ambient temperature and humidity without much sacrifice in the overall performance of the system. Thus it is the most suitable hybrid machine in regions that are liable to sudden variations in normally high temperatures and humidity. A maximum COP of 1.25 was obtained during the study and the unit was found to be best suited to hot and humid areas.

These are commonly operated in a hybrid mode with VCSs, to offer better IAQ and higher energy efficiencies especially for low sensible heat ratio and high humidity applications [43].

13.2.4 The photovoltaic–electrodialysis regeneration method for liquid desiccant cooling systems

13.2.4.1 Single-stage photovoltaic–electrodialysis regeneration for liquid desiccant cooling systems

Currently, the regeneration process is always fueled by thermal energy. Nevertheless, this regeneration pattern has some disadvantages in that its performance will become poor when the surrounding atmosphere is of high humidity, and the heat provided for regeneration will be unfavorable to the following dehumidification process. To ameliorate that, Xiu-Wei Li designed the single-stage PV–electrodialysis (PV–ED) regeneration method for liquid DCSs. That is, a membrane regenerator is employed to regenerate the liquid desiccant in an ED way, and a solar PV generator is adopted to supply electric power for this process [60].

The ED method is based on the transport of ions through selective membranes under the influence of an electrical field. Obviously, the adverse impact of high humidity in the surrounding atmosphere could be ignored in this regeneration style. More than that, there is no introduced heat that will be harmful to the dehumidification process, and no solution droplets can diffuse into the air of the environment.

The PV–ED regeneration process is described in Fig. 13.26. At the beginning, all the dilute cells are fed with the desiccant solution stream from solution storage tank 1

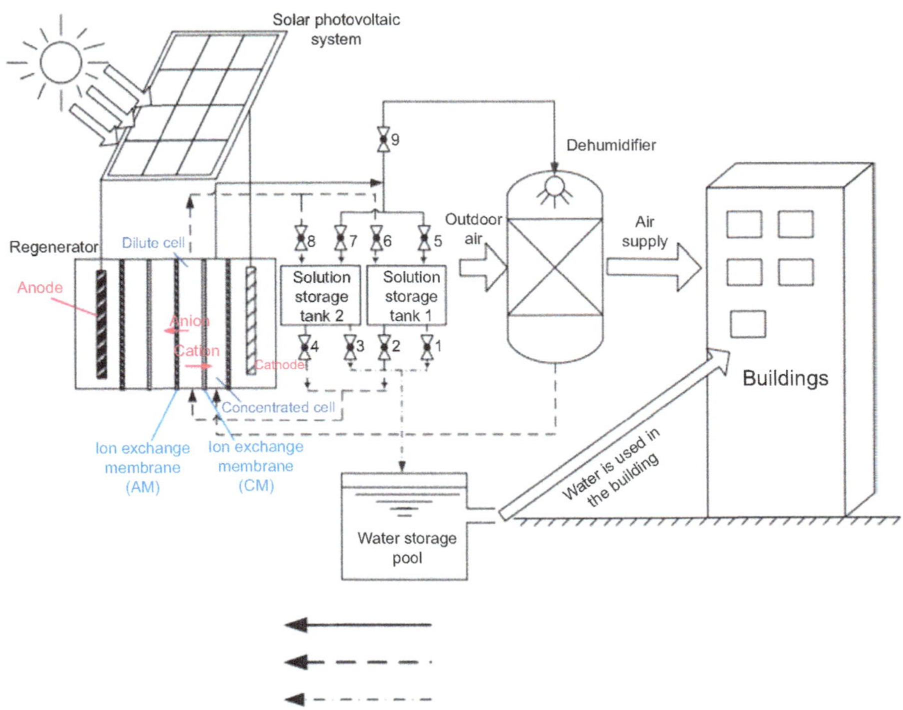

Figure 13.26 Photovoltaic–electrodialysis–liquid desiccant cooling system [60].

(valve 2 is open while valve 4 is closed; valves 1 and 3 are closed so that the water path is cut off); all the concentrate cells are fed with the weak desiccant solution stream from the dehumidifier. This diluted solution will be recycled from the regenerator to solution storage tank 1 (valve 6 is open while valve 8 is closed) and back to the dilute cells for the regeneration of a new round. This cycle will continue until its concentration falls to zero, which implies it has finally become purified water. Then, solution storage tank 2, which is full of strong desiccant solution, takes the place of tank 1 by feeding desiccant solution into the dilute cells of the regenerator (valve 4 is open while valve 2 is closed); the purified water in tank 1 will be removed (valve 1 is open while valve 3 stays closed) to the water storage pool for later, other use, and tank 1 comes to play the role that tank 2 played before (valves 5 and 8 are open while valves 7 and 6 are closed). Like this, tanks 1 and 2 keep on alternating their roles and both the strong desiccant solution and the purified water are acquired [60].

Compared to conventional thermal regeneration, this new method has the advantages in that it is immune to the instability caused by high humidity in the atmosphere; it gets rid of the added heat that is harmful to the dehumidification process, and thus saves the energy for producing a cooling effect; it is environmentally friendly by preventing the droplets of desiccant solution from sneaking into the surrounding atmosphere and introducing pollution; and it can also produce pure water in the regeneration process.

When the solar energy-conversion efficiency is taken into consideration, the PV–ED method is inferior to the solar thermal energy (TH) method, but the gap can be narrowed under certain working conditions like a smaller concentration change during the regeneration process or a very high relative humidity of the air. Nevertheless, when putting aside the poorer energy-conversion efficiency of a PV generator, the performance of the PV–ED method can be more than twice that of the conventional TH method. Therefore, under conditions without plenty of solar energy, the PV–ED method will be more efficient with electric power utilization. That also highlights that the PV–ED will be more appropriate for energy storage in the night operation mode when the electric power supply is at its valley; that electric power can be converted to a strong desiccant solution that can be used in the daytime. Beyond that, the PV–ED regeneration method can also be driven by a wind power generator, which gives more flexibility.

The total cost of a PV–ED system mainly rests on the cost of the PV component, which at present is much higher than that of the TH system. In today's Chinese market, the initial cost of a household PV system of 1 kW load is no less than $3000, whereas the initial cost of a typical TH system of the same load is no more than $200. So its development has received great limitation.

13.2.4.2 Double-stage photovoltaic–electrodialysis regeneration for liquid desiccant cooling systems

To further explore the PV–ED method, double-stage PV–ED regeneration systems have been presented and investigated [61]. Two regenerators are adopted; the configuration of a double-stage PV–ED regeneration system is described in Fig. 13.27: regenerators 1 and 2 respectively correspond to stages 1 and 2. In stage 1, the

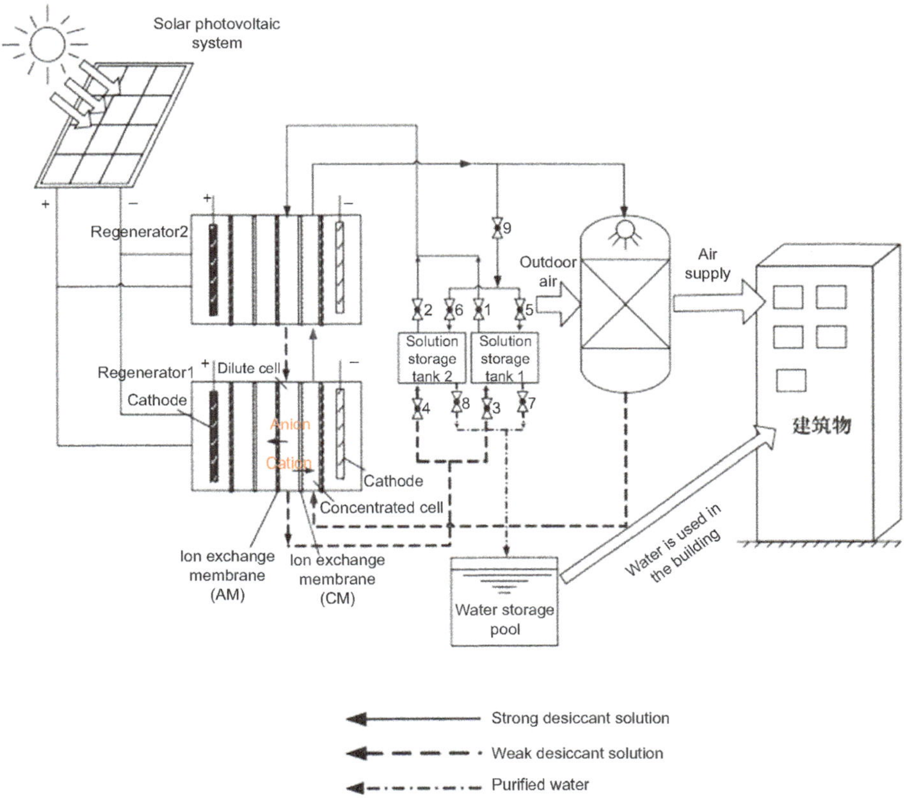

Figure 13.27 Double-stage photovoltaic−electrodialysis−liquid desiccant cooling system energy balance of the regeneration process [61].

weak desiccant solution from the dehumidifier feeds the concentrate cells of regenerator 1; the dilute cells of regenerator 1 are fed with the solution from the dilute cells of regenerator 2. In stage 2, the desiccant solution that preliminarily concentrates in regenerator 1 is then sent to the concentrate cells of regenerator 2 for further concentration; the dilute cells of regenerator 2 are fed with the desiccant solution stream from solution storage tank 1. The concentrations of the dilute and concentrate cells are both high for regenerator 2 and relatively lower for regenerator 1. In this way, the concentration difference between the dilute and the concentrate cells can be reduced.

The double-stage PV−ED regeneration system is more energy efficient than the single-stage PV−ED regeneration system. Many factors have impacts on the performance of the double-stage system, but the concentration change after regeneration is the most influential factor. Under appropriate working conditions, the double-stage system can save more than 70% energy compared to the single-stage system. These results provide useful information about the optimization of the PV−ED regeneration system.

13.2.5 Application and simulation

K. Gommed et al. [62] described a 16-kW system that was installed at the Energy Engineering Center at the Technion, in the Mediterranean city of Haifa. The system comprises a dehumidifier and a regenerator with their associated components operating together to dehumidify the fresh (ambient) air supply to a group of offices on the top floor of the building. LiCl—water is employed as the working fluid. The system is coupled to a solar collector field and employs two methods of storage—hot water and desiccant solution in the regenerated state. The performance of the system was monitored for five summer months under varying operating conditions.

The significance of this work lies in the potential to provide solar-powered cooling, dehumidification, and air conditioning for residential and commercial applications. A schematic description of the final design version of the system is given in Fig. 13.28. The system consists of six major components: an air dehumidifier or absorber, a solution regenerator or desorber, two water-to-solution HEs, a solution-to-solution HE, and an air-to-air HE. In Fig. 13.28, airflow is represented by thick solid lines, solution flow by thin solid lines, and water flow by thin dashed lines. Fig. 13.29 is a photograph of the liquid desiccant system, showing its main components. In the overall setup, the

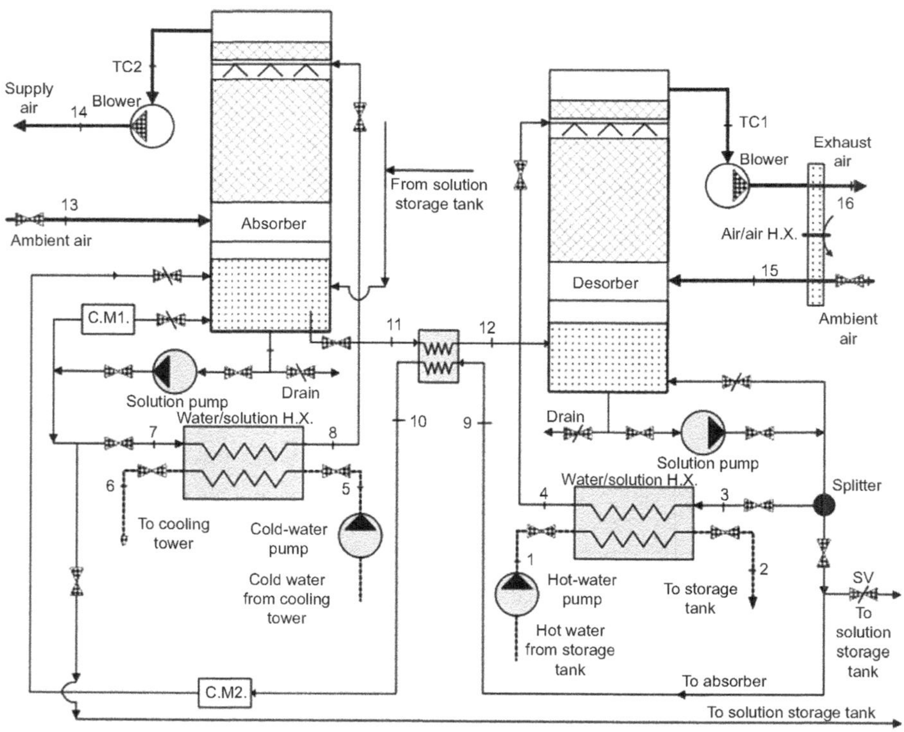

Figure 13.28 Schematic description of the liquid desiccant system. *Solid thick lines* indicate air flow; *solid thin lines* indicate solution flow; *dotted thin lines* indicate water flow [62].
H.X., Heat exchanger.

Figure 13.29 Photograph of the liquid desiccant system. (1) Absorber/dehumidifier, (2) desorber/regenerator, (3) air ducts, (4) fan, (5) rotary air/air heat exchanger, (6) control cabinet, (7) solar collector field, (8) hot water storage tank [62].

liquid desiccant system is connected in a flow arrangement allowing storage of concentrated solution and the capability to work in three different modes. The first is a manual mode used for testing individual components of the system. The other two modes are automatic, as may be selected by the user. One automatic mode is for full operation of the system and the second is for regeneration only.

The liquid desiccant system has been fully operational since April 2003, and its operation was monitored throughout the summer of 2003 (May 27 through November 6). The objective of this project has been to construct a solar-driven liquid desiccant system for cooling, dehumidification, and air conditioning, to test the concept, identify problems, and carry out preliminary design optimization and measure performance. The design process involved initial measurements to determine unknown parameters, along with extensive performance simulations. With many unknown factors, the initial design of the system underwent several changes during the development period. The characteristic performance of individual components, analyzed theoretically in an earlier simulation, was studied experimentally. The data analysis indicates a COP_{th} of about 0.8, with parasitic losses on the order of 10%. The COP calculations performed on the monitoring data have yielded satisfactory results, particularly with regard to the COP_{th}. By more elaborate design in the future, it is anticipated that parasitic losses could be minimized and better overall COPs could be achieved.

N. Audah et al. [63] studied the feasibility of using a solar-powered liquid desiccant system to meet both building cooling and freshwater needs in the Beirut humid climate using parabolic solar concentrators as a heat source for regenerating the liquid desiccant. The water condensate is captured from the air leaving the regenerator. An integrated model of a solar-powered calcium chloride liquid desiccant system for air dehumidification/humidification has been developed.

In the hot and humid climate of the Middle East, where clean fresh water is a scarce resource, and a comfortable surrounding is not naturally available, we have two major needs of occupants. These include the need to dehumidify air to meet comfort conditions and at the same time have a freshwater supply. Fig. 13.30 depicts the solar-powered liquid desiccant system (LDS) and condensing unit. The system is composed of a dehumidifier, a regenerator, parabolic solar collectors, HEs, and a desiccant cooler. The entering humid air moisture is absorbed by the desiccant. The air leaving the desiccant bed is used for household air conditioning. The liquid desiccant leaving the dehumidifier enters an HE where it is warmed up. It is further heated before entering the regenerator by another exchanger in the water storage tank heated indirectly by solar energy transmitted from the concentrators' closed fluid circuit. The hot low-concentration liquid desiccant enters the regenerator where counterflow ambient air absorbs the water accumulated from the desiccant. The regenerator-ejected humidified air is directed to a cooling coil submerged in cold seawater (temperature varying between 16°C and 20°C depending on the depth from which it is withdrawn) where condensation occurs and distilled water is collected.

To design the LDS for the identified water and air-conditioning needs, we have to calculate the space peak load conditions and identify the peak load design values for air supply temperature, humidity ratio, and air-flow rate. The maximum dehumidifier airflow rate should be equal to that design value. A case study of a typical residential space (80 m^2) in line with the practice of 40 W/m^2 maximum cooling is considered for assessing the feasibility of the solar-powered calcium chloride LDS in providing thermal comfort over the 9 h of operation and produce 15 L of freshwater over that period during the summer months of Beirut climate. The building envelope has conductance values for glass of 3.4 W/m^2 K, an overall wall conductance, $U = 1.3$ W/m^2 K, and roof conductance of 0.6 W/m^2 K. The window-to-wall ratio is 20% and the sensible heat factor is taken as 0.8.

As mentioned earlier, the critical need to be met is cooling in August (high temperature and humidity ambient air) and the most critical need to be met in June is freshwater because the air is relatively drier and at lower temperature. At the peak load of 40 W/m^2, the required supply air mass flow rate is 1.0 kg/s. The LDS simulations are run at an airflow rate of 1 kg/s for the months of June, July, August, and September based on the loads of each month and their weather data. It was found that for the month of June, with lowest ambient humidity, the amount of water produced is less than the set requirement of 15 L. A 1.205 kg/s rate is required for air mass flow rate through the LDS towers to produce the 15 L of water for the month of June. According to the design calculations for ensuring the daily delivery of 15 L during the summer season, six concentrator modules are needed as determined by the operation of the system in the month of June (the summer month of the least solar radiation) in Beirut's humid weather. The airflow rate needed for air conditioning in June is 1.18 kg/s, which is not enough for water production. So to meet both air-conditioning and water production needs, we increase our flow rate to 1.205 kg/s to achieve both needs in June.

G. Lychnos et al. [64] studied a solar-powered liquid DCS for greenhouse food production in hot climates. The concept of the cooled greenhouse system is shown

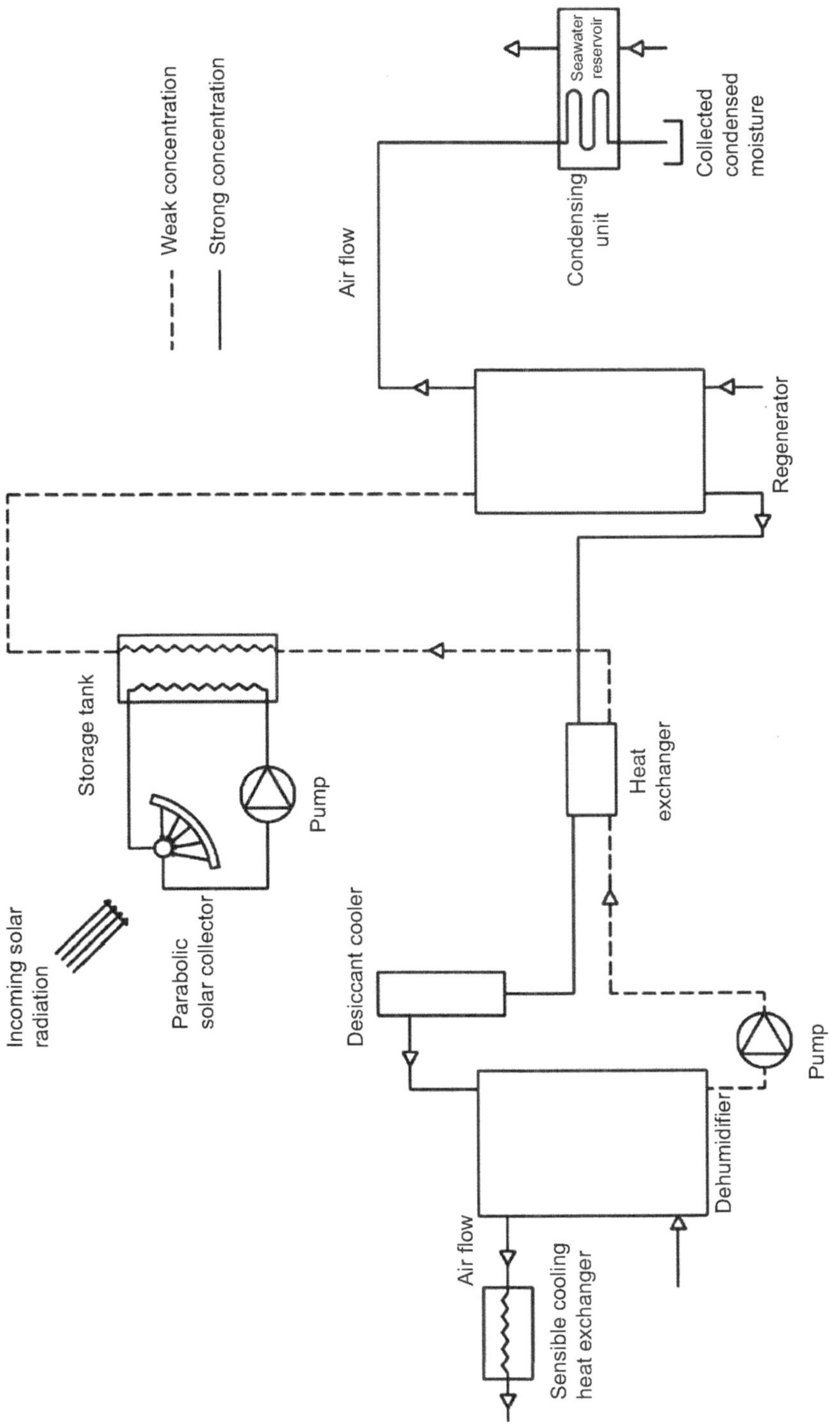

Figure 13.30 Schematic of the liquid desiccant system [63].

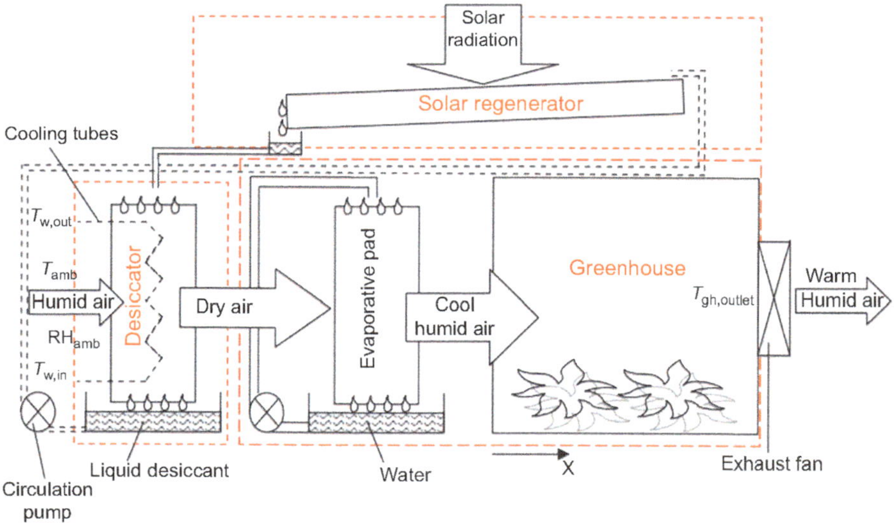

Figure 13.31 Schematic of the proposed liquid desiccant cooling system and greenhouse with evaporative pad. The desiccator pad cooling tubes are fed with water at the ambient wet bulb temperature $T_{w,in}$ [64].

schematically in Fig. 13.31. The essential parts of the system are the regenerator, the desiccator, and the greenhouse including the evaporative pad. There are three processed fluids: air, liquid desiccant, and cooling water. The desiccator and evaporative pad are both wetted porous media, permeable to air. Atmospheric air enters the system through the desiccator where it is dehumidified as it comes in contact with the liquid desiccant. The lowered humidity means that the cooling effect of the evaporative pad downstream is enhanced. As a result of absorbing water from the air, the liquid desiccant becomes less concentrated. The solar regenerator provides the necessary latent heat to drive off water from the liquid desiccant and thus restore its concentration and desiccant property. The solar regenerator may be placed on the greenhouse roof or on the ground. After leaving the regenerator, the concentrated desiccant is returned to the desiccant pad. To remove the latent heat of condensation and the heat of dilution of the liquid desiccant, the desiccant pad includes cooling tubes carrying cooling water provided at the ambient wet bulb temperature. This cooling water may be supplied from the sea or from a cooling tower. The air is drawn through the system by an exhaust fan. The aim of this study was to investigate the technical feasibility of these ideas with the help of experiments and modeling. The specific objectives were: (1) to build and test prototypes of the regenerator and desiccator in the laboratory; (2) to compare the results with the predictions of theories based on studies from the literature that used other liquid desiccants; (3) to use a computer model based on the theory, to predict the performance of the cooled greenhouse at a large scale in a range of hot climates represented by some coastal cities (Mumbai, India; Chittagong, Bangladesh; Muscat, Oman; Messina, Italy; Sfax, Tunisia; and Havana Cuba); and (4) to thus reach conclusions about what is technically possible with the system with regard to lowering of temperatures and the extension of the growing seasons of typical crops.

The regenerator is of an open type, consisting of a tilted black surface exposed to the sun with desiccant solution running over it. This is the simplest and cheapest arrangement and therefore considered potentially interesting for the greenhouse application for which large areas will be required. Based on the relevant inputs (ie, conditions and flow rates of the inlet air, liquid desiccant, and cooling water) the theoretical model predicts the mass flux of water absorption and the properties of the fluids at the outlet. The experimental regenerator consisted of a flat rectangular plate inclined at 2.5 degrees to the horizontal and measured 0.73 m long in the direction of flow, by 0.84 m wide, giving an active area of 0.61 m^2. Its surface consisted of a 6-mm-thick black neoprene foam sheet providing thermal insulation, covered by a woven black mulch sheet for better spreading of the solution over the surface. Magnesium chloride solution (ie, liquid desiccant) was pumped to a polythene distribution pile at the top of the regenerator. This pipe extended over the width of the regenerator with 21 holes 2 mm in diameter spaced at 30-mm intervals. At the regenerator outlet, the concentrated solution was collected in a rectangular transparent plastic tank. To monitor the concentration throughout the regeneration process, samples of the $MgCl_2$ solution were taken from the tank at predefined time intervals. The effectiveness of the regenerator was calculated as the ratio of the latent heat needed to evaporate water from the solution to the absorbed solar energy.

The model enables prediction of the temperature and humidity inside the greenhouse given environmental conditions (temperature, humidity, and solar irradiance) and design parameters including component sizes and flow rates of the fluids in the system. For convenience, results are presented here on the basis of a greenhouse of plan area 1000 m^2.

The climatic data required for the analysis were obtained from Meteonorm version 4.0 software. Cooling systems are usually employed in the hot seasons during the hottest hours of the day, which are normally between 1000 and 1700 h. Therefore, the hottest 3 months (with temperatures exceeding 30°C) were chosen for each location and then the average hourly data over each month of ambient temperature, relative humidity, and irradiance were fed into the model.

The simulations showed that the liquid DCS would lower the greenhouse temperature by 5.5–7.5°C in places such as Mumbai, Muscat, Chittagong, Havana, Messina, and Sfax, compared to evaporative cooling. They also showed that the required regenerator area can vary significantly (from 500 to 4000 m^2 per 1000 m^2 of greenhouse plan area) depending on the location and time of year, with larger variations at more humid places (Chittagong, Mumbai, and Havana). In some cases the large size of the open regenerator may not be practical and future work may need to consider alternative designs, eg, using glazed covers.

At times of year when the liquid DCS provides lower temperatures than needed, it could be switched off or run in evaporative cooling mode only. With appropriate control of this kind, the system could enable year-round cultivation of all the crops and at all the locations studied. At some locations the system has the capacity to cope with increased temperatures in the future.

13.3 Summary

Solar-powered DCSs have become a good alternative to conventional VCSs, and a lot of investigations have been conducted to evaluate the performance of the system. This work can be divided into two groups: separate solar rotary desiccant wheel cooling systems and hybrid systems. Within the first group, research can be subdivided based on the solar collector system. And within the second group, hybrid systems can be classified by whether a conventional VCS is adopted.

Though there has already been a large amount of work conducted on solar-powered desiccant wheel cooling systems, it is still possible to make progress especially in following aspects:

1. Almost 80% of the reviewed work was conducted based on simulation models. This is effective to obtain performance under various conditions and to analyze the feasibility in the initial phase.
2. Most of the work is done with constant structural parameters; there is rare work conducted on obtaining optimized performance and parameters with respect to different conditions.
3. For solar desiccant wheel air-conditioning systems, especially for the hybrid systems, there exist various operation modes: separate desiccant cooling mode, separate VC mode, hybrid desiccant and VC mode, solar regeneration mode, auxiliary regeneration mode, etc. Until now, simple control methods based on the temperature and humidity ratio have been proposed. Much more complex and accurate control methods linking to all parameters are needed especially for actual systems.

Acknowledgment

The authors acknowledge the financial support from the State Key Program of National Natural Science of China (Grant No. 51336004).

References

[1] Daou K, Wang R, Xia Z. Desiccant cooling air conditioning: a review. Renewable Sustainable Energy Rev 2006;10:55–77.
[2] La D, Dai YJ, Li Y, Wang RZ, Ge TS. Technical development of rotary desiccant dehumidification and air conditioning: a review. Renewable Sustainable Energy Rev 2010;14: 130–47.
[3] Ge TS, Li Y, Wang RZ, Dai YJ. A review of the mathematical models for predicting rotary desiccant wheel. Renewable Sustainable Energy Rev 2008;12:1485–528.
[4] Misha S, Mat S, Ruslan MH, Sopian K. Review of solid/liquid desiccant in the drying applications and its regeneration methods. Renewable Sustainable Energy Rev 2012;16: 4686–707.
[5] Preisler A, Brychta M. High potential of full year operation with solar driven desiccant evaporative cooling systems. Energy Procedia 2012;30:668–75.
[6] Joudi KA, Dhaidan NS. Application of solar assisted heating and desiccant cooling systems for a domestic building. Energy Convers Manag 2001;42:995–1022.

[7] Guidara Z, Elleuch M, Ben Bacha H. New solid desiccant solar air conditioning unit in Tunisia: design and simulation study. Appl Therm Eng 2013;58:656−63.
[8] Enteria N, Yoshino H, Mochida A, Satake A, Yoshie R, Takaki R, et al. Performance of solar-desiccant cooling system with silica-gel (SiO_2) and titanium dioxide (TiO_2) desiccant wheel applied in East Asian climates. Sol Energy 2012;86:1261−79.
[9] Enteria N, Yoshino H, Mochida A, Takaki R, Satake A, Yoshie R, et al. Construction and initial operation of the combined solar thermal and electric desiccant cooling system. Sol Energy 2009;83:1300−11.
[10] Enteria N, Yoshino H, Satake A, Mochida A, Takaki R, Yoshie R, et al. Development and construction of the novel solar thermal desiccant cooling system incorporating hot water production. Appl Energy 2010;87:478−86.
[11] Enteria N, Yoshino H, Takaki R, Yonekura H, Satake A, Mochida A. First and second law analyses of the developed solar-desiccant air-conditioning system (SDACS) operation during the summer day. Energy Build 2013;60:239−51.
[12] Enteria N, Yoshino H, Satake A, Mochida A, Yoshie R, Mizutani K. Numerical evaluation and optimization of the combined solar thermal and electric desiccant cooling system. In: ASME 2009 International Mechanical Engineering Congress and Exposition. American Society of Mechanical Engineers; 2009. p. 181−92.
[13] White SD, Kohlenbach P, Bongs C. Indoor temperature variations resulting from solar desiccant cooling in a building without thermal backup. Int J Refrig 2009;32: 695−704.
[14] Bourdoukan P, Wurtz E, Joubert P, Spérandio M. Potential of solar heat pipe vacuum collectors in the desiccant cooling process: modelling and experimental results. Sol Energy 2008;82:1209−19.
[15] Maclaine-Cross IL, Banks PJ. Coupled heat and mass transfer in regenerators—prediction using an analogy with heat transfer. Int J Heat Mass Transfer 1972;15:1225−42.
[16] Stabat P, Marchio D. Heat-and-mass transfers modelled for rotary desiccant dehumidifiers. Appl Energy 2008;85:128−42.
[17] Bourdoukan P, Wurtz E, Joubert P. Experimental investigation of a solar desiccant cooling installation. Sol Energy 2009;83:2059−73.
[18] Ge TS, Dai YJ, Wang RZ, Li Y. Experimental investigation on a one-rotor two-stage rotary desiccant cooling system. Energy 2008;33:1807−15.
[19] Ge TS, Li Y, Wang RZ, Dai YJ. Experimental study on a two-stage rotary desiccant cooling system. Int J Refrig 2009;32:498−508.
[20] Ge TS, Ziegler F, Wang RZ, Wang H. Performance comparison between a solar driven rotary desiccant cooling system and conventional vapor compression system (performance study of desiccant cooling). Appl Therm Eng 2010;30:724−31.
[21] Fong KF, Chow TT, Lin Z, Chan LS. Simulation−optimization of solar-assisted desiccant cooling system for subtropical Hong Kong. Appl Therm Eng 2010;30:220−8.
[22] Hatami Z, Saidi MH, Mohammadian M, Aghanajafi C. Optimization of solar collector surface in solar desiccant wheel cycle. Energy Build 2012;45:197−201.
[23] Kabeel AE. Solar powered air conditioning system using rotary honeycomb desiccant wheel. Renewable Energy 2007;32:1842−57.
[24] Li H, Dai YJ, Li Y, La D, Wang RZ. Case study of a two-stage rotary desiccant cooling/heating system driven by evacuated glass tube solar air collectors. Energy Build 2012;47: 107−12.
[25] Li H, Dai YJ, Köhler M, Wang RZ. Simulation and parameter analysis of a two-stage desiccant cooing/heating system driven by solar air collectors. Energy Convers Manag 2013;67:309−17.

[26] Elzahzby AM, Kabeel AE, Bassuoni MM, Abdelgaied M. A mathematical model for predicting the performance of the solar energy assisted hybrid air conditioning system, with one-rotor six-stage rotary desiccant cooling system. Energy Convers Manag 2014;77: 129–42.
[27] Mei L, Infield D, Eicker U, Loveday D, Fux V. Cooling potential of ventilated PV façade and solar air heaters combined with a desiccant cooling machine. Renewable Energy 2006; 31:1265–78.
[28] Beggs C, Sleigh P, Halliday S. Solar desiccant cooling model. In: DETR seminar (desiccant and solar desiccant systems). London: CIBSE; April 6, 2000.
[29] Mavroudaki P, Beggs CB, Sleigh PA, Halliday SP. The potential for solar powered single-stage desiccant cooling in southern Europe. Appl Therm Eng 2002;22:1129–40.
[30] Halliday SP, Beggs CB, Sleigh PA. The use of solar desiccant cooling in the UK: a feasibility study. Appl Therm Eng 2002;22:1327–38.
[31] Finocchiaro P, Beccali M, Nocke B. Advanced solar assisted desiccant and evaporative cooling system equipped with wet heat exchangers. Sol Energy 2012;86:608–18.
[32] Baniyounes AM, Liu G, Rasul MG, Khan MMK. Analysis of solar desiccant cooling system for an institutional building in subtropical Queensland, Australia. Renewable Sustainable Energy Rev 2012;16:6423–31.
[33] Baniyounes AM, Rasul MG, Khan MMK. Experimental assessment of a solar desiccant cooling system for an institutional building in subtropical Queensland, Australia. Energy Build 2013;62:78–86.
[34] Fong KF, Lee CK, Chow TT, Fong AML. Investigation on solar hybrid desiccant cooling system for commercial premises with high latent cooling load in subtropical Hong Kong. Appl Therm Eng 2011;31:3393–401.
[35] Khalid A, Mahmood M, Asif M, Muneer T. Solar assisted, pre-cooled hybrid desiccant cooling system for Pakistan. Renewable Energy 2009;34:151–7.
[36] Hürdoğan E, Büyükalaca O, Yılmaz T, Hepbasli A, Uçkan İ. Investigation of solar energy utilization in a novel desiccant based air conditioning system. Energy Build 2012;55: 757–64.
[37] La D, Dai Y, Li Y, Ge T, Wang R. Case study and theoretical analysis of a solar driven two-stage rotary desiccant cooling system assisted by vapor compression air-conditioning. Sol Energy 2011;85:2997–3009.
[38] Cejudo López JM, Hernández FF, Muñoz FD, Andrés AC. The optimization of the operation of a solar desiccant air handling unit coupled with a radiant floor. Energy Build 2013;62:427–35.
[39] Beccali M, Finocchiaro P, Nocke B. Energy performance evaluation of a demo solar desiccant cooling system with heat recovery for the regeneration of the adsorption material. Renewable Energy 2012;44:40–52.
[40] Fong KF, Chow TT, Lee CK, Lin Z, Chan LS. Solar hybrid cooling system for high-tech offices in subtropical climate – radiant cooling by absorption refrigeration and desiccant dehumidification. Energy Convers Manag 2011;52:2883–94.
[41] Beccali M, Finocchiaro P, Nocke B. Energy and economic assessment of desiccant cooling systems coupled with single glazed air and hybrid PV/thermal solar collectors for applications in hot and humid climate. Sol Energy 2009;83:1828–46.
[42] Fong KF, Chow TT, Lee CK, Lin Z, Chan LS. Advancement of solar desiccant cooling system for building use in subtropical Hong Kong. Energy Build 2010;42:2386–99.
[43] Jain S, Bansal PK. Performance analysis of liquid desiccant dehumidification systems. Int J Refrig 2007;30:861–72.

[44] Mei L, Dai YJ. A technical review on use of liquid-desiccant dehumidification for air-conditioning application. Renewable Sustainable Energy Rev 2008;12:662–89.
[45] Katejanekarn T, Kumar S. Performance of a solar-regenerated liquid desiccant ventilation pre-conditioning system. Energy Build 2008;40:1252–67.
[46] Abdel-Salam AH, Ge G, Simonson CJ. Thermo-economic performance of a solar membrane liquid desiccant air conditioning system. Sol Energy 2014;102:56–73.
[47] Tu M, Ren C-Q, Zhang L-A, Shao J-W. Simulation and analysis of a novel liquid desiccant air-conditioning system. Appl Therm Eng 2009;29:2417–25.
[48] Xiong ZQ, Dai YJ, Wang RZ. Development of a novel two-stage liquid desiccant dehumidification system assisted by $CaCl_2$ solution using exergy analysis method. Appl Energy 2010;87:1495–504.
[49] Xiong ZQ, Dai YJ, Wang RZ. Investigation on a two-stage solar liquid-desiccant (LiBr) dehumidification system assisted by $CaCl_2$ solution. Appl Therm Eng 2009;29:1209–15.
[50] Li Y, Lu L, Yang H. Energy and economic performance analysis of an open cycle solar desiccant dehumidification air-conditioning system for application in Hong Kong. Sol Energy 2010;84:2085–95.
[51] Niu X, Xiao F, Ge G. Performance analysis of liquid desiccant based air-conditioning system under variable fresh air ratios. Energy Build 2010;42:2457–64.
[52] Liu XH, Zhang Y, Qu KY, Jiang Y. Experimental study on mass transfer performances of cross flow dehumidifier using liquid desiccant. Energy Convers Manag 2006;47:2682–92.
[53] Dai YJ, Zhang HF. Numerical simulation and theoretical analysis of heat and mass transfer in a cross flow liquid desiccant air dehumidifier packed with honeycomb paper. Energy Convers Manag 2004;45:1343–56.
[54] Elsarrag E. Evaporation rate of a novel tilted solar liquid desiccant regeneration system. Sol Energy 2008;82:663–8.
[55] Alizadeh S, Saman WY. An experimental study of a forced flow solar collector/regenerator using liquid desiccant. Sol Energy 2002;73:345–62.
[56] Dai YJ, Wang RZ, Zhang HF, Yu JD. Use of liquid desiccant cooling to improve the performance of vapor compression air conditioning. Appl Therm Eng 2001;21:1185–202.
[57] Howell JR, Peterson JL. Preliminary performance evaluation of a hybrid vapor-compression/ liquid desiccant air conditioning system. New York, NY: American Society of Mechanical Engineers; 1987.
[58] Parsons BK, et al. Improving gas-fired heat pump capacity and performance by adding a desiccant dehumidification subsystem 1990.
[59] Khalid Ahmed CS, Gandhidasan P, Al-Farayedhi AA. Simulation of a hybrid liquid desiccant based air-conditioning system. Appl Therm Eng 1997;17:125–34.
[60] Li X-W, Zhang X-S. Photovoltaic–electrodialysis regeneration method for liquid desiccant cooling system. Sol Energy 2009;83:2195–204.
[61] Li X-W, Zhang X-S, Quan S. Single-stage and double-stage photovoltaic driven regeneration for liquid desiccant cooling system. Appl Energy 2011;88:4908–17.
[62] Gommed K, Grossman G. Experimental investigation of a liquid desiccant system for solar cooling and dehumidification. Sol Energy 2007;81:131–8.
[63] Audah N, Ghaddar N, Ghali K. Optimized solar-powered liquid desiccant system to supply building fresh water and cooling needs. Appl Energy 2011;88:3726–36.
[64] Lychnos G, Davies PA. Modelling and experimental verification of a solar-powered liquid desiccant cooling system for greenhouse food production in hot climates. Energy 2012;40: 116–30.

Other types of solar-powered cooling systems

14

X. Zheng, T.S. Ge
Institute of Refrigeration and Cryogenics, Shanghai Jiao Tong University, Shanghai, China

14.1 Introduction

In terms of solar thermal-driven air-conditioning cycles, in addition to the common solar-driven solid adsorption cycles, liquid absorption cycles, and desiccant cooling cycles, there are also some other types of solar-powered cooling systems in which solar-driven ejector cooling cycles and Rankine cycles are recently drawing more and more attention. This chapter aims to conclude the operation principles and recent development with respect to these two types of solar-powered cooling systems.

14.2 Other types of solar-powered cooling systems

14.2.1 Solar-driven ejector cooling technology

14.2.1.1 Operation principle

An ejector refrigeration system can be considered as a modification of a conventional vapor compression cycle (VCC). An ejector takes the place of a compressor to pressurize the refrigerant vapor flowing from an evaporator and discharge it to a condenser. The conventional ejector refrigeration cycle is shown in Fig. 14.1. Working fluid is heated at a high pressure and temperature in the generator. High-pressure refrigerant vapor enters the nozzle. Working fluid is then accelerated to a high velocity and entrains motive steam from the evaporator, resulting in a cooling effect. After that, mixed vapor steams are discharged from the nozzle to the condenser where they are cooled down and condensed to liquid fluids. A part of the liquid refrigerant returns to the evaporator through an expansion valve whereas the other part is pumped to the generator. Ejector cooling technology can be used for air conditioning in trains and large buildings [1]. However, ejector refrigeration systems always have a smaller coefficient of performance (COP) compared with that of vapor compression systems, but it can be more practical and economical when waste heat, solar energy, or exhaust heat are used to provide heat to the generator of an ejector system [2].

Advances in Solar Heating and Cooling. http://dx.doi.org/10.1016/B978-0-08-100301-5.00014-X

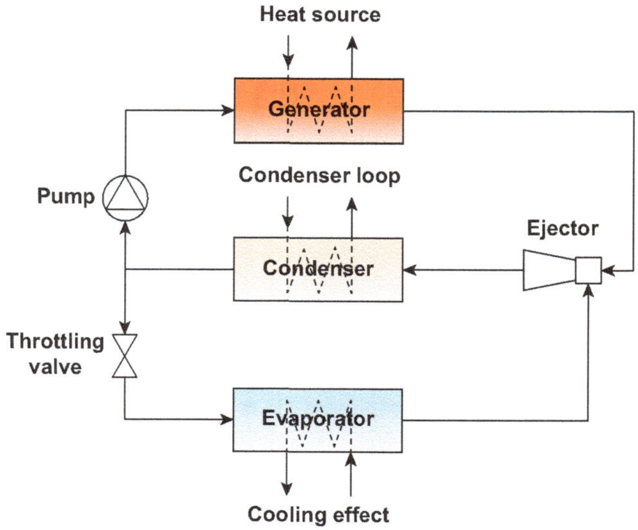

Figure 14.1 The schematic of an ejector refrigeration system.

14.2.1.2 Development of this technology

In recent years there has been a rapid growth in ejector cooling systems driven by solar heat. Solar-driven ejector refrigeration technologies cannot only satisfy cooling requirements such as air conditioning, ice-making, and medical or food preservation in remote areas, but they can also meet the demand for energy conservation and environment protection [3–6]. To improve the performance of solar-driven ejector refrigeration systems, various designs have been carried out, such as single-stage ejector systems, multistage ejector systems, or hybrid systems (in which the ejector technology is combined with other technologies). Numerous researches have concentrated on theoretical calculation, computer simulation, or experimental work to seek for a more efficient solar-driven ejector refrigeration system.

Single-stage solar-driven ejector refrigeration systems are one of the most investigated system types. Many researches and analyses have been focused on this single-stage type. Fig. 14.2 shows the diagram of a classic single-stage solar-driven ejector refrigeration system. It can be seen that the whole system consists of three subsystems: a solar collector–thermal storage subsystem, a thermal storage–generator subsystem, and an ejector subsystem. In the solar collector-thermal storage subsystem, water is often used as a heating medium. Working fluid is heated to become high temperature and pressure vapor in the generator by medium from the thermal storage–generator subsystem. The ejector subsystem is the same with the conventional ejector cooling system discussed in Fig. 14.1.

Many ejector refrigeration experimental rigs have been constructed and tested all over the world. Chlorofluorocarbon and hydrochlorofluorocarbon refrigerants were often used in the ejector subsystems [2,7–11]. Nowadays more environmentally friendly refrigerants are recommended, such as the hydrocarbon refrigerant [12] or

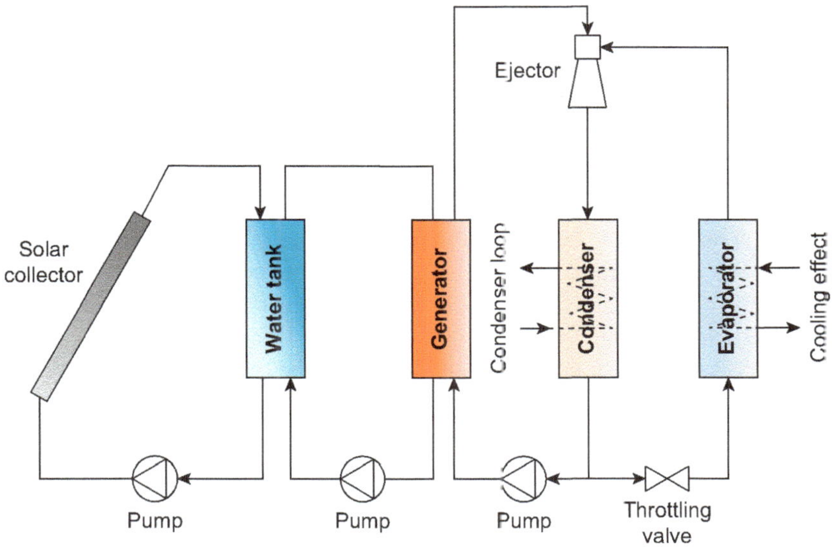

Figure 14.2 The schematic of a classic solar-driven ejector refrigeration system.

hydrofluoroether refrigerant [13]. Nguyen et al. performed experiments on a solar-powered passive ejector cooling system [6]. An evacuated-tube solar collector was used with water as its heating medium. This system was designed to provide a cooling capacity of 7 kW during summer and heating capacity of 20 kW during winter. Later, Alexis and Karayiannis [2] introduced R134a as a refrigerant for a solar-driven ejector system. The system was predicted to operate from May to September in conjunction with intermediate temperature solar collectors in Athens. Experimental results showed that the COP of the ejector cooling system varied from 0.035 to 0.199 when the temperature in the generator ranged between 82°C and 92°C, the temperature in the evaporator ranged from −10°C to 0°C, and the temperature in the condenser from 32°C to 40°C. In addition, the COP of the overall system was found to be in the range of 0.014−0.101 under the same operation conditions and total solar radiation of 536−838 W/m^2 in July. In 2007 Balaras et al. [14] reported a much greater COP of 0.85 for a pilot steam ejector plant. However, this desired high performance was only possible when the heat source temperature was no lower than 200°C. In addition, Noeres [15] investigated the possibilities for further development of combined heat, cooling, and power production with steam jet ejector chillers. A simple construction of these ejector systems can be considered as a great advantage, but their COP is still far from comparable with the other heat-driven technologies. More recently, Smierciew et al. [16] constructed a special prototype standing for a solar-driven ejector air-conditioning system in which isobutane was used as a working fluid. Experimental investigations confirmed that the ejector cycle may be effectively driven by a low-temperature heat source of approximately 80°C, which can be obtained from either solar panels or waste heat sources. In addition, the effects of operation parameters, ejector cycle performance coefficient, and loss coefficient for the ejector were thoroughly analyzed and discussed. Experimental results showed that under this temperature range the designed ejector cycles may effectively compete with absorption refrigeration systems. Likewise, Zhang et al. [17] designed a solar-driven ejector air-conditioning system with a nominal cooling capacity of 5 kW. Test results showed that a COP of up to 0.32 could be obtained for the ejector cooling system when the temperature in the boiler was approximately 90°C. Tashtoush et al. [18] proposed a simulation program based on TRNSYS and EES software to design solar collector subsystem components and to evaluate the performance of the solar ejector cooling system with R134a as a refrigerant. Under peak solar radiation and highest ambient temperature, the COP of the ejector cooling cycle was approximately 0.52–0.547 and the solar collector efficiency ranged between 0.52 and 0.92. In addition, the overall system efficiency was reported to be no less than 0.32, and the overall COP was in the range of 0.32–0.47.

Some researchers have concentrated on the ejector configuration design for an ejector playing a significant role in an ejector cooling system. Saito et al. [19] proposed a way to develop an appropriate design method of ejector configuration by applying computational fluid dynamics to study the behavior of the refrigerant in an ejector. Four types of configuration parameters in terms of a mixing section area and a nozzle exit area of an ejector were discussed in their work. Results showed that an appropriate mixing section area can increase cooling capacity but decrease condensing temperature whereas a smaller nozzle exit area can make the system more reliable but have lower cooling capacity. Dennis et al. [20] also pointed out that an ejector is a key to obtain high efficiency of the solar-driven ejector cooling system cycle, especially when operating conditions stray from designed points. A primary nozzle with variable throat diameter and variable nozzle exit diameter was proposed. Modeling results showed that the operating range of an ejector can be increased. Moreover, the ejector with a variable primary nozzle throat diameter can greatly enhance ejector performance via experimental tests.

On the basis of these studies, it seems that the main technical problem for solar refrigeration systems is that they highly rely on environmental factors such as air temperature, solar radiation, wind speed, and so on [21,22]. In particular, a severe restriction for a solar-driven ejector system is the heat rejection temperature. Although these single-stage solar-powered ejector cooling systems are simple, variation of working conditions such as ambient temperatures above design conditions or low sunlight will always have bad influences on these systems, making them deviate from optimum operating conditions [3,23]. To solve this problem, some researchers proposed multistage ejector systems [24,25]. In the 1990s Bejan et al. [24] placed several ejectors in parallel connection before the condenser. Each time, only one ejector operated according to the corresponding condenser pressure. Wang et al. [26] proposed a theoretical study on a solar-assisted two-stage ejector refrigeration system. Two injectors in series connections were designed to improve its COP. Different working fluids (ie, R134a, R152a, R245ca, and R227ea) were investigated in this two-stage system. Results showed that the total ejector coefficient was influenced by changing the two-stage compression ratio of working fluids. Later, Wang et al. [27] calculated the exergy loss rate of each component in this two-stage ejector refrigeration system and analyzed the influence of generator temperature and evaporation temperature on exergy. It was reported that maximum exergy loss existed in the solar collector with the loss rate of 47.67%. In addition, when other conditions were fixed, an optimal generator temperature could be found to minimize the total exergy loss.

However, these solar-driven ejector cooling systems cannot yet be ready to compete with the well-known vapor compression systems because of low energy efficiency. Intensive research still needs to be done for the replacement of conventional refrigeration machines and large-scale applications in industry. Thus many combined systems have been investigated to improve its performance by selectively combining different technologies or creation of new functions [28–31]. There are typically two kinds of combined ejector refrigeration systems: (1) an ejector refrigeration system combined with a booster or compressor and (2) an ejector refrigeration system connected to an adsorption or an absorption refrigeration system.

There are some research groups that have focused on the development of combined solar-assisted ejector cooling/heating systems (SACHs) for years. For example, Huang's group in the New Energy Center at National Taiwan University developed two types of SACH systems in 2010, named SACH-1 and SACH-2 [32,33]. Both SACH systems were made up of a usual inverter-type air conditioner, which is made of a variable-speed compressor. It can be in series or parallel connection with a solar-driven ejector cooling system. The SACH-1 system was in series configuration whereas the SACH-2 was in parallel configuration, as shown in Figs. 14.3–14.4. SACH-1 integrated a solar-driven pumpless ejector cooling system with an inverter-type air conditioner to provide a stable space cooling. The ejector cooling system was used to subcool the condenser in the inverter-type air conditioner. In this way COP can be increased and the power input of the compressor can be reduced by automatically regulating the rotational speed. Experimental results showed that with the use of the ejector cooling system in SACH-1, the condensing temperature of the heat pump can decrease by 7.3–12.6°C and the input power of the compressor can also decrease by 34.5–81.2%. In addition to the connection difference, a control

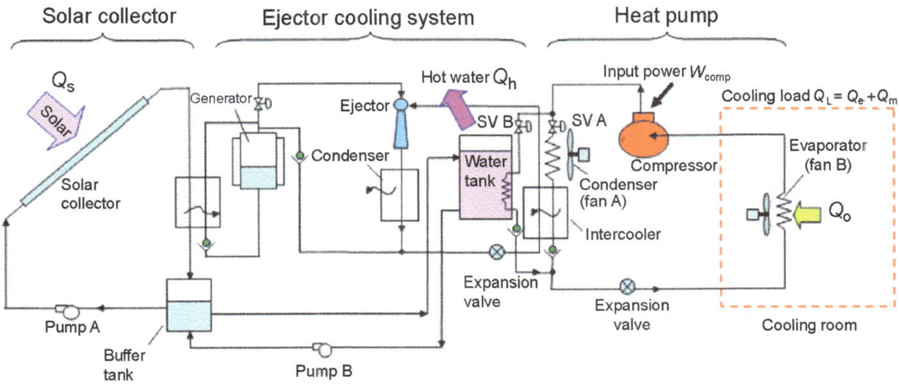

Figure 14.3 The schematic diagram of a solar-assisted ejector cooling/heating system in series configuration (SACH-1) [32].

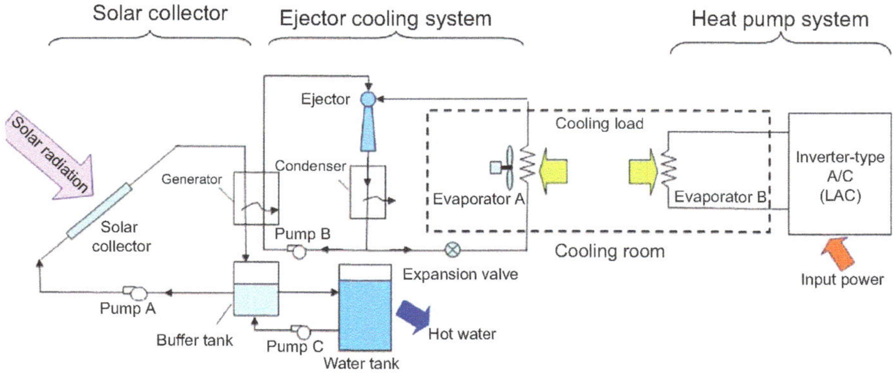

Figure 14.4 The schematic diagram of a solar-assisted cooling/heating system in parallel configuration (SACH-2) [33].

system was also added to SACH-2 to control the evaporating temperature of the ejector cooling system so that ejector failure can be avoided in variable weather. Long-term performance testing results displayed that the daily energy saving was approximately 30–70% compared with the energy consumption of an inverter-type air conditioner without a solar-driven ejector cooling system. However, in hot weather the parallel configuration in SACH-2 may cause the ejector cooling system to operating at a very low COP, resulting in very low system efficiency of SACH-2. Thus this group presented a further study on the improvement of SACH-2, named SACH-k2 [34]. SACH-k2 used series configuration and a refrigerant pump in the ejector cooling system, as shown in Fig. 14.5. In addition, advanced control technologies were also applied to ensure stable operation of the ejector cooling system and to minimize power consumption of peripherals. Test results showed that the overall system COP including power consumption of peripherals was approximately 33–43% higher that of a single inverter-type air conditioner, and the highest COP can be as high as 4.5.

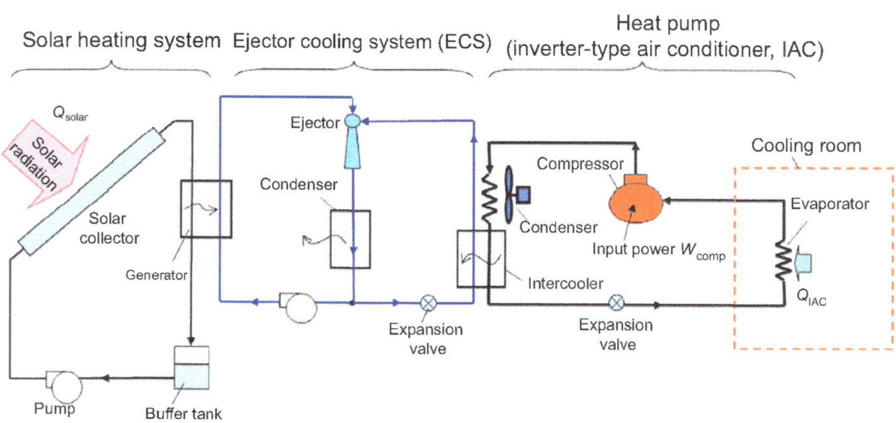

Figure 14.5 The schematic diagram of a modified solar-assisted cooling/heating system SACH-k2 [34].

In the early 21st century, Wang's group [35,36] investigated several combined ejector-sorption systems. Zhang and Wang [36] presented a theoretical analysis of solar-driven adsorption-ejector refrigeration and heating systems with zeolite–water as a working pair in the adsorption system. Simulation results showed that this combined hybrid system could reach the COP of 0.33 and could furnish 290 kg of hot water at 45°C for family use. Vereda et al. [37] presented a numerical model of an ejector-absorption refrigeration cycle with ammonia–lithium nitrate solution as working fluid. Simulation results showed that the use of an ejector was able to decrease the regeneration temperature by approximately 9°C in comparison to a conventional absorption cycle and to increase the COP for moderate temperatures. Ammonia–water is also a common used working pair in the absorption cooling cycle. Abdulateef et al. [38–40] reported a comprehensive investigation of a combined solar-driven ejector and ammonia–water absorption refrigeration system, including experimental study [38], economic analysis [39], and evaluation [40]. The schematic diagram of this combined ejector–absorption refrigeration cycle is shown in Fig. 14.6. Later, Sirwan et al. [41] made a modification in this combined solar-driven absorption-ejector cooling system by adding a removable flash tank between the condenser and the evaporator. The schematic diagram of this modified cycle is displayed in Fig. 14.7. Simulation results showed a significant improvement in the mean COP of the modified cycle. In addition, by adding the flash tank, the optimized ejector entertainment ratio and increased condenser pressure can be achieved, enabling the system to perform well in hot climates. Moreover, Abed et al. [42] made a further enhancement on the basis of Sirwan et al.'s [41] work by rearrangement of solution streamlines and adding a refrigerant heat exchanger (RHE), as displayed in Fig. 14.8. After this modification, the total COP of the system was found to be 12.2% higher, in which the COP increment by

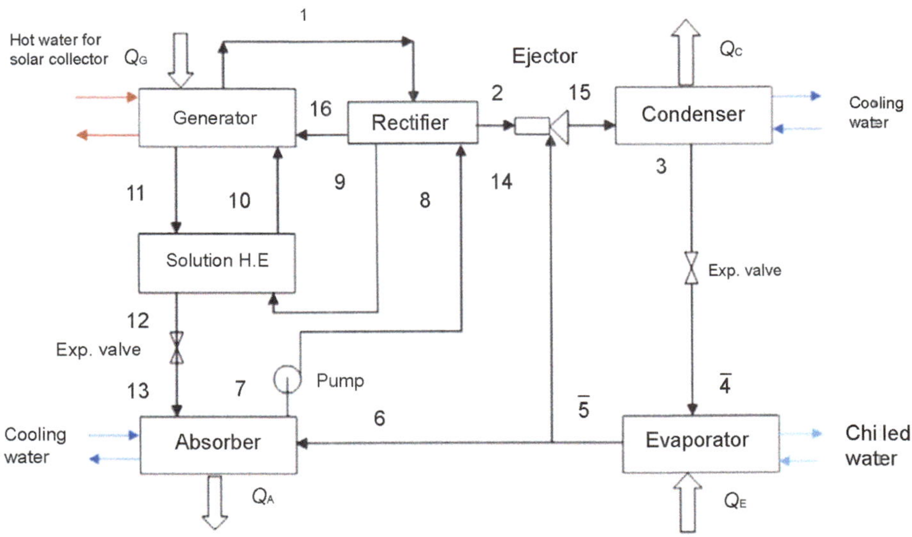

Figure 14.6 A combined ejector-absorption refrigeration cycle [38].

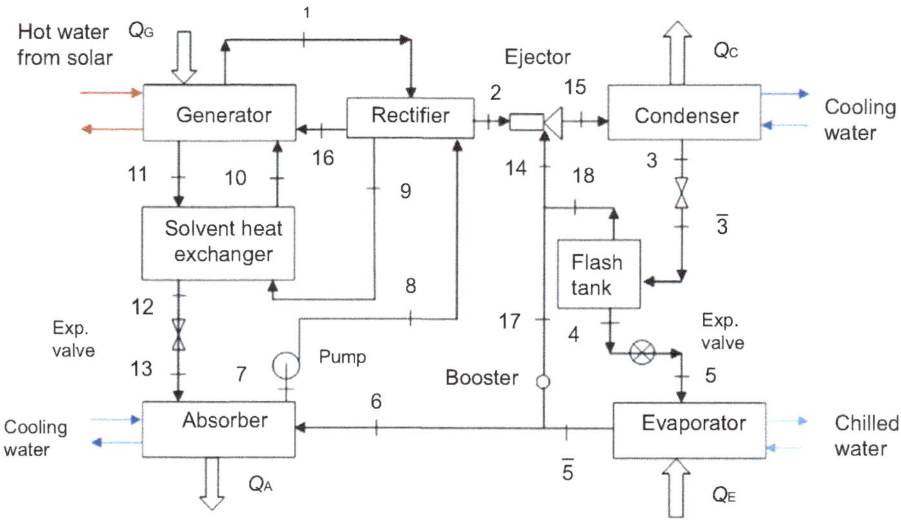

Figure 14.7 Combined flash tank—absorption—ejector refrigeration cycle [41].

rearrangement of the solution streamlines and by adding the RHE were 8% and 4.2%, respectively. In addition, cooling capacity was reported to increase by 4.85% with the adding of the RHE.

14.2.2 Solar-driven Rankine cycle

14.2.2.1 Operation principle

Solar-driven Rankine engines were first proposed in 1874 and offered an alternative means of electricity generation [43,44]. Classical solar-driven Rankine systems belong to solar thermomechanical refrigeration systems by converting solar power into a mechanical work through Rankine power machines to drive a conventional vapor compression refrigeration system [45]. The schematic diagram of a classical solar-powered organic Rankine system is presented in Fig. 14.9. Similar to a solar-driven ejector cooling system, a solar-powered organic Rankine system is also made up of three parts: a solar collector—thermal storage subsystem, a thermal storage—generator subsystem, and a Rankine subsystem. Heat was obtained from the solar collector to increase the temperature of heating medium in the solar collector subsystem. The solar collector subsystem could be combined with an auxiliary heater in low-lying places where solar energy is limited. Working fluid in the Rankine subsystem is heated to become high-temperature and -pressure vapor in the generator by means of the thermal storage—generator subsystem. The turbine is then driven by the high-pressure vapor that is connected to an electrical generator. Lower-pressure vapor is then released from the turbine into a condenser and cooled to liquid. Liquid refrigerant is finally pumped back into the vapor generator. In general, working fluids used in Rankine cycles are environmentally friendly refrigerants.

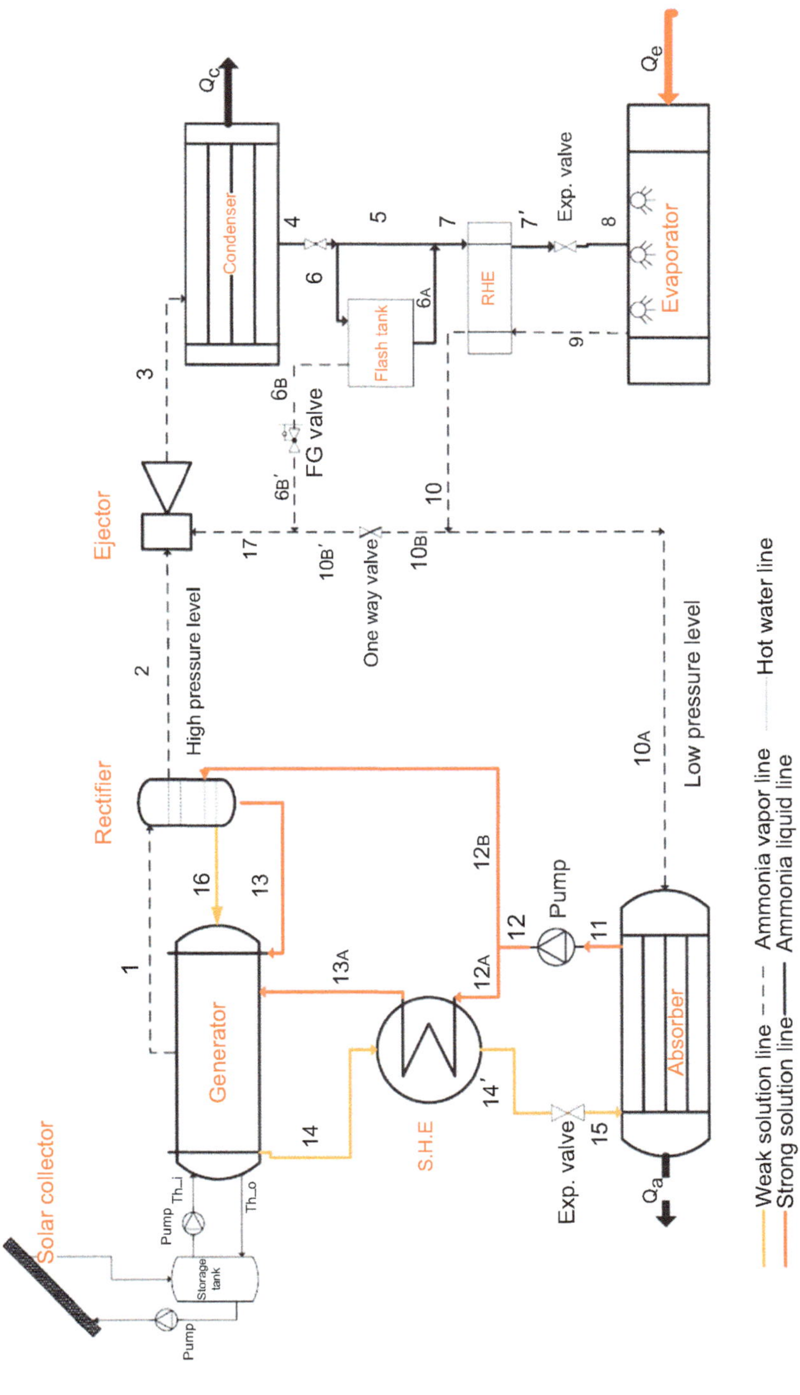

Figure 14.8 The schematic diagram of the further modified absorption–ejector refrigeration cycle [42].

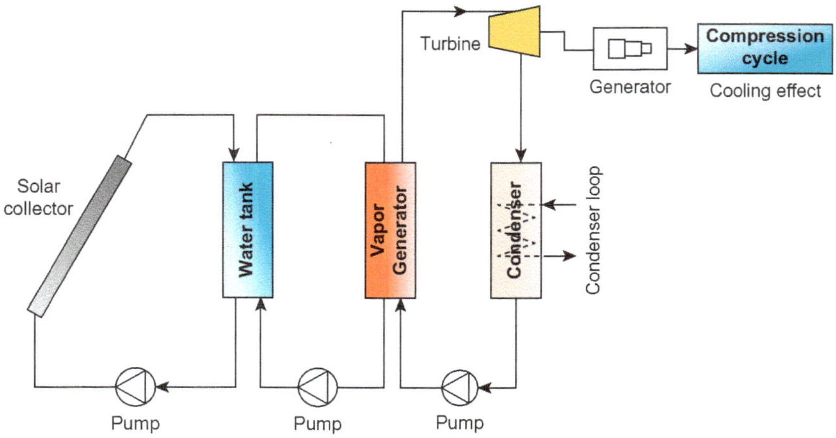

Figure 14.9 Solar-driven Rankine cooling cycle.

14.2.2.2 Development of this technology

A solar-powered organic Rankine cycle (ORC) with small-scale capacity has a lower working temperature, higher energy conversion efficiency, and less harmful impact on the environment when compared with a steam Rankine cycle. In the 1970s and 1980s, intensive research had been done to investigate this method of solar energy conversion for indoor cooling [46–50]. Among this research, Biancardi et al. [49] reported a design, development, and test of a solar-driven Rankine cycle heat pump prototype. R11 was used as the working fluid in the solar-powered chiller, and the COP of the heat pump ranged between 1.5 and 3.0. However, the development of these solar-powered Rankine cooling systems stagnated until the beginning of the 21st century, when environmental concerns and fossil energy depletion became more serious. Bu et al. [51] developed a solar energy–powered ORC/VCC ice maker. The system schematic was shown in Fig. 14.10. Four working fluids (ie, R123, R245fa, R600a, and R600) were used respectively in the system. The effects of parameters such as generation temperature and condensation temperature on overall performance of the system were analyzed. Results showed that R600 and R600a were more suitable working fluids for ORC considering power efficiency and expander size. R600a and R600 are also more appropriate for VCC in terms of pressure ratio and COP. However, if taking overall efficiency and ice production per square meter collector per day into consideration, R123 is the most suitable one for ORC/VCC. Later, Schimpf and Span [52] discussed the economic feasibility of a solar-assisted combined ground-source heat pump–ORC system and compared it to a conventional combined solar system coupled with a ground-source heat pump. Additional investments for the ORC system were evaluated to be approximately 400€ for a single-family house unit, including a pump, valves, and upgraded controls. Although isobutane as a working fluid can be more interesting because of lower global warming potential and smaller saturation pressures than R134a, the latter could cut down the additional investments for the ORC to 200€ because of a mass-produced rotary vane pump of

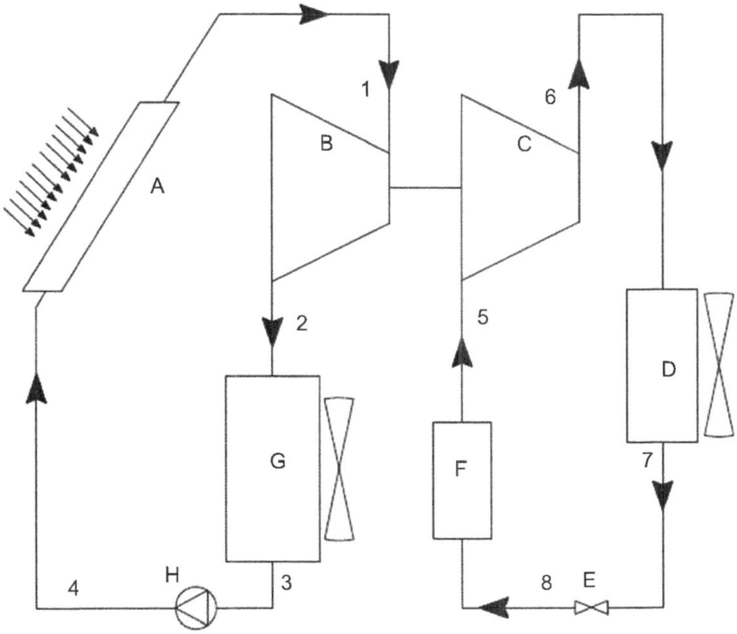

Figure 14.10 The system schematic of an organic Rankine cycle/vapor compression cycle. A, solar collector (generator); B, expander; C, compressor; D, condenser for cooling system; E, throttle valve; F, evaporator for ice maker; G, condenser for power system; H, working fluid pump [51].

R134a. Researchers believed that the ORC could be economically reasonable at all locations in Germany.

In addition to a conventional vapor compression refrigeration cycle, an ejector cooling cycle [53], an absorption condensation process [54,55], etc., have also been studied. For example, Oliveira et al. [53] proposed a novel combined system involving an ejector cycle and a Rankine cycle to provide cooling/heating and electricity generation for buildings, and environmentally friendly *N*-pentane was used as refrigerant. It was driven by solar energy and supplemented by a gas burner. The system was proven to be viable and reliable, and it could provide cooling/heating and electricity generation for buildings. Xu et al. [54] developed a combined Rankine power-absorption cooling cycle with an ammonia–water working pair. Later, Grosu et al. [56] discussed the exergy analysis of a combined solar-powered cycle: an ORC and an absorption cooling system using lithium bromide–water. Results displayed that by changing the connection mode between heat exchangers, minimum local exergy destruction ratios could be obtained; thus a maximum heat recovery was ensured. Likewise, Yamaguchi et al. [57] described a solar-driven Rankine cycle using supercritical carbon dioxide for combined electricity and thermal energy. The system comprises solar collectors, a turbine, a high-temperature heat recovery system, a low-temperature heat recovery system, and a feed pump, as shown in Fig. 14.11. Evacuated solar collectors were used to convert carbon dioxide into the high-temperature supercritical state so that the turbine could be

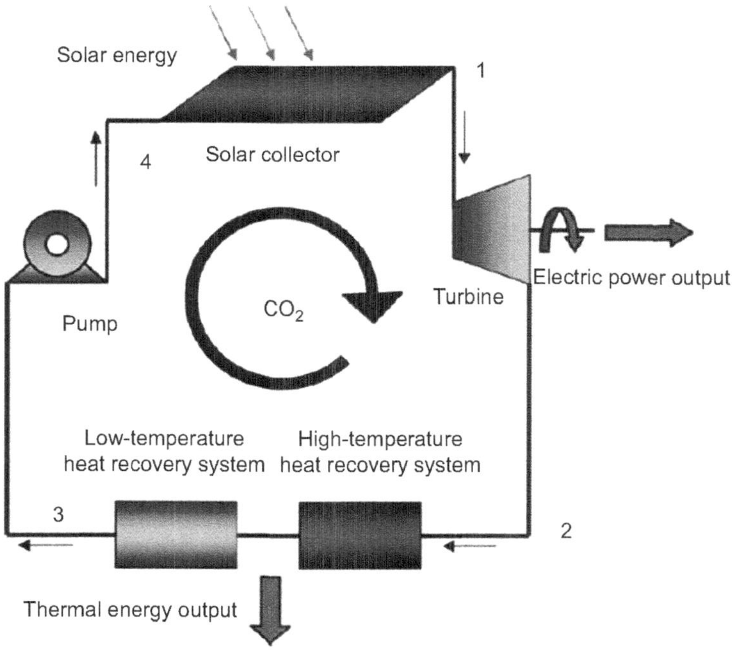

Figure 14.11 The scheme of a solar-powered Rankine cycle using supercritical carbon dioxide (Yamaguchi et al. [57]).

driven with the production of mechanical energy and hence electricity. The system also recovered both high-temperature heat and low-temperature heat for refrigeration, air conditioning, and hot water supply in buildings. Experimental results of this system proved that carbon dioxide is efficiently converted into the high-temperature supercritical state, and power generation efficiency and heat recovery efficiency were estimated to be 0.25 and 0.65, respectively. Later, Li and Zhang [58] also proposed a solar-powered transcritical carbon dioxide Rankine cycle prototype for simultaneous heating and cooling. Simulation results found that cooling COP for the system was above 0.3 whereas heating COP was more than 0.9. In addition, by optimizing parameters such as solar radiation, solar collector efficient area, the inlet water temperature of heat exchange components, etc., the performance of the system can be significantly improved. For example, with the increase of solar radiation, cooling capacity, heating capacity, and exergy loss rate increased whereas cooling and heating COPs decreased.

14.3 Conclusion

On the basis of the previous discussion, it can be seen that for solar-driven ejector refrigeration technology, including single-stage ejector systems, multistage ejector systems and hybrid systems can be regarded as an alternative way for conventional solar air conditioning. In addition, improved performance can be obtained by combining an ejector cooling system with other technologies. However, it should also be noticed that these hybrid systems are generally more complex and expensive. Thus more economical investigation and evaluation need to be studied to bring this technology into the market. It also can be seen that with a proper design of a solar-powered Rankine cooling system, it can produce electricity and cooling/heating at the same time. Nowadays, this popular system is known as combined cooling heating and power or trigeneration. This kind of system is so attractive that it makes zero-energy buildings possible. Numerous researches have been done on the design and evaluation of these cogeneration systems for domestic or commercial buildings [59–61]. Study on this area will be discussed in detail in Part V (*Advanced applications of solar heating and cooling systems*).

Although these technologies are not as popular as conventional adsorption, absorption, and desiccant cooling methods, they can provide a good supplement to the widely adopted technologies under some special conditions.

References

[1] C.A. Garris, W.J. Hong, C. Mavriplis, J. Shipman, A new thermally driven refrigeration system with environmental benefits, in: Proceedings of 33rd Intersociety Engineering Conference on Energy Conversion, Colorado Springs, USA, 1998.

[2] G.K. Alexis, E.K. Karayiannis, A solar ejector cooling system using refrigerant R134a in the Athens area, Renewable Energy 30 (9) (2005) 1457–1469.

[3] J. Chen, S. Jarall, H. Havtun, B. Palm, A review on versatile ejector applications in refrigeration systems, Renewable Sustainable Energy Rev. 49 (2015) 67–90.

[4] D.S. Kim, C.A. Infante Ferreira, Solar refrigeration options – a state-of-the-art review, Int. J. Refrig. 31 (1) (2008) 3–15.

[5] Z.C. Zhang, Y.Z. Li, X. Chen, Q. Wang, Application and development of solar refrigeration, Refrig. Air-Cond. 3 (1) (2003) 1–5 (in Chinese).

[6] V.M. Nguyen, S.B. Riffat, P.S. Doherty, Development of a solar-powered passive ejector cooling system, Appl. Therm. Eng. 21 (2) (2001) 157–168.

[7] S.S. Murthy, R. Balasubramanian, M.V.K. Murthy, Experiments on vapor jet refrigeration system suitable for solar energy applications, Renewable Energy 1 (5/6) (1991) 757–768.

[8] N. Al-Khalidy, Experimental investigation of solar concentrating collectors in a refrigerant ejector refrigeration machine, Int. J. Energy Res. 21 (12) (1997) 1123–1131.

[9] A. Bejan, J.V.C. Vargas, M. Sokolov, Optimal allocation of a heat-exchanger inventory in heat driven refrigerators, Int. J. Heat Mass Transfer 38 (16) (1995) 2997–3004.

[10] B.J. Huang, J.M. Chang, V.A. Petrenko, K.B. Zhuk, A solar ejector cooling system using refrigerant R141b, Sol. Energy 64 (4–6) (1998) 223–226.

[11] H.K. Ersoy, S. Yalcin, R. Yapici, M. Ozgoren, Performance of a solar ejector cooling-system in the southern region of Turkey, Appl. Energy 84 (9) (2007) 971–983.

[12] W. Pridasawas, P. Lundqvist, A year-round simulation of a solar-driven ejector refrigeration, in: Proceedings of the International Conference on Fluid and Thermal Energy Conversion, 2003.

[13] J.L. Wolpert, S.B. Riffat, S. Redshaw, Prototype for a novel solar powered ejector air conditioning system in Mazunte, Mexico, in: ISES Solar World Congress 2003. Solar Energy for a Sustainable Future, Göteborg, Sweden, O5 19, 2005.

[14] C.A. Balaras, G. Grossman, H.-M. Henning, C.A. Infante Ferreira, E. Podesser, L. Wang, et al., Solar air conditioning in Europe — an overview, Renewable Sustainable Energy Rev. 11 (2) (2007) 299—314.
[15] P. Noeres Thermische Kälteerzeugung mit Dampfstrahlkälte- maschinen KI Luft- und Kältetechnik, 11, 2006, pp. 478—483.
[16] K. Śmierciew, J. Gagan, D. Butrymowicz, et al., Experimental investigations of solar driven ejector air-conditioning system, Energy Build. 80 (2014) 260—267.
[17] W. Zhang, S.B. Riffat, X. Ma, S.A. Omer, Optimum Design of a Solar-driven Ejector Cooling System, Springer, Berlin Heidelberg, 2014, pp. 193—201.
[18] B. Tashtoush, A. Alshare, S. Al-Rifai, Hourly dynamic simulation of solar ejector cooling system using TRNSYS for Jordanian climate, Energy Convers. Manage. 100 (2015) 288—299.
[19] Y. Saito, T. Ito, A. Matsuo, H. Sato, Ejector configuration for designing a simple and high performance solar cooling system, Energy Procedia (2014) 2564—2571.
[20] M. Dennis, T. Cochrane, A. Marina, A prescription for primary nozzle diameters for solar driven ejectors, Sol. Energy 115 (0) (2015) 405—412.
[21] O. Ozgener, A. Hepbasli, A review on the energy and exergy analysis of solar assisted heat pump systems, Renewable Sustainable Energy Rev. 11 (3) (2007) 482—496.
[22] H.-M. Henning, Solar assisted air conditioning of buildings - an overview, Appl. Therm. Eng. 27 (10) (2007) 1734—1749.
[23] J.M. Abdulateef, K. Sopian, M.A. Alghoul, M.Y. Sulaiman, Review on solar-driven ejector refrigeration technologies, Renewable Sustainable Energy Rev. 13 (6—7) (2009) 1338—1349.
[24] T. Sun, B. Wan, C.J. Sun, Z.W. Ma, Multi-stage ejector design and numerical simulation for marine gas turbine, Adv. Mater. Res. 1078 (2015) 280—285.
[25] Z. Wang, H. Wang, G. Bao, Q. Li, Simulation and analysis on the two-stage ejector refrigeration system performance of solar refrigeration, Cryog. Supercond. 40 (8) (2012) (in Chinese).
[26] Z.J. Wang, H. Wang, H.T. Wang, The exergy analysis on the two-stage ejector refrigeration system performance of solar refrigeration, Energy Eng. 01 (2013) 44—48 (in Chinese).
[27] A. Arbel, M. Sokolov, Revisiting solar-powered ejector air conditioner — the greener the better, Sol. Energy 77 (2004) 57—66.
[28] L.M. Sun, Y. Feng, M. Pons, Numerical investigation of adsorptive heat pump systems with thermal wave heat regeneration under uniform-pressure conditions, Int J Heat Mass Transfer 40 (2) (1997) 281—293.
[29] S. Göktun, Performance analysis of a heat engine driven combined vapor compression—absorption-ejector refrigerator, Energy Convers. Manage. 41 (2000) 1885—1895.
[30] B.J. Huang, J.H. Wu, H.Y. Hsu, J.H. Wang, Development of hybrid solar-assisted cooling/heating system, Energy Convers. Manag. 51 (2010) 1643—1650.
[31] B.J. Huang, C.W. Yen, J.H. Wu, J.H. Liu, H.Y. Hsu, V.O. Petrenko, J.M. Chang, C.W. Lu, Optimal control and performance test of solar-assisted cooling system, Appl. Therm. Eng. 30 (2010) 2243—2252.
[32] B.-J. Huang, W.-Z. Ton, C.-C. Wu, H.-W. Ko, H.-S. Chang, H.-Y. Hsu, et al., Performance test of solar-assisted ejector cooling system, Int. J. Refrig. 39 (2014) 172—185.
[33] M. Li, R.Z. Wang, A study of the effects of collector and environment parameters on the performance of a solar powered solid adsorption refrigerator, Renewable Energy 27 (2002) 369—382.
[34] X.J. Zhang, R.Z. Wang, A new combined adsorption-ejector refrigeration and heating hybrid system powered by solar energy, Appl. Therm. Eng. 22 (2002) 1245—1258.

[35] C. Vereda, R. Ventas, A. Lecuona, M. Venegas, Study of an ejector-absorption refrigeration cycle with an adaptable ejector nozzle for different working conditions, Appl. Energy 97 (2012) 305–312.
[36] J.M. Abdulateef, N.M. Murad, M.A. Alghoul, A. Zaharim, K. Sopian, Experimental study on combined solar-assisted ejector absorption refrigeration system, in: Proceedings of the 4th WSEAS International Conference on Energy and Development – Environment – Biomedicine, 2011.
[37] J.M. Abdulateef, N.M. Murad, M.A. Alghoul, A. Zaharim, K. Sopian, Economic analysis of combined solar-assisted ejector absorption refrigeration system, in: Proceedings of the 4th WSEAS International Conference on Energy and Development – Environment – Biomedicine, 2011.
[38] K. Sopian, J. Abdulateef, M. Alghoul, K.S. Yigit, K.S. Yigit, Evaluation of a combined solar-assisted ejector absorption chiller, J. Mech. Eng. (1) (2013).
[39] R. Sirwan, M.A. Alghoul, K. Sopian, Y. Ali, J. Abdulateef, Evaluation of adding flash tank to solar combined ejector–absorption refrigeration system, Sol. Energy 91 (2013) 283–296.
[40] A.M. Abed, M.A. Alghoul, R. Sirawn, A.N. Al-Shamani, K. Sopian, Performance enhancement of ejector–absorption cooling cycle by re-arrangement of solution streamlines and adding RHE, Appl. Therm. Eng. 77 (2015) 65–75.
[41] C. Haddock, J.S.C. Mckee, Solar energy collection, concentration, and thermal conversion—a review, Energy Sources 13 (4) (1991) 461–482.
[42] S.B. R, J.L. Wolpert, Solar-powered Rankine system for domestic applications, Appl. Therm. Eng. 16 (1996) 281–290.
[43] A. Allouhi, T. Kousksou, A. Jamil, P. Bruel, Y. Mourad, Y. Zeraouli, Solar driven cooling systems: an updated review, Renewable Sustainable Energy Rev. 44 (2015) 159–181.
[44] D. Prigmore, R. Barber, Cooling with the sun's heat design considerations and test data for a Rankine cycle prototype, Sol. Energy 17 (1975) 185–192.
[45] H.M. Curran, M. Lokmanhekim, T. Alereza, Assessment of the Rankine Cycle for Potential Application to Solar Powered Cooling of Buildings, Final Report, Hittman Associates Inc., Columbia MD, 1974.
[46] E. Wali, Optimum working fluids for solar powered rankine-cycle cooling of buildings, Sol. Energy 25 (3) (1980) 235–241.
[47] F.R. Biancardi, J.W. Sitler, G. Melikian, Development and test of solar Rankine cycle heating and cooling systems, Int. J. Refrig. 5 (82) (1982) 351–360.
[48] N. Lior, K. Koai, Solar-Powered/Fuel-Assisted rankine cycle power and cooling system: sensitivity analysis, J. Sol. Energy Eng. 7 (4) (1982).
[49] X.B. Bu, H.S. Li, L.B. Wang, Performance analysis and working fluids selection of solar powered organic Rankine-vapor compression ice maker, Sol. Energy 95 (5) (2013) 271–278.
[50] S. Schimpf, R. Span, Techno-economic evaluation of a solar assisted combined heat pump – organic Rankine cycle system, Energy Convers. Manag. 94 (2015) 430–437.
[51] A.C. Oliveira, C. Afonso, J. Matos, S. Riffat, M. Nguyen, P. Doherty, A combined heat and power system for buildings driven by solar energy and gas, Appl. Therm. Eng. 22 (6) (2002) 587–593.
[52] F. Xu, D.Y. Goswami, S.S. Bhagwat, A combined power/cooling cycle, Energy 25 (3) (2000) 233–246.
[53] M.H. Ahmed, M.A. Rady, A.M.A. Amin, Multi applications of small scale solar power plant using organic rankine cycle and absorption chiller, in: Environmental Friendly Energies and Applications (EFEA), 2014 3rd International Symposium on, 2014.

[54] L. Grosu, A. Marin, A. Dobrovicescu, D. Queiros-Conde, Exergy analysis of a solar combined cycle: organic Rankine cycle and absorption cooling system, Int. J. Energy Environ. Eng. (2015).

[55] H. Yamaguchi, X.R. Zhang, K. Fujima, M. Enomoto, N. Sawada, Solar energy powered Rankine cycle using supercritical CO_2, Appl. Therm. Eng. 26 (17) (2006) 2345−2354.

[56] X.J. Li, X.R. Zhang, Preliminary investigation of a transcritical CO_2 heat pump driven by a solar-powered CO_2 Rankine cycle, Int. J. Energy Res. 37 (11) (2013) 1361−1371.

[57] W. Dan, A. Lu, P. Mendis, T. Ngo, Technical Feasibility of a Facade Integrated Solar Cooling System for Commercial Buildings, 2012. Solar.org.au.

[58] J. Freeman, K. Hellgardt, C.N. Markides, An assessment of solar-powered organic Rankine cycle systems for combined heating and power in UK domestic applications, Appl. Energy 138 (2015) 605−620.

[59] A. Hassoun, I. Dincer, Analysis and performance assessment of a multigenerational system powered by organic Rankine cycle for a net zero energy house, Appl. Therm. Eng. (2015) 25−36.

Part Four

Heat storage for solar heating and cooling applications

Sensible heat storage for solar heating and cooling systems

15

R. Velraj
Institute for Energy Studies, Anna University, Chennai, TamilNadu, India

15.1 Introduction

Thermal energy storage (TES) is an environmentally friendly technology that helps to economically bridge the gap between the energy supply and end-user demand. The concept of energy storage is very important in many industrial and commercial applications. However, this technology has received greater attention only in recent years because of the large promotion of renewable energy, particularly in the field of solar energy. TES is very essential in solar applications because there is a time mismatch between the availability and consumption of energy. This implies that a solar energy-based device or utility must be dynamically matched at the source point and the application point. Once the characteristics of the end-user demand and the nature of energy source options are known, the total demand and supply in the time domain have to be brought together through the integration of efficient energy storage and the distribution network. TES is an economically viable technology that aids in efficiently shaping the end-user demand.

Much development has been made in recent years on TES systems to make solar technology possible for several residential and industrial applications. The major applications that gained popularity in the field of solar thermal are solar water-heating, solar space-heating/cooling, solar thermal power generation, solar drying, and other process industrial application. TES enhances the renewable energy utilization and energy efficiency of conventional systems. It is understood from the thermodynamic principles that in any energy conversion or energy transfer process, the exergy degradation will be higher if the temperature difference associated with the process is higher. Thermal storage systems avoid the rapid or sudden disintegration of the energy available in the source with the environment (sink). This will reduce the exergy destruction and thereby reduce the environmental degradation as the operational temperature difference decreases. Hence, in developing an environmentally benign sustainable energy program, solar energy sources integrated with TES should be promoted and such activities form a basis for the short- and long-term policies of any country.

In a TES system, heat can be stored in sensible/latent form and by thermochemical reactions. In sensible thermal storage (STS), heat energy is stored by increasing the temperature of the solid/liquid storage medium. STS characterized by temperature variation is a simpler technique, and the amount of stored energy depends on the specific heat of the medium, the temperature change, and the mass of storage material whereas the latent heat storage is based on heat absorption or release when a phase

Advances in Solar Heating and Cooling. http://dx.doi.org/10.1016/B978-0-08-100301-5.00015-1

change material (PCM) undergoes the phase transition from solid to liquid or vice versa. In thermochemical heat storage systems, the heat is stored/retrieved in an endothermic/exothermic chemical reaction. Among these storage methods, sensible TES is dealt elaborately in this chapter, particularly in relevance to solar heating and cooling. The other types of thermal storage are dealt with in subsequent chapters.

15.2 Storage materials

Materials play a vital role in the thermal performance of any thermal storage system. The properties and other characteristics of a material influence the performance of any engineering component with which it is made of. Furthermore, to identify the suitable material with the highest performance for a given application, the criteria affecting the performance are to be well understood. Fernandez et al. [1] presented a methodology to identify the potential material to be used in a TES system with a case study that evaluates the materials for sensible TES in the temperature range of 150–200°C for short- and long-term storage with the objective of cost minimization.

An effective storage medium needs to satisfy the following criteria [2,3]:

- long life, noncorrosive, nontoxic, nonflammable, and large heat storage capacity;
- heat thermal diffusivity;
- capability to withstand charging/discharging cycles without loss in performance, storage capacity, or change in structure;
- wide availability and ease in handling and storage in simple containers;
- high thermal stratification;
- low thermal expansion coefficient;
- cost-effective; and
- reduced carbon footprint and environmental impacts.

15.2.1 Types of storage material

The STS materials are broadly classified into solid and liquid storage materials.

15.2.1.1 Liquid storage material

The commonly used sensible liquid storage materials are water, oils, inorganic molten salts, derivatives of alcohols, etc. For low-temperature requirements up to 100°C, water is the ideal choice for many applications. Oils are used for intermediate-temperature applications ranging from 100°C to 250°C. However, the degradation of these oils with respect to time is the major associated problem. Few molten inorganic salts have been considered for high-temperature storage applications (>300°C). Advantages of molten salts are high thermal stability, relatively low material costs, high heat capacity, high density, nonflammability, and low vapor pressure. Because of the low vapor pressure, pressurized vessels are not required.

For sensible heat storage in solar power plants, a noneutectic molten salt mixture consisting of 60 wt% sodium nitrate ($NaNO_3$) and 40 wt% potassium nitrate (KNO_3)

Table 15.1 Properties of liquid sensible heat storage materials [4–6]

Medium	Density ρ (kg/m^3)	Specific heat capacity c (J/kg K)	Heat capacity $\rho c*10^{-6}$ (J/m^3 K)
Butanol	809	2400	1.9416
Caloria HT43	–	2300	–
Dowtherms	867	2200	1.9074
Draw salt	1733	1550	2.6861
Engine oil	888	1880	1.6694
Ethanol	790	2400	1.8960
Ethylene glycol	1116	2382	2.6583
Hitec (53% KNO_3 + 7% $NaNO_3$ + 40% $NaNO_2$) Molten salt	1680	1560	2.6208
Isobutanol	808	3000	2.4240
Isopentanol	831	2200	1.8282
Lithium	510	4190	2.1369
Octane	704	2400	1.6896
Propanol	800	2500	2.0000
Sodium	960	1300	1.2480
Therminol 55	–	2400	–
Therminol 66	750	2100	1.5750
Water	1000	4190	4.1900
Water-ethylene glycol 50/50	1050	3470	3.6435

is used. This mixture is commonly known as "solar salt." Because of the increased amount of $NaNO_3$ as compared with the eutectic mixture, the material costs can be reduced. This noneutectic mixture has a melting temperature of approximately 240°C and the temperature limit for thermal stability is approximately 550°C. The most commonly used liquid sensible thermal materials are given in Table 15.1.

15.2.1.2 Solid storage materials

In general, solids exhibit a lower storing capacity than water. However, the cost of the storage media per unit energy stored is still acceptable for rocks. The commonly used solid storage materials are rocks, bricks, pebbles, refractory material/refractory bricks, concrete, dry and wet earth/soil, iron, wood, plasterboard, and corkboard. The properties of the most commonly used solid STS materials are given in Table 15.2.

Table 15.2 Properties of solid sensible storage materials [4–6]

Medium	Density (kg/m^3)	Specific heat capacity (J/kg K)	Heat capacity $\rho c \times 10^{-6}$ (J/m^3 K)
Aluminum	2707	896	2.4255
Aluminum oxide	3900	840	3.2760
Aluminum sulfate	2710	750	2.0325
Brick	1698	840	1.4263
Brick magnesia	3000	1130	3.3900
Calcium chloride	2510	670	1.6817
Cast iron	7900	837	6.6123
Clay	1458	879	1.2815
Concrete	2000	880	1.7600
Copper	8954	383	3.4294
Earth (dry)	1260	795	1.0017
Earth (wet)	1700	2093	3.5581
Glass	2710	837	2.2682
Granite	2750	892	2.4530
Graphite (solid)	2200	879	1.9338
Gravelly earth	2050	1840	3.7720
Lead	11,340	131	1.4855
Magnetite	5177	752	3.8931
Potassium chloride	1980	670	1.3266
Potassium sulfate	2660	920	2.4472
Pure iron	7897	452	3.5694
Sandstone	2200	712	1.5664
Slag	2700	836	2.2572
Sodium carbonate	2510	1090	2.7359
Soil (clay)	1450	880	1.2760
Soil (gravelly)	2040	1840	3.7536
Steel	7840	465	3.6456
Stone, granite	2640	820	2.1648
Stone, limestone	2500	900	2.2500
Stone, marble	2600	800	2.0800
Stone, sandstone	2200	710	1.5620
Wood	700	2390	1.6730

15.3 Classification of sensible thermal energy storage systems

Sensible TES systems are classified in various methods.

15.3.1 Short-term (diurnal)/long-term (seasonal) storage

In the short term, or diurnal TES, the heat energy is stored for a few hours to accommodate the fluctuation between energy availability and demand during a day. Storage systems used in water-heating systems, building air-conditioning applications, and solar power plant applications are generally short-term storage. The heat energy stored during the daytime to utilize in the nighttime for heating (building space-heating) or the cool energy stored during the nighttime for day-cooling requirements (free cooling of buildings) also come under this category. Water, pebbles, and oils are generally used for short-term energy storage.

In the long term, or seasonal thermal storage, the thermal energy stored during one season is utilized during the other season. Underground TES systems with boreholes, rock-bed systems (with gravel or solid rock), aquifers, soil (earth-to-air thermal storage system), caverns, and solar ponds are normally used in seasonal storage systems.

15.3.2 Cool/low/medium/high-temperature storage

On the basis of the operating temperature range of the storage medium, the storage systems are broadly classified as cool and hot storage. Cool storage is mainly used in passive space-cooling and active air-conditioning applications. In passive cooling, the building material itself is used as the storage medium. In active air-conditioning applications, the chilled water is used as the storage medium operating in the temperature range of approximately 4°C. In heating applications, the heat is utilized in low- (<100°C), medium- (100–250°C) and high-temperature (>250°C) applications such as space-/water-heating, process heating, and power generation.

15.3.3 Active and passive storage

In active storage systems, the storage medium is under circulation and performs all three functions of heat collection, storage, and retrieval. The solar water heater in which the water is circulated between the collector and the storage tank is an active storage system. In the passive storage system, the storage medium is stationary. An example of a passive storage system is the solar air heater, in which the hot air generated from the heater transfers its heat to any stationary storage medium such as water or pebbles. The concrete used in building structures acts as the storage medium and is an example of a passive storage system. Fig. 15.1 illustrates the examples of active and passive storage systems.

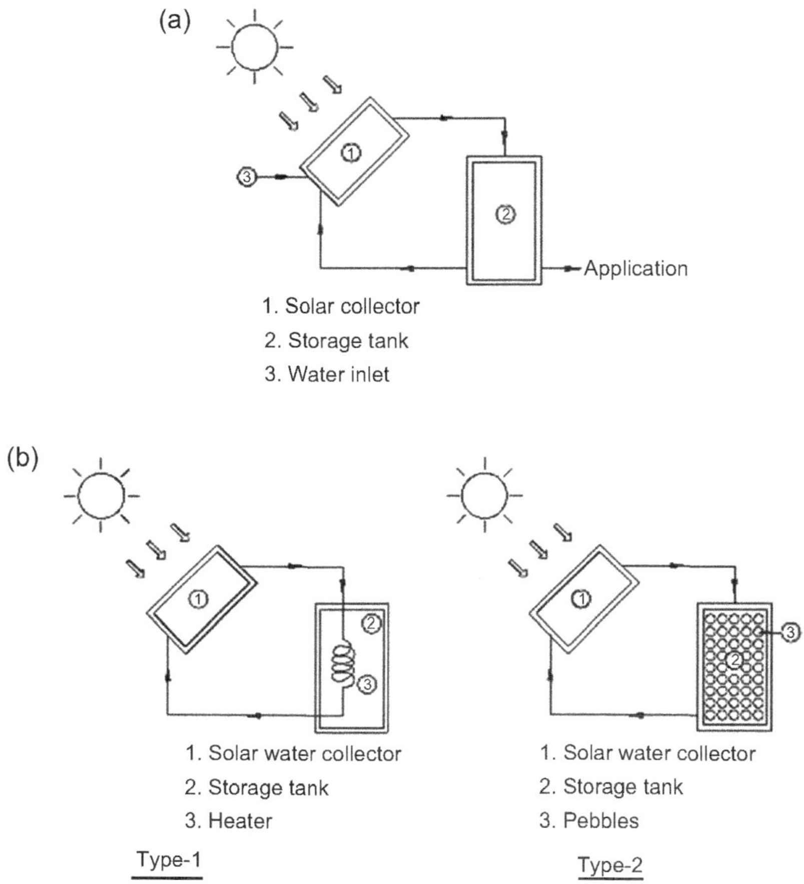

Figure 15.1 (a) Active storage and (b) passive storage.

15.4 Working principle

For any energy storage system, the energy from a source is supplied to its storage medium and this energy is stored in the storage medium for a period of time that depends upon the application and location to be implemented. This energy is retrieved for a desired application from the storage medium during the demand. There are three major processes involved in any storage system: charging, storing, and discharging. The heat transfer arrangement and the mechanism involved in these processes are different based on the type of applications (cooling/heating). Figs. 15.2 and 15.3 show the usual flow arrangement involved in the sensible heat and cooling energy storage systems, respectively.

In a heat storage system, for charging processes (Fig. 15.2), the inlet is normally kept in the top portion of the tank and the outlet is kept at the bottom of the tank to achieve better stratification. In the case of a solar thermal system, improved

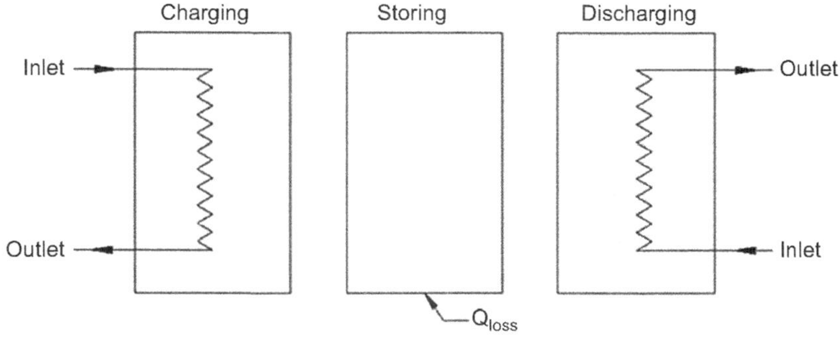

Figure 15.2 Flow arrangements in a sensible heat energy storage system.

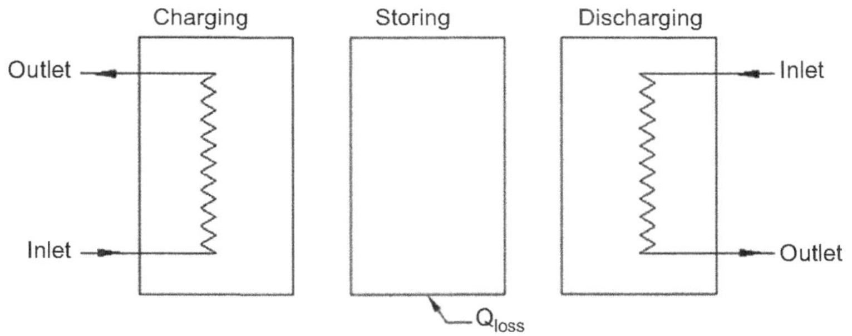

Figure 15.3 Flow arrangements in a sensible cool energy storage system.

stratification also increases the collector efficiency. Multiple inlets for charging process are used to effectively handle the varying temperature source available from the solar collectors at various times of a day. During the late afternoon, if the hot water available from the solar collector is lower in temperature than that of the hot water available in the top of the storage tank, the water may be sent to the intermediate inlet based on the temperature of the storage medium in the storage tank at various heights. The inlet to the storage tank for the discharging process is kept at the bottom of the tank and the outlet is kept at the top to supply the hot water at a higher temperature for applications.

In a sensible cool storage system as shown in Fig. 15.3, the inlet and outlet arrangements in the storage tanks are in the opposite way for charging and discharging compared with the sensible heat storage system.

15.5 Sensible thermal storage technologies

The storage concept plays a decisive role in the use of solar thermal systems, especially in the areas with considerable diurnal temperature variation and larger seasonal differences to accomplish the major residential thermal requirement. The implementation of

TES in water- and space-heating/-cooling systems is of great importance for effective use of intermittent solar radiation for residential applications. Efficient and economic storage technologies are intended to achieve high solar fraction and to meet the required operating conditions for specific application. In addition to the residential requirement, storage technologies are also used in solar-based power generation. The various storage technologies used in solar water-/space-heating, solar power generation, and solar-cooling are explained in detail in the following subsections.

15.5.1 Domestic hot water storage

Water can store more heat per unit volume compared with other substances; hence, it is considered an efficient storage medium. Furthermore, water can be used as both a thermal storage and heat transfer medium because of its high heat capacity and high heat transfer characteristics. Most of the residential heating needs can be easily fulfilled by the water available in the temperature range of 20–80°C.

Fig. 15.4 shows the schematic arrangement of a thermal storage tank integrated with a solar domestic water-heating system. The system consists of a solar collector, storage tank, auxiliary heater, and a series of tubes that connect the solar collector into the storage tank. The common materials used for the construction of storage tanks are steel, aluminum, reinforced concrete, and fiberglass. To prevent the thermal losses from the storage tank, the tanks are insulated with mineral wool, glass wool, or polyurethane. Achieving near-perfect insulation is the major requirement to reduce the overall heat loss coefficient in any storage tank.

Well-stratified and well-mixed storage systems are designed so as to ensure the maximum energy storage and uniform temperature, respectively, in a storage system. Maintaining the thermal gradient layer as shown in Fig. 15.5 with the warmer layers of fluid placed above the cooler ones because of the difference in fluid density is termed

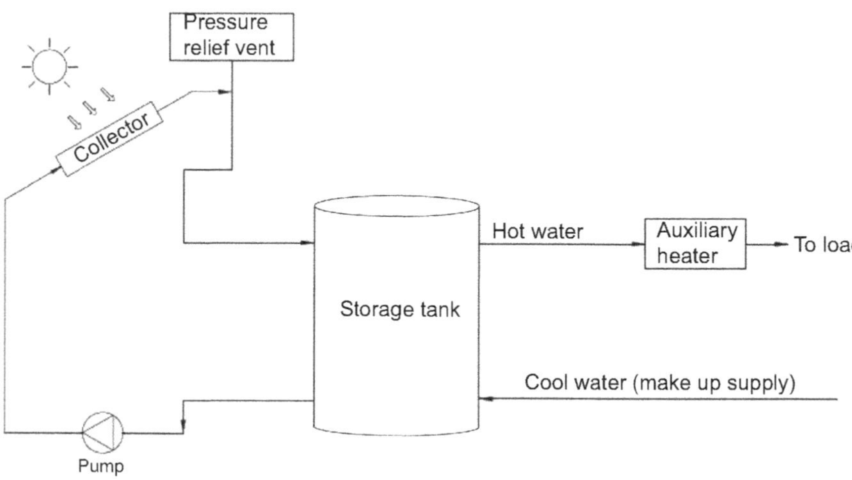

Figure 15.4 Basic schematic of domestic solar water-heating system with the storage tank.

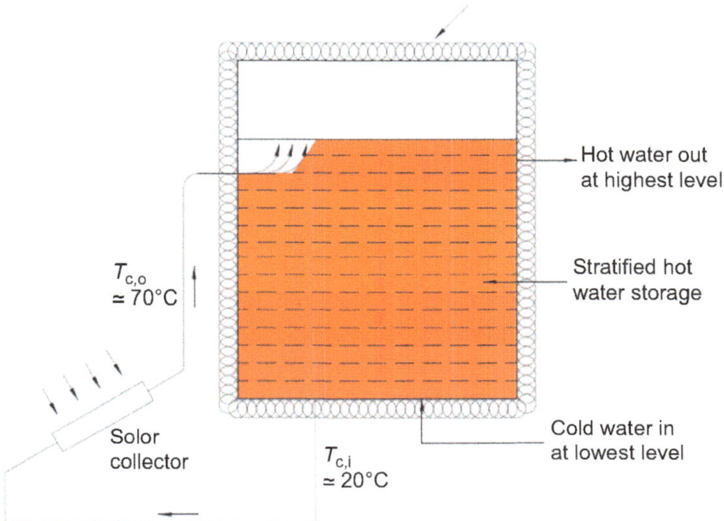

Figure 15.5 Thermal stratification in a storage tank [8].

as thermal stratification. The factors influencing the stratification performance in a storage tank are the aspect ratio (ratio of height to width/diameter), inlet/outlet locations in the storage tank, and the operating parameters such as inlet velocity and temperature of the heat transfer fluid. Further various baffles, geometry modifications, and stratifiers are adopted to enhance the stratification performance. The method of evaluating the stratification performance is given separately in the next section.

The storage tanks are integrated with the solar water-heating systems in several ways that are classified in the following subsections.

15.5.1.1 Direct/indirect storage

The storage tank configurations are broadly classified under two categories based on the heat exchange between the collector and the storage system: (1) direct storage (without heat exchanger) and (2) indirect storage (with heat exchanger).

In the direct storage method, the storage medium and heat transfer fluid are identical. Fig. 15.6(a) shows the plain vertical storage tank in which water from the collector directly flows in and out of the storage tank without any partition or heat exchanger.

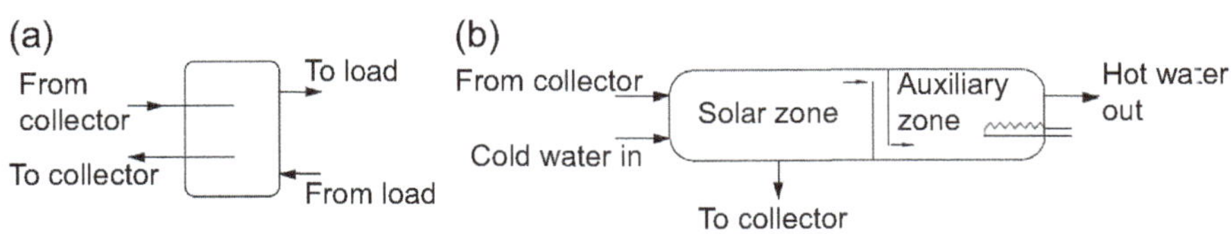

Figure 15.6 Direct storage (without heat exchangers): (a) vertical plain storage tank and (b) horizontal segmented storage tank.

This type of storage is widely used because it is simple in construction and less expensive. Better thermal stratification is possible in vertical storage tanks, where the height-to-diameter ratio is 2 or greater. This type of storage system is mostly used in thermosyphon-based solar water-heating systems because the thermal stratification favors the natural circulation of the water. In the horizontal tank system, usually the auxiliary heater is placed inside of the storage tank as shown in Fig. 15.6(b). Hence, the performance of the system is influenced by the conduction between the high-temperature auxiliary zone and the solar zone and by mixing the flow injection points. This problem is resolved in the segmented tank by introducing baffles in which the solar input cannot heat the auxiliary zone until there is a demand.

In the indirect storage method, the storage and heat transfer mediums are different. The heat transfer occurs through a heat exchanging surface and the storage medium does not enter into the collector. The heat exchange between the heat transfer fluid and the storage medium takes place either with internal or external heat exchangers. The configurations shown in Fig. 15.7 use different types of heat exchanger to transfer heat from the solar collector to the storage tank. The immersed coil heat exchanger as shown in Fig. 15.7(a) is generally located at the bottom of thermal storage tanks to take the advantage of the higher temperature differences between the heated fluid on the top of the storage tank and the incoming water from the load side. External heat exchangers as shown in Fig. 15.7(b) represent flexible options for indirect systems because they allow standard, less-expensive storage tanks and heat exchangers to be used. Most solar thermal systems used in North America are indirect systems with external heat exchangers and two circulation pumps: one to circulate an antifreeze

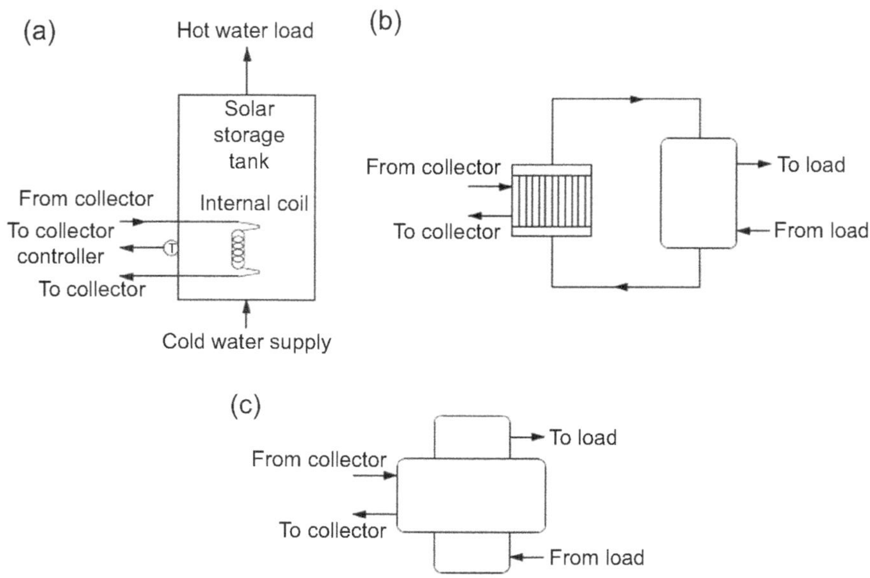

Figure 15.7 Indirect storage (with heat exchanger): (a) internal immersed coil type, (b) external coil and tube type, and (c) mantle heat exchanger.

solution through the solar collectors and the other to circulate water from the storage tank through the heat exchanger. Mantle heat exchangers as shown in Fig. 15.7(c) consist of a double-walled storage tank that allows the heat transfer fluid to be circulated through the storage mantle (ie, cavity formed by the two walls) transferring heat to the stored water. The mantle tank storage system tends to have a large heat transfer surface area that increases the performance but requires a specialized tank that makes it very expensive.

15.5.1.2 Open-/closed-loop storage tank

In the open-loop storage system, the storage medium receives the heat from the solar collector, and during the requirement, the same storage medium with heat energy is sent to the application. In the closed-loop system, the storage medium circulates between the source and the storage tank, and another heat transfer fluid indirectly absorbs the heat from the storage tank and is used for the application. The configurations used in the open-loop and closed-loop storage systems are shown in Fig. 15.8.

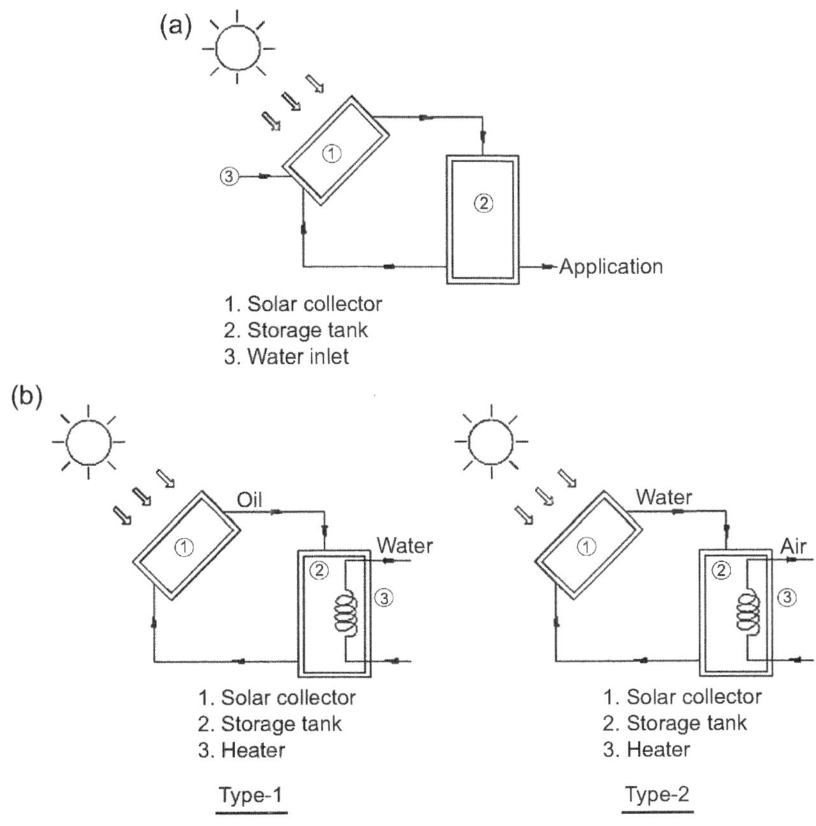

Figure 15.8 (a) Open-loop storage and (b) closed-loop storage.

15.5.1.3 Pressurized storage system

The boiling point of water restricts the use of water as heat transfer fluid in the solar collector beyond the 100°C requirement. Hence, to achieve higher temperature with water as heat transfer fluid in the solar collector, a pressurized water storage tank as shown in Fig. 15.9 is used in high-temperature applications that also reduce the evaporative losses. An external heat exchanger gives more flexibility because the tank and the heat exchanger can be independently selected. However, the additional pump used in the system increases the electrical energy consumption and hence the operating cost.

15.5.2 Storage methods in space-heating system

The storage technologies used in the previous subsection (solar water-heating system) can also be used as a "combi system" (ie, for combined solar water-heating and space-heating). The basic schematic arrangement of a solar-based space-heating system with a water storage unit is shown in Fig. 15.10. In this system, water from the storage tank is allowed to pass through a radiator located in the room through which the heat is transferred to the space to be heated.

The other types of storage technologies widely used for space-heating application include rock-bed storage, solar ponds, borehole thermal energy storage (BTES), gravel-water thermal energy storage (GWTES), and aquifer TES, which are particularly suitable for medium- and long-term storage and can also be used for water-heating applications. These technologies are explained in detail in the following sections.

15.5.2.1 Rock-bed thermal storage

In the rock-bed thermal storage system, the rock materials are loosely packed in a bed-like structure along the height of the tank. During the charging cycle, hot air

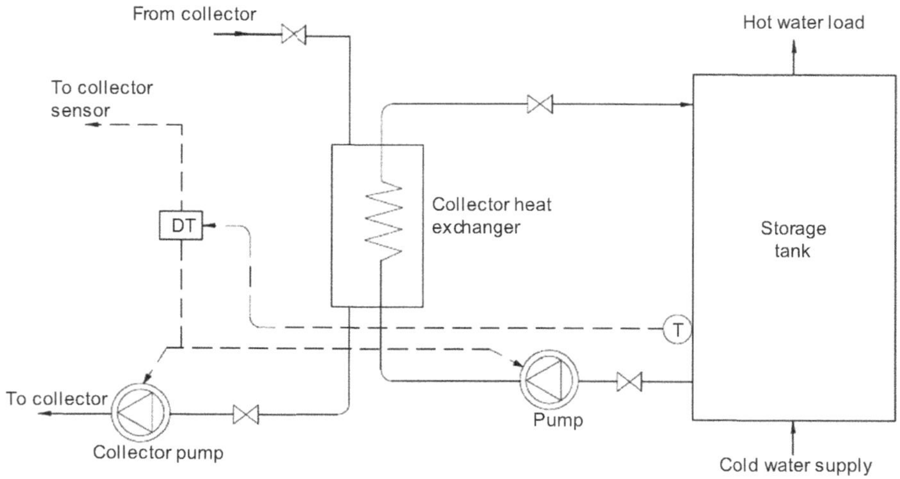

Figure 15.9 Pressurized storage system with heat exchanger [8].

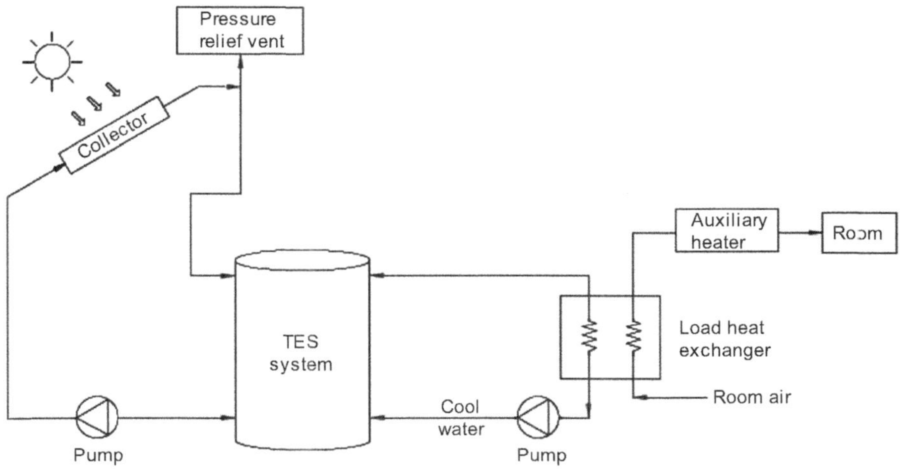

Figure 15.10 Basic schematic of solar-based space-heating system with water storage tank. *TES*, Thermal energy storage.

from the solar collector passes through the upper portion of the rock bed as shown in Fig. 15.11 and transfers the thermal energy to the rock bed. During the nighttime, cool air from the room flows through the bottom of the bed system and retrieves the heat energy stored from the bed, which is then utilized for space-heating. The efficiency and the viability of a rock-bed heat storage system strongly depend on its size and shape, porosity, heat transfer fluid flow pattern, etc. Heat storage performance of this kind of storage system is lower than that for a water storage system because the air and rock have low specific heat capacity compared with the water. However, the cost of a rock-bed storage system is less than that of a water-based system.

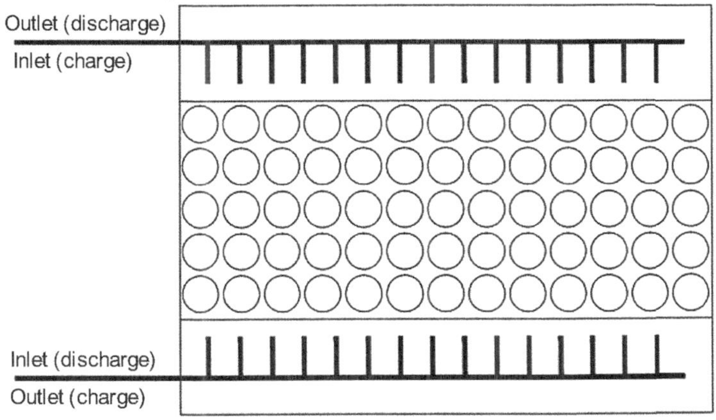

Figure 15.11 Illustration of rock-bed storage system [7].

15.5.2.2 Solar ponds

Solar ponds can be defined as large volumes of saline solution with a higher concentration of salts at the bottom layer than at the top. They are subdivided into three zones: the upper convective zone (UCZ), the nonconvective zone (NCZ), and the lower convective zone (LCZ), as shown in Fig. 15.12. The UCZ, which is located in the upper layers of the pond, is made up of low-salinity brine. The NCZ, which is located just below the UCZ, contains a salinity gradient such that the water at the top is always less salty than the water at the bottom of the pond. The LCZ has the greatest salinity density and is where the heat is stored and extracted. Solar ponds offer an economical method for collecting and storing the solar thermal energy in the temperature range of 50–95°C.

A solar pond operates as follows. Because of the salinity gradient within the pond, heat absorbed from solar irradiation remains trapped at the bottom. Indeed, the salinity gradient inhibits the natural convection within the pond and water at the surface remains cool and hence acts as an insulator for the deeper regions. The heat stored can easily be removed through a heat exchanger located in the LCZ. Solar ponds have 20% efficiency (the ratio between the total energy extracted from the pond and its estimated energy capacity) for a heat recovery temperature not higher than 65°C and 15% efficiency with a heat recovery at 85°C. Darkening the bottom of the pond results in more solar radiation being absorbed. The maximum possible efficiency of solar pond systems with heat extraction only from the LCZ is limited to 30%. However, this value can be enhanced up to 50% by combining the extraction of heat from the LCZ and the NCZ.

The desirable properties of salt to be used in solar ponds are as follows:

- a high value of solubility to allow for high solution densities, and the solubility should not vary appreciably with the temperature;
- transparency of the salt solution to solar irradiation;

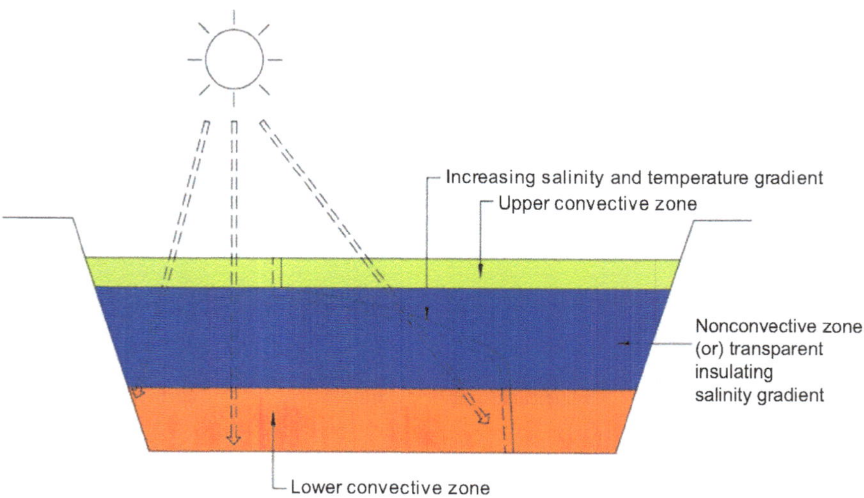

Figure 15.12 Solar pond [8].

- environmentally benign and safe to handle; and
- low cost.

Some commonly preferred salts for solar pond systems are as follows:

- sodium chloride;
- sodium carbonate;
- natural brine, magnesium, chlorine, etc.;
- magnesium chloride;
- potassium nitrate;
- ammonium nitrate; and
- urea.

15.5.2.3 Borehole thermal energy storage system

The BTES system is an underground structure for storing large quantities of solar heat collected in summer for later use in winter. It is basically a large, underground heat exchanger. The basic schematic arrangement of a BTES is shown in Fig. 15.13.

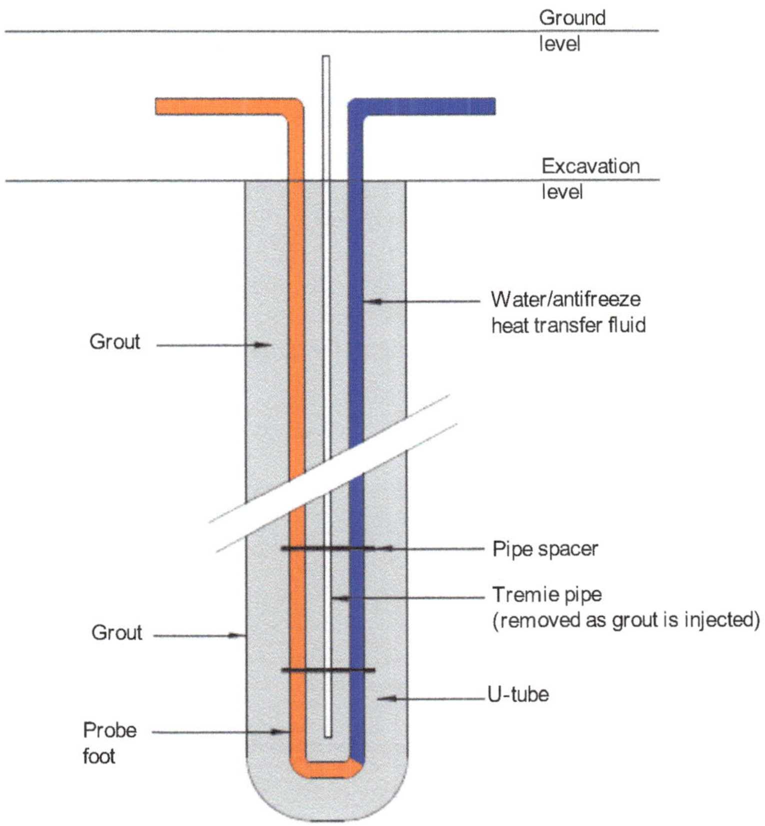

Figure 15.13 Borehole thermal energy storage system [8].

A BTES consists of an array of boreholes resembling the standard drilled wells. After drilling, a plastic pipe with a "U" bend at the bottom is inserted down the borehole. To provide good thermal contact with the surrounding soil, the borehole is then filled with a high thermal conductivity grouting material.

When the excess hot water from the solar collector is to be stored, it is pumped into the center of the BTES field through the U-pipe series. Heat is transferred to the surrounding soil and rock, and the water gradually cools as it reaches the outer edge of the BTES and returns to the source. Conversely, when the buildings require heat, cool water is pumped into the edges of the BTES field and as the water flows to the center, it picks up the heat. The heated water passes to the short-term storage tank and is then circulated to the buildings through district heating loops.

15.5.2.4 Gravel-water thermal energy storage

A GWTES system as shown in Fig. 15.14 is normally buried in the ground but closer to the ground surface in the range of 5–15 m deep to reduce excavation costs. GWTES needs to be insulated on the top and along the inclined walls. The top of the GWTES is usually covered with a load-carrying construction so that the surface area can be used for other purposes. Depending on the size and shape, insulation can also be provided at the bottom.

The storage material is usually a mixture of gravel and water, sand/water, or soil/water mixtures. The storage temperature can be up to a maximum of 95°C provided that the liner is made of either advanced polymer materials or metal. Heat is charged into and discharged out of the store either by direct water exchange or by polyvinyl chloride pipes installed in different layers inside of the store. Stratification should be supported by the charging devices. No load-bearing frame structure is required because forces are taken down to the side walls and to the bottom by the gravel. The heat transfer process in the storage is mainly due to convection. Because of the reduced specific heat capacity, the volume of the store has to be approximately 50%

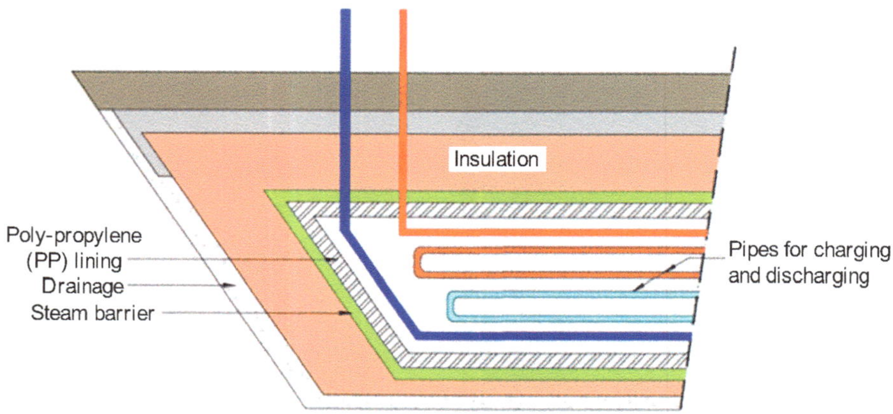

Figure 15.14 Gravel-water thermal energy storage [8].

higher compared with the water-based TES system to store the same amount of heat at the same temperature levels. The favorable geological requirement for the construction of such a storage system is stable ground condition and preferably no ground water.

15.5.2.5 Aquifer thermal storage system

Aquifers are geological formations containing ground water. Aquifer water is sometimes mixed with gravel or sand. The main advantages of aquifers for heat/cold storage are the allowable long storage periods and their low overall cost due to the large volumes that can be stored. Heat storage in aquifers is through the extraction of ground water from a well, heating this water with an available heat source (eg, solar collectors), and then re-injecting it back into the aquifer in the other well, as illustrated in Fig. 15.15.

Because of its bulk nature, aquifer heat storage systems are considered to be best suited for large-scale storage applications such as centralized urban heating systems. The energy costs can be reduced with increasing the storage volume in large-scale solar applications. Compared with classical water storage tanks, aquifer heat storage

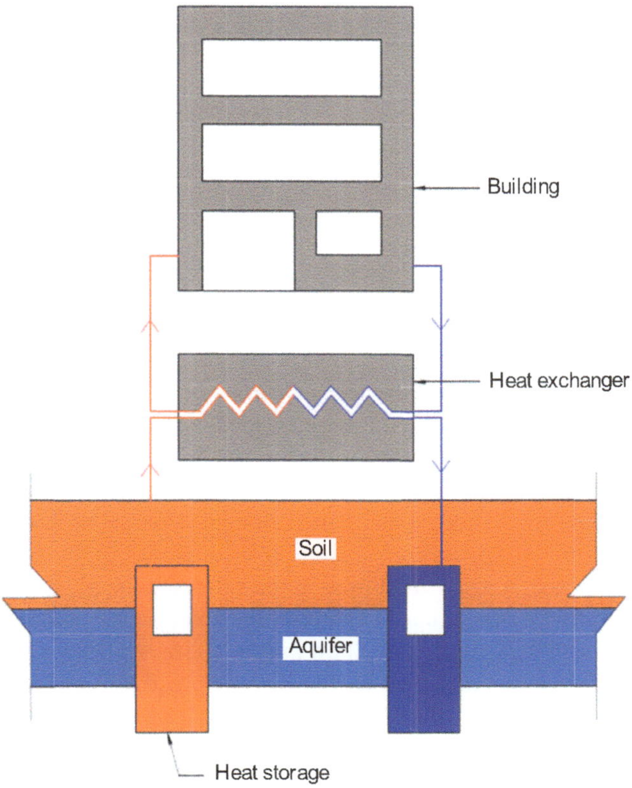

Figure 15.15 Aquifer thermal energy storage systems [8].

systems have lower investment and operational costs. In fact, use of the natural aquifer layer does not involve heavy excavation and the construction of a water tank that reduces the costs. However, specific hydrogeological conditions as given in the following are required for heat storage in aquifer sites:

- natural aquifer layer with high conductivity,
- confining layers above and below,
- no or low natural ground water flow,
- suitable water properties at sufficiently high temperatures, and
- aquifer thickness in the range of 20–50 m.

15.5.2.6 Building structure

The building structure itself can be used as a medium of energy storage in building thermal management. It plays a role as a thermal buffer to attenuate external heat flows and to balance the internal temperature swings. It uses the building inertia as a heat sink during the overheated periods and transfers the stored heat when needed through an appropriate ventilation system or by natural convection with the indoor air. The building materials generally have low specific heat capacity and low heat conductivity, which are the major limitations for heat storage. However, these materials have a relatively good density. Several recent studies have investigated toward the optimization of the building's mass and components to minimize the space-heating costs. Greater thermal effusivity, which characterizes the rate at which a material can absorb heat, will result in a greater amount of heat store. This will decrease the overall energy consumption of the building. Among the commonly available building materials, brick, granite, marble, steel slabs, or asphalt sheets have good potential in this context of space-heating.

15.5.3 Solar power plant with sensible thermal energy storage

The development of a short-term TES option that increases the annual load factor of the power plant is a major breakthrough in solar power plant technology. Integration of TES increases the investment cost and decreases the overall plant's efficiency because of the inevitable heat transfer losses involved during the charging and discharging period. Extending the operation of the solar power plant for 6 h beyond the sunshine hours is considered as optimal storage duration adopted from the economic point of view.

In general, the heat transfer fluid used in the solar field is oil or molten salts. The synthetic oil used as the heat transfer fluid in the solar field is a eutectic mixture of diphenyl oxide and biphenyl oxide. Few plants use a synthetic aromatic fluid as heat transfer fluid. This fluid is organic (benzene) based and as such cannot reach temperatures above 400°C with acceptable performance because of its degradation into unusable components at high temperatures. Using molten salt in both the solar field and TES system eliminates the need for expensive heat exchangers. It allows the solar field to be operated at higher temperatures than current heat transfer fluids allow. The major advantage of molten salt as a storage medium in central receiver technology is that it can be heated to 600°C, which allows high-energy steam to be generated and achieves high thermodynamic cycle efficiencies of approximately 40% in modern steam turbine systems. This high cycle efficiency is maintained while allowing the use of dry cooling towers, which is important in arid states with the best solar potential. Unfortunately, molten salts freezes in the temperature range of 120–220°C. This means that special care should be taken to ensure that the salt does not freeze in the solar field piping during the night.

The schematic representation of various storage configurations used in solar power generation is shown in Fig. 15.16. An indirect, two-tank TES system is shown in Fig. 15.16(a). This system requires two heat transfer fluids: one that captures energy from sunlight in the solar field and a second fluid that transfers the captured and stored energy to the traditional steam generation heat exchangers at the power block. The cost and performance of the indirect TES system are not independent because the size (cost) of the heat exchanger affects the physical size of the storage system and the performance of the turbine. Another option is a direct two-tank TES system that allows the same fluid to be used in the solar field and the thermal storage system, eliminating the need for the expensive heat exchanger involved in the solar field as shown in Fig. 15.16(b).

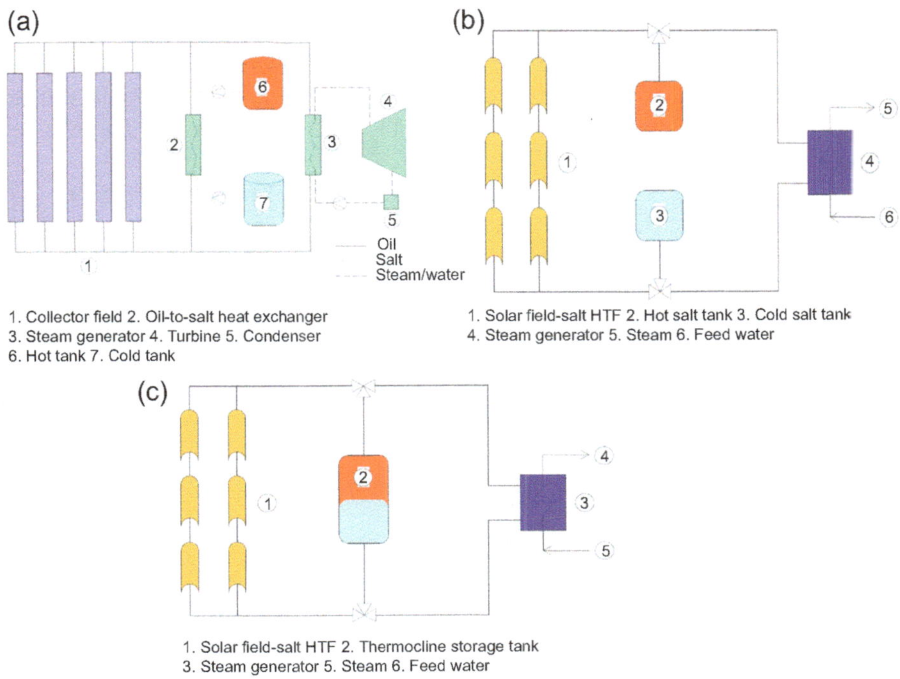

Figure 15.16 Storage configurations in solar power generation [9]: (a) indirect two-tank TES system, (b) direct two-tank TES system, and (c) thermocline setup. *HTF*, Heat transfer fluid; *TES*, thermal energy storage.

Figure 15.17 Concrete as solid thermal energy storage media [9].

A single tank for storing the hot and cold fluid provides one possibility for further reducing the cost of a direct two-tank storage system through a thermocline configuration that features the hot fluid on top and the cold fluid at the bottom as shown in Fig. 15.16(c). The zone between the hot and cold fluids is called the thermocline. The thermocline storage system has an additional advantage—most of the storage fluid can be replaced with a low-cost filler material. This filler material acts as the primary thermal storage medium and reduces the overall required quantities of the relatively higher cost molten salt heat transfer fluid.

The usage of solid TES media such as ceramic material or high-temperature concrete as shown in Fig. 15.17 is attempted by the German Aerospace Center for parabolic trough power plants. This system uses the standard heat transfer fluid in the solar field. The heat transfer fluid passes through an array of pipes imbedded in the solid medium to transfer the thermal energy to and from the media during plant operation. The primary advantage of this approach is the low cost of the solid media. Primary issues include maintaining good contact between the concrete and piping and the heat transfer rates into and out of the solid medium.

15.5.4 Storage for solar-cooling system

Solar cooling is one of the promising technologies given the fact that solar energy is the cheapest and most widely available renewable energy that matches the cooling load requirements. Storing cool energy produced during sunshine hours in a cool thermal storage tank is yet another option of storing energy. Cool thermal energy storage (CTES) has recently attracted increasing interest in building air-conditioning systems and industrial refrigeration applications such as process cooling and food preservation. CTES requires a proper insulation tank because the energy available in the cool state is expensive compared with the heat available in a hot storage tank. Solar cooling offers a wide variety of cooling techniques powered by solar collector-based thermally driven cycle operation, including vapor absorption refrigeration (VAR) systems, photovoltaic (PV)-based electrically operated mechanical cooling systems, and vapor compression refrigeration (VCR) systems.

15.5.4.1 *Storage concepts in vapor absorption refrigeration system*

A schematic diagram of a solar-operated absorption refrigeration system is shown in Fig. 15.18. The difference between this system and the conventional fossil fuel–fired units is that the energy supplied to the generator is from the solar collector. There are two different possible ways of storing the energy in the system as follows:

1. Storing the hot energy for heat supply to the generator.
2. Storing the produced cool energy.

In the first method, the heat transfer fluid from the solar collector is stored at a reasonably higher temperature in the hot storage tank, and the stored heat is supplied to the generator during the nonsunshine hours for the sustained operation of the VAR system to generate the cool energy. In the second method of storage, the cool energy is generated during the nonsunshine hours and the produced cool energy in the temperature range of $-10°C$ to $+10°C$ is stored in the cool thermal storage tank using brine as the storage medium. This stored energy is transferred to the application based on the demand during the nonsunshine hours.

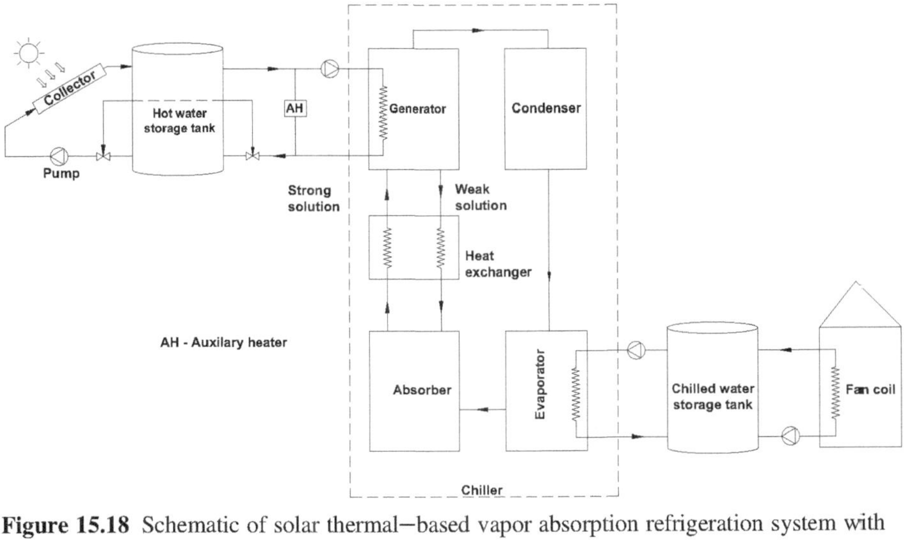

Figure 15.18 Schematic of solar thermal–based vapor absorption refrigeration system with chilled water storage system.

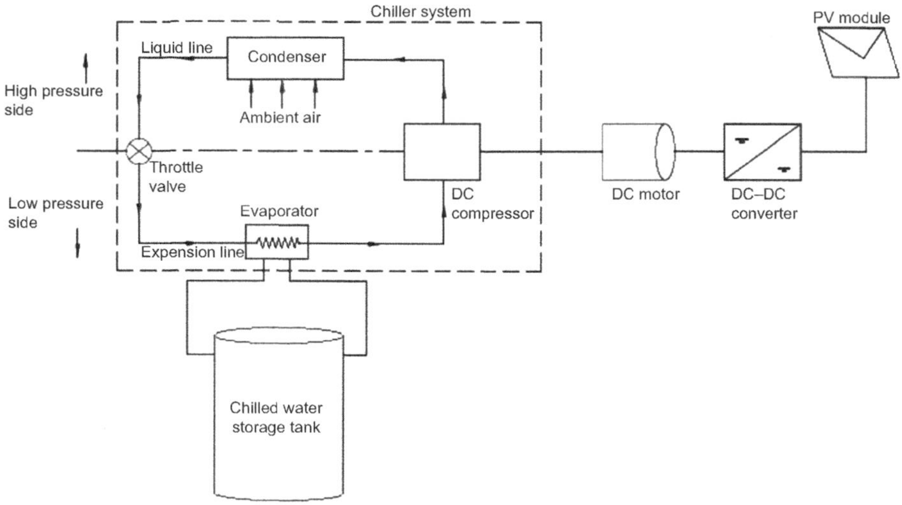

Figure 15.19 Schematic of solar PV-based vapor compression refrigeration system with chilled water storage system. *DC*, Direct current; *PV*, photovoltaic.

15.5.4.2 Cool storage in solar photovoltaic-operated vapor compression refrigeration system

The decreasing price of the solar PV module has made the solar PV-operated VCR technology more economical than the solar thermal–based VAR cooling system. In this technology the excess electricity generated during the sunshine hours will be used to produce cool energy, and this energy is stored in a cool thermal storage system, which can be used during the nonsunshine hours as shown in Fig. 15.19.

The major advantages of the CTES system used in the solar-based VCR system are as follows:

- The cool storage system allows for 50% reduction in the capacity of the chiller normally used in the central air-conditioning system.
- The compressor is allowed to run always under full-load condition irrespective of the load condition to achieve higher electrical efficiency of the motor because the excess cool energy generated can be delivered to the storage system.
- A high-efficiency direct current motor can be used to run the compressor.

15.6 Thermal performance evaluations

The important thermal performance characteristics of a sensible heat storage system are as follows:

1. heat capacity,
2. charging and discharging rates and efficiency,
3. heat loss rate, and
4. thermal stratification.

15.6.1 Heat capacity

Thermal energy can be stored in a material as sensible heat by raising its temperature.

The heat or energy storage can be calculated as

$$\begin{aligned} Q &= V\,\rho c_p\, dT \\ &= m\, c_p\, dT \end{aligned} \quad [15.1]$$

where Q is the sensible heat stored in the material (J), V is the volume of the storage medium (m^3); ρ is the density of the storage medium (kg/m^3), m is the mass of the storage medium (kg), c_p is the specific heat capacity of the storage medium (J/kg K), and dT is the temperature change (K).

When energy is stored as sensible heat, a high volumetric specific heat (ρc_p) of the material is required to make the storage system compact. Because packing densities of solid materials are often between 60% and 85%, the volumetric capacity (ρc_p) is proportionally reduced.

15.6.2 Charging and discharging efficiency

To charge/discharge a heat storage vessel, the heat transfer fluids (water, air, oil) take up the heat either by direct contract or through a heat exchanger. If the storage medium is a solid material and its thermal conductivity is very high, then the entire storage system will be of the same temperature and the lumped analysis concept can be used to evaluate the instantaneous rate of charging and discharging. Eq. [15.1] can be used to calculate the instantaneous charging and discharging by measuring the temperature of the storage system at any instant of time. If the thermal conductivity of the material is very low, then the conductivity of the storage material limits the heat transfer rate and a high temperature gradient exists in the storage system. Under such a situation, the heat transfer fluid temperatures at the inlet and outlet of the storage tank are normally used to evaluate the instantaneous rate of charging and discharging.

If the storage medium is liquid, then the lumped system analysis is possible if the storage tank is well mixed. Stratification normally exists in the liquid storage tanks; hence, the heat transfer fluid temperatures at the inlet and outlet of the storage tank are used to evaluate the instantaneous rate of charging and discharging. Eq. [15 2] is used to evaluate the instantaneous rate of charging and discharging:

$$Q_{ins} = mc_p(T_{in} - T_{out}) \quad [15.2]$$

where Q_{ins} is the instantaneous rate of charging/discharging (J/s), m is the mass flow rate of the heat transfer fluid (kg/s), c_p is the specific heat capacity of the heat transfer fluid (J/kg K), T_{in} is the temperature of the heat transfer fluid at the inlet, and T_{out} is the temperature of the heat transfer fluid at the outlet.

The positive value indicates the rate of charging and the negative value indicates the rate of discharging. It should be noted that for CTES the negative value indicates the rate of charging and the positive value indicates the rate of discharging.

The charging efficiency of the TES system is the ratio of instantaneous heat transfer to the maximum heat transfer at a given inlet temperature and flow rate of the heat transfer fluid. It is determined by using Eq. [15.3] [10]:

$$\eta_{\text{ch}}(t) = \frac{T_{\text{i}} - T_{\text{o}}(t)}{T_{\text{i}} - T_{\text{ini}}} \qquad [15.3]$$

where T_i and T_o are the inlet and outlet temperature of the heat transfer fluid, respectively, and T_{ini} is the initial temperature of the fluid in the storage tank.

Chan et al. [11] defined the thermal storage efficiency of the charging process as the actual energy change at time t divided by the maximum energy change after ideal plug flow replacement of the entire storage volume, given by Eq. [15.4]:

$$\eta_{\text{storage}}(t) = \frac{T_{\text{avg}}(t) - T_{\text{ini}}}{T_{\text{i}} - T_{\text{ini}}} \qquad [15.4]$$

where T_i and T_o and are the inlet and outlet temperature of the heat transfer fluid, respectively. T_{ini} and T_{avg} (t) are the initial and mass-weighted average temperature of fluid in the storage tank.

15.6.3 Heat loss evaluation

Heat losses of the storage vessel to the surroundings can severely diminish the heat storage efficiency. The overall heat loss coefficient of a storage system and the surface area of the storage tank are the major factors that decide the heat retention capacity of the storage system. The overall heat loss coefficient of a storage system is evaluated by conducting an experiment as detailed herein.

The storage medium in the tank is raised to a higher operating temperature (T_1) and left undisturbed. The temperature in the storage tank decreases slowly, and the average temperature inside of the storage tank is recorded at a regular interval of time and the variation of temperature with respect to time is observed. The experiment should be continued until the storage tank reaches the lower operating temperature (T_2) of the storage system. The overall heat loss coefficient is evaluated using Eq. [15.5]:

$$\begin{aligned} & mc_{\text{p}}(T_1 - T_2)/t = U_{\text{L}}A_{\text{s}}\,\text{LMTD} \\ & \text{LMTD} = ((T_1 - T_\infty) - (T_2 - T_\infty))/(\ln((T_1 - T_\infty) - (T_2 - T_\infty))) \end{aligned} \qquad [15.5]$$

where U_L is the overall heat loss coefficient (W/m K), m is the mass of the storage medium (kg), c_p is the specific heat capacity of the storage medium (J/kg K), T_1 is the temperature of the storage medium at the initial time (K), T_2 is the temperature of the storage medium at the end of the experiment (K), T_∞ is the average ambient temperature during the conduct of experiment (K), t is the time taken for the temperature to reach T_2 from T_1 (s), A_s is the surface area of the storage tank (m^2), and LMTD is the log mean temperature difference (K).

The heat loss from the storage tank in a given interval of time (Δt) is evaluated using Eq. [15.6]:

$$Q = (U_L A_s \mathrm{LMTD}) \times \Delta t \qquad [15.6]$$

15.6.4 Stratification performance parameters

Thermal stratification refers to the formation of the thermal gradient layer due to the difference in fluid density, with the warmer layers of the fluid placed above the cooler ones, resulting in a region of steep temperature gradient called a thermocline. With no physical barrier between the hot and cold regions of the TES tank, the buoyancy is the only mechanism separating them. A stable thermal stratification inside storage system lowers the energy degradation and thus maximizes the energy storage capacity with enhanced efficiency. For instance, the lower temperature of the flow returning from a stratified storage increases the solar collector's efficiency by reducing the convective and radiative heat losses from the collectors to the ambient. Because of fluid mixing inside of the storage tank, the quantity and quality of stored energy is lowered. However, with a stable stratification, loss in energy due to fluid mixing can be significantly reduced, thus enhancing the thermal performance of the TES system and the collector efficiency.

In a storage tank, because of the incoming forced circulation of hot water as in solar hot water storage tanks, a mixed region in the top layer of the tank is formed. The mixed layer moves down as time progresses, leading to a steep temperature gradient inside of the storage tank. This separates the hot and cold layers with an intermediate region termed as a thermocline region. Fig. 15.20 illustrates the differing degrees of stratification within a storage tank with the same amount of stored heat. The thermocline region with a large temperature gradient separating the hot and cold layers should be as minimal as possible. The selection of fluid parameters and appropriate stratification enhancing techniques need to match the energy supply, distribution, and utilization according to the temperature requirements, the required charge/discharge rates, and spatial limitations.

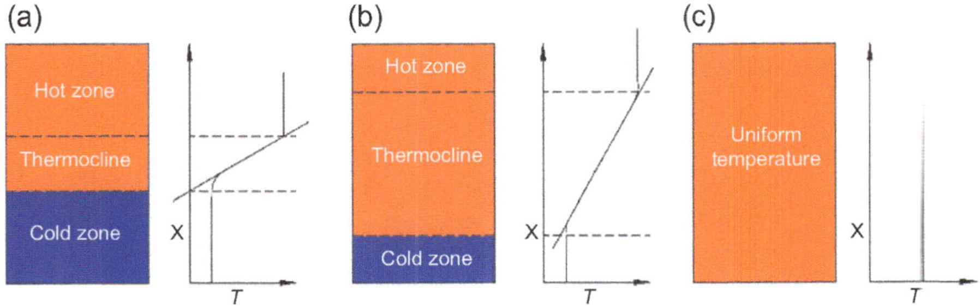

Figure 15.20 Stratification in a storage system: (a) highly stratified, (b) moderately stratified, and (c) fully stratified [12].

15.6.4.1 Design considerations for stratified thermal energy storage tanks

Thermal stratification in a storage tank improves the efficiency of the whole thermal system; hence, much research is performed to enhance the stratification in the thermal storage tanks. Inhibition of mixing between different temperature layers by reducing the flow turbulence is the key toward establishing a stable thermal stratification. The following criteria help in achieving a TES tank with enhanced thermal stratification:

- A deep water-storage container with high aspect ratio.
- Proper design of inlet diffusers that produces a uniform flow of water.
- Use of flow guiding baffles and fabric stratifiers.
- The velocity of the water flowing into and out of the tank should be low.
- The tank should be well insulated.

15.6.4.2 Performance evaluation

To quantify stratification in a TES system, various performance parameters such as stratification number, Richardson number, and mix number were introduced and are being used by researchers worldwide as the evaluating index. Among those, the stratification and Richardson number seemed to predict thermal stratification under all of the cases. Apart from this, temperature distribution along the height of the storage tank and the thermocline thickness were also used to characterize the thermal stratification in a solar thermal storage tank. The recent advancements in the computational analysis that predicts the temperature distribution in the entire storage tank paves the way for a detailed energy and exergy analysis of the storage system.

Stratification number

The stratification number is the ratio between the mean of the temperature gradients at each time interval and the temperature gradient at the start of the charging process ($t = 0$). During the charging process, as the charging proceeds the stratification number will increase and after reaching the peak, this value will start to decrease. The peak in stratification number signifies the time at which the outlet temperature of the storage tank starts to increase from its initial value. The decrease in stratification number after the peak represents the degradation of thermal stratification in the TES tank.

Fernandez-Seara et al. [13] proposed Eq. [15.7] to evaluate the stratification number based on the temperature measurements at equidistant positions in the TES tank:

$$Str = \frac{(\partial T/\partial y)_t}{(\partial T/\partial y)_{t=0}} \qquad [15.7]$$

where

$$(\partial T/\partial y)_t = \frac{1}{N-1}\left[\sum_{j=1}^{N-1}\left(\frac{T_j - T_{j+1}}{\Delta h}\right)\right]$$

$$(\partial T/\partial y)_{t=0} = \frac{T_i - T_{\text{ini}}}{(N-1)\Delta h}$$

where j corresponds to the nodal points where the temperature measurements are made, N is the number of nodal points, and Δh is the distance between the nodal points. T and T_{ini} represent the inlet and initial temperature of the heat transfer fluid, respectively.

These equations are only applicable when the temperature measurements are made at equidistant positions in the TES tank. However, if the temperature measurements are not made at some arbitrary positions, then these equations should be modified as in Eq. [15.8]:

$$(\partial T/\partial y)_t = \frac{1}{N-1}\left[\sum_{j=1}^{N-1}\left(\frac{T_j - T_{j+1}}{\Delta h_j}\right)\right] \quad [15.8]$$

$$(\partial T/\partial y)_{t=0} = \frac{T_i - T_{\text{ini}}}{\sum_{j=1}^{N-1} \Delta h_j}$$

where Δhj is the distance between the jth and $(j+1)$th nodal points.

Richardson number

The Richardson number is an effective indicator of stratification performance that considers the various aspects of storage systems such as the geometry of the storage tank, velocity, and the temperature at the top and bottom layers of the tank, and its higher value represents a better stratification. It is given by the ratio of buoyancy forces to mixing forces and is calculated by Eq. [15.9]:

$$Ri = \frac{g\beta H(T_{\text{tp}} - T_{\text{bt}})}{v_{\text{sf}}^2} \quad [15.9]$$

$$v_{\text{sf}} = \frac{\dot{V}}{\pi r^2}$$

where T_{tp} and T_{bt} represent the temperature of water measured at the top (T_1) and the bottom of the TES tank, respectively, and H is the distance between these locations. $\dot{V}$ and v_{sf} represent the volumetric flow rate and superficial velocity of heat transfer fluid entering the TES tank, and r is the radius of the inlet pipe.

Nomenclature and symbols

Abbreviations	
AH	Auxiliary heater
BTES	Borehole thermal energy storage
CTES	Cool thermal energy storage
GWTES	Gravel-water thermal energy storage
HTF	Heat transfer fluid
LCZ	Lower convective zone
LMTD	Log mean temperature difference
NCZ	Nonconvective zone
PCM	Phase change material
PV	Photovoltaic
STS	Sensible thermal storage
TES	Thermal energy storage
UCZ	Upper convective zone
VAR	Vapor absorption refrigeration
VCR	Vapor compression refrigeration
Symbols	
Q	Heat stored (J)
V	Volume (m^3)
ρ	Density (kg/m^3)
m	Mass (kg)
c_p	Specific heat capacity of the storage medium (J/kg K)
dT	Temperature change (K)
Q_{ins}	Instantaneous rate of charging/discharging (J/s)
m	Mass flow rate of heat transfer fluid (kg/s)
T_{in}	Temperature of the heat transfer fluid at the inlet (K)
T_{out}	Temperature of the heat transfer fluid at the outlet (K)
η_{ch}	Charging efficiency (%)
$\eta_{storage}$	Storage efficiency (%)
T_{ini}	Initial temperature of fluid in the storage tank (K)
T_{avg}	Mass weighted average temperature of fluid in the storage tank (K)

U_L	Overall heat loss coefficient (W/m·K)
T_1	Temperature of the storage medium at initial time (K)
T_2	Temperature of the storage medium at the end (K)
T_{co}	Average ambient temperature of experiment (K)
t	Time (s)
A_s	Surface area of the storage tank (m^2)
N	Number of nodal points
Str	Stratification number
Δh	Distance between the nodal points
Ri	Richardson number
g	Gravitational acceleration
β	Thermal expansion coefficient (1/K)
T_{ref}	Reference temperature (K)
r	Radius (m)
V_s	Superficial velocity (m/s)

References

[1] Fernandez AI, Martínez M, Segarra M, Martorell I, Cabeza LF. Selection of materials with potential in sensible thermal energy storage. Sol Energy Mater Sol Cells 2010;94: 1723−9.

[2] Abhat A. Short term thermal energy storage. Rev Phys Appl 1980;15:477−501.

[3] Khare S, Dell Amico M, Knight C, McGarry S. Selection of materials for high temperature sensible energy storage. Sol Energy Mater Sol Cells 2013;115:114−22.

[4] Ercan A. Energy storage systems − storage of thermal energy − Encyclopedia of life support systems. 2006.

[5] Hahne E. Energy storage systems − storage of sensible heat Encyclopedia of Life Support Systems. 1986.

[6] Dincer I, Rosen MA. Thermal energy storage: systems and applications. 2nd ed. 2010.

[7] Pinel P, Cruickshank CA, Morrison IB, Wills A. A review of available methods for seasonal storage of solar thermal energy in residential applications. Renew Sustain Energy Rev 2011;12:3341−59.

[8] Garg HP, Mullick SC, Bhargava AK. Solar thermal energy storage. Reidel Publishing; 1985.

[9] Kelly B. Lessons learned, project history, and operating experience of the solar two project. SAND2000-2598. Albuquerque, New Mexico: Sandia National Laboratories; November 2000.

[10] Ramana AS, Venkatesh R, Antony Aroul Raj V, Velraj R. Experimental investigation of the LHS system and comparison of the stratification performance with the SHS system using CFD simulation. Sol Energy 2014;103:378−89.

[11] Chan, et al. A numerical study of transient mixed convection flows in a thermal storage tank. J Sol Energy Eng 1983;105:246–53.
[12] Haller MY, Cruickshank CA, Streicher W, Harrison SJ, Andersen E, Furbo S. Methods to determine stratification efficiency of thermal energy storage processes – review and theoretical comparison. Sol Energy 2009;83:1847–60.
[13] Fernandez-Seara J, Uhia FJ, Sieres J. Experimental analysis of a domestic electric hot water storage tank. Part I. Static mode of operation. Appl Therm Eng 2007;27:129–36.

Latent heat storage for solar heating and cooling systems

16

C.A. Infante Ferreira
Delft University of Technology, Delft, The Netherlands

16.1 Introduction

This introductory section provides an overview of the possible system layouts and how latent heat storage can be integrated in the solar systems. Thermal energy storage systems are indispensable to minimize the discrepancy between the availability of solar energy and the demand for heating and/or cooling. Storage can be done at roughly two temperature levels: the temperature at which the thermally driven heating/cooling cycle is supplied or the temperature level at which heating and/or cooling are supplied. If the system has the purpose of delivering both heating and cooling, then even three temperature levels are relevant. Storage at the temperature level of the application (mid- and low-temperature level) has the advantage that a smaller amount of energy needs to be stored because most thermal losses will take place before storage takes place. The disadvantage is that it requires a lower (cooling) or higher (heating) temperature internally in the thermodynamic cycle, reducing the energetic performance of the cycle because a temperature driving force is needed to load/unload the storage. Storage at the supply (high) temperature level does not have this disadvantage. However, it requires a higher solar collector temperature to account for the temperature difference required to load/unload the latent heat storage. In addition, a larger amount of energy needs to be stored to compensate for the thermal energy losses involved with the energy conversion cycle. Fig. 16.1 illustrates the possible locations and temperature levels in the system when an absorption heat pump/chiller is used for the energy conversion.

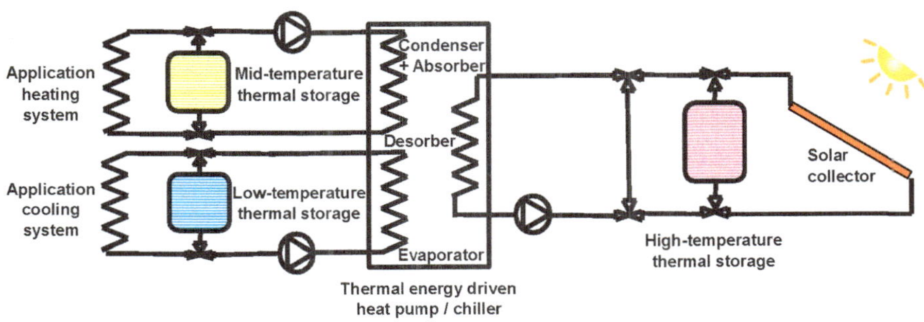

Figure 16.1 Schematic of a solar heating and cooling system illustrating the possible locations of thermal storage.

Advances in Solar Heating and Cooling. http://dx.doi.org/10.1016/B978-0-08-100301-5.00016-3

Table 16.1 Typical heating medium temperature required for the operation of sorption cycles for a heat rejection temperature of 30°C or 50°C and evaporating temperature of 2°C and corresponding coefficient of performance

Thermal conversion cycle	$T_{\text{heating fluid}}$ (°C)	$COP_{cooling}$	$T_{\text{heating fluid}}$ (°C)	$COP_{heating}$
Half-effect H_2O-LiBr	67	0.35	108	1.33
Single-effect H_2O-LiBr	92	0.72	NA	NA
Double-effect H_2O-LiBr	142	1.29	NA	NA
Single-effect NH_3-H_2O	86	0.60	140	1.52
Adsorption H_2O-silica gel	90	0.41	120	1.26

COP, Coefficient of performance; *H_2O*, water; *LiBr*, lithium bromide; *NH_3*, ammonia; *NA*, the cycle cannot operate under these specific conditions.

The required desorption temperature of thermally driven sorption heat pumps/chillers depends on the selected thermal conversion cycle. Infante Ferreira and Kim (2014) indicate the heating fluid temperature required to drive several types of thermally driven chillers. A summary is given in Table 16.1 for a heat rejection temperature of 30°C for cooling applications (Columns 2 and 3) and of 50°C for heating applications (Columns 4 and 5).

Table 16.1 makes it clear that, if evaporating temperatures of 2°C are required, then cooling applications require heating medium temperatures in the range 70–140°C associated with cooling coefficients of performance (COPs) from 0.35 to 1.30. Again assuming evaporating temperatures of 2°C (heat uptake from environmental ground source), application for heating requires higher temperatures of 110–140°C with heating COPs from 1.30 to 1.50. The high-temperature thermal storage requires storage at temperatures in the range 70–145°C.

16.2 Temperature level for latent heat storage design

This section discusses the impact of the temperature level of latent heat storage and how this affects the total system performance. Selection of the thermal conversion equipment requires that the cooling/heating effect is optimized in combination with solar thermal collectors (van Leeuwen, 2010). The required temperature level for the heating medium (and so the temperature level required for the high-temperature storage) can only be determined after this selection. The steps needed for the optimization are illustrated below for a specific day in a specific location with the radiation and environmental temperature given in Fig. 16.2.

Assume that the heat rejection temperature is 10K higher than the environmental temperature:

$$\overline{T}_{\text{sink}} = T_{\text{environment}} + \Delta T_{\text{rejection}} = T_{\text{environment}} + 10 \quad [16.1]$$

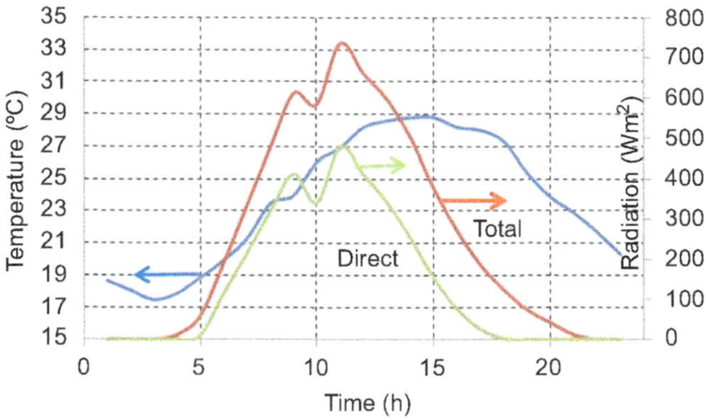

Figure 16.2 Environmental temperature and radiation (total and direct) for a specific day during the cooling period.

For cooling $\overline{T}_{\text{evaporation}} = 2°\text{C}$. Fitting polynomial equations to the relation between the heat rejection temperature and the required heating medium temperature reported by Infante Ferreira and Kim (2014), the following equation can be derived:

$$T_{\text{heating}} = a \times \overline{T}_{\text{sink}}^2 + b \times \overline{T}_{\text{sink}} + c \quad [16.2]$$

with the coefficients given in Table 16.2 when the temperature is expressed in Celsius.

Fig. 16.3 illustrates how this affects the required heating medium temperature.

The COP of the different cycles can be obtained from the second law efficiency, $\eta_{\text{2nd Law}}$: 0.44, 0.50, 0.52, 0.46, and 0.25, respectively, for half-, single-, and

Table 16.2 Coefficients in Eq. [16.2] that give the relation between the heat rejection temperature and the required temperature of the heating fluid

Thermal conversion cycle	*a*	*b*	*c*
Half-effect H_2O-LiBr	−0.005	2.45	−2
Single-effect H_2O-LiBr	0	1.2	55
Double-effect H_2O-LiBr	0	3.4	40
Single-effect NH_3-H_2O	−0.05	6.7	−70
Adsorption H_2O-silica gel	0	1.5	45

H_2O, Water; *LiBr*, lithium bromide; *NH_3*, ammonia.

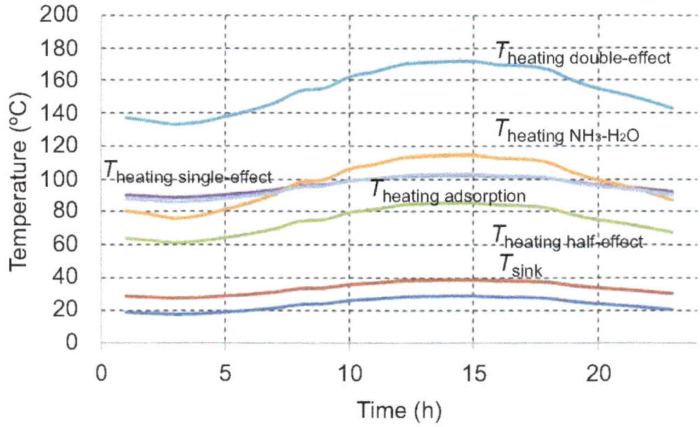

Figure 16.3 Example variation of the required heating medium temperature as a function of the environmental temperature.

double-effect, ammonia (NH_3)—water (H_2O), and adsorption systems (Infante Ferreira and Kim, 2014):

$$\begin{aligned} COP &= \eta_{2nd\ Law} \times COP_{Carnot} \\ &= \eta_{2nd\ Law} \times \left(1 - \frac{\overline{T}_{sink}}{T_{heating}}\right) \times \left(\frac{\overline{T}_{source}}{\overline{T}_{sink} - \overline{T}_{source}}\right) \end{aligned} \quad [16.3]$$

with $\overline{T}_{source}$ the average temperature of the chilled water (=9.3°C (282.5K), considering a chilled water glide from 12°C to 7°C in the evaporator). The cooling effect then follows from

$$\dot{q}_{cooling} = \dot{q}_{sun} \times COP \times \eta_{solar\ collector} \quad [16.4]$$

with the solar collector efficiency being approximated with

$$\eta_{solar\ collector} = d - e\frac{T_{heating} - T_{environment}}{\dot{q}_{sun}} \quad [16.5]$$

with *d* and *e* constants depending on the solar collector design. Table 16.3 provides values of these coefficients for some solar collector designs (Kalogirou, 2004). In the case of a compound concentrating collector (CPC) being applied, only the direct radiation should be taken into account. Fig. 16.1 also shows the direct radiation for the specific day selected for illustration of the calculation.

The results shown in Fig. 16.4 make clear that for the specific day considered the evacuated-tube solar collector always delivers the best results. The single-effect water—lithium bromide system gives the largest specific cooling capacity, closely followed by the single-effect ammonia—water and the double-effect water—lithium bromide systems. Fig. 16.3 shows that for both single-effect systems a maximum

Table 16.3 Coefficients in Eq. [16.5] that give the solar collector efficiency as a function of operating conditions

Solar collector type	*d*	*e*
Flat plate, FPC	0.79	6.67
Advanced flat plate, AFPC	0.80	4.73
Compound concentrating, CPC	0.72	1.56
Evacuated tube, ETC	0.82	2.19

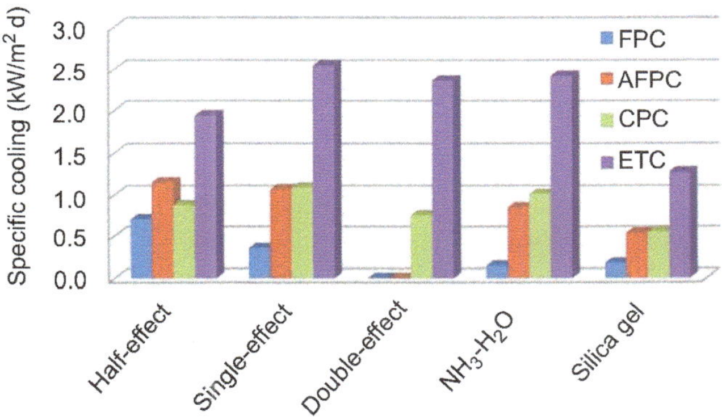

Figure 16.4 Effect of heating medium temperature, solar collector type, and thermal conversion system on the cooling delivered for an average summer day per square meter of collector surface. *FPC*, Flat-plate collector; *AFPC*, advanced flat-plate collector; *CPC*, compound concentrating collector; *ETC*, evacuated-tube collector.

temperature of 103°C is required whereas the double-effect cycle requires a temperature of 172°C.

The use of energy storage devices implies extra temperature driving forces and therefore lower performances. If a latent heat storage is applied at the solar collector side, then the solidification temperature must be at least 7.5K higher (110°C or 180°C) and the solar collector needs to operate again with at least a 7.5K higher temperature. This will increase the required solar collector temperature by 15K and will reduce the solar collector efficiency. This is illustrated in the left-hand side of Fig. 16.5, which indicates that this leads to a reduction of the cooling that can be generated per solar collector area. If the latent heat storage takes place at the application level, then, for cooling, a melting temperature of +2°C is required so that evaporation needs to take place at −5°C and only the NH_3-H_2O system can be used. This is illustrated in the right-hand side of Fig. 16.5, which indicates that storage at this temperature level has a higher effect on the total efficiency than storage at the solar collector temperature. This does not include the storage losses associated with heat transfer losses to the environment, which will be higher at higher temperatures.

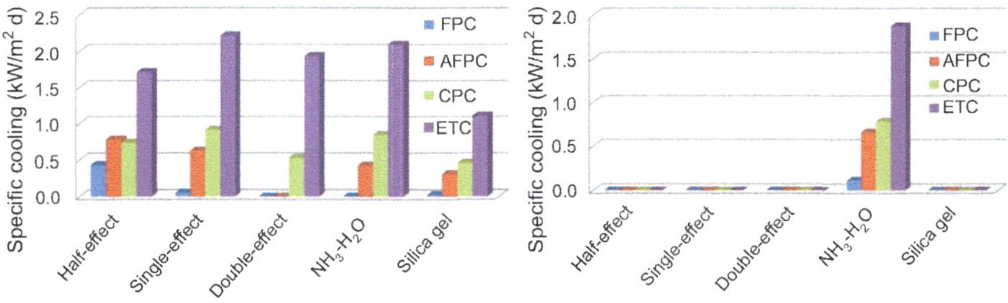

Figure 16.5 Effect of storage temperature level, solar collector type, and thermal conversion system into the cooling delivered for an average summer day per square meter collector surface. Left: storage at solar collector temperature level. Right: storage at cooling application level. *FPC*, Flat-plate collector; *AFPC*, advanced flat-plate collector; *CPC*, compound concentrating collector; *ETC*, evacuated-tube collector.

It should be remarked that the results shown here apply for very specific climate conditions and may be different for other conditions. Nevertheless, the method illustrates how the decision can be made about which temperature level is most suitable for latent heat storage application.

Fig. 16.6 illustrates the advantages of the application of pumpable phase change material (PCM) slurries. The use of this method of latent heat storage allows for higher evaporating temperatures in the sorption cycles giving a COP advantage and thereby increasing the specific cooling effect of the system. For this specific day the best solution is still the single-effect cycle combined with evacuated-tube solar collectors, and the highest possible evaporating temperature should be selected. The specific cooling is then higher than when high-temperature heat storage is applied.

Fig. 16.7 shows how the availability of cooling capacity varies with time for the considered day and location. Depending on the cooling load profile, the storage should be sized so that it can store the energy that is not directly needed to cover the required cooling capacity. In addition, the storage must be designed to deliver the cooling

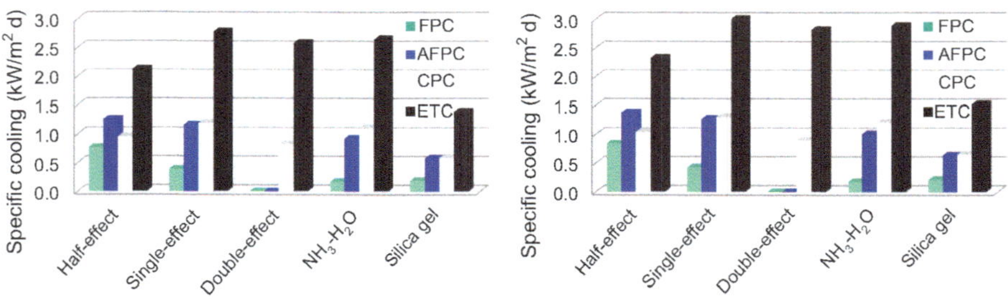

Figure 16.6 Effect of using pumpable phase change material slurries for storage at cooling application level. Left: evaporation takes place at 4°C (CO_2-hydrate slurry). Right: evaporation takes place at 6°C (tetra-*n*-butylammonium bromide–hydrate slurry). *FPC*, Flat-plate collector; *AFPC*, advanced flat-plate collector; *CPC*, compound concentrating collector; *ETC*, evacuated-tube collector.

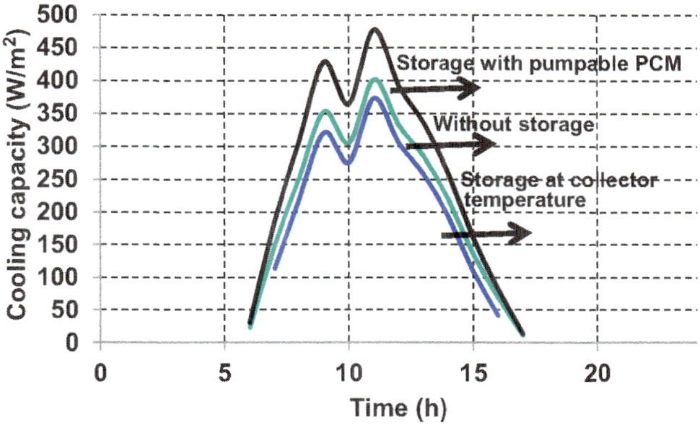

Figure 16.7 Cooling capacity per square meter evacuated-tube collector operating in combination with a single-stage lithium bromide–water absorption system. *PCM*, Phase change material.

capacity required in periods without (sufficient) insolation. Klein and Beckman (1979) studied the effect of system storage capacity per square meter of effective collector area and quantified the related solar fraction. As the name indicates, the solar fraction gives the fraction of useful energy derived from solar energy. These authors introduce a simple model for the thermal storage. A similar approach is also used in Tsoutsou et al. (2014). A more detailed model (Fan et al., 2014) will be introduced in Section 16.4.5, which allows derivation of the temperature of the heat transfer fluid (HTF) at the outlet of the latent heat thermal storage (LHTS) as a function of time. Optimization of the size of the solar collector field in combination with the volume of the storage should be the result of an economic evaluation. An optimization approach for the design of the latent heat storage taking economic aspects into account is given in Meewisse (2004).

16.3 Storage media

16.3.1 Required phase change temperature for collecting temperature level

This section provides an overview of high-temperature latent heat materials available for solar heating and cooling applications and its properties. As discussed earlier, the required phase change temperatures are approximately 110°C for single-effect systems and approximately 180°C for double-effect systems. The exact required temperature level will depend on the local climate conditions.

16.3.2 Double-effect H_2O/LiBr absorption systems

For any application it is required that the PCM has a melting temperature within the practical range of the application. Hence, as discussed previously, the selection of the PCM for the proposed application is derived from the temperature requirement of the high-temperature source for the double-effect H_2O/LiBr absorption system. This absorption system requires a high-temperature source in the range of 150–200°C depending on the local climate conditions.

Sharma et al. (2009) summarized the characteristics of PCMs suitable for thermal storage applications. These include

1. a suitable phase change temperature,
2. good thermal conductivity,
3. small volume change during the phase change process,
4. chemical stability,
5. safety, and
6. cost-effectiveness.

The literature introduces several PCMs for this operating range (Gallego et al., 2013; Gil et al., 2010; Hoshi et al., 2005; Kenisarin, 2010; Kenisarin and Mahkamov, 2007; Kumaresan et al., 2011; Pintaldi et al., 2015; Premkumar and Ramachandran, 2011; Sari et al., 2011; Takahashi et al., 1987).

These PCMs and their properties are listed in Table 16.4. The last column indicates the PCM price in relation to the price of hydroquinone. Inorganic materials (eg, salts) can be used as PCMs for applications with temperatures above 150°C. In general, the thermal conductivity of these materials is high. Takahashi et al. (1987) measured the heat capacity and the latent heat of fusion for a binary eutectic mixture composed of sodium hydroxide (NaOH) and potassium hydroxide (KOH). The authors concluded that this mixture is a useful PCM for LHTS applications because of the high latent heat value and the low cost. However, this mixture is very corrosive. Hoshi et al. (2005) reported a list of candidate materials for high-temperature LHTS applications. Lithium nitrate ($LiNO_3$) and sodium nitrate ($NaNO_3$) have a higher heat capacity compared with NaOH and KOH. However, both of these materials have higher costs, and $LiNO_3$ is almost twice as expensive compared with NaOH and KOH.

Sugar alcohols have been explored as candidate PCMs for applications in the temperature range between 150°C and 200°C. Sugar alcohols are often used as sweetening agents. They are safe and have a high latent heat of fusion value. Galactitol is a sugar alcohol and has a high latent heat storage capacity. The drawback of this material is the large supercooling degree, which is approximately 72K. Therefore this PCM is not suitable for LHTS applications (Sari et al., 2011). Kumaresan et al. (2011) examined D-mannitol as a PCM and concluded that D-mannitol is an excellent material because of the high melting temperature and the high latent heat of fusion value. Premkumar and Ramachandran (2011) chose D-mannitol as a potential PCM and measured the performance of D-mannitol as the storage material for a parabolic trough test unit. Gil et al. (2010) tested D-mannitol and hydroquinone in a pilot plant with different melting and solidification rates. The authors concluded that these materials are potential candidates for solar cooling applications. Furthermore, the authors found that D-mannitol is able to

Table 16.4 Candidate phase change materials for high-temperature application in combination with double-effect absorption systems

Material	Melting point (°C)	Heat of fusion (kJ/kg)	Sensible heat (s/L) (kJ/kgK)	Thermal conductivity (s/L) (W/mK)	Density (s/L) (kg/m^3)	Cost/ $Cost_{hydroquinone}$ (−)
Phenylhydrazone benzaldehyde	155	134.8				27.3
Salicylic acid	159	199		0.115	1443	8.9
$LiNO_3$-$NaNO_3$-KCl (55.4:4.5:40.1 wt%)	160	266				2.7
Benzanilide	161	162				3.2
E164 (EPS Ltd.)	164	306			1500	NA
$LiNO_3$-KCl (58.1:41.9 wt%)	166	272				2.8
D-Mannitol	166−168	261−340			1520	1.4
NaOH-KOH (50:50 mol%)	169−171	202−213				1.2
Hydroquinone	166−173	230	2.31/2.67	0.45/0.61	1358/1155	1.0
$LiNO_3$-$NaNO_3$-LiCl (47.9:50.7:1.4 wt%)	180	265				2.4
$LiNO_3$-$NaNO_3$-$Sr(NO_3)_2$ (45:47:8 wt%)	180	265				2.3
p-Aminobenzoic acid	187	153				16.9
Galactitol	188−189	351.8			1470	6.6
$LiNO_3$-$NaNO_3$ (57:43 wt%)	193	248−265				2.6
$LiNO_3$-$NaNO_3$ (49:51 wt%)	194	265				2.3

NA, not available.

store more energy compared with hydroquinone. On the other hand, D-mannitol shows supercooling during each experiment whereas hydroquinone shows only a little supercooling.

16.3.3 Single-effect absorption systems

The selection of the PCM for these systems is derived from the temperature requirement of the high-temperature source for the single-effect absorption systems. This absorption system requires a high-temperature source in the range of 85−125°C (see Fig. 16.3) depending on the local climate conditions.

The literature (Kenisarin and Mahkamov, 2007; Pintaldi et al., 2015) introduces several PCMs for this operating range. These PCMs and their properties are listed in Table 16.5.

16.3.4 Storage media for storage at cooling/heating application temperature level

This section provides an overview of latent heat materials available for solar heating and cooling applications and the properties to be integrated at the application temperature of the cooling/heating system. As discussed previously, if the latent heat storage takes place at application level then, for cooling, a melting temperature of +2°C is required.

Zalba et al. (2003) and Pintaldi et al. (2015) provide an overview of the PCMs in the cooling temperature range. Their data, added with manufacturers' data, are summarized in Table 16.6. Most of the products available in this range are commercial products.

If the latent heat storage takes place at the application level, then, for heating, a melting temperature between 25°C and 60°C is required.

Zalba et al. (2003) and Kenisarin and Mahkamov (2007) provide an overview of the PCMs in the heating temperature range. Their data, added with manufacturers' data, are summarized in Table 16.7. Most of the products available in this range are also commercial products.

16.4 Main materials for storage (Pumpable slurries and PCM)

This section explains how pumpable slurries can be applied for latent heat storage and how such systems can be designed. Pumpable slurries are mixtures of a solution (liquid) and crystals (solid). The properties of the solution and of the crystals need first to be determined. The properties of the slurry are then determined from the combination of solution and crystal properties (Pronk, 2006). It also explains how stationary PCM systems can be applied for latent heat storage and how such systems can be designed.

Table 16.5 **Candidate phase change materials for high-temperature application in combination with single-effect absorption systems**

Material	Melting point (°C)	Heat of fusion (kJ/kg)	Sensible heat (s/L) (kJ/kgK)	Thermal conductivity (s/L) (W/mK)	Density (s/L) (kg/m^3)	Cost/ $Cost_{D\text{-sorbitol}}$ (−)
TH89 (TEAP energy)	89	140−149	1.84/2.83	0.60/0.69	1640	8.0
SP90 (Rubitherm)	89	150	2.48	0.67		
$Mg(NO_3)_2\ 6H_2O$	89.5	167−175	2.51/1.84	0.67/0.49	1636/1550	2.5
RT90HC (Rubitherm)	91	163	1.80/2.40	0.20/0.20	930/770	4.0
Xylitol $C_5H_7(OH)_5$	94	263			1520	6.5
$(NH_4)Al(SO_4)$-$6H_2O$	95	269				4.2
D-Sorbitol $C_6H_8(OH)_6$	97	185			1500	1.0
KOH-H_2O/KOH	99					2.6
RT110 (Rubitherm)	112	213	1.80/2.40	0.20	930/770	NA
$MgCl_2$ -$6H_2O$	117	169	2.61	0.70/0.57	1570/1450	3.4
Erythritol	118	340	1.38/2.76	0.73/0.33	1480/1300	33.2

NA, not available.

Table 16.6 Candidate phase change materials for low-temperature application in combination with sorption systems

Material	Melting point (°C)	Heat of fusion (kJ/kg)	Sensible heat (s/L) (kJ/kgK)	Thermal conductivity (s/L) (W/mK)	Density (s/L) (kg/m^3)
RT-4 (Rubitherm)	−4	179	2.0/2.0	0.20/0.20	880/770
STL-3 (Mitsubishi)	−3	328			
AN.03 (Cristopia)	−2.6	174	4.3/2.9		
H_2O	0	333	2.1/4.2	2.24/0.61	917/1000
RT 0 (Rubitherm)	0	225	2.0/2.0	0.20/0.20	880/770
RT2HC (Rubitherm)	2	205	2.0/2.0	0.20/0.20	880/770
RT3HC_1 (Rubitherm)	3	200	2.0/2.0	0.20/0.20	880/770
RT4 (Rubitherm)	4	182	2.0/2.0	0.20/0.20	880/770
Latest 4T (TEAP)	4	105			1400
Paraffin C_{14}	4.5	165			
RT5 (Rubitherm)	5	180	2.0/2.0	0.20/0.20	880/770
RT5HC (Rubitherm)	5	240	2.0/2.0	0.20/0.20	880/760
RT6 (Rubitherm)	6	175	2.0/2.0	0.20/0.20	880/770
ClimSel C7	7	126	3.8/4.0	0.50/0.70	1400
Latest 7T (TEAP)	7	135			1400
Paraffin C_{15}−C_{16}	8	153			
Polyglycol E400	8	100		0.21/0.19	1128/1125
RT8 (Rubitherm)	8	180	2.0/2.0	0.20/0.20	880/770

Table 16.7 Candidate phase change materials for intermediate temperature application in combination with sorption systems

Material	Melting point (°C)	Heat of fusion (kJ/kg)	Sensible heat (s/L) (kJ/kgK)	Thermal conductivity (s/L) (W/mK)	Density (s/L) (kg/m^3)
$Mn(NO_3)_2-6H_2O$	25.8	126			1795/1728
Paraffin C_{18}	28.0	244		0.15/0.15	814/774
RT28HC (Rubitherm)	28.0	245	2.0/2.0	0.20/0.20	880/770
$CaCl_2-6H_2O$	29.0	191		1.09/0.54	1802/1562
$LiNO_3-3H_2O$	30.0	296			
ClimSel C32	32.0	162	3.6/3.6	0.50/0.70	1420
$Na_2SO_4-10H_2O$	32.4	254		–/0.54	1485/1458
$Na_2CO_3-10H_2O$	33.0	247			1442/1442
RT35HC (Rubitherm)	35.0	240	2.0/2.0	0.20/0.20	880/770
$Zn(N_3)_2$ $6H_2O$	36.2	147		–/0.47	1937/1828
Paraffin $C_{16}-C_{28}$	43.0	189		0.21/0.21	910/765
RT44HC (Rubitherm)	44.0	255	2.0/2.0	0.20/0.20	800/700
ClimSel C48	48.0	180	3.6/3.6	0.50/0.70	1360
$Mg(NO_3)_2$ $6H_2O$ (61.5%)+NH_4NO_3 (38.5%)	52.0	126		0.55/0.50	1596/1515
$Na(CH_3COO)-3H_2O$	58.0	264			1450/1450
ClimSel C58	58.0	288		0.50/0.70	1420
Paraffin C_{20}-C_{33}	49.0	189		0.21/0.21	912/769
Paraffin C_{22}-C_{45}	59.0	189		0.21/0.21	920/795
RT64HC (Rubitherm)	64.0	230	2.0/2.0	0.20/0.20	880/780
NaOH	64.3	228			1690/1690

16.4.1 Slurry properties

16.4.1.1 Density

The density of the slurry is obtained from the weighted average specific volumes of both phases:

$$\rho_{\text{slurry}} = \frac{1}{\frac{w_{\text{crystals}}}{\rho_{\text{crystals}}} + \frac{(1 - w_{\text{crystals}})}{\rho_{\text{sol}}}} \tag{16.6}$$

where w_{crystals} is the mass fraction of crystals.

16.4.1.2 Thermal conductivity

The thermal conductivity can be obtained from the model by Tareef (1940) for liquid–solid mixtures:

$$\lambda_{\text{slurry}} = \lambda_{\text{sol}} \left(\frac{2\lambda_{\text{sol}} + \lambda_{\text{crystals}} - 2\xi(\lambda_{\text{sol}} - \lambda_{\text{crystals}})}{2\lambda_{\text{sol}} + \lambda_{\text{crystals}} + \xi(\lambda_{\text{sol}} - \lambda_{\text{crystals}})} \right) \tag{16.7}$$

where ξ is the crystals volume fraction:

$$\xi = \frac{1}{1 + \frac{(1 - w_{\text{crystals}}) \times \rho_{\text{crystals}}}{w_{\text{crystals}} \times \rho_{\text{sol}}}} \tag{16.8}$$

16.4.1.3 Enthalpy and specific heat

The enthalpy of the slurry can be derived from the weighted average of the enthalpy of the solution and the enthalpy of the crystals:

$$h_{\text{slurry}} = w_{\text{crystals}} \times h_{\text{crystals}} + (1 - w_{\text{crystals}}) \times h_{\text{sol}} \tag{16.9}$$

Or, more explicitly,

$$h_{\text{slurry}} = w_{\text{crystals}} \times \left(\Delta h_{\text{crystals}} + \int_{T_{\text{phase change}}}^{T} c_{\text{p crystals}}\, dT \right) + (1 - w_{\text{crystals}}) \times \left[h_{\text{sol}}(T_{\text{phase change}}) + \int_{T_{\text{phase change}}}^{T} c_{\text{p sol}}\, dT \right] \tag{16.10}$$

The first term provides the enthalpy contribution of the solid phase with the latent heat at phase change temperature and its sensible heat contribution. The second term represents the enthalpy contribution of the solution and includes the enthalpy at phase change temperature and the sensible heat contribution of the solution. The derivative of Eq. [16.10] after rearranging shows a clear separation between latent and sensible heat contributions:

$$\frac{\partial h_{\text{slurry}}}{\partial T} = \frac{\partial w_{\text{crystals}}}{\partial T} \times \left[\Delta h_{\text{crystals}} + \int_{T_{\text{phase change}}}^{T} (c_{\text{p crystals}} - c_{\text{p sol}}) dT - h_{\text{sol}}(T_{\text{phase change}}) \right] + w_{\text{crystals}}\, c_{\text{p crystals}} + (1 - w_{\text{crystals}}) \times c_{\text{p sol}} \quad [16.11]$$

The apparent specific heat of the slurry takes both latent and sensible contributions into account:

$$c_{\text{p app slurry}} = \frac{\partial w_{\text{crystals}}}{\partial T} \times \left[\Delta h_{\text{crystals}} + \int_{T_{\text{phase change}}}^{T} (c_{\text{p crystals}} - c_{\text{p sol}}) dT - h_{\text{sol}}(T_{\text{phase change}}) \right] + w_{\text{crystals}}\, c_{\text{p crystals}} + (1 - w_{\text{crystals}}) \times c_{\text{p sol}} \quad [16.12]$$

whereas the sensible specific heat only takes the sensible contribution into account:

$$c_{\text{p sens slurry}} = w_{\text{crystals}}\, c_{\text{p crystals}} + (1 - w_{\text{crystals}}) \times c_{\text{p sol}} \quad [15.13]$$

16.4.1.4 *Dynamic viscosity*

The dynamic viscosity of slurries increases strongly with solid concentration. The model by Thomas (1965) is generally accepted to provide a good representation of this dependence at least up to volumetric solid fractions of 0.20 (Kauffeld et al., 1999):

$$\mu_{\text{slurry}} = \mu_{\text{sol}} \left(1 + 2.5\xi + 2.5\xi^2 + 0.00273 \times e^{16.6\xi}\right) \quad [16.14]$$

where again ξ is the crystals volume fraction.

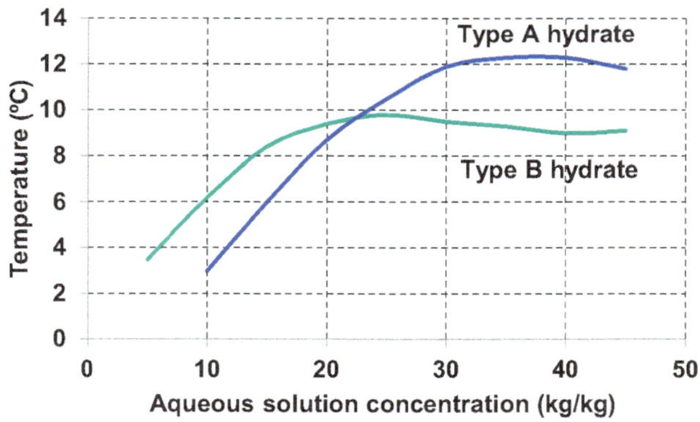

Figure 16.8 Phase diagram of tetra-*n*-butylammonium solutions in water at ambient pressure. Type A hydrates can reach phase change temperatures up to 12.5°C whereas the highest phase change temperature for type B hydrates is approximately 10°C. *PCM*, Phase change material.

16.4.2 Tetra-n-butylammonium–hydrate slurries

16.4.2.1 Phase diagram

Aqueous solutions of tetra-*n*-butylammonium (TBAB) can form two types of hydrate crystals depending on the initial solution concentration. Type A has a columnar shape and hydration number 26 and type B has an undefined form and hydration number 38. The solid-liquid equilibrium line at different solution concentrations at ambient pressure has been experimentally determined by several researchers (Darbouret et al., 2005; Hayashi et al., 2000; Ma et al., 2010; Oyama et al., 2005) and is, on average, as indicated in Fig. 16.8.

Pumpable slurries can be directly used to chill the air flow for cooling applications. For this reason a higher temperature would be more advantageous, and a phase change temperature of 12.5°C is generally preferred. From Fig. 16.8 it is clear that at this phase change temperature only type A hydrates can be formed. The required initial TBAB aqueous solution concentration is then 36.5 wt%.

16.4.2.2 Density

The density of type A crystals of TBAB-hydrate has been measured by Ma et al. (2010), Ogoshi and Takao (2004), and Oyama et al. (2005). All of these studies report a density of $\rho_{\text{crystals}} = 1080\ \text{kg/m}^3$.

The density of the aqueous solution of TBAB depends on concentration and temperature. Ma et al. (2010) and Ogoshi and Takao (2004) measured the density of the solution at various concentrations of TBAB and temperatures and concluded that the density is linearly dependent on temperature. From their experimental data the following equation, which applies for solution concentrations in the range 30–40%, can be derived:

$$\rho_{\text{sol}} = -0.2798 \times a_1 \times T_{\text{sol}} + 13.7 \times a_1 - 0.5357 \times T_{\text{sol}} + 1036.8 \qquad [16.15]$$

with

$$a_1 = \frac{w_{sol} - 0.30}{0.406 - 0.30} \quad [16.16]$$

and w_{sol} expressed in kg/kg, T in °C, and ρ_{sol} in kg/m^3. For a concentration of 36.5% $a_1 = 0.613$.

16.4.2.3 Thermal conductivity

Hayashi et al. (2000) reported that the thermal conductivity of type A TBAB-hydrate crystals is $\lambda_{crystals} = 0.42\ \text{W/mK}$. Ma et al. (2010) measured the thermal conductivity of the solution at 10°C for different concentrations, resulting in Eq. [16.17].

$$\lambda_{sol} = -0.45 \times w_{sol}^2 - 0.339 \times w_{sol} + 0.559 \quad [16.17]$$

16.4.2.4 Specific heat

Oyama et al. (2005) measured the specific heat of hydrate crystals at various temperatures below the melting temperature (solid phase) of TBAB-hydrate in a differential scanning calorimeter. Their data points have been fitted to a fourth-order polynomial function. The function is expressed as

$$c_{p\ crystals} = 2.3677 + 0.0232 \times T - 1.9697 \times 10^{-6} \times T^2 + 5.6768 \times 10^{-6} \times T^3 + 6.0606 \times 10^{-9} \times T^4 \quad [16.18]$$

where T is the temperature in degrees Celsius. The experimental data reported by Ogoshi and Takao (2004) is in agreement with Eq. [16.18].

The specific heat of TBAB solution has been measured by Asaoka et al. (2013) for different concentrations of TBAB in water at 25°C. Eq. [16.19] reproduces their experimental data.

$$c_{p\ sol} = 4.2255 - 0.41 \times w_{sol} - 1.3656 \times w_{sol}^2 \quad [16.19]$$

16.4.2.5 Latent heat

The latent heat of formation of type A TBAB-hydrate has been measured by several researchers. Ogoshi and Takao (2004) and Oyama et al. (2005) report a value of $\Delta h_{crystals} = 193\ \text{kJ/kg}$ while Kumano et al. (2006) report a value of 215 kJ/kg. Because Kumano et al. (2006) report the same value for both type A and type B hydrates whereas the hydration number of type B (38) is significantly higher than of type A (26), the value reported by the two first studies seems more reliable and should be adopted.

16.4.2.6 Dynamic viscosity

Kumano et al. (2011) experimentally determined the viscosity of aqueous TBAB solution and proposed the following dependence on temperature:

$$\mu_{sol} = \rho_{sol} \times \left(3.035 \times 10^{-6} - 1.539 \times 10^{-7} \times T_{sat} + 2.73 \times 10^{-8} T_{sat}^2\right) \quad [16.20]$$

where T_{sat} is the equilibrium temperature of the solution expressed in degrees Celsius.

16.4.3 Carbon dioxide hydrate slurries

Carbon dioxide (CO_2) hydrate slurry is considered to be one of the most promising PCMs for cold storage in space-cooling systems. It has a melting enthalpy that is even larger than the melting enthalpy of ice and the phase change takes place at temperatures above 0°C, mostly in the range 7–8°C, which matches the operating temperatures in space-cooling applications. Furthermore, the system only uses CO_2 and H_2O, natural fluids with limited environmental impact.

The properties of CO_2 hydrate slurry have been reviewed by de Sera et al. (2014). The most relevant properties are summarized here.

16.4.3.1 Phase diagram

Phase equilibrium data of CO_2 hydrate have been reported by many researchers: Vlahakis et al. (1972), Adisasmito et al. (1991), Ohgaki et al. (1993), Ng and Robinson (1985), Wendland et al. (1999), Yang et al. (2000), Mooijer-van den Heuvel et al. (2001), Yokozeki (2004), Sabil (2009), and Lirio and Pessoa (2013). These researchers show that the CO_2 hydrate formation temperature at 30 bar is approximately 7°C. Sabil's (2009) results are adopted because their data reproduce best the experimental data reported by Zhou et al. (2015). Eq. [16.21] provides the equilibrium temperature line as derived from Sabil's data.

$$T_{sat} = -6.800480 + 0.703765 \times p - 0.007446 \times p^2 \quad [16.21]$$

with p the operating pressure level expressed in bar and T_{sat} expressed in °C. Fig. 16.9 shows the equilibrium line between the liquid–vapor region and hydrate–liquid region of the CO_2–water solution.

16.4.3.2 Density

CO_2 clathrate hydrates, or simply CO_2 hydrates, are CO_2 guest molecules trapped inside of "cages," often referred to as cavities, of hydrogen-bonded water molecules, resulting in crystalline solids that look much like ice. If the cages, or cavities, in a hydrate structure are all filled with guest molecules, then the hydrate is called stoichiometric. In practice this is not often the case. CO_2 forms type I hydrates, which consist of 46 water molecules. The occupancy of CO_2 hydrate, which indicates how many

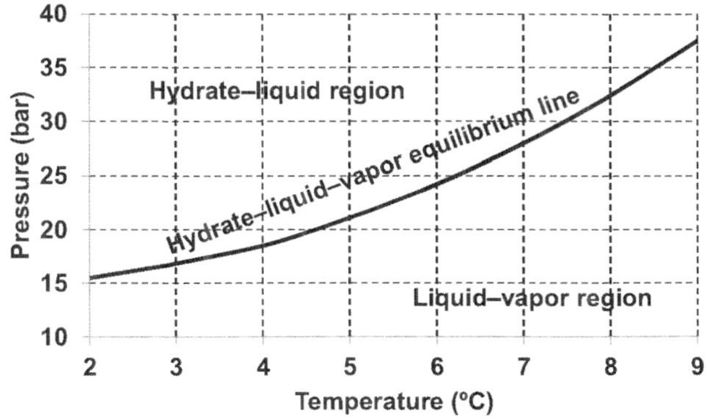

Figure 16.9 Phase diagram of CO_2 solutions in water at relevant pressure levels. With operating pressures in the range 24–33 bar, hydrates form at temperatures between 6°C and 8°C.

cages are occupied with CO_2 molecules, does not often exceed 75% in the case of hydrate structure type I. The chemical formula for CO_2 hydrate is $(x+y)CO_2{-}46$ H_2O, where $x(\leq 2)$ and $y(\leq 6)$ represent the numbers of the occupied 5^{12} (small) and $5^{1,2}6^2$ (large) cages, respectively. If all of the cages are occupied, then $x+y=8$, leading to the common expression of $8CO_2{-}46H_2O$ or $CO_2{-}5.75H_2O$ for CO_2 hydrate. However, because the ratio between small and large cages is 2–6, the practical expression for CO_2 hydrate should be $6CO_2{-}46H_2O$ or $CO_2{-}7.67$ H_2O; that is, CO_2 hydrate is a nonstoichiometric compound. The number of water molecules per CO_2 molecule, 5.75 in the stoichiometric case, is known as the hydration number. Lirio and Pessoa (2013) have shown that, depending on the formation conditions, the hydration numbers can vary between 6.0 and 8.3. The density of crystals of CO_2–hydrate is related to the hydration number. Teng et al. (1996) have proposed Eq. [16.22] to calculate this density:

$$\rho_{\text{hyd}} = \frac{46M_{CO_2}}{N_0 a^3}\left(0.409 + \frac{x_{CO_2}^{\text{hyd}}}{1 - x_{CO_2}^{\text{hyd}}}\right) \tag{16.22}$$

where M_{CO_2} is the molar mass of CO_2 in kg/kmol, N_0 is the Avogadro constant in molecules/kmol ($6.02214129 \times 10^{26}$), and $a = 12 \times 10^{-10}$ m is the lattice constant. Zhou et al. (2015) have determined from experimental data that on average, for CO_2 hydrates formed in fluidized beds, 70% of the hydrate cavities are occupied by CO_2 $\left(x_{CO_2}^{\text{hyd}} = 0.110;\ w_{CO_2}^{\text{hyd}} = 0.232\right)$, resulting in a hydration number of 8.06 and a hydrate density of $\rho_{\text{hyd}} = 1036\ \text{kg/m}^3$ when Eq. [16.22] is applied. In addition, Sari et al. (2008) report a value of $\rho_{\text{hyd}} = 1036\ \text{kg/m}^3$ for their CO_2 hydrates. With different occupancy and thus hydration number, the density of the produced hydrates will also differ. This number is highly sensitive to the experimental conditions.

Teng et al. (1996) report $\rho_{hyd} = 1049\ \text{kg/m}^3$ if only the large cavities are occupied (ie, $x_{CO_2}^{hyd} = 0.115$; $w_{CO_2}^{hyd} = 0.241$).

The density of the aqueous solution of CO_2 depends on concentration and temperature. Duan et al. (2008) have proposed the following relation:

$$\rho_{sol} = \frac{x_{H_2O} \times M_{H_2O} + x_{CO_2} \times M_{CO_2}}{\frac{x_{H_2O} \times M_{H_2O}}{\rho_{H_2O}} + \frac{x_{CO_2} \times M_{H_2O}}{\rho_{H_2O}}(1 + A_1 + A_2 p)} \quad [16.23]$$

with p expressed in MPa, A_1 and A_2 constants listed in Table 16.8, and A_i, with $i = 1{,}2$ given by

$$A_i = A_{i1}T^2 + A_{i2}T + A_{i3} + A_{i4}T^{-1} + A_{i5}T^{-2} \quad [16.24]$$

16.4.3.3 Thermal conductivity

The thermal conductivity of a solid hydrate has been reported by Sloan and Koh (2008) to be in the range 0.50–0.58 W/mK, closer to the thermal conductivity of liquid water than of ice. The thermal conductivity of CO_2 hydrates will be approximately $\lambda_{crystals} = 0.54\ \text{W/mK}$ for the relevant temperature range.

The thermal conductivity of the solution can, for different CO_2 concentrations, be obtained as proposed in Eq. [16.25].

$$\lambda_{sol} = \lambda_{l\ CO_2} \times w_{sol} + \lambda_{l\ H_2O} \times (1 - w_{sol}) \quad [16.25]$$

16.4.3.4 Specific heat

Youssef et al. (2014) report the specific heat of CO_2 hydrates to be $c_{p\ crystals} = 2.1\ \text{kJ/kgK}$ for the relevant temperature range.

The specific heat of the solution can, for different CO_2 concentrations, be obtained as proposed in Eq. [16.26].

$$c_{p\ sol} = c_{p\ l\ CO_2} \times w_{sol} + c_{p\ l\ H_2O} \times (1 - w_{sol}) \quad [16.26]$$

Table 16.8 Constants in Eq. [16.24]

j	A_{1j}	A_{2j}
1	$0.38384020 \times 10^{-3}$	$-0.57709332 \times 10^{-5}$
2	$-0.55953850 \times 10^{0}$	$0.82764653 \times 10^{-2}$
3	0.30429268×10^{3}	$-0.43813556 \times 10^{1}$
4	$-0.72044305 \times 10^{5}$	0.10144907×10^{4}
5	0.63003388×10^{7}	$-0.86777045 \times 10^{5}$

16.4.3.5 Latent heat

The latent heat of formation of CO_2-hydrate crystals has been reported by several researchers as indicated in Table 16.9. For a hydration number $n = 8.06$, as produced in fluidized bed heat exchangers (Zhou et al., 2015), a value of $\Delta h_{crystals} = 387.2$ kJ/kg seems most reasonable.

16.4.3.6 Dynamic viscosity

Zhou et al. (2015) have proposed to obtain the viscosity of the CO_2 water solution, μ_{sol}, as a function of both pressure and temperature using a linear dependence with the CO_2 concentration:

$$\mu_{sol} = \mu_{l\,CO_2} \times w_{sol} + \mu_{l\,H_2O} \times (1 - w_{sol}) \quad [16.27]$$

16.4.3.7 Diffusion coefficient

The diffusion coefficient of CO_2 in water has been determined by Lu et al. (2013) as a function of temperature. Zhou et al. (2015) have proposed to modify their equation to take into account the effect of the increased viscosity of the slurry:

$$D = \frac{\mu_{sol}(T)}{\mu_{slurry}(T)} e^{-\left[\frac{716.748}{T-126.9}\right]-15.9} \quad [16.28]$$

The temperature in this equation is expressed in K.

Table 16.9 Enthalpy of formation of carbon dioxide hydrate as reported by several sources

Source	T (°C)	$\Delta h_{crystals}$ (kJ/kg)	n
Vlahakis et al. (1972)	0.05	341.5	7.3
Bozzo et al. (1975)	0.05	336.3	7.3
Udachin et al. (2001)	2.85	398.5	6.2
Yoon et al. (2003)	0.05	370.1	6.21
Fournaison et al. (2004)	6.0	507.0	–
Sabil (2009)	0.45	386.7	8.25
Sabil (2009)	8.25	389.5	5.98
Lirio and Pessoa (2013)	2.15	380.2	7.9
Lirio and Pessoa (2013)	4.05	384.9	7.4
Lirio and Pessoa (2013)	6.05	387.6	6.9
Lirio and Pessoa (2013)	6.75	387.6	6.9

16.4.4 Stationary PCM storage designs

Agyenim et al. (2010) provide an overview of the most commonly used designs for PCM storage: cylindrical or rectangular containers. Because heat has to be added and removed in an efficient way, the container must be equipped with a heat exchanger. According to Agyenim et al. (2010), in 70% of the studied cases the shell-and-tube geometry has been preferred. In general, the HTF will flow inside tubes that are distributed in such way in the container that each tube serves approximately the same mass of PCM. Some studies report applications in which the PCM is inside of the tubes and the HTF flows outside of the tube (Lacroix, 1993; Esen et al., 1998; Bansal and Buddhi, 1992). Esen et al. (1998) have compared the two solutions and have concluded that the melting time is significantly reduced when the HTF is inside of the tubes. Fig. 16.10 illustrates how large these differences can be.

Gong and Mujumdar (1997) used a finite-element model to investigate the impact of introducing the hot (melting) and cold (solidification) HTF from the same or from different ends of the storage heat exchanger. They concluded that adding both hot and cold HTF from the same heat exchanger end leads to an enhancement of the charge/discharge rate of approximately 5%. This is because the penetration depth of the solid–liquid phase change interface during the charge/discharge is larger because of the higher temperature difference at the fluid inlet if the hot and cold fluids enter from the same end. Notice that if the geometry schematically given in Fig. 16.1 is applied, then the HTF will enter from opposite sides during melting and solidification and that this enhancement is limited in comparison to the location of the PCM in the shell-and-tube exchanger.

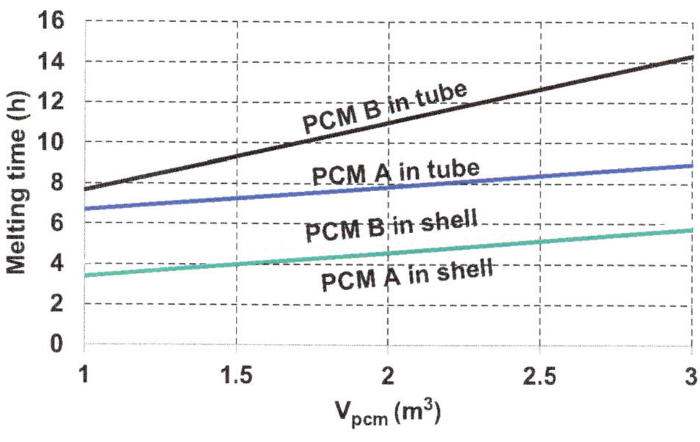

Figure 16.10 Comparison of the melting time of a shell-and-tube phase change material (PCM) storage with the heat transfer fluid in the tube side (PCM in shell) and in the shell side (PCM in tube) for two different PCMs with approximately the same melting temperature range.
Based on results by Esen, M., Durmus, A., Durmus, A., 1998. Geometric design of solar-aided latent heat storage depending on various parameters and phase change materials. Solar Energy 62, 19–28.

Because of the relatively low thermal conductivity of PCMs, many investigations have been performed to improve the heat transfer in LHTS. An overview of the various techniques adopted for enhancing the thermal performance of the LHTS units has been given by Agyenim et al. (2010) and more recently by Mosaffa et al. (2013):

- The dispersion of high-conductivity particles in the PCM (Mettawee and Assassa, 2007; Kumaresan et al., 2012; Oya et al., 2013),
- Using a high thermal conductivity porous matrix (Hoogendoorn and Bart, 1992; Py et al., 2001; Trelles and Duffy, 2003; Mesalhy et al., 2005),
- The addition of fins to the external surface of the storage (Sasaguchi and Takeo, 1994; Zhang and Faghri, 1996; Velraj et al., 1997; Horbaniuc et al., 1999; Ismail et al., 2001; Ermis et al., 2007; Agyenim et al., 2008; Talati et al., 2011; Mosaffa et al., 2012), and
- Using multiple families of PCMs in the unit to match as much as possible the temperature glide of the HTF (Shaikh and Lafdi, 2006; Brousseau and Lacroix, 1996; Wang et al., 1999; Cui et al., 2003; Kurnia et al., 2013; Mosaffa et al., 2013).

Agyenim et al. (2010) conclude that the general observations drawn from the various studies demonstrate that, irrespective of the PCM used, the heat transfer characteristics of the PCMs can be improved using all of the different enhancement techniques. Because different researchers used different experimental setups, different container materials, and different PCMs to investigate the heat transfer enhancement characteristics, it is not possible to compare the performance of the alternative enhancement methods. According to Agyenim et al. (2010), most enhancement techniques applied in practice make use of fins embedded in the PCM. This is probably because of the simplicity, ease in fabrication, and low cost of construction. This is followed by the introduction of a metal matrix into the PCM using high-conductivity materials such as carbon fiber and carbon nanotubes.

Mosaffa et al. (2013) have proposed the use of a unit consisting of several flat PCM slabs. This geometry appears to be favorable when the HTF is air. Because the HTF in solar latent heat storage applications will generally be a liquid, this geometry cannot directly be applied. However, similar geometries can be created when plate heat exchangers are used. One side of the plate heat exchanger will be filled with the PCM while the HTF flows in the other side. Medrano et al. (2009) have experimentally compared the performance of several commercially available heat exchangers for use as PCM storage systems and concluded that a typical evaporator/condenser copper tube—aluminum fin heat exchanger coil used in small air-conditioning units, which offers a large heat transfer area per unit volume, performs best. In this heat exchanger the PCM was filled in the space between tubes and fins and the heat exchanger was encapsulated in a rectangular container. The average thermal power values of this heat exchanger were 3 times higher than for the highest power levels achieved by the second best heat exchanger. These authors also concluded that because commercially available plate heat exchangers have the same volume on both heat exchanger sides, these heat exchangers have very low ratios of PCM heat capacity over HTF side capacity, making them not suitable for these applications.

16.4.5 Size of heat exchangers

The size of the required heat exchanger can be derived making use of the design approach for heat exchangers (Mills, 1999):

$$A_i = \frac{\dot{Q}_{lhts}}{U_i \times \Delta T_{ln}} \qquad [16.29]$$

with $\dot{Q}_{lhts}$ the heat transfer within the system, U_i the overall heat transfer coefficient based on the internal surface, ΔT_{ln} the logarithmic mean temperature difference, and A_i the HTF side surface of the heat exchanger. To prevent too large exergy losses between the HTF and the PCM, the ΔT_{ln} should be as small as possible and preferably not larger than 7K. The overall heat transfer coefficient will be determined mainly by the thermal conductivity of the PCM:

$$U_i = \frac{1}{\frac{1}{\alpha_i} + \frac{d_i \times \ln\left(\frac{d_o}{d_i}\right)}{2 \times \lambda_{tube}} + \frac{d_i \times \ln\left(\frac{d_{pcm}}{d_o}\right)}{2 \times \lambda_{pcm}}} \qquad [16.30]$$

If the internal heat transfer coefficient is $\alpha_i = 1000$ W/m^2 K, what can be easily attained, then the term related to the tube wall is negligible and the term related to the PCM is a factor 10 (for PCMs with thermal conductivity close to water) to 30 (for paraffin-based PCMs) times larger than the first term. Making use of external surface enlargement techniques, it is to some extent possible to compensate for this mismatch between the thermal resistances. The U_i value will be in the range 0.025–0.100 kW/m^2 K without external surface enhancement and can be increased up to 0.500 kW/m^2 K if enhancement techniques are used. Depending on the design choices, an internal heat exchanging area of 0.3–8.0 m^2 will be needed per kilowatt exchanged between PCM storage and HTF.

The liquid–solid and solid–liquid phase transitions of a PCM are moving boundary problems. The difficulty of predicting the behavior of phase change systems is due to its inherent nonlinear nature at the moving interface. In addition, the two phases have different thermophysical properties. A numerical model to simulate the transient behavior of a two concentric-tubes system will be introduced in the further lines. The melting and solidification processes can be described by the method of enthalpy formulation, and the differential equations can be numerically solved using a finite difference scheme. Implementation of the equations allows for a comparison with experimental data, after which more detailed heat exchanger designs can, numerically, be investigated.

The discussion applies for an LHTS system that is derived from the shell-and-tube heat exchanger design. An HTF flows through the tubes, whereas a PCM is located at the shell side. The dynamics of the system are induced by the fluctuation of the inlet conditions of the HTF. To simplify the simulation of the LHTS system, a model for two concentric tubes is preferred. The two concentric-tubes system is representative for one tube of the LHTS system that is surrounded with a PCM. A "bundle" of this system is a simple representation of the LHTS system. This is illustrated in Fig. 16.11.

For the two concentric-tubes system, an HTF flows through the inner tube and a PCM is located between the inner and outer tube. During charging, the HTF at a higher temperature is pumped through the tube and causes the PCM to melt. During discharging the PCM rejects heat to the HTF at a lower temperature and the PCM becomes solid. The system boundary of the two concentric-tubes system is shown in Fig. 16.12. Within the boundary, the system contains the HTF and the PCM as well as the tube wall.

The purpose of PCM models is generally to simulate the transient behavior of the system as a result of the conditions of the entering HTF. Therefore the mass flow rate $\dot{m}_{htf}$ and the temperature $T_{htf\ in}$ of the HTF at the inlet of the system are normally the input variables. The temperature of the HTF at the outlet of the system, $T_{htf\ out}$, the (average) temperature of the PCM, T_{pcm}, and the heat transfer within the

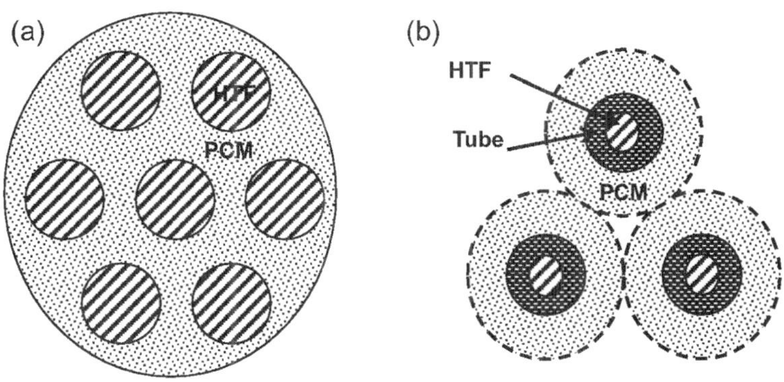

Figure 16.11 Simplified representation of an LHTS system: (a) LHTS system and (b) representation as a bundle of two concentric-tubes systems. An HTF flows in the tubes and a PCM is located at the shell side. The two concentric-tubes system represents a tube of the LHTS system. *LHTS*, Latent heat thermal storage; *HTF*, heat transfer fluid; *PCM*, phase change system.

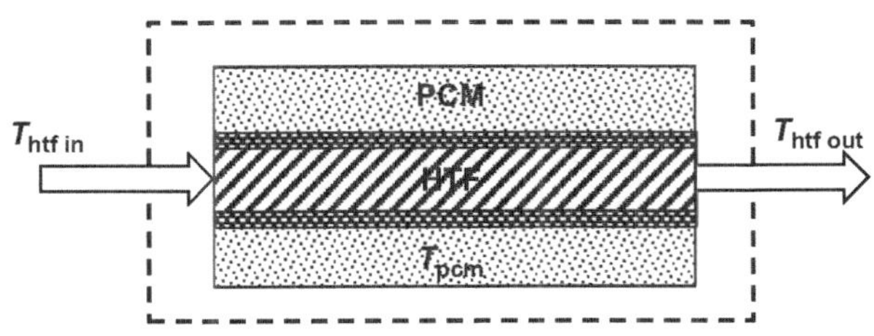

Figure 16.12 System boundary of the two concentric-tubes system. The system contains the HTF, the PCM, and the tube wall. *HTF*, Heat transfer fluid; *PCM*, phase change material.

system, $\dot{Q}_{\text{lhts}}$ are the output variables. The parameters of the geometry and the initial temperature of the system are the internal variables.

16.4.5.1 Model assumptions

The physics of the real process are simplified with assumptions and the mathematical model is derived under the following assumptions:

- The PCM is homogeneous and isotropic. The properties are identical in all directions.
- Both temperature variations of the tube and the PCM in the angular direction are neglected.
- The thermophysical properties of the PCM are constant with temperature but not with phase.
- The thermophysical properties of the tube are constant with temperature.
- The thermophysical properties of the HTF vary with temperature.
- The effect of natural convection of the PCM in the liquid phase is neglected.
- The system is well insulated; therefore it has no heat losses to the environment.

16.4.5.2 Constitutive equations

The heat transfer between the HTF and the tube wall is ascribed to convective heat transfer. The convective heat transfer coefficient is described by empirical relationships, which are a function of the Reynolds number, Re; the Prandtl number, Pr; and the geometry of the heat exchanger. The Reynolds number is expressed in Eq. [16.31]:

$$Re = \frac{4 \times \dot{m}_{\text{htf}}}{\pi \times d_{\text{i}} \times \mu_{\text{htf}}} \qquad [16.31]$$

where d_{i} (m) is the internal diameter of the inner tube, μ_{htf} (Pa s) is the dynamic viscosity of the HTF, and $\dot{m}_{\text{htf}}$ (kg/s) is the mass flow rate of HTF. The Prandtl number is given by Eq. [16.32]:

$$\text{Pr} = \frac{c_{\text{p htf}} \times \mu_{\text{htf}}}{\lambda_{\text{htf}}} \qquad [16.32]$$

where $c_{\text{p htf}}$ (J/kg K) is the specific heat capacity and λ_{htf} (W/m K) is the thermal conductivity of the HTF. The dynamic viscosity, the thermal conductivity, and the specific heat capacity depend on the temperature of the HTF. Once these values are known, the Nusselt number, Nu, can be calculated to estimate the convective heat transfer coefficient. However, the Nusselt number depends on the type of flow. According to Mills (1999), when the flow is fully developed and laminar, the Nusselt number can be expressed by Eq. [16.33]:

$$\text{Nu} = 3.66 \qquad [16.33]$$

The Nusselt number of fully developed flows with $3000 < Re < 10^6$ can be predicted using Gnielinski's correlation (Mills, 1999):

$$\mathrm{Nu} = \frac{\frac{f}{8} \times (Re - 1000) \times \mathrm{Pr}}{1 + 12.7 \times \left(\frac{f}{8}\right)^{0.5} \times (Pr^{0.667} - 1)} \qquad [16.34]$$

with the friction factor, f, given by

$$f = (0.79 \times \log(Re) - 1.64)^{-2}. \qquad [16.35]$$

The convective heat transfer coefficient, α_i (W/m^2 K), is then estimated with Eq. [16.36]:

$$\alpha_\mathrm{i} = \frac{\lambda_\mathrm{htf} Nu}{d_\mathrm{i}} \qquad [16.36]$$

16.4.5.3 Conservation equations

The HTF, the inner tube wall, and the PCM are described with the following energy conservation equations.

HTF:

$$\rho_\mathrm{htf} \times c_\mathrm{p\ htf} \times V_\mathrm{htf} \frac{\partial T_\mathrm{htf}}{\partial t} = \dot{m}_\mathrm{htf} \times c_\mathrm{p\ htf} \times dT_\mathrm{htf} - \alpha_\mathrm{i} \times A_\mathrm{i} \times (T_\mathrm{htf} - T_\mathrm{tube}) \qquad [16.37]$$

Internal tube wall:

$$\rho_\mathrm{tube} \times c_\mathrm{p\ tube} \times V_\mathrm{tube} \frac{\partial T_\mathrm{tube}}{\partial t} = \frac{\partial}{\partial z}\left(\lambda_\mathrm{tube} \frac{\partial T_\mathrm{tube}}{\partial z}\right) + \frac{1}{r}\left(r \times \lambda_\mathrm{tube} \times \frac{\partial T_\mathrm{tube}}{\partial r}\right) \qquad [16.38]$$

PCM:

$$\frac{\partial h_\mathrm{pcm}}{\partial t} = \frac{\partial}{\partial z}\left(\frac{\lambda_\mathrm{pcm}}{\rho_\mathrm{pcm} \times c_\mathrm{p\ pcm}} \frac{\partial h_\mathrm{pcm}}{\partial z}\right) + \frac{1}{r}\left(r \times \frac{\lambda_\mathrm{pcm}}{\rho_\mathrm{pcm} \times c_\mathrm{p\ pcm}} \times \frac{\partial h_\mathrm{pcm}}{\partial r}\right) \qquad [16.39]$$

where ρ (kg/m^3) is the density and h_pcm (J/m^3) is the volumetric enthalpy of the PCM. $\frac{\lambda_\mathrm{pcm}}{\rho_\mathrm{pcm} \times c_\mathrm{p\ pcm}}$ (m^2/s) is the thermal diffusivity, which is defined as the ratio of the thermal conductivity to the density and the specific heat capacity. Furthermore, A_i (m^2) and V (m^3) are the inner surface area of the internal tube and the volume, respectively. The distance in the axial direction is given by z (m) whereas the distance in the radial direction is given by r (m). The conservation equation of the inner tube wall and the

PCM are given in cylindrical coordinates. The melting and solidification processes of the PCM are described with the method of enthalpy formulation. Accordingly, the conservation equation of the PCM is written in the volumetric enthalpy form. The PCM state (solid, liquid, or undergoing phase change) depends on the volumetric enthalpy obtained from Eq. [16.39]. Hence:

- The PCM is in the solid phase if

$$h_{\mathrm{pcm}} < \rho_{\mathrm{s\ pcm}} \times c_{\mathrm{p\ s\ pcm}} \times T_{\mathrm{pcm}} \qquad [16.40]$$

- The PCM is undergoing a phase change if

$$\rho_{\mathrm{s\ pcm}} \times c_{\mathrm{p\ s\ pcm}} \times T_{\mathrm{pcm}} \leq h_{\mathrm{pcm}} \leq \rho_{\mathrm{s\ pcm}} \times (c_{\mathrm{p\ s\ pcm}} \times T_{\mathrm{m1}} + \Delta h_{\mathrm{pcm}}) \qquad [16.41]$$

- The PCM is in the liquid phase if

$$h_{pcm} > \rho_{s\ pcm} \times (c_{p\ s\ pcm} \times T_{m1} + \Delta h_{pcm}) \qquad [16.42]$$

where T_{m1} is the temperature at which the PCM starts to melt and Δh_{pcm} (J/kg) is the latent heat of melting. Once the phase of the PCM is determined, its corresponding temperature can be derived.

16.4.5.4 Initial and boundary conditions

The following initial and boundary conditions are applied to solve the differential equations:

- At time equals zero, the LHTS unit has a uniform temperature profile (the temperature of the HTF, the tube, and the PCM are equal to the ambient temperature).
- At time greater than zero, the HTF temperature of node 1 is equal to the inlet temperature of the HTF.
- There is a convective heat transfer process between the HTF and the tube wall.

$$\lambda_{tube}\frac{\partial T^i_{tube}}{\partial r} = \alpha_i \times \left(T^i_{tube} - T^i_{htf}\right) \qquad [16.43]$$

at $i = 0, \ldots, L$ and $r = r_i$.

- The heat transfer between the tube and the PCM is described by conductive heat transfer.

$$\lambda_{tube}\frac{\partial T^i_{tube}}{\partial r} = \lambda_{pcm}\frac{\partial T^i_{pcm}}{\partial r} = \overline{\lambda}\frac{\partial}{\partial r}\left(T^i_{pcm} - T^i_{tube}\right) \qquad [16.44]$$

at $i = 0, \ldots, L$ and $r = r_o$.

- There is no conduction along the tube and outer PCM walls. There is no heat transport through the outer walls of the PCM:

$$\frac{\partial T^i_{pcm}}{\partial r} = 0 \qquad [16.45]$$

The inner radius of the external tube is given by R (m) and the length of the system is defined by L (m). The average thermal conductivity between the tube and the PCM, $\overline{\lambda}$ (W/m K), is the distance weighted average thermal conductivity.

16.5 Examples

This section provides an overview of the solar cooling and heating systems that include latent heat storage that have been reported in the literature. Chidambaram et al. (2011) reviewed applications of thermal storage in solar cooling systems. Unfortunately all of the reported applications make use of sensible heat storage with hot (and cold) water. In addition, Pintaldi et al. (2015) give several examples of operating/demonstration units of absorption chiller solar cooling systems, but all of these examples use water as the storage medium. Helm et al. (2009, 2014) report experimental data obtained with four pilot installations between 7 and 90 kW nominal cooling capacity that were equipped with salt hydrate latent heat storages between 80 and 240 kWh energy content. These authors use the latent heat storage to support the heat dissipation system of the chiller and applied calcium chloride hexahydrate, which has a phase change temperature around 29°C and a melting enthalpy of approximately 150 kJ/kg or 240 kJ/dm^3 between 22°C and 36°C. This storage capacity comprises both sensible and latent heat. Together with Clina-Bionic Systems GmbH, the authors have developed two different latent heat storage module sizes with internal heat exchanger matrixes: one containing 1 m^3 of PCM providing 14 kW nominal thermal power and a storage capacity of 83 kWh and another one with 1.5 m^3 storage volume and 21 kW nominal thermal power at 124 kWh storage capacity. The heat exchanger consists of capillary tubes providing 36.7 m^2 heat exchanger surface per cubic meter PCM and generates low pressure loss in the external hydraulic circuit.

Noro et al. (2014) studied the energy and economy of a solar plant for heating and cooling of an office building, with the main objective of defining the size and the type (sensible vs. PCM) of thermal energy storage to be installed in the hot and in the cold side of the plant. From the energy point of view, the best solution was given by a 3000-L tank filled with PCM S44 (phase change at 44°C; storage at intermediate temperature) and a 2000-L tank filled with water (sensible heat at 7−15°C; cold storage). This combination provides the highest primary energy saving and solar ratio between the different scenarios considered because the mean operating temperature of the storage is near melting temperature of this PCM for a longer period during the year. The economic analysis reveals that the greater investment cost of PCM technology is justified in a solar cooling plant only in the case of high natural gas tariffs or lower PCM cost.

16.6 Conclusion

This chapter described the steps leading to the selection of the best latent heat storage for specific operating conditions. These steps can be finally summarized as follows:

- *Key components and storage strategy*: Solar cooling and heating systems must meet the specifications indicated by the local climate conditions and specific application. This includes

  - The amount of cooling/heating to be delivered by the system.
  - Sink and sources fluids and their respective inlet and outlet temperatures.
  - Identification of the most suitable combination of solar collector type, sorption system type, and temperature level of the latent thermal storage.
- Compatibility with fluids and operating conditions.
  - Materials of construction of the storage heat exchangers and other system components should be suitable for the latent heat storage material without excessive corrosion.
  - Ability for operation under fouling conditions of the process streams; ability to clean the equipment.
  - Ability to cope with fluid pressure stresses and stresses due to density changes between solid and liquid phases and due to temperature differences.
- Availability.
  - Delivery time of latent heat material.
  - Limitations of design methods in relation to size. Equipment setup in parallel leads to additional piping and installation costs and may lead to flow distribution problems.
- Economic factors.
  - Final choice should be based on economics.
  - Consider the economic advantages resulting from energy savings and emission cost reduction.
  - Consider capital, installation, and costs of solar collector field and sorption equipment.
  - Consider the economic advantages that can follow from governmental subsidies for implementation of heating/cooling systems with reduced energy use.
- Initial selection.
  - If one latent heat storage type shows much lower payback time than the others (say a difference in payback time of more than 1.5), then this type should be selected and detailed.
  - If several latent heat storage types show similar payback times, then estimate the three best types in more detail.

List of symbols

a	Lattice constant	m
A	Area	m^2
c_p	Specific heat	kJ/kg K
d	Tube diameter	m
D	Diffusion coefficient	m^2s
f	Friction factor	–
h	Enthalpy	kJ/kg
h_{pcm}	Volumetric enthalpy	kJ/m^3
L	Length of tube	m
$\dot{m}$	Mass flow	kg/s

M	Molar mass	kg/kmol
n	Hydration number	–
N_o	Avogadro constant	mol/kmol
Nu	Nusselt number	–
p	Pressure	kPa
Pr	Prandtl	–
$\dot{q}$	Heat flux	kW/m^2
r	Radial coordinate, radius	m
R	Inner tube radius	m
Re	Reynolds number	–
$\dot{Q}$	Heat flow	kW
t	Time	s
T	Temperature	K
$\overline{T}$	Equivalent temperature for gliding reservoirs	K
U	Overall heat transfer coefficient	kW/m^2 K
V	Volume	m^3
w	Mass concentration	kg/kg
x	Molar concentration	kmol/kmol
z	Axial coordinate	m

Greek symbols

α_i	Internal heat transfer coefficient	W/m^2 K
Δh	Enthalpy of melting	kJ/kg
ΔT	Temperature difference	K
η	Efficiency	–
λ	Thermal conductivity	W/m K
$\overline{\lambda}$	Distance weighted thermal conductivity	W/m K
μ	Viscosity	Pa s
ρ	Density	Kg/m^3
ξ	Volume fraction	m^3/m^3

Subscripts

app	Apparent
htf	Heat transfer fluid
hyd	Hydrate
i	Internal
in	At inlet conditions
l	Liquid phase
lhts	Latent heat thermal storage
ln	Logarithmic
m	Melting
o	External
out	At outlet conditions
pcm	Related to PCM
s	Solid
sat	Saturated conditions
sens	Sensible
sol	Solution

Abbreviations

AFPC	Advanced flat-plate collector
COP	Coefficient of performance
CPC	Compound concentrating collector
ETC	Evacuated-tube collector
FPC	Flat-plate collector
HTF	Heat transfer fluid
LHTS	Latent heat thermal storage
PCM	Phase change material

References

Adisasmito, S., Frank, R.J., Sloan, E.D., 1991. Hydrates of carbon dioxide and methane mixtures. Journal of Chemical and Engineering Data 36, 68–71.

Agyenim, F., Hewitt, N., Eames, P., Smyth, M., 2008. Numerical and experimental development of medium temperature thermal energy storage (Erythritol) system for the hot side of LiBr/ H_2O air conditioning applications. In: World Renewable Energy Congress.

Agyenim, F., Hewitt, N., Eames, P., Smyth, M., 2010. A review of materials, heat transfer and phase change problem formulation for latent heat thermal energy storage systems (LHTESS). Renewable and Sustainable Energy Reviews 14, 615–628.

Asaoka, T., Kumano, H., Serita, M., 2013. Measurement of latent heat of tetra-n-butylammonium bromide (TBAB) hydrate. International Journal of Refrigeration 36, 992–997.

Bansal, N.K., Buddhi, D., 1992. An analytical study of a latent heat storage system in a cylinder. Energy Conversion and Management 33, 235–242.

Bozzo, A.T., Chen, H.S., Kass, J.R., Barduhn, A.J., 1975. The properties of hydrate of chlorine and carbon dioxide. Desalination 1, 303–320.

Brousseau, P., Lacroix, M., 1996. Study of the thermal performance of a multi-layer PCM storage unit. Energy Conversion and Management 37, 599–609.

Chidambaram, L.A., Ramana, A.S., Kamaraj, G., Velraj, R., 2011. Review of solar cooling methods and thermal storage options. Renewable and Sustainable Energy Reviews 15, 3220–3228.

Cui, H., Yuan, X., Hou, X., 2003. Thermal performance analysis for a heat receiver using multiple phase change materials. Applied Thermal Engineering 23, 2353–2361.

Darbouret, M., Cournil, M., Herry, J.M., 2005. Rheological study of TBAB hydrate slurries as secondary two-phase refrigerants. International Journal of Refrigeration 28, 663–671.

de Sera, I.E.E., Zhou, H., Infante Ferreira, C.A., 2014. Experimental Study on the Continuous Production of CO_2 Hydrate Slurry. Food Future Project. Internal report. Delft University of Technology.

Duan, Z., Hu, J., Li, D., Mao, S., 2008. Densities of the CO_2 -H_2O and CO_2 -H_2O-NaCl systems up to 647K and 100 MPa. Energy and Fuels 22, 1666–1674.

Ermis, K., Erek, A., Dincer, I., 2007. Heat transfer analysis of phase change process in a finned-tube thermal energy storage system using artificial neural network. International Journal of Heat and Mass Transfer 50, 3163–3175.

Esen, M., Durmus, A., Durmus, A., 1998. Geometric design of solar-aided latent heat storage depending on various parameters and phase change materials. Solar Energy 62, 19–28.

Fan, Z., Infante Ferreira, C.A., Mosaffa, A.H., 2014. Numerical modelling of high temperature latent heat thermal storage for solar application combining with double-effect H_2O/LiBr absorption refrigeration system. Solar Energy 22, 398–409.

Fournaison, L., Delahaye, A., Chatti, I., Petitet, J.-P., 2004. CO_2 hydrates in refrigeration processes. Industrial and Engineering Chemistry Research 43, 6521–6526.

Gallego, A.J., Ruız-Pardo, A., Cerezuela-Parish, A., Sanchez, J., Martın-Macareno, C., Cabeza, L.F., Camacho, E.F., Oró, E., 2013. Mathematical modelling of a PCM storage tank in a solar cooling plant. Solar Energy 93, 1–10.

Gil, A., Medrano, M., Martorell, I., Lázaro, A., Dolado, P., Zalba, B., Cabeza, L.F., 2010. State of the art on high temperature thermal energy storage for power generation. part 1—concepts, materials and modellization. Renewable and Sustainable Energy Reviews 14, 31–55.

Gong, Z.X., Mujumdar, A.S., 1997. Finite-element analysis of cyclic heat transfer in a shell and tube latent heat energy storage exchanger. Applied Thermal Engineering 17, 583–591.

Hayashi, K., Takao, S., Ogoshi, H., Matsumoto, S., 2000. Research and development on high-density cold latent-heat medium transportation technology. In: Fifth Workshop, Final Proceedings, pp. 1–9.

Helm, M., Keil, C., Hiebler, S., Mehling, H., Schweigler, C., 2009. Solar heating and cooling system with absorption chiller and low temperature latent heat storage: energetic performance and operational experience. International Journal of Refrigeration 32, 596–606.

Helm, M., Hagel, K., Pfeffer, W., Hiebler, S., Schweigler, C., 2014. Solar heating and cooling system with absorption chiller and latent heat storage – a research project summary. Energy Procedia 48, 837–849.

Hoogendoorn, C.J., Bart, G.C.J., 1992. Performance and modelling of latent heat stores. Solar Energy 48, 53–58.

Horbaniuc, B., Dumitrascua, G., Popescub, A., 1999. Mathematical models for the study of solidification within a longitudinally finned heat pipe latent heat thermal storage system. Energy Conversion and Management 40, 1765–1774.

Hoshi, A., Mills, D.R., Bittar, A., Saitoh, T.T., 2005. Screening of high melting point phase change materials (PCM) in solar thermal concentrating technology based on CLFR. Solar Energy 79, 332–339.

Infante Ferreira, C.A., Kim, D.-S., 2014. Techno-economic review of solar cooling technologies based on location-specific data. International Journal of Refrigeration 39, 23–27.

Ismail, K.A.R., Alves, C.L.F., Modesto, M.S., 2001. Numerical and experimental study on the solidification of PCM around a vertical axially finned isothermal cylinder. Applied Thermal Engineering 21, 53–77.

Kalogirou, S.A., 2004. Solar thermal collectors and applications. Progress in Energy and Combustion Science 30, 231–295.

Kauffeld, M., Christensen, K.G., Lund, S., Hansen, T.M., 1999. Experience with ice slurry. In: Proceedings of the 1st IIR Workshop on Ice Slurries, 27–28 May 1999. International Institute of Refrigeration, Yverdon-les-Bains (Switzerland), Paris, pp. 42–73.

Kenisarin, M., 2010. High-temperature phase change materials for thermal energy storage. Renewable and Sustainable Energy Reviews 14, 955–970.

Kenisarin, M., Mahkamov, K., 2007. Solar energy storage using phase change materials. Renewable and Sustainable Energy Reviews 11, 1913–1965.

Klein, S.A., Beckman, W.A., 1979. A general design method for closed-loop solar energy systems. Solar Energy 22, 269–282.

Kumaresan, G., Velraj, R., Iniyan, S., 2011. Thermal analysis of d-mannitol for use as phase change material for latent heat storage. Journal of Applied Sciences 16, 3044–3048.

Kumaresan, V., Velraj, R., Das, S.K., 2012. The effect of carbon nanotubes in enhancing the thermal transport properties of PCM during solidification. Heat and Mass Transfer 48, 1345–1355.

Kumano, H., Saito, A., Okawa, S., Goto, Y., 2006. Study on fundamental characteristics of TBAB hydrate slurry. Transactions of the Japan Society of Mechanical Engineers, Part B 72, 3089–3095.

Kumano, H., Hirata, T., Kudoh, T., 2011. Experimental study on the flow and heat transfer characteristics of a tetra-n-butyl ammonium bromide hydrate slurry (first report: flow characteristics). International Journal of Refrigeration 34, 1953–1962.

Kurnia, J.C., Sasmito, A.P., Jangam, S.V., Mujumdar, A.S., 2013. Improved design for heat transfer performance of a novel phase change material (PCM) thermal energy storage (TES). Applied Thermal Engineering 50, 896–907.

Lacroix, M., 1993. Numerical simulation of a shell-and-tube latent heat thermal energy storage unit. Solar Energy 50, 357–367.

Lirio, C., Pessoa, F., 2013. Enthalpy of dissociation of simple and mixed carbon dioxide clathrate hydrate. Chemical Engineering Transactions 32, 577–582.

Lu, W., Guo, H., Chou, I.M., Burruss, R.C., Li, L., 2013. Determination of diffusion coefficients of carbon dioxide in water between 268 and 473K in a high-pressure capillary optical cell with in situ Raman spectroscopic measurements. Geochimica et Cosmochimica Acta 115, 183–204.

Ma, Z.W., Zhang, P., Wang, R.Z., Furui, S., Xi, G.N., 2010. Forced flow and convective melting heat transfer of clathrate hydrate slurry in tubes. International Journal of Heat and Mass Transfer 53, 3745–3757.

Medrano, M., Yilmaz, M.O., Nogués, M., Martorell, I., Roca, J., Cabeza, L.F., 2009. Experimental evaluation of commercial heat exchangers for use as PCM thermal storage systems. Applied Energy 86, 2047–2055.

Meewisse, J., 2004. Fluidized Bed Ice Slurry Generator for Enhanced Secondary Cooling Systems (Ph.D. thesis). Delft University of Technology.

Mesalhy, O., Lafdi, K., Elgafy, A., Bowman, K., 2005. Numerical study for enhancing the thermal conductivity of phase change material (PCM) storage using high thermal conductivity porous matrix. Energy Conversion and Management 46, 847–867.

Mettawee, E.S., Assassa, G.M.R., 2007. Thermal conductivity enhancement in a latent heat storage system. Solar Energy 81, 839–845.

Mills, A.F., 1999. Basic Heat & Mass Transfer, second ed. Prentice Hall, Inc.

Mooijer-van den Heuvel, M.M., Witteman, R., Peters, C.J., 2001. Phase behaviour of gas hydrates of carbon dioxide in the presence of tetrahydropyran, cyclobutanone, cyclohexane and methylcyclohexane. Fluid Phase Equilibria 182, 97–110.

Mosaffa, A.H., Talati, F., Basirat Tabrizi, H., Rosen, M.A., 2012. Analytical modeling of PCM solidification in a shell and tube finned thermal storage for air conditioning systems. Energy and Buildings 49, 356–361.

Mosaffa, A.H., Infante Ferreira, C.A., Talati, F., Rosen, M.A., 2013. Thermal performance of a multiple PCM thermal storage unit for free cooling. Energy Conversion and Management 67, 1–7.

Ng, H.J., Robinson, D.B., 1985. Hydrate formation in systems containing methane, ethane, propane, carbon dioxide or hydrogen sulphide in the presence of methanol. Fluid Phase Equilibria 21, 145–155.

Noro, M., Lazzarin, R.M., Busato, F., 2014. Solar cooling and heating plants: an energy and economic analysis of liquid sensible vs phase change material (PCM) heat storage. International Journal of Refrigeration 39, 104–116.

Ohgaki, K., Makihara, Y., Takan,o, K., 1993. Formation of CO_2 hydrate in pure and sea waters. Journal of Chemical Engineering of Japan 26, 558–564.

Ogoshi, H., Takao, S., 2004. Air-conditioning system using clathrate hydrate slurry. JFE GIHO 3, 1–5.

Oya, T., Nomura, T., Tsubota, M., Okinaka, N., Akiyama, T., 2013. Thermal conductivity enhancement of erythritol as PCM by using graphite and nickel particles. Applied Thermal Engineering 61, 825–828.

Oyama, H., Shimada, W., Ebinuma, T., Kamata, Y., Takeya, S., Uchida, T., Nagao, J., Narita, H., 2005. Phase diagram, latent heat and specific heat of TBAB semi-clathrate hydrate crystals. Fluid Phase Equilibria 234, 131–135.

Pintaldi, S., Perfumo, C., Sethuvenkatraman, S., White, S., Rosengarten, G., 2015. A review of thermal energy storage technologies and control approaches for solar cooling. Renewable and Sustainable Energy Reviews 41, 975–995.

Premkumar, M., Ramachandran, S., 2011. Investigation of solar thermal energy storage in therminol-55 with d-mannitol phase change material in brass cylindrical encapsulations with fins. European Journal of Scientific Research 80, 31–40.

Pronk, P., 2006. Fluidized Bed Heat Exchanger to Prevent Fouling in Ice Slurry Systems and Industrial Crystallizers (Ph.D. thesis). Delft University of Technology.

Py, X., Olives, R., Mauran, S., 2001. Paraffin/porous-graphite-matrix composite as a high and constant power thermal storage material. International Journal of Heat and Mass Transfer 44, 2727–2737.

Sabil, K., 2009. Phase Behaviour, Thermodynamics and Kinetics of Clathrate Hydrate Systems of Carbon Dioxide in Presence of Tetrahydrofuran and Electrolytes (Ph.D. thesis). Delft University of Technology.

Sari, O., Hu, J., Eicher, S., Egolf, P., Homsy, P., 2008. Thermo physical and flow properties of CO_2 hydrate slurry. In: Proceedings of the 8th IIR Gustav Lorentzen Conference on Natural Working Fluids, Copenhagen.

Sari, A., Boçer, A., Ladçi, Ö., Ceylan, M., 2011. Galaxitol hexa stearate and galactitol hexa palmitate as novel solid–liquid phase change materials for thermal energy storage. Solar Energy 85, 2061–2071.

Sasaguchi, K., Takeo, H., 1994. Effect of the orientation of a finned surface on the melting of frozen porous media. Heat and Mass Transfer 37, 13–26.

Shaikh, S., Lafdi, K., 2006. Effect of multiple phase change materials (PCMs) slab configurations on thermal energy storage. Energy Conversion and Management 47, 2103–2117.

Sharma, A., Tyagi, V.V., Chen, C.R., Buddhi, D., 2009. Review on thermal energy storage with phase change materials and applications. Renewable and Sustainable Energy Reviews 13, 318–345.

Sloan, E., Koh, C., 2008. Clathrate Hydrates of Natural Gases. CRC Press.

Takahashi, Y., Kamimoto, M., Abe, Y., Sakamoto, R., Kanari, K., Ozawa, T., 1987. Investigation of latent heat–thermal energy storage materials. IV. Thermoanalytical evaluation of binary eutectic mixtures of NaOH with LiOH or KOH. Thermochimica Acta 21, 193–202.

Talati, F., Mosaffa, A.H., Rosen, M.A., 2011. Analytical approximation for solidification processes in PCM storage with internal fins: imposed heat flux. Heat and Mass Transfer 47, 369–376.

Tareef, B.M., 1940. Thermal Conductivity of Colloidal Systems. Colloidal Journal USSR 6, 545.

Teng, H., Yamasaki, A., Shindo, Y., 1996. Stability of the hydrate layer formed on the surface of a CO_2 droplet in high-pressure, low-temperature water. Chemical Engineering Science 51, 4979–4986.

Thomas, D.G., 1965. Transport characteristics of suspension: VIII. A note on the viscosity of Newtonian suspensions of uniform spherical particles. Journal of Colloid Science 20, 267–277.

Trelles, J.P., Duffy, J.J., 2003. Numerical simulation of porous latent heat thermal energy storage for thermoelectric cooling. Applied Thermal Engineering 23, 1647–1664.

Tsoutsou, S., Infante Ferreira, C., Krieg, J., Ezzahiri, M., 2014. Building integration of concentrating solar systems for heating applications. Applied Thermal Engineering 70, 647–654.

Udachin, K.A., Ratcliffe, C.I., Ripmeester, J.A., 2001. Structure, composition, and thermal expansion of CO_2 hydrate from single crystal X-ray diffraction measurements. Journal of Physical Chemistry B 105, 4200–4204.

van Leeuwen, J., 2010. Feasibility Study Solar Refrigeration System for the Cooling Requirements of a Beach Bar. Internal report P&E 2430. Department Process & Energy. Delft University of Technology.

Velraj, R., Seeniraj, R.V., Hafner, B., Faber, C., Swharzer, K., 1997. Experimental analysis and numerical modelling of inward solidification on a finned vertical tube for a latent heat storage unit. Solar Energy 60, 281–290.

Vlahakis, J., Chen, H.S., Suwandi, M., Barduhn, A., 1972. The Growth Rate of Ice Crystals: Properties of Carbon Dioxide Hydrate, a Review of Properties of 51 Gas Hydrates. Technical report, Research and Development. Report 830 Prepared for US Department of Interior.

Wang, J., Chen, G., Zheng, F., 1999. Study on phase change temperature distributions of composite PCMs in thermal energy storage. International Journal of Energy Research 23, 277–285.

Wendland, M., Hasse, H., Maurer, G., 1999. Experimental pressure – temperature data on three- and four-phase equilibria of fluid, hydrate, and ice phases in the system carbon dioxide – water. Journal of Chemical and Engineering Data 44, 901–906.

Yang, S.O., Yang, I.M., Kim, Y.S., Lee, C.S., 2000. Measurement and prediction of phase equilibria for water + CO_2 in hydrate forming conditions. Fluid Phase Equilibria 175, 75–89.

Yokozeki, A., 2004. Solid-liquid-vapor phases of water and water-carbon dioxide mixtures using a simple analytical equation of state. Fluid Phase Equilibria 222-223, 55–66.

Yoon, J.H., Yamamoto, Y., Komai, T., Haneda, H., 2003. Rigorous approach to the prediction of the heat of dissociation of gas hydrates. Industrial and Engineering Chemistry Research 42, 1111–1114.

Youssef, Z., Hanu, L., Kappels, T., Delahaye, A., Fournaison, L., Zambrana, C., Pollerberg, C., 2014. Experimental study of single CO_2 and mixed CO_2 + TBAB hydrate formation and dissociation in oil-in-water emulsion. International Journal of Refrigeration 46, 207–218.

Zalba, B., Marín, J.M., Cabeza, L.F., Mehling, H., 2003. Review in thermal energy storage with phase change: materials, heat transfer analysis and applications. Applied Thermal Engineering 23, 251–283.

Zhang, Y., Faghri, A., 1996. Heat transfer enhancement in latent heat thermal energy storage system by using the internally finned tube. International Journal of Heat Mass Transfer 39, 3165–3173.

Zhou, H., de Sera, I.E.E., Infante Ferreira, C.A., 2015. Modelling and experimental validation of a fluidized bed based CO_2 hydrate cold storage system. Applied Energy 158.

Chemisorption heat storage for solar low-energy buildings

17

F. Kuznik
INSA Lyon, CETHIL, Villeurbanne, France

17.1 Introduction

Is it possible to store enough heat during the summer to be independent during the winter? Housing and tertiary buildings are responsible for the consumption of approximately 46% of all energy and approximately 19% of the total CO_2 emissions in France [1]. As the demand for thermal comfort in buildings rises, the energy consumption correspondingly increases. As a matter of fact, the energy consumption of buildings in France has increased by 30% since 1985. In Europe, 26% of the final energy consumption is related to the energy systems of households [3]. About 80% of this energy is needed for heating purposes [4], ie, heating and domestic hot water production.

The energy performance of buildings is key to achieving the "EU Climate & Energy" objectives, namely a reduction of 20% of the greenhouse gas emissions by 2020 and 20% energy savings by 2020. Improving the energy performance of buildings is a cost-effective way of fighting climate change and improving energy security while creating employment, especially in the building sector.

The thermal energy storage system (TESS) could make an important contribution to reducing our dependency on fossil fuels, but also to a more efficient and environmentally benign energy use [20]. Indeed, the primary role of these systems is to reduce the discrepancy between the availability of resources and their demand. In addition, designing efficient and inexpensive energy storage devices is vital because the energy supply will become more and more volatile as a consequence of the increasing share of renewable energies, which are inherently intermittent, in the energy mix. Furthermore, the use of a storage system during consumption peaks, by punctual discharges, contributes to smoothing the load curve and avoids the usage of the most polluting power plants throughout the duration of the peak.

The annual solar energy reaching the envelope of a low-energy building is far greater than the building needs for space heating and hot water, especially during the summer season. It is obvious that a long-term TESS is a way to rationalize the use of this renewable energy. Thermal energy storage can be accomplished by using sensible heat [46,50], latent heat [35,49], physical sorption [44], or chemical heat storage. Chemical heat storage has the highest potential for seasonal heat storage (Fig. 17.1): the storage density is high with few heat losses to the environment.

Advances in Solar Heating and Cooling. http://dx.doi.org/10.1016/B978-0-08-100301-5.00017-5

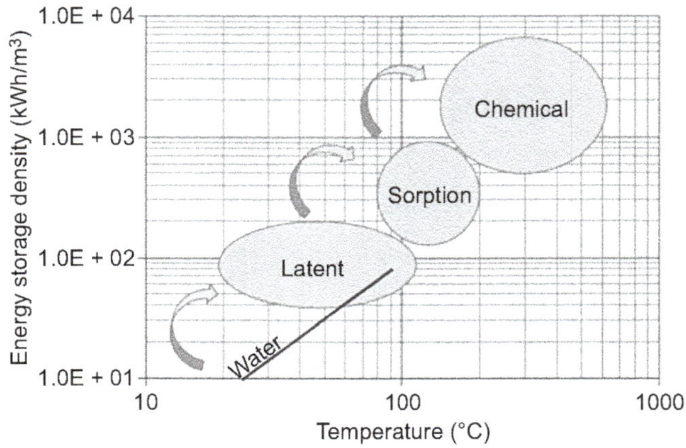

Figure 17.1 Storage density versus physical phenomena involved.

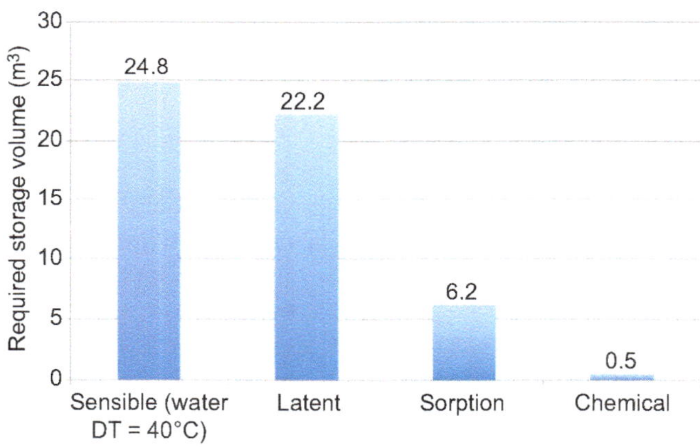

Figure 17.2 Storage volume depending on the material for Agen, France (oceanic climate). Extracted from V. Brick, F. Kuznik, K. Johannes, J. Virgone. Evaluation of thermal energy storage potential in low–energy buildings in France, in: ISES Solar World Congress 2011, Kassel, Germany, 28 August–2 September 2011.

In [14], a low-energy house (floor area of 98 m^2) located in France was numerically studied. The house was equipped with solar thermal collectors. The objective of the study was to evaluate the seasonal heat storage systems. Fig. 17.2 presents the storage volume required to heat the house using only energy from the solar collectors. This figure shows the high potential of chemical reactions for long-term heat storage.

In this chapter, chemisorption heat storage is first presented. Then, the applications to building specifications are described to understand the limits of the storage possibility. Pure chemical heat storage materials are then reviewed as well as composite salts inside porous matrices. Finally, applications are reviewed and conclusions are given concerning future research.

17.2 Basics of chemisorption

17.2.1 Basics of chemical reaction

The cycle of thermal energy storage/release can be carried out using a reversible chemical reaction:

$$\sum_i R_i \rightleftarrows \sum_i P_i$$

where R_i are the reactants and P_i the products. The creation of the products is the exothermic part of the reversible reaction, whereas the creation of the reactants is the endothermic part.

Each reversible reaction is a chemical equilibrium with a specific inversion temperature T^*. This temperature can be defined using the free energy at the equilibrium:

$$\Delta G_{T^*} = \Delta H - T^* \Delta S = 0 \Rightarrow T^* = \frac{\Delta H}{\Delta S}$$

where H is the enthalpy and S the entropy.

With T being the system temperature, if $T > T^*$ then the reaction is endothermic: it is the heat storage process. The products P_i generate the reactants R_i using heat. The reactants R_i can be separated, cooled, and stored at ambient temperature without thermal energy losses. This "ambient temperature storage" is thus suitable for long-term heat storage. If the reactants are mixed at a temperature $T < T^*$, then an exothermic reaction occurs: it is the heat release process.

The inversion temperature T^* corresponds to the minimal temperature for heat storage. The available heat source temperature T_S used for the storage must be at least equal to T^*. The potential candidates for chemical reaction require that $T^* < T_S$. Moreover, to have a high energy density (ie, high enthalpy difference $\Delta H = T^* \Delta S$), the candidates must have a high entropy difference, especially if T_S is low. High entropy variation can be obtained using the passage from a solid or liquid state (ordered state) to a gaseous state (disordered state). Thus, this chapter focuses on chemisorption, ie, a chemical reaction between a solid and a gas, or chemical adsorption.

17.2.2 Solid/gas reaction

The chemisorption reaction process can be expressed under the following general form:

$$A_{\text{solid}} + \nu B_{\text{gas}} \rightleftarrows A\nu B_{\text{solid}}$$

The reaction being monovariant, the equilibrium is given by the Clausius–Clapeyron relation:

$$\ln\left(\frac{p_e}{p_0}\right) = -\frac{\Delta h_r^0}{\nu R T_e} + \frac{\Delta s_r^0}{\nu R}$$

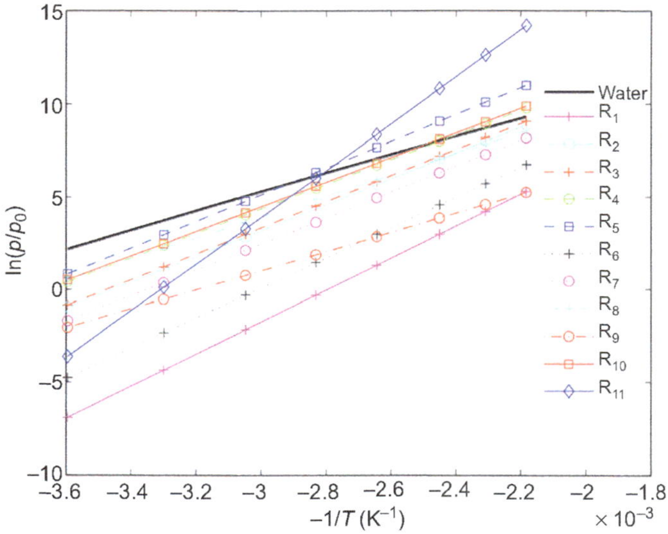

Figure 17.3 The $\ln(p/p_0)$ vs $-1/T$ curve. Reactions are: (R_1) $SrBr_2 + H_2O$; (R_2) $SrBr_2{\cdot}H_2O + 5H_2O$; ($R_3$) $MgSO_4 + 6H_2O$; (R_4) $MgSO_4{\cdot}H_2O + 5H_2O$; ($R_5$) $MgSO_4{\cdot}6H_2O + H_2O$; ($R_6$) $MgCl_2{\cdot}2H_2O + 2H_2O$; ($R_7$) $MgCl_2{\cdot}4H_2O + 2H_2O$; ($R_8$) $CaCl_2{\cdot}4H_2O + 2H_2O$; ($R_9$) $CaCl_2 + H_2O$; (R_{10}) $Na_2SO_3 + 5H_2O$; (R_{11}) $CaSO_4 + 2H_2O$.

where p_e is the equilibrium water vapor pressure (Pa) and T_e the equilibrium temperature (K). Examples of solid/gas reactions presented in Section 17.4 are given in Fig. 17.3. Fig. 17.3 shows the $\ln(p/p_0)$ versus $-1/T$ curve.

The schematic diagram of a perfect sorption thermal battery for energy storage using the solid/gas chemical reaction is presented in Fig. 17.4. The two curves present the solid/gas equilibrium of the adsorbent and the gas/liquid equilibrium of the adsorbate. Under the solid/gas line, the adsorbent is under the A form. Above the solid/gas line, the adsorbent is under the AvB form.

17.2.3 Chemisorption used in systems

Basically, the adsorbent can be integrated into two families of systems: the closed heat storage system and the open heat storage system.

The principle of a closed chemisorption heat storage system is given in Fig. 17.5. Initially, adsorbent is under the AvB form. During the storage phase, Q_{in} heat is transferred to the material at the temperature T_{in}. Then, the gas pressure increases and the gas moves from the material to the condenser where the gas pressure is p_h. Then, the gas condenses (in the condenser) and the heat of condensation is released, Q_{cond}. During the release phase, liquid evaporates (in the evaporator) at temperature T_e, requiring a quantity of heat, Q_{evap}. As the pressure is higher in the evaporator than in the material, a gas flow occurs. A quantity of heat Q_{out} is then released during the adsorption process in the material. Because of the principle described before, the closed system is

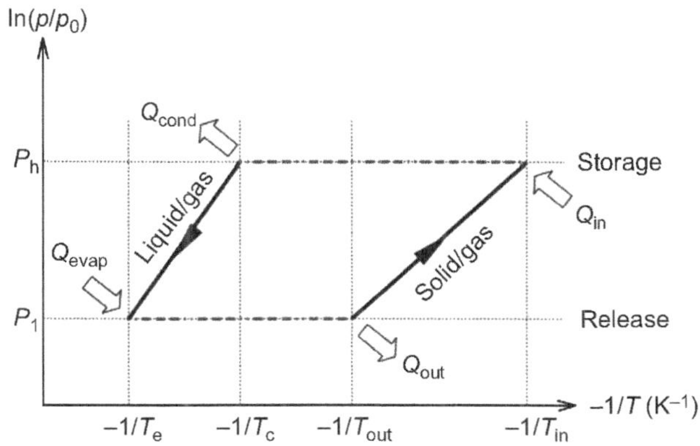

Figure 17.4 Theoretical Clausius–Clapeyron chemical adsorption cycle.

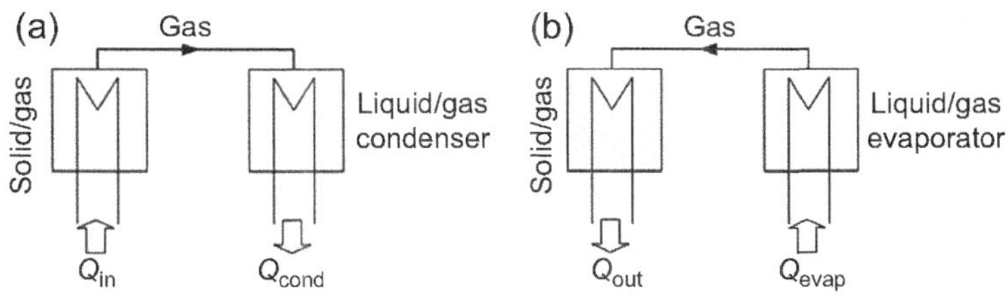

Figure 17.5 Closed heat storage system. Quantities refer to Fig. 17.4. (a) Storage–desorption. (b) Release–adsorption.

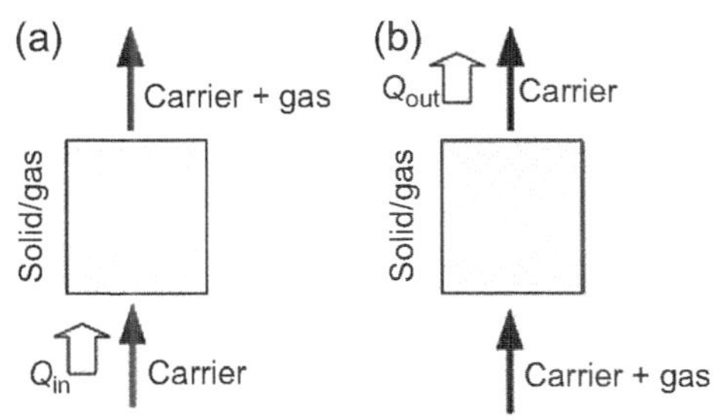

Figure 17.6 Open heat storage system. Quantities refer to Fig. 17.4. (a) Storage–desorption. (b) Release–adsorption.

also called a thermochemical heat pump. This system requires heat exchangers to transfer the heat in and from the material and in the condenser/evaporator.

The closed system requires a low gas pressure, which can limit the technical feasibility of the whole system. The other solution lies in the use of an open system, presented in Fig. 17.6. The idea is to use mixing of a carrier inert gas and the adsorbate, typically dry air and water vapor (ie, moist air). During the storage phase, the carrier (or the mix) passes through the material under the AvB form. Q_{in} heat is then transferred to the material at the temperature T_{in}, resulting in the desorption process. During the release phase, the mix passes through the material under the A form. The gas is adsorbed and heat Q_{out} is transferred to the carrier (or the mix). In such a process, the carrier gas transfers the heat.

For both open and closed systems, a low-temperature source of heat is required to produce the gas during the release phase. Typically, for water adsorbate, the energy required to produce vapor cannot be neglected for the calculation of the system's overall efficiency.

17.3 Important considerations concerning application to buildings

Designing a storage system requires one to know the specifications of the application: this is a *top-down methodology*, from the application to the system design. Thus, a large part of this section is dedicated to drawing guidelines for the determination of the storage system specifications. Quality criteria of a storage system are also developed in this section.

17.3.1 Where does the heat come from?

Thermal energy storage is a solution to rationalize the renewable energy use. Solar thermal collectors can be used to collect heat energy provided by the sun's radiation. A solution based on this concept is presented in [27], Fig. 17.7. It consists in two phases:

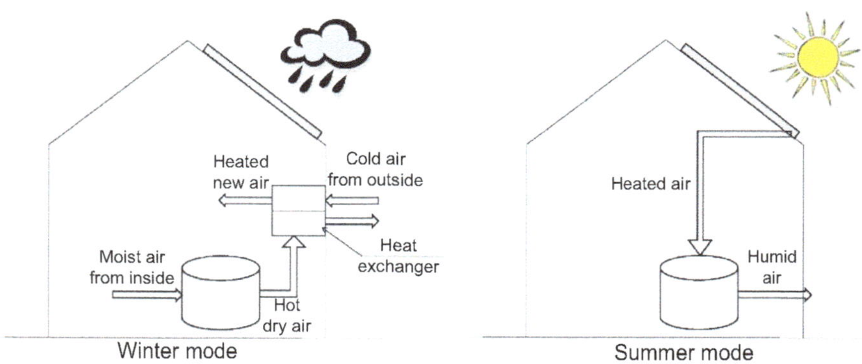

Figure 17.7 Presentation of the storage concept: heat release from the store (left) and heat storage (right).
Extracted from S. Hongois, F. Kuznik, P. Stevens, J.-J. Roux, M. Radulescu, E. Beaurepaire, Thermochemical storage using composite materials: from the material to the system, in: EuroSun Conference 2010, Graz (2010).

1. During the summer, the heat collected by the panels is used for the endothermic reaction.
2. During the winter, moist air passing through the system is used for the exothermic reaction (ie, release of heat).

Moreover, thermal energy storage can be used to rationalize the energy demand and shave the peak loads. An electricity grid or district heating grid can provide heat to storage systems during periods of low load. This heat is then released later, during periods of high demand. Finally, thermal energy storage can also be a way to recover waste heat from systems and thus increase their overall efficiency.

For all the above heat sources, except for electrical heating, the maximum temperature reached by the systems is about 200°C. The reaction temperature of the selected material must comply with this technical limit.

17.3.2 What are the specifications of a thermal energy storage system?

In buildings, the stored heat can be used for (1) space heating and (2) domestic hot water (DHW). Yet, both applications have different requirements concerning energy and power release. The objective of this section is to give an evaluation of the specifications of a TESS from the standpoint of material selection. More information about the data can be found in [13].

17.3.2.1 Chemical heat storage for space heating

The heat required for space heating depends on the building's location and construction mode. However, referring to standards, it is possible to evaluate, for an energy-efficient building, (1) the maximum peak load and (2) the heating demand. These specifications are indispensable to design the reactor and choose the appropriate material.

The main categories of energy-efficient buildings are:

- Low-energy buildings
- Ultralow-energy buildings
- Zero-energy houses
- Autonomous buildings
- Energy-plus houses

Different standards and guidelines exist concerning low- and ultralow-energy buildings. Focusing on Europe, it is possible to evaluate the mean values of the peak load for space heating and the annual energy demand (see Table 17.1) for low- and ultralow-energy buildings.

17.3.2.2 Chemical heat storage for domestic hot water

The evaluation of the energy and power requirements for DHW heating is more difficult because of the disparity of its use. In TASK 32 of IEA SHC [2], a reference building was designed to represent a typical four-person family house located in Sweden. For these four persons, the DHW is 200 L/day (ie, 50 L per day per person) with a DHW load profile given in [34]. Fig. 17.8 shows the power required to produce the

Table 17.1 Maximum peak load and annual heating demand in European standards and guidelines from [52]

Category	Maximum peak load (W/m^2)	Annual heating demand ($kW\ h/m^2$ per year)
Low-energy building	16 to 24	20 to 30
Ultralow-energy building	10 to 14	About 15

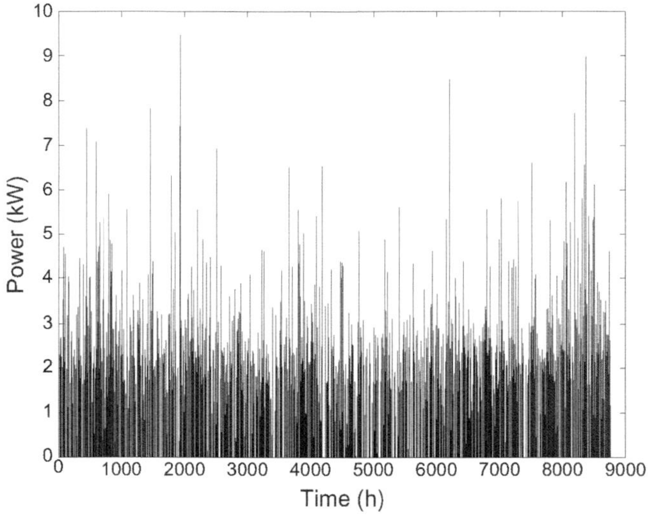

Figure 17.8 Instantaneous power required to produce the domestic hot water during 1 year (ie, 8760 h).
Evaluated using the data from U. Jordan, K. Vajen, Realistic Domestic Hot-Water Profiles in Different Time Scales. Technical report, IEA SHC, Task 26, Solar Combisystems (2001).

DHW, with the assumptions of a cold water supply temperature of 10°C and a DHW set-point temperature of 45°C.

From Fig. 17.9, a 2-kW water heating system will cover 92% of the demand, whereas a 3-kW system will cover 98% of the demand. The energy consumption for the DHW is about 3000 kW h per year for four persons.

17.3.3 What are the additional criteria?

The criteria used to evaluate the quality of a storage system are:

- Energy storage capacity and density (kW h, $kW\ h/m^3$)
- Release power (W)
- Efficiency ($E_{released}/E_{stored}$)

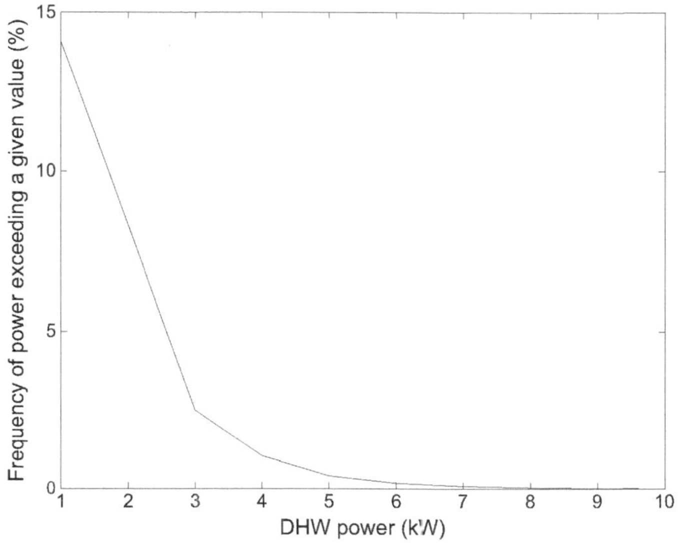

Figure 17.9 Frequency of instantaneous domestic hot water (DHW) power required exceeding a given value: % of time during 1 year (ie, 8760 h).
Evaluated using the data from U. Jordan, K. Vajen, Realistic Domestic Hot-Water Profiles in Different Time Scales. Technical report, IEA SHC, Task 26, Solar Combisystems (2001).

- Stability (chemical, mechanical)
- Reversibility/cyclability
- Cost (€/kW h) and then payback time
- Toxicity
- Life-cycle assessment

Of course, all the criteria must be seen within the scope of limits: physical limits, technological limits, and economical limits.

17.4 Chemical heat storage materials

17.4.1 Pure salts

Reversible chemical reactions of inorganic salts with water or other fluid reaction partners for the purpose of thermal energy storage have been studied for decades. Numerous reversible reactions can be used to store heat: oxide hydration or carbonation, hydration of hygroscopic salt, dehydrogenation of hydrocarbons, etc. A first comprehensive material screening was carried out at the University of Osnabrück, Germany [31]. The differential scanning calorimetry study showed promising results concerning energy storage density and reversibility of various compounds, but also some problems such as side reactions and agglomeration. Other researchers have also observed these undesired effects [51,54]. Workers at the Leuphana University of

Lüneburg have demonstrated which strategies can be used to avoid the mentioned difficulties: control of the temperature, control of the water amount, and the mix have a positive effect on the cycle stability of magnesium chloride and calcium chloride hydrates, respectively, at least at the milligram scale [47,48].

For building use, the material reaction temperature must be below 200°C (availability of energy source) and must not be dangerous for the occupants. The latter explains why most of the studies focus on water vapor adsorbate. Some potential candidates have been identified in [58] (see Table 17.2). These have been numerically evaluated and ranked in agreement with the following criteria:

- Energy storage density
- Reactor temperature for storage process
- Corrosiveness at storage and/or reaction
- Environmental impact and toxicity of the material
- Cost of the material
- Number of material components during synthesis reaction
- Reactor pressure

Among the potential candidates, $MgSO_4$, which is noncorrosive and nontoxic but expensive, seems to be the most promising material for an autonomous chemical storage system: the complete dehydration of $MgSO_4 \cdot 7H_2O$ generates the highest energy storage density (11 times the energy density of water) and the temperature of the synthesis reaction is consistent with the intended use in buildings (<200°C). Indeed, results of characterization studies conducted by ECN and TU Eindhoven [56] show that the dehydration of $MgSO_4 \cdot 7H_2O$ is carried out in at least three stages. The second phase has a high energy density and the reaction takes place in a range of temperatures at which vacuum thermal solar panels yield good performance.

Hydration experiments carried out at ambient pressure [53,56] show that the amount of water that reacts with $MgSO_4$ depends on the water vapor partial pressure and temperature and this vapor transport between the particles is the limiting factor for the hydration of $MgSO_4$. However, under practical conditions of use, $MgSO_4$ does not come up to expectations. Indeed, the usage of magnesium sulfate powder results in the

Table 17.2 Chemical heat storage candidates from [58]

Material	Reaction			Storage density (GJ/m³)	T^* (°C)	Rank
	C ⇔	B +	A			
Magnesium sulfate	$MgSO_4 \cdot 7H_2O$	$MgSO_4$	$7H_2O$	2.8	122	1
Iron carbonate	$FeCO_3$	FeO	CO_2	2.6	180	2
Iron hydroxide	$Fe(OH)_2$	FeO	H_2O	2.2	150	3
Calcium sulfate	$CaSO_4 \cdot 2H_2O$	$CaSO_4$	$2H_2O$	1.4	89	4

formation of a crust [7,26] and swelling of the grains, thus limiting mass transfer and causing reversibility issues. Also, the kinetics is not as fast as expected. However, increasing the water vapor pressure up to 5 kPa results in a higher reaction kinetics and could be an interesting alternative [23].

Both $MgSO_4 \cdot 7H_2O$ and $MgCl_2 \cdot 6H_2O$ have been studied in [22]. Their results are energy densities of 0.38 GJ/m^3 for $MgSO_4 \cdot 7H_2O$ and 0.71 GJ/m^3 for $MgCl_2 \cdot 6H_2O$. The assumption concerning the low-energy density of magnesium sulfate is the low water vapor pressure of 1.3 kPa.

Dehydration of $LiSO_4 \cdot H_2O$, $CuSO_4 \cdot 5H_2O$, and $CuSO_4 \cdot 3H_2O$ has been studied in [21] with a water vapor pressure of 1.3 kPa. The measured energy densities were, respectively, 0.71, 0.81, and 1.12 GJ/m^3.

In [55,57] $MgSO_4 \cdot 7H_2O$, $Al_2SO_4 \cdot 7H_2O$, $MgCl_2 \cdot 6H_2O$, and $CaCl_2 \cdot 2H_2O$ were studied to evaluate the temperature lift during hydration. The results showed that the temperature lift is higher for the chlorides than for the sulfates. However, the chlorides tend to form a gel-like substance due to melting or the formation of a solution. $Na_2S \cdot H_2O$ has also been studied in [19] as a potential candidate.

There are a lot of potential candidates with potential reactions but a systematic evaluation of salts was carried out in [45] for 125 materials. The discrimination procedure was based on three steps:

1. Discrimination based on toxicity, flammability, and well-known inconvenience
2. Discrimination based on Thermogravimetric Analysis (TGA) measurements and reversibility; phase change of the material was also regarded
3. Discrimination based on detailed TGA measurements

From this impressive work, the most promising candidates were $SrBr_2 \cdot 6H_2O$, $LaCl_3 \cdot 7H_2O$, and $MgSO_4 \cdot 6H_2O$. They also highlighted that the energy density was very low considering the water tank or the energy required for the vaporization of water.

On the whole, several drawbacks have been pointed out concerning the use of pure salts: low thermal conductivity and low mass transfer and consequently a low specific power during prototype experimentation. Scientists encountered problems such as the formation of a hydrated layer of material with a low porosity decreasing the access of the adsorbate molecules to the reaction sites [26].

17.4.2 Composite materials

Several drawbacks have been pointed out concerning the use of materials such as $MgSO_4$. The investigation of a novel family of composite materials is, therefore, under way so as to overcome these barriers. As of this writing, sorption material composites have been used for cooling (but can also be used for heating). In the literature, the composite materials are known as composite salts in porous matrix (CSPM):

- Selective water sorbents (SWS) [8–12,18,24,25].
- Chemical heat accumulators [37].
- Composite adsorbents [61,62].

The objectives governing the development of these materials are as follows:

- During hydration, the swelling of salt reduces heat transfer, whereas its agglomeration prejudices mass transfer. Dispersing the hygroscopic salt in a porous matrix with a high thermal conductivity is a way to overcome these difficulties and to enhance heat and mass transfer in the storage system.
- Another aim pursued is to increase the adsorption capacity of classic adsorbents and to obtain higher heat power release than that of the hygroscopic salt in its basic powder form. Indeed composite materials are a result of the combination of ordinary adsorbents (silica gel, zeolite, porous carbon, etc.) with hygroscopic salts ($CaCl_2$, LiBr, $MgSO_4$, etc.), thus improving the storage density.

$MgSO_4$ was dispersed in several porous matrices (zeolite and silica gel) in a project conducted by EDF R&D and CETHIL [26], so as to get a higher energy density than that of the salt alone, to obtain faster reaction kinetics, and to release more heat. Results show that the combination of $MgSO_4$ with zeolite renders the best performance [28]. Temperature lifts of 30°C and a maximum power of 28 mW/g^1 were measured during the hydration of this composite. Lower values were found with pure zeolite, silica gel, or a $MgSO_4$/silica gel composite under the same operating conditions. Similar zeolite-based CSPM were also tested in [59,60]. Because of the introduction of salt, the composite material is conferred new properties. It is therefore necessary to identify which parameters are influential on these properties [15].

The storage density of the composite material is higher than that of the bare matrix. The study of the water vapor adsorption kinetics for SWS-1L and SWS-1A showed that the adsorption process is faster for the matrix material than for the composite material [18]: because of the hygroscopic salt included in the matrix pores, the water vapor diffusion resistance is increased. The time necessary to reach 50% of the final value of adsorbed water is about four times higher for the composite than for the matrix alone. However, the sorption capacity of the composite is significantly superior to the capacity of the matrix alone (by a factor of about 2—3) so that this phenomenon is negligible for a reaction duration greater than 5 min.

It may be recalled that physisorption is a bivariant process. Consequently, when the adsorbate is a gas, the monovariant nature of the solid/gas reaction equilibrium can be used to adjust the sorption isotherms of the composite so that they comply with the application of interest. Moreover, the properties of the salt in the porous matrix differ from those of mere salt owing to:

- Dispersion at a small particle size, this being limited by the pore size. The hydration behavior of a composite material is closely related to the pore size distribution of the matrix material.
- Interactions between the salt and the matrix.

The behavior of the composite during hydration depends on the pore size distribution. For example, for lithium bromide included in a porous carbon, the sorption isobar curves have different shapes depending on mesoporosity (mean pore diameters about 8 nm) or macroporosity (mean pore diameters ranging from 10^3 to 10^4 nm). Fig. 17.10, extracted from [25], shows the isobars of macroporous and mesoporous carbon/LiBr composites. Whereas a plateau related to the formation of $LiBr \cdot H_2O$ (a crystal hydrate) appears for the macroporous composite, no crystal creation is detected in the mesoporous composite.

Further studies concerning calcium chloride composites present similar results: the greater the pore size, the closer the sorption is to the hygroscopic salt behavior with creation of one or more hydrates. No $CaCl_2$ hydrate has been revealed in the case of a microporous matrix with a mean porosity diameter inferior to 4 nm (silica gel [8], active carbon [32]). However, for meso- and macroporous materials, such as attapulgite (aluminum silicate and magnesium silicate) and vermiculite, the following successive products are created:

- $CaCl_2 \cdot 2H_2O$, $CaCl_2 \cdot 4H_2O$, $CaCl_2 \cdot 6H_2O$ in silica gel and attapulgite [32]
- $CaCl_2 \cdot H_2O$ and $CaCl_2 \cdot 2H_2O$ in vermiculite [10]

A vermiculite and $CaCl_2$ CSPM was tested in [16,17] and the results showed that it is a good alternative for an open system.

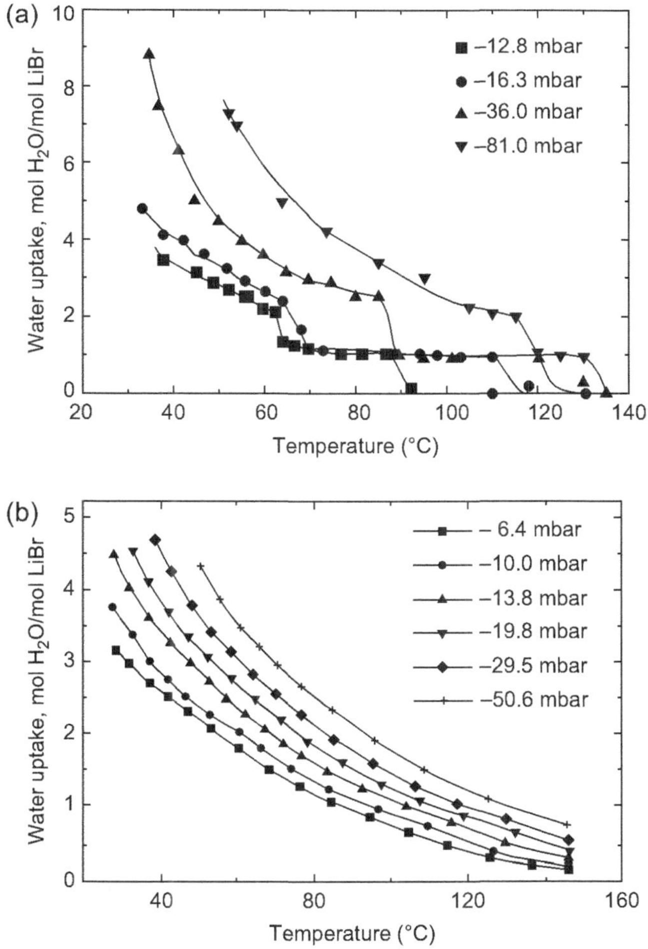

Figure 17.10 Sorption isobars for two composites of carbon/LiBr. (a) SWS−2EG macroporous carbon/LiBr and (b) SWS−2C mesoporous carbon/LiBr. *SWS*, Selective water sorbent. Extracted from L. Gordeeva, G. Restuccia, A. Freni, Y. Aristov, Selective water sorption on composites "LiBr in a porous carbon," Fuel Process. Technol. 79 (2002) 225−231.

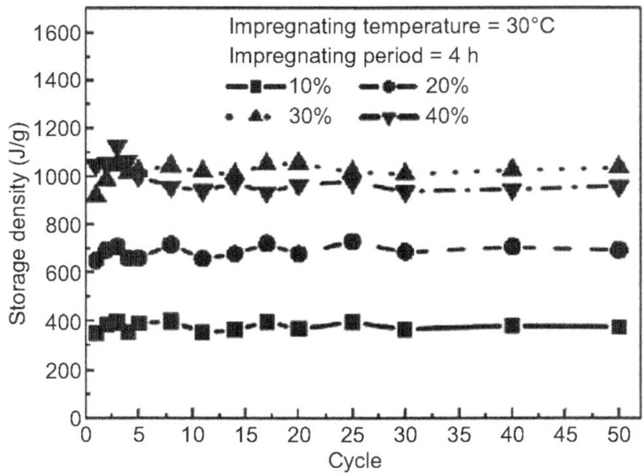

Figure 17.11 Energy density of sorption composites after 50 cycles.
Extracted from H. Wu, S. Wang, D. Zhu, Effects of impregnating variables on dynamic sorption characteristics and storage properties of composite sorbent for solar heat storage, Sol. Energy 81 (7) (2007) 864–871.

The addition of a hygroscopic salt increases the energy density of the material. The case of silica gel/$CaCl_2$ is an example of such an enhancement (Fig. 17.11).

Because the sorption capacity seems to increase with the salt content of the composite, the temptation to use matrices with a high pore volume and to insert important amounts of salt into the matrix is tempting. One must nevertheless pay attention to several issues:

- The pores must not be clogged so as not to hamper mass transfer and create unwanted phenomena.
- The sorption equilibrium can go from being monovariant to being bivariant, depending on the salt content of the composite.

The properties of the composite materials can be optimized by varying the solid matrix material, porosity, and hygroscopic salt material and quantity. Table 17.3 gives the characteristics of composite materials from the literature with silica gel sorption material (SWS type).

Fig. 17.11 shows that, during the first cycles, the energy density increases with $CaCl_2$ content: from 350 J/g (100 Wh/kg) for a weight ratio of $CaCl_2$ of 10% to 1045 J/g 290 W h/kg) for a weight ratio of $CaCl_2$ of 40%. From the sixth cycle on, the 30% $CaCl_2$ composite has the highest energy density. The composite made of 40% $CaCl_2$ absorbs a large quantity of water and reaches saturation: leaks of $CaCl_2$ solution from silica gel pores can be observed. Moreover, a good long-term cyclability of the composites can be observed in Fig. 17.11.

Because of the hygroscopic salt, the composite storage density is higher than that of the sorbent material. Because of the porous solid matrix, the hygroscopic salt is

Table 17.3 Characteristics of selective water sorbent materials from [12] and [18]

Matrix	Pore size (nm)	Salt	Salt mass fraction (%)	Name
KSK silicium	7.5	$CaCl_2$	33.7	SWS-1L
KSK silicium	7.5	LiBr	32.0	SWS-2L (32)
KSK silicium	7.5	LiBr	57.0	SWS-2L (57)
KSK silicium	1.8	$CaCl_2$	21.7	SWS-1S
Alumina	3.5	$CaCl_2$	28.0	SWS-1A

SWS, Selective water sorbent.

dispersed and the power heat release is higher than that of the hygroscopic salt in its basic powder form. Moreover, the solid matrix enables a large variety of shapes and forms, easy manipulation, and better cycling durability.

Instead of impregnation, another approach consists in synthesizing silica gel directly with a hygroscopic salt ($CaCl_2$) [25]. Using this method, the salt is dispersed more finely and uniformly in the porous matrix and generates nanocomposites. Thus, the energy density is relatively high, up to 3400 J/g for a $SiO_2/CaCl_2$ composite.

In the case of water vapor sorption, energy densities of some composites have been compared for the same operating conditions of material generation: heat production at 40°C, storage at 40°C or 160°C, and source of moist air at 15°C [30]. This study shows that the bond energy between salt and water is low compared to the condensation/evaporation energy of water. This bond energy is the main actor in chemical heat pumps. The sorption reaction causes the condensation of water vapor and its capture, enabling the chemical heat pump to function until material saturation.

Salt can also be integrated in a nonphysical adsorption matrix, for example, ceramics. In [33], a salt mixture of $MgCl_2-MgSO_4$ was impregnated in a ceramic porous matrix. Even if the first test on a honeycomb structure confirmed the suitability of the mixture, the storage capacity remained low.

17.5 Storage reactor developments

17.5.1 Closed system

A thermochemical heat pump based on $SrBr_2$ salt was developed for the aim of both heating and cooling in [40]. The prototype (with a volume of 1 m^3) is able to store 60 and 40 kW h/m^3, respectively, for heating and cooling. The specific powers for heating and cooling are in the range 2.5–4 kW/m^3. The main limitation identified by the authors was the low heat transfer at the interface between the reactive layer and the exchanger wall.

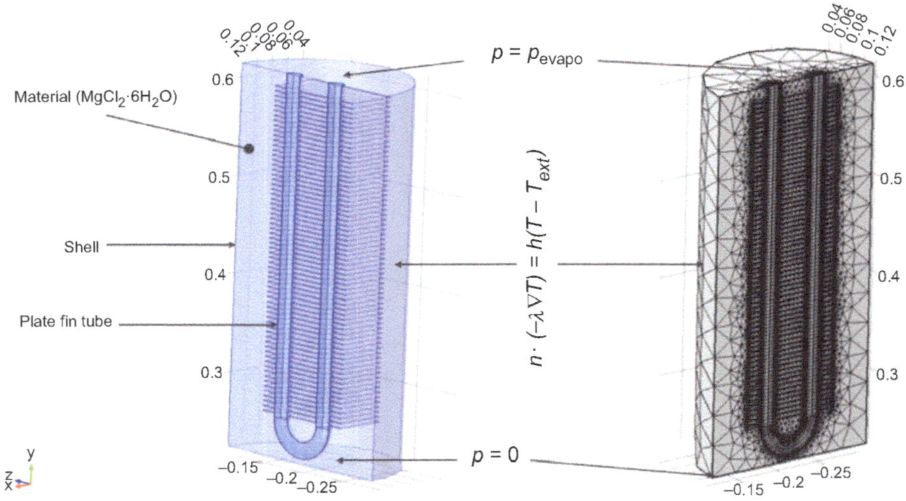

Figure 17.12 (Left) Geometry and (right) mesh of the reactor numerical modeling. Extracted from A.F. Lele, F. Kuznik, H.U. Rammelberg, T. Schmidt, W.K. Ruck, Thermal decomposition kinetic of salt hydrates for heat storage systems, Appl. Energy. ISSN: 0306-2619 154 (0) (2015) 447–458.

Na_2S was the salt used in [29] to develop a modular heat storage system. The average heating power is 3.79 kW and energy released about 1.8 MJ or 0.5 kW h. The calculations do not take into account the energy required for water vaporization during adsorption.

A closed system using $MgCl_2 \cdot 6H_2O$ was numerically studied in [36]. In this work, the material with the heat exchanger was modeled using COMSOL Multiphysics 4.3a software to evaluate the thermal and mass transport behavior of a salt–hydrate during the charging process. Fig. 17.12 shows the geometry and mesh of the model. From the model, it is also possible to evaluate the temperature field in the reactor (Fig. 17.13) and possible problems of melting.

17.5.2 Open system

Pure salt thermochemical heat storage has been studied in the laboratory with the design and test of a prototype using the hydration of strontium bromide ($SrBr_2$) [39,41–43]. References [41–43] report in their studies that hydration-specific powers from 0.75 to 2 W/kg (ie, 4.3 kW/m^3) and an energy density of 190 kW h/m^3 were reached for a theoretical bed salt energy density of 388 kW h/m^3.

$KAl(SO_4)_2 \cdot 12H_2O$ was tested in a reactor of 25 kg of salt [38]. The results were an energy density of 0.2 kW h/kg and a power density of 4 W/kg.

In [63], a selection procedure was realized to select the tested material, $MgCl_2 \cdot 6H_2O$. A prototype of 17 dm^3 was tested under laboratory conditions. The prototype was able to provide 50 W heating power at 60°C, with an instantaneous coefficient of performance of 12, ie, a density of about 3 kW/m^3. The authors also

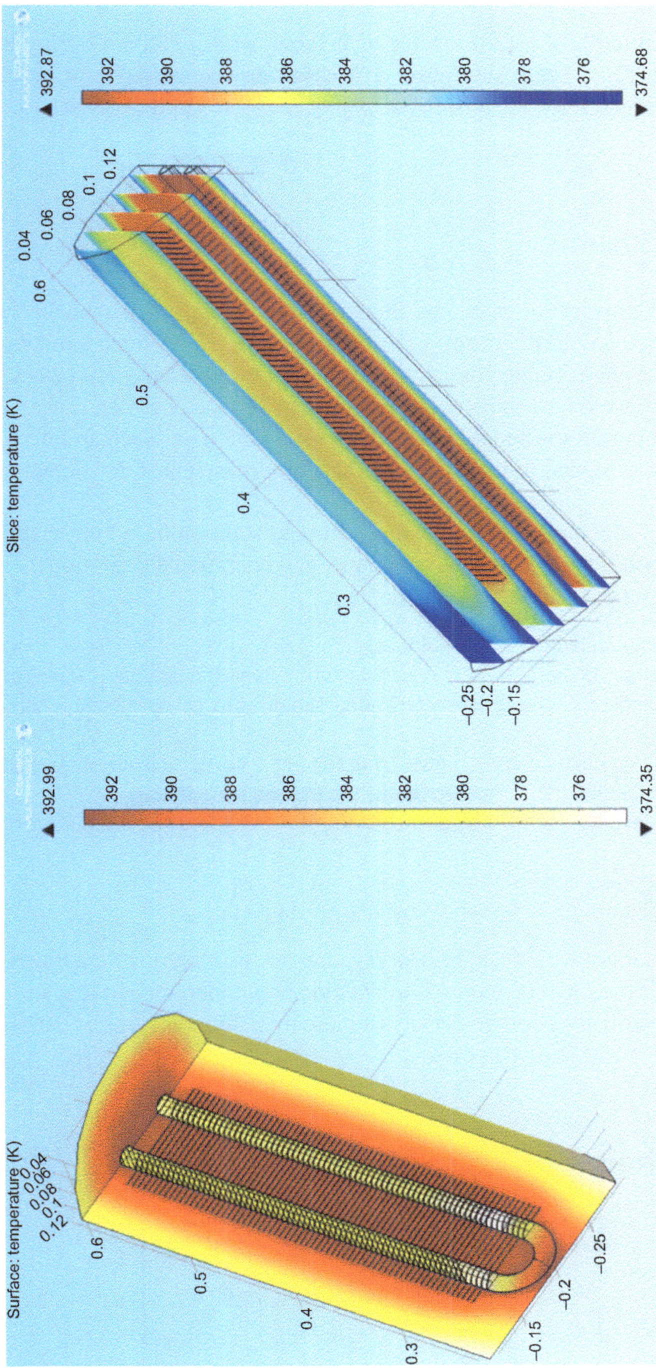

Figure 17.13 Temperature fields in the reactor during charging mode.
Extracted from A.F. Lele, F. Kuznik, H.U. Rammelberg, T. Schmidt, W.K. Ruck, Thermal decomposition kinetic of salt hydrates for heat storage systems, Appl. Energy. ISSN: 0306-2619 154 (0) (2015) 447–458.

highlighted that improvements in heat recovery and pressure drop could increase the performance.

A CSPM zeolite/$MgSO_4$ [28] reactor of 7.1 dm^3 was monitored during hydration and dehydration [26]. The results were a released power of 64 W for an energy of 636 W h, meaning a released power density of 9 kW/m^3 and an energy density of 90 kW h/m^3.

17.6 Conclusions

Theoretically, chemisorption materials allow high energy densities to be reached, up to 2.8 GJ/m^3, ie, 800 kW h/m^3. From a physical point of view, some limitations exist related to the salt phase change or salt dissolution in the condensation product. Moreover, gas diffusion and heat transfer are also barriers to reaching high power or energy densities. A solution lies in the design of composite materials. Such architectured materials should be designed to overcome the heat and mass transfers encountered with pure salts.

The design of a compact thermal energy storage solution based on chemisorption has to comply with several technical, economic, and operational limitations. Therefore, research and technical problems to tackle are:

- Finding materials suited to seasonal storage
- Characterizing the behavior of these materials during the different phases (heat storage/heat release); standards concerning characterization should also be developed to compare the candidates
- Enhancing heat and mass transfer, either by designing a more efficient heat exchanger or by improving the thermal transfer properties of the materials used
- Developing effective control strategies ensuring the stability of power during charge and discharge
- Conducting large-scale experiments
- Developing solutions for low-energy production vapor, Q_{evap}

On the whole, thermochemical heat storage has a low technology readiness level [6] and thus a lot of effort will be required to develop novel thermochemical storage technologies with an effective storage density of more than 300 kW h/m^3 by 2020 [5], with a payback time of less than 10 years.

List of symbols

H	Enthalpy	J
h	Reaction enthalpy	J/mol
P_i	Product	
p	Partial pressure	Pa

R	Gas constant	J/K mol
Q	Heat	W
R_i	Reactant	
S	Entropy	J/K
s	Reaction entropy	J/K mol
T	Temperature	K

Greek letters

Δ	Difference
ν	Stoichiometric coefficient

Subscripts

cond	Condensation
e	Equilibrium
evap	Evaporation
h	High
in	Inlet
l	Low
S	Source
out	Outlet

Superscript

0	Reference

References

[1] Technical report, Plan Climat 2004, Ministère de l'Écologie et du Développement durable, France, 2004.

[2] Technical report, Advanced Storage Concepts for Solar and Low Energy Buildings, IEA Solar Heating and Cooling Programme – Task 32, 2007.
[3] Technical report, EU Energy in Figures–Statistical Pocketbook, European Commissions, 2012.
[4] Technical report, Household Energy Consumption by End–Use in the EU–27, European Environment Agency, 2012.
[5] Technical report, Strategic Research Priorities for Solar Thermal Technology, European Technology Platform on Renewable Heating and Cooling, 2012.
[6] Technical report, Technology Roadmap – Energy Storage, IEA, 2014.
[7] Y. Aristov, L. Gordeeva, Salt in a porous matrix adsorbents: design of the phase composition and sorption properties, Kinet. Catal. 50 (2009) 65–72.
[8] Y. Aristov, M. Tokarev, G. Restuccia, G. Cacciola, Selective water sorbents for multiple applications, 2. $CaCl_2$ confined in micropores of silica gel: sorption properties, React. Kinet. Catal. Lett. 59 (1996) 335–342.
[9] Y. Aristov, G. Restuccia, G. Cacciola, M. Tokarev, Selective water sorbents for multiple applications, 7. Heat conductivity of $CaCl_2$-SiO_2 composites, React. Kinet. Catal. Lett. 65 (1998) 277–284.
[10] Y. Aristov, G. Restuccia, M. Tokarev, H.-D. Buerger, A. Freni, Selective water sorbents for multiple applications. 11. $CaCl_2$ confined to expanded vermiculite, React. Kinet. Catal. Lett. 71 (2000) 377–384.
[11] Y. Aristov, G. Restuccia, M. Tokarev, G. Cacciola, Selective water sorbents for multiple applications, 10. Energy storage ability, React. Kinet. Catal. Lett. 69 (2000) 345–353.
[12] Y. Aristov, G. Restuccia, G. Cacciola, V. Parmon, A family of new working materials for solid sorption air conditioning systems, Appl. Therm. Eng. 22 (2) (2002) 191–204.
[13] K. Belz, F. Kuznik, K. Werner, T. Schmidt, W. Ruck, Thermal energy storage systems for heating and hot water in residential buildings, in: L.F. Cabeza (Ed.), Advances in Thermal Energy Storage Systems, Woodhead Publishing Series in Energy, Woodhead Publishing, 2015, pp. 441–465.
[14] V. Brick, F. Kuznik, K. Johannes, J. Virgone, Evaluation of thermal energy storage potential in low-energy buildings in France, in: ISES Solar World Congress 2011, Kassel (Germany), 28 August–2 September 2011, 2011.
[15] E. Brouwer, C. Rindt, M. van Essen, W. van Helden, A. van Steenhoven, Hydration and dehydration of sorption materials: experiments in a small scale reactor, in: International Symposium on Convective Heat and Mass Transfer in Sustainable Energy, 2009.
[16] S.P. Casey, J. Elvins, S. Riffat, A. Robinson, Salt impregnated desiccant matrices for 'open' thermochemical energy storage–selection, synthesis and characterisation of candidate materials, Energy Buildings 84 (0) (2014) 412–425.
[17] S.P. Casey, D. Aydin, S. Riffat, J. Elvins, Salt impregnated desiccant matrices for 'open' thermochemical energy storage–hygrothermal cyclic behaviour and energetic analysis by physical experimentation, Energy Buildings. ISSN: 0378-7788 92 (0) (2015) 128–139.
[18] B. Dawoud, Y. Aristov, Experimental study on the kinetics of water vapor sorption on selective water sorbents, silica gel and alumina under typical operating conditions of sorption heat pumps, Int. J. Heat Mass Transfer 46 (2) (2003) 273–281.
[19] R. de Boer, W. Haije, J. Veldhuis, Determination of structural, thermodynamic and phase properties in the Na_2S–H_2O system for application in a chemical heat pump, Thermochim. Acta 395 (1–2) (2002) 3–19.
[20] I. Dincer, M. Rosen, Thermal Energy Storage – Systems and Applications, John Wiley and Sons, 2002.

[21] C.J. Ferchaud, H.A. Zondag, R. de Boer, C. Rindt, Characterization of the sorption process in thermochemical materials for seasonal solar heat storage application, in: 12th Proceedings of International Conference on Energy Storage, Innostock, 2012.
[22] C.J. Ferchaud, H.A. Zondag, J.B.J. Veldhuis, R. de Boer, Study of the reversible water vapour sorption process of $MgSO_4 \cdot 7H_2O$ and $MgCl \cdot 6H_2O$ under the conditions of seasonal solar heat storage, J. Phys. Conf. Ser. 395 (1) (2012) 012069.
[23] C.J. Ferchaud, R.A. Scherpenborg, H.A. Zondag, R. de Boer, Thermochemical seasonal solar heat storage in salt hydrates for residential applications—influence of the water vapor pressure on the desorption kinetics of $MgSO_4 \cdot 7H_2O$, Energy Procedia. ISSN: 1876-6102 57 (0) (2014) 2436—2440, 2013 ISES Solar World Congress.
[24] A. Freni, F. Russo, S. Vasta, M. Tokarev, Y. Aristov, G. Restuccia, An advanced solid sorption chiller using SWS-1L, Appl. Therm. Eng. 27 (13) (2007) 2200—2204.
[25] L. Gordeeva, G. Restuccia, A. Freni, Y. Aristov, Selective water sorption on composites "LiBr in a porous carbon", Fuel Process. Technol. 79 (2002) 225—231.
[26] S. Hongois, Stockage de chaleur inter—saisonnier par voie thermochimique pour le chauffage solaire de la maison individuelle (Ph.D. thesis), INSA, Lyon, 2011.
[27] S. Hongois, F. Kuznik, P. Stevens, J.-J. Roux, M. Radulescu, E. Beaurepaire, Thermochemical storage using composite materials: from the material to the system, in: EuroSun Conference 2010, Graz, 2010.
[28] S. Hongois, F. Kuznik, P. Stevens, J.-J. Roux, Development and characterisation of a new $MgSO_4$/zeolite composite for long-term thermal energy storage, Sol. Energy Mater. Sol. Cells. ISSN: 0927-0248 95 (7) (2011) 1831—1837.
[29] K. Iammak, W. Wongsuwan, T. Kiatsiriroj, Investigation of modular chemical energy storage performance, in: The Joint International Conference on "Sustainable Energy and Environment (SEE)", 2004, pp. 504—507.
[30] IEA Task 32, Thermal Energy Storage for Solar and Low Energy Buildings — State of the Art by IEA Solar Heating and Cooling Task 32, 2005.
[31] Z. Iyimen-Schwarz, M. Lechner, Energiespeicherung durch chemische Reaktionen. I. DSC-Messungen zur quantitativen Verfolgung der Enthalpieanderungen von Speicherstoffen fur die Hinund Reaktion, Thermochim. Acta 68 (2—3) (1983) 349—361.
[32] J. Janchen, D. Ackermann, E. Weiler, H. Stach, W. Brosicke, Thermochemical storage of low temperature heat by zeolite; Sapo's and impregnated active carbon, in: 7th Workshop of IEA/ECES Annex 17, 2004.
[33] J. Janchen, T. Brettschneider, S. Lauenroth, G. Fischer, U. Hellwig, Salt mixtures in porous ceramics for thermochemical storage of low temperature heat, in: Proceedings of ISES 2011, 2011, pp. 4858—4862.
[34] U. Jordan, K. Vajen, Realistic Domestic Hot-Water Profiles in Different Time Scales, Technical report, IEA SHC, Task 26, Solar Combisystems, 2001.
[35] F. Kuznik, D. David, K. Johannes, J.-J. Roux, A review on phase change materials integrated in building walls, Renewable Sustainable Energy Rev. 15 (1) (2011) 379—391.
[36] A.F. Lele, F. Kuznik, H.U. Rammelberg, T. Schmidt, W.K. Ruck, Thermal decomposition kinetic of salt hydrates for heat storage systems, Appl. Energy. ISSN: 0306-2619 154 (0) (2015) 447—458.
[37] E. Levitskij, Y. Aristov, M. Tokarev, V. Parmon, Chemical heat accumulator': a new approach to accumulating low potential heat, Sol. Energy Mater. Sol. Cells 44 (3) (1996) 219—235.
[38] F. Marias, G. Tanguy, J. Wyttenbach, S. Rouge, P. Papillon, Thermochemical storage first results of pilot storage with moist air, in: Proceedings of ISES 2011, 2011.

[39] F. Marias, P. Neveu, G. Tanguy, P. Papillon, Thermodynamic analysis and experimental study of solid/gas reactor operating in open mode, Energy 66 (2014) 757–765.
[40] S. Mauran, H. Lahmidi, V. Goetz, Solar heating and cooling by a thermochemical process. First experiments of a prototype storing 60 kW h by a solid/gas reaction, Sol. Energy. ISSN: 0038-092X 82 (7) (2008) 623–636.
[41] B. Michel, Procédé thermochimique pour le stockage intersaisonnier de l'énergie solaire: modélisation multi–échelles et expérimentation d'un prototype sous air humide (Ph.D. thesis), Université Via Domitia, Perpignan, 2012.
[42] B. Michel, N. Mazet, S. Mauran, D. Stitou, J. Xu, Thermochemical process for seasonal storage of solar energy: characterization and modeling of a high density reactive bed, Energy 47 (2012) 553–563.
[43] B. Michel, N. Mazet, P. Neveu, Experimental investigation of an innovative thermochemical process operating with a hydrate salt and moist air for thermal storage of solar energy: global performance, Appl. Energy 129 (2014) 177–186.
[44] K. N'Tsoukpoe, H. Liu, N. Le Pierres, L. Luo, A review on long–term sorption solar energy storage, Renewable Sustainable Energy Rev. 13 (9) (2009) 2385–2396.
[45] K.E. N'Tsoukpoe, T. Schmidt, H.U. Rammelberg, B.A. Watts, W.K. Ruck, A systematic multi–step screening of numerous salt hydrates for low temperature thermochemical energy storage, Appl. Energy. ISSN: 0306-2619 124 (0) (2014) 1–16.
[46] P. Pinel, C. Cruickshank, I. Beausoleil-Morrison, A. Wills, A review of available methods for seasonal storage of solar thermal energy in residential applications, Renewable Sustainable Energy Rev. 15 (7) (2011) 3341–3359.
[47] H. Rammelberg, J. Kollner, T. Schmidt, O. Opel, W. Ruck, Hydration and dehydration of salt hydrates and hydroxides for thermal energy storage – kinetics, energy release and cycleability, in: Proceedings of Heat Powered Cycles, 2012, pp. 65–66.
[48] H. Rammelberg, M. Myrau, T. Schmidt, W. Ruck, An optimization of salt hydrates for thermochemical heat storage, in: Chemical Science & Engineering Series: Innovative Materials for Processes in Energy Systems, 2013, pp. 550–555.
[49] A. Sharma, V. Tyagi, C. Chen, D. Buddhi, Review on thermal energy storage with phase change materials and applications, Renewable Sustainable Energy Rev. 13 (2) (2009) 318–345.
[50] H. Singh, R. Saini, J. Saini, A review on packed bed solar energy storage systems, Renewable Sustainable Energy Rev. 14 (3) (2010) 1059–1069.
[51] K. Sugimoto, R. Dinnebier, J. Hanson, Structures of three dehydration products of bischofite from in situ synchrotron powder diffraction data ($MgCl_2 \cdot nH_2O$; $n = 1, 2, 4$), Acta Crystallogr., Sect. B 63 (2) (April 2007) 235–242.
[52] K. Thullner, Low–Energy Buildings in Europe – Standards, Criteria and Consequences, Technical report, University of Lunds, 2010.
[53] I. van de Voort, Characterization of a Thermo Chemical Storage Material (Master's thesis), Technical University Eindhoven, 2007.
[54] V. van Essen, J. Gores, L. Bleijendaal, H. Zondag, R. Schuitema, M. Bakker, et al., Characterization of salt hydrates for compact seasonal thermochemical storage, in: Proceedings of the 3rd ASME International Conference on Energy Sustainability (ES2009), 2009.
[55] V. van Essen, J. Gores, L. Bleijendaal, H. Zondag, R. Schuitema, W. van Helden, Characterization of Salt Hydrates for Compact Seasonal Thermochemical Storage, Technical report, ECN, 2009.
[56] V. van Essen, H. Zondag, J. Gores, L. Bleijendaal, M. Bakker, R. Schuitema, et al., Characterization of $MgSO_4$ hydrate for thermochemical seasonal heat storage, J. Sol. Energy Eng. 131 (4) (2009).

[57] V. van Essen, J. Gores, L. Bleijendaal, H. Zondag, R. Schuitema, W. van Helden, Development of a compact heat storage system based on salt hydrates, in: International Conference on Solar Heating, Cooling and Buildings, 2010.
[58] K. Visscher, J. Veldhuis, Comparison of candidate materials for seasonal storage of solar heat through dynamic simulation of building and renewable energy system, in: Ninth International IBPSA Conference Proceedings, Montreal, Canada, August 15–18, 2005, 2005.
[59] G. Whiting, D. Grondin, S. Bennici, A. Auroux, Heats of water sorption studies on zeolite–$MgSO_4$ composites as potential thermochemical heat storage materials, Sol. Energy Mater. Sol. Cells. ISSN: 0927-0248 112 (0) (2013) 112–119.
[60] G.T. Whiting, D. Grondin, D. Stosic, S. Bennici, A. Auroux, Zeolite–$MgCl_2$ composites as potential long–term heat storage materials: influence of zeolite properties on heats of water sorption, Sol. Energy Mater. Sol. Cells. ISSN: 0927-0248 128 (0) (2014) 289–295.
[61] H. Wu, S. Wang, D. Zhu, Effects of impregnating variables on dynamic sorption characteristics and storage properties of composite sorbent for solar heat storage, Sol. Energy 81 (7) (2007) 864–871.
[62] D. Zhu, H. Wu, S. Wang, Experimental study on composite silica gel supported $CaCl_2$ sorbent for low grade heat storage, Int. J. Therm. Sci. 45 (8) (2006) 804–813.
[63] H. Zondag, B. Kikkert, S. Smeding, R. de Boer, M. Bakker, Prototype thermochemical heat storage with open reactor system, Appl. Energy. ISSN: 0306-2619 109 (0) (2013) 360–365.

Thermochemical heat storage for solar heating and cooling systems 18

T. Yan, T.X. Li, R.Z. Wang
Institute of Refrigeration and Cryogenics, Shanghai Jiao Tong University, Shanghai, China

18.1 Introduction

Solar energy is regarded as a promising renewable energy source and it is considered to be a complement to conventional fossil energy sources. Various solar thermal energy utilization technologies have been receiving more attention as the value of saving energy and reducing environmental pollution has been recognized. Solar thermal energy has a wide range of industrial and domestic applications, such as solar thermal electricity, solar space heating, solar desalination, solar industrial air heating, solar water heating, solar cooling, and so on. However, solar energy is always characterized by instability and discontinuity owing to the variations in solar insolation with the weather, location, time, and season of the year. This means that advanced energy storage is an essential key technology for adjusting the time discrepancy and instability between solar energy supply and energy demand. The common methods used for solar thermal energy storage include sensible heat energy storage, latent heat energy storage using phase-change materials (PCMs), and thermochemical energy storage.

The thermochemical energy storage method has been receiving more attention owing to its distinct advantage of higher energy storage density and low heat loss compared with the sensible and latent energy storage methods. Reversible thermochemical reactions have potential for both efficient conversion and storage of thermal energy. The endothermic reactions of the chemical compounds can be triggered by supplying thermal energy accumulated or captured from any heat source (surplus energy or solar energy). As a result, the chemical compounds (AB) are decomposed into individual ingredients (A, B) with the supply of heat. Meanwhile, the thermal energy is stored in the form of chemical potential energy. The individual reactive components can be stored separately, which makes it possible to store thermal energy for long periods. On the other hand, thermal energy can be retrieved again via the recombination of the reactive components in an exothermic reaction. The thermal energy stored and released is equivalent to the chemical reaction enthalpy.

$$AB + \text{heat input} \Leftrightarrow A + B \qquad [18.1]$$

Generally speaking, the reaction enthalpy is much larger than the transition heat of latent heat storage or sensible heat stored over a rational temperature span. Hence, the energy storage density of thermochemical heat storage is much larger than latent or sensible heat storage. In addition, latent or sensible heat storage loses thermal energy

Advances in Solar Heating and Cooling. http://dx.doi.org/10.1016/B978-0-08-100301-5.00018-7

gradually by heat conduction and radiation owing to the temperature difference between the stored materials and the ambient air; however, thermochemical heat storage can store thermal energy in the form of chemical potential energy with negligible heat loss.

18.2 Thermochemical heat storage

18.2.1 Thermochemical sorption heat storage

Solid–gas sorption working pairs and various cycles have already been used for refrigeration applications [1]. Among the various sorption working pairs, only those working pairs with low dissociation temperature below 100°C are suitable for solar cooling applications using the conventional flat plate solar collector, such as calcium chloride–ammonia, barium chloride–ammonia [2–6], and strontium chloride–ammonia [7].

A solar air-conditioning pilot plant was installed in the PROMES Laboratory in France in 2006 [8,9]. This pilot has a daily cooling capacity of 20 kWh and consists of four subsystems: a solar heating loop, a thermochemical unit using the working pair $BaCl_2$–NH_3, a ground cooling loop, and a chilled-water production loop (Fig. 18.1). The reactive sorbent, which consists of a compressed mixture of 140 kg of anhydrous $BaCl_2$ and 35 kg of expanded natural graphite, is loaded into a set of 19 tubes in the thermochemical sorption unit. The hot PCM storage enables the storage of excess solar heat if the desorption reaction in the reactor is completed, and the cold PCM storage makes a supply of cooling effect possible when the sorption reaction is not available. Two years of experimental operation illustrated an average yearly efficiency of solar collectors ranging from 0.4 to 0.5 and a coefficient of performance (COP) varying from 0.3 to 0.4. A daily cold storage capacity is about 0.8–1.2 kWh/m^2 solar collector plate at a cold temperature of 4°C.

Li et al. [10] proposed a dual-mode chemical sorption energy storage method used for seasonal storage of solar thermal energy. Fig. 18.2 shows a schematic diagram of a dual-mode chemical sorption thermal energy storage system. The overall system includes two units. Each unit is composed of a reactor and a refrigerant vessel. Solar thermal energy is stored in summer and released for use in winter. There are two operating modes depending on the ambient temperature during the energy release phase in winter. If the ambient temperature is high enough to make the solid–gas chemical reaction heat between the reactors and the evaporators satisfy the user heat demand, the generated reaction heat arising from the exothermic synthesis reaction is provided for the heat consumer. Each unit forms a chemical adsorption heat storage system and can supply the thermal energy to the end user. If the ambient temperature is relatively low in winter, the heat output temperature of both units cannot meet the user's heat demand. An internal heat recovery process is employed between the two energy storage units. The generated reaction heat of one of the units is used to heat the evaporator of another unit. This leads to the temperature-lift effect and therefore elevates the heat output temperature of the overall system to meet the requirements of the heat consumer. The theoretical analysis suggested that the proposed energy storage method is feasible for the seasonal storage of solar thermal energy. The energy density is

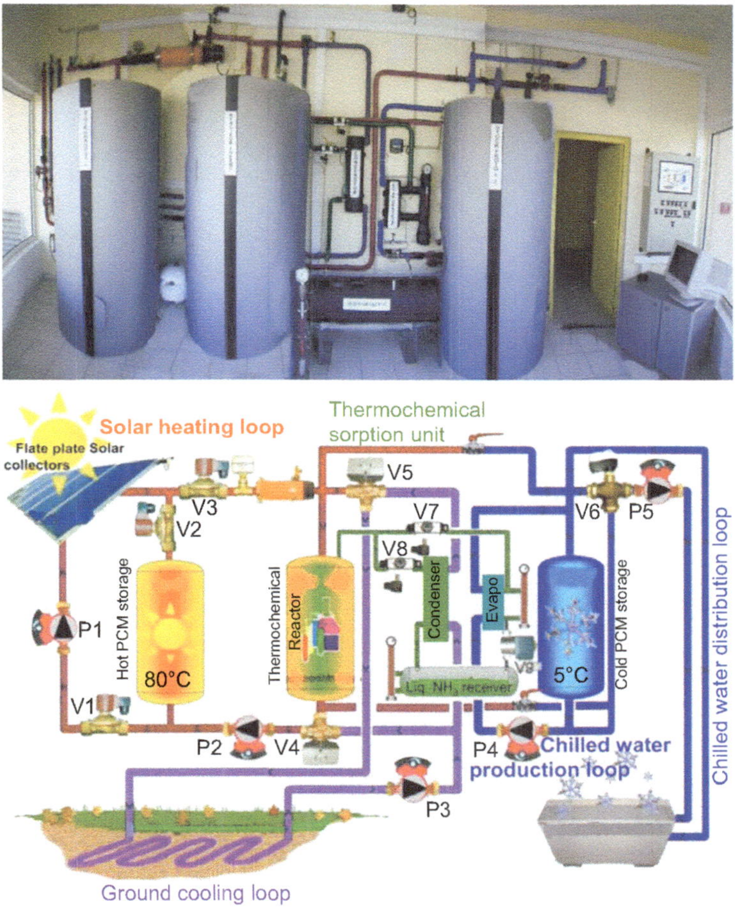

Figure 18.1 Solar sorption pilot plant for air conditioning [8]. *PCM,* phase-change material.

1043 and 579 kJ/kg salt, respectively, under the same conditions when the heat output temperature is 62°C with ambient temperatures of 0°C (without internal heat recovery) and −20°C (with internal heat recovery) in winter.

Li et al. [11] proposed and analyzed a solid−gas thermochemical sorption thermal battery for solar cooling and heating energy storage and heat transformer. The proposed sorption thermal battery can be used as short-term energy storage and long-term energy storage according to the length of storage period for various demands of the end user. The purpose of the former mode is to store solar thermal energy on sunny days to be used during cloudy days and/or at night, whereas the latter mode is to store solar thermal energy in summer to be used for heating in winter. Moreover, the advanced sorption thermal battery can also work as a heat transformer for integrated energy storage and energy upgrade, and the working temperature of stored solar thermal energy can be upgraded from a low heat input temperature to a high heat output temperature by using a solid−gas thermochemical sorption heat transformer

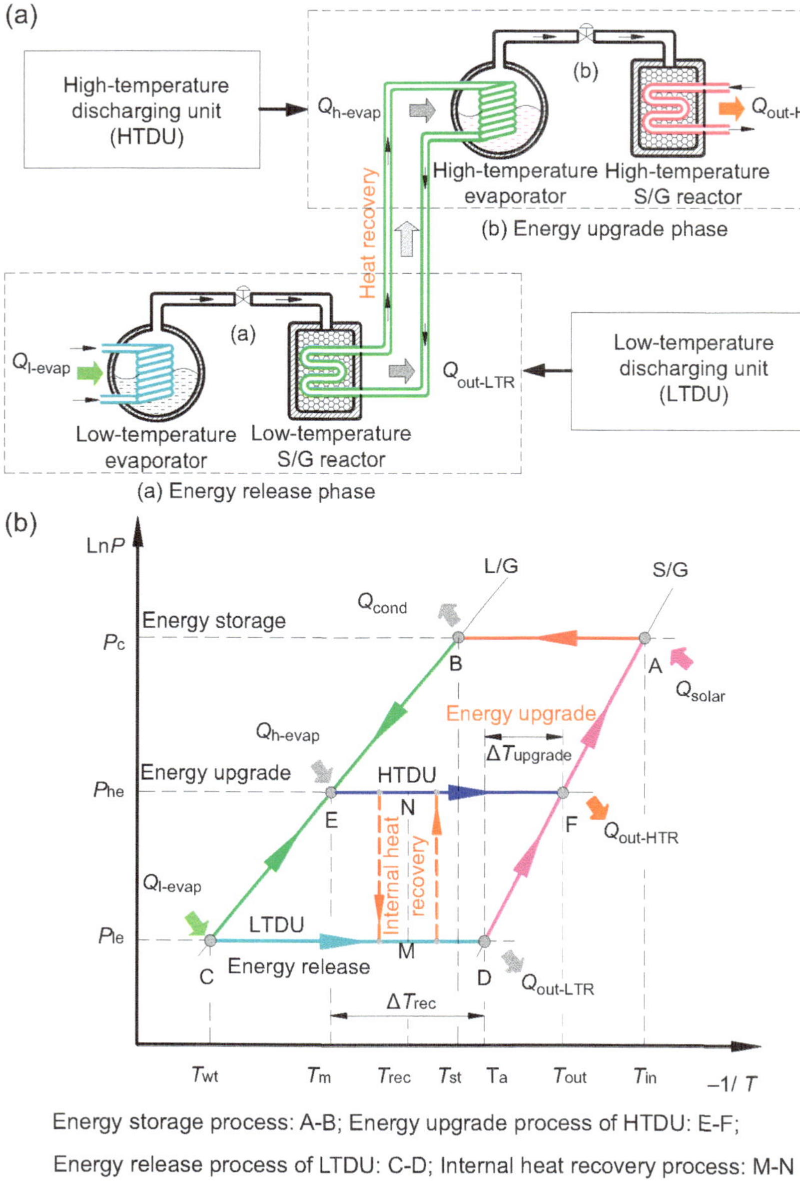

Figure 18.2 Schematic diagrams of the dual-mode solid–gas chemical sorption system for seasonal storage of solar thermal energy [10]. (a) Internal heat recovery process between reactor in LTDU and evaporator in HTDU. (b) Clausius–Clapeyron diagram. *HTDU*, high-temperature discharging unit; *LTDU*, low-temperature discharging unit.

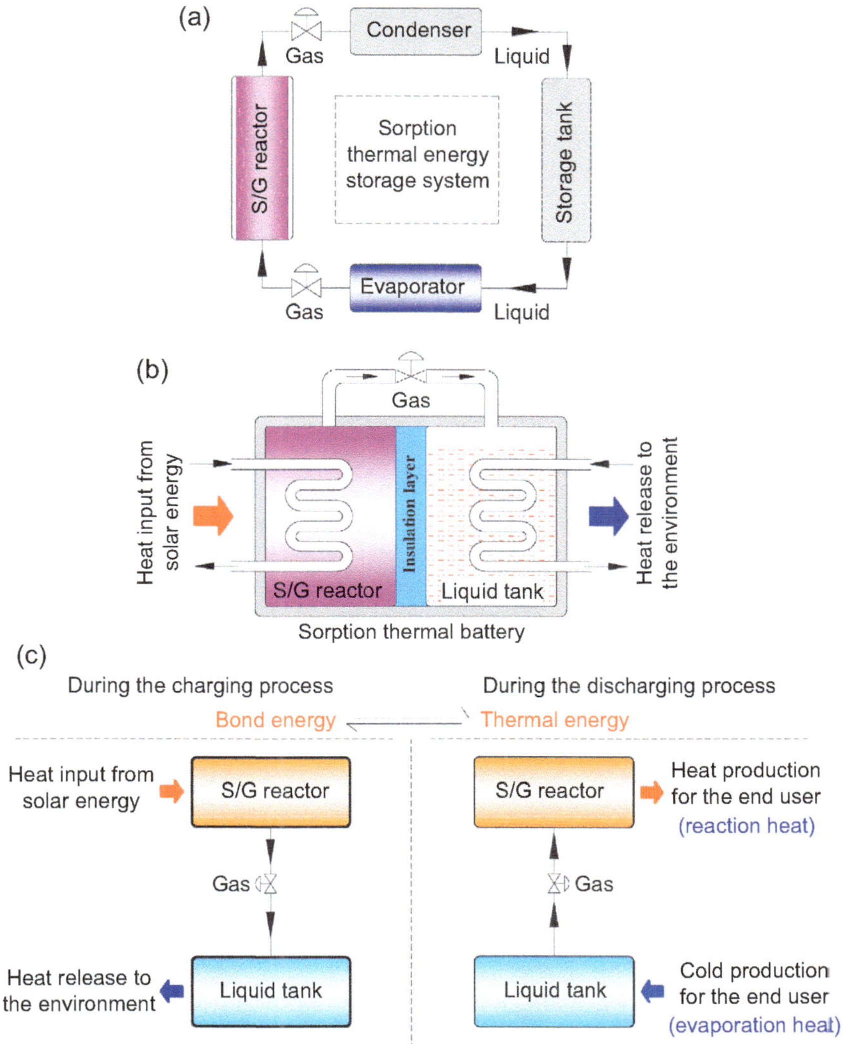

Figure 18.3 Schematic diagram of a solid–gas thermochemical sorption thermal battery for solar cooling and heating energy storage [11]. (a) Sorption thermal energy storage system. (b) Schematic diagram of sorption thermal battery. (c) Working processes of sorption thermal battery.

cycle to achieve high-temperature thermal energy. Fig. 18.3 shows a schematic diagram of the solid–gas thermochemical sorption thermal battery for solar cooling and heating energy storage [11]. The multimode performance of the proposed thermochemical sorption thermal battery for short-term and long-term energy storage and energy upgrade was evaluated and analyzed using as the sorption working pair strontium chloride–ammonia. The involved reaction is as follows:

$$SrCl_2 \cdot 8NH_3(s) \Leftrightarrow SrCl_2 \cdot 1NH_3(s) + 7NH_3(g) \quad [18.2]$$

For short-term energy storage operation mode, the advanced thermochemical sorption thermal battery has a heat energy storage density higher than 1600 kJ/kg with heat production temperatures of 52–87°C and a cold energy storage density higher than 640 kJ/kg with cold production temperatures of −20 to −35°C. For the long-term seasonal energy storage mode, the heat energy storage density is higher than 1300 kJ/kg with heat production temperatures of 52–70°C. For heat transformer operation mode, the working temperature of the stored energy can be upgraded from the heat input temperature of 96°C to a heat output temperature of 112–146°C with heat energy density higher than 1362 kJ/kg and energy upgrade efficiency higher than 0.65.

The desorption–adsorption reaction of hydrates of strontium bromide was chosen by H. Lahmidi [12]. It has an energy storage capacity of 0.35 kWh/kg anhydrous salt. The chemical reaction equation between strontium bromide and water is as follows:

$$SrBr_2 \cdot 6H_2O(s) \Leftrightarrow SrBr_2 \cdot 1H_2O(s) + 5H_2O(g) \quad [18.3]$$

The experimental results demonstrated that power levels in the range of 40 kW/m^3 can be reached with reactive composites ($SrBr_2$ and expanded graphite) and heating storage capacities higher than 250 kWh/m^3 can be achieved. To achieve the dual effects of heating and cooling, the sorption reaction is selected under high energy density [13]. The selection criteria involve the toxicity of the products, energy density, complexity, cost, satisfactory comfort, etc. Water was used as the reactive gas because of its environmental safety. In addition, thermodynamic constraints have been considered, such as the heat source or sink temperature and the external environment temperature. A prototype was designed and tested for achieving the heating and cooling functions, and a heat storage capacity of 60 kWh/m^3 and cooling storage capacity of 40 kWh/m^3 were obtained.

Magnesium sulfate ($MgSO_4$) is considered a potential thermochemical storage material, by means of the following reversible reaction:

$$MgSO_4 \cdot 7H_2O(s) \Leftrightarrow MgSO_4(s) + 7H_2O(g) \quad [18.4]$$

This material is very tempting for the solar seasonal heat storage. Posern and Kaps [14] demonstrated the working performance of $MgSO_4/H_2O$ system. The hydration and dehydration of the chemical material proceed in the reactor. The system can employ either a separate reactor or an integrated reactor, as shown in Fig. 18.4 [15].

A separate reactor requires a much smaller heat exchange area. It needs to heat up only the required quantity of storage material, making the charging process much more efficient. Similar benefits occur in the discharging process. Bales et al. [16] gave the main findings of studies on the storage material ($MgSO_4 \cdot 7H_2O$), as reported by ECN and TU Eindhoven. They proposed a prototype system using such hydration and dehydration reaction in TASK 32 of the IEA SHC program [17]. Dehydration of $MgSO_4 \cdot 7H_2O$ occurs in three steps. $MgSO_4 \cdot 6H_2O$ is formed as a result of the loss of 1 mol water first. Thenceforward, the removal of 5.8 water molecules leads to the generation of $MgSO_4 \cdot 0.2H_2O$. Finally, $MgSO_4$ is produced with the removal of residual water molecules. Of these steps the first and third are single-step reactions.

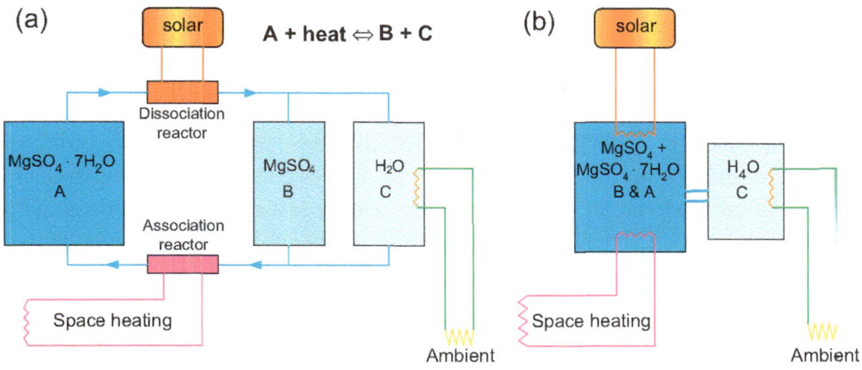

Figure 18.4 Chemical sorption heat storage system using $MgSO_4/H_2O$ [15]. (a) Separate reactor. (b) Integrated reactor.

The second dehydration step is the most promising for storing thermal energy. The corresponding energy density is about 420 kWh/m^3. The former two dehydration steps are endothermic processes. Moreover, the second dehydration step contains an endothermic conversion from crystal $MgSO_4 \cdot 6H_2O$ to an amorphous state. The third dehydration step is an exothermic process. This phenomenon contributes to the transition from amorphous precursor to crystalline $MgSO_4$, which leads to the heat release. The dehydration of $MgSO_4 \cdot 7H_2O$ can occur at temperatures below 150°C, which can be achieved by solar thermal collectors. $MgSO_4$ is formed when $MgSO_4 \cdot 7H_2O$ is heated up to 300°C. $MgSO_4$ absorbs water until $MgSO_4 \cdot 6H_2O$ is produced with a thermal energy of 200–400 kWh/m^3.

$MgSO_4 \cdot 7H_2O$ is neither toxic nor corrosive. The $MgSO_4$–H_2O pair has a theoretical storage density of 780 kWh/m^3 [18]. van Essen et al. [19] investigated the dehydration of $MgSO_4 \cdot 7H_2O$ by means of differential scanning calorimetry (DSC) and thermogravimetric analysis (TGA). They found that the hydration strongly depends on the water vapor pressure and this material cannot release all of the stored heat during dehydration. There are several hydrates of magnesium sulfate, so the intermediates from $MgSO_4$ to $MgSO_4 \cdot 7H_2O$ cannot be determined during the hydration reactions. There exist some problems that need to be solved for the application of magnesium sulfate as a thermochemical heat storage material, for example, $MgSO_4$ hydrates too slowly and the cycle performance of this material is poor. Magnesium sulfate powder will rapidly form a skin layer of the hydrated salt on the surface of the bulk powder when it reacts with humid air. This skin will impede the further solid–gas reaction and limit the energy power [20]. Therefore, magnesium sulfate power is difficult to use in a practical storage system because of its poor mass transfer ability. To overcome the aforementioned problems, composite materials have been developed by confining magnesium sulfate into a porous matrix, offering a large specific surface for the hydration reaction. Hongois et al. [20] prepared the form-stable composite materials by impregnating magnesium sulfate into the zeolite matrix for low-temperature heat storage applications in buildings. The magnesium sulfate was confined in the porous structure of the zeolite pellets with an optimum percentage of 15 wt%. This composite material has energy densities of 0.54—1.44 GJ/m^3 (150—400 kWh/m^3) at a storage temperature of 150°C. This temperature is sufficient for the release of 6 mol of water from $MgSO_4 \cdot 7H_2O$, whereas the thorough zeolite dehydration occurs at a temperature above 200°C. Characterization by TGA and DSC shows that the optimum dehydration temperature is about 150°C, corresponding to a dehydration level of about 89%. About 80% of the total energy density can be stored at 150°C although the material is not fully dehydrated. Moreover, this energy storage temperature is compatible with evacuated tube solar collectors. Other larger scale experimental results indicate that an energy density of 0.18 Wh/g (166 kWh/m^3) is achieved, which is 45% of the theoretical energy density of the form-stable zeolite—$MgSO_4$ composite sorbent under an airflow rate of 8 L/min and inlet air at 50% relative humidity. It appears that this composite material is a promising thermal energy storage material suitable for long-term solar heat storage owing to its good thermal properties and thermal and chemical reliability. Generally speaking, $MgSO_4$ is considered a potential solar heat storage material.

van Essen et al. [21] investigated the potential of four kinds of hydrated salts for seasonal heat storage of solar energy. All four salt hydrates can be dehydrated at 150°C and this temperature can be obtained through a medium-temperature collector. The theoretical study showed that three hydrated salts could be used as promising materials for compact seasonal heat storage in addition to $MgSO_4 \cdot 7H_2O$: aluminum sulfate ($Al_2(SO_4)_3 \cdot 18H_2O$), magnesium chloride ($MgCl_2 \cdot 6H_2O$), and calcium chloride ($CaCl_2 \cdot 2H_2O$). The chlorides ($MgCl_2 \cdot 6H_2O$ and $CaCl_2 \cdot 2H_2O$) have a larger temperature lift than the sulfates ($MgSO_4 \cdot 7H_2O$ and $Al_2(SO_4)_3 \cdot 18H_2O$) under similar conditions. However, the hygroscopic chlorides are inclined to produce a gel-like material. The dehydration of $Al_2(SO_4)_3 \cdot 18H_2O$ proceeds through three steps, even though at the temperature of 300°C, this material is not entirely dehydrated. First, $Al_2(SO_4)_3 \cdot 17.5H_2O$ is produced, followed by a gradual removal of 2.5 mol water leading to $Al_2(SO_4)_3 \cdot 15H_2O$, and in the final step $Al_2(SO_4)_3 \cdot 2H_2O$ is generated as a result of the loss of 13 mol water. The calcium chloride is hydrophilic and has already absorbed the water molecules at room temperature. Hence, it is observed that its hydration begins with the transformation from $CaCl_2 \cdot 2.3H_2O$ to $CaCl_2 \cdot 2H_2O$. Thereafter, $CaCl_2 \cdot H_2O$ is formed, accompanied by the loss of 1 mol water. At last, full dehydration occurs at 110°C and anhydrous $CaCl_2$ is produced as a result of the removal of residual water molecules. This material can be completely dehydrated under the temperature of 150°C. The dehydration of $MgCl_2 \cdot 6H_2O$ occurs in four steps. First, $MgCl_2 \cdot 4H_2O$ is produced, corresponding to the expulsion of 2 mol water. Thenceforth, $MgCl_2 \cdot 2H_2O$ is generated, followed by a gradual removal of 1 mol water resulting in $MgCl_2 \cdot H_2O$. MgOHCl is formed and gaseous HCl released when the temperature is higher than 138°C. The formation of both HCl and MgOHCl is undesired. HCl is harmful to metallic reactors, and MgOHCl is unable to take up water and thus no heat is released. Table 18.1 shows the four most promising reversible hydration/dehydration reactions of hydrated salts with corresponding energy density as identified in the theoretical investigation [22]. The formation of HCl and MgOHCl can be avoided when hydrolysis proceeds at 403K during the dehydration of $MgCl_2 \cdot 6H_2O$. The largest temperature lift can be obtained using $MgCl_2 \cdot 6H_2O$ under similar conditions. In addition, the

Table 18.1 Selected salt–water vapor reversible reactions with corresponding energy densities [22]

Reversible reactions	Solid density of salt hydrate (kg/m^3)	Theoretical energy density (GJ/m^3)
$MgSO_4 \cdot 7H_2O(s) \Leftrightarrow MgSO_4(s) + 7H_2O(g)$	1680	2.8
$Al_2(SO_4)_3 \cdot 18H_2O(s) \Leftrightarrow Al_2(SO_4)_3(s) + 18H_2O(g)$	1690	2.75
$MgCl_2 \cdot 6H_2O(s) \Leftrightarrow MgCl_2 \cdot H_2O(s) + 5H_2O(g)$	1560	2.48
$CaCl_2 \cdot 2H_2O(s) \Leftrightarrow CaCl_2(s) + 2H_2O(g)$	1710	1.44
Heating up water ($\Delta T = 60°C$)	—	0.25

agglomeration phenomenon occurs during hydration of the chlorides. Moreover, a hard impermeable layer is formed during hydration, which will prevent further hydration.

Bertsch et al. [23] investigated the hydration reactions of two different anhydrous salts ($MgSO_4$, $CuSO_4$) for low-temperature chemical heat storage. They found that magnesium sulfate monohydrate and copper sulfate monohydrate are good candidates for chemical heat storage materials. The theoretical storage densities of magnesium sulfate monohydrate and copper sulfate monohydrate are 2.3 GJ/m^3 heptahydrate (633 kWh/m^3 heptahydrate) and 2.1 GJ/m^3 pentahydrate (574 kWh/m^3 pentahydrate), including the condensation heat of water vapor. The thermal energy storage densities are about 11 and 10 times higher, respectively, than that of water under the same conditions. Zondag et al. [24,25] investigated the application of $MgCl_2 \cdot 6H_2O$ for chemical seasonal solar heat storage. The thermal energy storage density is 10 times higher than that of the water. The research indicated that $MgCl_2 \cdot 6H_2O$ is one of the most promising materials for compact seasonal solar heat storage. In addition, some hydrated salts have been selected for chemical heat storage as shown in Table 18.2 [26].

The chemical heat storage system using sodium sulfide is based on the following chemical equation:

$$Na_2S \cdot 5H_2O(s) \Leftrightarrow Na_2S(s) + 5H_2O(g) \quad [18.5]$$

Sodium sulfide (Na_2S) has high thermal energy storage density due to its large sorption capacity and high heat of sorption. Furthermore, the Na_2S/H_2O pair (pentahydrate $Na_2S \cdot 5H_2O$) has a high thermal power density as well as a large energy storage density (1980 kWh/m^3 for heating and 1300 kWh/m^3 for cooling) [27,28]. However, this material is very corrosive and might give off the effluvial hydrogen sulfide (H_2S), which therefore restricts its practical use. A vacuum pump is required to maintain the vacuum environment to extract unwanted intermediate gases. Fig. 18.5 shows the system diagram of a chemical heat storage system using sodium sulfide [29].

Table 18.2 Some potential thermochemical storage materials [26]

A ⇔	B +	C	GJ/m^3	*T* (°C)
$MgSO_4 \cdot 7H_2O$	$MgSO_4$	$7H_2O$	2.8	122
$MgSO_4 \cdot 7H_2O$	$MgSO_4 \cdot 1H_2O$	$6H_2O$	2.3	105
$MgSO_4 \cdot H_2O$	$MgSO_4$	H_2O	1.3	216
$CaSO_4 \cdot 2H_2O$	$CaSO_4$	$2H_2O$	1.4	89
$CaCl_2 \cdot 2H_2O$	$CaCl_2 \cdot 1H_2O$	H_2O	0.6	174
$FeCO_3$	FeO	CO_2	2.6	180
$ZnCO_3$	ZnO	CO_2	2.5	133

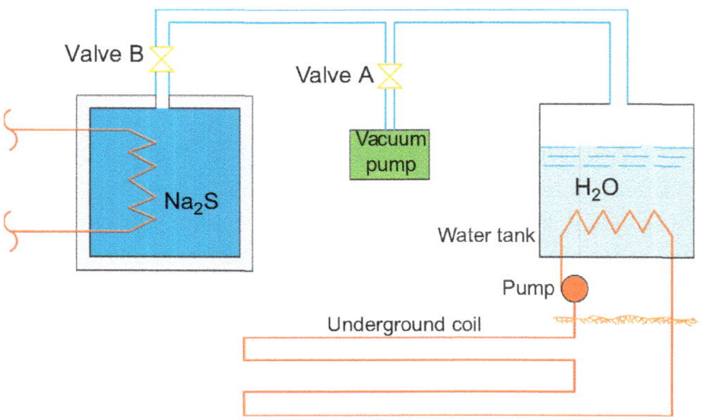

Figure 18.5 Diagram of chemical heat storage system based on the hydration of sodium sulfide [29].

This hygroscopic salt releases the sorption heat due to hydration when water vapor flows across it. In contrast, when heat is applied to the hydrated salt, crystal water is deprived because of the deterioration of hydrated salt as soon as the salt absorbs heat. The anhydrous salt regains the capacity to produce heat upon the addition of moisture. Ucar [29] investigated the sorption properties of Na_2S for solar thermal energy storage systems. The storage system can be driven by almost any low-grade heat, such as solar heat and waste heat. The system has an energy storage density of 3600 kJ/kg Na_2S, which is 10 times higher than that of the PCM. Nevertheless, the initial cost of this system is high owing to the operation of a vacuum pump for the purpose of maintaining the desired vacuum level. The hydrate of Na_2S mainly involves pentahydrate $Na_2S \cdot 5H_2O$ and nonahydrate $Na_2S \cdot 9H_2O$. Boer et al. [30] investigated the structure, thermodynamics, and phase properties of a Na_2S/H_2O system available for a chemical heat pump. They found that there exists a new phase of $Na_2S \cdot 2H_2O$ apart from the above known crystalline phases. The transition temperature from

$Na_2S \cdot 9H_2O$ to $Na_2S \cdot 5H_2O$ is 322K. The transformation temperature from $Na_2S \cdot 5H_2O$ to $Na_2S \cdot 2H_2O$ is 356K. $Na_2S \cdot 0.5H_2O$ is not a phase but rather a 3:1 mixture of Na_2S and $Na_2S \cdot 2H_2O$. The theoretical cold storage capacity of the Na_2S/H_2O working pair (pentahydrate $Na_2S \cdot 5H_2O$) is 2.54 MJ/kg Na_2S according to the evaporation heat of water.

18.2.2 Thermochemical heat storage without sorption

18.2.2.1 The $NH_3/N_2/H_2$ chemical heat storage system

The ammonia-based chemical reaction heat storage system uses reversible dissociation of ammonia to store thermal energy. The chemical reaction is:

$$2NH_3(g) \Leftrightarrow N_2(g) + 3H_2(g) \quad [18.6]$$

The reaction enthalpy is 66.8 kJ/mol ammonia at 20 MPa and 300K. This reaction can be made faster via introduction of an appropriate catalyst. The application of a ruthenium-on-carbon catalyst in place of the iron-based catalyst helps to lower the dissociation temperature to 673K and below. Carden initially investigated the ammonia reversible reaction at the Australian National University (ANU). The research work has been continued by Lovegrove et al. [31,32]. ANU devoted work to an ammonia dissociation and synthesis system for a period of approximately 40 years. They carried out a great deal of theoretical and experimental investigations on an ammonia-based chemical reaction heat storage system [33–41]. Various studies that were done revealed that the ammonia-based system for concentrating solar power is technically feasible. A 1500 *t*/day ammonia synthesis reactor can supply a 10-MW base-load plant with 400 large 400-m^2 dishes arranged in an array. The heat released by the synthesis reaction is used to drive the power cycle to produce superheated steam at a temperature and pressure of up to 793K and 10 MPa. Tower or dish solar collector systems enable the dissociation reaction to proceed more completely owing to the comparatively high temperature in the vicinity of 1273K. In addition, the increase in the system pressure can greatly accelerate ammonia synthesis. This makes it necessary for ammonia-based storage systems to operate at high pressures (10–30 MPa), although this will increase the costs due to material restrictions on the reactor and storage vessel. The ammonia dissociation and synthesis reaction has a number of obvious advantages [34,39]. One of the advantages is that it is easy to control the reactor because this reversible reaction is very simple with no unwanted side reactions. In addition, the temperature of this endothermic reaction is particularly suited to solar concentrators. Moreover, the phase separation of ammonia and hydrogen/nitrogen can proceed automatically at ambient temperature owing to the condensation of the ammonia fraction. In addition, the ammonia-based chemical reaction heat storage system can benefit from the well-developed experience of the mature ammonia synthesis industry. The basic principle is illustrated in Fig. 18.6 [32,34,41]. Dunn and Lovegrove [42] summarized ANU's work on ammonia-based chemical reaction heat storage for concentrating solar power.

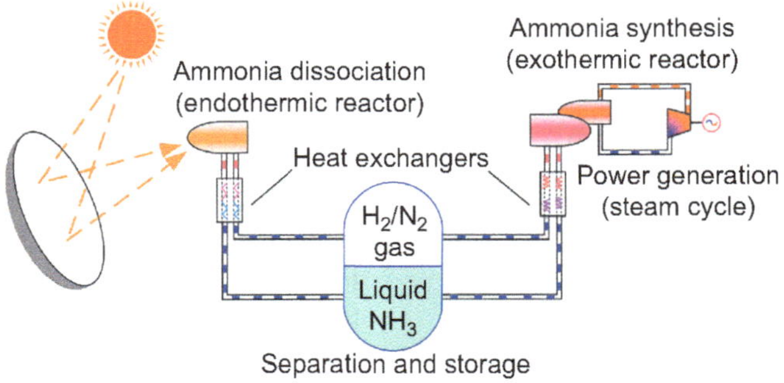

Figure 18.6 Thermochemical storage of solar heat using ammonia [32,34,41].

18.2.2.2 The $SO_3/O_2/SO_2$ chemical heat storage system

The sulfur trioxide (SO_3) chemical reaction heat storage system is based on the reversible dissociation of SO_3:

$$2SO_3(g) \Leftrightarrow 2SO_2(g) + O_2(g) \quad [18.7]$$

The decomposition reaction of SO_3 gas proceeds in the temperature range of 1073–1273K [43]. The dissociation and synthesis reaction of SO_3 has no undesirable side reactions and possesses high reactivity under the circumstances of suitable catalysts [44]. Fig. 18.7 shows a schematic diagram of a chemical reaction heat storage system using $SO_3/O_2/SO_2$.

In this storage system, the resultant gases of the dissociation reaction, SO_2/O_2 can be delivered via the pipes and used for the heating purposes. SO_3 can be generated by recombination of the SO_2/O_2 gas mixture in the presence of a catalyst at 773–873K and release the chemical reaction heat. This system can effectively achieve the transportation and storage of high-temperature thermal energy. Dayan [45] evaluated the technical feasibility of the $SO_3/O_2/SO_2$ thermal energy storage system. They integrated this kind of thermal energy storage system into a solar steam-cycle power plant. The heat storage efficiency of this storage system is about 58%. The overall conversion efficiency from thermal energy to electric energy is approximately 26%. For a similar steam-cycle power plant without the storage unit in the solar energy system, the overall efficiency is about 40%.

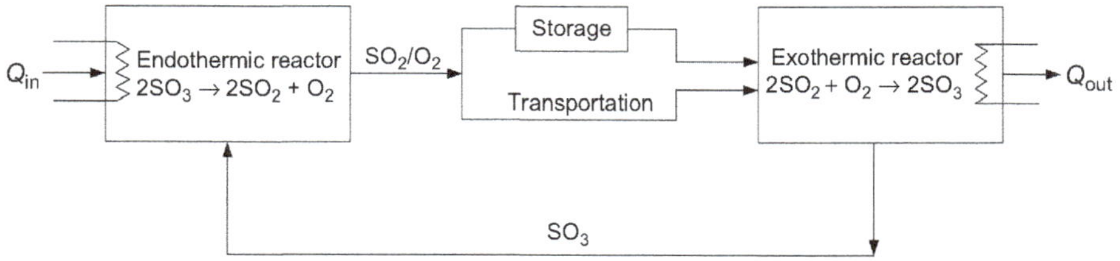

Figure 18.7 Schematic diagram of a chemical reaction heat storage system using the $SO_3/O_2/SO_2$ reversible thermochemical reaction.

18.2.2.3 The $CH_3OH/H_2/CO$ chemical heat storage system

The decomposition of methanol occurs at a relatively low temperature. Apart from being very clean and cheap, methanol can be stored and delivered conveniently. Hence, methanol has been considered as a potential candidate for thermal energy storage. The dissociation reaction is endothermic, whereas the reverse reaction is exothermic, and obeys the following equation:

$$CH_3OH(g) \Leftrightarrow 2H_2(g) + CO(g) \qquad [18.8]$$

Fig. 18.8 shows the thermal energy storage system using the decomposition and synthesis reaction of methanol. Liu et al. [46] proposed a methanol energy transport system integrating a two-step liquid-phase methanol synthesis process. The two-step methanol synthesis reaction, including the intermediate product methyl formate, is expressed as follows:

$$CH_3OH + CO \rightarrow HCOOCH_3 \qquad [18.9]$$

$$HCOOCH_3 + 2H_2 \rightarrow 2CH_3OH \qquad [18.10]$$

Compared to present gas-phase reactions, the reaction pressure and temperature of this system are decreased because of the use of the liquid-phase reaction of methanol. The heat transportation efficiency increases with the increase in the reaction conversion ratio. The corresponding transportation efficiency of 75% can be obtained with the reaction conversion ratio of 90%.

18.2.2.4 The cyclohexane/benzene/hydrogen chemical heat storage system

The cyclohexane/benzene/hydrogen system involves the reversible reaction $C_6H_{12}(g) \Leftrightarrow C_6H_6(g) + 3H_2(g)$ at atmospheric pressure in the temperature range of 478–589K. Both endothermic dehydrogenation and exothermic hydrogenation reactions

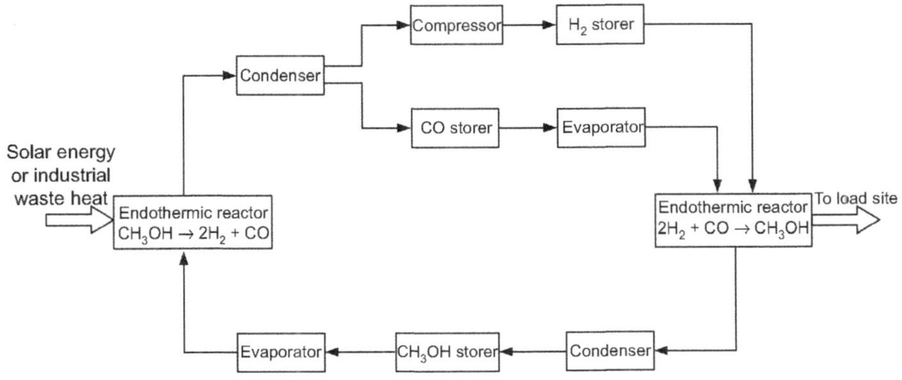

Figure 18.8 Schematic diagram of a chemical reaction heat storage system using the methanol reversible thermochemical reaction.

are catalytic reactions. The enthalpy of the reaction is very large. Moreover, the technology for the hydrogenation of benzene has already been formed on the industrial scale [47], which will promote the potential application of the cyclohexane/benzene/hydrogen system for chemical reaction heat storage. Therefore, this reversible reaction is considered as a promising candidate for chemical reaction heat storage. The existence of impurities in reagents may incur a side reaction. It is important to pretreat the reagents to eliminate impurities. Cacciola et al. [48] investigated the interrelationship between the COP and various influential factors for the cyclohexane/benzene/hydrogen system. For a dehydrogenation temperature of 473K and hydrogenation temperature of 573K, both thermal energy storage density and the COP reached the highest value when reaction conversions of the dehydrogenation and hydrogenation reactions were 30% and 70%, respectively.

18.2.2.5 Methane re-forming chemical heat storage system

The methane re-forming reaction is a promising candidate for the energy storage of solar heat and industrial waste heat because it is highly endothermic [49–53]. The synthesis gas produced by the re-forming reaction can achieve long-distance transportation by pipeline, via a reversible exothermic reaction. The methane re-forming reaction can realize energy conversion, storage, and transportation. Consequently, a methane re-forming reaction system, also known as a thermochemical heat pipe, is considered as an important technology for reasonable and effective utilization of existing energy as well as optimizing the use of renewable energy.

Methane with water or carbon dioxide is converted to synthesis gas at comparatively high temperatures. The endothermic catalytic reaction is reversible. Until now, the syngas has been mainly obtained by means of steam or CO_2 re-forming in industrial production [54]. Solar re-forming of methane has captured great interest. The two methane re-forming reactions can be applied to chemical heat pipes to achieve chemical heat storage and transmission of solar heat:

$$CH_4(g) + H_2O(g) \rightarrow CO(g) + 3H_2(g) \quad [18.11]$$

$$CH_4(g) + CO_2(g) \rightarrow 2CO(g) + 2H_2(g) \quad [18.12]$$

In theory, endothermic reactions via solar energy for this process, which produces syngas, can upgrade the caloric value of methane feedstock by 22–28% [49,55–57]. The CH_4/CO_2 re-forming reaction has an advantage over the steam re-forming [58,59]. Reaction [18.11], the steam re-formation of methane, is mainly concomitant with the side reaction of the water–gas shift [52,54], which is slightly exothermic:

$$CO + H_2O \rightarrow CO_2 + H_2 \quad [18.13]$$

The chemical reaction heat storage technology based on solar CH_4/CO_2 re-forming has a series of merits [60]. The solar CH_4/CO_2 re-forming reaction can provide a way to store and convey solar energy into a high-grade energy as well as reducing the CO_2 emissions. The CH_4/CO_2 re-forming reaction and its reverse CO/H_2 methanation reaction

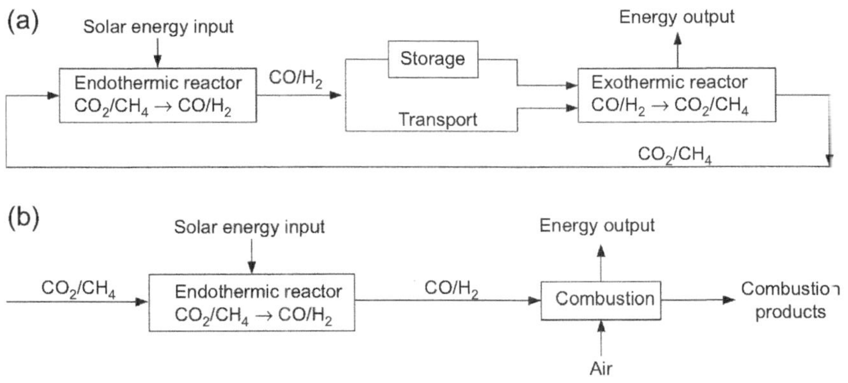

Figure 18.9 Concepts for (a) closed- and (b) open-loop chemical reaction heat storage system based on solar energy CH_4/CO_2 re-forming [58,60].

have been considered as the foregoers of thermochemical heat pipe applications in the storage and transmission of solar energy [58]. Solar re-forming of methane with CO_2 has been investigated for more than 20 years [61–64]. The reactor is operated under temperatures of 973–1133K with a pressure of 0.35 MPa. The power absorbed by the reactor ranges from 200 to 300 kW. The maximum methane conversion can be as high as 88%. A Ni–Cr–Al metallic foam absorber activated with a Ru/γ–Al_2O_3 catalyst is employed to measure the CO_2 re-forming of methane in a small-scale volumetric solar receiver–reactor [55]. Compared to a SiC foam absorber activated using the same Ru/γ–Al_2O_3 catalyst, the metallic foam absorber possesses better catalytic stability. Moreover, the metallic foam has superior plasticity, which can prohibit the emergence of cracks arising from mechanical or thermal shock. The solar-driven thermochemical heat pipe concepts using this reaction are shown in Fig. 18.9 [58,60].

18.2.2.6 The isopropanol/acetone/hydrogen chemical heat storage system

Prevost and Bugarel [65] first proposed the isopropanol (2-propanol)/acetone/hydrogen chemical heat pump system in 1980. Until now, many computational simulations of and experimental research on this system have been performed. The system involves a reversible reaction with proper catalysts. The endothermic reaction is dehydrogenation of isopropanol, producing acetone and hydrogen, as shown in the following chemical reaction:

$$(CH_3)_2CHOH(l) \rightarrow (CH_3)_2CO(g) + H_2(g) \qquad [18.14]$$

This liquid–gas endothermic reaction occurs at 353–363K in the presence of an appropriate catalyst. The exothermic reaction is the hydrogenation of acetone. This reaction occurs at 423–483K with a suitable catalyst [66]. The reaction equation is as follows:

$$(CH_3)_2CO(g) + H_2(g) \rightarrow (CH_3)_2CHOH(g) \qquad [18.15]$$

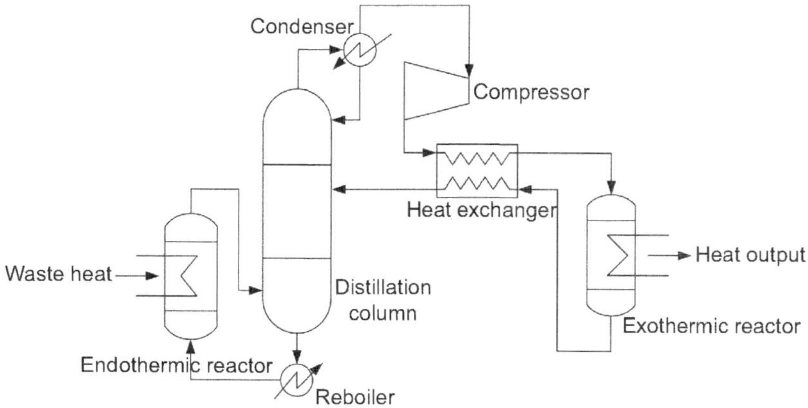

Figure 18.10 Isopropanol/acetone/hydrogen thermochemical reaction heat storage system [66,67].

Fig. 18.10 illustrates the basic principle of the isopropanol/acetone/hydrogen chemical reaction heat storage system [66,67]. The overall system mainly consists of two reactors: an endothermic reactor and an exothermic reactor. In the dehydrogenation process of isopropanol, isopropanol needs to be separated from the reaction products. The fractional condensation can easily be achieved at ambient temperature by including a distillation column in the system [68,69]. Kitikiatsophon and Piumsomboon [70] investigated the dynamic behavior of the isopropanol/acetone/hydrogen system by steady-state simulation. They found that the system stability hinged on the working pressure in the distillation column and at the outlet of the compressor. The simulation results showed that the corresponding thermal efficiency was 0.06 when the reaction conversions of the dehydrogenation and hydrogenation reactions were 7.9% and 17.9%, respectively. KlinSoda and Piumsomboon [66] developed an isopropanol/acetone/hydrogen chemical heat pump system. The conversions in the endothermic and exothermic reactors were about 8.5% and 11.6%, respectively. As a result, the fluid temperature in the exothermic reactor rose from 368 to 409K. Theoretically, the exit fluid temperature in the exothermic reactor could be as high as 473K. Improving the performance of the exothermic reactor could lift the level of thermal energy upgrading. The thermal efficiency of the system was around 38% and the exergy efficiency was 27%.

18.2.2.7 The ammonium hydrogen sulfate chemical heat storage system

Wentworth and Chen [71] evaluated many thermal decomposition reactions available for storing solar thermal energy. Considering the kinetics performance of the chemical reaction and the physical properties of the substances in the reaction, they constructed 10 criteria for choosing reactions for solar thermal energy storage. They found that the dissociation reaction of ammonium hydrogen sulfate could meet almost all the proposed criteria. Furthermore, they concluded that it is a very promising candidate reaction for stockpiling solar thermal energy.

The decomposition of ammonium hydrogen sulfate (NH_4HSO_4) into ammonia (NH_3), water (H_2O), and sulfur trioxide (SO_3) for the thermal energy storage stage is shown as follows:

$$NH_4HSO_4(l) \xrightarrow{T_1} NH_3(g) + H_2O(g) + SO_3(g) \qquad [18.16]$$

The reverse synthesis reaction is:

$$NH_3(g) + H_2O(g) + SO_3(g) \xrightarrow{T_2} NH_4HSO_4(l) \qquad [18.17]$$

The actual temperature required for energy storage stage (T_1) and energy release (T_2) depends on the reaction conditions. The thermal energy storage density provided by this process (storage of H_2O, NH_3, and SO_3 as liquids at ambient temperature) is between 3.098 GJ/m^3 (740 kcal/L) and 4.102 GJ/m^3 (980 kcal/L).

For the purpose of the effective separation of SO_3 from NH_3 and H_2O, Wentworth [72] explored a two-step reaction to realize the function of reaction [18.16]. The first step, endothermic reaction, employs metal sulfates (M_2SO_4) for forming intermediate pyrosulfates ($M_2S_2O_7$). The subsequent endothermic reaction is the decomposition of $M_2S_2O_7$. The gas products can be also condensed into liquid for long periods of storage and transportation at room temperature. The process is based on the following reaction equations:

$$NH_4HSO_4(l) + M_2SO_4(s) \xrightarrow{T_{1a}} NH_3(g) + H_2O(g) + M_2S_2O_7(l) \qquad [18.18]$$

$$M_2S_2O_7(l) \xrightarrow{T_{1b}} M_2SO_4(s) + SO_3(g) \qquad [18.19]$$

In addition, the use of metal oxides (MO) to form intermediate sulfates (MSO_4) can also realize the desired product separation in principle. The chemical reactions are described as:

$$NH_4HSO_4(l) + MO(s) \xrightarrow{T_{1c}} NH_3(g) + H_2O(g) + MSO_4(s) \qquad [18.20]$$

$$MSO_4(s) \xrightarrow{T_{1d}} MO(s) + SO_3(g) \qquad [18.21]$$

Additionally, the reactions of 25 different metal oxides and metal sulfates with NH_4HSO_4 have been examined experimentally using differential thermal analysis and TGA. The results indicated that only three sulfates (K_2SO_4, Rb_2SO_4, Cs_2SO_4) and six metal oxides (ZnO, Cu_2O, NiO, MnO, PbO, MgO) might react with NH_4HSO_4 to achieve the aforementioned two-step reaction. Among these different reagents, ZnO is the best choice owing to the achievement of complete product separation. The products can be released step by step (water vapor at around 436K, ammonia gas at

638−691K, and a mixture gas of SO_2 and SO_3 at about 1173K) as a result of the decomposition of NH_4HSO_4 by adding ZnO. The product separation can be achieved easily by controlling the reaction temperature. In addition, the thermal decomposition reaction of $ZnSO_4$ progresses rapidly when the reaction temperature exceeds 1123K.

18.2.2.8 Thermal dissociation reaction of metal oxides for chemical heat storage system

The reduction reaction of multivalent metal oxides and reverse oxidation reaction can be used for storing thermal energy. Using air as a reactant in the reverse exothermic reactions allows us to avoid storing product gases. Furthermore, metal oxides have good stability and thus are considered candidate materials for chemical reaction heat storage. The redox reaction can be given by the following equation:

$$MO_{2x+1} + \mathrm{Heat}(Q) \Leftrightarrow MO + xO_2 \quad [18.22]$$

Fahim and Ford [73] illuminated the selection criteria for reversible chemical reactions used for thermal energy storage. They studied the reversible reaction system of BaO_2/BaO for the purpose of examining its potential utilization for thermal energy storage. The activation energies of the forward dissociation reaction of BaO_2 and the backward oxidation reaction of BaO are 140.6 and 51.9 kJ/mol, respectively. Comparing with the other oxide systems, they concluded that the BaO_2/BaO system is a promising candidate reaction for thermal energy storage. This system has an energy storage density of 3.01 GJ/m^3 based on the solid product. The corresponding chemical equation is shown as:

$$BaO_2 \Leftrightarrow BaO + 1/2O_2 \quad [18.23]$$

In addition, the direct solar thermal dissociation of ZnO into its elements is an alluring process for solar energy storage [74]. The reversible chemical reaction is:

$$ZnO \Leftrightarrow Zn + 1/2O_2 \quad [18.24]$$

Möller and Palumbo [74] developed a solar chemical reactor for the direct thermal dissociation of zinc oxide using concentrated sunlight. The thermal decomposition of ZnO into its ingredients needs a temperature above 2200K, and the produced gases Zn(g) and O_2(g) need to be either quenched or separated at high temperature to avoid their recombination [75]. The utilization of carbonaceous materials as reducing agents can not only lower the reduction temperature significantly but also prevent recombination.

The direct thermal dissociation of metal oxides requires very high temperatures. When the temperature is up to 3000K, dissociation reactions can occur, such as for B_2O_3 or SnO_x (x representing 1 or 2) [76,77]. This means that the direct thermal dissociation of metal oxides is difficult to accomplish in practice for the chemical reaction heat storage of solar energy. Only a few metal oxides with high reaction enthalpy and suitable reaction temperature can be applied to chemical reaction heat storage for concentrated solar power. The mixture of metal oxides can provide a considerable range of appropriate thermal storage materials. Block et al. [78] employed the binary cobalt oxide/iron oxide system for high temperature heat storage. Cobalt oxide has a high reaction enthalpy and relatively low reaction temperature. These properties make it a very promising material for chemical reaction heat storage of solar energy. However, cobalt oxide is expensive and suspected to be carcinogenic. Compared to cobalt oxide, iron oxide is cheap, is nontoxic, and has a high reaction enthalpy as well. The conversion temperature of Fe_2O_3/Fe_3O_4 is 1665K and its reaction enthalpy is 599 J/g. The reaction temperature is too high for concentrated solar power applications. The transformation temperature of Fe_2O_3/Fe_3O_4 is 1188K with a reaction enthalpy of 576 J/g. The conversion temperature of a binary cobalt oxide/iron oxide system is similar to that of pure cobalt oxide, whereas the reaction enthalpy decreases with increased doped iron content. Binary cobalt oxide/iron oxide systems have higher structural stability and long duration reversibility of the chemical reaction compared to pure cobalt oxide. Moreover, the binary system with about 10% iron oxide has sufficient heat storage capability because of the appropriate enthalpy.

18.2.2.9 Decarbonation reaction of carbonate for chemical heat storage system

Kato et al. [79] surveyed the utility of several chemical reactions available for heat pump systems. They mainly focused on inorganic oxide/carbon dioxide reaction systems owing to their high chemical reaction heat. They found that inorganic oxides such as PbO, MgO, CuO, MnO, ZnO, NiO, and CoO can be selected as candidates. As for chemical reaction heat storage, the lead oxide/carbon dioxide (PbO/CO_2) reaction displays the most outstanding features.

The working principle of the PbO/CO_2 reaction system is based on the following chemical equation:

$$PbCO_3(g) \Leftrightarrow PbO(s) + CO_2(g) \quad [18.25]$$

The principle of the chemical reaction heat storage system using the PbO/CO_2 reaction is shown in Fig. 18.11 [79]. Kyaw et al. [80,81] investigated the calcium oxide/carbon dioxide reaction for chemical reaction heat storage, and the corresponding reversible chemical reaction can be written as:

Carbonation:

$$CaO(s) + CO_2(g) \rightarrow CaCO_3(g) \quad [18.26]$$

Decarbonation:

$$CaCO_3(g) \rightarrow CaO(s) + CO_2(g) \quad [18.27]$$

The CaO/CO_2 reaction system is very attractive for storing high-temperature thermal energy. The characteristics of this system are as follows [80,82]: (1) $CaCO_3$

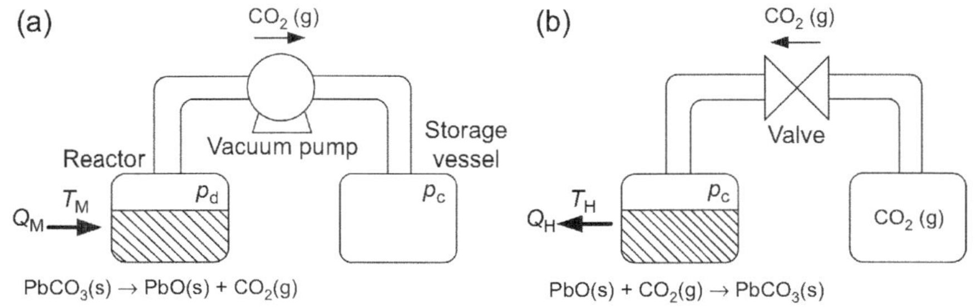

Figure 18.11 Principle of the PbO/CO_2 chemical reaction heat storage system [79]. (a) Heat storage mode. (b) Heat supply mode.

decomposes at 1100K under 101 kPa of CO_2 pressure. (2) The substances involved in this reversible reaction are nontoxic and noncorrosive. (3) There are no undesirable side reactions during this reversible reaction. (4) $CaCO_3$ is an abundant and relatively cheap raw material. (5) This reaction system has a relatively high energy storage density (about 3.26 GJ/m^3).

The major challenges of the CaO/CO_2 reaction system are: (1) CO_2 is rather difficult to liquefy. (2) The CO_2 pressure dominates the mode of the CaO/CO_2 thermal energy storage system, either the carbonation (heat release) process or the decarbonation (heat storage) process. Actually, a similar problem also arises in the PbO/CO_2 storage system.

Considering the aforementioned aspects, CO_2 gas resulting from the decomposition of carbonates must be stored. Storage of the produced CO_2 gas can be generally implemented in the following ways [80,82]:

1. The produced CO_2 gas is compressed and stored, which is the CaO/CO_2 compressor system.
2. A CaO/CO_2—metal oxide system is used by making CO_2 gas react with other metal oxides and storing it in the form of the metal carbonate.
3. A CaO/CO_2—adsorbent system is used by adsorbing CO_2 gas with a suitable adsorbent (ie, zeolite) and storing in the form of the adsorbed CO_2.

Kyaw et al. [80] investigated the corresponding operating efficiency and the upgraded temperature of the above three different methods that are used to store the produced CO_2. The CaO/CO_2 compressor system is the most suitable for storing and transporting thermal energy at the same temperature. This system, however, needs to expend mechanical work. The thermal efficiency of CaO/CO_2—zeolite is strongly influenced by the adsorptivity of the zeolite. With an increase in adsorptivity, the efficiency will be enhanced. Kyaw et al. [82] point out that the zeolite 13*X* adsorbent can be used as a CO_2 storage medium in a CaO/CO_2 high-temperature thermal energy storage system.

The crucial consideration in the development of an inorganic oxide/carbon dioxide system using these reversible reactions is how to handle and store the carbon dioxide gas-phase reactant [83]. Kato et al. [79] constructed a combination system by integrating the CaO/CO_2 and PbO/CO_2 reactions. The principle of the combination system is shown in Fig. 18.12 [79,83—85]. The system is composed of two reactors and the

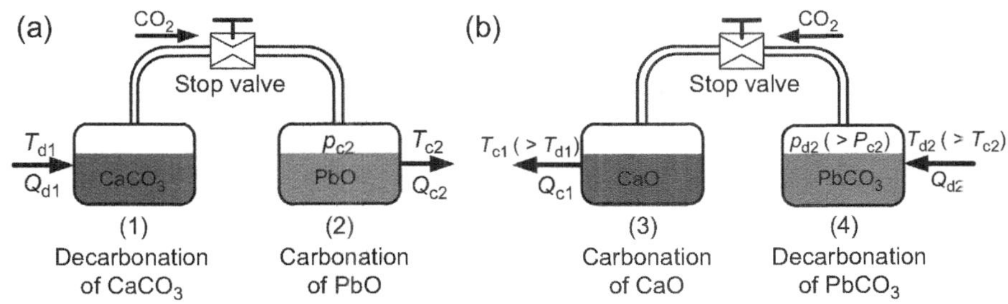

Figure 18.12 Principle of $CaO/PbO/CO_2$ chemical reaction heat storage system [79,83–85]. (a) Heat storage mode. (b) Heat output mode.

operation form comprises: (1) the heat storage stage and (2) the heat output stage. The CO_2 is transformed into the corresponding metal carbonates to achieve the storage of CO_2. Owing to the higher reaction pressure, the temperature T_{c1} of the generated heat is higher than T_{d1}. The $CaO/PbO/CO_2$ system is driven by thermal energy and no mechanical work is required, enabling ease in design as well as making the operation simple and stable. This kind of heat pump system is also called a heat transformer.

18.2.2.10 Dehydration reaction of hydroxide for chemical heat storage system

Darkwa and O'Callaghan [86] evaluated the potential of an inorganic oxide/hydroxide reaction system for chemical reaction heat storage. Thermal energy storage utilizing an inorganic oxide/hydroxide reaction system can produce high heat energy and possess much higher thermal energy storage capacities (about 1–3 GJ/m^3). A $CaO/Ca(OH)_2$ system, as well as a CaO/CO_3 system, has been claimed to be among the best candidates for thermal energy storage of high temperature heat [87].

The calcium hydroxide system exhibits good reversibility and cycling stability. Furthermore, the $CaO/Ca(OH)_2$ system possesses high reaction enthalpy and fast kinetics with an equilibrium temperature of around 773K depending on the H_2O partial pressure [88]. In addition, the price of this material is relatively low. The $CaO/Ca(OH)_2$ system for thermal energy storage utilizes the following reversible hydration/dehydration reaction:

$$Ca(OH)_2(s) \Leftrightarrow CaO(s) + H_2O(g) \qquad [18.28]$$

Fig. 18.13 shows a schematic diagram of a chemical reaction heat storage system using $CaO/Ca(OH)_2$ [89]. The $CaO/Ca(OH)_2$ system stores thermal energy through dehydration of calcium hydroxide and releases the stored energy on demand by means of the hydration of calcium oxide. Ogura et al. [89] investigated the performance of a $CaO/Ca(OH)_2$ chemical heat pump. The obtained thermal energy storage/release density was 1.86 MJ/kg CaO or 1640 MJ/m^3. Fujimoto et al. [90] conducted a dynamic simulation of a $CaO/Ca(OH)_2$ chemical heat pump using energy and exergy analyses. They stated that the $CaO/Ca(OH)_2$ chemical heat pump system can meet the demands

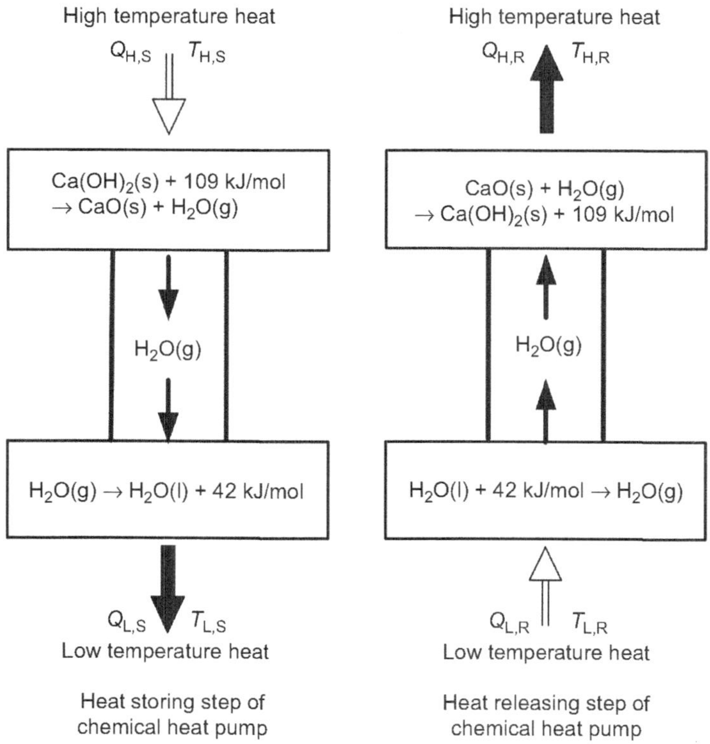

Figure 18.13 Schematic diagram of $CaO/Ca(OH)_2$ thermochemical reaction heat storage system [89].

of heating and cooling. The energy and exergy efficiencies were 58.7% and 61.6% for heating and 12.7% and 4.5% for cooling, respectively.

Many researchers have investigated thermal energy storage using a reversible magnesium oxide/water reaction [91–100]. To test the possibility of developing a magnesium oxide/water system, Kato et al. [95] performed a kinetic study of the hydration of magnesium oxide. The $MgO/Mg(OH)_2$ chemical reaction heat storage system is based on the following equation:

$$Mg(OH)_2(s) \Leftrightarrow MgO(s) + H_2O(g) \qquad [18.29]$$

Fig. 18.14 shows the working principle of the chemical reaction heat storage system using $MgO/Mg(OH)_2$ [94,96,97,99]. This system can be used for the storage of middle-temperature heat. The dehydration temperature is about 523K. The dehydration time highly depends on the temperature. Moreover, the reaction process highly relies on the heat and mass transfer. A relatively low dehydration temperature is more favorable for practical utilization. Enhancement of the heat conductivity in the reactor can reduce the dehydration temperature. The dehydration could occur at below 583K by enhancing the heat conductivity of the reactor. The more rapidly the

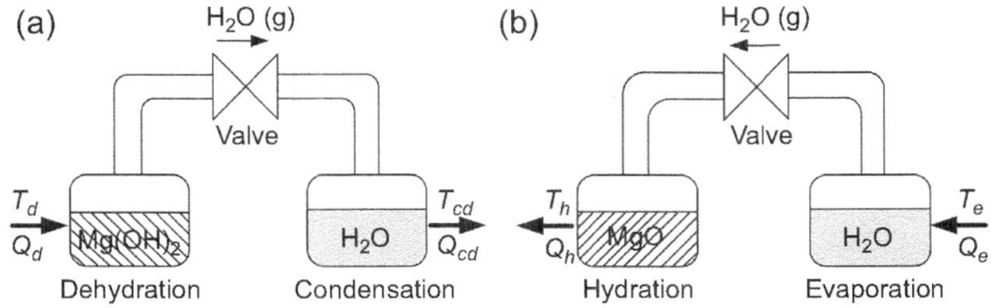

Figure 18.14 Schematic diagram of $MgO/Mg(OH)_2$ chemical reaction heat storage [94,96,97,99]. (a) Heat storage mode. (b) Heat output mode.

hydration proceeds, the higher is the vapor pressure. In addition, a higher-pressure hydration reaction is favorable for improving the heat output performance of the heat pump. Output heat at a temperature of around 473K was produced under a vapor pressure of 203 kPa [96]. The improvement of the thermal conductivity and mass diffusivity in the reactor is an important factor for upgrading the thermal performance of the heat pump. Higher temperature output can be realized by enhancement of the heat conductivity and mass diffusivity in the reactor. The presence of water on a reaction interface will significantly retard the decomposition reaction. Splitting the bulk reagents into tiny particles can improve the dehydration dynamics. Porous materials, such as expanded graphite and expanded vermiculite, can be employed for confining the hydroxide inside the pores, which enables the isolation of reagent particles. Dehydration relies on the heat conduction inside the reactor. The porous host matrix provides efficient transport of water vapor and heat as well, by accommodating the solid-phase swelling/shrinkage in the process of chemical transformation. Shkatulov et al. [101] synthesized a new composite material, "$MG(OH)_2$/expanded vermiculite," and investigated its properties relevant to chemical reaction heat storage. The decomposition temperature of the confined magnesium hydroxide was lower by about 50K than that of the bulk reagents. This fact might be ascribed to a significant acceleration of the dehydration reaction of magnesium hydroxide confined to the pores inside the expanded vermiculite pores. The in-depth reason for the decomposition temperature reduction needs further investigation. The maximal heat storage capacity is 540 kJ/kg of the composite materials when the magnesium hydroxide content is 67.4%. Ishitobi et al. [92] investigated the dehydration/hydration behavior of metal salt-modified materials. They concluded that LiCl-modified $Mg(OH)_2$ is a new potential material for chemical heat pumps. The dehydration of authentic $Mg(OH)_2$ behaves as a one-step reaction. However, the dehydration of LiCl-modified $Mg(OH)_2$ proceeds in two steps. $LiCl/Mg(OH)_2$ ($\alpha = 0.10$) has the fastest dehydration rate compared to other samples ($\alpha = 0, 0.02, 0.30$). The surplus LiCl might increase the diffusion resistance and therefore reduce the dehydration rate. The dehydration reaction rate increases owing to the effect of the LiCl modification, whereas the activation energy for the first-order dehydration reaction is reduced. Meanwhile, the hydration mechanisms for authentic MgO and LiCl/MgO ($\alpha = 0.10$) are different. Kato et al. [98] proposed chemical reaction heat storage materials mixed with metal hydroxides. $Mg_{\alpha}Ni_{1-\alpha}(OH)_2$, which is a hybrid hydroxide of $Mg(OH)_2$ and nickel hydroxide, $Ni(OH)_2$, on the atomic level, is prepared by a precipitation method. The decomposition of the hybrid hydroxide $Mg_{0.5}Ni_{0.5}(OH)_2$ and a physical mixture of $Ni(OH)_2$–$Mg(OH)_2$ was investigated. The physical mixture has a two-step dehydration because the hydroxides decompose independently. Only a single-step dehydration reaction occurs in the hybrid hydroxide. Hence, the hybrid hydroxide forms a new phase exhibiting a unique characteristic. The hybrid hydroxide can raise the operation's heat storage temperature by varying the composition of the addition of cation in the material. The dehydration temperature of $Mg(OH)_2$ can be reduced by the added cation in the hydroxide. $Mg_{0.5}Ni_{0.5}(OH)_2$ can be well decomposed, whereas only some of the $Mg(OH)_2$ is dehydrated at a temperature of 553K. Consequently, the hybrid hydroxide is capable of storing thermal energy at 473–573K, at which pure $Mg(OH)_2$ could hardly be decomposed and could not store heat. The second dehydration of the hybrid hydroxide is lower than that of the first dehydration because it embraces structural water. The thermal energy storage density obtained is 165 kJ/kg $Mg_{0.5}Ni_{0.5}(OH)_2$. So, the hybrid hydroxide materials are promising heat storage material candidates for chemically storing medium-temperature thermal energy.

Apart from the aforementioned chemical reaction heat storage using solid–gas reactions, Koutinas et al. [102] studied the chemical reaction heat storage system of carbon dioxide, ammonia, and ammonium carbamate (NH_2COONH_4). The chemical equation is:

$$NH_2COONH_4(s) \Leftrightarrow 2NH_3(g) + CO_2(g) \quad [18.30]$$

They investigated the kinetics and thermodynamics of this reversible reaction and found that the decomposition of NH_2COONH_4 proceeds very rapidly in the range of 353–383K. This chemical reaction heat storage system can be a solar thermal collector integrated into an organic Rankine cycle power plant to achieve the production of mechanical work and electricity.

18.2.2.11 Metal hydrides for chemical heat storage systems

The metal hydrides are widely used for many industrial purposes, such as hydrogen storage, thermal energy storage, etc. Many metals, alloys, or intermetallic compounds can react reversibly with hydrogen under certain conditions. This reaction is exothermic during the synthesis phase, whereas an endothermic reaction occurs when the hydrogen is emanated during the decomposition phase. The universal reaction can be described as follows:

$$M + x/2H_2 \Leftrightarrow MH_x + \Delta H \quad [18.31]$$

The relevant applications of metal hydrides have the common advantages of being environmentally benign, compact, and flexible for various operating conditions [103]. Because various alloys exhibit different properties, the chemical heat storage systems based on metal hydrides can cover a wide and attractive temperature range for industrial and domestic applications by choosing versatile materials. In addition, the system can be effectively driven by low-grade thermal energy, such as the solar heat energy and industrial waste heat.

Metallic magnesium is a very abundant and low-cost material. Magnesium hydride (MgH_2) plays a unique role among a number of hydrogen metal compounds and intermetallides [104]. Bogdanović et al. [105] demonstrated a process steam generator based on a high-temperature magnesium hydride/magnesium chemical heat storage system. Furthermore, the literature provides the exact temperature corresponding to the respective H_2 decomposition pressure (equilibrium pressure) of magnesium hydride. The theoretical storage density for chemically stored heat is about 37.2 MJ/kg H_2 and volume energy storage density is about 2.2464 GJ/m^3. The maximum power output is about 14.4 MJ when the evaporation quantity of water is about 7 kg/h. Felderhoff and Bogdanović [106] stressed that the H_2 decomposition pressure as a function of temperature is of fundamental importance for the operation of the MgH_2 heat and hydrogen storage system.

Mg-based metal hydrides are potential high-temperature heat storage media. The ternary hydride Mg_2FeH_6 is a very promising material for chemical heat storage in the vicinity of 773K [107,108]. This hydride is more economical for many applications compared to other hydrides. This is ascribed to its high volumetric hydrogen density, abundant availability, and low cost. Bogdanović et al. [109] reported the thermodynamic properties and the cyclic stability of the Mg-Fe-H system. The reversible Mg_2FeH_6 and the mixed Mg_2FeH_6/MgH_2 hydride systems proved to be highly suitable materials for chemical thermal energy storage at around 773K. The hydrogen dissociation pressure of Mg_2FeH_6 is lower than that of MgH_2 under the same temperature (for example, 6.6 MPa instead of 10 MPa at 773K). The relatively low hydrogen dissociation pressure is favorable in both the technical and the economical aspects. The volumetric heat storage density of Mg_2FeH_6 is considerably higher than that of MgH_2. Moreover, Mg_2FeH_6 has excellent cyclic stability. Herrich et al. [108] investigated the desorption behavior of Mg_2FeH_6 based on synthesis conditions and initial Fe content. The desorption temperatures of Mg_2FeH_6 decrease owing to the catalytic effect of Fe during the desorption process.

$NaMgH_3$ is considered one of the potential solar heat storage materials because of its high thermal stability and relatively low price [110]. The enthalpy and entropy change of the first step reaction are 86.6 kJ/mol and 132.2 J/(mol K), respectively. $NaMgH_3$ has a more advantageous hydrogen dissociation pressure of 1.1 MPa at 773K compared to 6.6 MPa for Mg_2FeH_6. Based on Mg compounds, Felderhoff and Bogdanović [106] found out that metal hydrides have both high heat storage capacities around 673K and excellent cycling stability for heat storage application. Reiser et al. [111] discussed the features and possibilities of thermal energy storage systems, such as Mg/MgH_2, Mg-Fe/Mg_2FeH_6, Mg-Ni/Mg_2NiH_4, and Mg-Co-H. The theoretical energy storage density is 2741 kJ/kg and the observed value is 2147 kJ/kg for the Ni-doped Mg/MgH_2 system when the hydrogen content is 6 wt%. The reaction enthalpy change of Mg_2NiH_4 is 62–64 kJ/mol. Compared to MgH_2, Mg_2NiH_4 is found to be less stable. The energy storage density is 916 kJ/kg with a hydrogen capacity of 3 wt%. They concluded that all the materials show cyclic

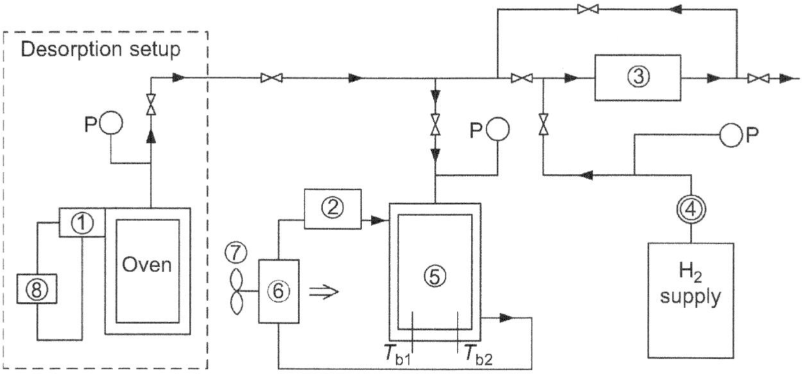

Figure 18.15 Test device of a chemical heat storage system based on metal hydrides [112]. 1, Proportion integration differentiation controller; 2, constant temperature bath; 3, mass flow meter; 4, pressure regulator; 5, reactor; 7, blower; 8, energy meter.

stability and their thermal energy storage densities can reach up to 2257 kJ/kg with the thermal output temperature between 523 and 823K. Satya Sekhar et al. [112] carried out performance tests on a Mg + 30% $MmNi_4$-based thermal energy storage device at different pressures (1−3 MPa) and temperatures (393−423K). The hydrogen storage capacity increased from 1.88 to 2.5 wt% when the pressure varied from 1.5 to 2.0 MPa. The thermal energy storage efficiency increased from 0.5 at 1 MPa to 0.74 at the pressure of 3 MPa when the temperature was 423K. For the given operating conditions under a pressure of 2 MPa and temperature of 423K, the maximum amount of stored thermal energy was about 0.714 MJ/kg and the corresponding thermal energy storage coefficient was 0.74. Fig. 18.15 shows the test device of the chemical heat storage system based on the metal hydrides [112].

18.3 Summary and perspective

Thermal energy storage is a promising environmentally friendly and energy-saving technology. Among the various technologies, it has been proved that thermochemical heat storage has a promising potential owing to its high thermal energy storage density and distinct advantage for long-term thermal energy storage with negligible heat loss. Since 2005, it has received great interest for the applications of renewable energy and low-grade thermal energy. A lot of theoretical and experimental studies have been carried out so as to promote the development of thermochemical heat storage. Based on the current situation of thermochemical heat storage technology, it is necessary to resolve numerous problems in the research and development stage before extensive application, such as the complexity of the reaction process, appropriate catalysts, special consideration of safety, the large initial investment, and the low efficiency. Thus, further work should be conducted to implement thermochemical heat storage in practice, such as the development of new composite reactants, chemical reaction kinetics, optimization designs, control strategy, and so on.

References

[1] Li TX, Wang RZ, Li H. Progress in the development of solid-gas sorption refrigeration thermodynamic cycle driven by low-grade thermal energy. Prog Energy Combust Sci 2014;40:1–58.

[2] Dueñas C, Pilatowsky I, Romero RJ, Oskam A, Finck PA. Dynamic study of the thermal behaviour of solar thermochemical refrigerator: barium chloride-ammonia for ice production. Sol Energy Mater Sol Cells 2001;70(3):401–13.

[3] Le Pierrès N, Driss S, Nathalie M. Design of a thermochemical process for deep freezing using solar low-grade heat. Chem Eng Process Process Intensif 2008;47(3):484–9.

[4] Le Pierrès N, Mazet N, Stitou D. Modelling and performances of a deep-freezing process using low-grade solar heat. Energy 2007;32(2):154–64.

[5] Le Pierrès N, Mazet N, Stitou D. Experimental results of a solar powered cooling system at low temperature. Int J Refrig 2007;30(6):1050–8.

[6] Rivera C, Pilatowsky I, Méndez E, Rivera W. Experimental study of a thermo-chemical refrigerator using the barium chloride–ammonia reaction. Int J Hydrogen Energy 2007; 32(15):3154–8.

[7] Erhard A, Hahne E. Test and simulation of a solar-powered absorption cooling machine. Sol Energy 1997;59(4–6):155–62.

[8] Stitou D, Mazet N, Mauran S. Experimental investigation of a solid/gas thermochemical storage process for solar air-conditioning. Energy 2012;41(1):261–70.

[9] Yu N, Wang RZ, Wang LW. Sorption thermal storage for solar energy. Prog Energy Combust Sci 2013;39(5):489–514.

[10] Li TX, Wang RZ, Kiplagat JK, Kang Y. Performance analysis of an integrated energy storage and energy upgrade thermochemical solid–gas sorption system for seasonal storage of solar thermal energy. Energy 2013;50:454–67.

[11] Li TX, Wang RZ, Yan T. Solid–gas thermochemical sorption thermal battery for solar cooling and heating energy storage and heat transformer. Energy 2015;84:745–58.

[12] Lahmidi H, Mauran S, Goetz V. Definition, test and simulation of a thermochemical storage process adapted to solar thermal systems. Sol Energy 2006;80(7):883–93.

[13] Mauran S, Lahmidi H, Goetz V. Solar heating and cooling by a thermochemical process. First experiments of a prototype storing 60 kWh by a solid/gas reaction. Sol Energy 2008; 82(7):623–36.

[14] Posern K, Kaps C. Humidity controlled calorimetric investigation of the hydration of $MgSO_4$ hydrates. J Therm Anal Calorim 2008;92:905–9.

[15] Zondag HA, Kalbasenka A, van Essen M, Bleijendaal L, Schuitema R, van Helden W, et al. First studies in reactor concepts for thermochemical storage. 2009. Available from: ftp://ftp.ecn.nl/pub/www/library/report/2009/m09008.pdf.

[16] Bales C, Gantenbein P, Jaenig D, Kerskes H, van Essen M, Weber R. Final report of Subtask B "chemical and sorption storage" The overview: A report of IEA solar heating and cooling programme-Task 32 Advanced storage concepts for solar and low energy buildings. 2008. Report B7 of Subtask B. Available from: http://archive.iea-shc.org/publications/downloads/task32-b7.pdf.

[17] Bales C, Gantenbein P, Jaenig D, Weber R. Laboratory prototypes of thermo-chemical and sorption storage units: A report of IEA solar heating and cooling programme-Task 32 Advanced storage concepts for solar and low energy buildings. 2007. Report B3 of Subtask B. Available from: http://archive.iea-shc.org/publications/downloads/task32-Laboratory_Prototypes_of_Thermo-Chemical_and_Sorption_Storage_Units.pdf.

[18] Bales C, Gantenbein P, Jaenig D, Kerskes H, Summer K, van Essen M, et al. Laboratory tests of chemical reactions and prototype sorption storage units: A report of IEA solar heating and cooling programme-Task 32 Advanced storage concepts for solar and low energy buildings. 2008. Report B4 of Subtask B. Available from: http://archive.iea-shc.org/publications/downloads/task32-b4.pdf.
[19] van Essen VM, Zondag HA, Schuitema R, van Helden WGJ, Rindt CCM. Materials for thermochemical storage: characterization of magnesium sulfate. 2009. Available from: ftp://ftp.ecn.nl/pub/www/library/report/2009/m09001.pdf.
[20] Hongois S, Kuznik F, Stevens P, Roux J-J. Development and characterisation of a new $MgSO_4$-zeolite composite for long-term thermal energy storage. Sol Energy Mater Sol Cells 2011;95:1831–7. http://dx.doi.org/10.1016/j.solmat.2011.01.050.
[21] van Essen VM, Gores JC, Bleijendaal LPJ, Zondag Ha, Schuitema R, Bakker M, et al. Characterization of salt hydrates for compact seasonal thermochemical storage. In: Proceedings of the ASME 3rd International Conference of Energy Sustainability; 2009;2: 3–8.
[22] van Essen VM, Bleijendaal LPJ, Kikkert BWJ, Zondag HA, Bakker M, Bach PW. Development of a compact heat storage system based on salt hydrates. 2010. Available from: ftp://ftp.ecn.nl/pub/www/library/report/2010/m10095.pdf.
[23] Bertsch F, Mette B, Asenbeck S, Kerskes H, Müller-Steinhagen H. Low temperature chemical heat storage – an investigation of hydration reactions. 2009. Available from: https://talon.stockton.edu/eyos/energy_studies/content/docs/effstock09/Session_6_2_Thermochemical_storage/49.pdf.
[24] Zondag HA, van Essen VM, Bleijendaal LPJ, Kikkert BWJ, Bakker M. Application of $MgCl_2 \cdot 6H_2O$ for thermochemical seasonal solar heat storage. 2010. Available from: ftp://ftp.ecn.nl/pub/www/library/report/2010/m10094.pdf.
[25] Zondag HA, Kikkert BWJ, Smeding S, Bakker M. Thermochemical seasonal solar heat storage with $MgCl_2 \cdot 6H_2O$: first upscaling of the reactor. 2011. Available from: ftp://ftp.ecn.nl/pub/www/library/report/2011/m11071.pdf.
[26] Zondag HA, Helden WGJ, Schuitema R, Bakker M. First systems studies and characterisation studies of $MgSO_4$ as a TCM for compact thermal storage. 2008. Available from: ftp://ftp.ecn.nl/pub/www/library/report/2008/m08061.pdf.
[27] N'Tsoukpoe KE, Liu H, Le Pierrès N, Luo L. A review on long-term sorption solar energy storage. Renew Sustain Energy Rev 2009;13(9):2385–96.
[28] Hou Y, Vidu R, Stroeve P. Solar energy storage methods. Ind Eng Chem Res 2011; 50(15):8954–64.
[29] Ucar M. Heat-of-adsorption solar thermal energy storage. J Therm Insulation 1982;6: 48–58.
[30] Boer RD, Haije WG, Veldhuis JBJ. Determination of structural, thermodynamic and phase properties in the Na_2S-H_2O system for application in a chemical heat pump. Thermochim Acta 2003;395:3–19.
[31] Lovegrove K, Luzzi A. Endothermic reactors for an ammonia based thermochemical solar energy storage and transport system. Sol Energy 1996;56(4):361–71.
[32] Kreetz H, Lovegrove K. Exergy analysis of an ammonia synthesis reactor in a solar thermochemical power system. Sol Energy 2002;73(3):187–94.
[33] Williams OM. Ammonia thermochemical energy transport in a distributed collector solar thermal power plant. Sol Energy 1981;27(3):205–14.
[34] Lovegrove K, Luzzi A, Soldiani I, Kreetz H. Developing ammonia based thermochemical energy storage for dish power plants. Sol Energy 2004;76(1–3):331–7.

[35] Carden PO. Energy corradiation using the reversible ammonia reaction. Sol Energy 1977; 19:365−78.
[36] Lovegrove K, Luzzi A, McCann M, Freitag O. Exergy analysis of ammonia-based solar thermochemical power systems. Sol Energy 1999;66(2):103−15.
[37] Lovegrove KM. High pressure ammonia dissociation experiments for solar energy transport and storage. Int J Energy Res 1996;20:965−78.
[38] Luzzi A, Lovegrove K. A solar thermochemical power plant using ammonia as an attractive option for greenhouse-gas abatement. Energy 1997;22(2−3):317−25.
[39] Lovegrove K, Luzzi A, Kreetz H. A solar-driven ammonia-based thermochemical energy storage system. Sol Energy 1999;67(4−6):309−16.
[40] Luzzi A, Lovegrove K, Filippi E, Fricker H, Schmitz-goeb M, Chandapillai M, et al. Techno-economic analysis of a 10 MW_e solar thermal power plant using ammonia-based thermochemical energy storage. Sol Energy 1999;66(2):91−101.
[41] Kreetz H, Lovegrove K. Theoretical analysis and experimental results of a 1 kW_{chem} ammonia synthesis reactor for a solar thermochemical energy storage system. Sol Energy 1999;67(4−6):287−96.
[42] Dunn R, Lovegrove K. A review of ammonia-based thermochemical energy storage for concentrating solar power. In: Proceedings of the IEEE; 2012.
[43] McCrary JH, McCrary GE, Chubb TA. An experimental study of SO_3 dissociation as a mechanism for converting and transporting solar energy. Sol Energy 1981;27(5):433−40.
[44] Hasatani M. Highly efficient conversion technologies for energy utilization. Energy Convers Manag 1997;38(10−13):931−40.
[45] Dayan J, Lynn S, Foss A. Evaluation of a sulfur oxide chemical heat storage process for a steam solar electric plant. 2011. Available from: http://escholarship.org/uc/item/1sr0w2p9.
[46] Liu Q, Yabe A, Kajiyama S, Fukuda K. A review of study on thermal energy transport system by synthesis and decomposition reactions of methanol. JSME Int J Ser B 2002;45: 473−80. http://dx.doi.org/10.1299/jsmeb.45.473.
[47] DeLancey GB, Kovenklioglu S, Ritter AB, Schneider JC. Cyclohexane dehydrogenation for thermochemical energy conversion. Ind Eng Chem Process Des Dev 1983;22(4):639−45.
[48] Cacciola G, Aristov YI, Restuccia G, Parmon VN. Influence of hydrogen-permeable membranes upon the efficiency of the high-temperature chemical heat pumps based on cyclohexane dehydrogenation-benzene hydrogenation reactions. Int J Hydrogen Energy 1993;18(8):673−80.
[49] Hatamachi T, Kodama T, Isobe Y. Carbonate composite catalyst with high-temperature thermal storage for use in solar tubular reformers. J Sol Energy Eng 2005;127(3): 396−400.
[50] Chang J-S, Park S-E, Chon H. Catalytic activity and coke resistance in the carbon dioxide reforming of methane to synthesis gas over zeolite-supported Ni catalysts. Appl Catal A General 1996;145:111−24. http://dx.doi.org/10.1016/0926-860X(96)00150-0.
[51] Gokon N, Inuta S-i, Yamashita S, Hatamachi T, Kodama T. Double-walled reformer tubes using high-temperature thermal storage of molten-salt/MgO composite for solar cavity-type reformer. Int J Hydrogen Energy 2009;34(17):7143−54.
[52] Berman A, Karn RK, Epstein M. A new catalyst system for high-temperature solar reforming of methane. Energy Fuels 2006;20(2):455−62.
[53] Kodama T, Isobe Y, Kondoh Y, Yamaguchi S, Shimizu KI. Ni/ceramic/molten-salt composite catalyst with high-temperature thermal storage for use in solar reforming processes. Energy 2004;29(5−6):895−903.

[54] Anikeev VI, Bobrin AS, Ortner J, Schmidt S, Funken KH, Kuzin N. Catalytic thermo-chemical reactor/receiver for solar reforming of natural gas: design and performance. Sol Energy 1998;63(2):97–104.

[55] Gokon N, Osawa Y, Nakazawa D, Kodama T. Kinetics of CO_2 reforming of methane by catalytically activated metallic foam absorber for solar receiver-reactors. Int J Hydrogen Energy 2009;34(4):1787–800.

[56] Gokon N, Yamawaki Y, Nakazawa D, Kodama T. Ni/MgO–Al_2O_3 and Ni–Mg–O catalyzed SiC foam absorbers for high temperature solar reforming of methane. Int J Hydrogen Energy 2010;35(14):7441–53.

[57] Gokon N, Yamawaki Y, Nakazawa D, Kodama T. Kinetics of methane reforming over Ru/γ-Al_2O_3-catalyzed metallic foam at 650–900°C for solar receiver-absorbers. Int J Hydrogen Energy 2011;36(1):203–15.

[58] Edwards JH, Maitra AM. The chemistry of methane reforming with carbon dioxide and its current and potential applications. Fuel Process Technol 1995;42(2–3):269–89.

[59] Xu Z, Zhen M, Bi Y. Catalytic properties of Ni modified hexaaluminates $LaNi_yAl_{12-y}O_{19-\delta}$ for CO_2 reforming of methane to synthesis gas. Appl Catal A General 2000:267–73.

[60] Edwards JH, Do KT, Maitra AM, Technology E, Ryde N, Schuck S, et al. The use of solar-based CO_2/CH_4 reforming for reducing greenhouse gas emissions during the generation of electricity and process heat. Energy Convers Manag 1996;37(6–8):1339–44.

[61] Levy M, Levitan R, Rosin H. Solar energy storage via a closed-loop chemical heat pipe. Sol Energy 1993;50(2):179–89.

[62] Kodama T, Kiyama A, Shimizu KI. Catalytically activated metal foam absorber for light-to-chemical energy conversion via solar reforming of methane. Energy Fuels 2003; 66(21):13–7.

[63] Kodama T, Koyanagi T, Shimizu T, Kitayama Y. CO_2 reforming of methane in a molten carbonate salt bath for use in solar thermochemical processes. Energy Fuels 2001;3(15): 60–5.

[64] Petrasch J, Osch P, Steinfeld A. Dynamics and control of solar thermochemical reactors. Chem Eng J 2009;145(3):362–70.

[65] Prevost M, Bugarel R. Chemical heat pumps: system isopropanol-acetone-hydrogen. In: Proceedings of the international conference on energy storage; 1980. p. 95–110.

[66] KlinSoda I, Piumsomboon P. Isopropanol–acetone–hydrogen chemical heat pump: a demonstration unit. Energy Convers Manag 2007;48(4):1200–7.

[67] Chung Y, Kim B-J, Yeo Y-K, Song HK. Optimal design of a chemical heat pump using the 2-propanol/acetone/hydrogen system. Energy 1997;22(5):525–36.

[68] Meng N, Shinoda S, Saito Y. Improvements on thermal efficiency of chemical heat pump involving the reaction couple of 2-propanol dehydrogenation and acetone hydrogenation. Int J Hydrogen Energy 1997;22(4):361–7.

[69] Mooksuwan W, Kumar SR. Study on 2-propanol/acetone/hydrogen chemical heat pump: endothermic dehydrogenation of 2-propanol. Int J Energy 2000;24:1109–22.

[70] Kitikiatsophon W, Piumsomboon P. Dynamic simulation and control of an isopropanol-acetone-hydrogen chemical heat pump. ScienceAsia 2004;30:135–47.

[71] Wentworth WE, Chen E. Simple thermal decomposition reactions for storage of solar thermal energy. Sol Energy 1976;18(3):205–14.

[72] Wentworth WE. Thermochemical cycles for energy storage: thermal decomposition of $ZnSO_4$ systems. NASA STI/Recon Technical Report N. April 1992.

[73] Fahim MA, Ford JD. Energy storage using the BaO_2-BaO reaction cycle. Chem Eng J 1983;27:21–8.

[74] Möller S, Palumbo R. The development of a solar chemical reactor for the direct thermal dissociation of zinc oxide. J Sol Energy Eng 2001;123(2):83–90.
[75] Osinga T, Olalde G, Steinfeld A. Solar carbothermal reduction of ZnO: shrinking packed-bed reactor modeling and experimental validation. Ind Eng Chem Res 2004;43(25):7981–8.
[76] Abu-Hamed T, Karni J, Epstein M. The use of boron for thermochemical storage and distribution of solar energy. Sol Energy 2007;81(1):93–101.
[77] Forster M. Theoretical investigation of the system SnO_x/Sn for the thermochemical storage of solar energy. Energy 2004;29(5–6):789–99.
[78] Block T, Knoblauch N, Schmücker M. The cobalt-oxide/iron-oxide binary system for use as high temperature thermochemical energy storage material. Thermochim Acta 2014; 577(0):25–32.
[79] Kato Y, Watanabe Y, Yoshizawa Y. Application of inorganic oxide/carbon dioxide reaction system to a chemical heat pump. In: Proceedings of the 31st Intersociety Energy Conversion Engineering Conference; 1996. p. 763–8.
[80] Kyaw K, Matsuda H, Hasatani M. Applicability of carbonation/decarbonation reactions to high-temperature thermal energy storage and temperature upgrading. J Chem Eng Jpn 1996;29:119–25. http://dx.doi.org/10.1252/jcej.29.119.
[81] Kyaw K, Kubota M, Watanabe F, Matsuda H, Hasatani M. Study of carbonation of CaO for high temperature thermal energy storage. J Chem Eng Jpn 1998;31:281–4.
[82] Kyaw K, Shibata T, Watanabe F, Matsuda H, Hasatani M. Applicability of zeolite for CO_2 storage in a CaO-CO_2 high temperature energy storage system. Energy Convers Manag 1997;38(10–13):1025–33.
[83] Kato Y, Yamada M, Kanie T, Yoshizawa Y. Calcium oxide/carbon dioxide reactivity in a packed bed reactor of a chemical heat pump for high-temperature gas reactors. Nucl Eng Des 2001;210(1–3):1–8.
[84] Kato Y, Saku D, Harada N. Utilization of high temperature heat using a calcium oxide/lead oxide/carbon dioxide chemical heat pump. J Chem Eng Jpn 1997;30: 1013–9.
[85] Kato Y, O-shima T, Yoshizawa Y. Thermal performance of a packed bed reactor for a high-temperature chemical heat pump. Int J Energy Res 2001;25(7):577–89.
[86] Darkwa K, O'Callaghan PW. Green transport technology (GTT): analytical studies of a thermochemical store for minimising energy consumption and air pollution from automobile engines. Appl Therm Eng 1997;17(7):603–14.
[87] Criado JM, Macias M. Analysis of the system CaO-CO_2-H_2O for storage of solar thermal energy. Sol Energy 1992;49(2):83–6.
[88] Fujii I, Tsuchiya K, Higano M, Yamada J. Studies of an energy storage system by use of the reversible chemical reaction: $CaO + H_2O \rightleftharpoons Ca(OH)_2$. Sol Energy 1985;34(4): 367–77.
[89] Ogura H, Yamamoto T, Kage H. Efficiencies of $CaO/H_2O/Ca(OH)_2$ chemical heat pump for heat storing and heating/cooling. Energy 2003;28(14):1479–93.
[90] Fujimoto S, Bilgen E, Ogura H. Dynamic simulation of $CaO/Ca(OH)_2$ chemical heat pump systems. Exergy Int J 2002;2(1):6–14.
[91] Kato Y, Yoshizawa Y. Application of a chemical heat pump to a cogeneration system. Int J Energy Res 2001;25:129–40.
[92] Ishitobi H, Uruma K, Takeuchi M, Ryu J, Kato Y. Dehydration and hydration behavior of metal-salt-modified materials for chemical heat pumps. Appl Therm Eng 2013;50(2): 1639–44.
[93] Kato Y, Nakahata J. Durability characteristics of the hydration of magnesium oxide under repetitive reaction. J Mater Sci 1999;4:475–80.

[94] Kato Y, Kobayashi K, Yoshizawa Y. Durability to repetitive reaction of magnesium oxide/water reaction system for a heat pump. Appl Therm Eng 1998;18(3−4):85−92.

[95] Kato Y, Yamashita N, Kobayashi K, Yoshizawa Y. Kinetic study of the hydration of magnesium oxide for a chemical heat pump. Appl Therm Eng 1996;16:853−62.

[96] Kato Y, Sasaki Y, Yoshizawa Y. Magnesium oxide/water chemical heat pump to enhance energy utilization of a cogeneration system. Energy 2005;30(11−12):2144−55.

[97] Kato Y, Minakami A, Li G. Operability of a thermally driven magnesium oxide/water chemical heat pump. Canadian J Chem Eng 2001;79:536−41.

[98] Kato Y, Takahashi R, Sekiguchi T, Ryu J. Study on medium-temperature chemical heat storage using mixed hydroxides. Int J Refrig 2009;32(4):661−6.

[99] Kato Y, Takahashi F-u, Watanabe A, Yoshizawa Y. Thermal analysis of a magnesium oxide/water chemical heat pump for cogeneration. Appl Therm Eng 2001;21(10): 1067−81.

[100] Kato Y, Takahashi F, Watanabe A. Thermal performance of a packed bed reactor of a chemical heat pump for cogeneration. Chem Eng Res Des 2000;78:745−8.

[101] Shkatulov A, Ryu J, Kato Y, Aristov Y. Composite material "$Mg(OH)_2$/vermiculite": a promising new candidate for storage of middle temperature heat. Energy 2012;44(1): 1028−34.

[102] Koutinas AA, Yianoulis P, Lycourghiotis A. Industrial scale modelling of the thermochemical energy storage system based on $CO_2 + 2NH_3 \leftrightarrow NH_2COONH_4$ equilibrium. Energy Convers Manag 1983;23:55−63. http://dx.doi.org/10.1016/0196-8904(83)90009-2.

[103] Yang FS, Wang GX, Zhang ZX, Meng XY, Rudolph V. Design of the metal hydride reactors − a review on the key technical issues. Int J Hydrogen Energy 2010;35(8): 3832−40.

[104] Klyamkin SN. Metal hydride compositions on the basis of magnesium as materials for hydrogen accumulation. Russ J General Chem 2007;77(4):712−20.

[105] Bogdanovic B, Ritter A, Spliethoff B. A process steam generator based on the high temperature magnesium hydride/magnesium heat storage system. Int J Hydrogen Energy 1995;20:811−22.

[106] Felderhoff M, Bogdanović B. High temperature metal hydrides as heat storage materials for solar and related applications. Int J Mol Sci 2009;10(1):325−44.

[107] Polanski M, Płociński T, Kunce I, Bystrzycki J. Dynamic synthesis of ternary Mg_2FeH_6. Int J Hydrogen Energy 2010;35(3):1257−66.

[108] Herrich M, Ismail N, Lyubina J, Handstein A, Pratt A, Gutfleisch O. Synthesis and decomposition of Mg_2FeH_6 prepared by reactive milling. Mater Sci Eng B 2004; 108(1−2):28−32.

[109] Bogdanović B, Reiser A, Schlichte K, Spliethoff B, Tesche B. Thermodynamics and dynamics of the Mg−Fe−H system and its potential for thermochemical thermal energy storage. J Alloys Compd 2002;345:77−89.

[110] Sheppard DA, Paskevicius M, Buckley CE. Thermodynamics of hydrogen desorption from $NaMgH_3$ and its application as a solar heat storage medium. Chem Mater 2011;23: 4298−300.

[111] Reiser A, Bogdanovic B, Schlichte K. The application of Mg-based metal-hydrides as heat energy storage systems. J Hydrogen Energy 2000;25:425−30.

[112] Satya Sekhar B, Muthukumar P, Saikia R. Tests on a metal hydride based thermal energy storage system. Int J Hydrogen Energy 2012;37(4):3818−24.

Part Five

Advanced applications of solar heating and cooling systems

Combined photovoltaic/thermal technology for building applications

19

T.T. Chow
City University of Hong Kong, Kowloon Tong, Hong Kong

19.1 Introduction

19.1.1 What is photovoltaic/thermal technology?

Solar energy utilization is one of the most encouraging means to deal with global environmental problems. There are currently two main solar product lines [1,2]. With 100 years of history, solar thermal (ST) collectors convert solar radiation into thermal energy through a transport medium and/or moving fluid. In less than a century, photovoltaic (PV) panels were introduced to convert sunlight into electric power by means of solar cells. These solar energy technologies have been well commercialized and are widely in use. More recently, a new area has emerged by integrating these two lines of products into single devices, generally called photovoltaic/thermal (PV/T) collectors. Through the hybrid solar collector system, thermal and electric power can be generated simultaneously in a controlled manner to support both demands, for example, in buildings.

In a PV panel, the solar cells absorb much of the incoming solar radiation, but convert only 4–18% of the incident level into electricity. The rest is converted into heat and consequently increases the PV working temperature. A PV/T system is able to facilitate the heat extraction from the solar cells through the coolant flow and derive useful means from this waste heat. This also favors electricity generation because solar cells suffer from a drop in efficiency at elevated temperature, which in most cases is within the low-to-medium range for building thermal applications [3,4]. An additional merit is its capture of cascading energy qualities, in that electric power has high exergy and heat energy has low exergy. Exergy represents the maximum quantity of work that can be produced in a given environment.

Solar power generation is commonly adopted in low-carbon buildings. In this era, there are more and more voices for and commitment to zero-carbon (or even positive-carbon) building developments. This sooner or later will call for the full use of available external facades to accommodate renewable energy devices. As long as carbon reduction remains a global concern, the market for solar products will grow continuously. Then collector overall efficiency, architectural uniformity, and space availability for housing of separate PV and ST systems are going to be important issues that ultimately favor the adoption of PV/T systems as an eco-friendly alternative [5].

Advances in Solar Heating and Cooling. http://dx.doi.org/10.1016/B978-0-08-100301-5.00019-9

19.1.2 What are the possible choices and applications?

While the concept of electric and heat power cogeneration was first documented in the 1950s, it was not until the 1970s that PV/T research really became active. The further increase in research activities since the 1990s was apparently a response to global environmental problems and the growing community interest in building-integrated solar device options [6]. At present, various types of practical PV/T collectors are available. However, their full-scale applications are not extensive. Fig. 19.1 gives a general classification of the types currently available in the commercial market, including those actively under research and expected to have good marketing potential.

There can be different approaches to the PV/T classification [7,8]. The main classification adopted here is first from the collector geometry and working principles, thus the flat plate technology and the concentrator technology demarcation. Fig. 19.2 shows the main features of a flat plate PV/T collector [6]. The key components include solar cells, encapsulated materials, thermal absorber, coolant flow channels, coolant inlet and outlet distributors, thermal insulation, electric wiring, and structural frame. The front glazing is optional. Then the various types of flat plate collectors can be categorized according to the type of cooling fluid in use, such as air, liquid, or refrigerant. Among the many subcategories, there can be selections among the various types of solar cells, natural or forced fluid flow, free-standing or building-integrated features, etc. Accordingly, the available system installations can range from simple devices such as PV/T air and water preheating systems to complex systems such as hot water supply through a PV-integrated heat pump, as well as PV/T solar tri-generation systems [9,10].

For solar cells, the selections can be among silicon-based (monocrystalline, polycrystalline, or amorphous) and thin-film, including dye-sensitized types. The electricity supply can be grid-connected or stand-alone in operation. For the thermal part, like in a liquid-based PV/T collector, the liquid coolant (water, glycol, or mineral oil) passes through flow channels that are in direct contact with the thermal absorber. The features are similar to those of an ST collector. In a closed-loop PV/T system, the heat absorbed by the circulating fluid will be transferred to its application means via a heat exchanger.

The PV and ST integration can be enriched further through solar concentration. Compared to the flat plate type, the concentrator-type PV/T (c-PV/T) technology is able to increase the intensity of solar radiation that falls on the cell surface [11,12]. High-efficiency solar cells can be used. This also makes possible the substitution of some of the expensive solar cells with low-cost reflector surfaces. The c-PV/T collectors can be categorized into three main groups: single cells, linear geometry, and densely packed modules. For all these, the use of either refractive lenses or reflector surfaces is common. Comparatively, a lens is lower in weight and material costs. For high-concentration systems, more concentrator material per unit cell/absorber area is necessary. Then the use of lenses is more appropriate. On the other hand, the reflector type is good for medium- to high-temperature hot water systems suitable for industrial applications. Examples are the production of steam for turbine running and the production of hot water for running an organic Rankine cycle generator.

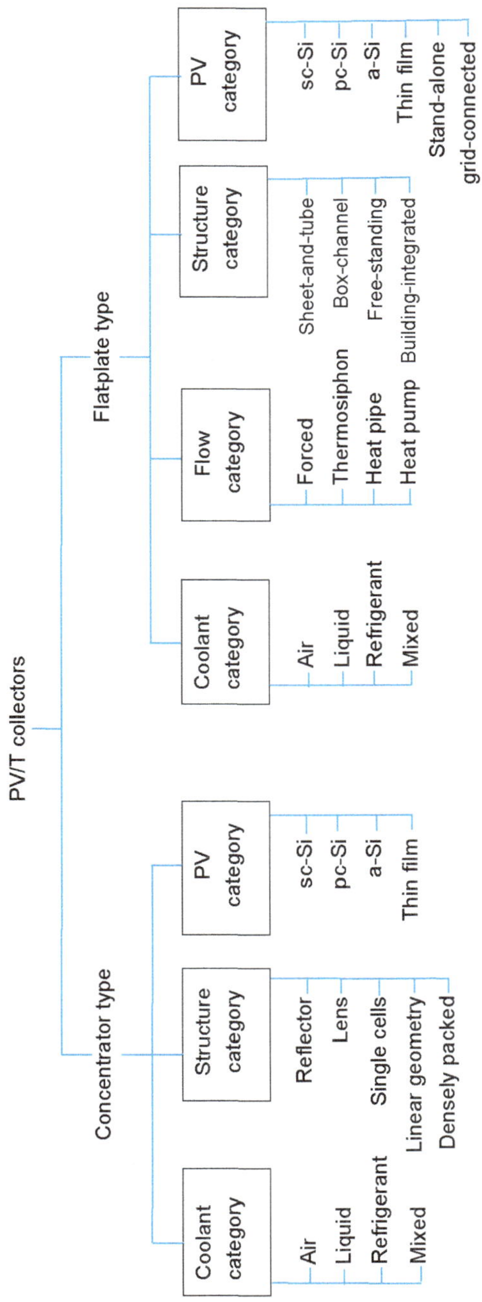

Figure 19.1 Classification of photovoltaic/thermal (PV/T) collectors. *a-Si*, Amorphous silicon; *pc-Si*, polycrystalline silicon; *sc-Si*, single-crystal silicon.

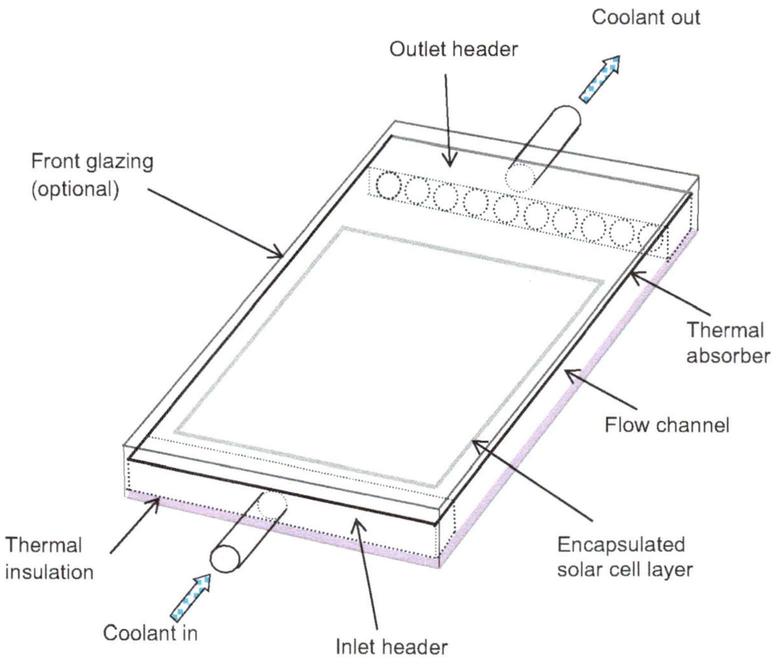

Figure 19.2 Main features of a flat plate photovoltaic/thermal collector.

The main concerns of the technological advancements are how to provide good uniformity in solar intensity and effective cooling to the lines of solar cells and a durable sun-tracking system. For low-temperature systems, flat plate collectors can achieve better energy efficiency. They are also good for water preheating when higher water temperature is needed. For these reasons, the c-PV/T collectors have very limited applications in buildings. The following discussions will be on the flat plate PV/T technology. More elaborations on the c-PV/T technology can be found in Refs [13,14].

19.2 Flat plate photovoltaic/thermal systems and equipment for building applications

19.2.1 Performance assessment criteria

Before we look into the various designs and engineering performance of a range of flat plate PV/T collectors, it is worthwhile to introduce the collector performance indicators that are often encountered in the equipment and/or system evaluation.

To assess the electrical performance of a PV/T system is straightforward since the electricity demand can be immediate and energy storage is optional. This is not the same for thermal performance. The PV/T collector system can be a minor part of a complete heat supply system that is composed of many subsystems, such as thermal storage, auxiliary heater, pumping station, or flow channels. The system designer determines the appropriate solar fraction and other design parameters in achieving the best overall benefits.

In principle, the electrical efficiency η_e and thermal efficiency η_t of a flat plate PV/T collector can be expressed as:

$$\eta_e = \frac{V_{mpp} I_{mpp}}{GA} \quad [19.1]$$

and

$$\eta_t = \frac{\dot{m}C(T_{out} - T_{in})}{GA} \quad [19.2]$$

where V_{mpp} and I_{mpp} are the voltage and electric current at maximum power point operation, G is the incident solar irradiance normal to the front surface, A is the collector aperture area, $\dot{m}$ and C are the mass flow rate and specific heat capacity of the coolant, respectively, and T_{in} and T_{out} are the coolant temperatures at the collector inlet and outlet.

The solar cells do not fully cover the aperture surface. The extent of coverage is defined by the term called packing factor, β, which is the ratio of the cell surface area A_{cell} to the aperture area. The electrical efficiency is then related to the cell efficiency η_{cell} by

$$\eta_e = \frac{A_{cell}\eta_{cell}}{A} = \beta\eta_{cell}. \quad [19.3]$$

Like in ST collectors, the PV/T thermal efficiency can be expressed as a function of reduced temperature [15], ie,

$$T^* \equiv \frac{T_{in} - T_a}{G} \quad [19.4]$$

where T_a is the ambient temperature. The Hottel–Whillier–Bliss model modified by Florschuetz [16] gives

$$\eta_t = F_R\left[(\tau\alpha)_e(1 - \eta_e) - U_L\left(\frac{T_{in} - T_a}{G}\right)\right] \quad [19.5]$$

where F_R is the modified heat removal factor, U_L is the overall heat loss coefficient, $(\tau\alpha)_e$ is the effective transmittance, and η_e is the electrical efficiency evaluated at T_a. F_R represents the ratio of the actual useful heat gain to the maximum possible useful heat gain. Alternatively, if T_m is the mean fluid temperature at the collector and F' is the collector efficiency factor, then

$$\eta_t = F'\left[(\tau\alpha)_e(1 - \eta_e) - U_L\left(\frac{T_m - T_a}{G}\right)\right]. \quad [19.6]$$

A linear expression of Eq. [19.6] gives

$$\eta_t = \eta_{t0} - aT^* \qquad [19.7]$$

where the ordinate intercept η_{t0} is the collector thermal efficiency at zero reduced temperature and the slope a relates to the collector heat loss factor.

For the overall performance evaluation, the concept of total system efficiency η_0 can be used. This is simply the sum of the two efficiency terms, ie,

$$\eta_0 = \eta_e + \eta_t \qquad [19.8]$$

Or considering that electricity is of a higher grade form than heat energy, the energy-saving efficiency term can then be used [17]. This is expressed as

$$\eta_{saving} = \frac{\eta_e}{\eta_{power}} + \eta_t \qquad [19.9]$$

where η_{power} is the electric power generation efficiency of a conventional power plant.

The third approach is from the viewpoint of exergy [18]. By definition, exergetic efficiency (ε) is the ratio of total exergy output to total exergy input. In a PV/T system, the exergetic efficiency within the time period from t_1 to t_2 can be expressed as

$$\varepsilon_{pvt} = \frac{\int_{t_1}^{t_2} \left(A\,\dot{E}x_t + A_{cell}\,\dot{E}x_{pv}\right)dt}{A\int_{t_1}^{t_2} \dot{E}x_{sun}\,dt} = \varepsilon_t + \beta\varepsilon_{pv} \qquad [19.10]$$

where $\dot{E}x_t$ is the thermal exergy output per unit collector area, $\dot{E}x_{pv}$ is the PV exergy output per unit solar cell area, $\dot{E}x_{sun}$ is the exergy input of solar radiation, and ε_t and ε_{pv} are the exergetic efficiencies of the thermal collector and solar cells, respectively. The exergy outputs are related to the energy outputs by

$$\dot{E}x_{pv} = \dot{E}_{pv} \qquad [19.11]$$

and

$$\dot{E}x_t = \dot{E}_t\left(1 - \frac{293}{293 + (T_2 - T_a)}\right) \qquad [19.12]$$

where $\dot{E}_t$ is the thermal output power per unit collector area, $\dot{E}_{pv}$ is the PV output power per unit cell area, and T_2 is the final temperature of the coolant.

On cost evaluation, life-cycle analysis can be performed [19]. The life-cycle cost of a PV/T system includes the total cost of investment and operation over its entire service life. As most of the related expenses will take place beyond the equipment acquisition date, an accurate evaluation should involve the time value of money, for instance, considering inflation, taxes, and/or company discount rates. A simplified approach is sometimes adequate. This is to ignore the time element so that the cost payback time (CPBT) can be used. This is done by adding together the cash inflows from successive years until the cumulative cash inflow is the same as the investment cost.

In analogy to cost evaluation, two environmental cost–benefit parameters, namely the energy payback time (EPBT) and greenhouse gas payback time (GPBT), can be used. These indicators are applied to evaluate the duration after which the real environmental benefit begins. EPBT is the ratio of embodied energy to annual energy output. Embodied energy refers to the quantity of energy required to produce the material in its product form. For a building-integrated PV/T (BiPV/T) system, for example,

$$\text{EPBT} = \frac{\Sigma_{\text{pvt}} + \Sigma_{\text{bos}} - \Sigma_{\text{mtl}}}{E_{\text{pv}} + E_{\text{t}} + E_{\text{ac}}} \quad [19.13]$$

where Σ_{pvt}, Σ_{bos}, and Σ_{mtl} are respectively the embodied energy terms of the PV/T system, of the balance of system, and of the replaced building materials; E_{pv} is the annual useful electricity output, E_{t} is the annual useful heat gain (equivalent), and E_{ac} is the annual electricity saving of the heating/ventilation/air conditioning system due to thermal load reduction.

Similarly, in terms of greenhouse gas (GHG) emission,

$$\text{GPBT} = \frac{\Omega_{\text{pvt}} + \Omega_{\text{bos}} - \Omega_{\text{mtl}}}{Z_{\text{pv}} + Z_{\text{t}} + Z_{\text{ac}}} \quad [19.14]$$

where Ω stands for the embodied GHG (or carbon dioxide equivalent) and Z is the reduction in annual GHG emissions from the local power plant owing to the BiPV/T operation.

EPBT and GPBT are therefore functions of the energy performance of the PV/T system under investigation as well as the environmental impacts, like those of the building systems, the power supply network, and the manufacturing, delivery, and on-site handling as a whole [20]. Also it should be noted that the energy performance of a given PV/T product or system design varies with the local environment, that is, the climatic conditions. Performance indicators derived for applications in different cities should not be compared directly.

19.2.2 Air-based design

The air-based PV/T (PV/T-a) collectors, either glazed or unglazed, provide a simple and economical means of energy cogeneration [21]. The solar cells can be in a mix of parallel and series connections to give the specific open-circuit voltage and closed-circuit current characteristics. They are laminated onto the absorber plate surface with encapsulated layers of EVA (ethylene–vinyl acetate) and TPT (Tedlar–polyester–Tedlar) in between. TPT is well known for its good electrical insulation and EVA for its adhesive properties.

The front glazing, when present, is usually a low-iron tempered glass plate that allows sunlight penetration but prevents thermal loss and the entry of dust and rain. But the reflection at the glass surface and the increase in solar cell working temperature because of the sealed air layer can be drawbacks. The thermal insulation layer at the back of the absorber plate is for reducing the thermal loss. A glazed collector favors heat collection and an unglazed one favors electricity generation. The cooling airstream can be warmed to various temperature levels through forced or natural flow. Forced circulation using a mechanical fan is more effective at heat extraction than natural circulation by buoyant flow, owing to the better thermal convection behavior. But the fan power consumption reduces the net electricity output. Its use is mostly to meet the demands of industrial hot air, indoor space heating, and/or agricultural dehydration. In the case of space heating, the building-integrated options can be more appropriate.

Fig. 19.3 shows longitudinal cross sections of four types of flat plate PV/T-a collectors. These include a flow channel above the solar cell encapsulation as mode 1, a channel

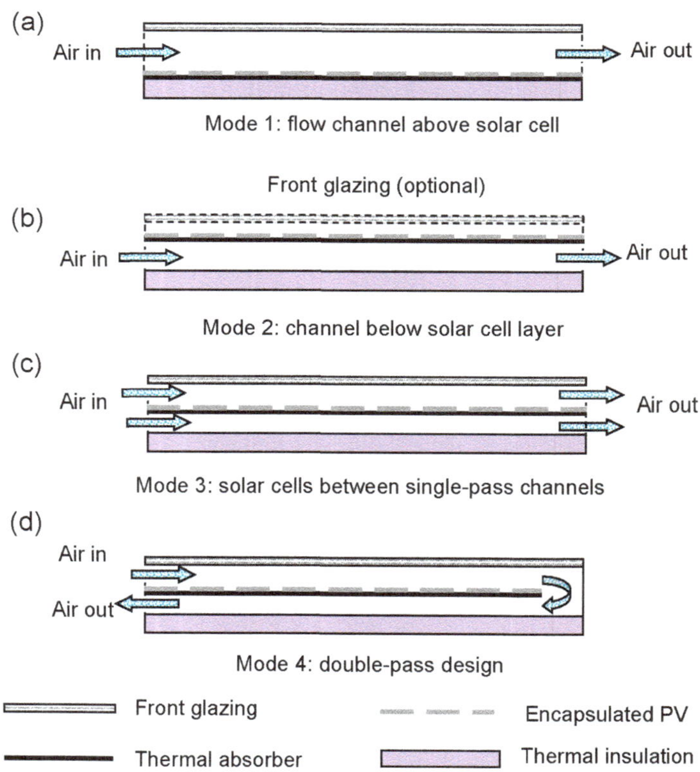

Figure 19.3 Longitudinal cross-sectional views of four modes of air-based photovoltaic/thermal collectors. (a) Mode 1: flow channel above solar cell. (b) Mode 2: channel below solar cell layer. (c) Mode 3: solar cells between single-pass channels. (d) Mode 4: double-pass design. *PV*, Photovoltaic cell.

underneath the thermal absorber as mode 2, an encapsulation layer between single-pass channels as mode 3, and finally the double-pass design as mode 4. Numerical analysis [22] shows that while mode 1 has the lowest performance, the other three have comparable energy outputs. On the whole, mode 3 requires the least fan power.

A smaller packing factor will lead to a higher ratio of thermal energy output to electricity output. When the surface area of the encapsulated solar cell layer is much smaller than the absorber surface, the encapsulated layer should be fixed as close to the coolant inlet end as possible, to achieve better overall energy performance.

19.2.3 Liquid (water)-based design

Generally speaking, liquid-based PV/T collectors have better overall energy performance than the air-based designs [23]. This is because of the higher heat capacitance and convective heat transfer characteristics of liquid compared to air. Water is by far the most popular liquid coolant because of its readiness and chemically neutral properties. Fig. 19.4 shows cross-sectional views of four common PV/T-w (water-based)

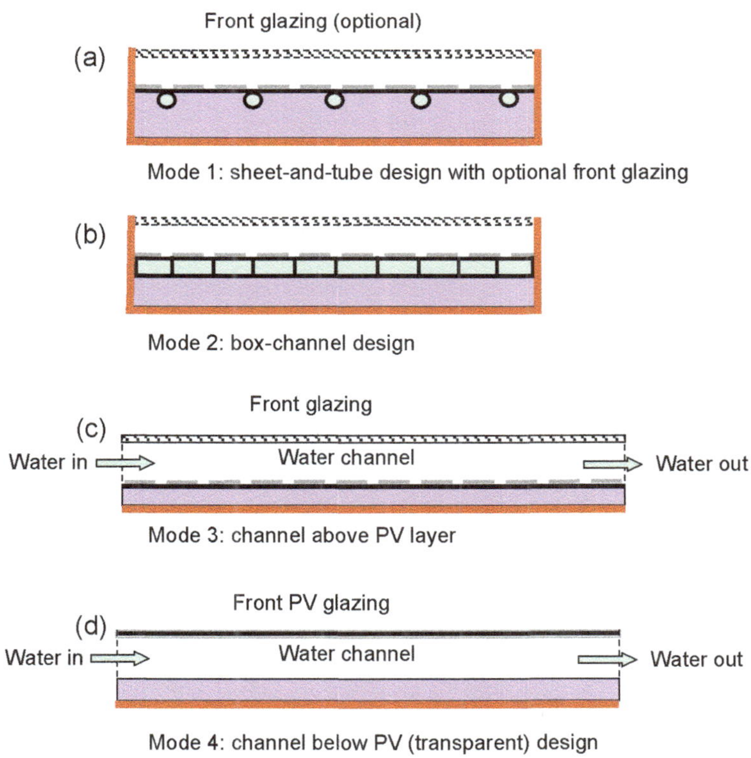

Figure 19.4 Transverse cross-sectional views of four common water-based photovoltaic/thermal collectors. (a) Mode 1: sheet-and-tube design with optional front glazing. (b) Mode 2: box-channel design. (c) Mode 3: channel above PV layer. (d) Mode 4: channel below PV (transparent) design. *PV*, Photovoltaic cell.

designs, ie, mode 1, sheet-and-tube design; mode 2, box-channel design; mode 3, channel above the solar cell encapsulation; and mode 4, channel below the PV glazing. Among these, the sheet-and-tube option is the one most economical and most commonly in use. When the solar cell encapsulation layer is removed, this mode 1 looks exactly like a conventional solar water-heating collector. But actually a stronger absorber plate is needed to hold the solar cells in position. Different patterns of tube running are possible, such as serpentine or parallel flow.

Because of the larger heat transfer surface area of the box-channel design in mode 2, it has better heat removal characteristics, with the value of fin effectiveness close to unity. Although this requires a higher production cost than mode 1, the difference is not obvious in PV/T collectors compared to ST collectors, because solar cells are the most expensive item. Among these four, the mode 4, channel-below-PV (transparent) configuration, may have the highest overall energy efficiency. However, its maintenance cost can be a concern in life-cycle service. Overall speaking, both mode 1 and mode 2 are promising for domestic hot water production.

A free-standing thermosiphon-flow design is good for single-family use. Fig. 19.5 shows such a mode 2 collector system in which the collector carries a water tank and the interconnecting pipes [24]. Water enters the collector at the lower header and leaves at the upper one. Table 19.1 lists some of its major design parameters for reference. When the solar radiation is rich, the buoyancy effect developed in the water

Figure 19.5 A box-channel photovoltaic/thermal collector designed for thermosiphon operation.

Table 19.1 Design parameters of the box-channel water-based photovoltaic/thermal collector

Glazing	4-mm-thick low-iron glass Surface area: 1.76 m^2
Air gap	25 mm deep
Photovoltaic cell encapsulation	Solar cell type: monocrystalline silicon Cell area: 1.11 m^2 Electrical efficiency at STC: 13% Solar cell temperature coefficient: 0.005/K Encapsulation: transparent TPT + EVA + solar cell + EVA + opaque TPT
Thermal absorber	Water holding capacity: 15 kg 14 numbers of aluminum-alloy flat box absorber module Absorber module size: 105 × 1380 × 22 mm (with three water channels)
Thermal insulation layer	20-mm-thick polystyrene
Water tank	Water holding capacity: 155 kg

EVA, Ethylene–vinyl acetate; *STC*, standard testing condition; *TPT*, Tedlar–polyester–Tedlar.

channels can effectively drive the water flow by natural circulation. The equipment was therefore very suitable for the warm and hot climate applications. Using a validated computer model and the typical weather conditions of Hong Kong, the overall efficiency of this PV/T-w system was found to be 44.2%, whereas the energy saving efficiency was 54.1%. Compared to the side-by-side system having the same surface areas of thermal absorber and PV cells, the combined energy yield of the two separate collector units is 8.6% higher than that of the PV/T-w system. However, in terms of combined energy yield per unit surface area, the PV/T-w collector is 50.1% higher than the side-by-side case. Also with water cooling, the daily variation in cell operating temperature is only around 30°C. The range can be doubled (around 60°C difference) in a similar plain PV module. The highest operating temperature of the solar cell in the latter case is above 80°C. For the PV/T-w case, the highest operating temperature is only 58.2°C. The annual electricity output is 2.2% higher than from the PV panel.

For a central system with arrays of PV/T-w collectors, pump circulation is generally required. On the other hand, the first law of thermodynamic evaluation indicates that the glazed design is always suitable if either the thermal or the overall energy output is to be maximized. On the other hand, the exergy analysis supports the use of an unglazed design under some conditions [25].

In some flat plate PV/T collectors, the configuration design allows the use of fins and reflectors to boost up the heat transfer [26], and some allow the use of either air or water (or even both) to work as the coolant at different times of the day. This provides the flexibility of changing the function of the solar collector to cater to the actual occupant's needs. Higher fluid temperature can be a result [27].

19.2.4 Refrigerant-based design

A refrigerant-based PV/T (PV/T-r) collector absorbs heat from the ambient air and brings about a phase change in the coolant fluid, which then becomes a refrigerant. So the solar device is called a PV/T evaporator [28]. This evaporation process takes place at low temperatures and the PV/T-r collector can serve either as a heat pipe component in the PV/T heat pipe (PV/T-pi) system or as a heat pump component in the PV/T heat pump (PV/T-pu) system. In buildings, the heat energy can be finally delivered to the domestic hot water (DHW) system. As the latent heat of the refrigerant is high, a large amount of heat can be transported through the small tubing even when the temperature difference is small. Unlike the PV/T-w systems, the PV/T-r systems can serve well in extreme cold climate zones without the frozen water problem.

Fig. 19.6 shows the structure of a PV/T-pi collector that houses a group of heat pipes within its metallic frame [29]. Inside each heat pipe there is a sealed vacuum partly filled with a refrigerant (like 134a or sometimes just pure water). The evaporator sections of the heat pipe array are fixed onto the back of a thermal absorber and the condenser sections are inserted into a water box that can be concealed. Under normal operation, heat energy is conducted through the absorber plate to the evaporator pipe sections. Then the refrigerant passes on this energy to the feed water flowing through the water box. The amount of useful thermal and electrical energy gains primarily depend on the available solar radiation on site.

Fig. 19.7 shows a simplified flow diagram of a PV/T-pu system [28]. The system comprises a water storage tank, an auxiliary gas heater, a circulation pump, an array

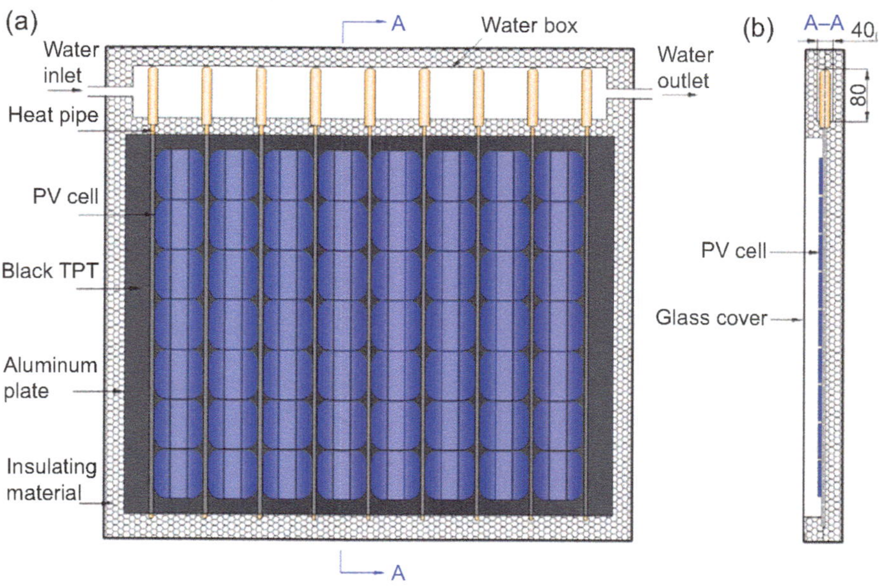

Figure 19.6 Features of a heat pipe photovoltaic/thermal collector for water heating. (a) Front view. (b) Section A–A. *PV*, Photovoltaic cell; *TPT*, Tedlar–polyester–Tedlar.

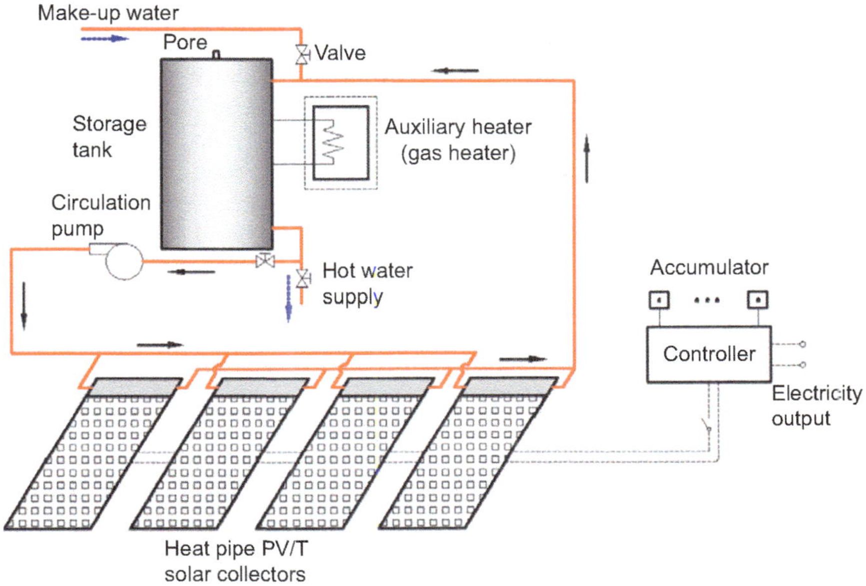

Figure 19.7 Flow diagram of a heat pump photovoltaic/thermal system.

of PV/T-r collectors, and the associated electrical and refrigeration circuits. The collectors are in parallel connection and work as the evaporators of the direct expansion refrigeration circuit, which is not shown in this figure. Solar heat absorbed by the collectors is delivered to the storage tank by the pumped water flow. With the auxiliary heater as backup, the water in the tank serves the DHW system. The ST efficiency mainly depends on the available solar radiation, the required pumping power, and the hot water load per unit collecting area. On the other hand, via a solar controller, the generated electricity first supports the variable-speed drive of the refrigerant compressor itself, and the surplus goes to the electricity consumer or the accumulator battery.

19.2.5 Building-integrated design

Building-integrated solar devices may be installed either at the building façade or on the roof. The system can be designed as invisible, as aesthetically appealing, or to appear as an architectural concept. The IEA TASK 41 working on solar energy and architecture set the following criteria for building-integrated solar devices [30]:

- Well composed and innovative design
- Architecturally pleasing design
- Good composition of colors and materials
- Size that suits the harmony and combination
- Consistent with the context of the building
- Integrating naturally

The air-based systems lend themselves to architectural integration. The outside air, before being drawn into an air distribution ductwork, can be preheated by successive PV/T elements that form parts of a roof or facade. Such integration may cut down on the material and installation costs of the building envelope.

The PV ventilated glazing technology for application in warm climate zones offers substantial energy saving opportunities through air-conditioning load reduction, more favorable daylight penetration, and renewable energy utilization [31]. As shown in Fig. 19.8, the double glazing consists of a PV glass pane at the outside and a clear glass pane at the inside. Different combinations of vent openings may come up with different ventilating flow schemes that can vary from buoyant-induced to forced-flow.

When the PV/T-w or PV/T-pi collectors are constructed outside the vertical wall, these are their building-integrated options, which can be multifunctional, such as electricity generation, water heat gain, and solar transmission reduction. The prime design considerations lie in fin effectiveness, lamination quality, thermal conductivity, and energy efficiency. These all have notable impacts on the electrical and thermal performance. The replacement of the expensive absorption material, like copper or aluminum, with precoated steel can be cost-effective. Nevertheless, for BiPV/T-w or BiPV/T-r system installations, the additional weight of the extra components on the building structure should not be overlooked in the life-cycle cost and/or energy evaluation.

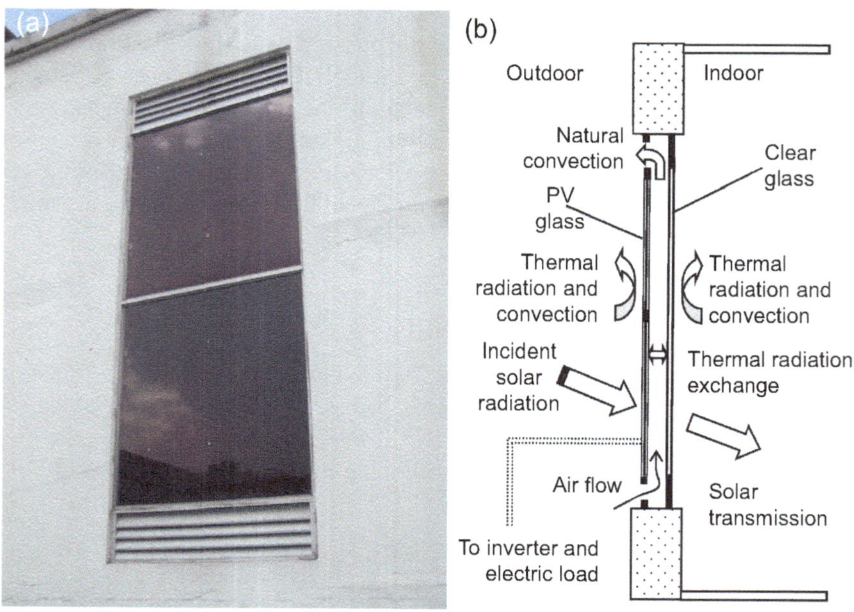

Figure 19.8 Features of natural ventilated photovoltaic glazing at test chamber. (a) Outside view. (b) Cross-sectional view and energy flow paths.

19.3 Advanced system design and performance analysis

19.3.1 On life-cycle analysis

Life-cycle analysis (LCA) is a technique that assesses the various aspects of a product in association with the product development and particularly the potential impact throughout its service life. Before the launch of a new energy technology, its environmental superiority over the competing alternatives can be evaluated by comparing their life-long consumption in terms of cost investments, energy uses, and GHG emissions. Although there have been many reported EPBT and GPBT studies on ST and PV systems, those on PV/T systems are very few. These studies are in fact essential when the main target of the technology is to combat climate change and carbon emission. Described below is one case study in Hong Kong, making use of the PV/T collector shown in Fig. 19.5 as the illustrating example. An evaluation of the CPBT, EPBT, and GPBT of a free-standing PV/T system and its building-integrated design option is shown.

In this case study, the annual energy outputs were the results of hourly dynamic simulation runs based on typical local weather data and a validated system computation model. Table 19.2 shows the technical data in the CPBT evaluation. Compared with the free-standing PV/T case, the BiPV/T system was found to have a lower capital demand on a unit collector area basis. This is because of the elimination of the steel structure for floor mounting and also the economy of scale in massive handling of the system components. As of this writing, aluminum thermal absorber parts are made available by raw material mining and extraction, ingot melting, mechanical extrusion, machining, and assembling into the whole piece. The major-component production and assembly processes include the front glass, PV laminated absorber, insulation material, and aluminum frame. As for the balance of the system (BOS), the electrical BOS components include inverters, electrical wiring, and electronic devices. The mechanical BOS includes the water storage tank, pipe work, supporting structure, and accessories. The embodied energy to be added on includes all the above items with respect to the manufacturing processes, the transportation from the suppliers to the building site, the installation and testing activities, and finally the decommissioning and equipment disposal measures.

For Hong Kong, the cumulative energy intensity of a monocrystalline-silicon PV panel was 976 kWh/m^2 [32]. That of the inverter and electrical parts together was approximately 5% of the PV panel, according to the Hong Kong government database. The evaluation showed that the PV/T-w collector itself accounts for around 80% of the embodied energy. For the BOS, the water tank accounts for 11.4%, the other mechanical components account for 7%, and the electrical accessories account for 1.8%. The BiPV/T option in this study included an array of PV/T collectors that are integrated into the external facade of an air-conditioned building. The water storage tank was on the rooftop and the water circulation was designed for thermosiphon means. The LCA showed that the BiPV/T collector itself accounts for 89% of the total embodied energy. For the BOS, the water tank accounts

Table 19.2 Evaluation of cost payback time based on the working conditions of Hong Kong in 2006

	Free-standing PV/T	BiPV/T
Investment (HK$)		
Water storage tank	400	750
Collector frame and support	400	1800
Modular thermal absorber	600	2700
Solar cells and encapsulation	4000	17,500
Inverter	700	1000
Piping, wiring, and accessories	300	900
Installation cost	1500	3000
Total system cost	7900	27,650
Useful energy savings (MJ (kWh))		
Thermal energy	2650.4 (736.2)	8127.5 (2257.6)
Electrical energy	473.2 (131.4)	1162.4 (322.9)
Space cooling load	–	742.6 (206.3)
Cost savings (HK$)		
Gaseous fuel @HK$0.2/MJ	530.1	1625.5
Electricity @HK$0.95/kWh	124.9	372.0
Annual savings	655.0	1997.5
Cost payback time	12.1 years	13.8 years

Note: US$1.00 is equivalent to HK$7.75. *PV/T*, Photovoltaic/thermal system; *BiPV/T*, building-integrated PV/T.

for 4.9%, the pipe and supporting components together account for 3.8%, and the remaining 1.8% goes to the electrical components. The findings show that based on a heat-to-electricity conversion factor of 0.38, the EPBT of the free-standing PV/T system is 2.4 years, which is an order of magnitude lower than the expected system working life. That of a BiPV/T system is 3.9 years, which is again considerably lower than the expected EPBT of comparable BiPV systems in Hong Kong. The corresponding GPBTs of the two PV/T options are 1.5 and 3.9 years, respectively. All these illustrate the environmental superiority of PV/T-w systems over the competing renewable energy system options based on the current engineering data. As time goes by, it is expected that the improvements in material and energy utilization and recycling will continue to improve the environmental profiles. Any increases in the electric-conversion efficiencies will also constitute a proportional improvement in the values of EPBT and GPBT.

For the system applications in real buildings, like the PV/T-w system options, the collector cooling effect is climate dependent and varies with the hot water usage pattern. Hence the merits of cogeneration at a specific location compared to the side-by-side system are best examined through real case studies [33]. The findings with respect to different climate zones and from a batch of building types having different functional usage will constitute an overall picture about the further research and development strategies needed.

19.3.2 On materials selection

In recent years, there have been increasing research efforts on the selection of coolant fluid [34–36]. Investigations on the cooling performance of a range of nanofluids have been reported. For example, there were comparisons among graphite/water, silver/water, and carbon nanotube/water in the performance of microscale direct absorption solar collectors. It was observed that the energy performance may increase substantially for those volume fractions less than 0.5%. Aluminum/water nanofluid at 1% volume fraction was found to improve solar absorption significantly. On the other hand, no significant increase in collector efficiency was found with an increase in the size of the nanoparticles from the volume fraction of 0.8%.

There have also been studies on the material selection for the thermal absorber. The possible use of copolymer absorbers as a substitute has been examined extensively [37]. In addition to the cost benefits, there are two additional advantages:

1. The weight reduction leads to less material wastage, easier material handling and site installation.
2. With the involvement of fewer collector components, the manufacturing process is simplified.

The use of copolymers may not be always advantageous. They are low in thermal conductivity, large in thermal expansion, and restrictive in service temperature. Overall, the copolymer for replacement must be chemically stable, good in physical strength, and UV-light protected.

19.3.3 On environmental impacts

Whereas conventional energy systems can damage climate, air, water, land, and the urban environment, leading to occupational and public health or fire risks, solar energy technologies can be considered as almost absolutely clean and safe. Their applications have significant socioeconomic benefits, such as security of energy supply and provision of job opportunities. To this end, the use of PV/T can have additional environmental benefits because of its multipurpose applications. Adverse effects are generally small and can be minimized by appropriate mitigation measures. The negative environmental impacts include the following areas [38]:

- Global warming emissions
- Land use
- Water use
- Hazardous materials

- Visual effect
- Occupational hazards
- Fire/health risk to public

Unlike fossil fuel–based technologies, PV/T does not lead to any harmful emissions during operation. However, there are emissions at other stages of the product life cycle, such as manufacturing, transportation, installation, maintenance, decommissioning, and dismantlement. The emissions associated with transportation and others are insignificant compared to those with manufacturing.

Generally speaking, minimum additional land will be required for the building applications because the PV/T system will usually be added onto the roof or the external facades of the building. Water resources are required particularly for water-based systems and also for general services. Both wind and solar power are intermittent energy sources. So all available outputs must be consumed when they are available and either stored or transported to where they can be used at once. Recyclable product components are to be encouraged. The solar cell production process includes a number of hazardous materials, such as those for cleaning and purifying the semiconductor surface. Batteries for electricity storage are responsible for most of the environmental impact, owing to their relatively short life span and their heavy metal content. Large amounts of energy and raw materials are required for their production. A module-recycling scheme can improve this situation.

The aesthetic impact of PV/T panels is evidently a matter of taste and may not be always negative. Advances in the development of multifunctional PV/T facades can provide an important stimulus for architectural expression. The accidental release of working fluids (water, oil, or nanofluid) could form a health hazard or lead to outbreak of fire. On the other hand, noise is insignificant in comparison to any other power option, such as conventional power plants, wind power generation, and gas turbines. As the long-term goal, all energy used to manufacture PV/T products should come from the PV/T technology itself or from other renewable energy options.

19.3.4 On technological advancements

PV/T collectors and systems have been a research and development focus for decades. The benefits of this hybrid solar technology originally were not well recognized in the commercial market although with time similar PV/T products were introduced by different suppliers. Experiences tell that the industrial demands were not stable. And because of the limited demand, further investments in the product development were not adequate to support the quality assurance and long-term reliability issues. For example, one technical problem remaining to be solved is the difference in coefficient of expansion between the thermal absorber and the encapsulated solar cells. Over the range of working temperatures, the cyclic stress tends to tear the two apart, leading to a degradation in heat transfer from the outer solar cell layer to the coolant in flow channels. More importantly, this may lead to a break in the electrical contact at some points along the line of solar cells that are in series connection. For these barriers, the manufacturers tend to keep putting marketing efforts into the PV and/or ST collector system equipment. In addition, there exists no popular and standardized laboratory test method to differentiate the energy performance and the overall quality of the PV/T products [39]. Likewise the lack of manufacturer guarantee on the service life of the individual product forces the client to face the risks of unexpected system breakdown as well as the financial burdens coming out of this.

Nevertheless, as long as the building industry is required to deal with the global environmental problems, the international protocol will continuously direct the industry toward technological breakthroughs and innovations that are able to maximize clean energy utilization.

19.4 Final remarks

The market for solar products is expected to grow continuously. For the building sector, solar power generation is commonly adopted in low-carbon buildings, and there is more and more interest in zero carbon or even positive carbon building designs. The target of maximizing renewable power outputs favors the use of hybrid solar systems.

Using liquid water as the coolant is more effective than using air for achieving better electrical output. The proper use of the hybrid solar technology requires a tactful balance between the low-temperature requirement of the water stream (to serve as the coolant) and the high-temperature requirement for its serving the hot water system. A PV/T system with flat plate thermal collector design is well applicable to a building with low-temperature fluid heating demands. Most large cities are now dominated by air-conditioned buildings in which space cooling demands are high in summer. In these buildings, the exposed facades provide very good opportunity for accommodating the building-integrated systems, hence favoring the BiPV/T options.

Through the active research and development activities of the past decades, a wide range of PV/T product designs are now available, which include the air-based, liquid-based, and refrigerant-based options, not to mention the various choices in solar cell types and thermal elements. In principle, there can be no straightforward rules in the selection of PV/T collector and/or system for building projects. The decision-making will depend on the geographical location and the actual application case by case. At locations with low levels of solar radiation and ambient temperatures, space heating is almost required all year and PV/T-a can be useful and cost-effective. At locations with high solar input as well as ambient temperature, PV/T-w can be useful for providing year-round water preheating services and on top with intermittent air heat extraction to provide space heating in winter and natural ventilation in summer. There are good opportunities for extension to solar cooling and heat pump integrations. The concentrator type can be used for elevating fluid service temperature from medium to high and therefore is more relevant to industrial applications.

A definite number of performance indicators are now available to evaluate and compare the energy performance of specific PV/T collectors. It should be cautioned that the performance indicators derived for applications in different cities should not be compared directly. As a matter of fact, the energy performance of a given product or system design varies with the local environment, such as climatic conditions. Despite the active research and development going on globally in the past decades, the numbers of commercially available PV/T collectors and systems are still very limited. Major barriers like product reliability and costs are yet to be overcome. Nevertheless, as long as there are international protocols with definite targets set on reductions in carbon emissions, the demand for solar devices that are able to maximize energy output will continue to increase. PV/T products are expected to have high market expansion potential in the near future.

References

[1] Tyagi VV, Kaushik SC, Tygai SK. Advancement in solar photovoltaic/thermal (PV/T) hybrid collector technology. Renewable Sustainable Energy Rev 2012;16:1384−98.
[2] Vera JT, Laukkanen T, Siren K. Performance evaluation and multi-objective optimization of hybrid photovoltaic-thermal collectors. Sol Energy 2014;102:223−33.
[3] Skoplaki E, Palyvos JA. On the temperature dependence of photovoltaic module electrical performance − a review of efficiency/power correlations. Sol Energy 2009;83:614−24.
[4] Gholampour M, Ameri M, Samani MS. Experimental study of performance of photovoltaic-thermal unglazed transpired solar collectors (PV/UTCs): energy, exergy, and electrical-to-thermal rational approaches. Sol Energy 2014;110:636−47.
[5] Agrawal S, Tiwari GN. Enviroeconomic analysis and energy matrices of glazed hybrid photovoltaic thermal module air collector. Sol Energy 2013;92:139−46.
[6] Chow TT. A review on photovoltaic/thermal hybrid solar technology. Appl Energy 2010; 87(2):365−79.
[7] Royne A, Dey CJ, Mills DR. Cooling of photovoltaic cells under concentrated illumination: a critical review. Sol Energy Mater Sol Cells 2005;86:451−83.
[8] Ibrahim A, Othman MY, Ruslan MH, Mat S, Sopian K. Recent advances in flat plate photovoltaic/thermal (PV/T) solar collectors. Renewable Sustainable Energy Rev 2011;15: 352−65.
[9] Fortuin S, Hermann M, Stryli-Hipp G, NItz P, Platzer W. Hybrid PV-thermal collector development: concepts, experiences, results and research needs. Energy Procedia 2014;48: 37−47.
[10] Calise F, d'Accadia MD, Vanoli L. Design and dynamic simulation of a novel solar trigeneration system based on hybrid photovoltaic/thermal collectors (PVT). Energy Convers Manag 2012;60:214−25.
[11] Nilsson J, Hakansson H, Karlsson B. Electrical and thermal characterization of a PV−CPC hybrid. Sol Energy 2007;81:917−28.
[12] Li M, Li GL, Ji X, Yin F, Xu L. The performance analysis of the trough concentrating solar photovoltaic/thermal system. Energy Convers Manag 2011;52:2378−83.
[13] Li G, Pei G, Yuehong S, Zhou X, Ji J. Preliminary study based on building-integrated compound parabolic concentrators (CPC) PV/thermal technology. Energy Procedia 2012;14:343−50.

[14] Tonui JK, Tripanagnostopoulos Y. Air-cooled PV/T solar collectors with low cost performance improvements. Sol Energy 2007;81:498–511.
[15] Florschuetz LW. Extension of the Hottel–Whiller model to the analysis of combined photovoltaic/thermal flat plate collectors. Sol Energy 1979;22:361–6.
[16] Coventry JS, Lovegrove K. Development of an approach to compare the 'value' of electrical and thermal output from a domestic PV/thermal system. Sol Energy 2003;75(1): 63–72.
[17] Raman V, Tiwari GN. A comparison study of energy and exergy performance of a hybrid photovoltaic double-pass and single-pass air collector. Int J Energy Res 2009;33:605–17.
[18] Kumar S, Tiwari GN. Life cycle cost analysis of single slope hybrid (PV/T) active solar still. Appl Energy 2009;86:1995–2004.
[19] Chow T-T, Ji J. Environmental life cycle analysis of hybrid solar photovoltaic/thermal systems for use in Hong Kong. Int J Photoenergy 2012;2012:9, 101968. http://dx.doi.org/10.1155/2012/101968.
[20] Garg HP, Adhikari RS. Transient simulation of conventional hybrid photovoltaic/thermal (PV/T) air heating collectors. Int J Energy Res 1998;22:547–62.
[21] Hegazy AA. Comparative study of the performances of four photovoltaic/thermal solar air collectors. Energy Convers Manag 2000;41:861–81.
[22] Aste N, del Pero C, Leonforte F. Water flat plate PV-thermal collectors: a review. Sol Energy 2014;102:98–115.
[23] Chow TT, He W, Ji J, Chan ALS. Performance evaluation of photovoltaic-thermosyphon system for subtropical climate application. Sol Energy 2007;81(1):123–30.
[24] Chow TT, Pei G, Fong KF, Lin Z, Chan ALS, Ji J. Energy and exergy analysis of photovoltaic-thermal collector with and without glass cover. Appl Energy 2009;86(3): 310–6.
[25] Tripanagnostopoulos Y, Nousia T, Souliotis M, Yianoulis P. Hybrid photovoltaic/thermal solar systems. Sol Energy 2002;72(3):217–34.
[26] Assoa YB, Menezo C, Fraisse G, Yezou R, Brau J. Study of a new concept of photovoltaic-thermal hybrid collector. Sol Energy 2007;81:1132–43.
[27] Ji J, Liu K, Tin-tai C, Pei G, Hanfeng H. Thermal analysis of PV/T evaporator of a solar assisted heat pump. Int J Energy Res 2007;31(5):525–45.
[28] Gang P, Huide F, Jie J, Tin-Tai C, Tao Z. Annual analysis of heat pipe PV/T systems for domestic hot water and electricity production. Energy Convers Manag 2012;56:8–21.
[29] Buker MS, Riffat SB. Building integrated solar thermal collectors – a review. Renewable Sustainable Energy Rev 2015;51:327–46.
[30] Chow T-T, Qiu Z, Li C. Potential application of "see-through" solar cells in ventilated glazing in Hong Kong. Sol Energy Mater Sol Cells 2009;93:230–8.
[31] Chow T-T, Tse K-K, Tse N. Energy performance of hybrid co-generation versus side-by-side solar water heating and photovoltaic for subtropical building application. In: Alternative energy and shale gas encyclopedia, solar 2. Wiley-Blackwell; September 2015, ISBN 978-0-470-89441-5.
[32] Otanicar TP, Phelan PE, Prasher RS, Rosengarten G, Taylor RA. Nanofluid-based direct absorption solar collector. J Renewable Sustainable Energy 2010;2:033102.
[33] Saidur R, Meng TC, Said Z, Hasanuzzaman M, Kamyar A. Evaluation of the effect of naonfluid-based absorbers on direct solar collector. Int J Heat Mass Transfer 2012;55: 5899–907.
[34] Taylor RA, Phelan PE, Otanicar TP, Walker CA, Nguyen M, Trimbie S, et al. Applicability of nanofluids in high flux solar collector. J Renewable Sustainable Energy 2011;3:023104.

[35] Mintsa Do Ango AC, Medale M, Abid C. Optimization of the design of a polymer flat plate solar collector. Sol Energy 2013;87:64–75.
[36] Tsoutsos T, Frantzerskaki N, Gekas V. Environmental impacts from the solar energy technologies. Energy Policy 2005;33:289–96.
[37] Hansen J, Sørensen H, Byström J, Collins M, Karlsson B. Market, modelling, testing and demonstration in the framework of IEA SHC Task 35 on PV/thermal solar systems. In: 22nd European photovoltaic solar energy conference and exhibition, DE2–5, Milan, Italy; September 3–7; 2007.

Future trends for solar energy use in nearly zero energy buildings

20

M. Moldovan, I. Visa, A. Duta
Transilvania University of Brasov, Brasov, Romania

20.1 Renewables in the built environment

20.1.1 Strategies

Energy is considered one of the main problems of humankind (Smalley, 2005). The energy demand is directly related to the world population increase and the level of development.

The world population growth can be divided into three periods of distinct trends: a very long age with very slow population growth until the 19th century, when the world population reached its first billion persons (Maddison, 2007); a second period, with rising standards of living allowing a faster increasing rate of growth, lasting until the 1960s when, according to the US Census Bureau International Data Base updated in July 2015 (USCB, 2015), a peak rate of 2.2% yearly growth was reached and the world population increased to 3.1 billion, thus, growing by three times in only one century; and a third period, with a still increasing world population (with 1 billion more each 12–15 years between 1962 and 2015) but with a slight decrease in the growth rate, which is estimated to lead to 9 billion by 2042 (Fig. 20.1).

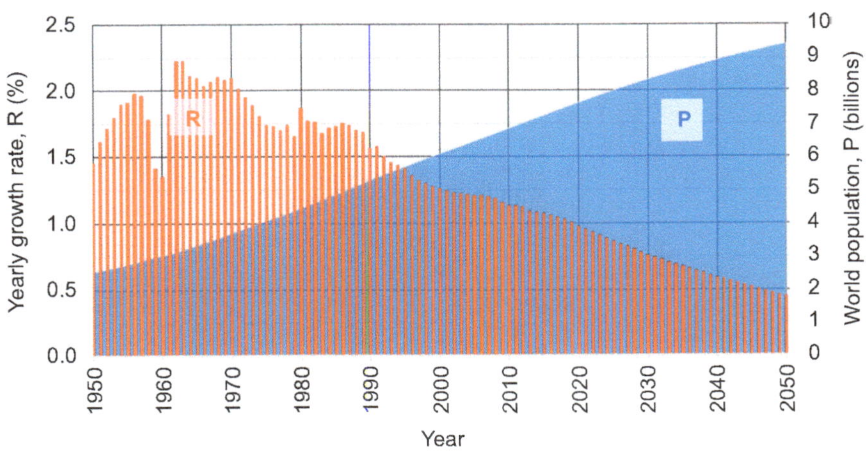

Figure 20.1 The yearly world population growth rate (R) and the world population (P) evolution from 1950 to 2050.

Advances in Solar Heating and Cooling. http://dx.doi.org/10.1016/B978-0-08-100301-5.00020-5

This increase in population at the global level in the 20th century raises the problem of growth limits, bearing in mind that the earth's resources are finite. In 1972 a research team at Massachusetts Institute of Technology forecast that if the growth trends in world population, industrialization, pollution, food production, and resource depletion continue unchanged, the limits to growth on the earth will be reached sometime within the next 100 years (Meadows et al., 1972). This conclusion was validated by Turner (2008), who found that after 30 years the changes are in line with the above "business as usual" scenario. In 1992 the same team argued that in many areas the capacity of the earth will be overpassed (Nørgård et al., 2010; Meadows et al., 1992). A peak followed by a decline is forecast in the next 40 years for the world population and resources (food, water, energy) if completely different long-term economic and environmental policies are not applied to developing highly resource-efficient and environmentally friendly advanced technologies, to achieve future economic growth (Meadow et al., 2004; Randers, 2012).

In addition to resource depletion, anthropogenic activity is responsible for climate changes. The exponential population growth over the past century increased emission levels at higher rates, the CO_2 emissions in this interval surpassing the emissions in all previous historical periods. From 1995, when the first Conference of the Parties of the United Nations Framework Convention on Climate Change (UNFCCC) took place, to 2012, greenhouse gas (GHG) emissions rose by more than 25% and the atmospheric concentration of these gases increased to 435 ppm carbon dioxide equivalents (EEA, 2015). Also the global distribution of the GHG emissions shifted: emissions originated almost exclusively from the United States and Europe at the beginning of the 20th century, but today these account for less than 30%, with China and other developing countries taking the lead (IEA, 2014). Efforts are being made to decouple GHG emissions from economic growth, through energy efficiency measures and the use of carbon-free sources of energy. Energy production and use account for two-thirds of the world's GHG emissions; for the other third agriculture and land use change (including deforestation) are mainly responsible (IEA, 2015).

Global, national, regional, and local strategies have been implemented to tackle these problems. At the global level, at the Rio Earth Summit in 1992, 195 countries adopted the UNFCCC, aiming at fighting global warming by reducing GHG atmospheric concentrations to "a level that would prevent dangerous anthropogenic interference with the climate system" (UNFCCC, 1992). The first reduction in GHG emissions targeted an average of 5%, having the 1990 level as reference, and represents the core statement of the Kyoto Protocol; the protocol was ratified by 37 industrialized countries and 15 member states of the European Union (EU) for the first commitment period (2008–12) (UNFCCC, 1997). During the second commitment period, the parties committed to reducing GHG emissions by at least 18% (vs the 1990 level) in the 2013–20 8-year period (UNFCCC, 2012).

In September of 2015, during the UN Sustainable Development Summit in New York, the post-2015 development agenda, "Transforming Our World: The 2030

Agenda for Sustainable Development," was adopted. Among the 17 sustainable development goals, the seventh aims at (UNGA, 2015):

- Ensuring, by 2030, universal access to affordable, reliable, sustainable, and modern energy for all; increasing substantially the share of renewable energy in the global energy mix; doubling the global rate of improvement in energy efficiency;
- Enhancing international cooperation to facilitate access to clean energy research and technology, including renewable energy, energy efficiency, and advanced and cleaner fossil fuel technology, and promoting investment in energy infrastructure and clean energy technology;
- Expanding infrastructure and upgrading technology for supplying modern and sustainable energy services for all in developing countries, in particular the least developed countries, small-island developing states, and landlocked developing countries, in accordance with their respective programs of support.

To secure the chance to limit global warming (below 2°C), the international community has decided to work toward an international climate agreement for the period following 2020, which should be applicable to all. The negotiation on this new global agreement was expected to be concluded at the UN Conference on Climate Change, COP21, in December 2015 in Paris. As of this writing, 123 parties of the UNFCCC had submitted their Intended Nationally Determined Contributions as follows:

- The United States pledged to cut net GHG emissions by 26–28% by 2025 (relative to 2005 levels);
- The European Union and its member states committed to a binding target of at least 40% domestic reduction in GHG emissions by 2030 (relative to 1990 levels);
- Russia pledged to limit anthropogenic GHGs to 70–75% of 1990 levels by the year 2030 as a long-term indicator;
- China aimed at achieving its peak of carbon dioxide emissions around 2030, to lower carbon dioxide emissions per unit of GDP by 60–65% (having as reference the 2005 level), to increase the share of non-fossil fuels in primary energy consumption to around 20%, and to increase the forest stock volume by around 4.5 billion cubic meters from the 2005 level;
- India, recognizing the climate change threat, declared a voluntary goal of reducing the emissions intensity of its GDP by 33–35% by 2030 (vs the 2005 level), achieving about 40% cumulative electric power installed capacity from non-fossil fuel–based energy resources, and creating an additional carbon sink of 2.5–3 billion tons of CO_2 equivalent through additional forest and tree cover.

The main contributors to the global energy demand are the built environment, economic activities (industry, agriculture, services), and transportation, with different shares depending on the regional development. Many studies are devoted to this aspect, with shares of around 40% of energy demand being presented for the building sector, exceeding the industrial and transportation sectors in developed countries (Abdellatif and Al-Shamma'a, 2015; Harvey, 2013; Nejat et al., 2015).

20.1.2 Net zero energy buildings

An ambitious target was formulated through the concept of net zero energy buildings (NZEBs) to mitigate, toward zero, GHG emissions and thus the energy demand from fossil fuels in the buildings sector. The literature mentions several definitions for NZEB, describing buildings that produce enough energy from renewable sources to fully meet their own yearly energy requirements (Marszal et al., 2011; Sartori et al., 2012). To reach this status, cost-effective energy efficiency measures should be implemented first and only afterward renewable energy systems should be installed, at the building level or nearby, to meet the energy needs (Deng et al., 2014).

According to the Intergovernmental Panel on Climate Change's Synthesis Report, the NZEB standard alone is not enough to limit global warming to 2°C; the carbon dioxide emissions from the energy sector need to continuously decrease, down to zero, between 2040 and 2070; afterward, complex and complementary measures are required to further decrease the already existent CO_2, by carbon-sequestration processes, reforestation, etc. (IPCC, 2014).

The main advantages of NZEBs at the beneficiary level is the low cost of utilities, whereas the strategic advantage is obviously related to the much lower environmental impact. Additionally, when implemented at the community level, the NZEB concept improves energy security and resilience against power outages and natural disasters (USDOE, 2015).

However, the initial investment is high, making this concept less economically affordable for many current building owners/users, all over the world. This has limited the spread of the NZEB concept being currently reported for pilot and demonstration cases (Lu et al., 2015). Therefore, transition concepts are being promoted, including by legal instruments; the nearly zero energy building (nZEB) represents such a transition concept that allows progress toward lowering GHG emissions and considers affordability, based on the current economic status of various countries worldwide.

20.1.3 Nearly zero energy buildings

In the EU, the large share of energy use in the built environment (almost 40% of the total energy demand) led to a new standard: nZEB design mandatory by 2021 for all new buildings or buildings undergoing major renovation; this standard becomes legally binding earlier, starting with 2019 for public buildings (EP, 2010). Each member state will formulate targets for the building energy demand (energy efficiency) and for the energy share covered by renewables implemented on or in close proximity to the building, under affordable and feasible conditions. To date, most member states have formulated these targets for the building energy demand (20–270 kWh/m^2/year, depending on the climatic zone and building destination) and for the renewable energy share (15–56%) (BPIE, 2015).

Site-specific methodologies are used for calculating the building's energy performance, considering the energy need and the energy provided from renewable sources. Because the nZEB standard refers to new buildings and buildings undergoing major renovation, one must consider that the stock renewal rate is only 1.94% in the EU (Uihlein and Eder, 2010) and the majority of existing buildings will last over the next 50 years. Therefore, both new and existing buildings must be addressed, the last category being a priority of the EU Horizon 2020 Research & Innovation Programme.

From a life-cycle perspective, energy efficiency measures must be taken first to reduce the high proportion (80–90%) of the building's energy demand during the utilization period (Ramesh et al., 2010), with highly different yearly values among the EU member states (depending on the climatic zone and building codes), ranging from 70 to 300 kWh/m^2 for residential buildings and from 200 to 500 kWh/m^2 for commercial buildings (ENTRANZE, 2014). Through integrated energy efficiency measures, up to 90% of the energy demand can be mitigated for new buildings and up to 50% for those undergoing major renovation (Broin et al., 2013; Urge-Vorsatz et al., 2013).

The most used energy efficiency measures consist of passive solar design principles, improvements in the insulation properties of the building, energy-saving measures, and use of energy-efficient equipment (Boyano et al., 2013; Tuominen et al., 2012). Attention must be paid to a problem caused by the use of a passive solar design principle, the orientation of the building with large, glazed facades toward the south; although efficient in reducing energy consumption for lighting, this positioning (if not optimized) can support thermal discomfort and increased cooling demand during sunny days in areas adjacent to large glazed facades, with temperature differences of up to 10°C between areas of a room (Hens, 2009; Moldovan et al., 2014b; Salvalai et al., 2013). Common solutions are to use shading systems placed outside, into, or inside these facades (Atzeri et al., 2014; Cho et al., 2014); electrochromic glassed surfaces (Kim and Todorovic, 2013); double-ventilated facades (Barbosa and Ip, 2014); natural ventilation (Campaniço et al., 2014); reversible heating/cooling systems (Arteconi et al., 2014); etc.; each of these has advantages and major limitations for existing buildings, as installing such systems can be difficult or impossible. These phenomena can be seasonal and require additional automation for control (Arteconi et al., 2014). Thus, in addition to thermal discomfort, unnecessary energy may be consumed if the building does not have a smart, selective heating/cooling system.

The second requirement of the nZEB standard refers to the use of renewable energy systems. The main barrier is the significant initial investment costs in heating/cooling systems, as in many climatic areas there is a need for backup systems based on conventional energy sources to compensate the diurnal and seasonal variability of the renewable energy sources (Azadian and Radzi, 2013; Lamnatou et al., 2015; Lang et al., 2015; Moldovan et al., 2014a). The next barrier is related to the operation of disparate renewable energy systems, owing to their sequential implementation following the technological development, the costs, and the support measures.

Thus, the optimal path (in the transition toward fully green energy) involves energy mixes consisting of conventional and renewable energy systems and their feasible operation (Visa et al., 2014b). The most implemented are the solar–thermal systems able to provide domestic hot water and contribute to covering the space heating and cooling demands. Their advantages are related to the rather high conversion efficiency (up to 85%), involving low amounts of energy during functioning, along with available aesthetic solutions for architectural acceptance (Munari Probst and Roecker, 2007; Visa et al., 2014a). Heat pumps are good candidates for renewable energy mixes; the use of one single piece of equipment with a high coefficient of performance (COP) is considered for heating (COP > 4), cooling (COP > 3), or both (COP > 7), for space or domestic hot water heating based on the heat recovered from the chilled space or from the exhausted air (Ellerbrok, 2014; Junghans, 2015). The main limitations to heat pump implementation are related to electricity consumption (mainly to drive the compressor), which, if not obtained from renewable sources, can pose a problem in reaching the nZEB status. To power the heat pump, photovoltaic systems represent a viable alternative as these can be integrated in all building elements (roof, facade, and glazing), with a multifunctional role: providing electricity, increasing the thermal resistance/insulation, shading to reduce excess solar energy, and improving aesthetics (Jelle and Breivik, 2012). The limitations to implementing photovoltaic systems are mainly linked to the rather large area required for implementation; alternatives to increase the photovoltaic output include: (1) the use of tracking systems that can increase yield energy up to 30% compared to fixed systems (Moldovan et al., 2011) and (2) the implementation of modules with large efficiencies in the working environment.

To conclude, the first steps in nZEB development should focus on decreasing the specific energy demand. Only in energy-efficient buildings can the renewables implementation meet the affordability requirements; even so, the initial investment costs to implement renewables depends on the climatic profile in the implementation location and on the actual market price of the equipment.

For an office building the correlation between the initial specific cost for two Renewable Energy Sources (RES) mixes and the yearly specific energy demand is plotted in

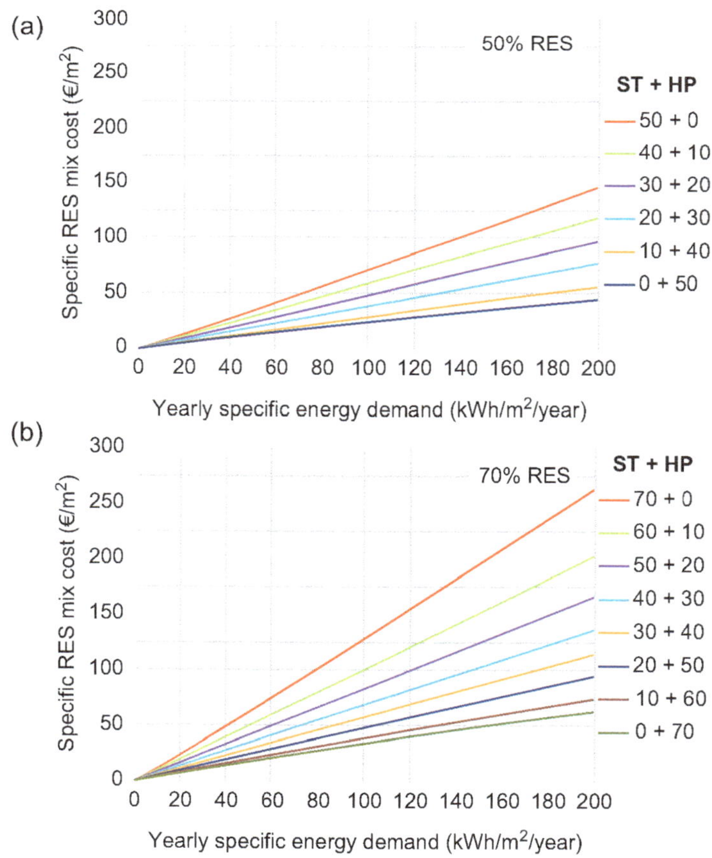

Figure 20.2 Correlation of the specific cost of a solar–thermal plus heat pump (ST + HP) energy mix with a yearly specific energy demand of: (a) 50% RES share and (b) 70% RES share covering the thermal energy demand. *RES*, Renewable Energy Systems.

Fig. 20.2. The energy mixes are designed to cover 50% and 70%, respectively, of the total thermal energy demand and consist of a solar–thermal (ST) system and a heat pump (HP) system, with various iterative shares of the ST system (50/70–0%, in 10% steps) and of the HP system (0–50/70%). The HP yearly electrical energy need is provided by a grid-connected photovoltaic system and its specific cost is included in the calculation. The lowest specific cost corresponds to the systems based on HP only, when considering a RES share of 50% and of 70%, owing to the lower specific energy demand for domestic hot water (DHW) compared to space heating. This represents a specific feature of office buildings, having a smaller DHW need (5 L/person/day) compared to residential buildings (90 L/person/day). A specific RES cost under €50/m^2 can be considered affordable for new buildings, considering the average office building specific investment of €500–750/m^2. This threshold can be respected if the specific energy demand is lower than 200 kWh/m^2/year, for energy mixes aiming at covering the thermal energy need using 50% of RES (Fig. 20.2a), and is lower than 140 kWh/m^2/year for a RES percentage of 70% (Fig. 20.2b).

20.2 Solar energy potential for thermal energy production in the built environment

20.2.1 General features

The solar energy potential is the main input data when designing a solar energy conversion system; thus it must be accurately evaluated. In addition to the daily and seasonal variations, directly linked to the geographical coordinates (latitude and longitude) and the on-site meteorological conditions, the available implementation area of the solar convertors is important. When evaluating this area, the building's roof and facades should be considered first, followed by the nearby ground areas. The built environment brings several constraints on the implementation of solar convertors (shadowing, wind shielding, competition with food production, etc.), which should be assessed to avoid the under- or overestimation of the renewable energy potential, with negative consequences on the final energy output and costs.

The best suited surfaces to install solar convertors in nZEBs are the roofs and rooftops. Among the roofs, those with optimal tilt and azimuth angles (south orientation) have the maximum output for fixed (and tilted) convertors. For rooftop mounting, an additional structure is required to optimally install the ST collectors or photovoltaic modules.

Solar tracking systems can be implemented to enhance the energy output; solar energy convertors installed on horizontal surfaces (rooftops, ground) can use full tracking; additionally, limited tracking (with limited strokes) can be used for convertors mounted on inclined roofs and vertical facades. The geometrical shape of the available surface imposes spatial restrictions: for nonrectangular shapes on buildings, the installed surface of a conventional solar convertor (with square or rectangle geometry) will be lower than the available surface. Therefore, for building integration, nonconventional shapes are proposed (triangle, trapezoid, and hexagonal) and are already on the market for photovoltaics; these allow a better coverage, also improving the acceptance (Visa et al., 2014a). Further restrictions are formulated for horizontal surfaces (rooftops or ground-mounted systems), imposed by the ratio between the available surface's lengths in the east—west and the north—south directions. The worst case scenario is represented by the situation in which the length of the available surface in the north—south direction is higher than its length in the east—west direction; in this case, a distance between two consecutive solar convertors is necessary to avoid reciprocal shadowing; thus an increased surface is required to install a given capacity. In contrast, the best case scenario occurs when the length of the available surface in the east—west direction is much higher than the north—south length, supporting the maximal use of the available surface.

20.2.2 Case study

An example of a renewable-based energy mix implemented in low-energy buildings (LEBs) to reach the nZEB status is further presented at the R&D Institute of Transylvania University in Brasov, Romania. This small community is located on the outskirts of Brasov City, Romania. The Institute is the result of a structural foundation project (R&D Institute for High-Tech Products, ID123, Contract No. 11/2009) with significant financial and logistic support from the university. The Institute was "in-house" designed by university teams (civil engineering, installations, renewables, IT, etc.) as a sustainable community. The R&D Institute has 11 new LEBs, which accommodate 27 research centers (15 with engineering R&D topics and 12 on informatics, humanities, health, music, sports, and law); the research community has its own water source and fresh- and wastewater treatment facilities (Fig. 20.3) and is connected to the power and gas public utilities grids. All the buildings are identical (for cost-efficiency reasons), incorporating state-of-the-art materials and systems. Through research and interactor cooperation, the R&D Institute aims at developing concrete, energy-efficient, and sustainable products and processes.

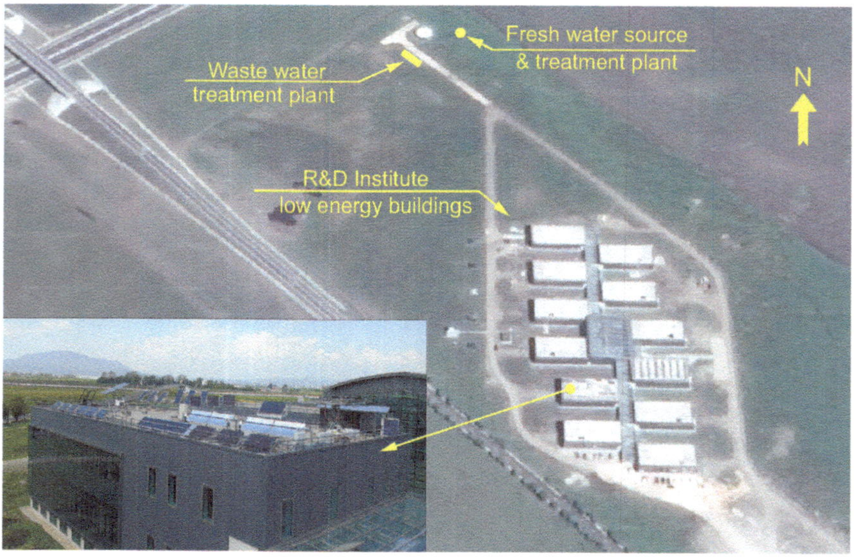

Figure 20.3 The R&D Institute of Transilvania University of Brasov, Romania.

Table 20.1 Energy consumption of each low-energy building of the community

Thermal energy in a building		Electrical energy in a building	
Appliances	Amount [kWh/year]	Appliances	Amount [kWh/year]
Domestic hot water	4244	Lighting and power appliances	13,224
Heating	78,027	Driving the circulating pumps and the heating and cooling systems	2322
Cooling	7759		
Total	*90,030*	*Total*	*15,546*
Total energy demand for one building			**105,576**

The total energy demand consists of the thermal and electrical energy demand of the 11 buildings and the electrical energy demand for external lighting and the fresh- and wastewater treatment facilities.

The steps followed were:

Step 1. The energy demand of each building was calculated on a monthly basis (Moldovan et al., 2014b), resulting in the data presented in Table 20.1.

Considering the yearly values, the total thermal energy demand of the entire community is 990,330 kWh/year and the total amount of electric energy is 219,022 kWh/year (thermal energy and electric energy for common appliances, for the water stations, etc.).

Step 2. Assessing the implementation area was done for each LEB and it consisted of 450 m^2 on the rooftop, 300 m^2 on the southern facade, 150 m^2 on the eastern or western facades, and 450 m^2 at the ground level (between buildings). In the close vicinity of the buildings, at the ground level, 7500 m^2 was available for geothermal heat exchangers and 25,000 m^2 for a photovoltaic park.

Step 3. Implementing the RES energy mix. The energy demand of the R&D Institute was partly covered by already implemented RES (Fig. 20.4):

The installed RES are:

- Ten ST systems for DHW: seven with flat plate ST collectors (FPSTCs) as 3.5-kW systems, two with evacuated tube ST collectors (ETCs) as 3.3-kW systems, and one with concentrated trough ST collectors as a 6-kW system
- Two geothermal HPs (22 kW) for heating and cooling two of the LEBs: one with a 1000-m^2 horizontal ground heat exchanger (HGHX) and another one with four vertical ground heat exchangers (VGHXs, at a depth of 90 m)
- Seven grid-connected photovoltaic systems with a total installed power of 27 kWp, for electrical energy production: a 12-kWp photovoltaic array of strings with a monoaxial tracking system installed on the rooftop of building L11, a total of 5 kWp from photovoltaic systems with mono- and biaxial solar tracking on the rooftop of building L7, and five 2-kWp photovoltaic platforms with mono- and biaxial solar tracking on the ground

Figure 20.4 The implemented RES in the R&D Institute of Transilvania University of Brasov and available surfaces for further RES development. *RES*, Renewable Energy Systems.

- Two wind turbine systems: one composed of three small (300-W) wind turbines installed on the rooftop of building L5 and the second composed of three small (600-W) wind turbines installed on the rooftop of building L6.

Further on, a photovoltaic (PV) system of 1.5 MWp and a system of five HGHXs and four VGHXs will be installed, aimed at ensuring a RES percentage above 90% for the community. Additional sustainability actions are planned for the midterm future, including small electric vehicles, powered by stationary stations that are fed by the PV system and by fuel cells.

As part of this community, one of the LEBs (laboratory building L7) has installed renewables to cover the entire yearly energy demand, for both thermal and electrical energy (Visa and Duta, 2015). However, seasonal peaks for heating during very cold periods with temperatures below −20°C (a maximum of 12 days in January and February), when the HP cannot perform at the rated efficiency, require backup sources (like natural gas).

20.3 New trends in increasing the use of solar energy conversion systems integrated in nearly zero energy buildings

20.3.1 Needs and trends

The worldwide effort toward the sustainable built environment, well supported by various financial and incentive schemes and by a legal frame with concrete milestones, will lead to efficient and affordable results only if implemented using concrete actions.

Solar energy conversion systems (mainly PV and ST systems) are key players in these actions, because solar radiation satisfies several major prerequisites:

- It is virtually inexhaustible, thus scarcity does not represent an issue.
- It is accessible all over the globe, therefore security aspects are not a real problem.
- Its conversion has no negative environmental side effects.
- There already are mature, market-accepted technologies and, in many countries, the future is expected to support economic development and limit unemployment.

Despite these features representing the pillars of sustainable development, solar energy conversion systems are far from being implemented at their true potential; the core reason is represented by the cost of the output energy, which is still not competitive compared to the energy produced using fossil fuels or other renewable sources (eg, wind). However, implementing nZEBs will require fast development by increasing the efficiency and the output, as presented in Fig. 20.5.

The input solar radiation represents a specific feature of the implementation location that has to be well characterized as yearly, seasonal, and peak (high/low) values. Accurate local data are preferable to modeled (simulated) data, particularly for extreme conditions (in terms of values and frequency).

Increasing the efficiency of the solar energy conversion systems is a general requirement, regardless of the application and should mainly focus on the convertors (PV modules and ST collectors), as these represent a significant share of the systems' cost. The financial aspects are important for market penetration, thus increasing the efficiency has to be done using upscalable, cost-effective solutions. The conversion system and the energy costs will obviously influence the payback time and the rate of return.

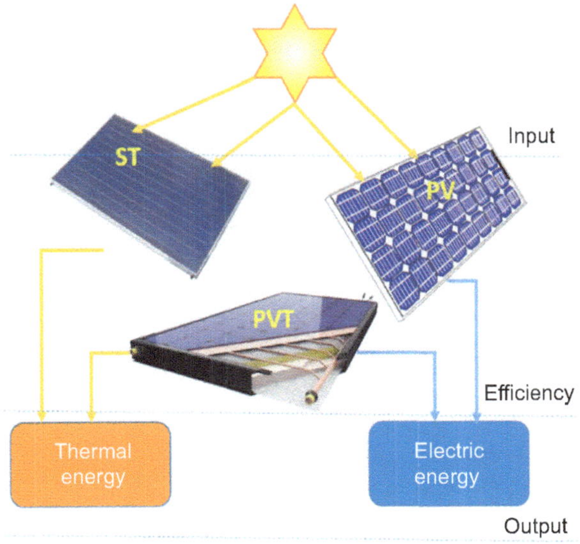

Figure 20.5 Input, efficiency, and output of solar energy conversion systems. *ST*, solar–thermal; *PV*, photovoltaic; *PVT*, photovoltaic–thermal.

For the PV modules, these financial indicators rank the wafer silicon modules on top, followed by the thin-film modules of types II—IV (Ismail et al., 2015); newcomers are already reported, such as the copper indium gallium selenide (CIGS), Cu2ZnSnS4 (CZTS), and other perovskite modules, that can be expected to reach higher infield efficiencies, by using low-cost and abundant raw materials (Abermann, 2013).

For ST collectors the current progress is mainly related to large concentrating systems, able to deliver steam for thermal and electric energy cogeneration (Zhao et al., 2015). Although efficient, these large systems are not well suited to the built environment, for which FPSTCs or ETCs are recommended. For these, commercial collectors reach efficiencies of 60—80% at acceptable costs (Kalogirou, 2004).

The combination of PV and FPSTC has started to be investigated for cogeneration convertors (photovoltaic—thermal, or PVT), particularly suited to the built environment, where the available implementation area is limited. The energy cost in regions with a large solar energy potential was reported for amorphous silicon building-integrated PVTs as being close to the grid-delivered power (Buker and Riffat, 2015).

Increasing the output of any solar energy conversion system at the end user is linked to the input radiation, the theoretical maximum efficiency, and the losses during conversion and energy storage/delivery. Whereas the last factors rely directly on the system's quality, the input solar energy can be maximized by positioning the convertors to fully collect the available solar radiation; thus the modules should always be able to collect the solar rays normal to their flat surface. As solar radiation varies throughout the year and during the day, the condition for maximizing the input radiation is fulfilled by continuously moving the modules, using tracking systems (Sharaf Eldin et al., 2016; Neagoe et al., 2014). The continuous motion is less cost-effective, therefore stepwise tracking algorithms are used and aim at meeting their main function (high tracking efficiency, usually collecting more than 94% of the available solar radiation) with low energy consumption and easy maintenance. In this respect tracking mechatronic systems are best at fulfilling the prerequisites, particularly for regions with large and sudden variations in solar energy input.

As already discussed, to implement solar energy convertors in the built environment (on or near the nZEB), additional limitations have to be considered, related to:

- *Available implementation area.* The solar energy conversion modules need to be oriented toward the south (with acceptable deviation of about ±10—15 degrees, which will decrease the input energy) and tilted at an optimal angle (usually 0.9 from the latitude angle), if not tracked. This leaves as suitable only part of the roof, the top terrace, part of the facade, and the architectural objects included (as balconies, shades, etc.). Because of this limitation, the convertors should have high conversion efficiencies (Visa et al., 2015b).
- *Architectural and aesthetic constraints.* The most effective PV, FPSTC, ETC, or PVT modules are dark colored (black or dark blue) because the best solar absorbers for ST conversion have to resemble the ideal blackbody and because the antireflective layers (eg, nanostructured TiO_2) are usually perceived as dark blue. However, these colors, exposed on large and visible surfaces, are considered rather unacceptable by most architects (Munari Probst and Roecker, 2007). Differently shaped and colored or "transparent" PVs are already reported on the market with efficiencies slightly lower than those of the conventional ones

Figure 20.6 The "PV Twin" tracked platform at Transilvania University of Brasov.

and with significantly higher cost, and this is a bottleneck that is the subject of intensive research (Bednar et al., 2015). For ST collectors research is ongoing to obtain various true colors of the absorber plate or superposed multiple reflections of the glazing, while preserving the efficiency (Mertin et al., 2014). An alternative path could be the implementation of the solar energy conversion systems in architectural objects with special design, well matching the implementation location and giving an identity to it; an example is the "PV Twin" tracked platform implemented at Transylvania University in Brasov (Fig. 20.6), which integrates German and Romanian architectural symbols in a system that functions as an outdoor PV testing laboratory and represents a tangible declaration on sustainability.

- *Working conditions.* The built environment is usually characterized by a higher atmospheric pollution level, with nano- or microsized suspended matter and/or aerosols combining oils and water that can stick to the glazed surface of the solar modules, thus reducing their transmittance and limiting the actual amount of solar radiation that reaches the photo-conversion assembly. This is why multifunctional glazing is under investigation, targeting high transmittance and self-cleaning properties (Midtdal and Jelle, 2013).

These general trends are differently approached and solved for the PV and ST systems in the built environment, and details on building integrated ST systems (BISTs) are discussed further on.

20.3.2 New trends in increasing the use of solar–thermal systems integrated in nearly zero energy buildings

A building uses thermal energy for heating, cooling, and DHW. During one year, the DHW need is almost constant, whereas the heating and cooling demands have a seasonal variation, depending on the implementation location (geographical coordinates and climatic profile) and on the building type, mainly its energy efficiency (directly correlated with the energy losses and thus to the insulation all around the building envelope). New buildings have a better energy performance (as a result of awareness and/or the legal constraints) but most buildings (built before 1980) need well-designed refurbishment.

ST systems can contribute, alone or in energy mixes, to fulfilling all the components in the thermal energy demand of an nZEB (Table 20.2). In addition to BISTs, presented in Table 20.2, other nonconventional solutions converting solar energy into heat are implemented, as massive ST collectors, using concrete, bricks, or gravel as heat storage systems (D'Antoni and Saro, 2012) or colored mild steel sheets integrated in the facade (Anderson et al., 2010; Bonhote et al., 2009). Each of these solutions has its advantages and drawbacks, being suitable for specific climatic profiles and, most of them, are still in the demonstration phase.

Considering the current situation and the real need for fast, efficient, and affordable solutions, a European network was developed under the COST Action TU1205 aiming at extending BISTs, through joint research, innovation, and technology transfer. An inventory of the worldwide situation outlines the current trends and focuses on the likelihood and challenges of BIST implementation (COST Action, 2015). Among these, promising developments are outlined for ST facades and for employing FPSTC and ETC tracking.

Table 20.2 Solar–thermal convertors in building-integrated solar–thermal systems

Type of collector	Thermal energy component	Functional options	Implementation options
Flat plate collector with liquid thermal fluid	Domestic hot water Space heating and cooling	Stand-alone system for a building or for a small group of buildings Part of an energy mix (eg, with heat pumps, biomass burners, etc.)	Roofs, terraces Facades, balconies Platforms near the building
Flat collector with air	Air heating	Stand-alone system for a building	Roofs, terraces Facades
Evacuated tube collector	Domestic hot water Space heating and cooling	Stand-alone system for a building or for a small group of buildings Part of an energy mix (eg, with heat pumps, biomass burners, etc.)	Roofs, terraces Facades, balconies, Staircases Platforms near the building
Small parabolic (dish, trough) collector	Domestic hot water Space heating and cooling	Stand-alone system for a building or for a small group of buildings	Terraces Platforms near the building

20.3.3 Case study 1: novel flat plate ST collector for solar–thermal facades

The novel concept of active ST facades (ASTFs) defines ST collectors (FPSTC, ETC, or PVT) integrated in the vertical envelope of a building, being able to produce thermal energy and to support additional features (insulation, shielding). The concept is still under development; therefore reports on roofs and terraces can also be found under this name; however, the most challenging issues related to ASTFs are raised by the highly visible parts of the buildings, and these should represent the focus (Zhang et al., 2015).

In addition to efficiency and cost-effectiveness, the aesthetic issues are very important in the ASTF; thus new forms and/or colors of the FPSTC are expected to increase the architects' and end users' acceptance.

Starting from this prerequisite, a novel ST collector was developed (Visa et al., 2015a). The design specifications include (Patent Proposal A/00156/18.02.2013) the following:

- The building block is an FPSTC.
- The FPSTC efficiency should be comparable with that of quality commercial collectors.
- The FPSTC must have nontraditional shape (other than rectangle or square).
- The FPSTC should be manufactured in a variety of colors: traditional (black or dark blue) and nontraditional.
- The FPSTC should have increased durability under working conditions (urban areas).
- The FPSTC manufacturing should employ market-competitive processes.
- The FPSTC has to act as a building block in developing ST arrays, with various geometries and surface areas.
- The FPSTC and the array must have light weight.
- The FPSTC should allow serial or parallel connection, in arrays.
- The FPSTC and the arrays should be connected to the current thermal installations already existent in the building.

Following these specification, a first concept was drafted: an FPSTC with an isosceles trapezoid shape and an overall area of 0.63 m^2. The shape was chosen to allow further development of various other geometries, by including new and traditional (rectangle) FPSTCs and/or PV and PVT modules (Visa et al., 2014a).

The optimized demonstrator has serpentine tubing and reaches a nominal efficiency of 65% (on the indoor testing rig, according to the International Organization of Standardization ISO 9806:2013 standard) and a maximum of 62% conversion efficiency under outdoor conditions (Visa and Duta, 2015).

Based on this optimized demonstrator, a prototype was developed, having the same dimensions but improved manufacturability and significantly lower weight. The nominal efficiency in standard indoor testing of this prototype is 68%, whereas using solar antifreeze causes the efficiency to slightly decrease (as expected) to 65.4% (Fig. 20.7).

Colored absorber plates were developed according to the design specifications. Various deposition techniques are reported for oxide, cermet, or metal nitride thin layers on the absorber plate or glazing, with colors ranging from red to green and violet (Schuller et al., 2005; Etherden et al., 2004). Most of these techniques are energy intensive, involving vacuum or rather high temperatures (Mertin et al., 2014). One approach

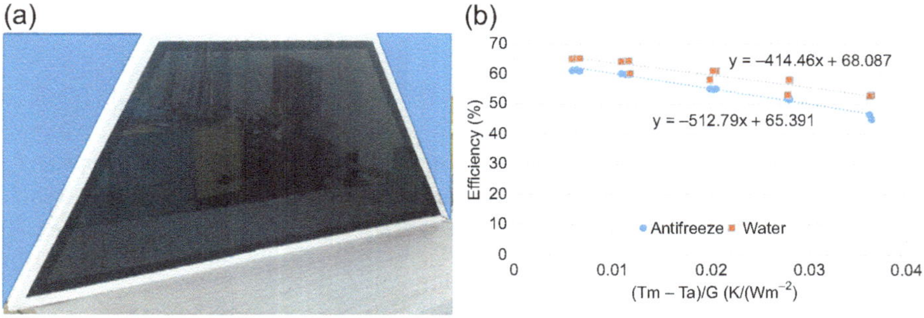

Figure 20.7 (a) The prototype of the trapezoid flat plate solar–thermal collector and (b) the characteristic curve.

relies on a low-cost, industrial technique, spray pyrolysis deposition, which allows one to obtain red absorber plates with spectral selectivity above 9, thus being highly efficient (Duta et al., 2014). In a further effort to increase the environmentally friendly features and the manufacturability, another deposition process is under development; it is based on sol-gel prepared powders dispersed as inks and further deposited by cold spraying. Red solar coatings with the chemical structure $Al/Al_2O_3/Fe_2O_3/Au(np)/TiO_2$ were deposited on large surfaces with a promising spectral selectivity (>5).

Additionally, multifunctional glazing, with very high transmittance (93%) and low reflectance (<10%), was obtained by depositing successive layers of SiO_2/TiO_2 on solar glass; these very thin, cold-sprayed multilayered films have superhydrophilic and photocatalytic properties, allowing self-cleaning of the glazing.

Two novel FPSTC prototypes have been assembled as an array and implemented outdoors, on the south facade of a testing building, in the R&D Institute of Transylvania University.

20.3.4 Case study 2: adaptive tracking to increase the output and durability of flat plate ST collectors

Solar tracking always increases the solar radiation input incident on the convertor. According to the functioning principle, tracking has to be highly accurate (deviation of less than 0.15 degree) for concentrating systems such as parabolic trough and dish ST collectors, concentrating PVs, and heliostat mirrors; for flat convertors (PV, FPSTC, ETC, or PVT), an average tracking accuracy is acceptable, thus the tracking motions can be done at time intervals of minutes (or hours).

The ST systems raise particular challenges: while the PV production is always used (injected into the grid or stored in rechargeable batteries), the thermal energy is required in large amounts in a building only during certain periods of the day (as DHW) or the year (eg, for heating during the cold seasons). Once the maximum heat storage capacity in the boiler is reached, the ST system enters the stagnation regime. During stagnation, the collector can be exposed to overheating, when very high temperatures are reached on the absorber plate (thus promoting cracks and, further on, accelerated corrosion); the casing, glazing, and polymeric sealing are also affected and destroyed much faster than expected. To increase durability by avoiding the overheating consequences, various solutions have been proposed: shades to cover the modules, drain-back systems, dumping the excess heat into the environment, or reversing the circuit during the night for cooling (Quiles et al., 2014). All these solutions have as a main drawback the rather long length of time required for reentering the system into function and/or the additional automation.

A novel alternative is represented by an adaptive tracking algorithm (Neagoe et al., 2014) that consists of four different programs, describing the most common working situations:

1. High thermal energy demand, supplied by increasing the output through direct (traditional) tracking
2. Stand-still position, during cloudy days (large amount of diffuse radiation), when the maximal output corresponds to the FPSTC horizontally positioned, facing the sky dome; the same position is highly recommendable during the night, as involving the lowest stress during high winds
3. Uneven thermal energy demand during the day, with periods of low/zero need alternating with high demand; this scenario can be well met through limited reversed tracking, when the collector is diverted by a few degrees from the incident radiation and can be quickly repositioned when the demand returns
4. Holiday regime, with no thermal energy consumption at all, when two extreme positions of the collector are recommended: extreme west from sunrise to noon and extreme east from noon to sunset.

These programs were investigated outdoors on the testing rig shown in Fig. 20.8 and proven to be efficient not only at avoiding overheating (thus supporting increased durability) but also at more evenly delivering the output energy, thus increasing the overall performance of the entire system.

20.3.5 Standardization

The durability tests of the ST collectors are standardized; steps toward a common international standard testing were made in a collaborative frame involving IEA-SHC TASK 43, Solar Keymark Network, SRCC, ISO/TC180, and CEN/TC312 targeting air heating, concentrated and PVT collectors, fixed or tracked (Carlsson et al., 2004). The standards regulate the indoor and outdoor testing conditions under various accelerated aging conditions (high-temperature resistance test, external thermal shock test, internal thermal shock test for liquid-heating collectors, rain penetration test, freezing test, impact resistance test, etc.).

However, one important aging factor is not standardized yet, the saline aerosols. Many FPSTCs are installed on buildings at seaside resorts, where the saline aerosols are common and permanent. The salt nanocrystals may thus adsorb to fine microcracks (eg, resulting from the sealing overheating) and expand them. After a first washout (rain), the tightness of the collector is no longer preserved, and the aerosols enter in the air gap between the absorber plate and the glazing and act as aggressive corrosion media.

Figure 20.8 Outdoor solar–thermal testing rig with biaxial tracking system.
Renewable Energy Systems and Recycling R&D Center, R&D Institute, Transilvania University of Brasov.

To outline these effects, a testing rig was developed for accelerated aging tests under saline aerosols and proved that, after 12 days of testing, large visible corrosion spots could be observed on the absorber surface of a performing commercial FPSTC (Ciobanu et al., 2014).

20.4 Instead of conclusions

The current trends in research and innovation in BISTs must be correlated with the trends in developing nZEBs. The development of isolated nZEBs will contribute to the progress but will not solve the sustainable energy problem.

Integratory concepts are required, targeting energy efficiency and renewables implementation as a distinct (a consistent) part of the sustainable economic, social, and environmental development in communities. Thus, the goal is the development of sustainable communities, with the use of all local resources (renewables, wastes) to meet the community's demands (household, local industry, agriculture, services, etc.) and involving the inhabitants from the design stage, through implementation and operation and in the economic and environmental benefits.

Sustainable communities in the true sense are very few, such as the Vauban district in Freiburg or the Kronsberg community near Hanover, in Germany (Fraker, 2013), and the R&D Institute of Transilvania University of Brasov (Visa and Duta, 2015).

These demonstration projects will have a good replication potential if well supported by the local authorities and by a coherent legal frame. Education and communication have an important role in sustainable communities' development and in increasing the acceptance of the systems (including BIST) that are now under development.

Acknowledgments

This work was done within the frame of the Program Cooperation in Priority Fields—PNII, developed with the support of ANCS, CNDI-UEFISCDI, Romania, in the project EST IN URBA, PN-II-PT-PCCA-2011-3.2-051.

References

Abdellatif, M., Al-Shamma'a, A., 2015. Review of sustainability in buildings. Sustainable Cities and Society 14, 171–177.

Abermann, S., 2013. Non-vacuum processed next generation thin film photovoltaics: towards marketable efficiency and production of CZTS based solar cells. Solar Energy 94, 37–70.

Anderson, T.N., Duke, M., Carson, J.K., 2010. The effect of colour on the thermal performance of building integrated solar collectors. Solar Energy Materials and Solar Cells 94, 350–354.

Arteconi, A., Costola, D., Hensen, J.L.M., 2014. Analysis of control strategies for thermally activated building systems under demand side management mechanisms. Energy and Buildings 80, 384–393.

Atzeri, A., Cappelletti, F., Gasparella, A., 2014. Internal versus external shading devices performance in office buildings. Energy Procedia 45, 463–472.

Azadian, F., Radzi, M., 2013. A general approach toward building integrated photovoltaic systems and its implementation barriers: a review. Renewable and Sustainable Energy Reviews 22, 527–538.

Barbosa, S., Ip, K., 2014. Perspectives of double skin façades for naturally ventilated buildings: a review. Renewable and Sustainable Energy Reviews 40, 1019–1029.

Bednar, N., Severino, N., Adamovic, N., 2015. A strategy for implementation of triangular thin-film photovoltaic modules. Solar Energy 120, 310–317.

Bonhote, P., Eperon, Y., Renaud, P., 2009. Unglazed colored solar absorbers on facade: modelling and performance evaluation. Solar Energy 83, 799–811.

Boyano, A., Hernandez, P., Wolf, O., 2013. Energy demands and potential savings in European office buildings: case studies based on energy plus simulations. Energy and Buildings 65, 19–20.

BPIE (Buildings Performance Institute Europe), 2015. Nearly Zero Energy Buildings Definitions Across Europe. Brussels, Belgium.

Broin, E., Mata, E., Goransson, A., Johnsson, F., 2013. The effect of improved efficiency on energy savings in EU-27 buildings. Energy 57, 134–148.

Buker, M.S., Riffat, S.B., 2015. Building integrated solar thermal collectors – a review. Renewable and Sustainable Energy Reviews 51, 327–346.

Campaniço, H., Hollmuller, P., Soares, P., 2014. Assessing energy savings in cooling demand of buildings using passive cooling systems based on ventilation. Applied Energy 134, 426–438.

Carlsson, B., Möller, K., Köhl, M., Heck, M., Brunold, S., Frei, U., Marechal, J.-C., Jorgensen, G., 2004. The applicability of accelerated life testing for assessment of service life of solar thermal components. Solar Energy Materials and Solar Cells 84, 255–274.

Cho, J., Yoo, C., Kim, Y., 2014. Viability of exterior shading devices for high-rise residential buildings: case study for cooling energy saving and economic feasibility analysis. Energy and Buildings 82, 771–785.

Ciobanu, D., Visa, I., Duta, A., 2014. Solar thermal collectors outdoor testing in saline environment. Energy Procedia 48, 707–714.

COST Action TU1205, 2015. Overview of BISTS State of the Art, Models and Applications. http://www.tu1205-bists.eu/wp-content/uploads/sites/13/2013/07/Book.pdf (lastly accessed on 19.10.15.).

D'Antoni, M., Saro, O., 2012. Massive solar-thermal collectors: a critical literature review. Renewable and Sustainable Energy Reviews 16, 3666–3679.

Deng, S., Wang, R.Z., Dai, Y.J., 2014. How to evaluate performance of net zero energy building – a literature research. Energy 71, 1–16, 1–16.

Duta, A., Isac, L., Milea, A., Ienei, E., Perniu, D., 2014. Coloured solar-thermal absorbers – a comparative analysis of cermet structures. Energy Procedia 48, 543–553.

EEA (European Environment Agency), 2015. Atmospheric Greenhouse Gas Concentrations. Copenhagen.

Ellerbrok, C., 2014. Potentials of demand side management using heat pumps with building mass as a thermal storage. Energy Procedia 46, 214–219.

ENTRANZE Consortium, 2014. Laying Down the Pathways to Nearly Zero-energy Buildings, a Toolkit for Policy Makers. Final Report.

EP (European Parliament), 2010. The Directive 2010/31/EU of the European Parliament and of the Council of 19 May 2010 on the energy performance of buildings. Official Journal of the European Union 53.

Etherden, N., Tesfamichael, T., Niklasson, G.A., Wackelgard, E., 2004. A theoretical feasibility study of pigments for thickness-sensitive spectrally selective paints. Journal of Physics D: Applied Physics 37, 1115–1122.

Fraker, H., 2013. The Hidden Potential of Sustainable Neighborhoods: Lessons from Low-Carbon Communities. Island Press, Washington, US.

Harvey, D.L.D., 2013. Recent advances in sustainable buildings: review of the energy and cost performance of the state-of-the-art best practices from around the world. Annual Review of Environment and Resources 38, 281–309.

Hens, H.S.L.C., 2009. Thermal comfort in office buildings: two case studies commented. Building and Environment 44, 1399–1408.

IEA (International Energy Agency), 2014. World Energy Outlook 2014. Paris, France.

IEA (International Energy Agency), 2015. World Energy Outlook Special Report – Energy and Climate Change. Paris, France.

IPCC (Intergovernmental Panel on Climate Change), 2014. Climate change 2014: synthesis report. In: Pachauri, R.K., Meyer, L.A. (Eds.), Contribution of Working Groups I, II and III to the Fifth Assessment Report of the Intergovernmental Panel on Climate Change. IPCC, Geneva, Switzerland.

Ismail, A.M., Ramirez-Iniguez, R., Asif, M., Munir, A.B., Muhammad-Sukki, F., 2015. Progress of solar photovoltaic in ASEAN countries: a review. Renewable and Sustainable Energy Reviews 48, 399–412.

Jelle, B., Breivik, C., 2012. State-of-the-art building integrated photovoltaics – Technoport RERC Research 2012. Energy Procedia 20, 68–77.

Junghans, L., 2015. Evaluation of the economic and environmental feasibility of heat pump systems in residential buildings, with varying qualities of the building envelope. Renewable Energy 76, 699–705.

Kalogirou, S.A., 2004. Solar thermal collectors and applications. Progress in Energy and Combustion Science 30, 231–295.

Kim, J., Todorovic, M., 2013. Tuning control of buildings glazing's transmittance dependence on the solar radiation wavelength to optimize daylighting and building's energy efficiency. Energy and Buildings 63, 108–118.

Lamnatou, C., Chemisana, D., Mateus, R., Almeida, M.G., Silva, S.M., 2015. Review and perspectives on life cycle analysis of solar technologies with emphasis on building-integrated solar thermal systems. Renewable Energy 75, 833–846.

Lang, T., Gloerfeld, E., Girod, B., 2015. Don't just follow the sun – a global assessment of economic performance for residential building photovoltaics. Renewable and Sustainable Energy Reviews 42, 932–951.

Lu, Y., Wang, S., Yan, C., Shan, K., 2015. Impacts of renewable energy system design inputs on the performance robustness of net zero energy buildings. Energy 93 (2), 1595–1606.

Maddison, A., 2007. Contours of the World Economy 1-2030 AD: Essays in Macro-Economic History Essays in Macro-Economic History. Oxford University Press.

Marszal, A.J., Heiselberg, P., Bourrelle, J.S., Musall, E., Voss, K., Sartori, I., Napolitano, A., 2011. Zero energy building – a review of definitions and calculation methodologies. Energy and Buildings 43, 971–979.

Meadows, D.H., Meadows, D.L., Randers, J., Behrens III, W.W., 1972. The Limits to Growth. Universe Books, New York, USA.

Meadows, D.H., Meadows, D.L., Randers, J., 1992. Beyond the Limits. Chelsea Green Publishing Company, Vermont, USA.

Meadows, D.H., Randers, J., Meadows, D.L., Behrens III, W.W., 2004. Limits to Growth: The 30-Year Update. Chelsea Green Publishing Company.

Mertin, S., Hody-Le Caër, V., Joly, M., Mack, I., Oelhafen, P., Scartezzini, J.-L., Schüler, A., 2014. Reactively sputtered coatings on architectural glazing for coloured active solar thermal façades. Energy and Buildings 68, 764–770.

Midtdal, K., Jelle, B.P., 2013. Self-cleaning glazing products: a state-of-the-art review and future research pathways. Solar Energy Materials and Solar Cells 109, 126–141.

Moldovan, M.D., Visa, I., Burduhos, B.G., 2011. Energetic autonomy for a solar house. Environmental Engineering and Management Journal 10, 1283–1290.

Moldovan, M.D., Visa, I., Ciobanu, D., 2014a. Towards nZEB—Sustainable solutions to meet Thermal energy demand in office buildings, sustainable energy in the built environment – steps towards nZEB Springer Proceedings in energy. In: Proceedings of the Conference for Sustainable Energy (CSE), pp. 115–133.

Moldovan, M.D., Visa, I., Neagoe, M., Burduhos, B., 2014b. Solar heating & cooling energy mixes to transform low energy buildings in nearly zero energy buildings. Energy Procedia 48, 924–937.

Munari Probst, M.C., Roecker, C., 2007. Towards an improved architectural quality of building integrated solar thermal systems (BIST). Solar Energy 81, 1104–1116.

Neagoe, M., Visa, I., Burduhos, B.G., Moldovan, M.D., 2014. Thermal load based adaptive tracking for flat plate solar collectors. Energy Procedia 48, 1401–1411.

Nejat, P., Jomehzadeh, F., Taheri, M., Gohari, M., Makid, M., 2015. A global review of energy consumption, CO_2 emissions and policy in the residential sector (with an overview of the top ten CO_2 emitting countries). Renewable and Sustainable Energy Reviews 43, 843–862.

Nørgård, J.S., Peet, J., Ragnarsdóttir, K.V., 2010. The history of the limits to growth. Solutions Journal 1 (2), 59–63.

Quiles, P.V., Aguilar, F.J., Aledo, S., 2014. Analysis of the overheating and stagnation problems of solar thermal installations. Energy Procedia 48, 172–180.

Ramesh, T., Prakash, R., Shukla, K., 2010. Life cycle energy analysis of buildings: an overview. Energy and Buildings 42, 1592–1600.

Randers, J., 2012. 2052: A Global Forecast for the Next Forty Years. Chelsea Green Publishing, Vermont, USA.

Salvalai, G., Pfafferott, J., Sesana, M.M., 2013. Assessing energy and thermal comfort of different low-energy cooling concepts for non-residential buildings. Energy Conversion and Management 76, 332–341.

Sartori, I., Napolitano, A., Voss, K., 2012. Net zero energy buildings: a consistent definition framework. Energy and Buildings 48, 220–232.

Schuler, A., Boudaden, J., Oelhafen, P., De Chambrier, E., Roecker, C., Scartezzini, J.-L., 2005. Thin film multilayer design types for colored glazed thermal solar collectors. Solar Energy Materials and Solar Cells 89, 219–231.

Sharaf Eldin, S.A., Abd-Elhady, M.S., Kandil, H.A., 2016. Feasibility of solar tracking systems for PV panels in hot and cold regions. Renewable Energy 85, 228–233.

Smalley, R.E., 2005. Future global energy: prosperity the terawatt challenge. MRS Bulletin 30, 412–417.

Tuominen, P., Klobut, K., Tolman, A., Adjei, A., de Best-Waldhober, M., 2012. Energy savings potential in buildings and overcoming market barriers in member states of the European Union. Energy and Buildings 51, 48–55.

Turner, G., 2008. A comparison of 'The Limits to Growth' with thirty years of reality, socio-economics and the environment in discussion. In: Commonwealth Scientific and Industrial Research Organization Working Paper Series, 09, pp. 52–59.

Uihlein, A., Eder, P., 2010. Policy options towards an energy efficient residential building stock in the EU-27. Energy and Buildings 42, 791–798.

UNFCCC, (United Nations Framework Convention on Climate Change), 1992. Draft Report of the Intergovernmental Negotiating Committee for a Framework Convention on Climate Change on the Work of the First Part of Its Fifth Session. New York.

UNFCCC, (United Nations Framework Convention on Climate Change), 1997. Kyoto Protocol.

UNFCCC, (United Nations Framework Convention on Climate Change), 2012. Doha Amendment to the Kyoto Protocol.

UNGA (United Nations General Assembly), 2015. Transforming Our World: The 2030 Agenda for Sustainable Development.

Urge-Vorsatz, D., Petrichenko, K., Staniec, M., Eom, J., 2013. Energy use in buildings in a long-term perspective. Current Opinion in Environmental Sustainability 5, 141–151.

USCB (United States Census Bureau), July 2015. International Data Base. Update, 2015.

USDOE (United States Department of Energy), 2015. A Common Definition for Zero Energy Buildings. National Institute of Building Sciences.

Visa, I., Comsit, M., Duta, A., 2014a. Urban acceptance of facade integrated novel solar thermal collectors. Energy Procedia 48, 1429–1435.

Visa, I., Moldovan, M.D., Comsit, M., Duta, A., 2014b. Improving the renewable energy mix in a building toward the nearly zero energy status. Energy and Buildings 68, 72–78.

Visa, I., Duta, A., 2015. The built environment in sustainable communities. In: Sustainable Energy in the Built Environment – Steps Towards NZEB. Springer Energy Series, pp. 3–38.

Visa, I., Jaliu, C., Duta, A., Neagoe, M., Comsit, M., Moldovan, M., Ciobanu, D., Burduhos, B., Saulescu, R., 2015b. The Role of Mechanisms in Sustainable Energy Systems. Transilvania University Publishing House.

Visa, I., Duta, A., Comsit, M., Moldovan, M., Ciobanu, D., Saulescu, R., Burduhos, B., 2015a. Design and experimental optimization of a novel flat plate solar thermal collector with trapezoidal shape for facades integration. Applied Thermal Engineering 90, 432–443.

Zhang, X., Shen, J., Lu, Y., He, W., Xu, P., Zhao, X., Qiu, Z., Zhu, Z., Zhou, J., Dong, X., 2015. Active solar thermal facades (ASTFs): from concept, application to research questions. Renewable and Sustainable Energy Reviews 50, 32–63.

Zhao, R., Zhao, L., Deng, S., Zheng, N., 2015. Trends in patents for solar thermal utilization in China. Renewable and Sustainable Energy Reviews 52, 852–862.

Index

'*Note*: Page numbers followed by "f" indicate figures and "t" indicate tables.'